T0189104

Lecture Notes in Computer Science   12533

More information about this subseries at http://www.springer.com/series/7407

Haiqin Yang · Kitsuchart Pasupa ·
Andrew Chi-Sing Leung ·
James T. Kwok · Jonathan H. Chan ·
Irwin King (Eds.)

Neural Information Processing

27th International Conference, ICONIP 2020
Bangkok, Thailand, November 23–27, 2020
Proceedings, Part II

 Springer

Editors
Haiqin Yang ⓘ
Department of AI
Ping An Life
Shenzhen, China

Andrew Chi-Sing Leung ⓘ
City University of Hong Kong
Kowloon, Hong Kong

Jonathan H. Chan ⓘ
School of Information Technology
King Mongkut's University
of Technology Thonburi
Bangkok, Thailand

Kitsuchart Pasupa ⓘ
Faculty of Information Technology
King Mongkut's Institute
of Technology Ladkrabang
Bangkok, Thailand

James T. Kwok ⓘ
Department of Computer Science
and Engineering
Hong Kong University of Science
and Technology
Hong Kong, Hong Kong

Irwin King ⓘ
The Chinese University of Hong Kong
New Territories, Hong Kong

ISSN 0302-9743   ISSN 1611-3349 (electronic)
Lecture Notes in Computer Science
ISBN 978-3-030-63832-0   ISBN 978-3-030-63833-7 (eBook)
https://doi.org/10.1007/978-3-030-63833-7

LNCS Sublibrary: SL1 – Theoretical Computer Science and General Issues

This Springer imprint is published by the registered company Springer Nature Switzerland AG
The registered company address is: Gewerbestrasse 11, 6330 Cham, Switzerland

Preface

This book is a part of the five-volume proceedings of the 27th International Conference on Neural Information Processing (ICONIP 2020), held during November 18–22, 2020. The conference aims to provide a leading international forum for researchers, scientists, and industry professionals who are working in neuroscience, neural networks, deep learning, and related fields to share their new ideas, progresses, and achievements. Due to the outbreak of COVID-19, this year's conference, which was supposed to be held in Bangkok, Thailand, was organized as fully virtual conference.

The research program of this year's edition consists of four main categories, Theory and Algorithms, Computational and Cognitive Neurosciences, Human-Centered Computing, and Applications, for refereed research papers with nine special sessions and one workshop. The research tracks attracted submissions from 1,083 distinct authors from 44 countries. All the submissions were rigorously reviewed by the conference Program Committee (PC) comprising 84 senior PC members and 367 PC members. A total of 1,351 reviews were provided, with each submission receiving at least 2 reviews, and some papers receiving 3 or more reviews. This year, we also provided rebuttals for authors to address the errors that exist in the review comments. Meta-reviews were provided with consideration of both authors' rebuttal and reviewers' comments. Finally, we accepted 187 (30.25%) of the 618 full papers that were sent out for review in three volumes of Springer's series of *Lecture Notes in Computer Science* (LNCS) and 189 (30.58%) of the 618 in two volumes of Springer's series of *Communications in Computer and Information Science* (CCIS).

We would like to take this opportunity to thank all the authors for submitting their papers to our conference, and the senior PC members, PC members, as well as all the Organizing Committee members for their hard work. We hope you enjoyed the research program at the conference.

November 2020

Haiqin Yang
Kitsuchart Pasupa

Organization

Honorary Chairs

Jonathan Chan   King Mongkut's University of Technology Thonburi, Thailand

Irwin King   Chinese University of Hong Kong, Hong Kong

General Chairs

Andrew Chi-Sing Leung   City University of Hong Kong, Hong Kong

James T. Kwok   Hong Kong University of Science and Technology, Hong Kong

Program Chairs

Haiqin Yang   Ping An Life, China

Kitsuchart Pasupa   King Mongkut's Institute of Technology Ladkrabang, Thailand

Local Arrangements Chair

Vithida Chongsuphajaisiddhi   King Mongkut University of Technology Thonburi, Thailand

Finance Chairs

Vajirasak Vanijja   King Mongkut's University of Technology Thonburi, Thailand

Seiichi Ozawa   Kobe University, Japan

Special Sessions Chairs

Kaizhu Huang   Xi'an Jiaotong-Liverpool University, China

Raymond Chi-Wing Wong   Hong Kong University of Science and Technology, Hong Kong

Tutorial Chairs

Zenglin Xu   Harbin Institute of Technology, China

Jing Li   Hong Kong Polytechnic University, Hong Kong

Proceedings Chairs

Xinyi Le	Shanghai Jiao Tong University, China
Jinchang Ren	University of Strathclyde, UK

Publicity Chairs

Zeng-Guang Hou	Chinese Academy of Sciences, China
Ricky Ka-Chun Wong	City University of Hong Kong, Hong Kong

Senior Program Committee

Sabri Arik	Istanbul University, Turkey
Davide Bacciu	University of Pisa, Italy
Yi Cai	South China University of Technology, China
Zehong Cao	University of Tasmania, Australia
Jonathan Chan	King Mongkut's University of Technology Thonburi, Thailand
Yi-Ping Phoebe Chen	La Trobe University, Australia
Xiaojun Chen	Shenzhen University, China
Wei Neng Chen	South China University of Technology, China
Yiran Chen	Duke University, USA
Yiu-ming Cheung	Hong Kong Baptist University, Hong Kong
Sonya Coleman	Ulster University, UK
Daoyi Dong	University of New South Wales, Australia
Leonardo Franco	University of Malaga, Spain
Jun Fu	Northeastern University, China
Xin Geng	Southeast University, China
Ping Guo	Beijing Normal University, China
Pedro Antonio Gutiérrez	Universidad de Córdoba, Spain
Wei He	University of Science and Technology Beijing, China
Akira Hirose	The University of Tokyo, Japan
Zengguang Hou	Chinese Academy of Sciences, China
Kaizhu Huang	Xi'an Jiaotong-Liverpool University, China
Kazushi Ikeda	Nara Institute of Science and Technology, Japan
Gwanggil Jeon	Incheon National University, South Korea
Min Jiang	Xiamen University, China
Abbas Khosravi	Deakin University, Australia
Wai Lam	Chinese University of Hong Kong, Hong Kong
Chi Sing Leung	City University of Hong Kong, Hong Kong
Kan Li	Beijing Institute of Technology, China
Xi Li	Zhejiang University, China
Jing Li	Hong Kong Polytechnic University, Hong Kong
Shuai Li	University of Cambridge, UK
Zhiyong Liu	Chinese Academy of Sciences, China
Zhigang Liu	Southwest Jiaotong University, China

Wei Liu	Tencent, China
Jun Liu	Xi'an Jiaotong University, China
Jiamou Liu	The University of Auckland, New Zealand
Lingjia Liu	Virginia Tech, USA
Jose A. Lozano	UPV/EHU, Spain
Bao-liang Lu	Shanghai Jiao Tong University, China
Jiancheng Lv	Sichuan University, China
Marley M. B. R. Vellasco	PUC of Rio de Janeiro, Brazil
Hiroshi Mamitsuka	Kyoto University, Japan
Leandro Minku	University of Birmingham, UK
Chaoxu Mu	Tianjin University, China
Wolfgang Nejdl	L3S Research Center, Germany
Quoc Viet Hung Nguyen	Griffith University, Australia
Takashi Omori	Tamagawa University, Japan
Seiichi Ozawa	Kobe University, Japan
Weike Pan	Shenzhen University, China
Jessie Ju Hyun Park	Yeungnam University, Japan
Kitsuchart Pasupa	King Mongkut's Institute of Technology Ladkrabang, Thailand
Abdul Rauf	Research Institute of Sweden, Sweden
Imran Razzak	Deakin University, Australia
Jinchang Ren	University of Strathclyde, UK
Hayaru Shouno	The University of Electro-Communications, Japan
Ponnuthurai Suganthan	Nanyang Technological University, Singapore
Yang Tang	East China University of Science and Technology, China
Jiliang Tang	Michigan State University, USA
Ivor Tsang	University of Technology Sydney, Australia
Peerapon Vateekul	Chulalongkorn University, Thailand
Brijesh Verma	Central Queensland University, Australia
Li-Po Wang	Nanyang Technological University, Singapore
Kok Wai Wong	Murdoch University, Australia
Ka-Chun Wong	City University of Hong Kong, Hong Kong
Raymond Chi-Wing Wong	Hong Kong University of Science and Technology, Hong Kong
Long Phil Xia	Peking University, Shenzhen Graduate School, China
Xin Xin	Beijing Institute of Technology, China
Guandong Xu	University of Technology Sydney, Australia
Bo Xu	Chinese Academy of Sciences, China
Zenglin Xu	Harbin Institute of Technology, China
Rui Yan	Peking University, China
Xiaoran Yan	Indiana University Bloomington, USA
Haiqin Yang	Ping An Life, China
Qinmin Yang	Zhejiang University, China
Zhirong Yang	Norwegian University of Science and Technology, Norway

De-Nian Yang	Academia Sinica, Taiwan
Zhigang Zeng	Huazhong University of Science and Technology, China
Jialin Zhang	Chinese Academy of Sciences, China
Min Ling Zhang	Southeast University, China
Kun Zhang	Carnegie Mellon University, USA
Yongfeng Zhang	Rutgers University, USA
Dongbin Zhao	Chinese Academy of Sciences, China
Yicong Zhou	University of Macau, Macau
Jianke Zhu	Zhejiang University, China

Program Committee

Muideen Adegoke	City University of Hong Kong, Hong Kong
Sheraz Ahmed	German Research Center for Artificial Intelligence, Germany
Shotaro Akaho	National Institute of Advanced Industrial Science and Technology, Japan
Sheeraz Akram	University of Pittsburgh, USA
Abdulrazak Alhababi	Universiti Malaysia Sarawak, Malaysia
Muhamad Erza Aminanto	University of Indonesia, Indonesia
Marco Anisetti	University of Milan, Italy
Sajid Anwar	Institute of Management Sciences, Pakistan
Muhammad Awais	COMSATS University Islamabad, Pakistan
Affan Baba	University of Technology Sydney, Australia
Boris Bacic	Auckland University of Technology, New Zealand
Mubasher Baig	National University of Computer and Emerging Sciences, Pakistan
Tao Ban	National Information Security Research Center, Japan
Sang Woo Ban	Dongguk University, South Korea
Kasun Bandara	Monash University, Australia
David Bong	Universiti Malaysia Sarawak, Malaysia
George Cabral	Rural Federal University of Pernambuco, Brazil
Anne Canuto	Federal University of Rio Grande do Norte, Brazil
Zehong Cao	University of Tasmania, Australia
Jonathan Chan	King Mongkut's University of Technology Thonburi, Thailand
Guoqing Chao	Singapore Management University, Singapore
Hongxu Chen	University of Technology Sydney, Australia
Ziran Chen	Bohai University, China
Xiaofeng Chen	Chongqing Jiaotong University, China
Xu Chen	Shanghai Jiao Tong University, China
He Chen	Hebei University of Technology, China
Junjie Chen	Inner Mongolia University, China
Mulin Chen	Northwestern Polytechnical University, China
Junying Chen	South China University of Technology, China

Chuan Chen	Sun Yat-sen University, China
Liang Chen	Sun Yat-sen University, China
Zhuangbin Chen	Chinese University of Hong Kong, Hong Kong
Junyi Chen	City University of Hong Kong, Hong Kong
Xingjian Chen	City University of Hong Kong, Hong Kong
Lisi Chen	Hong Kong Baptist University, Hong Kong
Fan Chen	Duke University, USA
Xiang Chen	George Mason University, USA
Long Cheng	Chinese Academy of Sciences, China
Aneesh Chivukula	University of Technology Sydney, Australia
Sung Bae Cho	Yonsei University, South Korea
Sonya Coleman	Ulster University, UK
Fengyu Cong	Dalian University of Technology, China
Jose Alfredo Ferreira Costa	Federal University of Rio Grande do Norte, Brazil
Ruxandra Liana Costea	Polytechnic University of Bucharest, Romania
Jean-Francois Couchot	University of Franche-Comté, France
Raphaël Couturier	University Bourgogne Franche-Comté, France
Zhenyu Cui	University of the Chinese Academy of Sciences, China
Debasmit Das	Qualcomm, USA
Justin Dauwels	Nanyang Technological University, Singapore
Xiaodan Deng	Beijing Normal University, China
Zhaohong Deng	Jiangnan University, China
Mingcong Deng	Tokyo University, Japan
Nat Dilokthanakul	Vidyasirimedhi Institute of Science and Technology, Thailand
Hai Dong	RMIT University, Australia
Qiulei Dong	Chinese Academy of Sciences, China
Shichao Dong	Shenzhen Zhiyan Technology Co., Ltd., China
Kenji Doya	Okinawa Institute of Science and Technology, Japan
Yiqun Duan	University of Sydney, Australia
Aritra Dutta	King Abdullah University of Science and Technology, Saudi Arabia
Mark Elshaw	Coventry University, UK
Issam Falih	Paris 13 University, France
Ozlem Faydasicok	Istanbul University, Turkey
Zunlei Feng	Zhejiang University, China
Leonardo Franco	University of Malaga, Spain
Fulvio Frati	Università degli Studi di Milano, Italy
Chun Che Fung	Murdoch University, Australia
Wai-Keung Fung	Robert Gordon University, UK
Claudio Gallicchio	University of Pisa, Italy
Yongsheng Gao	Griffith University, Australia
Cuiyun Gao	Harbin Institute of Technology, China
Hejia Gao	University of Science and Technology Beijing, China
Yunjun Gao	Zhejiang University, China

Xin Gao	King Abdullah University of Science and Technology, Saudi Arabia
Yuan Gao	Uppsala University, Sweden
Yuejiao Gong	South China University of Technology, China
Xiaotong Gu	University of Tasmania, Australia
Shenshen Gu	Shanghai University, China
Cheng Guo	Chinese Academy of Sciences, China
Zhishan Guo	University of Central Florida, USA
Akshansh Gupta	Central Electronics Engineering Research Institute, India
Pedro Antonio Gutiérrez	University of Córdoba, Spain
Christophe Guyeux	University Bourgogne Franche-Comté, France
Masafumi Hagiwara	Keio University, Japan
Ali Haidar	University of New South Wales, Australia
Ibrahim Hameed	Norwegian University of Science and Technology, Norway
Yiyan Han	Huazhong University of Science and Technology, China
Zhiwei Han	Southwest Jiaotong University, China
Xiaoyun Han	Sun Yat-sen University, China
Cheol Han	Korea University, South Korea
Takako Hashimoto	Chiba University of Commerce, Japan
Kun He	Shenzhen University, China
Xing He	Southwest University, China
Xiuyu He	University of Science and Technology Beijing, China
Wei He	University of Science and Technology Beijing, China
Katsuhiro Honda	Osaka Prefecture University, Japan
Yao Hu	Alibaba Group, China
Binbin Hu	Ant Group, China
Jin Hu	Chongqing Jiaotong University, China
Jinglu Hu	Waseda University, Japan
Shuyue Hu	National University of Singapore, Singapore
Qingbao Huang	Guangxi University, China
He Huang	Soochow University, China
Kaizhu Huang	Xi'an Jiaotong-Liverpool University, China
Chih-chieh Hung	National Chung Hsing University, Taiwan
Mohamed Ibn Khedher	IRT SystemX, France
Kazushi Ikeda	Nara Institute of Science and Technology, Japan
Teijiro Isokawa	University of Hyogo, Japan
Fuad Jamour	University of California, Riverside, USA
Jin-Tsong Jeng	National Formosa University, Taiwan
Sungmoon Jeong	Kyungpook National University, South Korea
Yizhang Jiang	Jiangnan University, China
Wenhao Jiang	Tencent, China
Yilun Jin	Hong Kong University of Science and Technology, Hong Kong

Chengchuang Lin	South China Normal University, China
Xinshi Lin	Chinese University of Hong Kong, Hong Kong
Jiecong Lin	City University of Hong Kong, Hong Kong
Shu Liu	The Australian National University, Australia
Xinping Liu	University of Tasmania, Australia
Shaowu Liu	University of Technology Sydney, Australia
Weifeng Liu	China University of Petroleum, China
Zhiyong Liu	Chinese Academy of Sciences, China
Junhao Liu	Chinese Academy of Sciences, China
Shenglan Liu	Dalian University of Technology, China
Xin Liu	Huaqiao University, China
Xiaoyang Liu	Huazhong University of Science and Technology, China
Weiqiang Liu	Nanjing University of Aeronautics and Astronautics, China
Qingshan Liu	Southeast University, China
Wenqiang Liu	Southwest Jiaotong University, China
Hongtao Liu	Tianjin University, China
Yong Liu	Zhejiang University, China
Linjing Liu	City University of Hong Kong, Hong Kong
Zongying Liu	King Mongkut's Institute of Technology Ladkrabang, Thailand
Xiaorui Liu	Michigan State University, USA
Huawen Liu	The University of Texas at San Antonio, USA
Zhaoyang Liu	Chinese Academy of Sciences, China
Sirasit Lochanachit	King Mongkut's Institute of Technology Ladkrabang, Thailand
Xuequan Lu	Deakin University, Australia
Wenlian Lu	Fudan University, China
Ju Lu	Shandong University, China
Hongtao Lu	Shanghai Jiao Tong University, China
Huayifu Lv	Beijing Normal University, China
Qianli Ma	South China University of Technology, China
Mohammed Mahmoud	Beijing Institute of Technology, China
Rammohan Mallipeddi	Kyungpook National University, South Korea
Jiachen Mao	Duke University, USA
Ali Marjaninejad	University of Southern California, USA
Sanparith Marukatat	National Electronics and Computer Technology Center, Thailand
Tomas Henrique Maul	University of Nottingham Malaysia, Malaysia
Phayung Meesad	King Mongkut's University of Technology North Bangkok, Thailand
Fozia Mehboob	Research Institute of Sweden, Sweden
Wenjuan Mei	University of Electronic Science and Technology of China, China
Daisuke Miyamoto	The University of Tokyo, Japan

Kazuteru Miyazaki	National Institution for Academic Degrees and Quality Enhancement of Higher Education, Japan
Bonaventure Molokwu	University of Windsor, Canada
Hiromu Monai	Ochanomizu University, Japan
J. Manuel Moreno	Universitat Politècnica de Catalunya, Spain
Francisco J. Moreno-Barea	University of Malaga, Spain
Chen Mou	Nanjing University of Aeronautics and Astronautics, China
Ahmed Muqeem Sheri	National University of Sciences and Technology, Pakistan
Usman Naseem	University of Technology Sydney, Australia
Mehdi Neshat	The University of Adelaide, Australia
Quoc Viet Hung Nguyen	Griffith University, Australia
Thanh Toan Nguyen	Griffith University, Australia
Dang Nguyen	University of Canberra, Australia
Thanh Tam Nguyen	Ecole Polytechnique Federale de Lausanne, France
Giang Nguyen	Korea Advanced Institute of Science and Technology, South Korea
Haruhiko Nishimura	University of Hyogo, Japan
Stavros Ntalampiras	University of Milan, Italy
Anupiya Nugaliyadde	Murdoch University, Australia
Toshiaki Omori	Kobe University, Japan
Yuangang Pan	University of Technology Sydney, Australia
Weike Pan	Shenzhen University, China
Teerapong Panboonyuen	Chulalongkorn University, Thailand
Paul S. Pang	Federal University Australia, Australia
Lie Meng Pang	Southern University of Science and Technology, China
Hyeyoung Park	Kyungpook National University, South Korea
Kitsuchart Pasupa	King Mongkut's Institute of Technology Ladkrabang, Thailand
Yong Peng	Hangzhou Dianzi University, China
Olutomilayo Petinrin	City University of Hong Kong, Hong Kong
Geong Sen Poh	National University of Singapore, Singapore
Mahardhika Pratama	Nanyang Technological University, Singapore
Emanuele Principi	Università Politecnica delle Marche, Italy
Yiyan Qi	Xi'an Jiaotong University, China
Saifur Rahaman	International Islamic University Chittagong, Bangladesh
Muhammad Ramzan	Saudi Electronic University, Saudi Arabia
Yazhou Ren	University of Electronic Science and Technology of China, China
Pengjie Ren	University of Amsterdam, The Netherlands
Colin Samplawski	University of Massachusetts Amherst, USA
Yu Sang	Liaoning Technical University, China
Gerald Schaefer	Loughborough University, UK

Rafal Scherer   Czestochowa University of Technology, Poland
Xiaohan Shan   Chinese Academy of Sciences, China
Hong Shang   Tencent, China
Nabin Sharma   University of Technology Sydney, Australia
Zheyang Shen   Aalto University, Finland
Yin Sheng   Huazhong University of Science and Technology,
                             China
Jin Shi   Nanjing University, China
Wen Shi   South China University of Technology, China
Zhanglei Shi   City University of Hong Kong, Hong Kong
Tomohiro Shibata   Kyushu Institute of Technology, Japan
Hayaru Shouno   The University of Electro-Communications, Japan
Chiranjibi Sitaula   Deakin University, Australia
An Song   South China University of Technology, China
mofei Song   Southeast University, China
Liyan Song   Southern University of Science and Technology, China
Linqi Song   City University of Hong Kong, Hong Kong
Yuxin Su   Chinese University of Hong Kong, Hong Kong
Jérémie Sublime   Institut supérieur d'électronique de Paris, France
Tahira Sultana   UTM Malaysia, Malaysia
Xiaoxuan Sun   Beijing Normal University, China
Qiyu Sun   East China University of Science and Technology,
                             China
Ning Sun   Nankai University, China
Fuchun Sun   Tsinghua University, China
Norikazu Takahashi   Okayama University, Japan
Hiu-Hin Tam   City University of Hong Kong, Hong Kong
Hakaru Tamukoh   Kyushu Institute of Technology, Japan
Xiaoyang Tan   Nanjing University of Aeronautics and Astronautics,
                             China
Ying Tan   Peking University, China
Shing Chiang Tan   Multimedia University, Malaysia
Choo Jun Tan   Wawasan Open University, Malaysia
Gouhei Tanaka   The University of Tokyo, Japan
Yang Tang   East China University of Science and Technology,
                             China
Xiao-Yu Tang   Zhejiang University, China
M. Tanveer   Indian Institutes of Technology, India
Kai Meng Tay   Universiti Malaysia Sarawak, Malaysia
Chee Siong Teh   Universiti Malaysia Sarawak, Malaysia
Ya-Wen Teng   Academia Sinica, Taiwan
Andrew Beng Jin Teoh   Yonsei University, South Korea
Arit Thammano   King Mongkut's Institute of Technology Ladkrabang,
                             Thailand
Eiji Uchino   Yamaguchi University, Japan

Nhi N.Y. Vo	University of Technology Sydney, Australia
Hiroaki Wagatsuma	Kyushu Institute of Technology, Japan
Nobuhiko Wagatsuma	Tokyo Denki University, Japan
Yuanyu Wan	Nanjing University, China
Feng Wan	University of Macau, Macau
Dianhui Wang	La Trobe University, Australia
Lei Wang	Beihang University, China
Meng Wang	Beijing Institute of Technology, China
Sheng Wang	Henan University, China
Meng Wang	Southeast University, China
Chang-Dong Wang	Sun Yat-sen University, China
Qiufeng Wang	Xi'an Jiaotong-Liverpool University, China
Zhenhua Wang	Zhejiang University of Technology, China
Yue Wang	Chinese University of Hong Kong, Hong Kong
Jiasen Wang	City University of Hong Kong, Hong Kong
Jin Wang	Hanyang University, South Korea
Wentao Wang	Michigan State University, USA
Yiqi Wang	Michigan State University, USA
Peerasak Wangsom	CAT Telecom PCL, Thailand
Bunthit Watanapa	King Mongkut's University of Technology Thonburi, Thailand
Qinglai Wei	Chinese Academy of Sciences, China
Yimin Wen	Guilin University of Electronic Technology, China
Guanghui Wen	Southeast University, China
Ka-Chun Wong	City University of Hong Kong, Hong Kong
Kuntpong Woraratpanya	King Mongkut's Institute of Technology Ladkrabang, Thailand
Dongrui Wu	Huazhong University of Science and Technology, China
Qiujie Wu	Huazhong University of Science and Technology, China
Zhengguang Wu	Zhejiang University, China
Weibin Wu	Chinese University of Hong Kong, Hong Kong
Long Phil Xia	Peking University, Shenzhen Graduate School, China
Tao Xiang	Chongqing University, China
Jiaming Xu	Chinese Academy of Sciences, China
Bin Xu	Northwestern Polytechnical University, China
Qing Xu	Tianjin University, China
Xingchen Xu	Fermilab, USA
Hui Xue	Southeast University, China
Nobuhiko Yamaguchi	Saga University, Japan
Toshiyuki Yamane	IBM Research, Japan
Xiaoran Yan	Indiana University, USA
Shankai Yan	National Institutes of Health, USA
Jinfu Yang	Beijing University of Technology, China
Xu Yang	Chinese Academy of Sciences, China

Feidiao Yang	Chinese Academy of Sciences, China
Minghao Yang	Chinese Academy of Sciences, China
Jianyi Yang	Nankai University, China
Haiqin Yang	Ping An Life, China
Xiaomin Yang	Sichuan University, China
Shaofu Yang	Southeast University, China
Yinghua Yao	University of Technology Sydney, Australia
Jisung Yoon	Indiana University, USA
Junichiro Yoshimoto	Nara Institute of Science and Technology, Japan
Qi Yu	University of New South Wales, Australia
Zhaoyuan Yu	Nanjing Normal University, China
Wen Yu	CINVESTAV-IPN, Mexico
Chun Yuan	Tsinghua University, China
Xiaodong Yue	Shanghai University, China
Li Yun	Nanjing University of Posts and Telecommunications, China
Jichuan Zeng	Chinese University of Hong Kong, Hong Kong
Yilei Zhang	Anhui Normal University, China
Yi Zhang	Beijing Institute of Technology, China
Xin-Yue Zhang	Chinese Academy of Sciences, China
Dehua Zhang	Chinese Academy of Sciences, China
Lei Zhang	Chongqing University, China
Jia Zhang	Microsoft Research, China
Liqing Zhang	Shanghai Jiao Tong University, China
Yu Zhang	Southeast University, China
Liang Zhang	Tencent, China
Tianlin Zhang	University of Chinese Academy of Sciences, China
Rui Zhang	Xi'an Jiaotong-Liverpool University, China
Jialiang Zhang	Zhejiang University, China
Ziqi Zhang	Zhejiang University, China
Jiani Zhang	Chinese University of Hong Kong, Hong Kong
Shixiong Zhang	City University of Hong Kong, Hong Kong
Jin Zhang	Norwegian University of Science and Technology, Norway
Jie Zhang	Newcastle University, UK
Kun Zhang	Carnegie Mellon University, USA
Yao Zhang	Tianjin University, China
Yu Zhang	University of Science and Technology Beijing, China
Zhijia Zhao	Guangzhou University, China
Shenglin Zhao	Tencent, China
Qiangfu Zhao	University of Aizu, Japan
Xiangyu Zhao	Michigan State University, USA
Xianglin Zheng	University of Tasmania, Australia
Nenggan Zheng	Zhejiang University, China
Wei-Long Zheng	Harvard Medical School, USA
Guoqiang Zhong	Ocean University of China, China

Jinghui Zhong	South China University of Technology, China
Junping Zhong	Southwest Jiaotong University, China
Xiaojun Zhou	Central South University, China
Hao Zhou	Harbin Engineering University, China
Yingjiang Zhou	Nanjing University of Posts and Telecommunications, China
Deyu Zhou	Southeast University, China
Zili Zhou	The University of Manchester, UK

Contents – Part II

Robotics and Control

Computational Intelligence

A Novel Mathematic Entorhinal-Hippocampal System Building Cognitive Map

Jianxin Peng[1]📷, Suogui Dang[1]📷, Rui Yan[2]([✉]), and Huajin Tang[3]

[1] College of Computer Science, Sichuan University, Chengdu, China
pjx1234567@foxmail.com, dangsuogui@foxmail.com
[2] College of Computer Science and Technology, Zhejiang University of Technology,
Hangzhou, China
yanrui2006@gmail.com
[3] College of Computer Science and Technology, Zhejiang University,
Hangzhou, China
huajin.tang@gmail.com

Abstract. Place cells and grid cells are crucial parts of the cognitive map, which shows a presentation of the real-world observation. However, the previous architecture, which uses CAN for simulating the activities of grid cells, is redundant. And it could not generate natural activities of place cells while it needs many computing resources and storage. In this paper, we proposed a simple novel mathematic entorhinal-hippocampal system to build an accurate cognitive map by combining the activities of head direction cells, grid cells, place cells, and visual cues. It has fewer parameters and could generate a natural pattern of place cells. Moreover, we could also perform a cognitive map building system with generated weight without training.

Keywords: Grid cell · Place cell · Cognitive map · SLAM

1 Introduction

Spatial cognition is an innate ability in humans and many animals. It involves the ability to understand and manipulate the environment, meaning that life can understand the environment and navigate through it. For a long time, researchers have been studying how animals perceive the environment and navigation.

In 1948, Thomas thought navigation was controlled or affected by the internal map-like representation, which is known as a cognitive map [16]. The concept explains the maze learning space layout of the mouse and then applied to other animals, including humans. Some specialists said that a cognitive map is a type

This work was supported by the National Natural Science Foundation of China under grant number 61773271 and the National Key Research and Development Program of China under grant 2017YFB1300201.

H. Yang et al. (Eds.): ICONIP 2020, LNCS 12533, pp. 3–14, 2020.
https://doi.org/10.1007/978-3-030-63833-7_1

of mental representation that serves an individual to acquire, code, store, recall, and decode information about the relative locations and attributes of phenomena in their everyday or metaphorical spatial environment. However, it was only a concept without concrete evidence until the discovery of the different types of spatial cells. O'Keefe and Dostrovsky first discovered place cells in the hippocampus and demonstrated place cells would fire whenever the rat was within a specific place in the environment [10,11]. HD cells are found in many areas in the brain, only when the animal heads to a specific direction, the firing rate will be higher than the baseline level of the neurons [14]. Then Moser et al. studied the activities from different levels of the entorhinal cortex and found that the cells in the dorsal part of the medial entorhinal cortex (MEC) had place fields similar to the pattern in the hippocampus but fired at multiple locations [5] and discovered the grid cells [7]. The activities of place cells are a linear summation of a subset of afferent grid cells with a range of spatial frequencies [4,12].

Many computational models were proposed to simulate the grid pattern. Additionally, they can be summed up in three types, which are continuous attractor network (CAN) [2,4], oscillatory interference [3,18], and sinusoidal gratings [12]. Also, there are some work searching for property of the grid cells [1,6,13]. It has been observed that rats can correct the accumulative errors of path integration when they meet the notable landmark [9]. And a few years ago, some works combining the grid cells and RatSLAM were done to build a cognitive map. Yuan et al. proposed an entorhinal-hippocampal model to build the cognitive map with CAN [17]. Hu et al. changed the learning rule between grid cells and place cells to produce a point centralized response [8]. Nevertheless, all of them are computationally intensive and complicated because they use many layers of grids to encode the grid cell. The activities of place cells in these models can be strange because they always have multiple responses, which is quite different from the single-peaked activities of real place cells.

In this work, we construct a simple novel mathematic entorhinal-hippocampal system to build a cognitive map by integrating visual cues, place cells, head direction cells, and velocity-coupled grid cells. This system provides an alternative method to build the cognitive map and could be used on the mobile robot to build an accurate cognitive map in the real world. The system might be faster for having fewer parameters than the system proposed before, and it can generate natural activities of place cells, which have only one central firing point. The discovery that the weight distribution is similar to the grid pattern confirms the distance measuring a property of the grid cells, and it helps us to build a system that does not require trained parameters.

Our future work may be improving the current grid cell model to solve the irregular grid cell pattern or use more powerful SLAM methods with the grid cell model to achieve a better cognitive map building result.

2   Model Description

2.1   Architecture of the System

The system can be divided into a few parts.

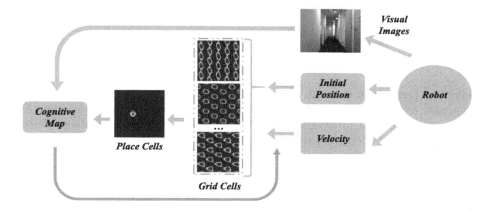

Fig. 1. The architecture of the system building cognitive maps

- Train the weight between grid cells and place cells. Here, we set up that a grid code was fully connected to place cells.
- The initial position and velocity of the robot are used to generate a grid code. The initial position was first provided to produce the first intermediate variable. Then the velocity information was used to generate the new intermediate variable, which will directly produce the grid code.
- Use the weight computed in the first step to compute the activities of the place cells, which could tell us the location where we are.
- Use the information combining with the visual cues to make sure whether it is the place we have ever been to.
- With the progress that the robot goes through the maze, the cognitive map will emerge into a more real trajectory.

2.2   Neuron Model

In our building system, three kinds of neurons are used: head direction cells, grid cells, and place cells. Head direction cells provide velocity and orientation. Grid cells use the data to compute a representation in a high dimension, which shows a hexagonal firing pattern in each dimension. Place cells are a population of neurons representing the location of the agent.

As to the grid gell, we should know that the grid cell spikes with a hexagonal pattern, which is described as Fig. 2a. The horizontal and vertical coordinates

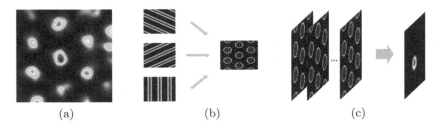

(a)   (b)   (c)

Fig. 2. (a) The natural grid pattern; (b) Grid cell pattern generating method; (c) Activities of place cells are generated by grid cells

indicate a position in the real world, and the value of these pixels indicates the firing rate of the grid neuron at the position. Besides, the pattern can be mathematically generated by overlaying periodic fringes with different orientations. Figure 2b showed the grid pattern generated by the model with orientation θ equaling 0, a scale ρ equaling 0.2 and a phase ϕ equaling (0, 0) in a $30\,\mathrm{m} * 30\,\mathrm{m}$ environment. The activities of place cells are thought to be computed by the activities of grid cells with different orientations, scales and phases as shown in Fig. 2c.

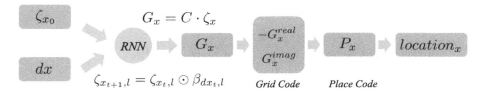

Fig. 3. Architecture of the model

The architecture of the grid cell and place cell in our system is shown as Fig. 3. In our grid cell model, we consider three important parameters—scale, orientation and phase, which decide the property of grid cells. We can compute the basis function $\zeta_{x,l}$ by Eq. 1.

$$\zeta_{x,l} = [e^{iw_{1,l}^T(x+\phi_l)}, e^{iw_{2,l}^T(x+\phi_l)}, e^{iw_{3,l}^T(x+\phi_l)}] \tag{1}$$

where $w_{j,l}$ can be computed by Eq. 2,

$$w_{j,l} = \rho_l[cos(\theta_l + \frac{2j\pi}{3}), sin(\theta_l + \frac{2j\pi}{3})]^T, j \in \{0,1,2\} \tag{2}$$

ρ_l is related to the scale of the grid cell, θ_l is related to the orientation of the pattern and ϕ_l is related to the phase of the pattern.

We will use the start position x_0 to initial the $\zeta_{x_0,l}$. And every time the agent moves, velocity and orientation will be provided to compute the $\beta_{dx_t,l}$ with Eq. 3.

$$\beta_{dx,l} = [e^{iw_{1,l}^T dx}, e^{iw_{2,l}^T dx}, e^{iw_{3,l}^T dx}] \tag{3}$$

And just like the computation described as $x_{t+1} = x_t + dx_t$ in Euclidean Space, the current $\zeta_{x_{t+1},l}$ can be iteratively computed by Eq. 4. Also, it could be directly got by Eq. 1.

$$\zeta_{x_{t+1},l} = \zeta_{x_t,l} \odot \beta_{dx_t,l} \tag{4}$$

For grid cells with different scales, orientations and phases, the grid code is separately computed by Eq. 5:

$$G_{x,l} = C \cdot \zeta_{x,l} \tag{5}$$

where C is required to be a positive definite unit matrix, which means:

$$C \cdot C^T = I \tag{6}$$

As mentioned above, $G_{x,l}$ is a subset of complex grid code with the same ρ, θ and ϕ. The best method to eliminate the complex number is to separate the complex number into two parts (the real part and the imaginary part) and combine them. So G_x^{final}, which is the final grid code of x, could be computed by Eq. 7:

$$G_x^{final} = [-G_x^{real}, G_x^{imag}] \tag{7}$$

Furthermore, we had known that the activities of place cells can represent the location and they are calculated by the output of grid cells, so a fully connected layer was taken to connect the grid cells and place cells. Thus the activities of place cells P_x can be computed by Eq. 8:

$$P_x = W^T G_x^{final} \tag{8}$$

where x is the position, and W is the weight of the fully connected layer.

Moreover, the activities of place cells will be much like a two-dimensional gaussian. Here, we use the WTA (winner take all) rule. That's to say that we choose the highest point as the center of the activities of the place cells to confirm the position $location_x$ from the by Eq. 9:

$$location_x = [argmax(P_x)/l_p, argmax(P_x)\%l_p] \tag{9}$$

Of course, some other methods can confirm the location through the activities around the highest point instead of the highest point only [17].

2.3 Learning Algorithm

Through the reproduction of the previous paper, we found that Hebb learning seems to be biological, but it always needs a good initial state and may cause a divergence of the place cells' activities. And if we train the network in the cognitive map building process, there might be a significant error of the place cells' activities at the beginning or in the whole process because it is a wrong firing pattern at first.

So we choose a supervised method to train the weight before the map building process. The target place cells' activities are generated by a combination of a radial function, which can be a gauss function, and a normalization:

$$P_m(x) = \frac{e^{-\frac{\|x-u_m\|_2^2}{2\sigma^2}}}{\sum_{j=1}^{M} e^{-\frac{\|x-u_j\|_2^2}{2\sigma^2}}} \tag{10}$$

where $P_m(x)$ present the current place cell activities, x indicates the current position and u_m is the place cell we choose, u_j indicates each of the place cells.

2.4   Visual Calibration

Although it is not clear how visual cues affect the internal cognitive map, what we confirmed was that visual cues are essential to the correction of the cognitive map. The Depth information is helpful to increasing the robustness of the model, because the information carried by 2D image may not be enough, which could cause a wrong loop closure detection or template match. Some work has been done to compare the image profiles between the RGB and Depth images for loop closure detection and new scene detection [15]. So we use the images including both the RGB and depth images as the visual cues to correct the path integration errors and detect the loop closure.

And Algorithm 1 describes the algorithm to build an accurate cognitive map in our system with more details.

Algorithm 1: The cognitive map building algrithm

Input : Raw odometry data from the robot wheel encoder and visual images from the RGB-D sensor
Out: Cognitive map
Begin:

- Pre-train the weight between grid cells and place cells with fake data.
- Compute the activities of the grid cells with the current speed.
- Compute the activities of the place cells.
- Compute the position of the highest value in the activities of place cell.
- Compare the current visual cues with the experience recorded.

    If Matched
    Then Perform map correction
    Else Create a new experience map point
    End if
End:

---

3 Experiment Result

3.1 Environment

The raw image data was collected on a real mobile robot platform, which consists of a Pioneer 3-DX mobile base, an RGB-D sensor, and a laptop. The front wheels of the base equipped with encoders provide raw odometry data, and the RGB-D sensor mounted on the front top of the base captures visual RGB-D images of environments [17].

3.2 Experiment for Cognitive Map Building

Before the formal experiment, it is crucial to make sure whether the model could perform path integration. We pretrain the weight from grid cells to place cells with the supervised learning method mentioned above. And it is obvious that the grid code could realize accurate path integration as shown in Fig. 4.

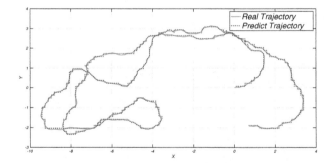

Fig. 4. The green solid line is a trajectory randomly generated from (0, 0), while the red dotted line shows the trajectory predicted by the model (Color figure online)

We had known that what we wanted to build a cognitive map of is about 35 m × 35 m office environment. First, 1600 place cells are used to cover an area, whose horizontal and vertical coordinates are both from −35 m to 35 m. Then, the proposed system is used to perform a cognitive map building. In this experiment, seventy-two different grid cells are used to encode and compute the position.

Figure 5a shows the trajectory calculated by the odometer, which has an accumulated error resulting in a shift in position. The subfigure is plotted in an environment with a fixed size (from −30 m to 20 m for each axis). Figure 5b is the cognitive map consisting of the corrected experience map points. We can see that the experience map detects the closure loop and corrects itself into a well-constructed cognitive map. Figure 5c show the activities of the place cell in our model.

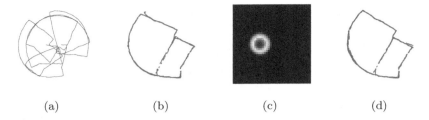

(a)	(b)	(c)	(d)

Fig. 5. (a) Trajectory computed from original odometer; (b) Cognive map building result; (c) Activities of place cells at the last time; (d) The cognitive map building result with the non-trained weight.

3.3   Experiment for Distribution of the Weight

The distribution of place cells' activities makes us pay attention to how the weight distributed and why the weight could result in a gauss distribution. So we visualized the weight W between the grid cells and all the place cells, which is shown as Fig. 6b. Each image shows the connection from a grid cell to all place cells. And it is a surprising discovery that the visualized weight images are similar to the pattern of the grid cells, which is visualized as Fig. 6a.

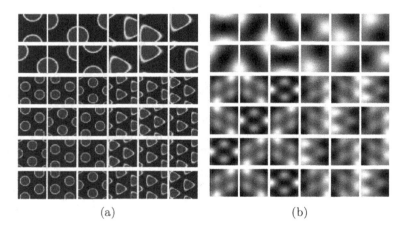

(a)	(b)

Fig. 6. (a) Some grid pattern sampled from cognitive map building process. (b) Visualized results of the parts of the weight W. Each of the images shows the connection from one grid cell to all of the place cells

We take that $W_{i,j}$ is the weight from the ith grid cell to the jth place cell, W_j is the weight connected to jth place cell, G_x is the grid code at position x, $P_{x,j}$ is the activity of the jth place cell, and X_{P_j} is the position whose center of the activities of place cells is just on the jth place cell. The weight connected to one place cell is similar to the grid code of the position corresponding to that place cell, which signifies $W_j \approx G_{X_{P_j}}$.

3.4 Experiment for Property Verification

Here, we test whether the inner product of the grid code can measure the relative distance. A five-star point and the other 100 are randomly chosen as shown in Fig. 7. In the picture below, we can see the two different lines almost overlap, which means the inner product of the two grid code could represent the relative distance. For visualization reasons, we take the reciprocal of the inner product to simulate that the closer two points have a smaller relative distance because the closer two points have a bigger inner product. But reciprocal operations won't affect relative relations.

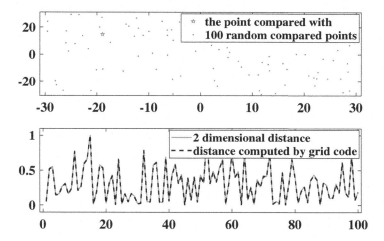

Fig. 7. In the picture above, the five-star is the point which compared to all of the others, and the dot indicates the other points we randomly chose; In the picture below: The green line is the relative distance in Euclidean Space, and the black dotted line represents a relative distance computed by grid code of that position. (Color figure online)

So we tried to replace the weight with the pattern of grid cells, which is computed by the model in our system. In this model, the scale parameter ρ is sampled with normal distribution, and the orientation factor θ is sampled with a uniform distribution, and the phase factor ϕ is randomly sampled under the environment scale. Experiments show it can generate a network that does not require training parameters. It can perform path integration very well and can be used in our cognitive map building system. The cognitive map building result with the non-trained weight is shown in Fig. 5d. And we could see that the system with the generated weight also detects the closure loop and builds a well-constructed cognitive map.

4   Discussion

4.1   Discussion on Cells and Parameters of the Model

The model in our system has fewer grid cells and parameters than the CAN model used in [17]. By analyzing the CAN model, we can see it is redundant that many grid cells with different phases in the same layer could have the same firing pattern for periodicity. While in our model, we can see that there are only a few neurons with different orientations and phases in each layer, each of them plays their own role.

If we set that the grid cells in each layer have the same scale. In the CAN model, each of the layers consists of $size^2$ neurons, where $size$ indicates the number of the neurons in each axis. For an environment with the same range, if we want to get a more accurate position, more place cells are needed. In the CAN model, the number of grid cells in each layer will rise up to be the same as the number of place cells. While in our model, the number of grid cells will be the same as before. And our model will have fewer parameters to have a faster speed, as shown in Table 1.

Table 1. Comparison of the Amount of Parameter

	CAN model	CAN model	Our model	Our model
Range of place cells	40	50	40	50
Number of place cells	1600	2500	1600	2500
Layer of grid cells	12	12	12	12
Number of grid cells	12×1600	12×2500	12×3	12×3
Number of weight	12×1600^2	12×2500^2	36×1600	36×2500

Through the reproduction of the previous paper, we have known that only a few layers of the grid cells dominate the activities of the place cell, and some of the weights are closer to zero in some of the CAN model, which means activities of the connected grid cells might be redundant to the activities of place cells. But in our system, we take all of the weights that could be useful.

4.2   Discussion on the Distribution of the Weight

From the experiments, we can see that the inner product of the grid code of two positions can measure the distance between them, and two positions with a larger inner product are thought to be closer. Furthermore, the computation of the inner product could be a radial function, which seems to be a function of distance. So we can take the forward computation $P_j = W_j \cdot G_x$ as the inner product computation of the grid codes representing the current position and the each of all target positions $\langle G_x, G_{X_{P_j}} \rangle$. Moreover, taking the max value of the forward computation results also conform to that two closer positions will have a larger inner product of their grid code, and two same grid code has the largest inner product.

4.3 Discussion on the Activities of Place Cells

Our system could generate a natural pattern of place cells. The activities of place cells should have only one spike center in order to get an unambiguous position in the real world. However, the previous model represents a strange pattern with more than one spike center, just as shown in Fig. 8a [17]. The activities of place cells that computed with weight, which is trained by the improved method called Pre-synaptically Gated Learning and Spatial Window [8], become more regular and have fewer centers. Nevertheless, the activities are more like a grid cell pattern with some center point weaken, as shown in Fig. 8b, whose pattern is not natural enough.

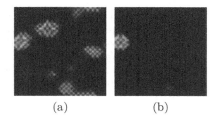

(a)   (b)

Fig. 8. Place patterns of some previous models. (a) The activities of place cell generated by the traditional CAN model [17]. (b) The activities of place cell generated by the CAN trained by the Pre-synaptically Gated Learning and Spatial Window in [8]

The activities of the place cells sampled from the map building process with our model are more natural than the above model.

The main reason might be that the Hebb learning algorithm in the CAN models is not robust enough. Also, most of the connections are from grid cells to grid cells in those models, and the number of the weight between grid cells and place cells is too small, which may cause a poor-fitting ability. However, in our model, the number of weights between grid cells and place cells is enough and the weight could present the activities of place cells very well.

For our model with generated weight, the activities of place cells are similar to a quadratic function with a negative coefficient of quadratic term, and the extent of the bump is mainly controlled by the scales of grid cells ρ.

5 Conclusion

In our system, we used a novel mathematic entorhinal-hippocampal model to build an accurate cognitive map by combining the activities of head direction cells, grid cells, place cells, and visual cues. Experiments and analysis show that the system might be simple for having fewer parameters than the system proposed before and it could generate natural activities of place cells. And generated grid patterns could be used as a weight to build a system that has no learned parameters to perform cognitive map building.

References

1. Anselmi, F., Murray, M.M., Franceschiello, B.: A computational model for grid maps in neural populations. J. Comput. Neurosci. **48**(2), 149–159 (2020)
2. Burak, Y., Fiete, I.R.: Accurate path integration in continuous attractor network models of grid cells. PLoS Comput. Biol. **5**(2), e1000291 (2009)
3. Burgess, N.: Grid cells and theta as oscillatory interference: Theory and predictions. Hippocampus **18**(12), 1157–1174 (2008)
4. Fuhs, M.C., Touretzky, D.S.: A spin glass model of path integration in rat medial entorhinal cortex. J. Neurosci. **26**(16), 4266–4276 (2006)
5. Fyhn, M., Molden, S., Witter, M.P., Moser, E.I., Moser, M.B.: Spatial representation in the entorhinal cortex. Science **305**(5688), 1258–1264 (2004)
6. Gao, R., Xie, J., Zhu, S.C., Wu, Y.N.: Learning grid cells as vector representation of self-position coupled with matrix representation of self-motion (2018)
7. Hafting, T., Fyhn, M., Molden, S., Moser, M.B., Moser, E.I.: Microstructure of a spatial map in the entorhinal cortex. Nature **436**(7052), 801 (2005)
8. Hu, J., Yuan, M., Tang, H., Yau, W.Y.: Hebbian learning analysis of a grid cell based cognitive mapping system. In: 2016 IEEE Congress on Evolutionary Computation (CEC), pp. 1212–1218. IEEE (2016)
9. Milford, M.J., Wyeth, G.F., Prasser, D.: Ratslam: A hippocampal model for simultaneous localization and mapping. In: IEEE International Conference on Robotics and Automation, 2004, Proceedings, ICRA 2004, vol. 1, pp. 403–408. IEEE (2004)
10. O'Keefe, J., Dostrovsky, J.: The hippocampus as a spatial map: Preliminary evidence from unit activity in the freely-moving rat. Brain Res. **34**(1), 171–175 (1971)
11. O'keefe, J., Nadel, L.: The Hippocampus as a Cognitive Map. Clarendon Press, Oxford (1978)
12. Solstad, T., Moser, E.I., Einevoll, G.T.: From grid cells to place cells: A mathematical model. Hippocampus **16**(12), 1026–1031 (2006)
13. Sorscher, B., Mel, G., Ganguli, S., Okko, S.: A unified theory for the origin of grid cells through the lens of pattern formation. In: NeurIPS 2019 : Thirty-third Conference on Neural Information Processing Systems, pp. 10003–10013 (2019)
14. Taube, J.S., Muller, R.U., Ranck, J.B.: Head-direction cells recorded from the postsubiculum in freely moving rats. I. Description and quantitative analysis. J. Neurosci. **10**(2), 420–435 (1990)
15. Tian, B., Shim, V.A., Yuan, M., Srinivasan, C., Tang, H., Li, H.: RGB-D based cognitive map building and navigation. In: 2013 IEEE/RSJ International Conference on Intelligent Robots and Systems, pp. 1562–1567. IEEE (2013)
16. Tolman, E.C.: Cognitive maps in rats and men. Psychol. Rev. **55**(4), 189 (1948)
17. Yuan, M., Tian, B., Shim, V.A., Tang, H., Li, H.: An entorhinal-hippocampal model for simultaneous cognitive map building. In: Twenty-Ninth AAAI Conference on Artificial Intelligence (2015)
18. Zilli, E.A., Hasselmo, M.E.: Coupled noisy spiking neurons as velocity-controlled oscillators in a model of grid cell spatial firing. J. Neurosci. **30**(41), 13850–13860 (2010)

Adaptive Risk-Return Control in Motor Planning

Qirui Yao$^{(\boxtimes)}$ and Yutaka Sakaguchi

Department of Mechanical Engineering and Intelligent Systems,
Graduate School of Informatics and Engineering,
University of Electro-Communications, Chofu, Tokyo, Japan
{qyao2017,yutaka.sakaguchi}@uec.ac.jp

Abstract. Bayesian decision-making theory presumes that humans can maximize the expected gains by trading off risk-returns in a predefined gain function. Recent findings from spatial reaching and coincident timing tasks have challenged this theory by revealing that humans exhibited risk-seeking or risk-aversive rather than risk-neutral tendency (i.e., failed to achieve Bayesian optimality) in asymmetric gain functions (the gain/loss asymmetric to the target time/position). The debate on why these participants' performances were sub-optimal remains unsettled. In the current paper, we argue that the abrupt change (i.e., gain volatility, a.k.a., risk magnitude) around the optimal point in the gain function, rather than its asymmetry, is a significant factor of this phenomenon, and that sub-optimality is resolved with an "adaptive risk control" where individual participants voluntarily adjust risk-return trade-off through a controllable task variable. We propose that the relationship between risk sensitivity and risk magnitude determines optimal motor planning.

Keywords: Bayesian decision-making · Movement planning · Adaptive risk-return control

1 Introduction

In daily life situations, we often find ourselves involved in planning motor actions intentionally or unintentionally. The available options of how to plan actions are often infinite. From which direction shall I kick the incoming ball? At which timing point shall I catch the falling book? At which speed shall I walk to cross the road? Each planned action associates with a certain risk. In economics, the expected utility theory was used to predict human decision-making behavior under risk [1]. This theory predicts that humans make decisions in a manner of maximizing the expected utility. In motor planning domains, maximizing the expected utility is equivalent to achieving the Bayesian optimality according to the Bayesian decision-making theory [2]. However, the optimality is not feasibly achieved given that the decision-making process is often susceptible to various cognitive factors, e.g., risk preference.

© Springer Nature Switzerland AG 2020
H. Yang et al. (Eds.): ICONIP 2020, LNCS 12533, pp. 15–24, 2020.
https://doi.org/10.1007/978-3-030-63833-7_2

Risk preference was often exhibited in binary decision-making scenarios, as described in Kahneman and Tversky's experiments [3]. Let us consider the following binary decisions:

Problem 1
A: 80% chance to win 4000 USD or B: win 3000 USD with certainty.
Problem 2
A: 20% chance to win 4000 USD or B: win 900 USD with certainty.

Participants were asked to make a one-shot decision over the two choices and chose a more desirable option. A large number of participants favored B in Problem 1 and A in Problem 2. Should human decision-making adhere to the expected utility hypothesis, participants would choose A in Problem 1 and B in Problem 2, because these choices yield higher expected returns. However, most participants accepted the unfavorable choice to secure 3000 USD (risk-aversive) in Problem 1, whereas they tended to be risk-seeking in hope of getting larger gains in Problem 2 [3, 4]. The contrasting results revealed that people express risk preferences as sensitivity to payoff variance. In any above cases, risk sensitivity leads to sub-optimal decisions as the overall gains are less than those brought by risk-neutral decisions.

Previous motor planning tasks often investigated the optimality of motor decisions without considering risk preference. A symmetric "two-circle" paradigm was adopted in the pioneering studies regarding spatial reaching [5–8]. "Two-circle" paradigm refers to two overlapping circles respectively defining gain/loss regions where the overlapping part separates gain and loss. Participants received points by touching the gain circle whereas losing points if the penalty circle was hit. Points gained on each trial were feed-backed and participants were asked to maximize the points received across trials. The "two-circle" paradigm allowed participants to adjust the mean end-point (MEP) to search for the optimal value. The corresponding findings supported the expected utility hypothesis, as most participants achieved Bayesian optimality (risk-neutral) [5–8]. However, Wu et al. argued that these participants might rely on the axis of symmetry as a shortcut to achieve optimality, and challenged the expected utility hypothesis by using a "three-circle" paradigm that formed a geometrically asymmetric gain/loss configuration [9]. Participants received the same reaching task to maximize the points received but most performed sub-optimally, leaving the unsolved question that whether humans could achieve optimality under asymmetric designs. Under the symmetric paradigm, the optimal value is cued by the axis of symmetry so that the risk sensitivity is overshadowed by the heuristic strategy [9]. On the other hand, Wu et al.'s asymmetric design, although ruling out the probability of implementing heuristic strategy, did not investigate the risk preference of individual participants that can affect decision-making.

A recent study by Ota et al. evaluated spatial reaching performance using a "cliff" paradigm (another type of asymmetric design) where a fixed risk-return control was implemented [10]. Under the "cliff" paradigm, the gain linearly increased along with reaching distance whilst the highest gain was located at the "cliff" edge (at a certain distance from the starting line); Overshooting the "cliff" incurred a fixed penalty. Ota et al. discovered that participants failed to achieve the optimal performance, and interestingly exhibited either risk-seeking or risk-aversive preference dependent on individual participants [10]. This indicates that the risk sensitivity does not simply stem from the task structure (i.e., the shape of gain function), but also depends on the characteristics of individuals. In other words, this type of risk preference is somewhat different from that of the prospect theory which is common to most people.

The same result was also found in motor planning under the temporal dimension. In Ota et al.'s experiments, participants were asked to reproduce the target interval in a coincident timing task [11]. The gains were proportional to the relative timing response if the responses were made before the target tone; otherwise, participants received no gains (i.e., temporal "cliff" paradigm). Participants were exposed to 9 d of practicing 2250 trials. Similar to the spatial task, they found that different risk preference (i.e., risk-seeking or risk-aversive) was exhibited at the initial phase of the trials, and this preference was mostly preserved to the final phase of the task sessions. However, examining their results in detail, we can see that the performance of most participants approached risk-neutral region as they repeated the experimental sessions (See Fig. 2 in the supplement material of [11]). This suggests that the asymmetry of gain structure is not the primary factor for the performance sub-optimality (if so, the risk sensitivity would persist even after many trials). In sum, these findings indicate that the risk sensitivity observed in an unfamiliar task can be resolved as people get familiar with the task (i.e., the statistical structure of the task is learned). In other words, long-term adaptation to the gain structure is required to reduce the individual component of the risk sensitivity.

In the current study, we discuss another approach, "adaptive risk-return control", to reduce the effect of the individual risk sensitivity instead of implementing long-term adaptation in various spatial and timing tasks. The idea of "adaptive" was firstly mentioned in speed-accuracy study by Nagengast et al. [12] but we are the first to term and applied the control to the motor planning tasks that evaluate performance optimality [13–15]. We demonstrated the Bayesian model under adaptive risk-return control, and compared it with the one mentioned in Ota et al.'s study [11,16]. We argue that risk sensitivity bringing the sub-optimality in human motor planning can be reduced by adaptive risk-return control, and through this control it is unnecessary to impose excessive practice to reach optimality.

2   Modeling

We take the coincident-timing task used in Ota et al. [11,16] as an example. Let us imagine that participants are asked to press a button after the presentation of a visual cue and receive a reward (gain) according to the response time t. Specifically, the reward is the highest when the response time t is equal to target time T (2.3 s, [11,16]), and linearly decreases as deviated from T while it is 0 if the responses are released before the target time (risk-before condition, Fig. 1a). Mathematically, the gain is defined as,

$$G(t) = \begin{cases} 0, \; t < T \\ a(t - T_m), \; T \leq t \leq T_m \\ 0, \; t > T_m \end{cases} \tag{1}$$

(where $a = -\frac{100}{T}$, $T_m = 4.6\,\mathrm{s}$, in Ota et al.'s experiment). Note that all participants receive the same gain function so that the gain function does not provide the flexibility on risk-return control.

Next, we introduce a linear gain function with an adjustable slope k,

$$G(t; \, k) = \begin{cases} 0, \; t < T \\ k(t - T_m), \; T \leq t \leq T_m \\ 0, \; t > T_m \end{cases} \tag{2}$$

Note that T_m can be adjusted together with k so that the amount of gain can satisfy task's demand (Fig. 1b). Gain functions under symmetric paradigm were plotted in Fig. 1c (symmetric linear function) and Fig. 1d (step gain function), respectively.

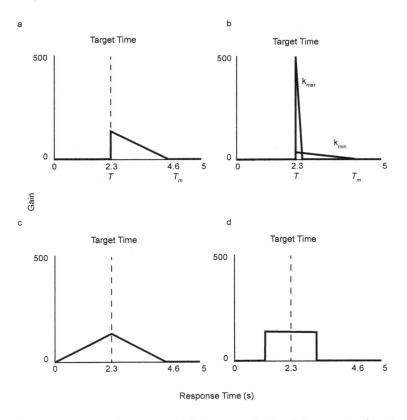

Fig. 1. Gain function configurations. (a) Asymmetric fixed linear gain function; (b) Asymmetric linear gain function with an adjustable slope where the two functions demonstrate two extreme situations of risk-return trade-off; (c) Symmetric linear gain function; (d) Step gain function; Figure a, c, d were retrieved and modified from [16].

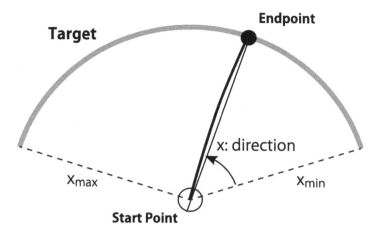

Fig. 2. Stimulus configuration in a planned design. Participants move a cursor from the start point to the green curve and click the mouse button at an appropriate timing. The slope of the gain functions is determined by the movement direction.

To access the adaptive control, several requirements need to be considered. Firstly, the key point of adaptive control lies in the sensitivity of the gain function (i.e., risk-return trade-off). To prepare the risk-return control, the selected gain function needs to have a spread parameter (such as k in the above example). For an example of using Gaussian gains, see [13,14]. Secondly, a task variable is required to allow direct control on the spread parameter. The variable needs to be measured simultaneously with the target variable (for example, movement direction while performing the reaching task where reaching distance is the target variable). Thirdly, a mapping between the spread parameter and the task variable is required to implement the control (see [13,14] for an example of linear mapping). In the above case, the mapping relation ensures that participants can choose a larger $|k|$ if they prefer higher risk sensitivity and by the same token for participants who prefer lower risk sensitivity can choose a smaller $|k|$.

Figure 2 shows an illustrative implementation of this control in a planned design. In this setting, participants perform a spatial reaching task by moving the cursor from the start point to the green curve and click the mouse button. The gain is determined by the reaching distance, whilst the spread parameter of the gain function is determined by the click position. Therefore, participants can control the risk-return trade-off of the task by choosing an appropriate movement direction. In the next section, we will argue that how the risk sensitivity could be diminished by this adaptive risk-return control.

3   Model Evaluation

The expected gain function in Ota et al.'s model is calculated by integrating the distribution of timing response $p(t|t_e)$ over Eq. (1)

$$EG(t_e) = \int G(t)p(t|t_e)dt \qquad (3)$$

The shape of the expected gain is determined by the mean timing response and the timing variance, expressed as (t_e, σ_t).

On the other hand, the expected gain function of the adaptive risk control is given by:

$$EG(t_e;\ x_e) = \int_0^\infty p\,(x|x_e)\int_0^\infty G\,(t;\ x)\,p\,(t|t_e)\,dtdx \qquad (4)$$

where x is the task variable determining the slope parameter of the gain functions. The adaptive risk-return model includes four parameters: t_e, σ_t, x_e, σ_x, where we assumed $p(x|x_e)$ obeys Gaussian distribution $N\,(x_e,\ \sigma_x)$, meaning the variable x fluctuates around the planned value x_e, and $p(t|t_e)$ also follows Gaussian distribution $N\,(t_e, \sigma_t)$, as modeled in previous studies [11, 16].

In the following demonstration regarding Eq. (2), we construct a linear mapping by letting $k = -\frac{5}{3}(mx+n)^{-2}$, $T_m = T - \frac{5}{k}(mx+n)^{-1}$, where m and n were constants ($m = 0.098$, $n = 0.01$). In order to visualize the model characteristics, we varied x_e from 0 to 5, (σ_t fixed at 0.2 s, σ_x at 0.1). As seen in Fig. 3, with a greater x_e, the contour intervals become wider (i.e., gain becomes less sensitive to the change in mean response time t_e). This indicates that participants can adjust the gain structure by choosing appropriate values of x_e. In other words, the adaptive control allows participants to either choose a smaller x_e to increase the variability of gains (high-risk high-return) or to choose a larger x_e if high variability of gains is unfavorable (low-risk low-return).

As Ota et al.'s results showed [10, 11, 16], the risk preference of individual participants was exhibited under the "cliff" paradigm, where the gain volatility is extremely high at the edge of the "cliff". To measure the volatility, we employed the standard deviation of gains as a proper index [17].

Regarding the gain functions in Ota et al.'s experiments and in the adaptive risk-return model, we set $\sigma_t = 0.05$ s, $\sigma_r = 0.05$ and $x_e = 0.1, 1.3, 2.5, 3.7, 4.9$ respectively, and simulated the corresponding standard deviations of gains (each SD computed from 1000 simulated trials) for every t_e ranging from 2 to 4.5 s.

The volatility under symmetric paradigm (Fig. 4c, d) remains extremely low at the axis of symmetry (i.e., $t_e = 2.3$ s). Particularly, even though gains are choppy at the edge of the step gain function, the gain volatility remains 0 around the axis of symmetry where the optimality value is located. In these conditions, no risk sensitivity could occur. Therefore, the optimality is easily to achieve, given low volatility around the optimality value.

On the other hand, the "cliff" replaces the axis of symmetry under the asymmetric paradigm. Note that the volatility (i.e., risk magnitude) abruptly changes around the "cliff" (Fig. 4a, b) and the potential location of optimal value, which, would lead to the risk sensitive to the volatility and cause sub-optimal performance. There are two possible approaches to reduce the risk sensitivity: One is through adaptation to the experimental task as indicated by Ota et al.'s design [11], whereas the other is to arbitrarily manipulate the gain volatility suggested by our adaptive model.

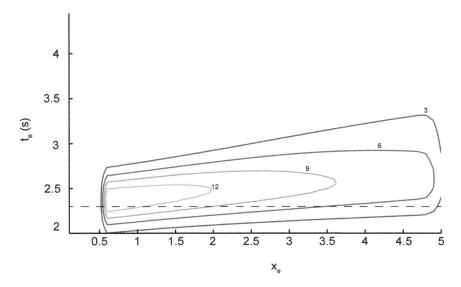

Fig. 3. The expected gains given $\sigma_t = 0.2\,\mathrm{s}$. The expected gains were computed taking Eq. (2) as gain function where the gain structure could be adjusted by x_e. The dashed line represents the "cliff".

Under Ota et al.'s model, the gain structure was fixed at a pre-experimentally defined shape [10,11,16]. When the gain structure is volatile at the cliff edge (Fig. 4a), participants are facing a high volatility and become risk sensitive. However, they are unable to exert direct control over the risk magnitude but can only adjust the response time, e.g., to shift the timing responses towards or away from the "cliff", resulting in risk-seeking or risk-aversive bias. This approach was proved to be less effective and required excessive trials of practice, in order to approximate the optimal performance [11].

On the other hand, the adaptive risk-return control model scales the gain variance on a spatial dimension, which forms a mapping relation between the gain structure and the risk sensitivity (so that participants can manipulate the risk magnitude). Under this model, the risk control is operated by integrating two statistical information: adjusting the response time to find the possible location of optimal value whilst adjusting the corresponding risk magnitude to modify the risk sensitivity. Therefore, the risk reference is not exhibited when the risk magnitude is consistent with the individual risk sensitivity. The adaptive risk-return control works faster and the optimality is easier to achieve, as we revealed that participants only needed around 150 trials (much less than 2250 trials) to find out the optimal value [13–15].

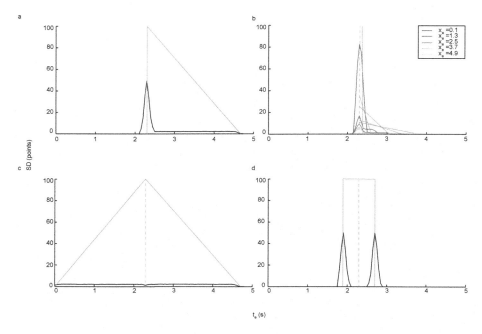

Fig. 4. The volatility of gain in Ota et al.'s experimental conditions [16] and adaptive risk-return model. The dashed line represents "cliff" in Graph a and b, and "the axis of symmetry" in Graph c and d. (a) The volatility around the "cliff" in asymmetric gain condition is highest, Fig. 1a; (b) The gain volatility under adaptive risk-return model. The volatility can be adjusted according to different values of x_e, Fig. 1b; (c) The volatility in symmetric gain condition is lowest as the curve tends almost flat, Fig. 1c; (d) The volatility remains 0 around the axis of symmetry but high at the edge of the step gain function, Fig. 1d. The corresponding gain functions were plotted by grey lines.

4   Summary and Discussion

The current paper reviewed the unresolved issue regarding risk control in various spatial and temporal decision-making tasks. Then, we proposed an adaptive risk-return control as a modification of the previous model. The adaptive control allows a direct manipulation of the gain structure by a task variable. The participant-dependent risk sensitivity could be adaptively matched by the corresponding payoff variance, and therefore the risk sensitivity is diminished and tends risk-neural. Our previous research has proved the effectiveness of adaptive risk-return control, as participants could achieve Bayesian optimality under an asymmetric paradigm in a coincident timing task [15].

The implementation of adaptive control requires repeated feedback. It was discovered that providing feedback could enhance task performance [18–20]. In Kahneman and Tversky's classical binary choice tasks, no feedback was provided [3]. Participants analyzed the expected returns and made the decisions with probability distortions. On the other hand, receiving repeated feedback reduced subjective distortion on low probability events, known as *decisions from experience* [21,22]. It might be arguable that whether it is the repeated feedback rather than the adaptive control that leads to performance optimality. However, so far to our best knowledge, all studies related to motor planning require repeated trial-by-trial feedback on the task performance, including our adaptive risk-return model [13–15]. Further researches are required to investigate the adaptive risk-return model under the feedback provided on different time schedules.

Acknowledgments. The research was supported by the fundamental fund for research and education of graduate students to Y. Sakaguchi from the University of Electro-Communications.

Author contributions. Q. Yao performed the simulations and analyzed the results. Q. Yao prepared the original draft. Y. Sakaguchi commented and revised the manuscript.
Conflict Interests. The authors declare no competing financial interest.

References

1. Schoemaker, P.J.: The expected utility model: Its variants, purposes, evidence and limitations. J. Econ. Lit. **20**, 529–563 (1982)
2. Weise, K., Woger, W.: A Bayesian theory of measurement uncertainty. Meas. Sci. Technol. **4**(1), 1–11 (1993). https://doi.org/10.1088/0957-0233/4/1/001
3. Kahneman, D., Tversky, A.: Prospect theory: An analysis of decision under risk. Econometrica **47**(2), 263 (1979). https://doi.org/10.2307/1914185
4. Kahneman, D., Tversky, A.: Prospect theory: An analysis of decision under risk. In: Handbook of the Fundamentals of Financial Decision Making: Part I, pp. 99–127 (2013)
5. Maloney, L., Landy, M., Trommershäuser, J.: Statistical decision theory and trade-offs in the control of motor response. Spat. Vis. **16**(3), 255–275 (2003). https://doi.org/10.1163/156856803322467527
6. Trommershäuser, J., Maloney, L.T., Landy, M.S.: Statistical decision theory and the selection of rapid, goal-directed movements. J. Opt. Soc. Am. A **20**(7), 1419 (2003). https://doi.org/10.1364/josaa.20.001419
7. Trommershäuser, J., Gepshtein, S., Maloney, L.T., Landy, M.S., Banks, M.S.: Optimal compensation for changes in task-relevant movement variability. J. Neurosci. **25**(31), 7169–7178 (2005)
8. Trommershäuser, J., Landy, M.S., Maloney, L.T.: Humans rapidly estimate expected gain in movement planning. Psychol. Sci. **17**(11), 981–988 (2006). https://doi.org/10.1111/j.1467-9280.2006.01816.x
9. Wu, S.W., Trommershäuser, J., Maloney, L.T., Landy, M.S.: Limits to human movement planning in tasks with asymmetric gain landscapes. J. Vis. **6**(1), 5 (2006). https://doi.org/10.1167/6.1.5
10. Ota, K., Shinya, M., Maloney, L.T., Kudo, K.: Sub-optimality in motor planning is not improved by explicit observation of motor uncertainty. Sci. Rep. **9**(1), 1–11 (2019). https://doi.org/10.1038/s41598-019-50901-x

11. Ota, K., Shinya, M., Kudo, K.: Sub-optimality in motor planning is retained throughout 9 days practice of 2250 trials. Sci. Rep. **6**(1), 37181 (2016). https://doi.org/10.1038/srep37181
12. Nagengast, A.J., Braun, D.A., Wolpert, D.M.: Risk sensitivity in a motor task with speed-accuracy trade-off. J. Neurophysiol. **105**(6), 2668–2674 (2011). https://doi.org/10.1152/jn.00804.2010
13. Yao, Q., Sakaguchi, Y.: Humans achieve bayesian optimality in controlling risk-return tradeoff of coincident timing task. In: Proceedings of JNNS2018, pp. 24–25 (2018)
14. Yao, Q., Sakaguchi, Y.: Humans achieve Bayesian optimality in controlling risk-return tradeoff of spatial reaching task. In: Proceedings of JNNS2019, pp. 59–60 (2019)
15. Yao, Q., Sakaguchi, Y.: Optimizing motor timing decision through adaptive risk-return control (2020, submitted)
16. Ota, K., Shinya, M., Kudo, K.: Motor planning under temporal uncertainty is suboptimal when the gain function is asymmetric. Front. Comput. Neurosci. **9**(9), 88 (2015). https://doi.org/10.3389/fncom.2015.00088
17. Aumann, R.J., Serrano, R.: An economic index of riskiness. J. Polit. Econ. **116**(5), 810–836 (2008). https://doi.org/10.1086/591947
18. Lejarraga, T., Gonzalez, C.: Effects of feedback and complexity on repeated decisions from description (2011). https://doi.org/10.1016/j.obhdp.2011.05.001
19. Neyedli, H.F., Welsh, T.N.: People are better at maximizing expected gain in a manual aiming task with rapidly changing probabilities than with rapidly changing payoffs. J. Neurophysiol. **111**(5), 1016–1026 (2014). https://doi.org/10.1152/jn.00163.2013
20. Neyedli, H.F., Welsh, T.N.: Optimal weighting of costs and probabilities in a risky motor decision-making task requires experience. J. Exp. Psychol. Hum. Percept. Perform. **39**(3), 638–645 (2013). https://doi.org/10.1037/a0030518
21. Hertwig, R., Barron, G., Weber, E.U., Erev, I.: Decisions from experience and the effect of rare events in risky choice. Psychol. Sci. **15**(8), 534–539 (2004). https://doi.org/10.1111/j.0956-7976.2004.00715.x
22. Jessup, R.K., Bishara, A.J., Busemeyer, J.R.: Feedback produces divergence from prospect theory in descriptive choice. Psychol. Sci. **19**(10), 1015–1022 (2008). https://doi.org/10.1111/j.1467-9280.2008.02193.x

Discrete Mother Tree Optimization for the Traveling Salesman Problem

Wael Korani⬤ and Malek Mouhoub$^{(\boxtimes)}$⬤

University of Regina, Regina, Canada
{wmk182,mouhoubm}@uregina.ca

Abstract. The Mother Tree Optimization (MTO) algorithm is a new swarm intelligence technique that we have recently proposed for solving continuous optimization problems. MTO is built on an offspring topology and a set of cooperating agents. In this paper, we first present a discrete version of MTO, that we call Discrete MTO (DMTO), for solving the Traveling Salesman Problem (TSP). DMTO is based on a new swap operation that is best suited for TSPs. We also used this swap operation to introduce an updated version of the Discrete Particle Swarm Optimization (DPSO) algorithm. With a careful application of our new swap operation, we will show that our Updated DPSO (UDPSO) is more effective than DPSO. In order to assess the performance of our proposed methods, DMTO and UPSO, we conducted several experiments comparing both to DPSO and an exact method (Branch and Bound), on ten TSP instances taken from the well-known TSPLIB dataset. The results clearly show that DMTO produces solutions of much better quality than in DPSO, and superior to those in UDPSO.

Keywords: Mother Tree Optimization · Particle Swarm Optimization · Combinatorial optimization · Traveling Salesman Problem

1 Introduction

Discrete optimization problems play a crucial role in many real-world applications, so that many scientific methods have been proposed to tackle them effectively [22]. These proposed methods are classified into exact and approximate techniques. Approximate methods include the following nature-inspired techniques: Particle Swarm Optimization (PSO) [15], Ant Colony Optimization (ACO) [8], Genetic Algorithms (GAs) [6], and Artificial Bee Colony (ABC) [16]. However, these methods have some drawbacks, such as complicated fitness function, premature convergence, and the challenge with tunable parameters. More recently, we have proposed the MTO algorithm, a population-based optimization method, that is inspired from the symbiotic relationship between Douglas fir trees and the mycorrhizal fungi network [18]. The fungi network allows plants to transfer nutrients between plants of same and different species, without any

© Springer Nature Switzerland AG 2020
H. Yang et al. (Eds.): ICONIP 2020, LNCS 12533, pp. 25–37, 2020.
https://doi.org/10.1007/978-3-030-63833-7_3

intervention. In addition, the fungi networks allow plants to supply fungi with carbohydrates that are essential for survival and growth. The fungi will then help their host plants to maximize the transfer of nutrients via their root system. The forest environment system includes different kinds of plants and fungi networks that are connected together in a way that allow them to exchange essential nutrients using what we call a cooperative system. This system allows plants from same and different species to communicate, exchange different essential nutrients, and protect each other [3]. Douglas fir trees communicate, protect, and help their kin and other trees by transferring different nutrients [1]. We claim that the cooperative system has three different subsystems: recognition, measurement, and defense [18].

The TSP is a well-know discrete optimization problem that consists of finding the shortest possible round trip through a given set of cities. Starting from a given city, the salesman should visit each other city once before returning to the starting city. The TSP is represented using a weighted graph $G = (V, E)$, where V is a set vertices (cities) and E is a set of edges that fully connects the nodes in the graph G. Each edge e in E has a weight $d(C_i, C_j)$ representing the distance between city i and city j. The distance between cities can either be symmetric or asymmetric. Finding the shortest tour for the TSP consists of looking for a sequence $\pi = C_1, C_2, C_3 \ldots, C_n$, where C_n corresponds to the nth selected city, minimizing the total distance between all cities. This total distance is expressed with the function, f, defined as follows.

$$f(\pi) = d(C_n, C_1) + \sum_{i=1}^{n-1} d(C_i, C_{i+1}) \tag{1}$$

There are many exact and nature-inspired techniques [29] that have been proposed for solving the TSP. Branch and bound [26,28], branch and cut [23], cutting plane [10], and dynamic programming [5] are examples of exact methods that have been developed to solve small and medium TSP instances. In [19,25], GAs are implemented to produce acceptable solution within a reasonable time frame as an example of evolutionary computation algorithms. Tabu search [11], simulated annealing [12], and neural networks [20] were also considered for solving TSPs. TSPs have also been tackled using Ant Colony Optimization (ACO) [2,14], Bee Colony Optimization [21] and ABC [17]. In [15,24], PSO has been adopted to solve the TSP using some basic operations. Some other approaches combining PSO with simulated annealing [9], PSO with ACO [13], and PSO with genetic simulated annealing and ACO [4] have also been considered.

These proposed methods are built on either a path construction or a path improvement strategies. Path construction consists of extending a partial solution into a complete and optimal one. More precisely, this method starts with one or more cities and adds other cities one by one until a complete tour is constructed. On the other hand, the path improvement approach starts with a complete (solution) tour and improves it at each iteration until no improvement can be done [30]. The path improvement techniques produce fast results, but do not often return a good solution in a reasonable time.

In this paper, we propose a discrete version of MTO (belonging to the path improvement category), that we call Discrete MTO (DMTO), for solving the TSP. DMTO is based on a new swap operation that is best suited for TSPs. We also used the same swap operation to introduce an updated version of the discrete variant of PSO (DPSO). The Updated DPSO (UDPSO) that we propose, is superior to DPSO, thanks to our new swap operation. In order to assess the performance in practice of our methods, we conducted several experiments comparing DMTO, DPSO, UDPSO, and Branch and Bound, on ten TSP instances taken from the well-known TSPLIB dataset. The results clearly show that DMTO produces solutions of much better quality than in DPSO and UDPSO. In case of Oliver TSP instance, the obtained relative error when using DMTO is decreased 194 times less than DPSO and 33 times less than UDPSO.

2   Discrete Mother Tree Algorithm (DMTO)

The MTO algorithm is based on a fixed-offspring topology [18], where agents update their positions in the search space according to the group to which they belong. The population is a set of Active Food Sources (AFSs) whose size is denoted as N_T, and it is divided into a the TMT (the agent receiving nutrients from a random source), the Partially Connected Trees (PCTs) group that has N_{PCTs} agents, and the Fully Connected Trees (FCTs) group that has N_{FCTs} agents. The numbers of agents within the PCTs and FCTs groups are given by the following equations.

$$N_{PCTs} = N_T - 4,$$
$$N_{FCTs} = 3,$$
$$N_T = N_{FCTs} + N_{PCTs} + 1. \quad (2)$$

2.1   Solution Representation

DMTO starts the search from a candidate solution (complete tour) corresponding to a given sequence of all the cities. The candidate solution corresponds to a vector of size n, where n is the total number of cities, and each entry corresponds to a different city. The fitness function corresponds to the total distance along the tour. Figure 1 shows a weighted graph (left) and a candidate solution (middle), with a fitness value of 25.

2.2   Swap Operation

During the DMTO search, a candidate solution moves from one position to another, in the search space, through swap operations. Each agent i corresponds to a given candidate solution (TSP tour). An agent is updated at each kin recognition signal (corresponding to an iteration) using a Swap Operation (Ⓢ), as shown in Fig. 1. In this regard, there are $\frac{n(n-1)}{2}$ possible swap operations that can be applied to any candidate solution. This number corresponds to the neighbourhood size (maximum number of possible neighbours) for each agent. The actual number of neighbours, to consider for each agent, is within the following range $[n : \frac{n(n-1)}{2}]$, and depends on the group that the agent belongs to. For instance, TMT has $\frac{n(n-1)}{2}$ neighbors, and the agent ranked 2nd (belonging to PCTs) has n neighbors only (as we see later in this Section). All possible swap operations are stored in a pool called Basic Swap Operation Pool (BSOP). In DMTO, the feeder influence and distance are represented by the number of swap operations. A swap operation removes four edges and adds four new ones in the fully connected graph. In our previous example, the swap operation results in adding two new edges $\{(A \rightarrow C), (E \rightarrow B)\}$ and removing two others, from the original solution: $\{(B \rightarrow C), (E \rightarrow A)\}$, as shown in Fig. 1. Thus, edge $(E \rightarrow A)$ is replaced with edge $(A \rightarrow C)$, and edge $(B \rightarrow C)$ is replaced with edge $(E \rightarrow B)$. The difference in fitness values (before and after performing the swap operation) can be calculated by considering those added and removed edges i.e. $(((3 + 7) + (3 + 4)) - ((3 + 3) + (3 + 4)))$, as shown in Fig. 1.

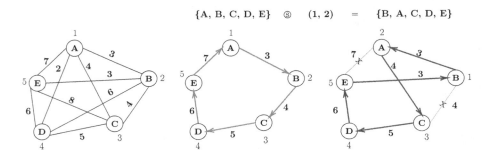

Fig. 1. An example of a TSP

There is a special swap operation pool called Supper Swap Operation Pool (SSOP) that has the capability to transform one solution to another using a successive number of SOs. The SSOP is used for all agents except the TMT to select a random SO to update an agent's position. A SSOP is a mediator swap operation pool between two solutions: the influencer solution and the influenced solution. Applying all the SOs, in order, of a given SSOP will transform the influenced solution into an influencer solution. For instance, assume we have two solutions {B, C, D, A, E} (Influencer) and {D, A, C, B, E} (influenced). The corresponding SSOP will be: $\{(1, 4), (2, 3), (3, 4)\}$. This pool can transform the influenced solution into the influencer solution by applying all SOs in order. Each DMTO run has a number of iterations called kin recognition K_{rs}. Note that the SSOP can have a maximum of n SOs (given that we need a maximum of n swaps to go from any sequence to another).

2.3   Updating the TMT

The TMT performs an exploration process at each K_{rs} iteration by searching among its neighbors for a better solution. The TMT has two levels of exploration: L_1 and L_2. Each level of exploration has N_{os} iterations. In the first iteration of L_1, the TMT computes the difference in the fitness values for each member of the BSOP. The resulted cost difference will be transferred to a range between [0 : 1] using the sigmoid function given by:

$$f(x) = \frac{1}{1 + e^{-x}},\tag{3}$$

where x is the difference in fitness values, and $f(x)$ represents the probability of a swap operation to be selected. The higher the sigmoid value the higher the chance of its associated swap operation to be selected. Then, the best swap operation is obtained as follows.

$$P^k_{L_1(n+1)} = (P^k_{L_1(n)} \; \text{Ⓢ} \; SOP^{best}_n)\tag{4}$$

where, $P^k_{L_1(n)}$ is the position of TMT at iteration k and at iteration n at level L_1, $P^k_{L_1(n+1)}$ is the updated solution at $n+1$ iteration , and SOP^{best}_n is the best swap operation in the current SOP_n at iteration n. If the updated solution of the TMT does not achieve a better fitness value, the TMT randomly picks and removes an SO from the current $BSOP_n$. Then, the TMT goes to the next iteration in L_1 (it will not pass though the second exploration level L_2). However, if the updated fitness value in L_1 is better than the current one, then this SOP^{best}_n will be removed from the current $BSOP_n$. Then, the TMT agent switches to the next exploration level L_2. The main reason to remove any SO after it is selected, is to avoid being trapped in a local minimum. In this regard, we have conducted several preliminary experiments, and the results show that removing every SO (after selection) improves significantly on the quality of the solution. In L_2, the same process is repeated for all members of the current $BSOP_n$ and the best SO that has the highest sigmoid value will be selected. This process is reflected by the following equation.

$$P^k_{L_2(n+1)} = (P^k_{L_2(n)} \; \text{Ⓢ} \; SOP^{best}_n)\tag{5}$$

The TMT agent continues repeating the same process in L_2 until the maximum number of iterations is reached. Then, the TMT witches again to the first level L_1 and repeats the same process until it reaches the maximum number of iterations.

There is a TMT special state called Sigmoid Deadlock (SD) situation. The SD state happens when several swap operations, form a pool called Swap Operation Deadlock Pool (SODP), achieve the same cost difference, and then the same sigmoid values. TMT resolves the deadlock situation by selecting the best agent in SODP that will achieve the highest sigmoid value for the next iteration.

2.4   Updating the Solutions of the FPCTs Group

The FPCTs group has a defense mechanism that helps any of its member to avoid being supported by a bad food source. Thus, if any member of this group gets an updated solution that has a worse fitness value than its current one, then it will pick a random SO from BSOP \ SSOP. The updated position is calculated as follows.

$$P_n^{k+1} = \sum_{i=1}^{n-1} \frac{1}{n-i+1}(P_n^k \, \text{\textcircled{S}} \, shuffle(SSOP_i)), \qquad (6)$$

where P_n^k represents the current solution of any member in range $[2 : \frac{N_T}{2} - 1]$, $shuffel$ is a function to change the order of the swap operations in the $SSOP_i$, $SSOP_i$ is the super swap operator pool of agent i to cause influence on agent n, and $\frac{1}{n-i+1}$ represents the probability of selecting a SO from $SSOP_i$. However, if any member of this group uses the defense mechanism, it picks a random SO as explained before.

Solution agent i is used to create the $SSOP_i$ by targeting solution n. Then, all members of $SSOP_i$ is implemented in a random way on solution n, and the sigmoid values are calculated as explained earlier. The probability of selecting a SO from $SSOP_i$ depends on its probability level. For example, the solution ranked 2nd is influenced only by the TMT with probability level $\frac{1}{2}$ [18]. Thus, $SSOP_1$ is generated using the TMT to influence the solution ranked 2nd. Each member of $SSOP_1$ is implemented on solution ranked 2nd in a random way, and then its sigmoid value is calculated. If the sigmoid value is higher than the probability level $(1 - \frac{1}{2})$, then the associate swap operation will be implemented and the solution will be updated; otherwise, the swap operation will be ignored. After implementing all good swap operation of $SSOP_1$ and the final updated solution is worse than the beginning solution, then the defense mechanism is implemented as explained before.

2.5   Updating the Solutions of FCTs Group

All members of the FCTs group are in the range $[\frac{N_T}{2} : \frac{N_T}{2} + 2]$. The updated solution is given by the following equation.

$$P_n^{k+1} = \sum_{i=n-N_{os}}^{n-1} \frac{1}{n-i+1}(P_n^k \, \text{\textcircled{S}} \, shuffle(SSOP_i)). \qquad (7)$$

For instance, in case of a population of size six, the 3rd solution is influenced by the TMT and the 2nd solution. Thus, two SSOPs are created. $SSOP_2$ is created using the solution ranked 2nd to influence the solution ranked 3rd with probability level $\frac{1}{2}$, and $SSOP_1$ is created to influence solution ranked 3rd with probability level $\frac{1}{3}$. When a sigmoid value resulted from $SSOP_2$ is higher than $(1 - \frac{1}{2})$ the associated SO will be selected and implemented on solution ranked 3rd. In addition, any swap operation associated with a sigmoid value higher than $(1 - \frac{1}{3})$ and located in $SSOP_1$ will be selected and implemented on the 3rd solution.

2.6   Updating the Solutions of LPCTs Group

LPCTs is the last group of candidate solutions in the population, and their members have the least number of nutrients in the population (worst fitness values). More precisely, LPCTs members are in the range $[\frac{N_T}{2} + 3 : N_T]$. Each LPCTs member is updated as follows.

$$P_n^{(k+1)} = \sum_{i=n-N_{os}}^{N_T-N_{os}} \frac{1}{n-i+1}(P_n^{(k)} \text{Ⓢ} shuffle(SSOP_i)). \tag{8}$$

2.7   DMTO Climate Change

The climate change is a diversification phoneme that helps DMTO find better solutions. The number of climate change events is denoted by Cl and each climate change happens once every cycle (a group of kin recognition signals) and corresponds to a distortion process. More precisely, the population is distorted by a certain level, called deviation level, that depends on the number of swap operations. The deviation level is adopted based on preliminary experiments to three random successive swap operation.

3   Discrete Particle Swarm Optimization for TSPs

3.1   Discrete Particle Swarm Optimization (DPSO)

In [15], Sarman et al. introduced a discrete version of the PSO (DPSO), and tested the algorithm for one instance (five cities). The proposed variant is built on swap operations. The population, of size n, is initialized in a random way, and the velocity corresponds to a random swap sequence. The particle position is updated as follows.

$$X_{id}^{k+1} = X_{id}^k + V_{id}^{k+1} \tag{9}$$

$$V_{id}^{k+1} = \omega V_{id}^k \oplus \alpha(P_{id} - X_{id}) \oplus \beta(Pgd - X_{id}) \tag{10}$$

α and β are random numbers in the range $[0 : 1]$. $(P_{id} - X_{id})$ produces a pool of swap operations (called the basic swap sequence) in the same way we produce the super swap operator. The cognitive term $\alpha(P_{id} - X_{id})$ means that each swap operation in the basic swap sequence is accepted or rejected based on a probability α. In other words, a random number is generated in the range $[0 : 1]$, and if this number is below α, it will be accepted otherwise it will be rejected, and its associated swap operation will not be used. The social term $\beta(Pgd - X_{id})$ is calculated in the same way as the cognitive term. The result of the velocity equation produces a pool of swap operations (by merging three pools). These swap operations are then applied to update the particle position. The cost function of each particle is calculated at each iteration, and the value of both P_{id} and P_{gd} are updated until the maximum number of iterations is reached [15].

DPSO has some drawbacks resulting in the production of very poor performance, when the number of cities is more than five. The authors explained the cognitive and social terms and how to calculate them. They did not however mention anything about the ωV_{id}^{k} term, and ignored this term in their implementation by assuming that it is equal to zero. In each iteration, the authors created a new velocity pool for each particle based on the updated cognitive and social terms. This contradicts with the basic concept of PSO where the velocity equation is updated by adding a part of the old velocity to the social and cognitive terms.

The authors basic idea is to apply a sequence of swap operations on a solution (particle) without any other selection criteria to improve the fitness value of the solution. Thus, after implementing their created velocity pool of the swap operations the result may produce a solution with lower fitness value [15]. In most cases, after applying all swap operations, the solution ends with a bad fitness value as every swap operator can have a significant effect on the solution. The velocity pool is updated in two cases: a solution is different than the local best and/or global best. The authors actually did not think about the global best solution that will not change, because it has the best local and the best global at the same time. This was an unwise decision to leave the global best solution without any updates.

3.2 Updated Version of Discrete Particle Swarm Optimization (UDPSO)

The DPSO limitations we listed in Sect. 3.1 have motivated us to introduce our UDPSO. In UDPSO, we assume that we have a velocity pool with $3n$ swap operations, where n is the number of cities. In the first iteration k, we create a random pool V_{id}^{k} (a group of $3n$ swap operations, that we call $Vpool$, we select the group size $3n$ after conducting preliminary experiments). In the next iteration, $k+1$, we firstly calculate the cognitive and social swap operations called Lpool and Gpool respectively. Lpool is a generated SSOP between the local best solution of the current agent and the current solution, and we calculate the SSOP in the same way that we use in our DMTO. Gpool is a generated SSOP between the global best solution and the current solution, and we calculate the SSOP in the same way that we use in our DMTO. The number of swap operations in any SSOP is equal to n, and we add only the unique items in the pool. Thus, the number of swap operations in Lpool and Gpool is less than or equal to $2n$. Then, we add ten random swap operations to V_{id}^{k+1}. Finally, we add the rest of the swap operations to reach $3n$ from V_{id}^{k}. More precisely, the velocity pool is calculated as follows.

$$V_{id}^{k+1} = Part(V_{id}^{k}) + Lpool + Gpool + SO10 \qquad (11)$$

where $SO10$ is a set of ten random swap operators. In iteration $k+1$, the new velocity is initially the sum of the number of swap operations in Lpool and Gpool. Then, we add random extra swap operations from $Part(V_{id}^{k})$ until the number of swap operations in the V_{id}^{k+1} reaches $(3n - 10)$. Finally, we add ten random swap operations to the V_{id}^{k+1} to reach $3n$. Note that all swap operators in Vpool are unique. We use the sigmoid values of all swap operations in the Vpool as an indicators to evaluate each swap operation. Then, the swap operation associated with the best sigmoid value is selected and applied on the current solution. Then, we calculate the sigmoid values for the same Vpool with the new updated solution to see if we can further improve the solution. We keep repeating this process until there is no way to improve this solution.

4   Experimentation

The timeout set for all the algorithms is set to five hours, for all the data sets. In the case of DPSO and UDPSO, the number of agents in the population is equal to the number of cities (n) of the TSP instance. In case of DMTO, the number of agents is set to 10; the kin recognition signal (K_{rs}) is set to 25; and the climate change, Cl, depends on the number of cities. More precisely, $Cl = \frac{S \cdot n}{10 \cdot 25}$. Branch and Bound (BAB) is implemented as a chronological backtrack search algorithm, that visits each node and explores it depending on the values of the Lower Bound (LB) and Upper Bound (UB). Here, LB corresponds to half of the sum of the 2 edges with the least cost, adjacent to each node. UB is the cost of the best solution found so far. The parameters of DPSO, UDPSO, and DMTO have been tuned after conducting extensive preliminary experiments. We have used the following TSP instances to conduct our experiments: Oliver30, Eil51, Berline52, St70, Pr76, Eil76, Kroa100, Eil101, Ch150, and Tsp225. These TSP instances were used in a recent research to compare ACO, ABC, and a hierarchic approach [14]. The optimum lengths of the TSP instances are obtained from TSPLIB [27]. The TSPLIB has been published since 1991 and includes a collection of TSP benchmark instances with different levels of difficulty. This library has been used in many research works for algorithm performance evaluation.

For each instance, we conducted 20 runs and report the best solution, the worse solution, the average (Avg.), the standard deviation (Std. dev.) and the relative error (RE), as shown in Table 1 and Fig. 2. RE is the ratio of the absolute error (difference between the best and the actual value) by the actual value, All the comparative results are reported in Fig. 2. The branch and bound method could only solve the first instance in the experiment allocated time. As expected, RE of DPSO is very high in all tested TSP instances. This is explained by the drawbacks of this method, as we reported earlier. Figure 2 shows that DPSO has the highest relative error among all three algorithms.

Table 1. Comparative Results

Instance	Method	Best	Worse	Avg.	Std. dev.	RE%
Oliver30	DPSO	685	866	767.4	48.41	81.84
	UDPSO	462	506	481.7	11.18	14.14
	DMTO	420	422	420.2	0.509	0.42
	BAB	420	420	420	0.00	**0.0**
Eil51	DPSO	962	1126	1046.7	39.743	145
	UDPSO	519	559	542.15	10.65	27.43
	DMTO	434	452	442.85	4.574	**3.95**
Berlin52	DPSO	15482	20040	18120.35	1242.55	140.25
	UDPSO	8644	9768	9325.2	304.07	23.64
	DMTO	7676	8194	7941.85	161.9	**5.3**
St70	DPSO	2077	2634	2297.35	149.764	240.34
	UDPSO	989	1057	1023.8	20.05	51.67
	DMTO	694	731	710.15	10.35	**5.20**
Eil76	DPSO	1564	1931	1724.25	98.89	220.49
	UDPSO	740	817	785.4	18.44	45.98
	DMTO	557	589	573.9	9.142	**6.67**
Pr76	DPSO	333033	419861	372852.55	24125.48	244.73
	UDPSO	148438	165467	158975.85	3906.65	46.98
	DMTO	114329	122206	118554.15	2333.01	**9.61**
Kroa100	DPSO	93629	122484	108342.4	7705.67	409.0
	UDPSO	37726	42611	39952.5	1183.87	87.72
	DMTO	22033	26350	23906.3	997.82	**12.33**
Eil101	DPSO	2180	2859	2421.7	169.99	285.0
	UDPSO	1017	1091	1055.75	20.81	67.84
	DMTO	658	703	681.95	10.753	**8.41**
Ch150	DPSO	33847	44681	38909.35	2555.98	496.04
	UDPSO	13994	15185	14532.35	262.540	122.61
	DMTO	7056	8183	7524.5	305.48	**15.26**
Tsp225	DPSO	26128	33431	30953.7	1895.78	690.44
	UDPSO	10520	11468	11069.5	208.04	182.67
	DMTO	4546	4959	4754.75	135.60	**21.41**

Table 1 lists all the resutls, with the lowest RE marked in bold. UDPSO is significantly superior to DPSO. In general, RE is reduced between 3.8 times to 6 times. For instance, the relative error is reduced six times when solving Berlin52. However, when we increase the number of cities to 225, the relative error is reduced from 690.44% to 182.67%, which is 3.8 times. DMTO achieves much

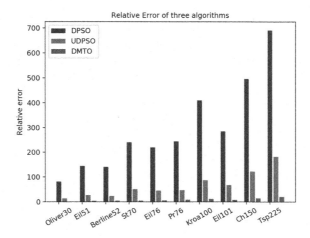

Fig. 2. Relative errors of DPSO, UDPSO, and DMTO

better performance in terms of RE. For instance, the results show that RE is decreased 194.85 times when compared to DPSO for solving Oliver30. However, when the number of cities increases to 225, RE for DMRO decreases to 32.24 times. Figure 2 shows the significant RE improvement when using DMTO with respect to the other two algorithms.

5   Conclusion and Future Works

We propose an effective nature-inspired method to solve the TSP, called DMTO, and an improvement of DPSO (that we call UDPSO). In order to assess the performance of the proposed algorithms, we conducted several experiments on known TSP instances. The results show that UDPSO has better performance than DPSO in terms of quality of solutions. Also, DMTO achieves much better performance than DPSO and UDPSO over all tested TSP instances. DMTO can therefore be an alternative tool for solving TSPs. These promising results have motivated us to consider, in the near future, other discrete optimization problems, using DMTO. These problems will include: vehicle routing, knapsack, set covering, and maximum clique. We also plan to tackle the Constraint Satisfaction Problem (CSP) [7] which is a general framework for representing and solving combinatorial problems.

References

1. Beiler, K.J., Durall, D.M., Simard, S.W., Maxwell, S.A., Kretzer, A.M.: Architecture of the wood-wide web: Rhizopogon spp. genets link multiple douglas-fir cohorts. New Phytologist. **185**(2), 543–553 (2010)

2. Bontoux, B., Feillet, D.: Ant colony optimization for the traveling purchaser problem. Comput. Oper. Res. **35**(2), 628–637 (2008)
3. Books, W.: Summary and Analysis of The Hidden Life of Trees: What They Feel, How They Communicate Discoveries from a Secret World: Based on the Book by Peter Wohlleben. Smart Summaries, Worth Books (2017). https://books.google.ca/books?id=OWmZDgAAQBAJ
4. Chen, S.M., Chien, C.Y.: Solving the traveling salesman problem based on the genetic simulated annealing ant colony system with particle swarm optimization techniques. Expert Syst. Appl. **38**(12), 14439–14450 (2011)
5. Chentsov, A., Korotayeva, L.: The dynamic programming method in the generalized traveling salesman problem. Math. Comput. Model. **25**(1), 93–105 (1997)
6. Davis, L.: Handbook of Genetic Algorithms. CUMINCAD (1991)
7. Dechter, R., Cohen, D.: Constraint Processing. Morgan Kaufmann, Burlington (2003)
8. Dorigo, M., Maniezzo, V., Colorni, A., et al.: Ant system: optimization by a colony of cooperating agents. IEEE Trans. Syst. Man Cybern. Part B Cybern. **26**(1), 29–41 (1996)
9. Fang, L., Chen, P., Liu, S.: Particle swarm optimization with simulated annealing for TSP. In: Proceedings of the 6th WSEAS International Conference on Artificial Intelligence, Knowledge Engineering and Data Bases, pp. 206–210 (2007)
10. Fleischmann, B.: A cutting plane procedure for the travelling salesman problem on road networks. Eur. J. Oper. Res. **21**(3), 307–317 (1985)
11. Gendreau, M., Laporte, G., Semet, F.: A tabu search heuristic for the undirected selective travelling salesman problem. Eur. J. Oper. Res. **106**(2–3), 539–545 (1998)
12. Geng, X., Chen, Z., Yang, W., Shi, D., Zhao, K.: Solving the traveling salesman problem based on an adaptive simulated annealing algorithm with greedy search. Appl. Soft Comput. **11**(4), 3680–3689 (2011)
13. Gomez-Cabrero, D., Armero, C., Ranasinghe, D.N.: The travelling salesman's problem: a self-adapting PSO-ACS algorithm. In: 2007 International Conference on Industrial and Information Systems, pp. 479–484. IEEE (2007)
14. Gündüz, M., Kiran, M.S., Özceylan, E.: A hierarchic approach based on swarm intelligence to solve the traveling salesman problem. Turk. J. Electr. Eng. Comput. Sci. **23**(1), 103–117 (2015)
15. Hadia, S.K., Joshi, A.H., Patel, C.K., Kosta, Y.P.: Solving city routing issue with particle swarm optimization. Int. J. Comput. Appl. **47**(15) (2012)
16. Karaboga, D., Basturk, B.: A powerful and efficient algorithm for numerical function optimization: artificial bee colony (ABC) algorithm. J. Global Optim. **39**(3), 459–471 (2007)
17. Karaboga, D., Gorkemli, B.: A combinatorial artificial bee colony algorithm for traveling salesman problem. In: 2011 International Symposium on Innovations in Intelligent Systems and Applications, pp. 50–53. IEEE (2011)
18. Korani, W., Mouhoub, M., Spiteri, R.J.: Mother tree optimization. In: 2019 IEEE International Conference on Systems, Man and Cybernetics (SMC), pp. 2206–2213. IEEE (2019)
19. Larranaga, P., Kuijpers, C.M.H., Murga, R.H., Inza, I., Dizdarevic, S.: Genetic algorithms for the travelling salesman problem: A review of representations and operators. Artif. Intell. Rev. **13**(2), 129–170 (1999)
20. Leung, K.S., Jin, H.D., Xu, Z.B.: An expanding self-organizing neural network for the traveling salesman problem. Neurocomputing **62**, 267–292 (2004)

21. Marinakis, Y., Marinaki, M., Dounias, G.: Honey bees mating optimization algorithm for the Euclidean traveling salesman problem. Inf. Sci. **181**(20), 4684–4698 (2011)
22. Mouhoub, M., Wang, Z.: Ant colony with stochastic local search for the quadratic assignment problem. In: 2006 18th IEEE International Conference on Tools with Artificial Intelligence (ICTAI 2006), pp. 127–131. IEEE (2006)
23. Padberg, M., Rinaldi, G.: Optimization of a 532-city symmetric traveling salesman problem by branch and cut. Oper. Res. Lett. **6**(1), 1–7 (1987)
24. Pang, W., et al.: Modified particle swarm optimization based on space transformation for solving traveling salesman problem. In: Proceedings of 2004 International Conference on Machine Learning and Cybernetics (IEEE Cat. No. 04EX826), vol. 4, pp. 2342–2346. IEEE (2004)
25. Qu, L., Sun, R.: A synergetic approach to genetic algorithms for solving traveling salesman problem. Inf. Sci. **117**(3–4), 267–283 (1999)
26. Radharamanan, R., Choi, L.: A branch and bound algorithm for the travelling salesman and the transportation routing problems. Comput. Ind. Eng. **11**(1–4), 236–240 (1986)
27. Reinelt, G.: Tsplib discrete and combinatorial optimization (1995). http://comopt. ifi.uniheidelberg.de/software/TSPLIB95
28. Sanches, D., Whitley, D., Tinós, R.: Improving an exact solver for the traveling salesman problem using partition crossover. In: Proceedings of the Genetic and Evolutionary Computation Conference, pp. 337–344 (2017)
29. Talbi, E.G.: Metaheuristics: From Design to Implementation, vol. 74. Wiley, New York (2009)
30. Zweig, G.: An effective tour construction and improvement procedure for the traveling salesman problem. Oper. Res. **43**(6), 1049–1057 (1995)

Dynamic Cloud Workflow Scheduling with a Heuristic-Based Encoding Genetic Algorithm

Jian-Ping Xiao[1,2], Xiao-Min Hu[3(✉)], and Wei-Neng Chen[1,2]

[1] School of Computer Science and Engineering, South China University
of Technology, Guangzhou 510006, China
[2] State Key Laboratory of Subtropical Building Science, South China University
of Technology, Guangzhou 510006, China
[3] School of Computers, Guangdong University of Technology,
Guangzhou 510006, China
xiaomin.hu@aliyun.com

Abstract. Cloud computing is a powerful and scalable computing platform that enables the virtualization, share and on-demand use of computing resources. Scientific workflows on clouds are promising for handling computational-intensive and complex scientific computing tasks. The scientific workflow scheduling problem has been regarded as an intractable optimization problem that determines the performance of a scientific cloud workflow management system. The problem becomes even more challenging if the dynamic and heterogeneous characteristics of cloud workflows are taken into account. In order to adapt to the dynamic environment, this paper proposes a hybrid genetic algorithm (HGA) algorithm. Different from the traditional evolutionary algorithms for workflow scheduling that uses a direct encoding scheme, the proposed HGA uses an indirect encoding scheme, i.e., a schedule is encoded as a sequence of heuristic rules. Since there have been some widely-studied heuristic information for scheduling on a directed acyclic graph, this heuristic information is adopted by HGA to improve performance. In addition, under the dynamic batch-processing environment, it is found that the results returned by HGA in the form of heuristic-based can still adaptive to the changes. The experimental results validate that HGA is promising.

Keywords: Workflow scheduling · GA · Heuristic rules · CPM · Batch processing

1 Introduction

With the burgeoning demand for computing power, the cloud computing industry is experiencing an explosive growth. Cloud computing refers to that cloud service providers build data centers or supercomputers through distributed computing and virtualization technology [14], and provide data storage, analysis,

© Springer Nature Switzerland AG 2020
H. Yang et al. (Eds.): ICONIP 2020, LNCS 12533, pp. 38–49, 2020.
https://doi.org/10.1007/978-3-030-63833-7_4

scientific computing and other services to technology developers or enterprise customers in a free or on-demand manner, such as Amazon data warehouse leasing business. Cloud computing usually involves providing dynamic, scalable and often virtualized resources through the Internet. Due to its strong scalability, elasticity and efficiency [1], tons of tasks are uploaded to cloud. How to effectively schedule these projects and allocate reasonable resources to tasks becomes a significant problem.

For collaborative scientific projects in domains such like structural biology and neuroscience, they usually involve the distributed data resource and small tasks. These resources and tasks are usually presented and structured as a scientific workflow. A workflow is widely used in cloud scheduling problems, and it considers different tasks as a group to achieve a particular result. The tasks in a workflow have parent and child relationships, a child node can be executed only after all of its parent nodes is done. Generally, a workflow can be presented by a weighted directed acyclic graph (DAG) [2] with nodes presenting tasks and weighted edges presenting the data transferred between tasks. As a commonly used model in scheduling problem, workflow can clearly indicate the relationships of tasks in a large project, and the aim of scheduling problem turns to generate an optimal scheduling solution to minimize the makespan of workflow. Scheduling problem in cloud environment is a NP-hard problem. With the increasing number of tasks in cloud environment, or even with the addition of VMs, the solution search space is proportionally increased [3]. Hence, it is impossible to find an optimal solution in a huge search space. Besides, some cloud service providers prefer that the cloud can finish the tasks in demanded time under extra constrains, such as energy cost minimization. In this case, how to find a scheduling solution in extra constrains makes scheduling problem more complex.

However, researchers and scholars have proposed many excellent algorithms to find an effective scheduling solution in a reasonable time. There are two main categories. One is based on greedy strategies, such like heterogeneous earliest finish time (HEFT), first come first serve (FCFS), shortest job first (SJF), critical path method (CPM) [16]. Nitish et al. [4] proposed a scheduling algorithm based on HEFT, and through different greedy strategies such as choosing the task with the largest deadline violation, the highest priority, the latest finish time, etc., to minimize the makespan and cost. Faragardi et al. [5] promoted a greedy strategy named Greedy Resource Provisioning (GRP) and modified the HEFT algorithm to avoid budget violation. Based on GRP policy, this algorithm greedily chooses the most effective VMs, making this algorithm standout. Li et al. [2] constructed the initial solution with three greedy rules, minimum average cost first, maximum cost ascending ratio first and earliest finish time first, and used the greedy improvement heuristic and fair improvement heuristic to improve the solution. The other category is based on heuristic algorithms, such as genetic algorithm (GA) [6], particle swarm optimization (PSO), ant colony optimization (ACO), etc. GA is a branch of evolutionary algorithm proposed by Holland. By applying the principle of evolution, GA provides robust search ability to allow high-quality solution to be derived from a large search space in polynomial time [7]. Therefore, GA is a powerful algorithm in dealing with NP-hard search problem. Wu et al. [8] proposed an algorithm named MOELS with a list of schedule heuristic embedded into GA, and it was compared with MOHEFT and EMS•C, and the results showed that with the expansion of the problem scale, the makespan of MOELS has not increased much as the other algorithms did. Ahmad et al. [9] proposed an algorithm named heuristic-based GA-PSO algorithm, this algorithm improves its initial population with GA for half of its max iterations, and uses PSO to improve for the other half.

Though scholars have proposed many excellent algorithms, most of them are in the condition that the problem only contains one workflow with fixed number of tasks. In another way, most algorithms are aiming at static scheduling, but are not suitable for dynamic scheduling. Dynamic problem means that the tasks are arriving at the cloud at unknown time. In this paper, the tasks arrive at the cloud in the form of workflows, which contains a series of organized tasks. Therefore, the algorithm needs to adjust the scheduling solution when new workflows arrive. Some researchers have proposed to use heuristic rules and the greedy algorithm to deal with the dynamic scheduling problem [10–12]. However, the algorithm with a single rule can easily get trapped in local optima. Hence, the performance of pure-rules-based algorithm is not good as expected.

In this paper, we intend to propose a hybrid genetic algorithm (HGA) algorithm. Different from the traditional evolutionary algorithms for workflow scheduling, the proposed HGA encodes a schedule as a sequence of heuristic rules. The heuristics, e.g., the earliest start time, the latest finish time of each task, etc., are some effective heuristic information proposed in the existing studies of workflow scheduling. Most of these heuristics are calculated based on the CPM method. CPM is an important management concept based on human experience. By embedding the heuristic information yielded by the CPM method, the proposed GA can quickly evolves to promising search areas, and thus achieves fast convergence. To handle dynamic scheduling problems, we further consider the situation of dynamic batch workflow scheduling. It is found that by using the proposed HGA, the solution yielded by the algorithm in the form of a list of heuristics can easily adapt to the dynamic environment.

The rest of this paper is organized as follows. Section 2 gives a detailed description of cloud workflow scheduling problem. In Sect. 3, the framework of proposed algorithm HGA is presented. Comparison studies are shown in Sect. 4, and the conclusions are summarized in Sect. 5.

2 Definition and Background

2.1 Workflow Modeling

In cloud, the workflows are uploaded at different times, and usually the upload time conforms to the Poisson distribution. Figure 1 is a sample of workflow. These workflows are indicated as w_i, and each w_i can be presented by a tuple, $w_i = a_i, G_i$, where a_i indicates the arrival time of w_i, and G_i presents the structure. The structure of workflow are usually presented by a Direct Acyclic Graph

(DAG), where $G_i = \{T_i, E_i\}$, $T_i = \{t_{i1}, t_{i2}, \ldots, t_{iN}\}$, $E_i = \{e_j k^i, \ldots\}$, where T_i is the set of tasks, t_{ij} presents the j-th task of the i-th workflow, and N presents the number of tasks in workflow; E_i presents the set of edges in the i-th workflows and $e_j k^i$ presents the data transferred from t_{ij} to t_{ik}. $\text{pred}(t_{ij}) = \{t_{ip} \mid e^i_{pj} \in E_i\}$ is the set of predecessor tasks of the task t_{ij}. $\text{succ}(t_{ip}) = \{t_{ij} \mid e^i_{pj} \in E_i\}$ is the set of successor tasks of the task t_{ip}. It is defined that a task can be executed only if all of its predecessor tasks have been done.

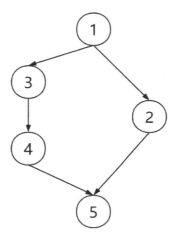

Fig. 1. A simple workflow example with 5 nodes and 5 edges.

2.2 Resource Modeling

A widely used technology in cloud computing is virtualization. It indicates that the cloud platform offers multiple virtual machine (VM) resources R = $\{r_1, r_2, r_3, \ldots, r_M\}$ for different resource types, and M presents the number of VMs. Each type of VM has its own specified attributes such as computing speed and prices, which can be indicated as 3-tuple $r_i = \{ocp_i, speed_i, price_i\}$, where ocp_i presents the occupied time of VM r_i, this indicator is for calculating the utilization, and $speed_i$ presents the processor speed of VM r_i and $price_i$ means the computing cost of VM r_i per unit time. Higher performance virtual machines have higher price. From Barton's experiment [15], it can be seen that the processor cost increases logarithmically with the increasing of performance. It can be assumed that:

- Consumption of a machine includes calculating consumption and system consumption.
- The faster the machine runs, the more energy it costs, while system consumption remains stable.
- Faster machines have higher energy utilization.

Based on the above assumptions, it can be known that mapping the tasks to faster machines can increase the utilization of energy, so this paper takes a greedy strategy that tasks are allocated to the fastest machine when scheduling.

2.3 Extended CPM

Heuristic rules are a set of rules based on human experience. According to Ozdamar [16], heuristic rules can be embedded into GA, guiding the evolution direction and reach a reasonable solution faster. These heuristic rules are mainly based on critical path method, also named CPM. This method defines a series of concept such as earliest start time (EST), latest finish time (LFT) to describe the start time and finish time of a task theoretically, and these theoretical values provides a judgement basis for task arrangement.

However, these rules didn't take data transmission time into account. And the problem this paper discusses doses consider the data transmitted between tasks. So the original CPM rules are utilized and they are defined as follows:

1) MINSK:
$$P_{ij} = lft_{ij} - est_{ij} - d_{ij} + dt_{ij}$$

where lft_{ij} and est_{ij} are the LFT and EST of j-th task in workflow w_i, and d_{ij} is the duration of j-th task in workflow w_i, dt_{ij} is the time of data transmission before executing t_{ij}, the aim of rule MINSK is to find out the slack time of each task, and the task with least slack time has higher priority.

2) MIN LFT:
$$P_{ij} = lft_{ij} + dt_{ij}$$

This rule makes the tasks that finishes earliest executed earliest.

3) MIN SPT:
$$P_{ij} = d_{ij} + dt_{ij}$$

This rule gives the lightest task highest priority. Lightest tasks executes earliest, reducing the total waiting time.

4) MIN LST:
$$P_{ij} = lft_{ij} - d_{ij} + dt_{ij}$$

This rule adopts delaying strategy. The tasks that can be delayed for a long time will be given a small priority.

5) MIN EFT:
$$P_{ij} = est_{ij} + dt_{ij}$$

This rule is similar to the rule first-in-first-out, since the task with earliest start time has the highest priority.

When scheduling, the task with least P_{ij} will be scheduled at once. Compared to the original rules, these polished rules all plus an additional dt_{ij}, dt_{ij} presents the data transmission time from t_i to t_j, since it is very likely that the predecessor task and posterity task may not be executed on the same machine. As for machine choose strategy, the scheduler firstly choose the fastest machine

r_i, and then choose a task t_{ij} to be executed on r_i. If none of its predecessor tasks is executed on that machine, the scheduler calls other machines to send the data of its predecessor tasks produced. If some of its predecessor tasks are executed on the same machine, the data does not need to be transmitted. Let $data_L$ present the largest data that transmitted from other machines among these data, then $dt_{ij} = data_L / bw$, bw is the bandwidth of cloud platform.

All of these rules are based on human experience that urgent tasks execute as soon as possible. But these rules judge the urgency of tasks by different criteria, there is not strong evidence says that which one is better or worse. Therefor this paper embeds all these heuristic rules into GA and takes full advantage of various rules.

2.4 Objective Functions

The aim of this paper is to provide a solution for cloud tasks scheduling, and for a optimal solution, it should finish the task as soon as possible and keep the consumption as low as possible at the same time, put it in another way, the makespan should be as small as possible and the cost should be in a reasonable range. Thought in terms of technology, the cloud platform takes divided tasks as basic units to execute, the users expect their whole projects can be finished in expected time. Hence the target of the research should focus on the makespan of single project. The makespan of workflow wi can be calculated as:

$$makespan_i = max\{ft_{ij}\} - min\{st_{ij}\}$$

where ft_{ij} is the finish time of task t_{ij}, and st_{ij} is the start time of task t_{ij}, $max\{ft_{ij}\}$ presents the finish time of the last task in w_i, it also means the finish time of w_i, and $min\{st_{ij}\}$ presents the start time of first tasks in w_i, it also means the start time of w_i.

After calculating the makespan of each workflow, our goal is to reduce the total computing time of all workflows in cloud platform, and the total time is concerned with the number of workflows N. Therefore, in this paper, it takes the average makespan of all workflows as the final criterion.

$$makespan_{avg} = \frac{\sum_{i=1}^{N} makespan_i}{N}$$

3 Algorithm

3.1 Batch Processing

Since the dynamic situation of cloud computing platform, workflows are uploaded and added to computing queue at any time. These workflows need to be rescheduled in real time depending on the current situation. Genetic algorithm is a effective method for workflow scheduling but not for real time processing, since GA can only search in fixed-size space. To make GA more suitable under

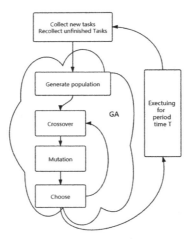

Fig. 2. Framework of batch processing.

dynamic cloud environment, this paper combines GA with the batch processing technique.

Figure 2 is the frame of batch processing. It explains how GA works in batch processing. At the beginning of batch, the scheduler collects all the newly-arrived workflows during the last batch, and re-collects all the unfinished tasks in last batch to re-unite as a new batch. Then the GA generates a new population for the new batch scheduling and the population iterates for generations. When a new scheduling solution is generated, it will be implemented on the platform, after the platform running for a period of time T, the scheduler collects the newly-arrived worfklows, re-collects unfinished tasks, and enters the next batch. Giving the period time T, the cloud platform can schedule the tasks in a near real-time way.

3.2 Scheduling Generator

The scheduling generator is defined to find an optimal scheduling solution. Algorithm 1 depicted the generating process in one batch.

In the final generations, an optimized solution is generated. This solution can be implemented on the platform during this batch.

3.3 Heuristic-Based Encoding GA

The traditional GA generates hundreds or thousands of chromosomes to present the solutions in search space. After generations of evolution, including crossover and mutations, GA may reach a right spot in the search space and find a optimal solution, the process of evolution is long, but with some human experience, this long process may speed up. GA embedded with heuristic rules may speed up the evolution and get a satisfied solution faster.

Algorithm 1 Heuristic-based GA.

Input: population size *Pop_size*, maximum generation number *MaxGen*, crossover
   possibility P_c, mutation possibility P_m
1: Collect unfinished tasks as UFset;
2: Initialize Population P_{Gen-1} of size *Pop_size* and initialize chromosome based on
UFset;
3: Calculate objective function and select top individuals with least makespan;
4: **While** *Gen < MaxGen*
5:   Use binary tournament selection, two-point crossover, mutation operations on
       population P_{Gen-1} to generate offspring Q_{Gen-1};
6:   Using the the offsprings in Q_{Gen-1} to replace the individual with largest
       makspan to form a new population of size *Pop_size*;
7:   *Gen = Gen + 1*;
8: **End While**
Output: Final population P_{Gen}

The traditional GA generates hundreds or thousands of chromosomes to present the solutions in search space. After generations of evolution, including crossover and mutations, GA may reach a right spot in the search space and find a optimal solution, the process of evolution is long, but with some human experience, this long process may speed up. GA embedded with heuristic rules may speed up the evolution and get a satisfied solution faster.

Chromosomes Encoding. Chromosomes encoding. For the traditional GA, a chromosome may conclude all tasks on cloud platform, and the priority of each task is presented by its position on the chromosome, however, CPM contains a series rule to define the priority of each task, and what the chromosome needs to do is to determine when and which rule to use. Suppose there are 5 rules and their sequence numbers are 1-MINSK, 2-MINLFT, 3-MINSPT, 4-MINLST, 5-MINEST, respectively, and there are 6 tasks to be scheduled. Then, the chromosome can be defined as Fig. 3:

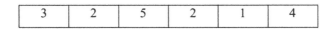

3	2	5	2	1	4

Fig. 3. An example of the chromosome in HGA

The number of the first position on this chromosome is 3, which means the first time to choose next task should use rule 3. The number of second position on this chromosome is 2, which means the generator should use rule 2 when the second time to choose next task, and so on. Through this chromosome encoding, the generator can choose tasks based on experience, instead of random order.

Crossover and Mutation. For crossover, here we use the traditional two-point crossover, we randomly choose two parents from the population, and then randomly choose two pair of same spot on both parent. The genes between the two spots will be exchanged to deliver two new offsprings. For mutation, we randomly choose one spot on chromosome according to the mutation rate and mutate that spot.

4   Experiments and Analysis

In order to fit the reality situation that many workflows are uploaded to the cloud computing platform, we randomly create several groups of workflows in the experiment. The number of tasks in each workflow varies from 10 to 30. The edges between tasks are produced randomly. The total number of workflow that program generated is 50. The workflow uploaded to the cloud platform is randomly selected. The parameters of the machine resources are shown in the Table 1.

Table 1. Parameter of machine resource

Machine type	CPUs	Num	Price
M1	2	2	0.48
M2	4	4	0.7
M3	8	9	0.95
M4	16	6	1.23
M5	32	3	1.52

In this experiment, the focus is to test and verify that the batch process and heuristic-based GA can effectively schedule the tasks on cloud platform. For comparison, the traditional GA with the pure encoding scheme is applied. Table 2 defines the parameter of population in both GA-related algorithms. Since the experience focuses on heuristic rules, the parameter of population is set as a constant.

Table 2. Parameter of population

Parameter Name	Value
Pop_size	1000
Crossover rate	0.7
Mutation rate	0.05
Selection rate	0.6
Iterations	50

4.1 Efficiency of Single Heuristic Rules

Heuristic method is essentially a greedy algorithm. To validate the effectiveness the heuristics, we first compare the efficiency of each single heuristic. In addition, two more rules are used in the comparison, e.g., the first-in-first-out (FIFO) rule and the random selection rule.

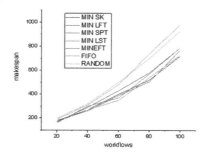

Fig. 4. Makespan increase with the number of workflows

Figure 4 presents the growth of makespan when the number of workflow grows. From the figure, it can be inferred that with the increasing scale of workflows, the makespan of each workflow also increases. The y-axis presents the makespan of each workflow. The lower the curve is, the better the rule is. It can be directly seen that the curves of FIFO and RANDOM are mostly above the other curves, which means heuristic rules with human experiences are better. From Fig. 4, it can also be deduced that when the number of workflows reaches 60, there is a sharp increasing on all curves. This means the execution power of machine resources has reach saturation. Tasks begin to compete for machine resources. Some tasks need to wait for the resources. Thus there is extra waiting time apart from the executing time of the tasks.

4.2 Heuristic-Based GA vs. Pure GA

The HGA based on heuristics and the Pure GA are both GA algorithms. To adapt to the dynamic environment. GA needs to combined with batch processing. This method provides GA a near real-time way to solve the dynamic cloud workflow scheduling problem. In this experiment, the period time T of batch processing is set as 500, which means every 500 unit time, the scheduler reschedules all of the tasks. The experiment runs 3 periods for observation.

Figure 5 shows 3 experiment results during 3 periods executing. In 3 situations, there are 30, 60, and 100 workflows uploaded to the cloud. Hence in each experiment, there are 2 jumps in the graph. For the first period, the efficiency of heuristic-based GA and Pure GA makes no difference, since the number of workflows is small. There is no pressure for the scheduler to handle the workflows since most machines are idle. But for the second period and when $N \geq 60$,

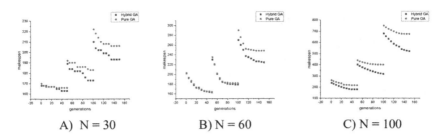

A) N = 30   B) N = 60   C) N = 100

Fig. 5. The least makespan in each generation under different circumstances N = 30, 60, 100.

the competition between tasks begins to emerge. So HGA begins to show its advantage. Its convergence rate and the final results yielded are better than pure GA. According to Fig. 5, when the workflow number grows to 100, the final makespan of HGA turns out to be much shorter than that of the Pure GA. So this experiment shows that HGA is promising for dynamic workflow scheduling.

5   Conclusion

This paper provides a heuristic-based GA to solve the dynamic cloud workflow scheduling problem. Though under most circumstances, GA are usually used to solve fixed-sized problem, to take advantages of the powerful search ability of GA, this paper combines GA with batch processing. It allows GA to solve dynamic problems in a near real-time way. Besides, to speed up GA, some heuristic rules based on CPM are embedded into GA. A special encoding scheme that regards a solution as a list of selected heuristics is developed. In the experiment, it can be seen that though HGA makes no difference with the Pure GA when the workflow number is small. But as the number of workflows increases, the advantage of HGA becomes significant.

Acknowledgement. This work was supported in part by the National Natural Science Foundation of China under Grants 61976093 and 61772142, in part by the Guangdong Natural Science Foundation Research Team No. 2018B030312003 and No. 2019A1515011270, and in part by Pearl River Science and Technology Nova Program of Guangzhou No. 201806010059.

References

1. Lehrig, S., Eikerling, H., Becker, S.: Scalability, elasticity, and efficiency in cloud computing: a systematic literature review of definitions and metrics. In: 2015 11th International ACM SIGSOFT Conference on Quality of Software Architectures (QoSA), Montreal, QC, pp. 83–92 (2015). https://doi.org/10.1145/2737182. 2737185

2. Li, X., Qian, L., Ruiz, R.: Cloud workflow scheduling with deadlines and time slot availability. IEEE Trans. Serv. Comput. **11**, 329–340 (2016)
3. Bilgaiyan, S., Sagnika, S., Mishra, S., et al.: Study of task scheduling in cloud computing environment using soft computing algorithms. Int. J. Mod. Educ. Comput. Sci. **7**(3), 32–38 (2015)
4. Chopra, N., Singh, S.: HEFT based workflow scheduling algorithm for cost optimization within deadline in heuristic-based clouds. In: 2013 4th International Conference on Computing, Communications and Networking Technologies (ICCCNT). IEEE (2013)
5. Faragardi, H.R., Sedghpour, M.R.S., Fazliahmadi, S., et al.: GRP-HEFT: a budget-constrained resource provisioning scheme for workflow scheduling in IaaS clouds. IEEE Trans. Parallel Distrib. Syst. **31**(6), 1239–1254 (2020)
6. Goldberg, D.E.: Genetic algorithms in search, optimization, and machine learning. Ethnographic Praxis Ind. Conf. Proc. **9**(2) (1988)
7. Yu, J., Buyya, R.: Scheduling scientific workflow applications with deadline and budget constraints using genetic algorithms. Sci. Program. **14**, 217–230 (2006)
8. Wu, Q., Zhou, M., Zhu, Q., Xia, Y., Wen, J.: MOELS: multiobjective evolutionary list scheduling for cloud workflows. IEEE Trans. Autom. Sci. Eng. **17**(1), 166–176 (2020)
9. Manasrah, A.M., Hanan, B.A.: Workflow scheduling using heuristic-based GA-PSO algorithm in cloud computing. Wirel. Commun. Mob. Comput. **2018**, 1–16 (2018)
10. Nazia, A., Huifang, D.: A heuristic-based Metaheuristic for multi-objective scientific workflow scheduling in a cloud environment. Appl. Sci. **8**(4), 538 (2018)
11. Rehani, N., Garg, R.: Meta-heuristic based reliable and green workflow scheduling in cloud computing. Int. J. Syst. Assur. Eng. Manage. **9**, 811–820 (2018). https://doi.org/10.1007/s13198-017-0659-8
12. Kaur, A., Kaur, B., Singh, D.: Meta-heuristic based framework for workflow load balancing in cloud environment. Int. J. Inf. Technol. **11**(1), 119–125 (2019)
13. Kohler, W.H.: A preliminary evaluation of the critical path method for scheduling tasks on multiprocessor systems. IEEE Trans. Comput. **C−24**(12), 1235–1238 (1975)
14. Xing, Y., Zhan, Y.: Virtualization and cloud computing. In: Zhang, Y. (ed.) Future Wireless Networks and Information Systems. LNEE, vol. 143, pp. 305–312. Springer, Heidelberg (2012). https://doi.org/10.1007/978-3-642-27323-0_39
15. Barton, M.L., Withers, G.R.: Computing performance as a function of the speed, quantity, and cost of the processors. In: Proceedings of the 1989 ACM/IEEE Conference on Supercomputing, Supercomputing 1989, Reno, NV, USA, pp. 759–764 (1989)
16. Ozdamar, L.: A genetic algorithm approach to a general category project scheduling problem. IEEE Trans. Syst. Man Cybern. Part C (Appl. Rev.) **29**(1), 44–59 (1999)

Multi-strategy Evolutionary Computation for Automated Jigsaw Puzzles

Senhua Zhao, Yue-Jiao Gong$^{(\boxtimes)}$, and Xiaolin Xiao

South China University of Technology, Guangzhou 510006, China
gongyuejiao@gmail.com

Abstract. Solving jigsaw-puzzles has been of increasing importance in many real-world applications. The existing methods endure the problem of local or premature convergence, which perform inefficiently on some challenging images. For an efficient optimizer of jigsaw puzzles, this paper utilizes the powerfulness of the global optimization technique and develops a multi-strategy evolution algorithm. The algorithm constantly generates jigsaw puzzle solutions by mimicking the process of natural evolution, while adopting a new objective function to evaluate the solutions. An elite-based crossover operator is designed to exploit the historically good patterns for generating competitive solutions. Then, a new mutation operator consisting of four perturbation strategies is developed to handle different puzzle situations. Experimental results verify the promising performance of the proposed algorithm that it outperforms the state-of-the-art methods on various image datasets.

Keywords: Jigsaw puzzle · Evolutionary computation · Global optimization · Shredded piece assembly

1 Introduction

The automatic solver of jigsaw puzzle has received increasing attention since it plays an important role in many scientific and engineering fields, such as reassembling archaeological artifacts [17] and recovering shredded documents [2]. Specifically, given non-overlapping and disordered square pieces of an image, the automatic solver needs to reconstruct the original image without any prior knowledge.

Assembling jigsaw puzzles is a technically challenging problem that has been proved to be NP-complete [4]. Recently, many efforts have been devoted to this domain to develop different solutions, such as greedy algorithms [13,14], hierarchical loop constraints [18], and the genetic algorithm [16]. The greedy methods

Supported in part by the Key Project of Science and Technology Innovation 2030, Ministry of Science and Technology of China, under Grant 2018AAA0101300; in part by the National Natural Science Foundation of China under Grant 61873095 and Grant U1701267; and in part by the China Postdoctoral Science Foundation under Grant 2019M662913.

© Springer Nature Switzerland AG 2020
H. Yang et al. (Eds.): ICONIP 2020, LNCS 12533, pp. 50–62, 2020.
https://doi.org/10.1007/978-3-030-63833-7_5

have a great risk of converging to local optima, resulting in unsatisfactory solution accuracy. The approach of hierarchical loop constraints [18] merges the pieces in a bottom up manner based on the basic loops of assembling 2×2 pieces, which may however endure the error propagation problem that the mismatched patterns propagate to the rest. The evolutionary computation (EC) algorithms, such as the genetic algorithm, are widely known for the powerful global optimization capability and the flexibility in dealing with arbitrary objective formulations [7,19]. This paper focuses on utilizing the EC paradigm for solving jigsaw puzzle. However, this type of algorithm faces the "curse of dimensionality" problem, whose performance degrades fast when the number of pieces increases.

To address the above-mentioned issues, we develop a multi-strategy evolutionary approach (MSEA) for the efficient assembling of image pieces. Typically, an evolutionary algorithm (EA) is composed of a main loop of the reproduction and selection operations to evolve the candidate solutions and approach to the optimum. On the other side, there are two crucial components of a jigsaw puzzle solver: an estimation function to evaluate the compatibility of adjacent pieces and an assemble strategy to place the pieces. Note that the first component determines the fitness evaluation step in the selection operation of EC, while the second component relies on the reproduction operations. Following this line of thinking, for an efficient jigsaw puzzle solver, we specifically develop novel reproduction mechanisms for generating high-quality solutions, and we also design a comprehensive fitness function to improve the effectiveness of the selection operation.

The novelties of MSEA are summarized in the following. 1) An elite-based crossover (EBC) operator is proposed to synthesize the genetic information of the parents to generate the offspring. EBC incorporates the historical guidance of elite individuals (i.e., solutions) to assign the adjacency patterns with different priorities. Then, it applies heuristic rules based on the priority hierarchy to reproduce solutions. The mechanism is good at identifying promising patterns and passing them to the offspring. 2) A four-strategy mutation is developed, which consists of an unfit point exchange (UPE) strategy, a random point exchange (RPE) strategy, an unfit line exchange (ULE) strategy, and a random line exchange (RLE) strategy. The mutation not only fixes some particularly unsuitable placement of pieces but also enhances the search diversity to avoid the premature convergence. 3) A comprehensive fitness function named distance and gradient compatibility (DGC) is designed to evaluate the solution quality and guide the selection operation that maintains good solutions and discards inferior ones. Experiments and the comparisons with existing algorithms validate that our MSEA is a powerful and stable algorithm for solving jigsaw puzzle problem.

2 Related Work

The first attempt of jigsaw puzzle solver has been made in [5]. The early works were mostly based on the shape of puzzles. Then, the research direction turned

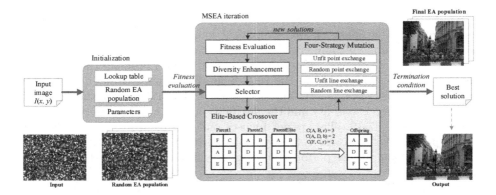

Fig. 1. The pipeline of MSEA.

towards the content-based jigsaw puzzle problems [10,20] and identical square pieces [3,6,8,13–16,18]. In this paper, we only discuss the content-based jigsaw puzzle problem with square pieces.

Greedy Methods: Pomeranz *et al.* [14] presented an automatic square puzzle solver based on a greedy method. The algorithm tackles three subproblems, namely, placement, segmentation, and reconstruction, iteratively. To improve the work of [14], Paikin *et al.* [13] presented a greedy algorithm with more careful initial configuration, which can solve jigsaw puzzles with additional challenges.

Hierarchical Methods: Gallagher [6] proposed a tree-based assembly approach using a new piece compatibility score focusing on the gradients near the boundary of adjacent pieces. Hammoudeh [8] introduced an agglomerative hierarchical clustering-based solver for mix-bag jigsaw puzzle problem. Son [18] proposed an algorithm based on hierarchical cyclic constraints, which uses a bottom-up loop puzzle fragment construction method to reconstruct images. However, for hierarchical methods, the existing matching error in the early fragments will propagate to the rest.

Advanced Approaches: Sholomon *et al.* [16] shown the potential of the genetic algorithm (GA) that can solve a very large jigsaw puzzle. Rika [15] combined GA and deep learning to reconstruct Portuguese tile panels. The evolutionary algorithms such as the GA are powerful global optimizers specialized in solving NP-complete problems, and they have seen successful applications in solving the jigsaw puzzles. However, there are several important issues that still remained unexplored. For example, how to fully exploit the historical information during the evolution in order to inherit promising patterns of piece matching and produce high-quality offspring, and how to specifically address targeted unfix situations in order to enhance the global optimization capability. In this paper, we apply multiple strategies to evolution algorithm and propose our MSEA for solving the jigsaw puzzle problem.

3 The Multi-strategy Evolution Algorithm

This paper develops a multi-strategy evolution algorithm, the MSEA, which is a global optimizer to solve jigsaw puzzle. Figure 1 depicts the pipeline of MSEA. The details are illustrated in Sect. 3.1–Sect. 3.5. The problem input is a set of $M \times N$ disordered pieces of an image. In MSEA, each individual is represented as an $M \times N$ matrix, where each element stores the ID of a jigsaw piece.

3.1 Compatibility Between Pieces and the Objective Function

In EC, the objective function is utilized to evaluate the fitness of each individual. We use (x_i, x_j, R) to denote the relation between pieces x_i and x_j, where $R \in \{l, r, a, b\}$ represents the space relationships (left/right/above/below) of the two pieces. Each piece is represented by a $K \times K$ matrix of pixels, where each pixel is a triple in the normalized CIELAB color space.

Our proposed measure combines the pixel distances of the connecting boundaries of the two adjacent pieces and the gradient changes across the boundaries. The study in [11,12] have shown the effectiveness of processing color components reasonably. By integrating the gradient information between adjacent pixel points, the pairwise compatibility measure between pieces can be more informative. Suppose that a piece x_j is placed on the right side of x_i, the compatibility between the two pieces is calculated as

$$V_{p,q}(x_i, x_j, r) = D_{p,q}(x_i, x_j, r) + G_{p,q}(x_i, x_j, r) \tag{1}$$

$$D_{p,q}(x_i, x_j, r) = \left(\sum_{k=1}^{K} \sum_{d=1}^{3} (|x_i(k, K, d) - x_j(k, 1, d)|)^p \right)^{\frac{q}{p}} \tag{2}$$

$$G_{p,q}(x_i, x_j, r) = \left\{ \sum_{k=1}^{K} \sum_{d=1}^{3} [|\delta_r^{ij}(k, d) - E_r^{ij}(k, d)|]^p \right\}^{\frac{q}{p}} \tag{3}$$

where,

$$\delta_r^{ij}(k, d) = x_j(k, 1, d) - x_i(k, K, d) \tag{4}$$

$$E_r^{ij}(k, d) = \frac{1}{2}(x_i(k, K, d) - x_i(k, K-1, d) + x_j(k, 2, d) - x_j(k, 1, d)) \tag{5}$$

In the equations, $x_i(k, K, d)$ represents the pixel value in the kth row, Kth column, and dth channel of the pixel matrix (jigsaw puzzle piece). Note that the gradient change calculation in Eq. (3) follows the work in [18]. We use the asymmetric dissimilarity with L_q^p norm [14]. To maximize the compatibility between two pieces, their $V_{p,q}$ values should be minimized. The overall cost function, distance and gradient compatibility (DGC), is defined as below:

$$DGC(p, q) = \sum_{i=1}^{N} \sum_{j=1}^{M-1} (V_{p,q}(x_i, x_j, r)) + \sum_{i=1}^{N-1} \sum_{j=1}^{M} (V_{p,q}(x_i, x_j, b)) \tag{6}$$

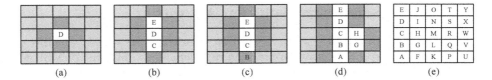

Fig. 2. Illustration of the construction manner in the crossover. The slots around the kernel are marked in blue, and the unsettled slots are marked in gray. The operator joins piece into blue slots that are neighboring the kernel. (a) initial stage with only one piece D. (b)–(d) allocation of selected pieces.

3.2 Selector

In MSEA, we use the tournament selection, which is a useful and robust selection mechanism [1]. The selector first randomly chooses a group of individuals from the population, and then it chooses the best one from the group of individuals for population reproduction. The above procedure involves a parameter, namely the group ratio s that decides the number of individuals participating in the competition, which controls the convergence rate of MSEA.

3.3 Elite-Based Crossover

We design an EBC that absorbs the promising genetic materials from both parents and elite individuals. Particularly, three individuals participate in the crossover, two of which are chosen by the selector, and the remaining one is selected randomly from the preserved elite individuals. The elite individuals are the individuals whose fitness values are ranked at the former e place in each generation. Using elite individuals to participate in crossover, the operator can make full use of historical experience information to improve the search efficiency.

The next issue becomes how to inherit the valid genetic information from the three parent individuals. The proposed crossover operator counts the pairwise adjacency patterns in the parent individuals and divides them into "3A", "2A", and "other" categories to obtain an agreement table. The "3A" relationship indicates that all three parent individuals contain the pattern (agree to the relationship), the "2A" relationship means that two of the parent individuals contain the pattern, and the "other" means that the pattern only occurs once. Our crossover operator will be executed in a hierarchical manner that it gives the highest priority to preserve the "3A" patterns and then the "2A" patterns. The basic idea is to maintain the previously found good patterns and explore the others. Details of the proposed EBC are presented below.

Constructive Manner: As shown in Fig. 2, given the selected parent individuals, the crossover operator first chooses a piece randomly as a "seed" (a kernel) and place it in the center of the solution matrix. Then, the kernel is grown up by gradually filling the empty slots around it with the other available pieces selected using some heuristic rules. When the placement of a piece exceeds the

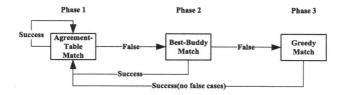

Fig. 3. The state transition diagram of EBC.

image boundary, the entire kernel is shifted one place to make a space for the piece. This makes the location of pieces be not certain before the offspring being completed, which ensures the position independence property when constructing the solutions. The above kernel grow-up and shift mechanisms repeat until the entire solution matrix has been filled up. Note that this constructive manner follows the approach developed by Sholomon [16].

Elite-Based Heuristic Rules: The elite-based heuristic rules are applied to select the appropriate piece when needed , by which the state transition diagram of EBC is shown in Fig. 3. The placer repeats the below three phases until it completely reproduces an offspring.

Phase 1 Agreement-Table Match: Given all the existing boundary pieces of the current kernel, the placer checks checks the agreement table. When there is only one available "3A" boundary pattern, the placer directly implements it. When there are multiple "3A" boundary patterns, one of these is selected randomly. After a piece x_j has been placed, the piece becomes no longer available that all the other patterns that involves the piece x_j will be ignored in the following procedure. Repeat this process until there is no more available "3A" boundary pattern existing. Then, the placer turns to check "2A" boundary patterns, and places the piece in a similar manner. This phase terminates when there is no more available "2A" boundary patterns to realize.

Phase 2 Best-Buddy Match: Two pieces are regarded as the best buddies [14] if each of them considers the other as its most appreciate piece. Piece x_i and piece x_j are the best buddies if they hold:

$$\forall x_k, \forall x_p, C(x_i, x_j, r_1) \geq C(x_i, x_k, r_1) \wedge C(x_j, x_i, r_2) \geq C(x_j, x_p, r_2) \quad (7)$$

where r_1 and r_2 are the opposite spatial direction. Given existing boundaries, the placer checks whether one of the parents contains best buddies. If so, the corresponding buddy piece will be assigned. As before, when multiple best-buddy pieces are available, the placer chooses one to assign at random.

Phase 3 Greedy: The third phase begins when there is no available best-buddy piece. The placer chooses a boundary at random, selects the most compatible piece of it, and place the piece to the proper slot.

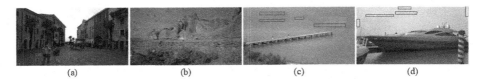

Fig. 4. Unfit points [(a) and (b)] and unfit lines [(c) and (d)] in MSEA iteration. The unfit points and lines in the figure have been marked by red boxes. (Color figure online)

3.4 Four-Strategy Mutation

Mutation is used to simulate a random perturbation on the genotype of individuals and alter the product of genes. In MSEA, we develop four solution variation strategies for mutation: UPE, RPE, ULE and RLE. Shown in Fig. 4, the UPE strategy is used to address the situation that some pieces appearing in the MSEA iteration are not suitable for the surrounding, and the ULE strategy is mainly for the case where a line of consecutive pieces is placed improperly. Besides, the RPE and RLE are random strategies as the traditional mutation does, but they are performed at different levels of granularity.

Point Exchange: the point exchange strategies are used to find those unfit pieces and try to place them in more appropriate positions. The number of points to be exchanged, denoted as enp, is determined by a probability P_{mt}.

UPE: the mutation operator sorts all patterns involved in the individual based on their pairwise compatibility. For the top $2 \times enp$ worst pieces, enp times of random exchanges are performed.

RPE: each piece with $random(0,1) < P_{mt}$ will be exchanged with a random piece once. The strategy increases the diversity of individuals in the population.

Line Exchange the line exchange strategies attempt to correct a line of L consecutive pieces that shifts slightly from its appropriate position, where the length L is generated randomly and validly in length. Since we need to consider both column exchange and row exchange, one of them is chosen at random.

ULE: taking row exchange as an example, the operator calculates the local fitness LF of every line segment, and chooses the most inappropriate one.

$$LF(L_r, L_c, L) = \begin{cases} f_{down}(L_r, L_c, L), & L_r = 0 \\ \frac{1}{2}(f_{down}(L_r, L_c, L) + f_{up}(L_r, L_c, L)), & 0 < L_r < M \\ f_{up}(L_r, L_c, L), & L_r = M \end{cases} \quad (8)$$

$$f_{down}(L_r, L_c, L) = \sum_{j=L_c}^{L} D_{p,q}(x_{L_r,j}, x_{L_r+1,j}, b)$$

$$f_{up}(L_r, L_c, L) = \sum_{j=L_c}^{L} D_{p,q}(x_{L_r,j}, x_{L_r-1,j}, a) \quad (9)$$

Algorithm 1: Pseudocode of MSEA.

Input: $I(M, N)$: a set of $M \times N$ disordered pieces of an image, Population size P ,
   Maximum generation $T_{\max}$, Tournament rate s , Elite number e, Random rate
   parameters: P_{mt}, P_{mb}, P_{mp}, Mutation parameters: ω_1, ω_2

Output: a jigsaw puzzle solution

1 population ← generate P random individuals;
2 evaluate all individuals of population using the objective function;
3 **repeat**
4 | new population ← ∅ ;
5 | copy e best individuals to new population;
6 | **while** *size(new population)*≤ P **do**
7 | | parent1 ← select individual;
8 | | parent2 ← select individual;
9 | | parentElite ← select elite individual;
10 | | child ← crossover (parent1, parent2, parentElite);
11 | | child ← mutation(child);
12 | | add child to new population;
13 | **end**
14 | population ← new population ;
15 | evaluate all individuals of population using the objective function;
16 | diversity enhancement strategy;
17 **until** *the maximum generation T_{max} is reached*;

---

where L_r represents the row of the piece x_{L_r, L_c}, L_c indicates the column of the piece x_{L_r, L_c}, and (L_r, L_c) is the start index of the line segment L. After the exchange segment has been selected, the operator checks the local fitness in condition of assigning it to all slots of the puzzle. If there exists one slot that is more appreciate to assign the segment, the operator exchanges the chosen segment and the one in the slot.

RLE: the operator randomly generates two start points to determine the exchange lines. Then, two selected lines of pieces are exchanged in position.

The proposed four-strategy mutation merges the above UPE, RPE, ULE, RLE strategies in a frame with two parameters. The parameter ω_1 controls the ratio of point exchange and line exchange, and the parameter ω_2 adjusts the proportion of random strategy participation.

3.5   Diversity Enhancement Strategy

We further incorporate a diversity enhancement strategy into MSEA to avoid the premature convergence. This strategy checks whether there is no significant change in the population fitness of four successive generations, which is called stagnation. When stagnation occurs, for each individual except the elite ones, the MSEA executes either a block exchange strategy or a point exchange strategy at random. The block exchange strategy simply generates two blocks and swaps them, in which the block size is determined by a probability parameter P_{mb}. The point exchange strategy is the same as the RPE in the mutation with a parameter P_{mp} , but they are executed in different situations.

Table 1. Comparison of different jigsaw puzzle solvers under the NC metric.

Image Set	Pomeranz *et al.* (%)		Sholomon *et al.* (%)		Paikin *et al.* (%)		MSEA (%)	
	Best	Avg	Best	Avg	Best	Avg	Best	Avg
MIT-432	94.81	88.65	94.16	93.19	94.48	-	**95.96**	**95.61**
McGill-540	89.82	81.72	88.61	87.51	92.50	-	**96.19**	**95.42**
BGU-805	89.73	81.74	90.97	89.70	93.50	-	**96.87**	**96.07**
BGU-2360	93.89	86.43	75.29	74.37	93.32	-	**97.09**	**96.52**
BGU-3300	86.39	83.50	73.65	72.56	90.73	-	**94.10**	**93.39**
SR-805	87.70	74.62	80.16	78.20	95.19	-	**98.71**	**98.14**
SC-3300	83.62	76.61	81.08	79.76	97.05	-	**98.58**	**98.29**

* The numerical results reported in [18] on MIT-432, McGill-540, BGU-805, -2360 and -3300 are 95.60, **97.00**, 95.50, 96.00 and **97.70**.

4 Experiments

4.1 Experimental Setup and Test Sets

Experiments are carried on seven datasets of different scales, including the set of images supplied by Cho [3] (MIT-432) and the four image sets supplied by Pomeranz [14] (McGill-540, BGU-805, BGU-2360 and BGU-3300). The image sets above are the most commonly used dataset to examine the performance of jigsaw puzzle solvers, which contains 20 images of 432-, 540-, 805-pieces puzzles and 3 images of 2360- and 3300-pieces puzzles. Following [14,16], all images are composed of 28×28-pixel patches. In addition, we also test the proposed MSEA on the 805-pieces partial images of Urban100 (SR-805) [9] and our generated image set consisting of 10 images with 3300 pieces (SC-3300).[1]

The proposed MSEA algorithm is compared with three jigsaw puzzle solvers: two are based on local greedy methods [13,14] and the other is based on the genetic algorithm (GA) of Sholomon *et al.* [16]. Each algorithm is independently executed ten times to obtain the statistic results, except for the greedy algorithm that is deterministic [13]. Besides, we also compare with the numerical results reported in [18] on the same datasets tested. The parameters p and q in the objective function are set to 0.5 and 1.0 after experiment. The pseudo code of MSEA is shown in Algorithm 1, in which the parameter values are set as: the population size $P = 1000$, maximum generation $T_{max} = 100$, group radio $s = 0.01$, elite number $e = 4$, random rate parameters: $P_{mt} = 0.01, P_{mb} = 0.8, P_{mp} = 0.1$, mutation parameters: $\omega_1 = 0.8, \omega_2 = 0.5$.

4.2 Performance Metrics

To evaluate the image reconstruction quality of different algorithms, standard performance metrics are used, namely, the direct comparison metric and the neighbor comparison metric [3]. **Direct Comparison (DC)**: the ratio between

---

[1] Link to the dataset: https://github.com/SenhuaZhao/SC-3300.

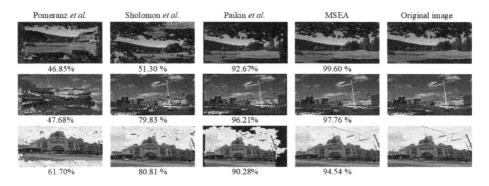

Pomeranz *et al.*   Sholomon *et al.*   Paikin *et al.*   MSEA   Original image

46.85%   51.30 %   92.67%   99.60 %

47.68%   79.83 %   96.21%   97.76 %

61.70%   80.81 %   90.28%   94.54 %

Fig. 5. Solved jigsaw puzzles on images 'shade', 'harbor' and 'station' of SC-3300.

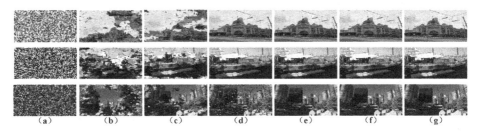

(a)   (b)   (c)   (d)   (e)   (f)   (g)

Fig. 6. Reconstruction process of MSEA on images 'station', 'boat' and 'buildings' of SC-3300. (a) The input images. (b)–(e) The best solutions in the first,second,tenth and twentieth generation. (f) Output images. (g) Original images.

the number of pieces that are placed in their correct position and the total number of pieces. **Neighbor Comparison (NC)**: the ratio between the number of pairwise piece adjacencies that are correct in the origin image and the total number of pairwise piece adjacencies. When comparing the performance of different algorithms, we mainly focus on the NC metric, since the DC metric is very sensitive to a slight shift of some patterns found in the solution.

4.3   Comparison Results

Table 1 presents the results achieved by the comparison algorithms mentioned above and our MSEA on each dataset. The data reported are the average of the mean and best NC scores for all images in each dataset. As can be seen from Table 1, the proposed MSEA obtains the best value on all datasets, and its performance is very stable for different scales of jigsaw puzzles. Note that Paikin *et al.*'s algorithm [13] is deterministic. In addition, the final assembly results obtained by different algorithms are shown in Fig. 5. The selected images are challenging because they contain smooth contents or narrow textures. It can be seen that the previous algorithms return the assembly with some false configurations. In contrast, the proposed MSEA performs much better than the others,

Table 2. The average accuracy of MSEA, where the best, worst, and standard deviation in each dataset are presented after ten executions.

Image Set	Avg. Best (%)		Avg. Worst (%)		Avg. Avg. (%)		Avg. Standard Deviation (%)	
	NC	DC	NC	DC	NC	DC	NC	DC
MIT-432	95.96	96.33	95.15	90.73	95.61	94.56	0.33	2.64
McGill-540	96.19	94.19	94.79	91.80	95.42	92.96	0.45	0.79
BGU-805	96.87	96.29	95.27	93.34	96.07	94.88	0.50	0.97
BGU-2360	97.09	94.92	96.25	94.35	96.52	94.61	0.24	0.16
BGU-3300	94.10	91.18	92.63	87.85	93.39	89.50	0.50	1.26
SR-805	98.71	98.80	97.58	89.24	98.14	94.08	0.39	4.65
SC-3300	98.58	98.82	97.82	97.71	98.29	98.50	0.24	0.36

owing much to its powerful global optimization capability and the specifically designed new evolution strategies.

4.4   In-Depth Performance of MSEA

The optimization process of MSEA is illustrated in Fig. 6. Three images are from the SC-3300 dataset, where each puzzle consists of 3,300 pieces. Note that the third image of column (d) triggers the diversity enhancement strategy to enhance population diversity, so that it is a little bit disordered. Generally, as the iteration progresses, the fitness value of the entire population becomes higher, and adjacent segments (or relationships) in individuals are gradually linked. Eventually, these segments and the adjacency relationship are shifted to a possible correct absolute position to generate the final solution. It can be seen that the final solutions are very close to the original images. The output NC accuracy from top to bottom is 94.5%, 99.5% and 99.9%.

For each image set, the averages of the best, worst, and mean results for all images obtained by MSEA, as well as the standard deviations, are recorded in Table 2. Note that each dataset contains one or more images that contain smooth contents, e.g.., a region of blue sky. The lack of sufficient texture information for these smooth regions challenges the algorithm to restore the pieces into the correct absolute positions. Nevertheless, the MSEA's NC accuracy on each dataset is above 94% and its DC accuracy is above 90%. At the same time, the standard deviation of the algorithm is small, and the best, worst value of the algorithm is similar to the average value. In addition, MSEA exhibits a good performance for the dataset with a large number of pieces. Thus, although the algorithm is non-deterministic with some random decisions, the results of multiple tests indicates that MSEA is stable and robust under different situaions.

5   Conclusion

In this paper, a multi-strategy evolution algorithm named MSEA is designed to tackle the jigsaw puzzle problem. The promising performance of MSEA owes much to the comprehensive objective function and the global optimization method it applies. The EBC operator generates offspring by making full use of historically promising patterns in a hierarchical manner, accelerating the algorithm to find high-quality solutions. The four-strategy mutation adapts the algorithm to different conditions of jigsaw puzzles, performing a local search on the solution generated by the algorithm. Besides, the diversity enhancement strategy prevents the algorithm from premature convergence. With the EBC and all these strategies, MSEA can handle different puzzle situations and perform well on various types of images. The experimental results show that the MSEA outperforms the state-of-the-art algorithms on different datasets. Owing to the good performance of MSEA, it is appealing to extend the algorithm for other types of jigsaw puzzles or piece assembly problems in the future.

References

1. Blickle, T.: Tournament selection. Evolut. Comput. **1**, 181–186 (2000)
2. Cao, S., Liu, H., Yan, S.: Automated assembly of shredded pieces from multiple photos. In: ICME, pp. 358–363 (Jul 2010)
3. Cho, T.S., Avidan, S., Freeman, W.T.: A probabilistic image jigsaw puzzle solver. In: CVPR, pp. 183–190. IEEE (2010)
4. Demaine, E.D., Demaine, M.L.: Jigsaw puzzles, edge matching, and polyomino packing: connections and complexity. Graphs and Combinatorics **23**(1), 195–208 (2007)
5. Freeman, H., Garder, L.: Apictorial jigsaw puzzles: the computer solution of a problem in pattern recognition. IEEE Trans. Electron. Comput. **EC−13**(2), 118–127 (1964)
6. Gallagher, A.C.: Jigsaw puzzles with pieces of unknown orientation. In: CVPR, pp. 382–389 (Jun 2012)
7. Gong, Y., Zhou, Y.: Differential evolutionary superpixel segmentation. IEEE Trans. Image Process. **27**(3), 1390–1404 (2018)
8. Hammoudeh, Z., Pollett, C.: Clustering-based, fully automated mixed-bag jigsaw puzzle solving. In: Felsberg, M., Heyden, A., Krüger, N. (eds.) CAIP 2017. LNCS, vol. 10425, pp. 205–217. Springer, Cham (2017). https://doi.org/10.1007/978-3-319-64698-5_18
9. Huang, J.B., Singh, A., Ahuja, N.: Single image super-resolution from transformed self-exemplars. In: CVPR, pp. 5197–5206 (2015)
10. Kosiba, D.A., Devaux, P.M., Balasubramanian, S., Gandhi, T.L., Kasturi, K.: An automatic jigsaw puzzle solver. In: ICPR, vol. 1, pp. 616–618 vol 1 (Oct 1994)
11. Lan, R., Lu, H., Zhou, Y., Liu, Z., Luo, X.: An LBP encoding scheme jointly using quaternionic representation and angular information. Neural Comput. Appl. **32**(9), 4317–4323 (2019). https://doi.org/10.1007/s00521-018-03968-y
12. Lan, R., Zhou, Y., Liu, Z., Luo, X.: Prior knowledge-based probabilistic collaborative representation for visual recognition. IEEE Trans. Cybern. **50**(4), 1498–1508 (2020)
13. Paikin, G., Tal, A.: Solving multiple square jigsaw puzzles with missing pieces. In: CVPR, pp. 4832–4839 (Jun 2015)
14. Pomeranz, D., Shemesh, M., Ben-Shahar, O.: A fully automated greedy square jigsaw puzzle solver. In: CVPR, pp. 9–16 (Jun 2011)

15. Rika, D., Sholomon, D., David, E., Netanyahu, N.S.: A novel hybrid scheme using genetic algorithms and deep learning for the reconstruction of portuguese tile panels. In: GECCO, pp. 1319–1327 (2019)
16. Sholomon, D., David, O.E., Netanyahu, N.S.: An automatic solver for very large jigsaw puzzles using genetic algorithms. Genet. Program. Evolvable Mach. **17**(3), 291–313 (2016). https://doi.org/10.1007/s10710-015-9258-0
17. Son, K., Almeida, E.B., Cooper, D.B.: Axially symmetric 3D pots configuration system using axis of symmetry and break curve. In: CVPR, pp. 257–264 (Jun 2013)
18. Son, K., Hays, J., Cooper, D.B.: Solving square jigsaw puzzle by hierarchical loop constraints. IEEE Trans. Pattern Anal. Mach. Intell. **41**(9), 2222–2235 (2019)
19. Xie, L., Yuille, A.: Genetic CNN. In: ICCV, pp. 1379–1388 (2017)
20. Yao, F.H., Shao, G.F.: A shape and image merging technique to solve jigsaw puzzles. Pattern Recogn. Lett. **24**(12), 1819–1835 (2003)

Real Valued Card Counting Strategies
for the Game of Blackjack

Mózes Vidámi[1], László Szilágyi[1,2,3], and David Iclanzan[1(✉)]

[1] Computational Intelligence Research Group, Sapientia Hungarian University
of Transylvania, Târgu-Mureş, Romania
iclanzan@ms.sapientia.ro
[2] Physiological Controls Research Center, Obuda University, Budapest, Hungary
[3] Department of Control Engineering and Information Technology, Budapest
University of Technology and Economics, Budapest, Hungary

Abstract. Card counting is a family of casino card game advantage
gambling strategies, in which a player keeps a mental tally of the cards
played in order to calculate whether the next hand is likely to be in the
favor of the player or the dealer. A card counting system assigns point
values (weights) to the cards. Summing the point values of the already
played cards gives a concise numerical estimate of how advantageous the
remaining cards are for the player. In theory, any assignment of weights is
permissible. Historically, card counting systems used integers and rarely
the 1/2 and 3/2 fractions, as computation with these are easier and more
tractable for the human memory.

In this paper we investigate how much advantage would a system
using real valued weights provide. Using a blackjack simulator and a
simple genetic algorithm, we evolved weights vectors for ace-neutral and
ace-reckoned balanced strategies with a fitness function that indicates
how much a given strategy empirically under or outperforms a simple
card counting system. After convergence, we evaluated the systems in
the three efficiency categories used to characterize card counting strategies: playing efficiency, betting and insurance correlation. The obtained
systems outperform classical integer count techniques, offering a better
balance of the efficiency metrics. Finally, by applying rounding and scaling, we transformed some real valued strategies to integer point counts
and found that most of the systems' extra edge is preserved. However,
because of the large weight values, it is unlikely that these systems can
be played quickly and accurately even by professional card counters.

Keywords: Card counting strategies · Evolutionary computation.

1 Introduction

Blackjack is unique among casino games as it affords to an observant player
an opportunity to have an advantage over the house. There is ample statistical

This project was supported by the Sapientia Foundation Institute for Scientific
Research.

H. Yang et al. (Eds.): ICONIP 2020, LNCS 12533, pp. 63–73, 2020.
https://doi.org/10.1007/978-3-030-63833-7_6

evidence that high cards benefit the player, while the low cards are advantageous to the dealer. In his 1962 book, Beat the Dealer [17], Edward O. Thorp described a system and proved that it gave a blackjack player an edge over the house. While Thorp is considered the father of card counting, even before the publication of its seminal work, professional card counters were already exploiting casino blackjack games for a profit. Since the early days of card counting, a plethora of other systems were proposed with the aim of offering a better ease-of-use vs. profitability balance, or as responses and adaptations to the counter-measures taken by casinos to curb the profitability of card counting. The documentary film "The Hot Shoe"[1] provides a nice overview of the card counting history.

Blackjack in general [19] and optimal strategies (when to hit, double, stand or split) and count systems in special [2,3,5,8,12], have received considerable attention from the AI community. Most approaches use evolutionary algorithm (EA) to optimize the strategies over simulated hands, while others use neural network to develop complete blackjack players.

Historically, the manually developed count systems or the ones obtained via artificial evolution [5] were targeted for use by humans. Therefore, these systems restrict the point counts to only integers (and rarely simple fractions) so people can perform the calculations mentally, relatively simply. In this paper we investigate i) if a count strategy that use real valued weights offers any meaningful edge over the integer restricted ones; ii) and if it does, can the system be transformed into an integer point count system that preserves (part of) the additional advantage.

2   The Game of Blackjack

Blackjack, also known as 21, is a card game in which a player or players compete against the dealer or "house", by obtaining a sum of cards that is as close to 21 as possible, without exceeding that value (busting). The game is played with one to eight decks of 52 French cards. The rules of Blackjack can vary by country and even by casino.

First, the dealer shuffles the card, while the players make their bets that are in-between a minimum and maximum bet size and can not be changed once or taken back the first card is dealt. The house deals cards from left to right, one by one. Players start with two cards, both face up, while only one of the dealers card is visible. The values of the cards between two and ten are their pip value (2 to 10), Jacks, Queens and Kings are all worth ten while the Ace has two values: one or eleven. The value of a hand equals sum of the card values. While pursuing the goal of getting as close to twenty-one as possible, every player can draw, request as many cards as they wish, an action called Hit. The player can also choose to Stand - take no more cards, Double - double the bet and draw one last card or Split, to obtain two separate hands from an initially dealt set of pairs. The dealer cannot double down. Some casinos let the player Surrender after seeing the first two cards, for a portion of the bet.

---

[1] https://www.imdb.com/title/tt9414698/.

The most valuable hand is an Ace paired with a ten value card. This is called a Blackjack. It depends on the casino whether or not this beats, draws or loses to the Blackjack of the dealer. A player automatically loses, if they draw more than twenty-one. Once all players completed their hands, the dealer turns over his hidden card and is obligated to draw until his hand values is at least seventeen, where they stop and they compare their hand to the ones obtained by the players. A player is considered a winner if is closer to twenty-one than the house. In case their hand value is equal, it's a draw, otherwise the player loses.

When the dealer's up card is an Ace, the players are allowed to take an "insurance" bet. If they two, the price is half the bet. If the dealer face down card is a ten, the insurance bet pays 2:1. The maximum size of the insurance bet is half of the current bet size. The odds of the dealer making a Blackjack is 9:4, therefore insurance can become profitable only if the player counts the cards and knows that there are proportionally above average ten-point cards still left in the shoe.

Edward O. Thorp used computer simulations to test each distinct situation in a blackjack game and derive the best action the player can take. This collection of rules is called Basic Strategy and when strictly followed, it decreases the edge of the house from 4% to 0.5% [9]. Since then the game has changed, now it is usually played with more than one deck of cards (to reduce the efficiency of card counting systems). Nevertheless, each blackjack game still has a Basic Strategy[2], which describes the optimal method of playing any hand against whatever the dealers up card is. Rarely, casino promotions such as limited 2:1 blackjack payouts enables players to have an edge over the house just by playing the basic strategy.

2.1 Card Counting Principles

Card counting strategies are built upon the observation that high cards benefit the player more than the dealer, while the opposite is true for the low cards. 5s help the dealer the most, thus many such cards remaining in the shoe is very disadvantageous for the player. Higher concentration of high cards benefit the player because it increases the player's chances of hitting a Blackjack, which pays out at a 3:2 rate, while the dealers Blackjack is valued at 1:1. When the deck is stacked in such way the player has the option of Doubling down on additional hands, to increase the expected profit, while the dealer is restricted from Doubling. It also leads to more splitting opportunities for the player, while the dealer, again, is restricted from Splitting. Also, a high enough concentration of 10's increases the probability of the dealer making a blackjack from 4/9 to over 0.5, making the insurance bet profitable.

A concentration of low cards benefit the dealer, since according to the rules the dealer must continue Hitting until he reaches 17. For the common hand values of 12–16, the dealer would bust for every 10-valued card, while low cards provide safety, and hand values close to or spot-on 21.

---

[2] https://en.wikipedia.org/wiki/Blackjack#%23Basic_strategy.

Casinos have implemented many changes to the game rules and casino policies, in an effort to combat bleeding money to professional card counters. While the edge of the card counter player can be severely reduced, it can not be completely eliminated. Countermeasures include increasing the number of decks or shoe count, preferential shuffling - shuffling when the remaining cards are deemed to favour the player, decreasing deck penetration by reshuffling early, no mid-shoe entry into the game, continuous shuffling etc.

Card counting systems assign a positive, negative, or zero point value to each card value available. Once a card is dealt, the so called running count, which starts from 0, is adjusted by that card's point value. Low cards usually have positive point values and raise the value of the count, signaling the increased percentage of high cards in the remaining decks. Conversely, high cards have negative values and they decrease the count for the opposite reason. System that assign 0 point values to cards (usually 7–9, sometimes aces) consider them neutral and they do not affect the running count.

2.2   Efficiency Metrics

Good card counting strategies must perform well several objectives and metrics, that gauge different aspects of the game. Following the work of Peter A. Griffin [6], strategies aim to achieve a balance of efficiency in three categories:

1. Playing Efficiency (PE) or Strategic Efficiency. This metric indicates how well a counting system can be used to vary playing strategy, according to the actual composition of the remaining cards in the shoe. PE is particularly important in hand-held games that only use one or two decks of cards. Approximately 0.70 is the cap on the highest possible PE [6] for a single parameter counting system, that does not use side counts. PE is not relevant to unbalanced counting systems (the running count does not equal zero after all cards are dealt), therefore in this paper we only develop balanced strategies.
2. Betting Correlation (BC) gauges how well the system detects the player advantage based on the remaining undealt cards. Effective card counting system assign point values to each card that correlates well with the card's "effect of removal" as computed in [17], enabling a good estimation of the edge provided by the composition of cards still to be dealt. The player advantages in percentages, when removing card types from Aces, 2, 3 ... to ten-valued cards are: -2.42, 1.75, $+2.14$, $+2.64$, $+3.58$, $+2.40$, $+2.05$, $+0.43$, -0.41, $+1.62$. Larger ratios between the point values permit a higher correlation but they also result in an increased complexity, mentally more taxing computations of the count. By taking the ratio between the highest and lowest assigned point values of a system, counting strategies may be referred to as "level 1", "level 2" etc. The correlation value computed by the BC can approach 1.00.
3. Insurance Correlation (IC) expresses how well a counting strategy indicates whether an Insurance bet should be taken. A high IC offers additional value to a card counting system, as the expected gain from counting cards also comes from taking the insurance bet, when the count is high. A point value of -9 for tens and +4 for all other cards provides a maximal IC value.

To obtain a single valued overall metric that permits an easy comparison of strategies, in this paper we will use the Unified Performance Metric (UPM), that for a point value count vector w, simply computes the normalized averages of the above mentioned metrics:

$$UPM(w) = \frac{PC(w)/0.7 + BC(w) + IC(w)}{3} \tag{1}$$

2.3   Card Counting Strategies

Table 1 illustrates a few famous balanced card counting systems and their respective performance metrics.

Table 1. Comparison of different balanced card counting strategies. The first 10 columns after the strategy name describe the card values used in counting, while the last 4 contain different performance metrics, namely the Playing Efficiency, Betting Correlation, Insurance Correlation and Unified Performance Metric defined in Eq. 1

Strategy	A	2	3	4	5	6	7	8	9	10JQK	PE	BC	IC	UPM
Hi-Lo	-1	1	1	1	1	1	0	0	0	-1	0.51	0.97	0.76	**0.8195**
Hi-Opt I	0	0	1	1	1	1	0	0	0	-1	0.61	0.88	0.85	**0.8671**
Hi-Opt II	0	1	1	2	2	1	1	0	0	-2	0.67	0.91	0.91	**0.9257**
Mentor	-1	1	2	2	2	2	1	0	-1	-2	0.62	0.97	0.8	**0.8852**
Omega II	0	1	1	2	2	2	1	0	-1	-2	0.67	0.92	0.85	**0.9090**
Revere Point Count	-2	1	2	2	2	2	1	0	0	-2	0.55	0.99	0.78	**0.8519**
Revere RAPC	-4	2	3	3	4	3	2	0	-1	-3	0.53	1	0.71	**0.8223**
Revere 14 Count	0	2	2	3	4	2	1	0	-2	-3	0.65	0.92	0.82	**0.8895**
Wong Halves	-1	0.5	1	1	1.5	1	0.5	0	-0.5	-1	0.56	0.99	0.72	**0.8366**
Zen Count	-1	1	1	2	2	2	1	0	0	-2	0.63	0.96	0.85	**0.9033**
Averages											**0.599**	**0.9454**	**0.8009**	**0.8674**

The Hi-Lo or the "Complete Point-Count System" balanced card counting strategy was first introduced by Harvey Dubner in 1963 at the Fall Joint Computer Conference in Las Vegas and was later refined by Julian Braun and discussed by Edward Thorp's famous book, Beat the Dealer [17] (pp. 93–101). The Hi-Lo is the most commonly used card counting strategy and the majority of simulations and studies are based on this count. Hi-Lo has a high BC of 0.97 but its PE is the smallest and the IC is also below average.

Hi-Opt I and Hi-Opt II are strategies developed by Lance Humble and its collaborators [7]. Because of their high PE they are very suited for single deck games. Hi-Opt II has the highest UPM and its still used by many professional blackjack players as it works very well in shoe games, outperforming many other systems [15].

Mentor [13] is a strategy developed with the aim of being suitable for both hand-held and shoe games. It achieves this balance with an above average PE and BC, and slightly below average IC.

Omega II [1] is a more complex counting system created by Bryce Carlson, with one of the highest PE and above average IC. It has the second highest UPM of the strategies from Table 1.

Revere Advanced Plus-Minus, Revere Point Count, Revere RAPC and Revere 14 Count are balanced strategies developed by Lawrence Revere and described in its book Playing Blackjack as a Business [14]. Revere was originally a blackjack dealer, and trained many players to count cards with his advanced systems and also shared techniques meant to avoid detection by casinos. His counting systems have very high BC, thus are very suited for shoe games.

Wong Halves [20] is special counting system that also uses fractions, not just integers. It has a near perfect BC of 0.99, however the PE and IC is way below the average. In practice, many players double the tag values to remove the fractions.

Zen Count [16] is an advanced balanced counting strategy, with all 3 metrics well above average. Similarly to Mentor, it was designed to provide a balance between single-deck and multi-deck strategies.

3   Methods

3.1   Genetic Algorithm

For evolving the strategies we use a Genetic Algorithm [18] implemented with the help of the Distributed Evolutionary Algorithms in Python (DEAP[3]) [4] framework.

The solution are encoded as vectors of float numbers of length 9 in case of ace-reckoned strategies, one float weight for the cards ranging from Ace, 2, 3 to 9. In the case of ace-neutral strategies, the first weight is always zero, therefore the method optimizes the remaining 8 point counts.

As observed, the genetic algorithm does not search for the point value for the 10-J-Q-K cards. Instead, this value is computed from the other weights, in order to ensure that the strategies are all balanced, the count after playing all cards is zero:

$$w_{10} = -\frac{\sum_{i=1}^{9} w_i}{4} \qquad (2)$$

The method uses a population size of 100 individuals, tournament selection of size 2 [10], crossover probability of 0.8. After crossover, individuals are mutated with a probability of 0.5; if mutations occurs, each allele is perturbed with a probability of 0.2 using a Gaussian mutation with $\mu = 0$ and $\sigma = 0.1$.

Objective Function. Blackjack is a nonlinear potentially chaotic game[5], therefore attempts to actually calculate the expected gain from a particular system often rely on simulation techniques [11,17].

---

[3] https://deap.readthedocs.io/en/master/.

In this paper we also asses the quality of the evolved strategies according to bench–marks obtained from a simple but efficient blackjack simulator. The simulator takes a given strategy expressed as an array of 10 elements as input and simulates that strategy over a user selected shoe count (number of decks used for playing), allowed shoe penetration before reshuffling and number of random hands played. The bet sizes are adjusted according to the current count and the supplied bet spread. If the input is comprised only of zeros (no card counting), the simulator executes the basic strategy. The simulator returns a net overall result summing up all the game results.

The fitness function for a strategy characterized by a weight vector w is defined as the difference between its benchmarks results and the net Hi-Lo strategy, divided by the number of played hands:

$$F(x) = \frac{sim(w, hands) - sim([-1, 1, 1, 1, 1, 1, 0, 0, 0, -1], hands)}{length(hands)} \qquad (3)$$

In our experiments, the number of hands used in the bench-marks is 10e6. Due to the high variability of blackjack game, these hands are re-sampled every generation. Therefore, the fitness function is noisy, an individual fitness can vary slightly from generation to generation. $F(w)$ is positive if the strategy encoded by w outperforms the Hi-Lo strategy on the actual *hands*. We have chosen the Hi-Lo system as the baseline as this is the most commonly used card counting strategy and the majority of previous studies and simulations were also based on this count. However, other strategies can also be used to provide a baseline.

3.2 Expert Advisor Mobile Application

For testing and training purposes we also developed an Expert Advisor that scans and recognizes the played cards and indicates what to play next, according to the actually loaded strategy. While card counting with the mind is legal, the use of an automatic card counting device in a casino game would be illegal in most jurisdictions.

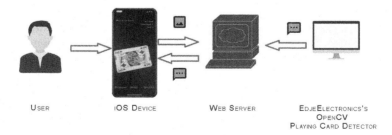

USER   iOS DEVICE   WEB SERVER   EDJEELECTRONICS'S
                                                          OPENCV
                                                     PLAYING CARD DETECTOR

Fig. 1. Expert Advisor for testing and training purposes.

The flowchart of the card processing is depicted in Fig. 1. To identify the cards we used EdjeElectronics's OpenCV Playing Card Detector [4]. The method first detects the card object in the image frame. Then it processes the card image determining its corner points and corrects for perspective and obtains a flattened 200×300 pixels sized image of the card. In the last step, the method isolates the card's suit and rank from the flattened image. The detector works best if it is provided with sample rank and suit images generated from the actual playing cards.

4   Experiments and Results

Some strategies count the ace (ace-reckoned strategies) while others do not (ace-neutral strategies). Counting the aces usually improves betting correlation since the ace is the highest value card in the deck for betting purposes. However, since the ace can either be counted as one or eleven, is both a small and a high card. Including it in the count decreases playing efficiency, therefore many experts prefer to assign a value of zero to the ace. To obtain strategies with emphasis on BC (more important in shoe games) but also ones that emphasize PE (more important in single- and double-deck games) more, we searched for both flavours of strategies with 150 runs of the genetic algorithm, each run spanning 50 generations. The simulator was configured with a shoe count of 6 and 10e6 played hands, the minimum bet size was 1 while the maximum was set to 100.

The average, minimum and maximum fitness values obtained from the runs is depicted in Fig. 2. The runs show a great variability in the range of fitness values, which can be attributed to the noisy nature of the fitness function (the hands are re-sampled every generation) and also to the inherent variability of the game. The average fitness curve shows a steep increase in the first generations then a steady but small growth in the later ones.

Next we computed the PE, BC and IC performance metrics for each one of the 300 evolved strategies using `blackjackinfo.com`'s the free Card Counting Efficiency Calculator[5] then we computed the normalized averages per Eq. 1 to obtain the UPM. The distribution of these values is depicted in Fig. 3.

For both ace-neutral and ace-reckoned strategies the average values of 0.9261 and 0.87342 significantly exceeds the 0.8674 average value of the strategies summarized in Table 1. The average is much higher in the case of ace-neutral strategies, even exceeding the 0.9257 maximum value from Table 1, belonging to the Hi-Opt II strategy.

The average metric values for the ace-neutral strategies were PE=0.6652 $\pm$ 0.0136, BC=0.8910 $\pm$ 0.0155, IC=0.9370 $\pm$ 0.0197 and for the ace-reckoned ones PE = 0.5784 $\pm$ 0.02369, BC=0.9610 $\pm$ 0.0168, IC=0.83291 $\pm$ 0.0330. As expected, considering the ace to be neutral leads to strategies with a high PE and lower BC, while the opposite happens when aces are also counted. Surprisingly,

---

[4] https://github.com/EdjeElectronics/OpenCV-Playing-Card-Detector.

[5] https://www.blackjackinfo.com/card-counting-efficiency-calculator/.

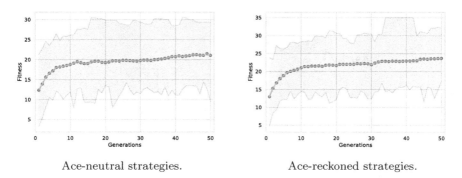

Ace-neutral strategies.   Ace-reckoned strategies.

Fig. 2. Average, minimum and maximum fitness values observed over the 2×150 runs of the Genetic Algorithm.

the evolved ace-neutral strategies have a very high IC, the biggest observed one being 0.9735, while the average IC value of the strategies from Table 1 is just 0.8. The average again exceeds even the maximum IC value of 0.91 belonging to Hi-Opt II.

Table 2. Interesting counting strategies obtained by the Genetic Algorithm.

	A	2	3	4	5	6	7	8	9	10JQK	PE	BC	IC	UPM
AN	0	2.8	2.68	3.55	4.67	3.52	2.99	1.32	-0.13	-5.35	0.6850	0.9012	0.9404	**0.9400**
AR	-2.36	2.7	3.9	4.62	4.93	4.51	3.1	0	-1.16	-5.06	0.6401	0.9702	0.8485	**0.9110**

Table 2 contains the best ace-neutral (AN) and ace-reckoned (AR) strategies, with the point counts truncated to two decimal points, an the value of 10-J-Q-K adjusted to maintain a balanced strategy. The AN outperforms all strategies from Table 1 on PE, IC and UPM. The AN strategy provides a great balance between PE and BC, outperforming strategies like Mentor and Zen Count that were especially developed to be suitable for both hand-held and shoe games. The only metric where the evolved strategies did not beat classical ones is BC, the highest obtained value being 0.98902, while several published strategies have a BC of 0.99.

4.1 Integer Weights

We also tested if the strategies can be converted to integer counts while also retaining their advantageous properties. For this we rounded each weight to the nearest quarter an then scaled each value by 4. Finally, we slightly increased-decreased some point counts, until the re-balancing for the 10-J-Q-K also yielded an integer value.

The resulting weights and the corresponding performance metrics can be found in the first two rows of Table 3. As can be observed from the last column,

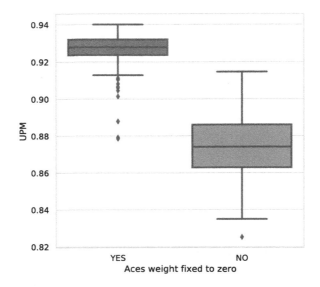

Fig. 3. Distribution of the evolves strategies UPM values

Table 3. Integer counting strategies derived from the real weights obtained by the Genetic Algorithm.

Strategy	A	2	3	4	5	6	7	8	9	10JQK	PE	BC	IC	UPM
AN_I	0	11	11	14	19	14	12	7	0	-22	0.6833	0.8953	0.9461	**0.9391**
AR_I	-9	11	16	18	20	18	12	0	-6	-20	0.6414	0.9704	0.8386	**0.9084**
Combined	-4	12	12	16	20	16	12	6	-6	-21	0.6725	0.9451	0.8907	**0.9321**

the remapping to integers leads to only a slight decrease in UPM and other metrics, the edge of the original real valued systems are retained.

The strategy from row three was obtained as a combination of the first two and provides a great balance between PE, BC, and IC, outperforming the best PE strategies Hi-Opt II and OMEGA II not only on PE but also on BC.

Unfortunately, the integer point counts are very high, making the mental counting of the running count difficult.

5   Conclusions

We have shown that by using real valued weights in card counting strategies offers a significant extra edge in the game of blackjack. Many of the developed systems have a very high Insurance Correlation value while also matching and outperforming the classical systems on Performance Efficiency and Betting Correlation.

We also found, that when re-scaling these strategies to integer point counts, most of the edge is preserved. However, this leads to high level counting strategies, that are harder to mentally operate with, and may detract players from their ability to count cards quickly and accurately. Experts suggest, that the return of a simpler and less advantageous system that can be played flawlessly for hours, typically outperforms the return of more complex systems that are prone to user error.

Future work will experiment with unbalanced and suite aware counting strategies and the application of other intelligent search algorithms.

References

1. Carlson, B.: Blackjack for Blood. Pi Yee Press (2001)
2. Coltin, K.: Optimal strategy for casino blackjack: a markov chain approach. Tech. rep. (2012)
3. Curran, D., O'Riordan, C.: Evolving blackjack strategies using cultural learning in multi-agent systems. Tech. rep., National University of Ireland, Galway
4. De Rainville, F.M., Fortin, F.A., Gardner, M.A., Parizeau, M., Gagné, C.: Deap: A python framework for evolutionary algorithms. In: Proceedings of the 14th Annual Conference Companion on Genetic and Evolutionary Computation, pp. 85–92 (2012)
5. Fogel, D.B.: Evolving strategies in blackjack. In: Proceedings of the 2004 Congress on Evolutionary Computation (IEEE Cat. No. 04TH8753), vol. 2, pp. 1427–1434. IEEE (2004)
6. Griffin, P.A.: The Theory of Blackjack: the Compleat Card Counter's Guide to the Casino Game of 21. Huntington Press, Las Vegas (1999)
7. Humble, L., Cooper, K., Cooper, C.: The world's greatest blackjack book. Main Street Books (1987)
8. Kendall, G., Smith, C.: The evolution of blackjack strategies. In: The 2003 Congress on Evolutionary Computation CEC 2003, vol. 4, pp. 2474–2481. IEEE (2003)
9. Mezrich, B.: 21: Bringing Down the House-Movie Tie-In: the Inside Story of Six MIT Students Who Took Vegas for Millions. Simon and Schuster, New York (2008)
10. Miller, B.L., Goldberg, D.E., et al.: Genetic algorithms, tournament selection, and the effects of noise. Complex Syst. 9(3), 193–212 (1995)
11. Millman, M.H.: A statistical analysis of casino blackjack. Am. Math. Mon. 90(7), 431–436 (1983)
12. Pérez-Uribe, A., Sanchez, E.: Blackjack as a test bed for learning strategies in neural networks. In: 1998 IEEE International Joint Conference on Neural Networks Proceedings. IEEE World Congress on Computational Intelligence (Cat. No. 98CH36227), vol. 3, pp. 2022–2027. IEEE (1998)
13. Renzey, F.: Blackjack Bluebook II the Simplest Winning Strategies Ever Published. Blackjack Mentor (2017)
14. Revere, L.: Playing Blackjack as a Business: a Proffessional Player's Approach to the Game of "21". Lyle Stuart, New York (2000)
15. Schlesinger, D.: Blackjack Attack: playing the Pros' Way. Huntington Press, Las Vegas (2018)
16. Snyder, A.: Blackbelt in Blackjack. Cardoza Publishing, Las Vegas (1998)
17. Thorp, E.O.: Beat the Dealer: a Winning Strategy for the Game of Twenty One, vol. 310. Vintage (1966)
18. Whitley, D.: A genetic algorithm tutorial. Stat. Comput. 4(2), 65–85 (1994)
19. Widrow, B., Gupta, N.K., Maitra, S.: Punish/reward: learning with a critic in adaptive threshold systems. IEEE Trans. Syst. Man Cybern. 5, 455–465 (1973)
20. Wong, S.: Professional blackjack. Pi Yee Press (1994)

Machine Learning

A Feature Selection Approach to Visual Domain Adaptation in Classification

Rakesh Kumar Sanodiya[1], Debdeep Paul[2], Leehter Yao[1(✉)], Jimson Mathew[2], and Aparna Juhi[3]

[1] National Taipei University of Technology, Taipei 10608, Taiwan
`rakesh.pcs16@gmail.com, ltyao@ntut.edu.tw`
[2] Indian Institute of Technology Patna, Patna, India
[3] National Institute of Technology Patna, Patna, India

Abstract. In machine learning, we presume datasets to be labeled while performing any operation. But, is it true in real-life scenarios? To its contrary, we have an enormous amount of unlabeled datasets available in the form of images, videos, audios, articles, and many more. The major challenge we face is to train our classification model with primitive machine learning algorithms because these algorithms only expect labeled data. To overcome these limitations visual domain adaptation algorithms such as MEDA (Manifold Embedded Distribution Alignment) have been introduced. The main motto of MEDA is to minimize the distribution difference between the source domain (an application that contains enough labeled data) and the target domain (an application that contains only unlabeled data). In this way, the source domain labeled data can be utilized to enhance the performance of the target domain classifier. Though MEDA (Manifold Embedded Distribution Alignment) approach shows remarkable improvement in classification accuracy, but still there is considerable scope of improvement. There are plenty of irrelevant features in both domains. These irrelevant features create a hole for this algorithm and prevent the target domain classifier from becoming more robust. Therefore, for the purpose of filling this hole, we propose a new feature selection based visual domain adaptation (FSVDA) method which uses particle swarm optimization (PSO), where the MEDA method is considered as a fitness function that leads to automatically select a good subset of features over both the domains. Extensive experimental results on two real-world domain adaptation (DA) data sets such as object recognition and digit recognition demonstrate that our proposed method outperforms state-of-the-art primitive and domain DA algorithms. It is a big challenge to train the classifier for a new unlabeled image dataset in image classification and computer vision. The two magnificent solutions to this challenge are transfer learning and domain adaptation. By transfer learning, we can use our knowledge from previously trained models for training newer models.

Keywords: Transfer learning · Feature selection · Domain adaptation · Classification.

© Springer Nature Switzerland AG 2020
H. Yang et al. (Eds.): ICONIP 2020, LNCS 12533, pp. 77–89, 2020.
https://doi.org/10.1007/978-3-030-63833-7_7

1   Introduction

Machine learning (ML) is one of the most compelling recent technology which attempts to imitate how the human brain learns. Some of the ML applications which are being frequently used in our day to day life are image recognition, speech recognition, effective web search etc. [4]. The ML algorithm intends to discover and exploit the hidden patterns accessible in the training data. And those patterns can be used to identify new or unknown patterns in test data. The main constraint with the primitive ML algorithm is that both training and test data must follow the same distribution. Therefore, they were incapable to withstand any shifts from the training and test data. But recent technological advances such as Transfer Learning (TL) and Domain Adaptation (DA) [5,8,15] have made these algorithms flexible and thus increased its appropriateness in real-world applications.

Domain adaptation is a form of transfer learning, which intends to transfer information from the source domain to enhance the performance of the target classifier. If a distributional change happens after learning a classifier, it can lead to the degradation of performance during test time. The DA technique attempts to decrease the distribution difference between the source domain labeled data and the target domain unlabeled data. In the literature, DA method is compiled into two major divisions:(a) feature-based DA [8]- It aims to discover a universal feature space between both the domains; and (b) instance re-weighting based DA [6]- It attempts to re-weight the source domain instances to reduce the distribution difference between domains. Since our proposed work is concerned to feature learning, we concentrate on methods of feature learning. In the existing feature-based DA methods, the two most important challenges are unevaluated distribution alignment and degenerated feature transformation. As the name suggests, 'Unevaluated distribution alignment' means not evaluating the relative importance of marginal and conditional distributions i.e treating both the distributions with equal importance. It may result in degraded model performance because we can encounter scenarios where assigning equal weights to both the distribution may not help, like- if both domains are very different from each other, more weight should be assigned to the marginal distribution alignment. Similarly, if the distance between both the domains is marginally distributed, then more weight should be assigned to the conditional distribution. 'Degenerated feature transformation' means that existing distribution alignment [8] and subspace learning [1] methods are only able to reduce, but are not able to eliminate, the distribution differences between domains. For example, subspace learning methods only attempts to transform the subspaces of both domains for good feature representation but fails to eliminate feature deviations as it only deals with subspace structure. Whereas the distribution alignment methods only minimize the distribution distance in the original feature space. It is quite difficult to minimize divergence between domains due to the presence of distortion in the majority of the features. Therefore we need to take advantage of both subspace learning and distribution alignment methods to enhance the performance of DA methods.

To overcome the threat of degenerated feature transformation, most existing work like Joint Geometrical and Statistical Alignment (JGSA)[14] acknowledges both subspace learning and distribution alignment. However, these methods suffer from some drawbacks like- (i) They fail to assess the significance of both marginal and conditional distributions, (ii) They do not consider the Laplacian regularization term, which is required to conserve the original similarity of the data samples. Therefore, to address all these challenges, Wang et al. [12] propose a Manifold Embedded Distribution Alignment (MEDA) DA method. MEDA first uses original features of both domains for learning manifold features to eliminate the threat of degenerated feature transformation. Then, it performs dynamic distribution alignment to quantitatively measure the relative significance of both the distributions to overcome the limitation of unevaluated distribution alignment. However, if key features in both domains are too irregular (or there are plenty of irrelevant features), even after performing manifold feature learning, the risk of degenerated feature transformation may not be eliminated.

Fortunately, in the literature, there are some feature selection approaches [7], which can overcome the risk of degenerated feature transformation by selecting a good subset of features across both the domains. The task of feature selection is quite difficult when there are a huge number of features because a) Search space varies exponentially with respect to the number of features;b) The interaction between the features is complicated. Some of the features in our domain are invariant i.e they have no change in characteristics under different circumstances, while some possess high distinguishing capacity (relevant features). It is equitably crucial to select both types of features. Selecting only highly relevant features can't serve the purpose, to its contrary it can lead to redundant features. So, weakly relevant features should also be considered to increase the classification performance. To address this issue, we need to do a global search, and one of the most simple EC (Evolutionary Computation) technique is PSO [13].

The principal contributions of this work can be listed as follows:

- Our proposed method FSVDA is the first framework, which crosses the limits of all existing state-of-the-art methods, by selecting a good subset of features and considering MEDA's proposed functions as our fitness function.
- To prove the strength of our proposed methodology, we have considered two broadly utilized real-world visual domain adaptation problem datasets- Office + Caltech10 and Digit Recognition datasets. We have done a comparison between them and several state-of-the-art primitive and domain adaptation methods.
- To make fair comparisons, we have taken the outcomes of previous methods straight from previous papers for both datasets and also explained why our method is superior to others.

2 Related Work

Although much research has been done in the domain of DA, we have only discussed those methods that are more related to the current work.

Transfer Component Analysis (TCA) [8] method attempts to get transfer components across domains in a Reproducing Kernel Hilbert Space (RKHS) with the help of Maximum Mean Discrepancy (MMD). Joint Distribution Adaptation (JDA) [5] advances TCA by considering both marginal and conditional distribution for reducing the distribution gap between both domains. Transfer Joint Matching (TJM) [5] extends JDA by considering both feature learning and instance re-weighting into a common framework. Correlation Alignment (CORAL) [11], reduces domain shift by aligning the second-order statistics of source and target distributions, without demanding any target labels. Geodesic Flow Kernel (GFK) [3] minimizes the distribution gap by integrating an infinite number of subspaces from the source domain to domain. Scattered component analysis (SCA) [2] explores such a representation that weighed between maximizing classes isolation, lessening mismatch between domains, and maximizing data isolation; each is determined by scattering. JGSA [14] succeeds the shortcomings of TCA and JDA by acknowledging the following objectives into a collaborative framework: (i)preserving source domain discriminative information (SDI) using linear discriminant analysis(LDA),(ii) preserving both marginal and conditional distributions (MCD), (iii) subspace alignment (SA),(iv) maximizing target domain variance (TV). The most recent method MEDA eliminates the drawbacks of JGSA by evaluating the significance of both marginal and conditional distributions. In order to preserve the fundamental similarity of data samples, it considers the Laplacian regularization(LR) term.

3   A Feature Selection Approach to Visual Domain Adaptation in Classification

3.1   Particle Swarm Optimization

Particle swarm optimization (PSO) algorithm is a stochastic optimization technique which was proposed by Eberhart and Kennedy (1995) [13]. It is originated from the simulation of flock of birds, fish schooling or ant colonies seeking food. Birds are acknowledged as particles without mass and volume in the algorithm. The search space is parallelly explored using a swarm of particles. The position and velocity of each particle moving in search space are unique. During the searching process, they share their best positions with each other to guide the swarm towards the optimal solution.A particle's moving state is influenced by the speed and direction of- its neighbors and the whole particle swarm. Particles having good position and direction have a greater tendency to approach the optimal solution. We represent the position and velocity by numeric vectors, whose lengths are equal to the number of dimensions in the search space. Finally, it employs the fitness function to estimate the quality of the solution.

In this work, PSO aims to select good subsets of common features over the source and target domains. It will eradicate the risk of degenerated feature transformation. Each particle consists of a position field p (to select subsets of features in both domains) and a fitness value field v (which includes accuracy corresponding to its position field). Initially, the PSO generates a collection of random particles S(or solutions) and then seeks for an optimal particle by renewing the position field with respect to velocity. In each iteration, the velocity of each particle is renewed by two values, namely, Particle Best (PB) and Global Best (GB). PB is the local best particle in each iteration, while GB is the global best particle obtained until the current iteration. After finding both GB and PB, the velocity of each particle (i^{th} particle) is updated with the help of the following equation.

$$\mathcal{V}'(i) = \omega\mathcal{V}(i) + a_1 * d_1 * (PB.p - S.S(i)) + a_2 * d_2 * (GB.p - S.S(i)) \quad (1)$$

where, a_1 and a_2 are acceleration coefficients, ω is adaptation factor, d_1 and d_2 are random numbers ranging between 0 and 1, $PB.p$ represents position field of particle PB, $\mathcal{V}(i)$ and $S.S(i)$ are the current (or present) velocity and position of ith particle, and $\mathcal{V}'(i)$ is updated velocity of ith particle.

Based on the updated velocity $\mathcal{V}'(i)$ of ith particle, its next position field is updated as follows:

$$\mathcal{S}'(i).p = \mathcal{V}'(i) + \mathcal{S}(i).p \quad (2)$$

where $\mathcal{S}(i).p$ is the present position of ith particle.

3.2   Fitness Function

To address both the challenges, we consider MEDA algorithm's stated function $f(\cdot)$ as a fitness function for our proposed framework. This function $f(\cdot)$ takes subsets ($X'_s \in X_s$ and $X'_t \in X_s$, where X_s and X_t are the source and target domain data) of the selected common features in both domains based on the position field of each particle and gives us accuracy as a fitness value. The two important steps of MEDA are :(i)manifold feature learning; (ii)dynamic distribution alignment.

As per the first step, we need to eradicate the threat of degenerated feature transformation. It can be done by manifold feature learning. The manifold feature learning classifier $g(\cdot)$ is learnt in the Grassmann manifold $G(d)$. Geodesic Flow Kernel (GFK) [3] is used to learn $g(\cdot)$. Let S_s and S_t denote the Principle Component Analysis (PCA) subspaces for the source and target domains respectively and G denote collection of all d-dimensional subspaces. Finding a geodesic flow $\phi(t) : 0 \leq t \leq 1$ from S_s to S_t where $S_s = \phi(0)$ and $S_t= \phi(1)$, can be seen as an incremental way of walking from $\phi(0)$ to ϕ (1).The new features can be represented as $z = g(x) = \phi(t)^T x$. The inner product of transformed features z_i and z_j gives rise to a PSD geodesic flow kernel:

$$< z_i, z_j > = \int_0^1 (\phi(t)^T x_i)^T (\phi(t)^T x_j) dt = x_i^T G x_j \quad (3)$$

where G Grassmann manifold and can be computed by SVD decomposition.

After application of the first step, we get transformed features Z_s and Z_t of source domain and the target domain, respectively. Then, according to the second step, we need to evaluate the importance of aligning marginal (P) and conditional (Q) distributions in domain adaptation. Many methods assume that the importance of both distributions is same, but this assumption doesn't fit for real applications. Therefore, Wang et al.[12] proposed an adaptation factor δ to tackle this problem.In order to measure the distribution divergence between domains empirically, MMD [8] method is widely adopted. The MMD distance between distributions q and r can be calculated as $d^2(q,r) = \|E_q[\phi(z_s)] - E_r[\phi(z_t)]\|^2_{\mathcal{H}_K}$, $\mathcal{H}_K$ is the reproducing kernel Hilbert space (RKHS) induced by feature map $\phi(\cdot)$ and $E[\cdot]$ denotes the mean of the embedded samples. After associating MMD with function $f(\cdot)$, the marginal distribution between both domains can be computed as $\mathcal{D}_f(P_s, P_t) = \|E[f(z_s)] - E[f(z_t)]\|^2_{\mathcal{H}_K}$. Similarly, the conditional distribution between both domains can be computed as $\mathcal{D}_c^f(Q_s, Q_t) = \|E[f(z_s^{(c)})] - E[f(z_t^{(c)})]\|^2_{H_K}$, where c denotes c^{th} class samples. Thus, according to [12], both the distribution alignment terms can be added by using dynamic adaptation factor δ as follows:

$$\mathcal{D}_f^d(Z_s, Z_t) = (1 - \delta)\|E[f(z_s)] - E[f(z_t)]\|^2_{\mathcal{H}_K} + \delta \sum_{c=1}^{C} \|E[f(z_s^{(c)})) - E[f(z_t^{(c)})]\|^2_{\mathcal{H}_K} \tag{4}$$

When $\delta \to 0$ the second term of the equation will become 0, which suggests that the marginal distribution alignment is more significant. It suggests that the distribution distance among both domains is huge. Whereas, when $\delta \to 1$, the first term of the equation becomes 0, which suggests that the conditional distribution alignment is more significant. It means that the distribution of each class is dominant. When $\delta = 0.5$, both distributions have equal significance. Parameter sensitivity testing is one way to find its proper value, but it is a time-consuming process. Therefore, Wang et al.[12] developed an idea to find out its appropriate value automatically.

The overall learning function ($f(\cdot)$) of MEDA method can be learned by summarizing these two steps with the function ($l(\cdot)$) of SRM (Structural risk minimization) [10] and a Laplacian regularization term [9] as follows:

$$f = \underset{f \in \sum_{i=1}^{n} \mathcal{H}_K}{\arg\min} \; l(f(z_i), y_i) + \eta\|f\|^2_K + \lambda\mathcal{D}_f^d(Z_s, Z_t) + \rho R_f(Z_s, Z_t) \tag{5}$$

where η, λ, and ρ are regularization parameters, n is the total number samples available in both domains, $\|f\|^2_K$ is the squared norm of f, and finally $\mathcal{D}_f^d(\cdot, \cdot)$ and $R_f(\cdot, \cdot)$ represent dynamic distribution alignment term and Laplacain regularization term, respectively. The optimization procedure of this function f is given in [12].

Algorithm 1: FSVDA

    Input   : Input data:X_s, X_t, Y_s, Y_t

    Output: Selected Positions:Pos, Accuracy:Acc

    /* Initialize Constant: */

1   $i = 0, m_f = 800, n_f = 750, p_s = 100, a_1 = 1, a_2 = 1, \omega = 0.5, T = 10$;

    /* Initialize Particles: */

2   $GB.v(i) = 0, \mathcal{V} = ones(p_s, n_f), \mathcal{P}_o = zeros(p_s, n_f), S = $ struct

    /* Initializing each row of $\mathcal{P}_o$ with the feature index or position
       randomly generated between 1 to m_f */

3   **for** $j \leftarrow 1$ **to** p_s **do**

4 | $\mathcal{P}_o(j, n_f) = $ randperm(m_f);

5   **end**

6   **while** $i < T$ **do**

      /* Generate p_s particles for population set matrix S by MEDA's
        function $f(\cdot)$ */

7   **for** $j \leftarrow 1$ **to** p_s **do**

8 | $S(i).p(j,:) = \mathcal{P}_o(j,:)$;

9 | $S(i).v(j) = f(\mathcal{P}_o(j,:))$;

10   **end**

      /* Find a particle with maximum fitness value in matrix P and
        then it will become PB particle   */

11      $[PB.v(i), ind] = max(S(i).v)$;/* where ind is the index of best
        particle in P   */

12      $PB.p(i,:) = S(i).p(ind,:)$;

      /* Find GB particle after comparing with PB particle   */

13   **if** $(GB.v(i) < PB.v(i))$ **then**

14 | $[GB.v(i), GB.v(i+1)] = PB.v(i)$;

15 | $GB.p(i,:) = PB.p(i,:)$;

16   **else**

17 | $GB.v(i+1) = GB.v(i)$;

18 | $GB.p(i+1,:) = GB.p(i,:)$;

19   **end**

      /* Update the velocity matrix $\mathcal{V}'$ and the position matrix $\mathcal{P}'_o$ based
        on PB and GB particles   */

20   **for** $j \leftarrow 1$ **to** p_s **do**

21 | $\mathcal{V}'(j,:) = \omega * \mathcal{V}(j,:) + a_1 * d_1(PB.p(j,:) - \mathrm{P}_o(j,:)) + a_2 * d_3(GB.p(j,:) - \mathrm{P}_o(j,:))$

22 | $\mathcal{P}'_o(j,:) = \mathcal{P}_o(j,:) + \mathrm{V}'(j,:)$

23   **end**

24   i=i+1;

25   **end**

    /* Select the global best particle   */

26 Pos=$GB.p(i,:)$, Acc:$GB.v(i)$

27 Return (Pos, Acc)

3.3   Main Idea

The goal of our proposed feature selection technique based on PSO in the DA framework is to find out an optimal position vector p for deciding good subsets of common features across both domains so that the risk of degenerated feature transformation(Even after features in both domains are too much distorted) is eliminated. The pseudo-code of the introduced framework FSVDA is shown in Algorithm 1.

The working steps of proposed FSVDA Algorithm is as follows:

- **Step 1 (or line no. 1)**:Initialize all constant parameter values such as maximum number of features available for selection (m_f), number of features for selection (n_f), population size (p_s), current iteration (i), maximum number of iterations (T), a_1, a_2, ω etc.
- **Step 2 (or line no. 2)**: As the given population size is (p_s), we have total (p_s) particles in population set S,i.e. $(S = s_1, s_2, \ldots, s_{p_s})$ and each particle has two fields, namely position (p) and value (v), where the position field contains the positions for selecting subset of features across both domains, while the value field includes corresponding accuracy calculated from MEDA algorithm's proposed function $f(\cdot)$. $\mathcal{V}$ is the velocity parameter matrix that contains the velocity of each particle present in the population set S and is of size $(p_s \times n_f)$. Similarly, $\mathcal{P}_o$ is the position parameter matrix that contains positions of each particle present in the population set S for selecting subsets of common features across both the domains and is of size $(p_s \times n_f)$. In this step, the value field of GB particle is initialized to zero, velocity matrix $(\mathcal{V})$ is initialized with one(i.e., $ones(p_s, n_f)$), and the position matrix $(\mathcal{P}_o)$ is initialized with zero (i.e., $zeros(p_s, n_f)$).
- **Step 3 (or line no. 3–5)**: In this step, we initialize each row of the position matrix $(\mathcal{P}_o)$ with permutation of numbers from 1 to m_f for selecting subsets of features across both domains.
- **Steps 4 (or line no. 7–10)**: In this step, based on each entry (or positions) in the position matrix $(\mathcal{P}_o)$, the subsets of features across both both domains are selected and then the fitness value corresponding to these features is calculated by using MEDA's function $(f(\cdot))$. Later, all the entries and computed corresponding fitness values are assigned to respective fields in the particle set S.
- **Step 5 (or line no. 11–12)**: A particle that has the highest fitness value or accuracy is chosen among all the particles in the population set S. And then this particle will become the PB particle.
- **Step 6 (or line no. 13–19)**: If the value field of PB particle is greater than the value field of GB particle (i.e., $PB.v > GB.v$), then GB particle is assigned with the PB particle.
- **Step 7 (or line no. 20–23)**: First compute new velocity of each particle by using Eq. (1) and update the velocity matrix $(\mathcal{V})$. After that, we find new velocity matrix $(\mathcal{V}')$. Similarly, new velocity matrix $(\mathcal{V}')$, we find new position matrix $(\mathcal{P}'_o)$ by using Eq. (2).

- **Step 8 (or line no. 20–23):** Repeat steps 3–7 until the maximum number of iterations are completed or until there is no optimal value adjustment.
- **Step 9 (or line no. 27):** Return best particle position vector and corresponding accuracy.

After the application of FSVDA approach, we obtain best position vector (p) for choosing best subsets of features across both domains and corresponding accuracy (v).

Table 1. Comparison of the proposed method with other state-of-the-art methods on the Office + Caltech10 and digit recognition data-sets

	Primitive Algorithms			Domain Adaptation Algorithms								
Office+Caltech10 datasets using SURF features.												
Task	1NN	PCA	SVM	GFK	TCA	JDA	CORAL	TJM	SCA	JGSA	MEDA	Proposed
C_A	23.7	39.5	53.1	46.0	45.6	43.1	52.1	46.8	45.6	51.5	56.5	**58.04**
C_W	25.8	34.6	41.7	37.0	39.3	39.3	46.4	39.0	40.0	45.4	53.9	**56.45**
C_D	25.5	44.6	47.8	40.8	45.9	49.0	45.9	44.6	47.1	45.9	50.3	**60.51**
A_C	26.0	39.0	41.7	40.7	42.0	40.9	45.1	39.5	39.7	41.5	43.9	**45.86**
A_W	29.8	35.9	31.9	37.0	40.0	38.0	44.4	42.0	34.9	45.8	53.2	**58.98**
A_D	25.5	33.8	44.6	40.1	35.7	42.0	39.5	45.2	39.5	47.1	45.9	**50.96**
W_C	19.9	28.2	28.8	24.8	31.5	33.0	33.7	30.2	31.1	33.2	34.0	**35.26**
W_A	23.0	29.1	27.6	27.6	30.5	29.8	36.0	30.0	30.0	39.9	42.7	**44.47**
W_D	59.2	89.2	78.3	85.4	91.1	92.4	86.6	89.2	87.3	90.5	88.5	**92.36**
D_C	26.3	29.7	26.4	29.3	33.0	31.2	33.8	31.4	30.7	29.9	34.9	**37.67**
D_A	28.5	33.2	26.2	28.7	32.8	33.4	37.7	32.8	31.6	38.0	41.2	**43.84**
D_W	63.4	86.1	52.5	80.3	87.5	89.2	84.7	85.4	84.4	91.9	87.5	**90.51**
Avg.	31.4	43.6	41.1	43.1	46.2	46.8	48.8	46.3	45.2	50.0	52.7	**56.45**
Office+Caltech10 datasets using DeCaf6 features												
Task	1NN	PCA	SVM	GFK	TCA	JDA	CORAL	TJM	SCA	JGSA	MEDA	Proposed
C_A	87.3	88.1	91.6	88.2	89.8	89.6	92.0	88.8	89.5	91.4	93.4	**94.11**
C_W	72.5	83.4	80.7	77.6	78.3	85.1	80.0	81.4	85.4	86.8	95.6	**96.23**
C_D	79.6	84.1	86.0	86.6	85.4	89.8	84.7	84.7	87.9	93.6	91.1	**95.51**
A_C	71.7	79.3	82.2	79.2	82.6	83.6	83.2	84.3	78.8	84.9	87.4	**88.35**
A_W	68.1	70.9	71.9	70.9	74.2	78.3	74.6	71.9	75.9	81.0	88.1	**93.22**
A_D	74.5	82.2	80.9	82.2	81.5	80.3	84.1	76.4	85.4	88.5	88.1	**95.90**
W_C	55.3	70.3	67.9	69.8	80.4	84.8	75.5	83.0	74.8	85.0	93.2	**94.18**
W_A	62.6	73.5	73.4	76.8	84.1	90.3	81.2	87.6	86.1	90.7	99.4	**99.42**
W_D	98.1	99.4	100.0	100.0	100.0	100.0	100.0	100.0	100.0	100.0	99.4	**100.0**
D_C	42.1	71.7	72.8	71.4	82.3	85.5	76.8	83.8	78.1	86.2	87.5	**88.18**
D_A	50.0	79.2	78.7	76.3	89.1	91.7	85.5	90.3	90.0	92.0	93.2	**94.02**
D_W	91.5	98.0	98.3	99.3	99.7	99.7	99.3	99.3	98.6	99.7	97.6	**99.66**
Avg.	71.1	81.7	82.0	81.5	85.6	88.2	84.7	86.0	85.9	90.0	92.8	**94.89**
USPS+MNIST												
Task	1NN	PCA	SVM	GFK	TCA	JDA	CORAL	TJM	SCA	JGSA	MEDA	Proposed
U_M	44.7	45.0	62.2	46.5	51.2	59.7	30.5	52.3	48.0	68.2	72.1	**74.3**
M_U	65.9	66.2	68.2	61.2	56.3	67.3	49.2	63.3	65.1	80.4	89.5	**90.5**
Avg.	55.3	55.6	65.2	53.85	53.75	63.5	39.85	57.8	56.55	74.3	80.8	**82.4**

4   Experiment Section

This section deals with the thorough experiments which we have conducted for image classification problems to estimate the performance of the FSVDA strategy on various primitive and DA methods.

4.1   Data Preparation

Office+Caltech10 and Digit Recognition [14] are the widely chosen public image datasets for evaluating the performance of visual DA algorithms. Office+Caltech10 dataset consists of two data-sets such as Office-31 and Caltech10. Office-31 dataset comprises of three real-world object domains: Amazon (A), DSLR (D), and Webcam (W). It contains 4,652 images with 31 categories. Caltech-256 (C) comprises of 30,607 images and 256 categories. The objects in Office and Caltech datasets hold distinct distributions which aid to implement cross-domain recognition. For the experiment, we have considered 10 common classes available in both the data-sets. Since there are 4 domains, a total of 12 tasks will be possible such as $C \rightarrow A, \ldots, D \rightarrow W$, where ($W \rightarrow D$ denotes transfer of knowledge from source domain Webcam to target domain DSLR). Similar to previous work [12,15], we have also considered both 800 SURF and 4,096 DeCaf6 features for these datasets. We have extended our experiment to Digital Recognition Data-sets also, like- USPS and MNIST [14]. It consists of handwritten digits between 0 and 9. USPS is comprised of 7,291 training images and 2,007 test images of size 16 ×16 whereas MNIST is comprised of 60,000 training images and 10,000 test images of size 28 × 28. As there are two domains, a total of 2 tasks will be possible such as $M \rightarrow U$ and $U \rightarrow M$. Here, for experiment, we have considered 256 SURF features of both domains.

4.2   State-of-the-Art Comparison Methods

We have compared here, the performance of our proposed algorithm to various state-of-the-art traditional approaches like 1NN, SVM, and PCA. The domain adaptation methods are TCA[8] (performs marginal distribution), GFK [6](performs manifold feature learning), JDA[6] (assign equal weights to marginal and conditional distribution), CORAL [1](performs second-order subspace alignment), TJM [5] (adapts marginal distribution with source sample selection by instance re-weighting), SCA [2] (adapts scatters in subspace), JGSA[14](considers both distribution and subspace alignment), MEDA[12] (addresses degenerate feature transformation and unevaluated distribution alignment).

4.3   Parameter Sensitivity and Experimental Setup

Since the performance of an algorithm depends on the appropriate value of each parameter, we need to implement parameter sensitivity tests manually to determine them. As our proposed method uses MEDA's proposed function as a fitness function, we consider appropriate values of all its parameters reported in [12] for the experiment. However, for the parameters of our proposed PSO method, we have performed parameter sensitivity tests and found that the proposed method is performing remarkably well for all the considered data-sets for the parameter values like $T = 10, a_1 = 1, a_2 = 2, p_s = 100$, and $w = 0.5$. But for the parameter m_f, the proposed method outperforms at $m_f = 750$ for Office+Caltech10 using SURF features, at $m_f = 3000$ for Office+Caltech10 using DeCaf6 features, and finally at $m_f = 220$ for digit recognition using SURF features datasets.

After finding the appropriate value of each parameter, we experimented with the proposed method with those values for all tasks in both data-sets and reported the performance as accuracy in Table 1. The results reported in Table 1 for other methods, we have directly taken them from previous domain adaptation papers [5,12,14].

4.4   Comparative Analysis

By comparing the experimental results given in Table 1, the following observations can be drawn:

1. The performance of primitive machine learning algorithms such as 1NN, PCA, and SVM are not notable due to the distribution differences between both source and target domain data.
2. Earlier methods such as TCA, GFK, TCA, JDA, CORAL, TJM, and SCA are not performing well due to the lack of one or more import objectives (as discussed above) such as SA, MCD, TV, LR, and SDI, in their proposed functions.
3. The performance of JGSA is remarkable compared to other DA approaches except for MEDA and the proposed methodology. This is because of considering all the above important objectives except LR, in a common framework. However, it still suffers from degenerated feature transformation and unevaluated distribution alignment problems.
4. MEDA improves JGSA by overcoming the limitations of degenerated feature transformation and unevaluated distribution alignment problems. It also considers the Laplacian regularization term for preserving the original similarity of data samples. Therefore, it outperforms the other DA methods(except ours). However, if the original features itself are distorted, it suffers from degenerated feature transformation problem.
5. Our proposed method eliminates the problem of distorted features by selecting appropriate good subsets of features across both domains. Therefore, in Table 1, it is clearly shown that the average accuracies (i.e., 55.45%, 94.89%, and 82.4%) for all tasks of the various datasets of our proposed method are excellent over all other primitive and domain adaptation methods.

5   Conclusions

In this work, we have proposed a novel feature selection based visual domain adaptation (FSVDA) method for real-world visual domain adaptation problems.

In contrast to previous studies, the proposed method selects appropriate subsets of features across both domains and considers a more robust fitness function for guiding the PSO algorithm. Each time we select a set of features and then perform dynamic distribution alignment for manifold domain adaptation until we get a set of good features. MEDA has successfully tackled both the challenges of degenerated feature transformation and unevaluated distribution alignment. It has also preserved the relative significance of marginal and conditional distribution in domain adaptation. Experimental results evaluation on two real-world domain adaptation problems demonstrates the superiority of our proposed method over the state-of-the-art methods.

References

1. Fernando, B., Habrard, A., Sebban, M., Tuytelaars, T.: Unsupervised visual domain adaptation using subspace alignment. In: Proceedings of the IEEE International Conference on Computer Vision, pp. 2960–2967 (2013)
2. Ghifary, M., Balduzzi, D., Kleijn, W.B., Zhang, M.: Scatter component analysis: a unified framework for domain adaptation and domain generalization. IEEE Tran. Pattern Anal. Mach. Intell. **39**(7), 1414–1430 (2016)
3. Gong, B., Shi, Y., Sha, F., Grauman, K.: Geodesic flow kernel for unsupervised domain adaptation. In: Proceedings of the IEEE Conference on Computer Vision and Pattern Recognition (CVPR 2012), pp. 2066–2073. IEEE (2012)
4. Ionescu, B., Lupu, M., Rohm, M., Gînsca, A.L., Müller, H.: Datasets column: diversity and credibility for social images and image retrieval. ACM SIGMultimedia Rec. **9**(3), 7 (2018)
5. Long, M., Wang, J., Ding, G., Pan, S.J., Philip, S.Y.: Adaptation regularization: a general framework for transfer learning. IEEE Trans. Knowl. Data Eng. **26**(5), 1076–1089 (2014)
6. Long, M., Wang, J., Ding, G., Sun, J., Yu, P.S.: Transfer feature learning with joint distribution adaptation. In: Proceedings of the IEEE International Conference on Computer Vision, pp. 2200–2207 (2013)
7. Nguyen, B.H., Xue, B., Andreae, P.: A particle swarm optimization based feature selection approach to transfer learning in classification. In: Proceedings of the Genetic and Evolutionary Computation Conference, pp. 37–44 (2018)
8. Pan, S.J., Tsang, I.W., Kwok, J.T., Yang, Q.: Domain adaptation via transfer component analysis. IEEE Trans. Neural Networks **22**(2), 199–210 (2011)
9. Sanodiya, R.K., Mathew, J.: A framework for semi-supervised metric transfer learning on manifolds. Knowl.-Based Syst. **176**, 1–14 (2019)
10. Shawe-Taylor, J., Bartlett, P.L., Williamson, R.C., Anthony, M.: Structural risk minimization over data-dependent hierarchies. IEEE Trans. Inf. Theory **44**(5), 1926–1940 (1998)
11. Sun, B., Saenko, K.: Subspace distribution alignment for unsupervised domain adaptation. BMVC. **4**, 24–1 (2015)
12. Wang, J., Feng, W., Chen, Y., Yu, H., Huang, M., Yu, P.S.: Visual domain adaptation with manifold embedded distribution alignment. In: 2018 ACM Multimedia Conference on Multimedia Conference, pp. 402–410. ACM (2018)
13. Zeugmann, T., et al.: Particle swarm optimization, Encyclopedia of machine learning, Springer, US, pp. 760–766 (2011)

14. Zhang, J., Li, W., Ogunbona, P.: Joint geometrical and statistical alignment for visual domain adaptation. In: Proceedings of the IEEE Conference on Computer Vision and Pattern Recognition, pp. 1859–1867 (2017)
15. Zhang, J., Li, W., Ogunbona, P.: Transfer learning for cross-dataset recognition: a survey. arXiv preprint arXiv:1705.04396 (2017)

A Framework for Reinforcement Learning with Autocorrelated Actions

Marcin Szulc, Jakub Łyskawa, and Paweł Wawrzyński[(✉)] [iD]

Institute of Computer Science, Warsaw University of Technology, Warsaw, Poland
{marcin.szulc.stud,jakub.lyskawa.stud,pawel.wawrzynski}@pw.edu.pl

Abstract. The subject of this paper is reinforcement learning. Policies are considered here that produce actions based on states and random elements autocorrelated in subsequent time instants. Consequently, an agent learns from experiments that are distributed over time and potentially give better clues to policy improvement. Also, physical implementation of such policies, e.g. in robotics, is less problematic, as it avoids making robots shake. This is in opposition to most RL algorithms which add white noise to control causing unwanted shaking of the robots. An algorithm is introduced here that approximately optimizes the aforementioned policy. Its efficiency is verified for four simulated learning control problems (Ant, HalfCheetah, Hopper, and Walker2D) against three other methods (PPO, SAC, ACER). The algorithm outperforms others in three of these problems.

Keywords: Reinforcement learning · Actor-Critic · Experience replay · Fine time discretization

1 Introduction

The usual goal of Reinforcement Learning (RL) to optimize a policy that samples an action on the basis of a current state of a learning agent. The only stochastic dependence between subsequent actions is through state transition: The action moves the agent to another state which determines the distribution of another action. Main analytical tools in RL are based on this lack of other dependence between actions. E.g., for a given policy, its value function expresses the expected sum of discounted rewards the agent may expect starting from a given state. The sum of rewards does not depend on actions taken before the given state was reached. Hence, only the given state and the policy matter.

Lack of dependence between actions beyond state transition leads to several difficulties. In physical implementation of RL, e.g. in robotics, it usually means that white noise is added to control actions. However, that makes control discontinuous and rapidly changing all the time. This is often impossible to implement since electric motors that are to execute these actions can not operate this way. Even if it is possible, it requires a lot of energy, makes the controlled system shake, and exposes it to damages.

© Springer Nature Switzerland AG 2020
H. Yang et al. (Eds.): ICONIP 2020, LNCS 12533, pp. 90–101, 2020.
https://doi.org/10.1007/978-3-030-63833-7_8

It is also questionable if the lack of dependence between actions beyond state transition does not reduce efficiency of learning. Each action is an experiment that leads to policy improvement. However, due to limited accuracy of (action-)value function approximation, consequences of a single action may be difficult to recognize. The finer the time discretization, the more serious this problem becomes. Consequences of a random experiment distributed over several time instants could be more tangible thus easier to recognize.

The contribution of this paper may be summarized in the following points:

- A framework is introduced in which a policy produces actions on the basis of states and values of a stochastic process. That enables relation between actions that is beyond state transition.
- An algorithm is introduced that approximately optimizes the aforementioned policy.
- The above algorithm is tested on four benchmark learning control problems: Ant, Half-Cheetah, Hopper, and Walker2D.

The rest of the paper is organized as follows. Section 2 overviews related literature. Section 3 introduces a policy that produces autocorrelated actions along with tools for its analysis. Section 4 introduces an algorithm that approximately optimizes that policy. Section 5 presents simulations that compare the presented algorithm with state-of-the-art reinforcement learning methods. The last section concludes the paper.

2   Related Work

2.1   Stochastic Dependence Between Actions

The idea of introducing stochastic dependence between actions was analyzed in [16] as a remedy to problems with application of RL in fine time discretization. The control process was divided there into "non-Markov periods" in which actions were stochastically dependent. A policy with autocorrelated actions was analyzed in [18] with a standard RL algorithm applied to its optimization that did not account for the dependence of actions.

In [5] a policy was analyzed whose parameters were incremented by the autoregressive stochastic process. Essentially, this resulted in autocorrelated random components of actions. In [8] a policy was analyzed that produced an action being a sum of the autoregressive noise and a deterministic function of state. However, no learning algorithm was presented in this paper that accounted for specific properties of this policy.

2.2   Reinforcement Learning with Experience Replay

The Actor-Critic architecture of reinforcement learning was introduced in [1]. Approximators were applied to this structure for the first time in [7]. In order to boost efficiency of these algorithms, they were combined with experience replay for the first time in [17].

Application of experience replay to Actor-Critic encounters the following problem. The learning algorithm needs to estimate quality of a given policy on the basis of consequences of actions that were registered when a different policy was in use. Importance sampling estimators are designed to do that, but they can have arbitrarily large variance. In [17] that problem was addressed with truncating density ratios present in those estimators. In [15] specific correction terms were introduced for that purpose.

Another approach to the aforementioned problem is to prevent the algorithm from inducing a policy that differs too much from the one tried. That idea was first applied in Conservative Policy Iteration [6]. It was further extended in Trust Region Policy Optimization [12]. This algorithm optimizes a policy with the constraint that the Kullback-Leibler divergence between that policy and the tried one should not exceed a given threshold. The K-L divergence becomes an additive penalty in Proximal Policy Optimization algorithms, namely PPO-Penalty and PPO-Clip [13].

A way to avoid the problem of estimating quality of a given policy on the basis of the tried one is to approximate the action-value function instead of estimating the value function. Algorithms based on this approach are Deep Q-Network (DQN) [11], Deep Deterministic Policy Gradient (DDPG) [10], and Soft Actor-Critic (SAC) [4]. In the original version of DDPG the time-correlated OU noise was added to action. However, this algorithm was not adapted to this fact in any specific way. SAC uses white noise in actions and it is considered one of the most efficient in this family of algorithms.

3   Policy with Autocorrelated Actions

Let an action, a_t, be computed as

$$a_t = \pi(s_t, \xi_t; \theta) \tag{1}$$

where π is a deterministic transformation, s_t is a current state, θ is a vector of trained parameters, and $(\xi_t, t = 1, 2, \dots)$ is a stochastic process. We require this process to have the following properties:

- Stationarity: The distribution of ξ_t is the same for each t.
- Zero mean: $E\xi_t = 0$ for each t.
- Autocorrelation decreasing with growing lag:

$$E\xi_t^T \xi_{t+k} > E\xi_t^T \xi_{t+k+1} \geq 0 \text{ for } k \geq 0. \tag{2}$$

  Essentially that means that values of the process are close to each other when they are in close time instants.
- Markov property: For any t and $k, l \geq 0$, the conditional distributions

$$(\xi_t, \dots, \xi_{t+k} | \xi_{t-1}, \dots, \xi_{t-1-l}) \quad \text{and} \quad (\xi_t, \dots, \xi_{t+k} | \xi_{t-1}) \tag{3}$$

  are the same. In words, dependence of future values of (ξ_t) on its past is entirely carried over by ξ_{t-1}.

Consequently, if only π (1) is continuous for all its arguments, and subsequent states s_t are close to each other, then the corresponding actions are close, although random. In words, they create a consistent, distributed in time experiment that can lead to policy improvement.

Example: Auto-Regressive (ξ_t). Let $\alpha \in [0, 1)$ and

$$\epsilon_t \sim N(0, C), \quad t = 1, 2, \dots$$

$$\xi_1 = \epsilon_1$$

$$\xi_t = \alpha \xi_{t-1} + \sqrt{1 - \alpha^2} \epsilon_t, \quad t = 2, 3, \dots \tag{4}$$

Figure 1 demonstrates a realization of both the white noise (ϵ_t) and (ξ_t). Let us analyze if (ξ_t) has the required properties.

Both ϵ_t and ξ_t have the same distribution $N(0, C)$. Therefore (ξ_t) is stationary and zero-mean. A simple derivation reveals that

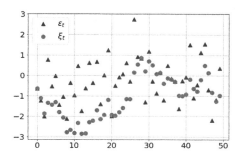

Fig. 1. A realization of the normal white noise (ϵ_t), and the auto-regressive process (ξ_t) (4).

$$E\xi_t \xi_{t+k}^T = \alpha^{|k|} C \quad \text{and} \quad E\xi_t^T \xi_{t+k} = \alpha^{|k|} \text{tr}(C)$$

for any t, k. Therefore, (ξ_t) is autocorrelated, and this autocorrelation decreases with growing lag. Consequently, the values of ξ_t are closer to one another for subsequent t than the values of ϵ_t, namely

$$E\|\epsilon_t - \epsilon_{t-1}\|^2 = E(\epsilon_t - \epsilon_{t-1})^T(\epsilon_t - \epsilon_{t-1}) = 2\text{tr}(C)$$

$$E\|\xi_t - \xi_{t-1}\|^2 = E\left((\alpha-1)\xi_{t-1} + \sqrt{1-\alpha^2}\epsilon_t\right)^T\left((\alpha-1)\xi_{t-1} + \sqrt{1-\alpha^2}\epsilon_t\right)$$

$$= (\alpha - 1)^2 \text{tr}(C) + (1 - \alpha^2)\text{tr}(C) = (1 - \alpha)2\text{tr}(C).$$

The Markov property of (ξ_t) directly results from how ξ_t (4) is computed.

In fact, marginal distributions of the process (ξ_t), as well as its conditional marginal distributions are normal, and their parameters have compact forms. We shall not present derivations of these parameters due to lack of space, but we shall denote them for further use. Namely, let as consider

$$\bar{\xi}_t^n = [\xi_t^T, \dots, \xi_{t+n-1}^T]^T. \tag{5}$$

The distribution of $\bar{\xi}_t^n$ is normal

$$N(0, \Omega_0^n), \tag{6}$$

where Ω_0^n is a matrix dependent on n, α, and C. The conditional distribution $(\bar{\xi}_t^n | \xi_{t-1})$ is also normal,

$$N(B^n \xi_{t-1}, \Omega_1^n), \tag{7}$$

where both B^n and Ω_1^n are matrices dependent on n, α, and C.

The neural-normal policy. A simple and practical way to implement π (1) is as follows. A feedforward neural network,

$$A(s; \theta), \tag{8}$$

has input s and weights θ. An action is computed as

$$a_t = \pi(s_t, \xi_t; \theta) = A(s_t; \theta) + \xi_t, \tag{9}$$

for ξ_t in the form (4). While the discussion below can be extended to the general formulation (1), in order to make it simpler we will further assume that a policy is of the form (9).

Let us consider

$$\bar{s}_t^n = [s_t^T, \ldots, s_{t+n-1}^T]^T,$$
$$\bar{a}_i^n = [a_t^T, \ldots, a_{t+n-1}^T]^T,$$
$$\bar{A}(\bar{s}_i^n; \theta) = [A(s_t; \theta)^T, \ldots, A(s_{t+n-1}; \theta)^T]^T,$$

and fixed θ. With (9) the distributions $(\bar{a}_t^n | \bar{s}_t^n)$ and $(\bar{a}_t^n | \bar{s}_t^n, \xi_{t-1})$ are both normal, namely $N(\bar{A}(\bar{s}_t^n; \theta), \Omega_0^n)$, and $N(\bar{A}(\bar{s}_t^n; \theta) + B^n \xi_{t-1}, \Omega_1^n)$, respectively (see (6) and (7)). The algorithm defined in the next section updates θ to manipulate the above distributions. Density of the normal distribution with mean μ and covariance matrix Ω will be denoted by

$$\varphi(\cdot; \mu, \Omega). \tag{10}$$

Noise-value function. In policy (1) there is a stochastic dependence between actions beyond the dependence resulting from state transition. Therefore, the traditional understanding of policy as distribution of actions conditioned on state does not hold here. Each action depends on the current state, but also previous states and actions. Analytical usefulness of the traditional value function and action-value function is thus limited.

As a valid analytical tool we propose *noise-value function* defined as

$$W^\pi(\xi, s) = E_\pi \left(\sum_{i \geq 0} \gamma^i r_{t+i} \Big| \xi_{t-1} = \xi, s_t = s \right). \tag{11}$$

The course of events starting in time t depends on the current state s_t and the value ξ_{t-1}. Because of Markov property of ξ_t (3), the pair (ξ_{t-1}, s_t) is a proper condition for the expected value of future rewards.

The *value function* $V^\pi : S \mapsto \mathbb{R}$ is slightly redefined, namely

$$V^\pi(s) = E(W(\xi_{t-1}, s_t) | s_t = s). \tag{12}$$

The random value in the above expectation is ξ_{t-1} and its distribution is conditional with the condition $s_t = s$. The distribution of ξ_{t-1} may differ for different s_t. However, being in the state s_t and not knowing ξ_{t-1} the agent may expect the sum of future rewards equal to $V^\pi(s_t)$.

4   ACERAC: Actor-Critic with Experience Replay and Autocorrelated aCtions

The algorithm presented here has Actor-Critic structure. It optimizes a policy of the form (9) and uses Critic,

$$V(s; \nu),$$

which is an approximator of the value function (12) parametrized by a vector, ν.

For each time instant of the agent-environment interaction the policy (9) is applied and a tuple, $\langle s_t, A_t, a_t, r_t, s_{t+1} \rangle$, is registered, where $A_t = A(s_t; \theta)$.

The general goal of training is to maximize $W^\pi(\xi_{i-1}, s_i)$ for each state s_i registered during the agent-environment interaction. In this order previous time instants are sampled, and sequences of actions that follow these instants are made more/less probable depending on their return. More specifically, i is sampled from $\{1, \ldots, t-1\}$ and the conditional density of the sequence of actions $(a_i, \ldots, a_{i+n-1})$ is being increased/decreased depending on the return

$$r_i + \cdots + \gamma^{n-1} r_{i+n-1} + \gamma^n V(s_{i+n}; \nu)$$

this sequence of actions yields. At the same time adjustments of the same form are performed for several sequences of actions starting from a_i, namely for $n = 1, \ldots, \tau$, where $\tau \in \mathbb{N}$ is a parameter.

4.1   Actor and Critic Training

The following procedure is repeated several times at each t-th instant of agent–environment interaction:

1. A random i is sampled from the uniform distribution over $\{1, \ldots, t-1\}$.
2. If i is the initial instant of a trial, then consider for $n = 1, \ldots, \tau$

$$\mu_{i+j} = E(\xi_{i+j}) = 0, \ j = 0, \ldots, n-1$$
$$\eta_{i+j} = E(\xi_{i+j}) = 0, \ j = 0, \ldots, n-1$$
$$\Omega_2^n = \Omega_0^n.$$

Otherwise, consider

$$\mu_{i+j} = E(\xi_{i+j} | \xi_{i-1} = a_{i-1} - A_{i-1}), \ j = 0, \ldots, n-1$$
$$\eta_{i+j} = E(\xi_{i+j} | \xi_{i-1} = a_{i-1} - A(s_{i-1}; \theta)), \ j = 0, \ldots, n-1$$
$$\Omega_2^n = \Omega_1^n.$$

3. Consider the following vectors for $n = 1, \ldots, \tau$

$$\bar{\mu}_i^n = [\mu_i^T, \ldots, \mu_{i+n-1}^T]^T,$$
$$\bar{\eta}_i^n = [\eta_i^T, \ldots, \eta_{i+n-1}^T]^T,$$
$$\bar{s}_i^n = [s_i^T, \ldots, s_{i+n-1}^T]^T,$$

$$\bar{a}_i^n = [a_i^T, \ldots, a_{i+n-1}^T]^T,$$
$$\bar{A}_i^n = [A_i^T, \ldots, A_{i+n-1}^T]^T,$$
$$\bar{A}(\bar{s}_i^n; \theta) = [A(s_i; \theta)^T, \ldots, A(s_{i+n-1}; \theta)^T]^T.$$

4. Temporal differences are computed for $n = 1, \ldots, \tau$

$$d_i^n(\theta, \nu) = \left(r_i + \cdots + \gamma^{n-1} r_{i+n-1} + \gamma^n V(s_{i+n}; \nu) - V(s_i; \nu)\right) \times$$
$$\times \psi_b\left(\frac{\varphi(\bar{a}_i^n; \bar{A}(\bar{s}_i^n; \theta) + \bar{\eta}_i^n, \Omega_2^n)}{\varphi(\bar{a}_i^n; \bar{A}_i^n + \bar{\mu}_i^n, \Omega_2^n)}\right), \tag{13}$$

where ψ_b is a soft-truncating function, e.g. $\psi_b(x) = b \tanh(x/b)$, for a certain $b > 1$.

5. Actor and Critic are updated. The improvement directions for Actor and Critic are

$$\Delta\theta = \frac{1}{\tau} \sum_{n=1}^{\tau} \nabla_\theta \ln \varphi(\bar{a}_i^n; \bar{A}(\bar{s}_i^n; \theta) + \bar{\eta}_i^n, \Omega_2^n) d_i^n(\theta, \nu) - \nabla_\theta L(s_i, \theta) \tag{14}$$

$$\Delta\nu = \frac{1}{\tau} \sum_{n=1}^{\tau} \nabla_\nu V(s_i; \nu) d_i^n(\theta, \nu), \tag{15}$$

where $L(s, \theta)$ is a loss function that penalizes Actor for producing actions that do not satisfy conditions e.g., they exceed their boundaries. $\Delta\theta$ is designed do increase/decrease the likelihood of the sequence of actions $\bar{a}_i^n$ proportionally to $d_i^n(\theta, \nu)$. $\Delta\nu$ is designed to make $V(\cdot; \nu)$ approximate the value function (12) better. The improvement directions $\Delta\theta$ and $\Delta\nu$ are applied to update θ and ν, respectively, with the use of either ADAM, SGD, or other method of stochastic optimization.

In Point 1 the algorithm selects an experienced event to replay. In Points 2 and 3 it determines the parameters the distribution of the sequence of subsequent actions, $\bar{a}_i^n$. In Point 4 it determines the relative quality of $\bar{a}_i^n$. The temporal difference (13) implements two ideas. Firstly, θ is changing due to being optimized, thus the conditional distribution $(\bar{a}_i^n | \xi_{i-1})$ is now different than it was at the time when the actions $\bar{a}_i^n$ were happening. The density ratio in (13) accounts for this discrepancy of distributions. Secondly, in order to limit variance of the density ratio, the soft-truncating function ψ_b is applied. In Point 5 the parameters of Actor, θ, and Critic, ν, are being updated.

5   Empirical Study

This section presents simulations whose purpose has been to compare the algorithm introduced in Sect. 4 to state-of-the-art reinforcement learning methods. We compared the new algorithm (ACERAC) to ACER [17], SAC [4] and PPO [13]. We used the rllib implementation [9] of SAC and PPO in the simulations.

Our implementation of ACERAC is available at https://github.com/mszulc913/acerac.

We used four control tasks, namely Ant, Hopper, HalfCheetah, and Walker2D (see Fig. 2) from PyBullet physics simulator [2] to compare the algorithms. A simulator that is more popular in the RL community is MuJoCo [14].[1] Hyperparameters that assure optimal performance of ACER, SAC, and PPO applied to the considered environments in MuJoCo are well known. However, PyBullet environments introduce several changes to MuJoCo tasks, which make them more realistic, thus more difficult. Additionally, physics in MuJoCo and PyBullets slightly differ [3], hence we needed to tune the hyperparameters. Their value can be found in appendix A.

For each learning algorithm we used Actor and Critic structures as described in [4]. That is, both structures had the form of neural networks with two hidden layers of 256 units each.

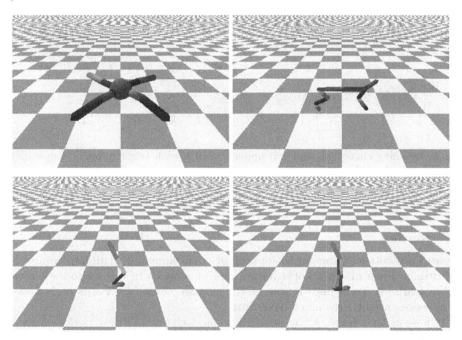

Fig. 2. Environments used in simulations: Ant (left upper), HalfCheetah (right upper), Hopper (left lower), Walker2D (right lower).

5.1 Experimental Setting

Each learning run lasted for 3 million timesteps. Every 30000 timesteps of a simulation was made with frozen weights and without exploration for 5 test

---

[1] We chose PyBullet because it is a freeware, while MuJoCo is a commercial software.

episodes. An average sum of rewards within a test episode was registered. Each run was repeated 5 times.

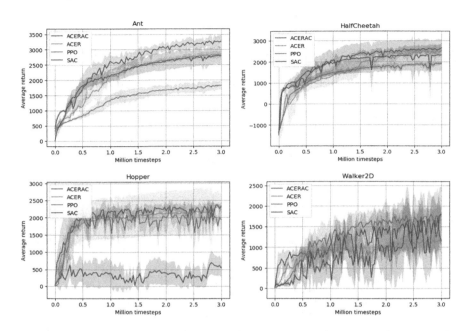

Fig. 3. Learning curves for Ant (left upper), HalfCheetah (right upper), Hopper (left lower) and Walker2D (right lower) environments: average sums of rewards in test trials.

5.2 Results

Figure 3 presents learning curves for all four environments and all four compared algorithms. Each graph reports how a sum of rewards in test episodes evolves within learning. Solid lines represent the average sums of rewards and shaded areas represent their standard deviations.

It is seen that for Ant the algorithm that achieve the best performance is ACERAC, then ACER and SAC, then PPO. For HalfCHeetah, the best performance is achieved by ACERAC which is slightly better than ACER, then SAC, then PPO. For Hopper the algorithms to win are ACERAC *ex aequo* with ACER, then PPO, then SAC; actually SAC fails in this task. Eventually, for Walker2D, PPO achieves the best performance, then ACERAC and SAC, and then ACER.

5.3 Discussion

It is seen in Fig. 3 that ACERAC is the best performing algorithm for three environments out of four (in one ACER preforms equally well). ACERAC extends

Table 1. ACERAC hyperparameters

Parameter	Value
Action std. dev. for Hopper	0.3
Action std. dev. for other envs.	0.4
α	0.5
Critic step-size for Walker2D	10^{-4}
Critic step-size for other envs.	$6 \cdot 10^{-5}$
Actor step-size for Walker2D	$5 \cdot 10^{-5}$
Actor step-size for other envs	$3 \cdot 10^{-5}$
τ	4
b	2
Memory size	10^{6}
Minibatch size	256
Target update interval	1
Gradient steps	1
Learning start	10^{3}

Table 2. ACER hyperparameters

Parameter	Value
Action std. dev	0.3
Critic step-size	10^{-5}
Actor step-size	10^{-5}
λ	0.9
b	2
Memory size	10^{6}
Minibatch size	256
Target update interval	1
Gradient steps	1
Learning start	10^{3}

Table 3. SAC hyperparameters

Parameter	Value
Step-size for Hopper	0.0001
Step-size for other envs	0.0003
Replay buffer size	10^{6}
Minibatch size	256
Target smoothing coef. τ	0.005
Target update interval	1
Gradient steps	1
Learning start for Ant	10^{4}
Learning start for HalfCheetah	10^{4}
Learning start for Hopper	10^{3}
Learning start for Walker2D	10^{3}
Reward scale for Ant	1
Reward scale for HalfCheetah	0.1
Reward scale for Hopper	30
Reward scale for Walker2D	30

Table 4. PPO hyperparameters

Parameter	Value
GAE parameter (λ)	0.95
Minibatch size	64
Step-size	0.0003
Horizon	2048
Number of epochs	10
Policy clipping coef.	0.2
Value function clipping coef.	10
Target KL	0.01

ACER in two directions. Firstly, it admits autocrrelated actions. This enables exploration distributed in many actions instead in one. Secondly, in order to mimic learning with eligibility traces [7], ACER estimates improvement directions with the use of a sum whose limit is random. This increases variance of these estimates. Instead, for each state ACERAC computes an improvement direction as an average of increments similar to those ACER selects on random. Hence smaller variance of improvement direction estimates in ACERAC which enables larger step-sizes and faster learning.

It is important to note that the algorithm introduced here, ACERAC, has been designed for fine time discretization and real life control problems. However, here it has been tested on simulated benchmarks in which time discretization was not particularly fine and control could be arbitrarily discontinuous. Its relatively good performance is a desirable result. It allows to expect that this algorithm will perform relatively even better in real life control problems. That remains to be confirmed experimentally in further studies.

6   Conclusions and Future Work

In this paper a framework was introduced to apply reinforcement learning to policies that admit stochastic dependence between subsequent actions beyond state transition. This dependence is a tool to enable reinforcement learning in physical systems and fine time discretization. It can also yield better exploration thus faster learning.

An algorithm based on this framework, Actor-Critic with Experience Replay and Autocorrelated aCtions (ACERAC), was introduced. Its efficiency was verified in simulations of four learning control problems: Ant, HalfCheetah, Hopper, and Walker2D. The algorithm was compared with PPO, SAC, and ACER. ACERAC outperformed the competitors in Ant and HalfCheetah. For Hopper ACERAC was the best *ex aequo* with ACER. For Walker2D the best results was obtained by PPO.

It is desirable to combine the framework proposed here with adaptation of dispersion of actions by introducing reward for the entropy of their distribution, as it is done in PPO. The framework proposed here is specially designed for applications in robotics. An obvious step of our further research is to apply it in this field, obviously much more demanding than simulations.

Acknowledgement. This work was partially funded by a grant of Warsaw University of Technology Scientific Discipline Council for Computer Science and Telecommunications.

A   Algorithms' Hyperparameters

This section presents hyperparameters used in simulations reported in Sect. 5. All algorithms used the discount factor equal to 0.99. The rest of hyperparameters for ACERAC, ACER, SAC, and PPO, are depicted in Tables 1, 2, 3, and 4, respectively.

References

1. Barto, A.G., Sutton, R.S., Anderson, C.W.: Neuronlike adaptive elements that can learn difficult learning control problems. IEEE Trans. Syst. Man Cybern. B **13**, 834–846 (1983)
2. Coumans, E., Bai, Y.: Pybullet, a python module for physics simulation for games, robotics and machine learning (2016–2019). http://pybullet.org
3. Erez, T., Tassa, Y., Todorov, E.: Simulation tools for model-based robotics: comparison of Bullet, Havok, MuJoCo, ODE and PhysX. In: 2015 IEEE International Conference on Robotics and Automation (ICRA), pp. 4397–4404 (2015)
4. Haarnoja, T., Zhou, A., Abbeel, P., Levine, S.: Soft Actor-Critic: offpolicy maximum entropy deep reinforcement learning with a stochastic actor (2018). arXiv:1801.01290
5. van Hoof, H., Tanneberg, D., Peters, J.: Generalized exploration in policy search. Mach. Learn. **106**, 1705–1724 (2017). https://doi.org/10.1007/s10994-017-5657-1
6. Kakade, S., Langford, J.: Approximately optimal approximate reinforcement learning. In: Proceedings of the Nineteenth International Conference on Machine Learning, ICML'02, pp. 267–274 (2002)
7. Kimura, H., Kobayashi, S.: An analysis of actor/critic algorithms using eligibility traces: reinforcement learning with imperfect value function. In: ICML (1998)
8. Korenkevych, D., Mahmood, A.R., Vasan, G., Bergstra, J.: Autoregressive policies for continuous control deep reinforcement learning. In: Proceedings of the Twenty-Eighth International Joint Conference on Artificial Intelligence (IJCAI-19), pp. 2754–2762 (2019)
9. Liang, E., et al.: RLlib: abstractions for distributed reinforcement learning. In: Dy, J., Krause, A. (eds.) Proceedings of the 35th International Conference on Machine Learning. Proceedings of Machine Learning Research, vol. 80, pp. 3053–3062. PMLR, Stockholmsmässan, Stockholm, Sweden, 10–15 July 2018
10. Lillicrap, T.P., et al.: Continuous control with deep reinforcement learning (2016). arXiv:1509.02971
11. Mnih, V., et al.: Playing atari with deep reinforcement learning (2013). arXiv:1312.5602
12. Schulman, J., Levine, S., Moritz, P., Jordan, M.I., Abbeel, P.: Trust region policy optimization (2015). arXiv:1502.05477
13. Schulman, J., Wolski, F., Dhariwal, P., Radford, A., Klimov, O.: Proximal policy optimization algorithms (2017). arXiv:1707.06347
14. Todorov, E., Erez, T., Tassa, Y.: Mujoco: a physics engine for model-based control. In: 2012 IEEE/RSJ International Conference on Intelligent Robots and Systems, pp. 5026–5033. IEEE (2012)
15. Wang, Z., et al.: Sample efficient actor-critic with experience replay (2016). arXiv:1611.01224
16. Wawrzyński, P.: Learning to control a 6-degree-of-freedom walking robot. In: Proceedings of EUROCON 2007 the International Conference on Computer as a Tool, pp. 698–705 (2007)
17. Wawrzyński, P.: Real-time reinforcement learning by sequential actor-critics and experience replay. Neural Networks **22**(10), 1484–1497 (2009)
18. Wawrzyński, P.: Control policy with autocorrelated noise in reinforcement learning for robotics. Int. J. Mach. Learn. Comput. **5**(2), 91–95 (2015)

A Motif-Based Graph Neural Network to Reciprocal Recommendation for Online Dating

Linhao Luo[1], Kai Liu[2](✉), Dan Peng[1], Yaolin Ying[1], and Xiaofeng Zhang[1](✉)

[1] Harbin Institute of Technology, Shenzhen, China
{luolinhao,pengdan,yaolinying}@stu.hit.edu.cn, zhangxiaofeng@hit.edu.cn
[2] Ping An Life, Shenzhen, China
kennethkliu@foxmail.com

Abstract. Recommender systems have been widely adopted in various large-scale Web applications. Among these applications, online dating application has attracted more and more research efforts. Essentially, online dating data is a bipartite graph with sparse reciprocal links. Reciprocal recommendations consider bi-directional interests of service and recommended users, not merely the service user's interest. This paper proposes a motif-based graph neural network (MotifGNN) for online dating recommendation task. We first define seven kinds of motifs and then design a motif based random walk algorithm to sample neighbor users to learn feature embeddings of each service user. At last, these learned feature embeddings are used to predict whether a reciprocal link exists or not. Experiments are evaluated on two real-world online dating datasets. The promising results demonstrate the superiority of the proposed approach against a number of state-of-the-art approaches.

Keywords: Recommender system · Graph convolutional networks · Online dating · Reciprocal recommendation

1 Introduction

Nowadays there exist various web-scale social applications attracting a huge amount of users. Among these popular applications, online dating applications could even have over hundred millions of registered users. This kind of application aims to match users who are mutually interested in each other. To better serve such a large volume of users, effective reciprocal recommender systems (RRS) are urgently needed, which is seldom studied in the literature. Different from traditional recommender systems [18,19] which recommend items of interest to users, the reciprocal recommender systems try to make reciprocal recommendations by simultaneously matching mutual interests between *service user* (receiving recommendations) and *target user* (recommended user) [6,13,16].

Apparently, it is more challenging to make RRS-type recommendations than making conventional recommendations. For conventional recommendation, it

© Springer Nature Switzerland AG 2020
H. Yang et al. (Eds.): ICONIP 2020, LNCS 12533, pp. 102–114, 2020.
https://doi.org/10.1007/978-3-030-63833-7_9

only cares whether the *service user* is interested in recommended items or not. However, for RRS recommendation, the successful recommendation should consider bi-directional interests, not merely the service user's interest. In the literature, there exist a few conventional techniques based RRS approaches [6,13,16]. Most of these approaches simply employ conventional user-item recommendation system twice and then match the directional recommendation results. On top of this, several attempts have been made to simultaneously model users' mutual interests. To summarize, most of these existing approaches build their recommendation models directly based on users' features such as profile data or behavior records. As is known, online dating data is a directed attributed graph data, and graph representation approaches like Graph Neural Network (GNN) is more suitable for this task. However, this task is still challenging due to the following practical difficulties. First, online dating data is essentially a bipartite graph which means an edge only appears between heterosexual users. Second, a service user sends messages to other users and only a few messages have been replied, and thus the reciprocal edges are seriously imbalanced. Last, such a bipartite graph is relatively sparse when compared with other social graphs like WeChat or Facebook. Therefore, how to learn a highly predictable GNN-based model from such large sparse graph data has become a challenging research issue.

In this paper, we cast this reciprocal recommendation task into a reciprocal link prediction problem. Practically, there are no sufficient reciprocal link data for model training. Intuitively, we seek the help from neighbor users who have similar dating preferences. How to select such users especially in sparse bipartite graph data is a challenging task. To this end, the motivating GraphSAGE [4] first samples certain k-hop neighbors for each given node, then aggregates neighbors' features. Unfortunately, these approaches cannot be directly applied on sparse bipartite graph data. Network motifs, i.e., small subgraphs, could well preserve users interactive behaviour information [8,9], and thus is suitable for our problem. Consequently, this paper proposes a motif-based graph neural network model for reciprocal recommendation task. The major contributions of this paper can be summarized as follows.

- We define seven kinds of motifs to capture mutual interests or attractiveness among users. Then, a motif-based random walk algorithm is designed to sample more informative neighbor users based on the defined network motifs, and then their node features are learned to represent each service user.
- We propose a motif-based Graph Neural Network (MotifGNN) approach for reciprocal online dating recommendations. Particularly, MotifGNN adopts a motif-based neighbor sampling method to sample the neighbors users and an attention aggregation mechanism is employed to aggregate users features.
- We collect a real-world dataset from one of the most popular online dating application. Then, extensive experiments are evaluated on this private dataset and one public dataset. The proposed approach achieves superior performance against a number of baseline and state-of-the-art approaches with respect to several widely adopted evaluation metrics.

2   Related Work

As is well known, recommender systems have long been studied in the history literature. However, reciprocal recommender systems (RRS) are seldom investigated. We briefly review *conventional approaches*, *deep learning based approaches*, and *graph neural network based approaches* for RRS in this section.

2.1   Conventional RRS Approaches

As one of the benchmark RRS models, the RECON [13] is proposed for online dating recommendations. It measures mutual interests between the service user and the target user. Particularly, the authors propose a content-based algorithm that calculates the reciprocal scores of each pair of users to be matched. Unfortunately, only users' attributes are considered to calculate the score and thus users' interaction history is bypassed by the model. In addition to RECON, [16] computes such reciprocal scores by using the similarities between users' attributes and the similarities of mutual interests and attractiveness. The similarity of interest and attractiveness for a pair of users is calculated based on the intersected set of users approached by service and target users. Alternatively, there exists another line of research efforts that focus on social relations among users to make the recommendations. One common assumption of these approaches is that users' preferences are susceptible to their cliques (nearest neighbors) which is theoretically supported by general principles in social sciences.

2.2   Deep Learning Based RRS Approaches

There exist some recently proposed deep learning based approaches for recommender systems [3]. These approaches consider the recommendation system as a CTR prediction problem, which use deep learning algorithms to learn deep latent features and then predict the probability that a service user would be interested in the recommended users. For reciprocal recommendation, a hybrid model is proposed in [6] which combines collaborative filtering and deep neural network to predict the probability that a target user is interested in a service user. It first calculates users' similarity according to their interaction history. Then, the features of service users and target user are concatenated as the input of the proposed deep model to predict the probability that the target user would reply to the service user's messages.

2.3   Graph Neural Networks Based Approaches

With the popularity of Graph Neural Network (GNN) model [5], researchers attempt to adapt GNN to solve the recommendation problem [1,4,17]. Graph-SAGE [4] is proposed to learn feature representations via sampling and aggregating strategies. In [17], PinSAGE is proposed for Web-scale recommender systems. It learns node representations via selecting neighbor nodes by a revised random walk algorithm and then aggregating representations of neighbor nodes by GCN model. [1] proposes to learn representations of users and items based on their social relations. Alternative to these random walk based approaches, "motif" is proposed to capture graph structural information [10,11,14]. In [10], a spectral motif convolution approach is proposed for convolutional filters. Motif-CNN [14] defines several kinds of motifs to design the receptive fields around the target node of interest, and then motif-based spatial convolution operations are performed to extract local connectivity features. For graph node classification, [11] proposes a motif-level self-attention model to differentiate the importance of different motifs.

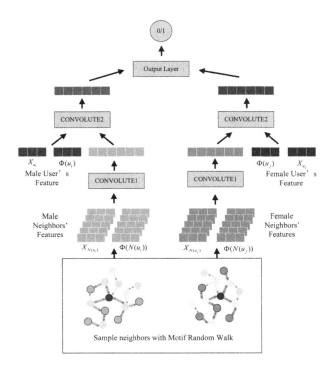

Fig. 1. The framework of the proposed MotifGNN approach. Note that we learn the representations of male user and female user in parallel. If a male user is chosen as service user, the recommended female users are target users.

3 The Proposed MotifGNN Approach

In this section, we propose the motif-based graph neural network (MotifGNN) for the reciprocal recommendation task. The overall framework of MotifGNN is illustrated in Fig. 1. We first introduce the designed motif-based random walk algorithm. Then, we employ the motif-based graph convolution operations to learn feature representations of service and target user in parallel. These feature representations of service and target users are then concatenated to predict whether there exists a reciprocal link or not between them.

3.1 Preliminaries

Generally, online dating network data could be represented as a bipartite graph $G = \{U_s, U_t, E\}$, where $U_s = \{u_1, u_2, \cdots, u_N\}$ and $U_t = \{u_1, u_2, \cdots, u_M\}$ respectively denote the service user set and target user set, and $E = \{f(u_i, u_j)\}$, $u_i \in U_s, u_j \in U_t$ denotes the edge set. If u_i sends a message to u_j, there exists a directed edge from u_i to u_j. "REPLY" is a special kind of "SEND" action. The proposed motif-based random walk algorithm is to sample neighbor user sets $N(u_i)$ and $N(u_j)$ for a pair of service user u_i and target user u_j. A mapping function $\Phi : U \rightarrow \mathbb{R}^d$ is to embed features of a user $u \in U$ into d-dimensional feature space. The purpose of the proposed approach is to predict whether there exists a link between u_i and u_j.

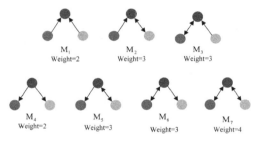

Fig. 2. The defined seven kinds of motifs for online dating graph. The dark blue nodes denote source nodes u_s, the light blue nodes denote neighbor nodes u_n and dark red nodes denote intermediate nodes u_m. (Color figure online)

3.2 Defined Motifs

Generally, social graphs like *Facebook* or *Twitter* assume that densely connected users may have similar preferences, and thus have similar behavior patterns. Consequently, their feature representations should be close to each other. However, online dating network G is a bipartite graph, where users can only send "SEND" or "REPLY" messages to heterosexual users. Intuitively, the node features of these heterosexual users as well as their behavior patterns would be fundamentally different. Hence, we further define "neighbor user" to be "homosexual neighbor user", which means that two homosexual users could be neighbors if they "SEND" or "REPLY" to the same heterosexual users. This motivates us to find high-order connected homosexual users instead of randomly sampling adjacent neighbors.

In this paper, we propose to employ network motifs to capture this high-order connectivity patterns to sample homosexual neighbors. According to the

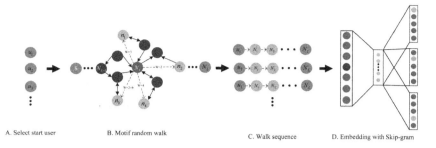

A. Select start user   B. Motif random walk   C. Walk sequence   D. Embedding with Skip-gram

Fig. 3. The working process of the proposed motif based random walk algorithm. For each service user in (a), a motif based random work is applied in (b), and then the walking paths are extracted in (c) and we embed selected users using skip-gram model as illustrated in (d).

characteristics of online dating data, we only consider 3-node network motifs, which are plotted in Fig. 2. As is observed, there are only 7 network motifs for bipartite graph instead of 13 network motifs for conventional homogeneous directed graphs [8]. We denote the defined motifs as $\mathbf{M}_S = \{M_1, M_2, \ldots, M_7\}$. Let $M_\tau = \{u_s, u_n, u_m, e_s, e_n\}, M_\tau \in \mathbf{M}_S$ denotes a 3-node motif where u_s, u_n and u_m respectively denote source node, neighbor node and intermediate node, and e_s and e_n respectively denote the edge between u_s, u_m and u_n, u_m. Then, we have $\forall u_s, u_n, u_m \in U$ and $\forall e_s, e_n \in E$.

Although there are only 3 nodes in the defined motifs, the directional edges represent different interaction patterns among these three users, and each pattern could be used to measure closeness of two homosexual users u_s and u_n. Taking M_1 as an example, both u_s and u_n send messages to the same intermediate user u_m, and thus M_1 simply shows that u_m has the same attractiveness to u_s and u_n. The u_m of M_2 and M_3 have the same attractiveness to u_s and u_n, but not the same interest in u_s and u_n. Alternatively, M_4 represents that u_s and u_n have the same attractiveness to u_m, and so does M_5 and M_6. But u_s and u_n have different interests in u_m for M_5 and M_6. Obviously, M_7 represents mutual attractiveness between u_m and $\{u_s, u_n\}$. From above observations, it is well noticed that different network motifs can well differentiate local neighbors' contribution to attractiveness or interests of a service user. To further enhance the contributions of motifs having reciprocal links, we set the edge's weight w to be 1 for non-reciprocal edge, i.e., "SEND", and 2 for reciprocal edge. Therefore, the total weight W of motif can be calculated as $W = w_{sm} + w_{nm}$ where w_{sm} and w_{nm} denote the weight of edges between u_s, u_m and u_n, u_m.

3.3   Motif Based Random Walk Algorithm

In this subsection, we propose a motif based random walk algorithm as conventional graph embedding techniques like DeepWalk [12] and Node2Vec [2] cannot be directly applied on bipartite graph. The proposed motif based random walk algorithm is to sample homosexual neighbor nodes within 2-hops and we sample L times forming a L length neighbor user sequence. Let n_k denote the k-th node in a sequence and n_l denote the next node to walk. Let M_{kl}^τ denote a 3-node motif containing n_k and n_l, where $u_s = n_k$ and $u_n = n_l$, W_{kl}^τ denote the weight of motif M_{kl}^τ. The transition probability from node n_k to n_l can be calculated as

$$
P(n_l|n_k) = \begin{cases} \dfrac{E[k][\tau] * W_{kl}^\tau}{\sum\limits_{u_t \in M_+(n_k)} W_{kt}}, & M_{kl}^\tau \in M_S, M_+(n_k) \in M_S, \\ 0, & M_{kl} \notin M_S, \end{cases} \tag{1}
$$

where $M_S = \{M_1, M_2, \ldots, M_7\}$ is the set of 3-node motifs, $M_+(n_k)$ is the set of adjacent motifs of node n_k, and $\forall M_{kt}^\tau \in M_+(n_k) \rightarrow M_{kt}^\tau \in M_S$, $E[k][\tau]$ denote the total number of τ type motif which W_{kl}^τ belongs to among $M_+(n_k)$. After sampling a sequence of neighbor nodes, the Skip-Gram model is adopted to embed each node into low-dimensional feature space. The corresponding feature representation is learned by maximizing the co-occurrence probability of contextual nodes in the generated random walk path, calculated as

$$
\begin{aligned}
\underset{\Phi}{\text{Min}}\, E &= -\log\, \Pr(n_{k-w}, n_{k-w+1}, \ldots, n_{k+w}|\Phi(n_k)) \\
&= -\log\, \prod_{l=k-w, l\neq k}^{k+w} \Pr(n_l|\Phi(n_k)),
\end{aligned} \tag{2}
$$

where w is the window size of contextual nodes in the path, Φ is mapping function which embeds u_i into a $d-$dimensional space. By adopting the negative sampling strategy, the new loss function can be written as

$$
\underset{\Phi}{\text{Min}} \sum_{l \in N(u_k)} \log\, \sigma(\Phi(n_l)^\top \Phi(n_k)) + \sum_{l' \in N(u_k)'} \log\, \sigma(-\Phi(n_{l'})^\top \Phi(n_k)), \tag{3}
$$

where $N(u_k)$ is the set of sampling neighbors, $N(u_k)'$ is the set of negative samples for u_k and $\sigma()$ is sigmoid function, and we set $|N(u_k)'| = |N(u_k)|$ in the experiments. The working process of the proposed motif based random walk algorithm is illustrated in Fig. 3.

3.4   Embedding with Attentive Graph Convolution

After embedding user features with skip-gram model, these feature representations are treated as the input of the graph neural network component. The proposed graph neural network component is to convolute users features and their neighbor users' features using a two-layer graph convolution operation. The convolution operation is performed by iteratively aggregating features of all neighboring nodes. The *CONVOLUTE1* operation differentiates the importance of user extracted from different motifs. Instead of concatenating neighbor users' features, the *CONVOLUTE2* operation learns the high-order feature interactions between each given user and his (her) neighbor users.

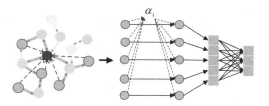

Fig. 4. The structure of the proposed CONVOLUTE1 operation with attention mechanism.

For any user u, motif-based random walk first samples a set of neighbor users $N(u)$ with its size to be $|N(u)| = L$. To differentiate the importance of each user in $N(u)$, we adopt an attention mechanism to aggregate features of each neighbor user, as illustrated in Fig. 4. For each user u_t in $N(u)$, the weight α_t is calculated as

$$h_{u_t} = X_{u_t} \oplus \Phi(u_t) \tag{4}$$

$$\mu_t = \boldsymbol{W}_2^\top \sigma_1(\boldsymbol{W}_1^\top h_{u_t} + \boldsymbol{b}_1) + b_2 \tag{5}$$

$$\alpha_t = \frac{e^{\mu_t}}{\sum_{u_l \in N(u)} e^{\mu_{tl}}}, \tag{6}$$

where $\sigma_1()$ is the tanh function, $\boldsymbol{W}_1^\top \in R^{(m+d) \times n_1}, \boldsymbol{W}_2^\top \in R^{n_1 \times 1}$ are learnable parameters. The *CONVOLUTE1* operation is calculated as

$$h_{N(u)}^1 = \boldsymbol{W}_3^\top (\alpha_1 h_{u_1} \oplus \alpha_2 h_{u_2} \cdots \oplus \alpha_L h_{u_L}) + b_3, \tag{7}$$

where $\boldsymbol{W}_3^\top \in R^{L \times (m+d) \times n_3}$ and b_3 are the parameters of the convolution layer. The CONVOLUTE2 operation combines the embeddings of neighbor users as well as the given user. It contains a fully connected layer and a batch normalized layer, computed as

$$\mu^2 = \sigma_2(\boldsymbol{W}_4^\top (X_u \oplus \Phi(u) \oplus h_{N(u)}^1) + b_4) \tag{8}$$

$$h_u^2 = \frac{\mu^2 - \mathrm{E}[\mu^2]}{\sqrt{\mathrm{Var}[\mu^2] + \epsilon}} * \gamma + \beta, \tag{9}$$

where $\sigma_2()$ is the ReLU function, $\boldsymbol{W}_4^\top \in R^{(m+d+n_3) \times n_4}$, γ and β are learnable parameter vectors. The mean and standard deviation are calculated within each mini-batch.

3.5 Reciprocal Recommendation Component

The purpose of this component is to predict whether there exists a reciprocal link or not. Given a pair of users (u_i, u_j) and their learnt feature representations

$h_{u_i}^2, h_{u_j}^2$, a feature vector $\boldsymbol{Z}$ is acquired by concatenating these feature representations, and then is fed into the reciprocal recommendation component for prediction task. This prediction component consists of two fully connected layers and a dropout layer, calculated as

$$\boldsymbol{Z} = h_{u_i}^2 \oplus h_{u_j}^2 \tag{10}$$

$$\mu^3 = \sigma_2(\boldsymbol{W}_5^\top \boldsymbol{Z} + b_5) \tag{11}$$

$$\hat{\mu}^3 = Dropout(\mu^3) \tag{12}$$

$$y = \sigma(\boldsymbol{W}_6^\top \hat{\mu}^3 + b_6), \tag{13}$$

where dropout rate is set to 0.5 and $y \in [0,1]$ indicating whether there exists a reciprocal link or not. The cross-entropy loss function is adopted to optimize the proposed model, given as

$$\mathcal{L} = \min\{-(\overline{y} \log y + (1 - \overline{y}) \log(1 - y))\}, \tag{14}$$

where $\overline{y}$ is the ground truth reciprocal link between u_i and u_j.

4   Experiments

4.1   Dataset and Experimental Settings

Two real-world datasets are chosen to evaluate the proposed model called "Dating dateset 1" and "Dating dataset 2". "Dating dataset 1" is a public online dating dataset provided by a competition[1]. "Dating dataset 2" is collected by ourselves from one of the most popular online dating Applications. "Dating dataset 1" contains 34 user attributes and their interaction history, i.e. "message" and "click". We treat these action equally between users to generate both reciprocal links and non-reciprocal links. "Dating dataset 2" contains 228,470 registered users with 25,168,824 links, where 14.5% of the links are reciprocal. Each registered user has 28 different attributes. Statistics of these two datasets are summarized in Table 1. Before experiments, we respectively extract pre-defined motifs from each dataset and the statistics are reported in Table 2.

For evaluation criteria, several widely adopted metrics are adopted in the experiments which are precision, Recall, F1-measure and AUC. Following baseline model as well as state-of-the-art approaches are evaluated which are DeepFM [3], xDeepFM [7], DeepWalk [12], Node2vec [2], GraphSage [4], PinSage [17] and Social GCN [15]. To prepare the experiments, datasets are randomly partitioned into training and testing dataset at the ratio of 90% to 10%.

---

[1] https://cosx.org/2011/03/1st-data-mining-competetion-for-college-students/.

Table 1. Statistics of experimental datasets.

Dataset	# Users	# Messages	# Reciprocal links	Percentage of reciprocal links
Dating dataset 1	59,921	232,954	9,375	0.0402
Dating dataset 2	228,470	25,168,824	1,592,945	0.1449

Table 2. Statistics of the defined motifs.

Dataset	M1	M2	M3	M4	M5	M6	M7
Dating dataset 1	4.57%	2.76%	2.76%	27.50%	14.23%	14.23%	33.97%
Dating dataset 2	4.16%	12.94%	12.94%	15.13%	19.24%	19.24%	16.09%

4.2 Model Performance Evaluation

We evaluate all models and report the corresponding experimental results in Table 3. From this table, it is well noticed that deep neural network based approaches, e.g., deepFM and xDeepFM, are generally better than those graph convolution based approaches, e.g., GraphSage and PinSage. This indicates that node attributes play a more important role in online dating prediction task. This is consistent with our common intuition that users may be interested in different attributes of dating users. The performance of graph neural network based approaches are slightly better than node embedding based methods, e.g., Deep-Walk and Node2vec, in "dataset 1" but not in "dataset 2". The possible reasons might be as follows. Both graph neural network based approach and node embedding approach utilize graph structural information. The neighbor users sampled by these approaches might not contain sufficient information due to the data sparsity issue. The node embedding based approach could achieve comparably good model performance in "dataset 2", and the reason is that the adopted walk-

Table 3. Results of reciprocal recommendations on two experimental datasets.

Methods	Dating dataset 1				Dating dataset 2			
	Precision	Recall	F1	AUC	Precision	Recall	F1	AUC
DeepWalk	.5177	.3544	.4208	.7646	.8801	.7579	.8144	.9026
Node2vec	.4865	.4380	.4610	.7752	.8138	.8558	.8343	.9135
DeepFM	.7004	.4852	.5732	.8609	.8533	.7477	.7970	.8849
xDeepFM	.7714	.5094	.6136	.8877	.9357	.8605	.9047	.9506
GraphSage	.7151	.3383	.4593	.8679	.6829	.6643	.6735	.7126
PinSage	.6428	**.7493**	.6920	.9020	.7220	.7549	.7991	.8891
SocialGCN	.4667	.4434	.4547	.7669	.8588	.8314	.7991	.8886
MotifGNN (M7)	**.8576**	.6900	**.7647**	**.9206**	**.9476**	**.8713**	**.9078**	**.9684**

Table 4. Evaluation results about how the defined motifs affect model performance.

Methods	Dating dataset 1				Dating dataset 2			
	Precision	Recall	F1	AUC	Precision	Recall	F1	AUC
MotifGNN (M7)	**.8576**	**.6900**	**.7647**	**.9206**	.9476	.8713	.9078	**.9684**
MotifGNN (M1,M4)	.7975	.5997	.6846	.8627	.9880	.8484	.9129	.9660
MotifGNN (M1,M2,M3)	.8339	.6496	.7303	.8993	.8483	**.9338**	.8890	.9670
MotifGNN (M4,M5,M6)	.6908	.6806	.6857	.8890	.9682	.8605	0.9112	.9676
MotifGNN (M1–M7)	.8055	.6752	.7346	.8933	**.9986**	.8999	**.9468**	.9467

ing strategy can well sample neighbor nodes and extract higher order information to avoid the over-smoothing problem usually occurred in graph representation learning related tasks. While the proposed MotifGNN significantly outperforms all compared methods w.r.t. all evaluation criteria except the Recall in "dataset 1". This verifies that effectiveness of the proposed approach.

4.3   Evaluation Results on Motif Effect

This experiment is to investigate how the proposed motifs could affect the reciprocal recommendation performance. To recall that, the motif $M7$ is the desired one which can best capture the reciprocal relationships among bipartite users. The combination of $M1$ and $M4$ could be used to approximate the results of motif $M7$. For motifs $M1$, $M2$ and $M3$, they are considered to model the interests of the service user, whereas the motifs $M4$, $M5$ and $M6$ capture the attractiveness of service users. Thus, we respectively sample homosexual neighbor users by using these combinations of motifs and evaluate the corresponding model performance which are reported in Table 4.

From this table, it is well noticed that in "dataset1", MotifGNN with $M7$ is the best model w.r.t. most evaluation criteria. But in "dataset2", the combination of $M1$, $M2$ and $M3$ achieves the best performance w.r.t. *recall* criterion and the combination of all motifs achieves the best *precision* and *F1* score. As for recommendation task, the AUC score is the most important evaluation criterion. In terms of the AUC score, we still can conclude that the MotifGNN(M7) is the best model on "dataset 2". Furthermore, different combinations of motifs could achieve quite different model performance and this needs further investigation.

5   Conclusion

Recommender systems have long been studies in various applications. However, the reciprocal recommender systems have seldom been investigated. This paper proposes a motif based graph neural network model for this task to address existing research challenges in RRS problem. Experiments are evaluated on two real-world datasets. The promising empirical evaluation results demonstrate the superiority of the proposed approach against a number of state-of-the-art approaches w.r.t several widely adopted evaluation criteria. In the near future, we will further explore a better way to utilize the defined motifs.

Acknowledgments. This work was supported in part by the National Key R&D Program of China under Grant no. 2018YFB1003800, 2018YFB1003804, the National Natural Science Foundation of China under Grant No. 61872108, and the Shenzhen Science and Technology Program under Grant No. JCYJ20170811153507788.

References

1. Fan, W., et al.: Graph neural networks for social recommendation. In: WWW, pp. 417–426 (2019)
2. Grover, A., Leskovec, J.: node2vec: scalable feature learning for networks. In: ACM SIGKDD, pp. 855–864 (2016)
3. Guo, H., Tang, R., Ye, Y., Li, Z., He, X.: DeepFM: a factorization-machine based neural network for CTR prediction. arXiv preprint arXiv:1703.04247 (2017)
4. Hamilton, W., Ying, Z., Leskovec, J.: Inductive representation learning on large graphs. In: NeurIPS, pp. 1024–1034 (2017)
5. Kipf, T.N., Welling, M.: Semi-supervised classification with graph convolutional networks. arXiv preprint arXiv:1609.02907 (2016)
6. Kleinerman, A., Rosenfeld, A., Ricci, F., Kraus, S.: Optimally balancing receiver and recommended users' importance in reciprocal recommender systems. In: ACM Conference on Recommender Systems, pp. 131–139 (2018)
7. Lian, J., Zhou, X., Zhang, F., Chen, Z., Xie, X., Sun, G.: xDeepFM: combining explicit and implicit feature interactions for recommender systems. In: ACM SIGKDD, pp. 1754–1763 (2018)
8. Liu, K., Cheung, W.K., Liu, J.: Detecting multiple stochastic network motifs in network data. Knowl. Inf. Syst. **42**(1), 49–74 (2013). https://doi.org/10.1007/s10115-013-0680-4
9. Milo, R., Shen-Orr, S., Itzkovitz, S., Kashtan, N., Chklovskii, D., Alon, U.: Network motifs: simple building blocks of complex networks. Science **298**(5594), 824 (2002)
10. Monti, F., Otness, K., Bronstein, M.M.: Motifnet: a motif-based graph convolutional network for directed graphs. In: IEEE Data Science, pp. 225–228 (2018)
11. Peng, H., Li, J., Gong, Q., Wang, S., Ning, Y., Yu, P.S.: Graph convolutional neural networks via motif-based attention. arXiv preprint arXiv:1811.08270 (2018)
12. Perozzi, B., Al-Rfou, R., Skiena, S.: Deepwalk: online learning of social representations. In: ACM SIGKDD, pp. 701–710 (2014)
13. Pizzato, L., Rej, T., Chung, T., Koprinska, I., Kay, J.: RECON: a reciprocal recommender for online dating. In: RecSys, pp. 207–214 (2010)
14. Sankar, A., Zhang, X., Chang, K.C.C.: Motif-based convolutional neural network on graphs. arXiv preprint arXiv:1711.05697 (2017)
15. Wu, L., Sun, P., Hong, R., Fu, Y., Wang, X., Wang, M.: SocialGCN: an efficient graph convolutional network based model for social recommendation. In: SIGIR (2019)
16. Xia, P., Liu, B., Sun, Y., Chen, C.: Reciprocal recommendation system for online dating. In: ASONAM, pp. 234–241 (2015)
17. Ying, R., He, R., Chen, K., Eksombatchai, P., Hamilton, W.L., Leskovec, J.: Graph convolutional neural networks for web-scale recommender systems. In: ACM SIGKDD, pp. 974–983 (2018)

18. Zhang, X., Liu, H., Chen, X., Zhong, J., Wang, D.: A novel hybrid deep recommendation system to differentiate user's preference and item's attractiveness. Inf. Sci. **519**, 306–316 (2020)
19. Zhang, X., Zhong, J., Liu, K.: Wasserstein autoencoders for collaborative filtering. Neural Comput. Appl. 1–10 (2020)

A Spiking Neural Architecture for Vector Quantization and Clustering

Adrien Fois[1,2] and Bernard Girau[1,2(✉)]

[1] Université de Lorraine, LORIA, UMR 7503, 54506 Vandoeuvre-lès-Nancy, France
{adrien.fois,bernard.girau}@loria.fr
[2] CNRS, LORIA, UMR 7503, 54506 Vandoeuvre-lès-Nancy, France

Abstract. Although a couple of spiking neural network (SNN) architectures have been developed to perform vector quantization, good performances remains hard to attain. Moreover these architectures make use of rate codes that require an unplausible high number of spikes and consequently a high energetical cost. This paper presents for the first time a SNN architecture that uses temporal codes, more precisely first-spike latency code, while performing competitively with respect to the state-of-the-art visual coding methods. We developed a novel spike-timing-dependent plasticity (STDP) rule able to efficiently learn first-spike latency codes. This event-based rule is integrated in a two-layer SNN architecture of leaky integrate-and-fire (LIF) neurons. The first layer encodes a real-valued input vector in a spatio-temporal spike pattern, thus producing a temporal code. The second layer implements a distance-dependent lateral interaction profile making competitive and cooperative processes able to operate. The STDP rule operates between those two layers so as to learn the inputs by adapting the synaptic weights. State-of-the art performances are demonstrated on the MNIST and natural image datasets.

Keywords: Neural network models · Self-organizing map · Vector quantization · Temporal coding · Representation learning

1 Introduction

Spiking Neural Networks (SNNs) have gained an increasing attention in the recent years and have been used to perform supervised and unsupervised learning [4,17] tasks. From a neuromorphic perspective, SNNs offer the advantages of energetic and communication efficiency [15]. Indeed, they don't need to produce and send real-valued outputs at each iteration. Instead they sparsely emit binary events, the so-called spikes. This event-driven message passing scheme greatly alleviates the communication channels as compared to the clock-driven communication of traditional neural network implementations. Further energetic and communication improvements can be obtained by making the memory storage (the synapses) locally accessible to the computational unit (the neurons). This

© Springer Nature Switzerland AG 2020
H. Yang et al. (Eds.): ICONIP 2020, LNCS 12533, pp. 115–126, 2020.
https://doi.org/10.1007/978-3-030-63833-7_10

hardware architectural scheme can then be fully exploited at the algorithmic level by the use of local learning rules such as STDP (spike timing dependent plasticity). Local learning rules are leveraging information available at the presynaptic terminal (or synaptic input) and the postsynaptic cell membrane (or synapse output). This paper follows this algorithmic line of work. We propose a SNN architecture that incorporates a novel STDP rule operating locally in space and time to perform vector quantization and clustering.

The paper is organized as follows. In Sect. 2, we rapidly situate this work with respect to vector quantization, self-organizing maps and spiking neural networks. The architecture and learning method of our model is described in Sect. 3, and the experimental results are presented in Sect. 4.

2   Background

Vector quantization (VQ, see [19]) consists in approximating the probability density of a possibly highly multi-dimensional input space with a finite set of prototype vectors, often called codewords, the set of codewords being called the codebook. Many VQ algorithms exist such as k-means [10], self-organizing maps (SOM) [8], neural gas (NG) [12], growing neural gas (GNG) [5] or growing when required (GWR) [11]. The well-known k-means method provides a codebook without any internal structure, meaning that no additional knowledge is provided with respect to the similarities between codewords. Algorithms such as NG, GNG or GWR create a codebook as a network of codewords and connections between codewords. These connections stand for significant similarities between codewords, thus inducing the codebook structure that derives from the learned data. On the contrary, SOM are based on a static underlying topology and a fixed number of codewords (neuron weight vectors). Each codeword has an associated position in the underlying map that is usually a 2D lattice, and the learning process captures the topographic relationships between the inputs: similar input samples are represented by the weights of spatially close neurons in the map.

All above mentioned neural models for VQ use analog neurons. More biologically inspired computational paradigms recently attract more and more attention. The so-called spiking neurons mimic the neurons in the brain that communicate thanks to action potentials. Their success is tightly linked to the recent trend towards neuromorphic processors (such as IBM TrueNorth or Intel Loihi) or sensors (such as DVS event cameras) that make use of spikes. Our work aims at defining an efficient spiking model for VQ inspired by SOM. Other works already tackles the problem of vector quantization in SNN with local rules [3,7,18]. However, they make use of rate coding. Rate coding is energetically expensive as it encodes a continuous variable in the spike rate of a neuron. The higher the value the higher the rate. Furthermore, rate-coding is not an efficient coding scheme for non-stationary data, that are the norm rather than the exception in real-world applications. Indeed, the firing rate is defined as a limit that theoretically involves an infinite number of spikes [2]. Getting a reliable firing rate estimate then requires to integrate enough spikes either with a large temporal

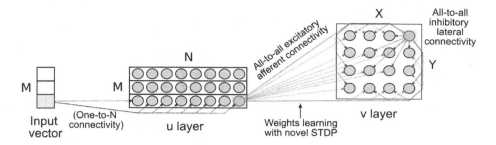

Fig. 1. Spiking neural network architecture.

window or with a high number of spiking neurons [4]. Consequently temporal codes are often preferable due to their biological plausibility and energetical efficiency. Contrary to rate codes, temporal codes leverage the precise spike times to encode continuous variables. Among temporal codes we can cite first-spike latency code that uses the precise spike time and rank-order-code, which can be considered as a special case as it only leverages the rank of the spikes.

Our model performs vector quantization on the basis of a temporal coding of data. It also addresses other common limitations of SNN architecture such as the lack of scalability: changing input data dimensionality often requires an additional fine tuning of parameters. To sum up, we propose here a new SNN model for fast and efficient vector quantization and clustering, that differs from previous works by making use of a new STDP rule for temporal codes, more precisely first-spike latency code, and by being easily scalable to N input data dimensions, without additional parametric fine-tuning.

3   Methods

Our SNN architecture is composed of two layers of LIF neurons (leaky integrate and fire). The first layer u encodes a vector of continuous variables into spike times, thus producing a temporal code. This layer is fully connected to the second layer v that we call the spiking SOM. Layer v performs vector quantization in a way that is conceptually similar to the initial SOM algorithm [8]. The first neuron that spikes with respect to an input vector is considered to be the best matching unit (BMU), i.e. the neuron that is best tuned to the current input vector. This spike triggers the learning of the input vector by its afferent synaptic weights with our new STDP rule. The spike produced by the BMU also influences the other neurons in layer v via a distance-dependent lateral interaction profile that mimics the usual neighbourhood kernel in the usual 2D underlying topology of a Kohonen SOM. Spatially close neurons will tend to spike in close temporal proximity and thus learn the input vector, while more distant neurons will not spike and therefore not learn it. By iterating this process, the map gets gradually topologically ordered, i.e. vectors that are close in the input space will be represented by spatially close neurons in layer v.

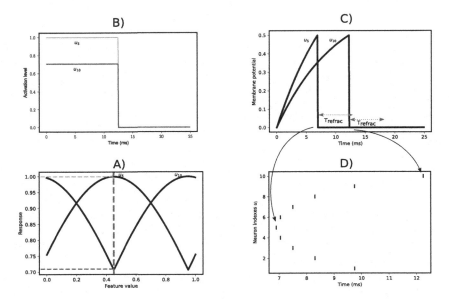

Fig. 2. A feature value of 0.45 is encoded by a spatio-temporal spike pattern. A) A bank of 10 neurons represent this feature. Each neuron u_i of that bank has a gaussian tuning curve centered aroung a preferential value. u_5 and u_{10} have 0.45 and 0.95 as preferential values, respectively. The tuning curves of u_5 and u_{10} and their resulting activation levels for an input value of 0.45 are plotted. B) These activations levels are constant for an interval of time of 12.5 ms and are followed by a quiescent period of the same duration. C) The activation levels are integrated by the membrane potentials of u_5 and u_{10}. The membrane potential of u_5 grows faster than for u_{10} as the input is higher. Consequently u_5 reaches its threshold faster, is resetted and enters in a refractory period. D) Each neuron u_i emits a spike when it crosses it's threshold, thereby producing a spatio-temporal spike pattern at a mesoscopic level.

The architecture was simulated using the BRIAN2 simulator. The following sections more precisely describe the different components of our model.

3.1   Neuron and Synapse Model

The network only contains leaky integrate-and-fire (LIF) neurons. This one-variable model captures the basic behavior of a biological neuron while maintaining a low computational cost and analytical tractability. The dynamic of a LIF neuron is given by

$$\tau_m \frac{dV}{dt} = -V(t) + E(t)$$
$$V(t) \leftarrow V_{\text{reset}}, \text{ if } V(t) \geq \theta \tag{1}$$

where τ_m is the membrane time constant, $V(t)$ is the membrane potential, and $E(t)$ is the input to the neural membrane at time t. When the neuron's membrane potential $V(t)$ crosses its membrane threshold θ, the neuron instantaneously emits a spike and the membrane potential is reset to V_{reset}. The neuron then enters an absolute refractory period T_{refrac} during which it cannot emit a new spike as the membrane potential stops being integrated.

The synaptic transmission is modeled with an instantenous jump of the postsynaptic potential (PSP) by the synaptic weight when a spike is emitted by a presynaptic neuron, followed by an exponential decay. This exponential decay can be expressed by a linear differential equation. Rather than modeling the synaptic transmission with one differential equation per synapse, only one differential equation is needed for the whole set of synapses, thanks to the superposition principle of linear systems. This reduces the computational load significantly.

$$E_j(t) \leftarrow E_j(t) + \sum_{i=1}^{I} s_i(t)w_{ij}(t), \qquad \text{if a presynaptic neuron } i \text{ spikes}$$

$$\tau_f \frac{dE_j}{dt} = -E_j(t), \qquad\qquad\qquad\qquad \text{otherwise} \qquad (2)$$

where i and j index the pre- and post-synaptic neurons, respectively. $E_j(t)$ is the input potential of postsynaptic neuron j. $s_i(t)$ is an indicator function returning the value 1 when a presynaptic neuron i spikes at time t, 0 otherwise. $w_{ij}(t)$ is the synaptic weight between pre-synaptic neuron i and post-synaptic neuron j, and τ_f is the fall time constant.

3.2   Input Encoding

Continuous input variables need to be encoded into spike times in layer u. This can be achieved by distributing a variable x over a bank of N neurons (see Fig. 1), where each of the $i \in 1, ..., N$ neuron of that bank has a receptive field size σ and is tuned to a preferential value μ_i in a periodic space. In our experiments $N = 10$ and all neurons used an identical $\sigma = 0.6$ while the prefential values μ_i were equally spaced between 0.05 and 0.95. A wrapped gaussian activation function $G(\sigma, \mu_i, x)$ is then used to compute the input potential of each neuron. The resulting periodic tuning curves cover the entire interval of variation of the normalized input variable. The neuron whose preferential value is the closest to the current input value will get the highest input, and will thus spike first. The other neurons of the bank then spike in an order that is determined by the distance between the input value and their preferential values (see Fig. 2). In this way a continuous variable is encoded in a spatio-temporal spike pattern. This coding scheme can be easily generalized for an input vector of M dimensions, where each dimension is independently encoded by a corresponding bank of N neurons in layer u, as inspired by [1].

3.3   Learning Rule

STDP rules that are not weight-dependent result in the steady-state in a bimodal distribution, where the synaptic weights are either fully potentiated, or fully depressed, as shown by Rossum [14]. With our rule, the learned distribution is unimodal and thus the interval of variation of the synaptic weights is fully leveraged to learn the inputs. This makes possible an easy subsequent decoding step.

To learn the inputs encoded in first-spike latency codes, we developed an event-based plasticity rule derived from a vector-quantization criterion. It takes the form of a novel STDP rule operating 1) locally in space with a single pair of pre and postsynaptic neuron i and j, and 2) locally in time within a temporal learning window. STDP rules are characterized by a weight change Δw_{ij} that is a function of the time difference $\Delta_t = t_{post} - t_{pre}$ between the post and presynaptic spikes, respectively. The magnitude of the weight change Δw_{ij} usually depends on the term $e^{-\Delta_t/\tau}$. Two different time constants τ_+ and τ_- account for $\Delta_t > 0$ and $\Delta_t \leq 0$, respectively and determine the width of the temporal learning window. To keep computation efficient and local, we implemented an online version of STDP. Each synapse has two local state variables x_{ij} and y_{ij}. These variables serves as memories of recent pre and postsynaptic activities, respectively. When a pre (post) synaptic spike is emitted, $x_{ij}(t)$ $(y_{ij}(t))$ is set to 1 and then decays exponentially with a time constant τ_+ (τ_-) to 0. In this way we reproduce the behavior of $e^{-\Delta_t/\tau}$.

$$
\begin{aligned}
x_{ij}(t) &\leftarrow 1, && \text{if neuron } i \text{ spike} \\
\tau_+ \frac{dx_{ij}}{dt} &= -x_{ij}(t), && \text{otherwise} \\
y_{ij}(t) &\leftarrow 1, && \text{if neuron } j \text{ spike} \\
\tau_- \frac{dy_{ij}}{dt} &= -y_{ij}(t), && \text{otherwise}
\end{aligned}
\tag{3}
$$

The weight change Δw_{ij} of a synapse is then computed as a function of $x_{ij}(t)$ and $y_{ij}(t)$. The weight is updated by that change such that $w_{ij}(t) \leftarrow w_{ij}(t) + \Delta w_{ij}$. The final learning rules read:

$$
\Delta w_{ij} = \begin{cases} \alpha_+ \big(1 - x_{ij}(t) - w_{ij}(t) + w_{\text{offset}}\big), & \text{if neur. } j \text{ spikes and } x_{ij}(t) > \theta_{\text{stdp}} \\ \alpha_- \big(1 - y_{ij}(t)\big), & \text{if neur. } i \text{ spikes and } y_{ij}(t) > \theta_{\text{stdp}} \end{cases}
\tag{4}
$$

and are combined with the hard bounds $0 \leq w_{ij} \leq 1$. α_+ and α_- are the learning rates. w_{offset} is a positive offset that shifts the attracting fixed point up. We will discuss about the fixed point in the next paragraph. If $x_{ij}(t)$ $(y_{ij}(t))$ is smaller than the threshold θ_{stdp}, then it does not lead to a change. This is done to account for the time elapsed between the presentation of two successive input patterns. Without this condition, the memories $x_{ij}(t)$ $(y_{ij}(t))$ decaying to 0 after a first input presentation would automatically lead to a maximal weight change when a new input pattern is presented.

In the first adaptation case - that corresponds to $\Delta_t > 0$ - we can observe that the learned weight is an attractive fixed point by analyzing the equilibrium solution:

$$\Delta_{w_{ij}} = 0$$
$$\Leftrightarrow\ 0 = 1 - x_{ij}(t) - w_{ij}(t) + w_{\text{offset}}$$
$$\Leftrightarrow\ w_{ij}(t) = 1 - x_{ij}(t) + w_{\text{offset}} \tag{5}$$

Consider a pattern that is repeatedly presented as an input. w_{ij} will converge to a value that depends on the memory $x_{ij}(t)$ and thereby on Δ_t. The higher Δ_t, the higher the learned value, as the memory $x_{ij}(t)$ relaxes to 0 for large Δ_t values. Thus the first presynaptic neuron that spike will have the highest weight and the last one that spike will have the lowest weight (see Fig. 3). This process happens for each pair of pre and postsynaptic neuron that meet the conditions of the first adaptation case. This behavior is reminiscent of first-spike latency code or of rank-order coding which can be considered as a special case of the former.

The second adaptation case - which corresponds to $\Delta_t \leq 0$ - implements depression. The weight gets depressed by a magnitude that grows with Δ_t, as the memory $y_{ij}(t)$ relaxes to 0 for large Δ_t. Thus in the limit, the weights w_{ij} takes the value 0, as the lower bound for the weight is set to 0.

3.4   Lateral Interactions

We describe in this section the lateral interaction profile in layer v, where the neurons are arranged in a grid. Instead of using a high global inhibition to implement a Winner-Take-All (WTA) behavior, we use a distance-dependent lateral interaction profile, similar to the Kohonen SOM [8]. This profile implements weak short-range inhibition and strong long-range inhibition [6]. In this way, spatially close neurons will tend to spike together and represent close input vectors, while distant neurons will not spike together and thus will learn uncorrelated codes. This also tends to accelerate the learning process, as a population of neurons spike in response to an input vector rather than a single neuron. The lateral interaction profile in layer v reads

$$w_{ij} = \begin{cases} c_{\min}\|i - j\|, \text{ if } \|i - j\| \leq r \\ c_{\max}, \qquad\qquad \text{otherwise} \end{cases} \tag{6}$$

where w_{ij} is the lateral connection weight between the neurons at locations i and j in the grid. $c_{\min}$ scales the inhibitory level for neurons whose distance is less than a radius r, otherwise w_{ij} takes a maximum value $c_{\max}$.

In the learning phase of the Kohonen SOM, the interaction radius progressively decreases so as to first organize the map and then make the neurons learn more and more individualized receptive fields. When the interaction radius is equal to zero, we get a WTA behavior. We mimic this behavior by increasing the lateral connection weights w_{ij} during the training phase. By increasing the inhibition level, we approach a WTA behavior. The growing inhibition during the training phase is implemented by the following homeostatic mechanism:

$$\tau_w \frac{dw_{ij}}{dt} = c_{\max} - w_{ij} \tag{7}$$

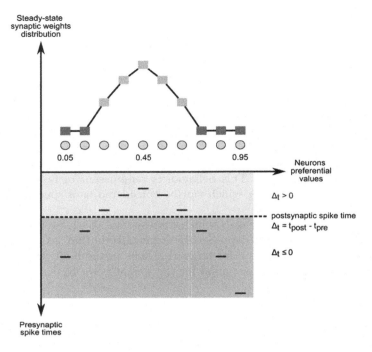

Fig. 3. Qualitative sketch of the steady-state learned weights when presenting an input value of 0.45. The value gets encoded in a spatio-temporal spike pattern with a bank of 10 neurons in layer u. The neuron whose preferential value is the closest to the input spikes first, followed by the other neurons of that bank. At some time, a postsynaptic spike is emitted by a neuron in layer v, here after the 5th spike and before the 6th spike. The sign of the time difference Δ_t between a post and presynaptic spike delimits two learning regimes for the STDP rule. If $\Delta_t > 0$, then the learned weight is an attracting fixed point. If $\Delta_t \leq 0$ the weight gets depressed by a magnitude that grows with Δ_t, and finally the weight tends to 0.

Thus w_{ij} will exponentially converge to the target value $c_{\max}$ with a timescale τ_w. In our experiments τ_w is equal to the period of the training phase.

4   Experiments and Results

This section presents the results of the conducted experiments to evaluate the performance of our architecture in terms of reconstruction error. The data are reconstructed from the weights with a center of mass with periodic boundary conditions. This is a consequence of the circular nature of the gaussian receptive fields of neurons in layer u. For the two datasets, data were normalized in range [0.05, 0.95] to prevent the wrapping effect of the circular gaussian receptive fields covering the range [0,1]. Each input vector is encoded by the receptive fields in a vector of activation levels for the neurons of u layer. Activation levels are kept

Table 1. Network parameters used in all simulations

Neuronal parameters, used in (1)							
dt	τ_m^u	τ_m^v	$V_{reset}^{u,v}$	θ^u	θ^v	T_{refrac}^u	T_{refrac}^v
0.1 ms	10.0 ms	1.3 ms	0.0	0.5	1.3 n_{dim}	6 ms	4 ms
Synaptic parameters, used in (2)							
τ_f (u to v)	τ_f (v to v)						
2.8 ms	0.7 ms						
Learning parameters, used in (3) and (4)							
τ_+	τ_-	α_+	α_-	w_{offset}	θ_{stdp}		
2.2 ms	5.5 ms	0.005	0.045	0.2	0.35		
Neighborhood parameters, used in (6) and (7)							
r	c_{min}	c_{max}					
0.3	4 θ_v	37 θ_v					

constant for a period of 12.5 ms, followed by a quiescent period of 12.5 ms (see Fig. 2B). Synaptic weights between layers u and v are randomly initialized in range [0.4,0.6]. Layer v is initialized with 100 neurons arranged in a 10*10 grid.

Table 1 show the network parameters and gives a simple strategy to scale the architecture to higher input data dimensions n_{dim}. Only three parameters need to be scaled: the threshold level θ_v and the inhibitory levels c_{min} and c_{max}.

4.1 Quality Assessment of the Reconstructed Images

We used the root mean squared (RMS) error to quantify the difference between an original image patch $\mathbf{y}_p$ and a reconstructed patch $\hat{\mathbf{y}}_p$.

$$RMS = \frac{1}{P}\sum_{p=1}^{P}\sqrt{\frac{1}{D}\sum_{i=1}^{D}(y_{i,p}-\hat{y}_{i,p})^2} \qquad (8)$$

where P is the number of extracted patches from the images and D is the number of pixels of the patch.

4.2 Results on MNIST and Natural Images

We performed two experiments on two real datasets. For the testing phase and for both experiments we fixed the lateral inhibition in layer v to c_{max} and disabled the plasticity.

The first experiment was conducted on the MNIST dataset [9]. We randomly selected a subset of 15 000 and 1000 training and testing digits, respectively, from the MNIST dataset. The 28 * 28 digits were splitted in 4 * 4 patches. We scaled θ_v, c_{min}, c_{max} with the strategy presented in Table 1 for 16 dimensions. 150,000

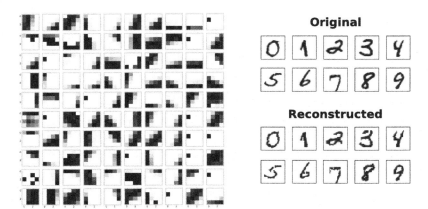

Fig. 4. Left: learned receptive fields of the neurons of v layer after the training phase. Right: original digits and reconstructed digits from the receptive fields.

Table 2. RMS reconstruction error values reported by Tavanei et al. [18]

Dataset 1	Tavanei [18]	Our model
MNIST	0.17	0.13
Natural images	0.24	0.10

randomly selected patches were used to train the network. Figure 4 shows the learned receptive fields and the reconstructed images for visual assessment.

The second experiment was performed on natural images [13]. 512 * 512 images were splitted in 16 * 16 patches. Again, we scaled θ_v, c_{min}, c_{max} with the strategy presented in Table 1 for 256 dimensions. 150,000 randomly selected patches were used so as to train the network.

The RMS reconstruction error for the two datasets is reported in Table 2 and is compared to the state-of-the art performances of Tavanei et al. [18]. Our architecture shows better performances for the two datasets.

5   Discussion

Related works make use of rate codes [3, 7, 18]. Our work makes use of first-spike latency code which brings significant advantages compared to rate code and rank-order code. Consider a bank of N neurons in which each neuron only spike once to represent a feature. Rank-order codes used in conjunction with gaussian receptive fields will lead to a resolution of $1/N$ for the encoded feature. Thus increasing the resolution requires to increase the number of neurons and consequently the number of spikes. This is costly but the situation is getting worse for a rate code, as it requires N^2 neurons to achieve the same resolution. It is possible to be more efficient: by using the spike latencies we can theoretically decode a feature value with a resolution that is only limited by the time step resolution. What is then missing is a mechanism able to efficiently decode these latencies. This is provided by our novel STDP rule that learns a representation of the latencies in the synaptic weights. In our experiments only 10 neurons in layer u, generating 10 spikes, are able to represent a feature. The first neurons that spike in response to a feature value convey a maximal amount of information - as they are best tuned to the feature value - thus the resulting latencies are encoded by non-zero synaptic weights. The neurons that spike later convey few information, thus the resulting latencies induce weight depression and make the weights decay to 0. In this way a postsynaptic neuron in layer v learns a filter in its afferent synaptic weights that makes it able to quickly detect an input pattern. In other words the postsynaptic neuron spikes without having to wait for the whole set of presynaptic neurons to spike. Rumbell et al. [16] also developped a spiking SOM that uses temporal codes and mimics the original Kohonen SOM [8]. However, this architecture only performs categorization and not vector quantization.

A limitation of our architecture is the costly all-to-all lateral connectivity in layer v. The scalability of the architecture also needs further investigations, though preliminary results show that a simple strategy to adapt parameters to higher dimensions already provides satisfactory performances.

To conclude, we presented in this paper a SNN architecture able to perform for the first time vector quantization using temporal codes, more precisely first-spike latency code. This code brings a good theoretical compromise between the amout of neurons and/or spikes required to encode a continuous variable compared to rate code or rank-order code. What was missing was a mechanism able to decode the latencies. We solved this issue by developing a novel STDP rule able to efficiently learn a representation of the latencies in the synaptic weights. The locality in time and space of the computation brought by the STDP rule makes it neuromorphic-friendly. The learned synaptic weights fall in range [0,1] and the distribution is unimodal, allowing an easy subsequent decoding step. A soft competition is implemented in the lateral connections of layer v so that neighboring neurons of the BMU have a chance to spike and thus to represent close vectors in the input space. The lateral inhibition increases during the training phase with an homeostatic mechanism, so as to first organize the map and then make the neurons learn more and more individualized receptive fields.

Acknowledgments. This work has been supported by ANR project SOMA ANR-17-CE24-0036.

References

1. Bohte, S., La Poutre, H., Kok, J.: Unsupervised clustering with spiking neurons by sparse temporal coding and multilayer RBF networks. IEEE Trans. Neural Netw. **13**(2), 426–435 (2002)
2. Brette, R.: Philosophy of the spike: rate-based vs. spike-based theories of the brain. Front. Syst. Neurosci. **9**, 151 (2015)

3. Burbank, K.S.: Mirrored STDP implements autoencoder learning in a network of spiking neurons. PLOS Comput. Biol. **11**(12), e1004566 (2015)
4. Diehl, P.U., Cook, M.: Unsupervised learning of digit recognition using spike-timing-dependent plasticity. Front. Comput. Neurosci. **9**, 99 (2015)
5. Fritzke, B.: A growing neural gas network learns topologies. In: Advances in Neural Information Processing Systems, vol. 7, pp. 625–632. MIT Press (1995)
6. Hazan, H., Saunders, D.J., Sanghavi, D.T., Siegelmann, H., Kozma, R.: Lattice map spiking neural networks (LM-SNNs) for clustering and classifying image data. Ann. Math. Artif. Intell. 1–24 (2019). https://doi.org/10.1007/s10472-019-09665-3
7. King, P.D., Zylberberg, J., DeWeese, M.R.: Inhibitory interneurons decorrelate excitatory cells to drive sparse code formation in a spiking model of V1. J. Neurosci. Official J. Soc. Neurosci. **33**(13), 5475–5485 (2013)
8. Kohonen, T.: Self-organized formation of topologically correct feature maps. Biol. Cybern. **43**(1), 59–69 (1982)
9. LeCun, Y., Cortes, C., Burges, C.: The MNIST database. http://yann.lecun.com/exdb/mnist/
10. MacQueen, J.B.: Some methods for classification and analysis of multivariate observations. In: Cam, L.M.L., Neyman, J. (eds.) Proceedings of the fifth Berkeley Symposium on Mathematical Statistics and Probability, vol. 1, pp. 281–297. University of California Press (1967)
11. Marsland, S., Shapiro, J., Nehmzow, U.: A self-organising network that grows when required. Neural Netw. Official J. Int. Neural Netw. Soc. **15**(8–9), 1041–1058 (2002)
12. Martinetz, T.M., Berkovich, S.G., Schulten, K.J.: Neural-gas network for vector quantization and its application to time-series prediction. IEEE Trans. Neural Netw. **4**(4), 558–569 (1993)
13. Olshausen, B.A., Field, D.J.: Emergence of simple-cell receptive field properties by learning a sparse code for natural images. Nature **381**(6583), 607–609 (1996)
14. Rossum, M.C.W., Bi, G.Q., Turrigiano, G.G.: Stable Hebbian learning from spike timing-dependent plasticity. J. Neurosci. **20**(23), 8812–8821 (2000)
15. Roy, K., Jaiswal, A., Panda, P.: Towards spike-based machine intelligence with neuromorphic computing. Nature **575**(7784), 607–617 (2019)
16. Rumbell, T., Denham, S.L., Wennekers, T.: A spiking self-organizing map combining STDP, oscillations, and continuous learning. IEEE Trans. Neural Netw. Learn. Syst. **25**(5), 894–907 (2014)
17. Tavanaei, A., Ghodrati, M., Kheradpisheh, S.R., Masquelier, T., Maida, A.: Deep learning in spiking neural networks. Neural Netw. **111**, 47–63 (2019)
18. Tavanaei, A., Masquelier, T., Maida, A.: Representation learning using event-based STDP. Neural Netw. **105**, 294–303 (2018)
19. Vasuki, A., Vanathi, P.: A review of vector quantization techniques. IEEE Potentials **25**(4), 39–47 (2006)

A Survey of Graph Curvature and Embedding in Non-Euclidean Spaces

Chandni Saxena[✉], Tianyu Liu, and Irwin King

Department of Computer Science and Engineering, The Chinese University of Hong Kong, Shatin, NT, Hong Kong
{csaxena,tyliu,king}@cse.cuhk.edu.hk

Abstract. Interest has been growing lately towards learning representations for non-Euclidean geometric data structures. Such kinds of data are found everywhere ranging from social network graphs, brain images, sensor networks to 3-dimensional objects. To understand the underlying geometry and functions of these high dimensional discrete data with non-Euclidean structure, it requires their representations in non-Euclidean spaces. Recently, graph embedding in Riemannian spaces has been explored to successfully capture the geometric properties of networks and achieve the state-of-the-art quality in graph representation learning tasks. In this survey, we provide an overview on graph embeddings based on Riemannian geometry with different curvature spaces. We further present recent developments in various application areas using graph embedding models in non-Euclidean domains.

Keywords: Graph curvature · Geometric data · Graph embeddings · Hyperbolic spaces · Applications

1 Introduction

Graphs are universal data structures and have been perceived as empirical models for representing complex relational data in terms of nodes and edges. In multiple domains (e.g., social networks, recommender systems, knowledge graphs, molecular fingerprints, traffic networks, etc.), the datasets usually contain millions of samples as well as very rich interaction information. It is difficult to store and query relationships between the data in traditional relational databases due to their fixed schema. Graph, as a very flexible data structure captures essential topological properties and is regarded as a powerful analysis paradigm across fields as divergent as biology, neuroscience, linguistics, engineering, finance, marketing, and social sciences.

In order to learn the underlying structural information and graph features, many graph representation learning methods have been proposed. After obtaining the graph as input, representation learning (also called network embedding) provides dense representation of features, i.e., low dimensional vectors for nodes, edges or subgraphs (shown in Fig. 1). This conversion benefits the downstream

© Springer Nature Switzerland AG 2020
H. Yang et al. (Eds.): ICONIP 2020, LNCS 12533, pp. 127–139, 2020.
https://doi.org/10.1007/978-3-030-63833-7_11

applications including node classification, link prediction and graph categorization, as the learning models can only understand and deal with numerical data. There are roughly three classes of graph embedding methods: factorization-based, random walk-based, and deep learning-based algorithms [12]. The goal of these methods is that the semantics of the discrete data is captured by distances in the embedding space, i.e., two similar nodes should be located nearly to each other. Recently, there has been a growing interest in deep learning-based models where Graph Neural Networks (GNN) have achieved state-of-the performance [12]. Due to powerful simplicity and efficiency of Euclidean geometry, these models exploit conventional Euclidean spaces to learn graph representation. While deep learning-based models show high performances on

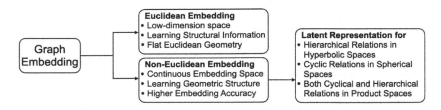

Fig. 1. Illustration of graph embedding in Euclidean and non-Euclidean spaces.

non-geometric data such as images, speech signals, or videos, Euclidean spaces have limitations for geometric data such as graphs and three-dimensional (3D) objects. However, in recent works it has been validated that flat Euclidean space is not an appropriate isometric space to embed geometric-structured graphs [10, 22, 28]. The tree-like properties of graphs suffer from substantial distortion upon embedding in such spaces. Moreover, when the number of nodes increases in graphs, it is inappropriate to position nodes in such embedding spaces as the distances get more distorted towards leaves of the tree [2].

Over the last few years, there has been a surge of interest in trying to exploit non-Euclidean space for graphs, since the underlying geometry of embedding space is favourable for representation of such geometric data [9]. Riemannian geometry provides a mathematical framework to manipulate non-Euclidean geometries and leads to the fundamental theory of manifold learning and information geometry [2]. A space of constant sectional curvature assigns an interesting trade-off between Euclidean space and Riemannian manifolds and defines hyperbolic, spherical, and Euclidean geometries. As presented in Fig. 1, Non-Euclidean spaces exploit structural properties and show efficient learning representation capacity. For instance, a non-Euclidean hyperbolic space continuously grows as analogous to a tree and even provides nearly isometric embedding when a tree grows exponentially. Additionally, hyperbolic embeddings in the case of complex networks, scale-free graphs, and hierarchical data exhibit low distortion and naturally unfold their properties, such as strong clustering, hierarchical community structure, and heterogeneous degree distributions [22, 29, 33, 39].

In this survey we introduce a mathematical generalization of constant curvature spaces and provide classes of graph embedding models in canonical Riemannian manifolds with constant curvature. We then review a taxonomy of graph embedding models in non-Euclidean spaces. Finally, we briefly present the recent advances in graph embedding applications and tasks. This unified formation of the geometric background, recent development in embedding methods, and applications would help the reader to gain grasp of the foundation and insight into recent advancements in graph representation learning.

2   Mathematical Background

This section serves as background of concepts and mathematical definitions of basic notions of manifolds and description of various curvatures in Riemannian geometry to understand non-Euclidean spaces.

2.1   Preliminaries

We briefly introduce the concepts and terminologies which we shall use in the following parts.

Curvature. Curvature is a measure which defines how much a geometric object deviates from being flat. Graphs are discrete data and for a given graph the "flatness" is to be understood to manifest connectivity and interdependence between distant nodes.

Metric Spaces and Embeddings. Metric space is an ordered pair $(\mathcal{X}, d)$, where $\mathcal{X}$ is the underlying space and $d : \mathcal{X} * \mathcal{X} \to \mathbb{R}$ is a distant function (metric) measuring distance between pairs of elements of $\mathcal{X}$. An embedding of $(\mathcal{X}_1, d_1)$ into $(\mathcal{X}_2, d_2)$ is a mapping of $f : \mathcal{X}_1 \to \mathcal{X}_2$. Embedding of graph metric (V, d_G) of a graph $G = (V, E)$, with V nodes and E edges into k-dimensional target space $(\mathcal{T}^k, d_{\mathcal{T}})$, where d_G measures the length of shortest paths between two nodes in G and $d_{\mathcal{T}}$ is a standard metric of $\mathcal{T}^k$ [41]. The worst-case distortions over all the distances in the metric determine the quality of the embedding.

Riemannian Manifold. Manifolds belong to the branches of mathematics of topology and differential geometry, which can be considered as collections of points that locally but not globally resemble Euclidean space [5]. For instance, a manifold can be considered as a continuous approximation of a discrete graph. The shortest paths on manifolds are termed as geodesics. Riemannian manifolds $(\mathcal{M}, g)$ are smooth manifolds $\mathcal{M}$ equipped with Riemannian metrics g, which is described by smoothly varying choices of inner products on tangent spaces denoted by $T_p\mathcal{M}$ for tangent vectors T_p at each point $p \in \mathcal{M}$. Riemannian metric g is a function to measure geometric quantities such as geodesic distances, angles and curvatures [24].

2.2   Curvatures of Riemannian Geometry

In Riemannian geometry, curvature is defined in terms of first and second order derivatives of the metric tensor [26]. There are different notions of curvatures to describe geometry objects: Principal curvature, Gaussian curvature [24], sectional curvature, Ricci curvature [26], etc. Table 1 lists different curvature notions and their properties. For further investigation of the two discretizations of Ricci curvature [34] may be referred.

Principal Curvature. Principal curvature is the simplest curvature, which is described by two values k_{max} and k_{min}, are the maximum and minimum values of the curvature at each point p of a differential manifold.

Gaussian Curvature. The product of the principal curvatures is known as Gaussian curvature: $\mathcal{K} = k_{max} * k_{min}$. The Gaussian curvature assigns a scalar for each point on the surface in 3D Euclidean space and it also equals the Jacobian determinant of the Gaussian map. This curvature notion is helpful to be able to distinguish different types of manifolds from each other.

Sectional Curvature. Sectional curvature is a full invariant and encodes all information about a Riemannian metric. It further generalizes Gaussian curvature to high dimension and assigns a scalar for 2-dimensional (2D) linear subspace of tangent space at each point on a Riemannian manifold by encoding all the information about Riemannian metric.

Ricci Curvature. The Ricci curvature is the average of sectional curvature. The Ricci curvature assigns a scalar for each unit tangent vector at each point on the manifold. Sectional curvature only involves one 2D subspace $\mathcal{P}$ in the tangent space $T_p\mathcal{P}$ at point p, while Ricci curvature further involves each 2D subspace and takes the average as the curvature metric at p. Generalization to the Ricci curvature have been explored under different notions including Ollivier's Ricci curvature (ORC) [30] and Forman's Ricci curvature (FRC) [16]. These discrete formulations on curvature provide important tools to study geometric and topological properties of graphs in terms of curvatures of nodes and edges.

Ollivier Ricci Curvature. Ollivier reformulated Ricci curvature in the view of an optimal transportation plan between two points in a metric space. In fact, ORC generalizes Ricci curvature to general metric space with probability measure and this gives intuition that ORC discretizes Ricci curvature to undirected graphs with nodes having probability measure attributes.

Forman Ricci Curvature. Forman Ricci curvature generalizes Ricci curvature which measures how fast distance volume grows between points and in graph analogy this can be inferred to measure dispersion rate of geodesics. Based on Forman curvature of edges, Forman curvature (a scalar value) for nodes are defined.

All above variants of curvatures are useful to understand and define the notion of graph curvature and representation learning in Riemannian manifolds. It is beyond the scope of this paper to define them in canonical forms.

Table 1. Various curvature notions on manifolds

Curvature	Output	Definition	Geometric nature
Principal	2 scalars	Extremums of curvature of $3D$ line	Extrinsic
Gaussian	1 scalar	Product of principal curvatures	Intrinsic
Sectional	1 scalar	Gaussian curvature in one tangent space	Intrinsic
Ricci	1 scalar	Average of Gaussian curvatures	Intrinsic
Ollivier Ricci	1 scalar	Minimal transportation expense	Intrinsic
Forman Ricci	1 scalar	Dispersion rate of geodesic	Intrinsic

Table 2. Properties of constant curvature spaces

Property	Euclidean	Spherical	Hyperbolic
Curvature	zero	positive	negative
Parallel lines	1	0	∞
Sum of angles in triangle	180°	>180°	<180°
Characteristic Graph			

2.3   Spaces of Constant Curvature

We now present an intuitive description of curvatures and the geometry of homogeneous spaces, also called "spaces of constant curvature", in this case curvature is constant everywhere in the space and deforms to a constant curvature space. According to the different conditions, there are three kinds of connected Riemannian manifolds with constant sectional curvature: Euclidean space with zero curvature, spherical space with constant positive curvature and hyperbolic space with constant negative curvature. Table 2 summarizes the properties of the three geometries of constant curvatures. The notion of curvature plays a central role due to correlation between structure of the data (hierarchical, cyclical) and the geometry of non-Euclidean embedding space [19]. The geometry of hyperbolic spaces is suited to embed a tree structured graph and that of spherical spaces is appropriate to embed a cyclic graph. We define (1) **Hyperbolic Spaces** and (2) **Spherical Spaces** with related models in the remaining part of this section.

(1) Hyperbolic Spaces. In Hyperbolic spaces projections preserve angles but massively distort distances. Another interesting property regarding hyperbolic spaces is that the area and volume of a ball grows exponentially with the radius, which makes the hyperbolic space a perfect model to embed tree-like data. In hyperbolic geometry distances can grow exponentially towards the edges of the disk (shown in Fig. 2(a)) and a set of hyperbolic straight lines can pass through a single point (shown in Fig. 2(b)). These properties contribute infinite trees to have nearly isometric embeddings in hyperbolic space [10].

We denote a Euclidean space as $\mathcal{R}$, hyperbolic space as $(\mathcal{H}, g)$, and g is a Riemannian metric. The three commonly used models for hyperbolic space in Riemannian geometry are (i) **Poincaré Ball Model** [28], (ii) **Klein Model**, and (iii) **Lorentz Model** [29].

Fig. 2. Properties of hyperbolic spaces: (a) Each tile is a constant area in hyperbolic plane, however vanishes at the Euclidean space boundary and (b) Number of straight lines passing through a point, parallel to the blue line [10]. (Color figure online)

(i) Poincaré Ball Model. Let $\mathcal{B}^n = \{x \in \mathcal{R}^n \, \|x\| < 1\}$ be an n-dimensional hyperbolic space and $\|.\|$ denotes Euclidean norm. The Poincaré ball model corresponds to the Riemannian manifold $(\mathcal{B}^n, g_x)$ and Riemannian metric tensor defined on $\mathcal{B}^n$ is defined as:

$$g_x = \left(\frac{2}{1 - \|x\|^2} \right)^2 g^E, \tag{1}$$

where $x \in \mathcal{B}^n$ and g^E stands for the Euclidean metric tensor. Furthermore, hyperbolic distance between points $Z_i, Z_j \in \mathcal{B}^n$ in Poincaré ball model is calculated as:

$$g_p(Z_i, Z_j) = \operatorname{arcosh} \left(1 + 2 \frac{\|Z_i - Z_j\|^2}{(1 - \|Z_i\|^2)(1 - \|Z_j\|^2)} \right), \tag{2}$$

where $\operatorname{arcosh}(w) = \ln(w + \sqrt{w^2 - 1})$ is the inverse of hyperbolic cosine function.

(ii) Klein Model. The metric space $(\mathcal{B}^n, g_k)$ is the Klein model of n dimensional hyperbolic space, in this case $\mathcal{B}^n$ is a Euclidean unit ball, and the distance is given as:

$$g_k(Z_i, Z_j) = \operatorname{arcosh} \left(\frac{1 - (Z_i | Z_j)}{\sqrt{1 - \|Z_i\|^2} \sqrt{1 - \|Z_j\|^2}} \right), \tag{3}$$

(iii) Lorentz Model. The metric space $(\mathcal{L}, g_l)$ is the Lorentz model (also called upper sheet hyperboloid model) of n dimensional hyperbolic space where distance is defined as:

$$g_l(Z_i, Z_j) = \operatorname{arcosh}(-\langle Z_i | Z_j \rangle) \in [o, \infty], \tag{4}$$

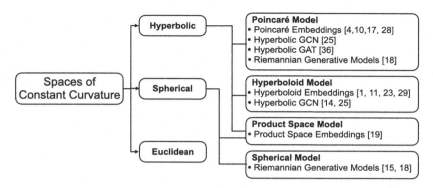

Fig. 3. Taxonomy of graph embedding models in non-Euclidean spaces.

where $\langle Z_i|Z_j\rangle$ is a Minkowski bilinear form and $\langle Z_i|Z_j\rangle = -Z_{i0}Z_{j0} + \sum_{k=1}^{n} Z_{ik}Z_{jk}$. It is experimentally proven that among all the three models, the Lorentz model has the best performance in graph analysis since it contains simple formulas and avoids numerical instabilities that arise from the fraction in the Poincaré distance [29]. It should be noted that the Poincaré Model exhibits conformal behaviour in view of the fact that the Euclidean angles between hyperbolic lines are equal to their hyperbolic angles in the model [42]. Due to this property of the model it is well-suited for gradient-based optimization.

(2) Spherical Space. In the case of spherical space, the spherical model $\mathcal{S}_K^n$ is most easily defined when embedded in $\mathcal{R}^{n+1}$ Euclidean space. The spherical manifold for any K is defined on the subset $\{x \in \mathcal{R}^{n+1} : ||x||_2 = K^{1/2}\}$ with metric g_s originated by the Euclidean metric on $\mathcal{R}^{d+1}$. The spherical distance on $\mathcal{S}^n$ is defined as: $g_s(Z_i, Z_j) = \arccos(\langle Z_i, Z_j\rangle)$

With above background and description of Riemannian geometry and hyperbolic space, following sections provide graph embedding models and summarization on recent development in the wide range of applications concerning graph structured data embedded in non-Euclidean spaces.

3 Graph Embeddings in Non-Euclidean Spaces

Non-Euclidean spaces are well-suited for graph embeddings to explore underlying geometry and function and also have been considered for various network science problems [1,35,41]. In this section we present taxonomy (shown in Fig. 3) of graph embedding models in non-Euclidean geometry with a constant curvature for downstream machine learning tasks and applications. We discuss these models in the following part of this section.

Poincaré Embeddings. The Poincaré ball model of hyperbolic space is specifically suitable to represent hierarchies present in graph data. The Poincaré embedding model [28] learns hyperbolic space embeddings via Poincaré distance function defined in Eq. (2). The points $\{Z_i\}$, $i \in V$ represented on hyperbolic spaces denote embeddings for node i, where graph properties (local and global) are well preserved. The smooth change in distances within the Poincaré ball with respect to the points Z_i and Z_j is the key for continuous hierarchical embeddings in hyperbolic spaces. The embeddings are then learned by computing loss function which minimizes distances between connected nodes while maximizes distance between disconnected nodes given as:

$$\mathcal{L}(\theta) = \sum_{Z_i, Z_j} \log \frac{e^{-g_p(Z_i, Z_j)}}{\sum_{Z'_j \in \mathcal{N}(Z_i)} e^{-g_p(Z_i, Z'_j)}}, \tag{5}$$

where $\mathcal{N}(Z_i)$ is the set of negative examples for Z_i (including Z_i). The distance function for Poincaré in Eq. (2) is differentiable and pertinent for gradient-based optimization. This requires Riemannian optimization techniques, for example, the Riemannian stochastic gradient descent (Riemannian SGD), a generalization of the SGD [8] in Euclidean space, except that there is no concept of vector addition. In hyperbolic space, the similar concept is exponential map [7]. Other variants of Poincaré embedding model have been proposed with relevant loss functions to offer isometric embedding in hyperbolic space. For example, Chamberlan et al. [10] extend Poincaré embedding and define learning by optimising objective function which integrates hyperbolic inner product in skip-gram losses with negative sampling. The model has different adaptations including multi-relational Poincaré model [4], Poincaré Glove for word embeddings [39], and Hyperbolic Entailment Cone to embed symbolic objects in hyperbolic space [17].

Hyperboloid Embeddings. This model aims to learn non-Euclidean graph representation based on distance function of Lorentz model as defined in Eq. (3). The model provides improved stability and efficient learning than Poincaré ball model. Nickel and Kiela [29] define Hyperboloid (Lorentz) embeddings in hyperbolic spaces with Riemannian SGD optimization. The model preserves pairwise similarity and defines the loss function based on similarity semantic approach. As stated in [29], the basic idea of the model is to define embedding of concept c_i, where $\{c_i\}_{i=1}^m \in \mathcal{C}$ (concepts set) and $X \in \mathbb{R}^{m \times m}$ denotes the similarity scores of these concepts. According to our notations Z_i be the embeddings for c_i and $\mathcal{N}(i, j)$ denotes the concepts set that are less similar to c_i then c_j as negative sampling cases. The embeddings $\theta = \{Z\}_{i=1}^m$ are learned by optimizing the loss function given as:

$$Prob(\phi(i, j)|\theta) = \sum_{Z_i, Z_j} \log \frac{e^{-g_l(Z_i, Z_j)}}{\sum_{Z'_j \in \mathcal{N}(Z_i)} e^{-g_l(Z_i, Z'_j)}}, \tag{6}$$

where $\phi(i, j) = \arg\min_{k \in \mathcal{N}(i,j)} (g_l(Z_i, Z_k))$. It is significant to note that due to equivalence of both models, points in Lorentz model and Poincaré ball model can be mapped onto each others' spaces via diffeomorphism (ψ) [23], such that $\psi : \mathcal{B}^n \to \mathcal{L}^n$, where $\psi(p_0, p_1, .., p_n) = \frac{(p_0, p_1, ..., p_n)}{p_0 + 1}$ maps the Lorentz model into a

Poincaré ball. Inheriting hyperboloid embedding, there are various modifications for embeddings in hyperbolic spaces [1,11,23,38].

Product Space Embeddings. Graph data with hierarchical and cyclical nature of local regions need embedding spaces with heterogeneous curvature. Product Space (also known as mixed spaces) provides higher quality of representation in such a framework. Gu et al. [19] define product space, where the mixed curvature embeddings are characterised with Riemannian product manifolds of hyperbolic, spherical and Euclidean components and provide a decomposable non-constant curvature. The curvature is directly learned from geometry of underlying data for each component space with the embedding via Riemannian optimization. Authors define distance-based loss function to learn optimal embedding on product spaces.

Hyperbolic GCN. Graph Convolutional Networks (GCNs) are GNN based models. GNNs leverage node features and provide deep learning frameworks for graph embedding in Euclidean geometry and associated vector spaces [12]. Furthermore, GCNs bring up inductive graph embeddings and define layer-wise propagation for neural network models on GNN. However, GNN embeddings in non-Euclidean geometry lead to a large distortion due to hierarchical and scale-free property of graphs. On the other hand, non-Euclidean frameworks extend GNN to enable learning on hyperbolic geometry and perform convolution in hyperbolic spaces. We present hyperbolic GCNs which introduce and define core GCN operations to perform convolution and next define attentions based graph attention networks (GAT) model in hyperbolic spaces with curvature as trainable parameters. GCNs compute node representations by aggregating messages from neighbors over multiple steps and define information propagation operations on graphs. A new representation for node u at propagation step $k + 1$ is given as:

$$h_u^{k+1} = \sigma(\sum_{v \in \mathcal{I}(u)} \tilde{A}_{uv} W^k h_v^k), \tag{7}$$

where σ is an activation function, h_v^k is representation of node v at step k, $W^k \in \mathbb{R}^{h \times h}$ represents a trainable parameter at time t, $\tilde{A}_{uv}$ is derived from adjacency matrix A and it captures connectivity of node u and set $\mathcal{I}(u)$ of neighbors of u. Recently, HGCN [14] and HGNN [25] extend GNN in hyperbolic spaces with learnable curvatures. These models propose new set operations for aggregation to perform graph convolution and provide a first-order approximation of the hyperbolic manifold at a point. Furthermore, Bechmann et al. [2] introduce an extension of hyperbolic graph neural networks to the stereographic model of a trainable and unified hyperbolic and spherical curvatures. Ye et al. [44] define curvature graph network (CurvGraph) and formulate weighted aggregation parameter in update step of GCN to incorporate advanced structural information of Ricci curvature of edges connecting set of neighbors $\mathcal{I}(u)$.

Hyperbolic GAT. Attention operation in GCN computes a notion of neighbors' importance to the center node and learns attention weights to construct weighted neighborhood aggregation. Hyperbolic attention-based aggregation uses hyperbolic embeddings to compute attention weights. HGCN [14] extends graph convolutions to hyperbolic geometric space with attention based architecture for neighborhood propagation. Shimizu et al. [36] define hyperbolic multi-head attention mechanism perceived in the Poincaré ball model.

Riemannian Generative Models. In recent times, generative models have been generalized to hyperbolic and spherical latent spaces. Following the same analogy, Davidson et al. [15] explored hyperspherical latent representation for variational auto-encoders. Grattarola et al. [18] define adversarial autoencoders framework for graph embedding on manifolds with constant curvatures.

Other Models. Recently, several models of deep networks extend the hyperbolic geometric representation in terms of generalizing operations defined on Lorentz and Klein models for graph embeddings. Following that, Gülçehre et al. [20] define hyperbolic attention mechanism using hyperbolic distance between nodes and generate hyperbolic aggregation weights. Zhang et al. [45] investigate GNN in hyperbolic spaces and define graph operation for attention mechanism based on hyperbolic proximity.

4   Applications and Tasks

Graph embedding in a non-Euclidean space can be applied to a wide range of tasks and applications. We review some recent developments in this section.

Knowledge Graph Embedding. Recently, knowledge graphs (KG) embedding models in the hyperbolic planes have shown to be effective as they can represent model topological structure. Recent work presented in [4,13,21] exploited hyperbolic embedding to learn relationship-specific hyperbolic transformations.

Community Analysis. Ni et al. [27] considered the problem of community detection in geometric view. Bakker et al. [3] considered Riemannian geometry for community detection tasks where graphs change over the time.

Hyperbolic Embedding. The geometry view of graph embedding has generated considerable attention and offered a new perspective on the topology of complex networks structure. Most recent advancements in this direction are [1,6,22,28,29,37].

Hyperbolic Graph Neural Networks. Hyperbolic graph neural networks offer low distortion and enable to learn embedding for graph structures like hierarchical and scale-free graphs with hyperbolic distance metric [14,32,36].

Recommender Systems. Lately, hyperbolic spaces have gained significant attention to investigate the notion of training recommender systems. Recent works in [11,31,40,43] offer effective performances in this direction.

5   Conclusion

This survey highlights the significance of graph curvature and the representation of graph-structured data in non-Euclidean spaces. We present an overview of models of curvature spaces and description of Riemannian hyperbolic geometry. We further provide categories of graph embedding models in geometric spaces and summarize the survey on recent works in various application domains using non-Euclidean geometry. We explore the use of hyperbolic representation on graph-structured data and encourage future research in deep geometric models for graph learning methods and applications. In the future we may conduct a more comprehensive coverage of learning graph embedding on different hyperbolic models with challenges in optimization methods, further studies may investigate complex Riemannian manifolds and exploit architectures on non-Euclidean spaces for graph learning.

Acknowledgements. The work described in this paper was supported by the Research Grants Council of the Hong Kong Special Administrative Region, China (CUHK 2410021, Research Impact Fund, No. R5034-18).

References

1. Asta, D., Shalizi, C.R.: Geometric network comparison. arXiv preprint arXiv:1411.1350 (2014)
2. Bachmann, G., Bécigneul, G., Ganea, O.E.: Constant curvature graph convolutional networks. arXiv preprint arXiv:1911.05076 (2019)
3. Bakker, C., Halappanavar, M., Sathanur, A.V.: Dynamic graphs, community detection, and Riemannian geometry. Appl. Netw. Sci. **3**(1), 3:1–3:30 (2018)
4. Balazevic, I., Allen, C., Hospedales, T.: Multi-relational poincaré graph embeddings. In: NIPS, pp. 4465–4475 (2019)
5. Berger, M., Gostiaux, B.: Differential Geometry: Manifolds, Curves, and Surfaces, vol. 115. Springer Science & Business Media, New York (2012). https://doi.org/10.1007/978-1-4612-1033-7
6. Boguná, M., Papadopoulos, F., Krioukov, D.: Sustaining the internet with hyperbolic mapping. Nat. Commun. **1**(1), 1–8 (2010)
7. Bonnabel, S.: Stochastic gradient descent on Riemannian manifolds. IEEE Trans. Autom. Control **58**(9), 2217–2229 (2013)
8. Bottou, L.: Stochastic gradient descent tricks. In: Montavon, G., Orr, G.B., Müller, K.-R. (eds.) Neural Networks: Tricks of the Trade. LNCS, vol. 7700, pp. 421–436. Springer, Heidelberg (2012). https://doi.org/10.1007/978-3-642-35289-8_25
9. Bronstein, M.M., Bruna, J., LeCun, Y., Szlam, A., Vandergheynst, P.: Geometric deep learning: going beyond Euclidean data. IEEE SPM **34**(4), 18–42 (2017)
10. Chamberlain, B.P., Clough, J.R., Deisenroth, M.P.: Neural embeddings of graphs in hyperbolic space. arXiv preprint arXiv:1705.10359 (2017)
11. Chamberlain, B.P., Hardwick, S.R., Wardrope, D.R., Dzogang, F., Daolio, F., Vargas, S.: Scalable hyperbolic recommender systems. arXiv preprint arXiv:1902.08648 (2019)
12. Chami, I., Abu-El-Haija, S., Perozzi, B., Ré, C., Murphy, K.: Machine learning on graphs: A model and comprehensive taxonomy. arXiv preprint arXiv:2005.03675 (2020)

13. Chami, I., Wolf, A., Sala, F., Ré, C.: Low-dimensional knowledge graph embeddings via hyperbolic rotations. In: NIPS Workshop (2019)
14. Chami, I., Ying, Z., Ré, C., Leskovec, J.: Hyperbolic graph convolutional neural networks. In: Advances in Neural Information Processing Systems, pp. 4869–4880 (2019)
15. Davidson, T.R., Falorsi, L., De Cao, N., Kipf, T., Tomczak, J.M.: Hyperspherical variational auto-encoders. arXiv preprint arXiv:1804.00891 (2018)
16. Forman, R.: Bochner's method for cell complexes and combinatorial Ricci curvature. Discrete Comput. Geom. **29**(3), 323–374 (2003)
17. Ganea, O., Bécigneul, G., Hofmann, T.: Hyperbolic entailment cones for learning hierarchical embeddings. arXiv preprint arXiv:1804.01882 (2018)
18. Grattarola, D., Zambon, D., Livi, L., Alippi, C.: Change detection in graph streams by learning graph embeddings on constant-curvature manifolds. IEEE Trans. Neural Netw. Learn. Syst. **31**(6), 1856–1869 (2019)
19. Gu, A., Sala, F., Gunel, B., Ré, C.: Learning mixed-curvature representations in product spaces. In: International Conference on Learning Representations (2018)
20. Gülçehre, Ç., et al.: Hyperbolic attention networks. arXiv preprint arXiv:1812.08434 abs/1805.09786 (2018)
21. Kolyvakis, P., Kalousis, A., Kiritsis, D.: Hyperkg: hyperbolic knowledge graph embeddings for knowledge base completion. arXiv preprint arXiv:1908.04895 (2019)
22. Krioukov, D.V., Papadopoulos, F., Kitsak, M., Vahdat, A., Boguñá, M.: Hyperbolic geometry of complex networks. Phys. Rev. E **82**(3), 036106 (2010)
23. Le, M., Roller, S., Papaxanthos, L., Kiela, D., Nickel, M.: Inferring concept hierarchies from text corpora via hyperbolic embeddings. arXiv preprint arXiv:1902.00913 (2019)
24. Lee, J.M.: Riemannian Manifolds. GTM, vol. 176. Springer, New York (1997). https://doi.org/10.1007/b98852
25. Liu, Q., Nickel, M., Kiela, D.: Hyperbolic graph neural networks. In: NIPS, pp. 8230–8241 (2019)
26. Najman, L., Romon, P. (eds.): Modern Approaches to Discrete Curvature. LNM, vol. 2184. Springer, Cham (2017). https://doi.org/10.1007/978-3-319-58002-9
27. Ni, C., Lin, Y., Luo, F., Gao, J.: Community detection on networks with Ricci flow. Sci. Rep. **9**(1), 1–12 (2019)
28. Nickel, M., Kiela, D.: Poincaré embeddings for learning hierarchical representations. In: Advances in Neural Information Processing Systems, pp. 6338–6347 (2017)
29. Nickel, M., Kiela, D.: Learning continuous hierarchies in the Lorentz model of hyperbolic geometry. arXiv preprint arXiv:1806.03417 (2018)
30. Ollivier, Y.: Ricci curvature of Markov chains on metric spaces. J. Funct. Anal. **256**(3), 810–864 (2009)
31. Papadis, N., Stai, E., Karyotis, V.: A path-based recommendations approach for online systems via hyperbolic network embedding. In: IEEE. ISCC, pp. 973–980 (2017)
32. Pei, H., Wei, B., Chang, K.C., Lei, Y., Yang, B.: Geom-GCN: geometric graph convolutional networks. In: ICLR. OpenReview.net (2020)
33. Sala, F., Sa, C.D., Gu, A., Ré, C.: Representation tradeoffs for hyperbolic embeddings. In: ICML, pp. 4457–4466 (2018)
34. Samal, A., Sreejith, R., Gu, J., Liu, S., Saucan, E., Jost, J.: Comparative analysis of two discretizations of Ricci curvature for complex networks. Sci. Rep. **8**(1), 1–16 (2018)

35. Shavitt, Y., Tankel, T.: Hyperbolic embedding of internet graph for distance estimation and overlay construction. IEEE/ACM Trans. Netw. **16**(1), 25–36 (2008)
36. Shimizu, R., Mukuta, Y., Harada, T.: Hyperbolic neural networks++. arXiv preprint arXiv:2006.08210 (2020)
37. Sun, Z., Deng, Z., Nie, J., Tang, J.: Rotate: knowledge graph embedding by relational rotation in complex space. CoRR abs/1902.10197 (2019)
38. Suzuki, R., Takahama, R., Onoda, S.: Hyperbolic disk embeddings for directed acyclic graphs. In: ICML, pp. 6066–6075 (2019)
39. Tifrea, A., Bécigneul, G., Ganea, O.: Poincaré glove: hyperbolic word embeddings. CoRR abs/1810.06546 (2018)
40. Tran, L.V., Tay, Y., Zhang, S., Cong, G., Li, X.: HyperML: a boosting metric learning approach in hyperbolic space for recommender systems. In: WSDM (2020)
41. Verbeek, K., Suri, S.: Metric embedding, hyperbolic space, and social networks. In: Proceedings of the Thirtieth Annual Symposium on Computational Geometry, pp. 501–510 (2014)
42. Wang, X., Zhang, Y., Shi, C.: Hyperbolic heterogeneous information network embedding. AAAI **33**, 5337–5344 (2019)
43. Xu, D., Ruan, C., Korpeoglu, E., Kumar, S., Achan, K.: Product knowledge graph embedding for e-commerce. In: Web Search and Data Mining, pp. 672–680 (2020)
44. Ye, Z., Liu, K.S., Ma, T., Gao, J., Chen, C.: Curvature graph network. In: 8th International Conference on Learning Representations, ICLR. OpenReview.net (2020)
45. Zhang, Y., Wang, X., Jiang, X., Shi, C., Ye, Y.: Hyperbolic graph attention network. arXiv preprint arXiv:1912.03046 (2019)

A Tax Evasion Detection Method Based on Positive and Unlabeled Learning with Network Embedding Features

Lingyun Mi[1,2], Bo Dong[3,4(✉)], Bin Shi[1,2], and Qinghua Zheng[1,2]

[1] School of Computer Science and Technology,
Xi'an Jiaotong University, Xi'an, China
`mly1015@stu.xjtu.edu.cn`, {`shibin,qhzheng`}`@mail.xjtu.edu.cn`
[2] SPKLSTN Lab, Xi'an Jiaotong University, Xi'an, China
[3] School of Continuing Education, Xi'an Jiaotong University, Xi'an, China
[4] National Engineering Lab for Big Data Analytics,
Xi'an Jiaotong University, Xi'an, China
`dong.bo@mail.xjtu.edu.cn`

Abstract. Tax evasion detection has a crucial role in addressing tax revenue loss. In the real world, an accessed tax dataset only contains a small number of labeled taxpayers who evade tax (positive samples) and a large number of unlabeled taxpayers who either evade tax or do not evade tax. It is difficult to address this issue due to this nontraditional dataset. In addition, the basic features of taxpayers designed according to tax experts' domain knowledge and experience are very limited to determining whether taxpayers evade tax. These limitations motivate the contribution of this work. In this paper, we argue that the tax evasion detection task in the real world should be formalized as a positive unlabeled (PU) learning problem. We propose a novel tax evasion detection method based on PU learning with Network Embedding features (PUNE). PUNE effectively detects tax evasion based on basic features and transaction network features that are extracted by a network embedding algorithm. Moreover, PUNE can work well even under label noise. To evaluate the effectiveness of PUNE, we conduct experimental tests on a real-world tax dataset. The results demonstrate that PUNE can significantly improve the performance of tax evasion detection.

Keywords: Positive and unlabeled learning · Tax evasion · Network embedding · Label noise

1 Introduction

Tax evasion has always been a crucial issue to both governments and academic research. Billions to trillions of dollars in revenue are lost every year due to various tax evasion means [7,18]. Tax evasion violates tax-related laws and causes a large amount of tax loss, which not only reduces the national fiscal revenue but also seriously undermines the economic order in society [9,25].

© Springer Nature Switzerland AG 2020
H. Yang et al. (Eds.): ICONIP 2020, LNCS 12533, pp. 140–151, 2020.
https://doi.org/10.1007/978-3-030-63833-7_12

To detect tax evasion, numerous detection methods have been proposed, which can be divided into two categories: whistle-blowing-based methods and manual-select-based methods. The former method mainly relies on the internal or mutual supervision of taxpayers, while the latter method relies on the experience of tax inspectors to judge whether taxpayers have abnormal behaviors related to tax evasion. Although these methods have been employed to identify many tax evasion taxpayers, it is still very difficult to manually process rapidly growing tax data by a limited number of tax inspectors. Therefore, in recent studies, machine learning techniques have been utilized to help tax inspectors detect suspicious taxpayers, extract taxpayer-related features and train detection models for mining vast amounts of tax data [2,4,8]. However, due to the complexity of tax data and the concealment of tax evasion means, two main challenges still exist for detecting tax evasion.

First, there are limited features for describing transactions. Traditional machine learning based models are mainly built on the basic features of taxpayers, which are designed according to tax experts' domain knowledge and experience. However, with exponentially growing tax transaction data, the measurement of trading characteristics in large-scale transaction data using manually designed features, such as the amount and number of transactions, is difficult.

To extract additional features that are beneficial for improving the performance of detecting tax evasion by machine learning, the network embedding (NE) algorithm is applied to describe transaction characteristics based on a transaction network that is built by business records between taxpayers. The combination of network features and basic features forms the feature space of taxpayers, which is subsequently employed for training the tax evasion detection model.

Second, there are a small number of labeled samples and a vast number of unlabeled data. In the real world, tax inspectors randomly select several taxpayers and analyze their income and expenses to determine whether they have evaded tax. Thus, it is a common case in practical classification applications, especially in tax evasion detection scenarios in which the available data only contain labeled positive samples and a vast number of unlabeled samples that could be positive or negative, that is, only a small number of taxpayers have been confirmed as evading tax. The vast amount of taxpayer data and transaction data provided by tax authorities are unlabeled, which include both normal tax evasion taxpayers and suspicious tax evasion taxpayers.

Thus, we are motivated to use positive unlabeled (PU) learning [1], which is a kind of semisupervised learning, to learn a binary classifier from a training set of only positive and unlabeled samples. Compared to the previously discussed machine learning methods, PU learning methods can only utilize labeled and unlabeled samples so that negative samples are not necessary in the training process, which matches our tax evasion detection scenario.

We propose a novel tax evasion detection method based on PU learning with Network Embedding features (PUNE), which can fully utilize the large-scale unlabeled samples in training process. Moreover, in order to improve accuracy of tax evasion detection, we extract transaction features of taxpayers by utilizing network embedding algorithms in our method.

To evaluate the effectiveness of PUNE, we conduct experimental tests on a real-world tax dataset obtained from our collaborated local tax authority. The results demonstrate that PUNE can significantly improve the performance of tax evasion detection. The main contributions of this work are summarized as follows:

- We build a transaction network according to the business records between taxpayers. Based on the network, we apply three network embedding algorithms that are employed to extract the transaction features of taxpayers to verify that transaction features can improve the tax evasion detection performance.
- We formalize the tax evasion detection task as a positive and negative learning problem. In PU training, we can fully utilize not only the positive samples but also the large-scale unbalanced samples. Moreover, we give each sample an individual weight by rank labeling to weaken the influence of label noise, which is significant in tax evasion detection.
- We conduct an experimental test on a real-world tax dataset to evaluate the effectiveness of our tax evasion detection method.

The remainder of this paper is organized as follows: Sect. 2 introduces the preliminary of this paper. Section 3 explains the details of our approach of tax evasion detection. Our approach mainly consists of extracting transaction network features and PU training. Section 4 evaluates PUNE and demonstrates the experiment. Section 5 concludes this work.

2   Preliminary

In the PU problem, let $X \in \mathbb{R}^d$ be a d-dimensional feature space and $Y \in \{-1, 1\}$ be a label space as input and output random variables, respectively. Another random variable $S \in \{-1, 1\}$ is applied to represent whether a sample is labeled.

Assuming that the number of labeled positive samples in the training set is n_l, we can obtain the labeled positive sample set X_l and the unlabeled sample set X_u as follows:

$$X_l = \{x_1, x_2, \ldots, x_{n_l}\}$$

$$X_u = \{x_{n_1+1}, x_{n_l+2}, \ldots, x_N\}$$

We can represent the datasets by X and S as:

$$DS_l = \{(x_1, s_1), (x_2, s_2), \ldots, (x_{n_l}, s_{n_l})\}$$

$$DS_u = \{(x_{n_l+1}, s_{n_l+1}), (x_{n_l+2}, y_{n_l+2}), \ldots, (x_N, s_N)\}$$

Where $s_1, s_2, \ldots, s_{n_l} = 1$ and $s_{n_l+1}, s_{n_l+2}, \ldots, s_N = -1$, which means that labeled samples are considered "positive" and unlabeled samples are considered "negative". In PU learning, the class prior π refers to the proportion of positive samples in the training set; it is represented by $\pi = P(Y = 1)$, π is assumed to be known throughout the paper and can be estimated from positive and unlabeled data [3, 13, 16, 20].

In addition, our basic assumption is that the labeled positive samples are chosen completely randomly from all positive samples (SCAR) [1]. Under the SCAR, X and S are conditionally independent given Y. Stated formally, that is,

$$P(S = 1|X, Y = 1) = P(S = 1|Y = 1) \tag{1}$$

In addition, $c = P(S = 1|Y = 1)$ indicates the probability that a positive sample is labeled, which is regarded as a constant on a definite dataset.

In the scenario of tax evasion detection, suspicious tax evasion taxpayers are labeled by tax inspectors. Tax inspectors always randomly inspect taxpayers. Thus, the labeled positive suspicious taxpayers are selected randomly in all taxpayers that exactly satisfy the SCAR assumption.

3 PUNE Method

Our tax evasion detection method PUNE contains three process, as shown in Fig. 1. In the first process, a network embedding algorithm is utilized to extract network features that can describe the characteristics of businesses based on transaction networks. In the rank labeling process, it utilizes class prior to rank samples and gives each sample an individual weight to weaken the influence aroused by label noise. In the last process, the final tax evasion detection model is trained by using weighted samples.

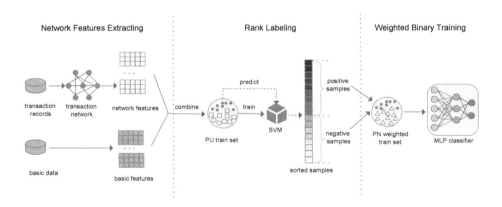

Fig. 1. Framework of PUNE method

3.1 Problem Formulation

To help tax inspectors detect tax evasion, many machine learning based techniques have been explored in previous literature [21,23,24]. These works disregard the notion that transaction characteristics can distinguish tax evading taxpayers more accurately than basic features because tax evasion is often hidden in transactions. Thus, network embedding algorithms are utilized to extract network features based on transaction networks that are built according to business records. In our method, we apply three famous network embedding algorithms to verify the effectiveness of transaction features in the task of tax evasion detection.

Moreover, these works disregard the notion that tax inspectors tend to randomly select several taxpayers and analyze their income and expenses to determine whether they have evaded tax. Thus, this case is common in practical classification applications, especially in tax evasion detection scenarios, in which the available data only contain labeled positive samples (suspicious tax evasion taxpayers) and a vast number of unlabeled samples that could be positive or negative. The special dataset reduces the effectiveness of these works.

This PU problem can be solved by PU learning, which belongs to semisupervised learning. PU learning is useful in many real-world problems. Many PU methods have been proposed in the literature [5,6,14,15,17]. The bagging SVM [17] and biased SVM [15] train binary classifiers using labeled samples and unlabeled samples. They directly regard all unlabeled samples as negative, which causes poor classifier performance because of bias.

The weighted SVM [7] simultaneously regards unlabeled data as weighted positive and negative samples. In addition, this SVM gives unit weights to all labeled positive samples. UPU [6] and NNPU [14] are both unbiased methods; they propose unbiased risk estimators that are state-of-the-art. However, when tax inspectors label data, they often make a mistake, that is, there is label noise in labeled positive samples, which will influence the performance of tax evasion detection. Thus, when these methods are applied to detect tax evasion, their performance will be poor because of the inevitable noise.

Based on this claim, in PUNE, first, we build a transaction network according to business records between taxpayers and then extract transaction features by a network embedding algorithm. Second, based on the transaction features and basic features, we weight both labeled positive samples and unlabeled samples to solve the label noise problem and PU problem. Last, a multilayer perceptron (MLP) is employed to train the final tax evasion detection classifier based on the weighted samples.

3.2 Extract Network Features

In this section, we illustrate how to construct the transaction network from raw transaction data. The feature space is built based on basic features and transaction network features extracted by a network embedding algorithm.

Transaction Network Construction. The raw transaction record describes details such as the name of the buyer and seller taxpayers, the category and price of trading commodities and other identification information. After data preprocessing, such as removing irrelevant information, merging records of the same buyers and sellers over time, and encoding commodity categories, we obtain a dataset that contains transaction relationships of all taxpayers.

Based on the information of the transaction relationship, the transaction network $G = (V, E, W)$ can be built, where $V = v_1, v_2, \cdots, v_n$ is a set of taxpayers, n is the number of taxpayers, and $E = e_{ij}$ denotes the transaction relationships between taxpayers and $W = w_{ij}$ represents the weights of the edges which are according to the transaction amount.

Network Embedding. In recent years, a significant amount of progress has been made in studies of network embedding; nodes in a network are assigned to low-dimensional representations and the local and global network structure is preserved [22]. In these studies, many network embedding methods, such as DeepWalk [19], LINE [22] and Node2vec [11], have been proposed. For example, the Deepwalk [19] algorithm learns latent social representations of vertices using local information from truncated random walks as input. The LINE [22] algorithm designs an objective function that preserves first-order and second-order proximities, which are complementary to each other. [11] proposed Node2vec, which learns a mapping of nodes to a low-dimensional space of features that maximizes the likelihood of preserving network neighborhoods of nodes.

These three network embedding algorithms are applied to for extract the transaction network features based on the transaction network and verify whether the transaction network is beneficial for tax evasion detection.

3.3   PU Training

After extracting network features, the feature space of the taxpayer has been built based on the basic features and network features of the taxpayer. In this stage, we mainly employ positive and unlabeled data to train a binary classifier to detect tax evasion. The stage consists of two processes: rank labeling process and binary classifier training process. First, in the rank labeling process, a biased classifier is trained by datasets DS_l and DS_u to obtain $P(s = 1|x)$ of all PU data. Second, according to the value of $P(s = 1|x)$, the weight of each samples can be calculated. In binary classifier training, we train the final model by an MLP based on weighted samples.

Rank Labeling Process. With the SCAR assumption, [7] has indicated that

$$
\begin{aligned}
P(S = 1|X) &= P(Y = 1 \ \wedge \ S = 1|X) \\
&= P(S = 1|Y = 1, X)P(Y = 1|X) \\
&= P(S = 1|Y = 1)P(Y = 1|X) \\
&= c \cdot P(Y = 1|X)
\end{aligned}
\tag{2}
$$

It shows that a classifier trained on positive and unlabeled samples predicts probabilities that differ by only a constant factor from the true conditional probability of being positive. Although, the true conditional probability $P(Y = 1|X)$ is not equal to $P(S = 1|X)$, the rank relationship between samples based on the conditional probability is never changed. Stated formally,

$$P(s_i = 1|x_i) > P(s_j = 1)|x_j) \Rightarrow P(y_i = 1|x_i) > P(y_j = 1)|x_j) \tag{3}$$

This conclusion motivates us to perform the classification by class prior instead of a threshold as employed by other methods.

As shown in Fig. 1, the basic learning algorithm is an support vector machine(SVM). We train a biased SVM where DS_l and DS_u are regarded as the training set. Then, we use this classifier, which works on all train samples X_l and X_u to obtain the probability estimates $P(s = 1|x)$, which indicates the confidence degree that the samples are positive. As previously mentioned, the rank relationship between samples based on $P(s = 1|x)$ is the same as that based on $P(y = 1|x)$. Thus, we sort $P(s = 1|x)$ and label the remaining samples as negative, where n_p can be calculated by $n_p = N * \pi$ After rank labeling, D_p and D_u can be obtained by reorganizing the dataset:

$$D_p = \{(x_1, y_1), (x_2, y_2), \ldots, (x_{n_p}, y_{n_p})\}$$

$$D_n = \{(x_{n_p+1}, y_{n_p+1}), (x_{n_p+2}, y_{n_p+2}), \ldots, (x_N, y_N)\}.$$

where $y_1, y_2, \ldots, y_{n_p}$ is equal to 1 and $y_{n_p+1}, y_{n_p+2}, \ldots, y_N$ is equal to -1.

Weight Training Process. After the rank labeling process, we obtained rough labels of the training set and the probabilities that indicate the confidence degree that the samples are positive. Thus, the positive and unlabeled problem has been transformed to an ordinary binary classification problem that is supervised. Supervise learning is easier than PU learning because all samples in the training set are labeled. For our binary classifier training, the employed neural network type is the standard layered neural network type, which is referred to as called the "multilayer perceptron".

An MLP is a supervised learning algorithm that learns a nonlinear function approximator. An MLP consists of an input layer of source nodes, one or more hidden layers of neurons and an output layer [12, 26]. Each layer is fully connected to the next layer. The hidden layers are nonlinear layers that are situated between the input layer and output layer. Because it can have one or more hidden layers, an MLP renders it different and flexible from other algorithms. The number of nodes in the input layer and the number of nodes in the output layer depend on the dimensions of the feature space and the dimensions of the label space, respectively. The input signal propagates through the network layer-by-layer. In the training of ordinary binary classification, the risk estimator of MLP is

$$R(f) = \pi R_1(f) + (1 - \pi)R_{-1}(f) \tag{4}$$

where $R_{-1}(f)$ and $R_1(f)$ denote the expected risk of positive samples and the expected risk of negative samples, respectively. π represents $P(Y = 1)$, which is the proportion of positive samples in the training set. It is obvious that every sample has the same weight in training.

In PUNE, to further reduce the influence of noise, we give each sample an individual weight rather than the same weight as given in ordinary classification.

First, we treat the positive samples in D_p with weight $W_p = \{p_1, p_2, \ldots, p_{n_p}\}$ and negative samples in D_n with weight $W_n = \{1 - p_{n_p+1}, 1 - p_{n_p+2}, \ldots, 1 - p_N\}$. Second, a three-layer MLP g is taken into account because it has been shown that a single hidden layer is sufficient to approximate any continuous function [10]. The optimization objective of our MLP is

$$R_{min}(g) = \frac{\pi}{\sum_{w_i \in W_p} w_i} \cdot \sum_{(x_i, y_i) \in D_p, w_i \in W_p} w_i \cdot l(g(x_i), y_i) + \frac{1 - \pi}{\sum_{w_j \in W_n} w_j} \cdot \sum_{(x_j, y_j) \in D_n, w_j \in W_n} w_j \cdot l(g(x_j), y_j) \tag{5}$$

The first term of R_{min} represents the risk of positive samples; it is divided by the sum of the weights of the positive samples because it is necessary to normalize the weights of the positive samples. The same situation applies to the second term.

Compared with the risk estimator of the ordinary binary MLP, PUNE assigns each sample an individual weight so that PUNE can weaken the influence of the label noise. The details of the PU training algorithm are described in Algorithm 1.

4 Experiments

In this section, we perform exhaustive experiments using real-world tax data. We compare PUNE with state-of-the-art PU learning methods, such as the weighted SVM [7], biased SVM [15], bagging SVM [17] and unbiased PU (UPU) [6] and non-negative PU (NNPU) [14]. In addition, the traditional linear SVM is also trained, where the unlabeled samples are directly treated as negative (PN-SVM). What's more, we conduct experiment on TEDM-PU, the state of art method that used to detect tax evasion by PU learning. Among these methods, the kernel function of SVM is linear function, and the base model of Unbiased PU and NNPU is the same three-layer MLP. Precision (Pre), recall (Rec), F1 score (F1) and accuracy (Acc) are employed as performance measurements in our experiments.

Our tax dataset is obtained from collaborated local tax authorities because of the lack of a standard dataset for a tax evaluation. This dataset contains tax information about 85,791 taxpayers, which consists of 15,420 tax evasion taxpayers (positive samples) and 70,371 normal taxpayers (negative samples). For each taxpayer, there are two categories of data, basic data and transaction data. The basic data include some inherent attributes of taxpayers, such as the

---

Algorithm 1: PU training algorithm

Input: samples X_l and X_u, dataset DS_l and DS_u, number of train set N, loss function l, class prior π, max iteration T

Output: final model g^*

1　**Rank Labeling Process**
2　Train an SVM such that $SVM(DS_l, DS_u)$;
3　Obtain predicted scores of train samples such that scores=predict(SVM, $X_l \cup X_u$);
4　Rank the scores in descending order;
5　Label the first $n_p = \pi * N$ samples as positive and label the remainder as negative to obtain the labeled data sets D_p and D_n after reorganizing;
6　Record $W_p = predict(SVM, x_i), x_i$ such that x_i in D_p and $W_n = 1 - predict(SVM, x_j)$ such that x_j in D_n;
7　**Binary Classifier Training Process**;
8　Initialized MLP g_0,and let t=0;
9　**while** $t < T$ **do**
10　　　Input training sets D_p and D_n into g_t;
11　　　Obtain output and compute the empirical risk $R(g_t)$;
12　　　Backward propagation of the empirical risk $R(g_t)$;
13　　　Update the parameters of MLP to obtain g_{t+1};
14　　　$t = t + 1$;
15　**end**
16　$g^* = g_T$;
17　**return** g^*

---

names of taxpayers, business codes, register types, addresses and other company-related indicators. We extract 74 basic features from the basic data to describe each taxpayer, including industry features (e.g., register time, industry category, and location), capital features (e.g., registered capital and top 3 shareholding ratios), employee features (e.g., total number of registered employees) and tax features (tax amount and ratio in last 3 years). The transaction data include transaction information that is based on invoices, such as the names of both sides of trade, commodity names, quantity, commodity unit price, amount, trade date, etc. After data preprocessing, we compress the raw transaction records from $854,625$ to $54,914$.

Before the classifier task, we use three famous network embedding algorithms DeepWalk, LINE and Node2vec to extract 128-dimensional network features for each sample to describe the transaction characteristics.

For this dataset, we randomly choose approximately 80% of the samples for training; the remaining 20% are utilized for testing, that is, the training set has 70,791 samples, and the test set has 15,000 samples. The training set contains 12,300 abnormal taxpayers (positive samples) and 58,491 normal taxpayers (negative samples). To simulate the PU problem, we choose 5,000 abnormal taxpayers (positive samples) of the training set as the labeled set. In addition, we randomly select 1,000 normal taxpayers (negative samples) as label noise with a labeled set, that is, the labeled set of train data contains 6,000 samples. The remainder of the positive samples and normal taxpayers (negative samples) are considered the unlabeled set.

To verify that the transaction network is beneficial for tax evasion detection, we conduct additional experiments based on only basic features of taxpayers. The first line in Table 1 is the baseline, in which the feature space only contains basic features, and the classifier model is trained by the traditional PN method, which directly treats unlabeled samples as negative samples. The results in Table 1 show that the performances of tax evasion detection are improved to different degrees regardless of which network embedding algorithm is applied (compared to PN-SVM and TEDM-PU). Moreover, PUNE performs better than other PU methods when it is employed to detect tax evasion.

Table 1. Performances of various methods for the tax dataset

PU Method	NE Method	Pre	Rec	F1	Acc
PN-SVM	–	0.409	0.584	0.481	0.580
TEDM-PU	–	0.833	0.525	0.644	0.729
W-SVM	Deepwalk	0.714	0.836	0.770	0.796
	LINE	0.719	0.807	0.751	0.760
	Node2vec	0.721	0.877	0.791	0.799
B-SVM	Deepwalk	0.693	0.672	0.682	0.710
	LINE	0.703	0.651	0.676	0.704
	Node2vec	0.642	0.652	0.647	0.721
Bg-SVM	Deepwalk	0.791	0.680	0.731	0.750
	LINE	0.832	0.530	0.707	0.701
	Node2vec	0.769	0.691	0.728	0.741
UPU	Deepwalk	0.778	0.878	0.825	0.820
	LINE	0.712	0.873	0.784	0.797
	Node2vec	0.747	0.881	0.808	0.805
NNPU	Deepwalk	0.778	0.978	0.867	0.850
	LINE	0.712	0.973	0.792	0.807
	Node2vec	0.747	0.981	0.848	0.825
PUNE	Deepwalk	0.957	**0.844**	**0.897**	**0.896**
	LINE	0.934	0.741	0.826	0.857
	Node2vec	**0.961**	0.788	0.866	0.878

5   Conclusion

This paper proposes a tax evasion method that is based on positive and unlabeled learning and network embedding. Before classification, we use network embedding techniques to extract transaction network features, which significantly improved the accuracy of tax evasion detection. In the PU training process, first, we use the SVM to predict the scores of all training samples. Second, we label them according to the class prior and sorted rank scores and assign them individual weights during the rank labeling process. The rank labeling process transforms the original PU problem into a binary classification problem. Last, an MLP is employed to train a binary classifier with weighted samples based on minimizing the empirical risk. The experimental results with the real-world tax dataset show that our method is superior to the state-of-the-art PU methods in terms of tax evasion detection.

Acknowledgments. This research was partially supported by "The Fundamental Theory and Applications of Big Data with Knowledge Engineering" under the National Key Research and Development Program of China with Grant No. 2018YFB1004500, the MOE Innovation Research Team No. IRT_17R86, the National Science Foundation of China under Grant Nos. 61721002 and 61532015, and Project of XJTU-SERVYOU Joint AI Lab.

References

1. Bekker, J., Davis, J.: Learning from positive and unlabeled data: a survey. arXiv preprint arXiv:1811.04820 (2018)
2. Chen, Y.S., Cheng, C.H.: A Delphi-based rough sets fusion model for extracting payment rules of vehicle license tax in the government sector. Expert Syst. Appl. **37**(3), 2161–2174 (2010)
3. Christoffel, M., Niu, G., Sugiyama, M.: Class-prior estimation for learning from positive and unlabeled data. In: Asian Conference on Machine Learning, pp. 221–236 (2016)
4. DeBarr, D., Eyler-Walker, Z.: Closing the gap: automated screening of tax returns to identify egregious tax shelters. ACM SIGKDD Explor. Newslett. **8**(1), 11–16 (2006)
5. Du Plessis, M., Niu, G., Sugiyama, M.: Convex formulation for learning from positive and unlabeled data. In: International Conference on Machine Learning, pp. 1386–1394 (2015)
6. Du Plessis, M.C., Niu, G., Sugiyama, M.: Analysis of learning from positive and unlabeled data. In: Advances in Neural Information Processing Systems, pp. 703–711 (2014)
7. Elkan, C., Noto, K.: Learning classifiers from only positive and unlabeled data. In: Proceedings of the 14th ACM SIGKDD International Conference on Knowledge Discovery and Data Mining, pp. 213–220. ACM (2008)
8. Junqué de Fortuny, E., Stankova, M., Moeyersoms, J., Minnaert, B., Provost, F., Martens, D.: Corporate residence fraud detection. In: Proceedings of the 20th ACM SIGKDD International Conference on Knowledge Discovery and Data Mining, pp. 1650–1659. ACM (2014)
9. Fung, G.P.C., Yu, J.X., Lu, H., Yu, P.S.: Text classification without negative examples revisit. IEEE Trans. Knowl. Data Eng. **18**(1), 6–20 (2005)
10. Greenwood, P.E., Nikulin, M.S.: A Guide to Chi-Squared Testing, vol. 280. Wiley, New York (1996)

11. Grover, A., Leskovec, J.: node2vec: scalable feature learning for networks. In: Proceedings of the 22nd ACM SIGKDD International Conference on Knowledge Discovery and Data Mining, pp. 855–864. ACM (2016)
12. Hornik, K., Stinchcombe, M.: Halbert: multilayer feedforward networks are universal approximators. Neural Netw. **2**(5), 359–366 (1989)
13. Jain, S., White, M., Radivojac, P.: Estimating the class prior and posterior from noisy positives and unlabeled data. In: Advances in Neural Information Processing Systems, pp. 2693–2701 (2016)
14. Kiryo, R., Niu, G., du Plessis, M.C., Sugiyama, M.: Positive-unlabeled learning with non-negative risk estimator. In: Advances in Neural Information Processing Systems (2017)
15. Liu, B., Dai, Y., Li, X., Lee, W.S., Philip, S.Y.: Building text classifiers using positive and unlabeled examples. In: ICDM, vol. 3, pp. 179–188. Citeseer (2003)
16. Menon, A., Van Rooyen, B., Ong, C.S., Williamson, B.: Learning from corrupted binary labels via class-probability estimation. In: International Conference on Machine Learning, pp. 125–134 (2015)
17. Mordelet, F., Vert, J.P.: A bagging SVM to learn from positive and unlabeled examples. Pattern Recogn. Lett. **37**, 201–209 (2014)
18. Pérez López, C., Delgado Rodríguez, M.J., de Lucas Santos, S.: Tax fraud detection through neural networks: an application using a sample of personal income taxpayers. Future Internet **11**(4), 86 (2019)
19. Perozzi, B., Al-Rfou, R., Skiena, S.: Deepwalk: Online learning of social representations. In: Proceedings of the 20th ACM SIGKDD International Conference on Knowledge Discovery and Data Mining, pp. 701–710. ACM (2014)
20. Ramaswamy, H., Scott, C.: Mixture proportion estimation via kernel embeddings of distributions. In: International Conference on Machine Learning, pp. 2052–2060 (2016)
21. Ruan, J., Yan, Z., Dong, B., Zheng, Q., Qian, B.: Identifying suspicious groups of affiliated-transaction-based tax evasion in big data. Inf. Sci. **477**, 508–532 (2019)
22. Tang, J., Qu, M., Wang, M., Zhang, M., Yan, J., Mei, Q.: Line: large-scale information network embedding. In: Proceedings of the 24th International Conference on World Wide Web, pp. 1067–1077. International World Wide Web Conferences Steering Committee (2015)
23. Tian, F., et al.: Mining suspicious tax evasion groups in big data. IEEE Trans. Knowl. Data Eng. **28**(10), 2651–2664 (2016)
24. Wu, R.S., Ou, C.S., Lin, H.y., Chang, S.I., Yen, D.C.: Using data mining technique to enhance tax evasion detection performance. Expert Syst. Appl. **39**(10), 8769–8777 (2012)
25. Zhu, X., Yan, Z., Ruan, J., Zheng, Q., Dong, B.: IRTED-TL: an inter-region tax evasion detection method based on transfer learning. In: 2018 17th IEEE International Conference On Trust, Security And Privacy In Computing And Communications/12th IEEE International Conference On Big Data Science And Engineering (TrustCom/BigDataSE), pp. 1224–1235. IEEE (2018)
26. Zurada, J.M.: Introduction to Artificial Neural Systems, vol. 8. West publishing company St, Paul (1992)

Adversarial Rectification Network for Scene Text Regularization

Jing Li[1], Qiu-Feng Wang[1(✉)], Rui Zhang[2], and Kaizhu Huang[1]

[1] Department of Intelligent Science, School of Advanced Technology,
Xi'an Jiaotong-Liverpool University, Suzhou, China
`Jing.Li19@student.xjtlu.edu.cn`, {`Qiufeng.Wang,Kaizhu.Huang`}`@xjtlu.edu.cn`
[2] Department of Foundational Mathematics, School of Science,
Xi'an Jiaotong-Liverpool University, Suzhou, China
`Rui.Zhang02@xjtlu.edu.cn`

Abstract. Scene text recognition with irregular layouts is a challenging yet important problem in computer vision. One widely used method is to employ a rectification network before the recognition stage. However, most previous rectification methods either did not consider recognition information or were integrated into end-to-end recognition models without considering rectification explicitly. To overcome this issue, we propose an adversarial learning-based rectification network that integrates transformation (from irregular texts to regular texts) with recognition information into a unified framework. In this framework, we optimize the rectification network with an extended Generative Adversarial Network that competes between rectifier and discriminator, together with the results of a recognizer. To evaluate the rectification performance, we generated a regular-irregular pair set from the benchmark datasets, and experimental results show that the proposed method can achieve significant improvement on the rectification performance with comparable recognition performance. Specifically, the PSNR and SSIM are improved by 0.81 and 0.051, respectively, which demonstrates its effectiveness.

Keywords: Rectification network · Scene text recognition · Generative adversarial networks · Irregular text.

1 Introduction

Scene text recognition (STR) has attracted much attention recently. It can be applied in a wide range of text-related applications, from tools including text translation and license plate recognition, to an essential part of integrated systems, such as intelligent transportation, surveillance and product search.

With the advance of deep neural networks [1], regular text recognition methods [2,3] have achieved tremendous advances. However, reading irregular text is still not satisfied in practice due to their various shapes and distorted patterns. Figure 1 presents some examples of regular and irregular text images. Therefore,

© Springer Nature Switzerland AG 2020
H. Yang et al. (Eds.): ICONIP 2020, LNCS 12533, pp. 152–163, 2020.
https://doi.org/10.1007/978-3-030-63833-7_13

recent recognition methods [4–6] have concentrated on dealing with irregular texts. ASTER [4] and MORAN [5] adopt rectification networks before recognition networks, which can rectify irregular text to be a canonical form and ease latter recognition task. The rectification module of ASTER [4] predicts concrete transformation matrices to rectify images. MORAN [5] adopts a simple CNN to predict offset maps for every pixel. However, these methods are usually overtrained for recognition performance while rectification networks may not learn well.

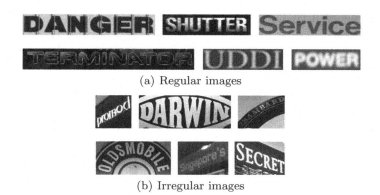

(a) Regular images

(b) Irregular images

Fig. 1. Examples of regular and irregular images.

Inspired by the adversarial process from GANs [8–10], we propose an adversarial learning-based rectification network to facilitate rectification ability. Following the structure of GAC [10], our method integrates rectification, recognition, and discrimination networks together, where a rectifier is used for the rectification, a recognizer is utilized to integrate recognition information, and a discriminator is used to distinguish real regular images and fake rectified images.

Through adversarial learning of the rectifier and discriminator, the rectification network can extract more precise spatial features and output more regular images. However, with such training, the rectifier may concentrate on rectifying the geometry of images while ignoring text details. Therefore, in addition to the normal generator loss, we also design a recognition loss to constrain the rectifier and keep more text information on rectified images. To our best knowledge, this is the first network that trains a text rectification network with a discriminator via an adversarial process. We test the proposed method on a regular-irregular pair set and achieve encouraging rectification performance without losing much recognition performance. The PSNR and SSIM are improved from 9.05 to 9.86 and from 0.4562 to 0.5072, respectively.

2 Related Works

2.1 Generative Adversarial Networks

Generative Adversarial Network (GAN) is proposed by Goodfellow et al. [8] based on adversarial learning. It consists of two competing modules, a generator and a discriminator. The generator aims at producing fake samples close to training data by capturing the real data distribution, while the discriminator tries to classify if a sample is from training data or generator. Due to its powerful ability to generate realistic-looking samples, various extension methods [9–11] have been proposed and successfully used in many areas, such as image synthesis, video processing, and image super-resolution. SRGAN [9] is able to reconstruct high-resolution images from heavily low-resolution ones via a perceptual loss including an adversarial loss and a content loss. Besides a generator and a discriminator, Triple-GAN [11] adds a classifier to learn how to label images, thereby producing more training data. Similar to [11], GAC [10] proposed by Qian et al. also consists of three parts. It is based on SRGAN [9] with removing the perceptual loss and adding the classifier in order to recognize handwriting character images reconstructed by the generator. Extending the above methods, the proposed network also has three components. Differently, our generator is not for recovering data, but to rectify images. In addition, the third component is a recognizer used to read text instances in images. The purpose is thus different among them and our network.

2.2 Scene Text Rectification and Recognition

Scene text recognition aims to recognize text sequences from cropped natural scene images. Recently, deep learning-based works [2–5] have focused on directly recognizing text from scene images. They regard scene text recognition as a sequence-to-sequence problem. Existing methods [2] for reading regular scene text images have achieved state-of-the-art performance. However, most of those methods fail in recognizing irregular text images (perspectively distorted and curved) that are more prevalent in daily life. Therefore, researchers have started to propose novel networks for irregular text images. Rectification-based methods are extremely effective and intuitive, which introduce rectification networks before recognition networks. Rectifications of irregular texts mostly based on morphological analysis or registration techniques have been previously studied in well-structured documents OCR. At present, [4] utilizes CNNs to predicts a set of control points of upper and lower text edges and utilizes Thin-Plate Spline transformation [7] to rectify images via these points. Different from those methods using transformation methods, MORAN [5] is more flexible. It predicts displacement maps for pixels and moves pixels to transform images, which is free of geometric constrains and generate more complicated transformation. Due to its elasticity, we choose MORAN as the baseline model. In the adversarial learning stage, the rectifier competes with a discriminator as well as considering recognizer results simultaneously.

3   Methodology

In this section, we first describe the architecture of our network, which contains three parts: a generative network G (i.e. Rectifier), a discriminative network D, and a recognition network R as shown in Fig. 2. We then detail the adversarial learning framework in the system and finally introduce the specific loss function of G.

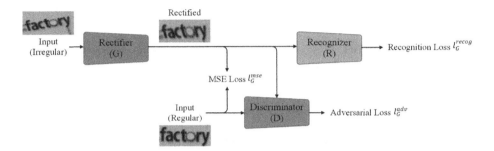

Fig. 2. The overall structure of the proposed method.

3.1   Network Architecture

Initially, basic G and R are obtained through end-to-end training. Next, D participates in the adversarial learning phase with pretrained G and R. It is noted that G and D are trained alternatively in an adversarial manner, but R is frozen. Therefore, to make R match with the well-learned G, R is fine-tuned for the final recognition. During the recognition inference, we only use G and R. Here, our G and R are based on MORAN [5] and D follows SRGAN [9].

Rectification Network (G). A multi-object rectification network (MORN) from MORAN [5] is adopted as G. As demonstrated in Fig. 2, irregular images are rectified by G to be more regular. According to the MORAN [5], we only consider vertical transformation. It predicts offset maps $offset$ which record vertical displacement of every pixel from the original position. Then, $offset$ are summed with original pixel coordinates $grid$ with the following:

$$offset'_{r,c} = offset_{r,c} + grid_{r,c}, \tag{1}$$

where r, c represent the r-th row and c-th column respectively, $offset'$ denotes the mapping from input image I to the rectified image I'. The final rectified image I' can be sampled using bilinear interpolation as follows:

$$c' = offset'_{r,c} \Rightarrow I'_{r,c} = I_{r,c'}, \tag{2}$$

Discriminative Network (D). The same architecture as in SRGAN [9] is adopted. G contains eight convolutional blocks that each of them consists of from 64 to 512 3×3 filter kernels, a batch normalization layer and a LeakyReLU layer with 0.2 negative slopes. Strided convolutions are adopted to reduce feature size when the number of features is doubled. The last resulting 512 feature maps are followed by an adaptive average pooling layer, two dense layers and a final sigmoid activation. In Fig. 2, we can see that input of D is a pair of a regular image and the corresponding rectified image from G, and its output is the probability of being the real regular text.

Recognition Network (R). We exploit the Attention-based Sequence Recognition Network (ASRN) from MORAN as R to recognize text instances. It employs a ResNet-BLSTM encoder and an attentional bidirectional GRU decoder. For the encoder, ResNet is used to extract spatial features from rectified images. Then, spatial features are reshaped and fed into a two-layer BLSTM. BLSTM analyzes the feature sequence and outputs the contextual features. The attention-based decoder translates the contextual features and outputs the target sequence. More details can be seen in [5].

In the pretraining phase, G and R both aim to minimize the recognition loss as follows:

$$L_{recog} = -\sum_{i=1}^{N}\sum_{t=1}^{T}\log(y_{i,t}|I_i), \tag{3}$$

where $y_{i,t}$ is the ground-truth of the t-th character in I_i, T is the length of sequence per image, and N is the number of training images. G needs to compete with D in the following adversarial process. Therefore, its loss function is modified and will be introduced in Sect. 3.3.

3.2 Adversarial Learning

Traditional adversarial learning-based methods [8,9] train G and D simultaneously. G captures a vector d' sampled from a noise distribution $P_{noise}(d')$ and outputs a fake data $G(d')$. Given a fake sample $G(d')$ or a real training sample d from the distribution $P_{data}(d)$, D identifies whether the sample is from training data or not and outputs the corresponding probability. This kind of adversarial min-max problem can be optimized as follows:

$$\min_{G}\max_{D}\mathbb{E}_{d\sim P_{data}(d)}[\log(D(d))] + \\ \mathbb{E}_{d'\sim P_{noise}(d')}[\log(1 - D(G(d')))], \tag{4}$$

We propose to play the adversarial game among G, D and R. Our G learns a mapping of z' from the distribution of irregular images $P_{irregular}(z')$ to the regular distribution $P_{regular}(z)$ over real regular images z. Given the rectified images, R outputs the conditional distribution $P(s|z')$ that should be like the training data distribution $P(s|z)$, where s represents the text ground-truth of the input image. Therefore, our model is defined by:

$$\min_{G} \max_{D} \mathbb{E}_{z \sim P_{regular}(z)}[\log(D(z))] +$$
$$\mathbb{E}_{z' \sim P_{irregular}(z')}[-\log(D(G(z')))] + \qquad (5)$$
$$\mathbb{E}_{I' \sim P_{irregular}(z')}[l_G^{recog}(R(G(z')), s)],$$

In our framework, both R and D are linked to G, trying to guide G to produce more regular but recognizable rectified images. We alternatively train G and D in an adversarial manner, while fixing R. Initially, G generates rectified images $G(z')$ which are relatively friendly to R but not enough to deceive D. With the progress of adversarial learning, it learns strategies to obtain more precise intermediate offset maps so that it becomes hard for D to identify. Also, G considers recognition results $R(G(z'))$ so that R still can recognize rectified images. At the same time, we seek to use labeled samples to train D in a supervised manner, while encouraging it to accurately classify real and fake images, despite the existence of corrective measures.

3.3 Loss Function for Rectifier

In the pretraining stage, G is guided by R, which is trained implicitly and only focuses on the recognition performance. However, in the adversarial training process, G aims to generate more regular images that can fool D. Therefore, the initial loss function of G is unsuitable. We design a weighted combination of several loss functions for G, integrating information from D and R:

$$\hat{G} = \arg \min_{G} \frac{1}{N} \sum_{n=1}^{N} l_G(G(z'_n), z_n), \qquad (6)$$

$$l_G = \lambda \cdot l_G^{mse} + \mu \cdot l_G^{adv} + l_G^{recog}, \qquad (7)$$

where N is the number of training regular-irregular image pairs. l_G^{mse} is the content loss, l_G^{adv} is the adversarial loss, and l_G^{recog} is the recognition loss. λ, μ are trade-off parameters for the content and adversarial loss functions, respectively. We use $\lambda = 100$ and $\mu = 0.1$ in our experiments.

Content Loss. We utilize the pixel-wise MSE loss as the content loss for G:

$$l_G^{mse} = \frac{1}{WH} \sum_{x=1}^{W} \sum_{y=1}^{H} (z_{x,y} - G(z')_{x,y})^2, \qquad (8)$$

Regular images z and rectified images $G(z')$ have the same width W and height H. Irregular images z' are transformed from z through a series of rotation, cropping and affine transformations.

Here, we use MSE loss to evaluate the difference between regular images and the rectified images at the pixel level, which makes G extract more accurate transform features. However, [9] raises that MSE loss usually leads to a lack of high-frequency content while realizing high PSNR. The edges of text usually belong to the high-frequency part which are important elements for text recognition. Therefore, we follow [9] and utilize an adversarial loss.

Adversarial Loss. The solution of adversarial loss is designed to make D regard the $G(z')$ as a regular image. The adversarial loss can be formulated as:

$$l_G^{adv} = -\log(D(G(z'))),\tag{9}$$

$D(G(z'))$ is the probability from D that the rectified image is classified as a regular image.

Recognition Loss. G is to ease R for reading irregular text images. To make the output of G suitable for the recognition, we also minimize the recognition loss of rectified image $G(z')$ for G as follows:

$$l_G^{recog} = -\frac{1}{T}\sum_{t=1}^{T}\log(y_t|G(z')).\tag{10}$$

where y_t is the t-th character in ground-truth.

4   Experiments

In this section, we begin by specifying datasets and implementation details. Then, we conduct extensive experiments on regular and irregular benchmarks. The evaluation criteria are Peak Signal to Noise Ratio (PSNR) and Structural Similarity Index (SSIM) [23] for rectification, and sequence accuracy for recognition under lexicon-free.

4.1   Datasets

When pretraining G and R, we merely use two synthetic datasets [14,15]. In the adversarial learning stage, pairs of regular and irregular images are required, but such pairs are not available in public datasets. Therefore, we generate **PairSet** for training and testing, more details are described in the following. Examples of **PairSet** are shown in Fig. 3. We also test the model on 4 regular benchmarks and 3 irregular benchmarks for general recognition performance. Datasets are described in details as follows:

Synth90K [14] is the synthetic scene text dataset containing 8-million images with relevant ground-truth.
SynthText [15] is the synthetic dataset for scene text detection. We crop around 5-million images for training according to bounding boxes.

(a) Training pairs (rotation, curved, affine transformation)

(b) Testing pairs (rotation, curved, affine transformation)

Fig. 3. Examples of training and testing pairs on **PairSet**. For every two rows, the first shows original regular images and the second row shows transformed irregular images.

IIIT5K-Words (IIIT5K) [16] contains 3,000 test images collected from the Internet. Most of them are nearly horizontal.

Street View Text (SVT) [17] has 647 cropped word images from 249 street images. Most are horizontal and many of them are severely corrupted by noise, blur and low resolution.

ICDAR 2003 (IC03) [18] contains 867 cropped horizontal images for testing, removing the images with non-alphanumeric characters or having few than three characters.

ICDAR 2013 (IC13) [19] is mostly from IC03 and add some new images. The total number of horizontal text images is 1015.

ICDAR 2015 Incidental Text (IC15) [20] exists more than 200 perspective and oriented word images. It totally contains 2077 images for testing.

SVT-Perspective (SVTP) [21] contains 645 cropped text images for testing. Most of them are highly perspective distorted.

CUTE80 (CUTE) [22] consists of 80 high-resolution scene text images. According to the bounding box, there are 288 cropped word images for recognition testing.

PairSet contains 5 groups of regular-irregular pairs, one group for training and four groups for evaluation. The training regular images are picked up from **Synth90K** by a simple CNN classifier (output regular or irregular) and the other four test groups are all samples of **IIIT5K**, **SVT**, **IC03** and **IC13**, respectively. The corresponding irregular images are obtained by the rotation, curved and affine transformations on the regular images. Finally, we get 100 k and 5529 pairs of regular-irregular scene text samples for training and testing, respectively.

4.2 Implementation Details

Our model follows a three-stage training. In the first stage, we pretrain G and R as our baseline (same as [5]), adopting ADADELTA optimizer [12]. The learning rate is set to 1.0 for the first 3 epochs and decreases to 0.1 for another 3 epochs. In the second stage, G and D compete through adversarial learning strategy with Adam optimizer [13] and 0.001 learning rate for 1 epoch. In the final stage, R is finetuned to match with the well-learned G, and the training strategy is the same as the first stage. The output of R contains 37 classes including 26 letters, 10 digits and a terminator. The batch size is set to 64. All models are implemented with the Pytorch framework on an NVIDIA GTX-1080Ti GPU.

4.3 Rectification Performance

In order to explain the rectification improvement, we compare our method with the baseline from PSNR and SSIM on the **PairSet**. Different from PSNR and SSIM on super-resolution tasks, these criteria are more strict in rectification tasks due to the movement of every pixel. Therefore, values of PSNR and SSIM are relatively lower, and the results are reported in Tables 1 and 2. The value of **PS**-All is calculated by all samples in the **PairSet**. As can be seen, our model outperforms the baseline in all datasets, particularly on PSNR of SVT (1.39) and IC03 (1.05) and SSIM of SVT (0.0727) and IC03 (0.0703). Therefore, the proposed method presents great rectification performance in irregular text images.

Table 1. The PSNR comparison of our model with the baseline on **PairSet**

Method	**PS**-IIIT5K	**PS**-SVT	**PS**-IC03	**PS**-IC13	**PS**-ALL
Baseline	7.85	11.66	9.43	10.45	9.05
Our model	**8.50**	**13.05**	**10.48**	**11.29**	**9.86**

Table 2. The SSIM comparison of our model with the baseline on **PairSet**

Method	**PS**-IIIT5K	**PS**-SVT	**PS**-IC03	**PS**-IC13	**PS**-ALL
Baseline	0.4477	0.4892	0.4341	0.4735	0.4562
Our model	**0.4915**	**0.5619**	**0.5044**	**0.5203**	**0.5072**

For further comparison, Fig. 4 visualizes the rectification results on some examples from irregular datasets. We can see that the rectified images from the baseline (the second column in Fig. 4) are adjusted slightly, which are not consistent with the results from [5]. This may be caused by unstable weak supervised learning [5], where G is constrained by the over-trained R, thus leading to inferior rectification results. Through adversarial learning with D, G not only serves for R but also tries to deceive D. Therefore, the rectifier can extract precise offset mappings for pixels, finally resulting in more regular rectified images as shown in the third column in Fig. 4. Although some of the rectified images are still distorted (merely vertical rectification), they are more flat and recognizable than the original and baseline ones.

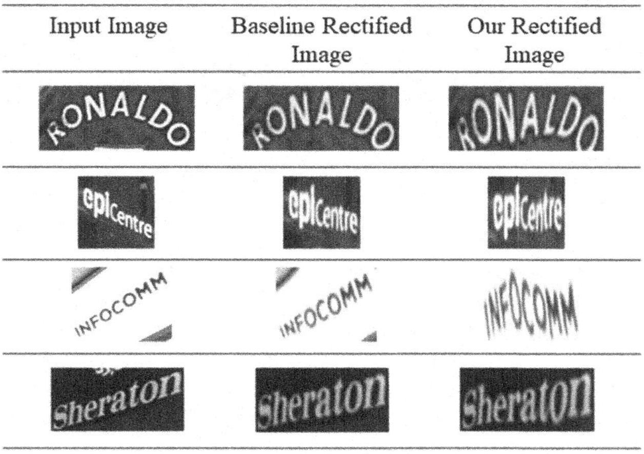

Fig. 4. Results of rectified images from irregular datasets. The first column shows the original images, the second column shows the rectified images from the baseline, and the last column presents the results from our proposed method.

4.4   Recognition Performance

To show the effect of the proposed adversarial rectification network on the recognition performance, we evaluate it on the benchmark datasets (both regular and irregular datasets), and the results are shown in Table 3. We can see that the recognition performance is comparable with the baseline system, which is an end-to-end recognition framework only focusing on the recognition performance. In the irregular datasets of CUTE80 and SVTP, the recognition performance is improved slightly from 78.47% to 79.51% and from 78.60% to 79.05%, respectively. However, for those highly curved text images, the proposed method usually cannot rectify them well, which results in low recognition performance. Some failure examples are shown in Fig. 5.

Table 3. Recognition accuracy (%) on regular and irregular benchmarks.

Method	Regular datasets				Irregular datasets			
	IIIT5K	SVT	IC03	IC13	SVTP	CUTE80	IC15_1811	IC15_2077
Baseline	**92.16**	**86.70**	93.31	**91.62**	78.60	78.47	**75.70**	**71.73**
Our model	91.83	86.08	**93.42**	90.04	**79.05**	**79.51**	73.49	69.61

Fig. 5. Failure examples with highly curve text. The first and third column are original images and the rest two columns are rectified images.

5   Conclusion

In this paper, we propose an adversarial rectification network to rectify scene texts, which contains a generator to rectify the regular scene texts, a discriminator to improve the rectification performance of the generator, and a recognizer to help the generator capture the text property. Compared with those end-to-end recognition models, the proposed method optimizes the rectification of scene texts explicitly. Extensive experiments on the generated regular-irregular **PairSet** from benchmark datasets show that the proposed method can improve the rectification performance significantly and achieve the comparable recognition performance with the end-to-end recognition model. In the future, we will improve the rectification performance on those severely curved scene texts.

Acknowledgements. This study was funded by National Natural Science Foundation of China under no. 61876154 and 61876155; Natural Science Foundation of Jiangsu Province BK20181189 and BK20181190; Key Program Special Fund in XJTLU under no. KSF-A-10, KSF-A-01, KSF-P-02, KSF-E-26 and KSF-T-06; and XJTLU Research Development Fund RDF-16-02-49 and RDF-16-01-57.

References

1. Huang, K., Hussain, A., Wang, Q., Zhang, R. (eds.): Deep Learning: Fundamentals, Theory and Applications. Cognitive Computation Trends, vol. 2. Springer, Cham (2019). https://doi.org/10.1007/978-3-030-06073-2
2. Shi, B., Bai, X., Yao, C.: An end-to-end trainable neural network for image-based sequence recognition and its application to scene text recognition. IEEE Trans. Pattern Anal. Mach. Intell. **39**(11), 2298–2304 (2016)
3. Bissacco, A., Cummins, M., Netzer, Y., et al.: PhotoOCR: reading text in uncontrolled conditions. In: Proceedings of the IEEE International Conference on Computer Vision, pp. 785–792 (2013)

4. Shi, B., Yang, M., Wang, X., et al.: ASTER: an attentional scene text recognizer with flexible rectification. IEEE Trans. Pattern Anal. Mach. Intell. **41**(9), 2035–2048 (2018)
5. Luo, C., Jin, L., Sun, Z.: MORAN: a multi-object rectified attention network for scene text recognition. Pattern Recogn. **90**, 109–118 (2019)
6. Guo, Z., Xu, H., Lu, F., et al.: Improving irregular text recognition by integrating gabor convolutional network. In: 2019 IEEE 31st International Conference on Tools with Artificial Intelligence (ICTAI), pp. 286–293. IEEE (2019)
7. Bookstein, F.L.: Principal warps: thin-plate splines and the decomposition of deformations. IEEE Trans. Pattern Anal. Mach. Intell. **11**(6), 567–585 (1989)
8. Goodfellow, I., Pouget-Abadie, J., Mirza, M., et al.: Generative adversarial nets. In: Advances in Neural Information Processing Systems, pp. 2672–2680 (2014)
9. Ledig, C., Theis, L., Huszár, F., et al.: Photo-realistic single image super-resolution using a generative adversarial network. In: Proceedings of the IEEE Conference on Computer Vision and Pattern Recognition, pp. 4681–4690 (2017)
10. Qian, Z., Huang, K., Wang, Q., et al.: Generative adversarial classifier for handwriting characters super-resolution. Pattern Recogn. **107**, 107453 (2020)
11. Li, C.X., Xu, T., Zhu, J., et al.: Triple generative adversarial nets. In: Advances in Neural Information Processing Systems, pp. 4088–4098 (2017)
12. Zeiler, M.D.: Adadelta: an adaptive learning rate method. arXiv preprint arXiv:1212.5701 (2012)
13. Kingma, D.P., Ba, J.: Adam: a method for stochastic optimization. arXiv preprint arXiv:1412.6980 (2014)
14. Jaderberg, M., Simonyan, K., Vedaldi, A., et al.: Synthetic data and artificial neural networks for natural scene text recognition. In: Workshop on Deep Learning, NIPS (2014)
15. Gupta, A., Vedaldi, A., Zisserman, A.: Synthetic data for text localisation in natural images. In: Proceedings of the IEEE Conference on Computer Vision and Pattern Recognition, pp. 2315–2324 (2016)
16. Mishra, A., Alahari, K., Jawahar, C.V.: Scene text recognition using higher order language priors. In: British Machine Vision Conference, pp. 1–11 (2012)
17. Wang, K., Babenko, B., Belongie, S.: End-to-end scene text recognition. In: 2011 International Conference on Computer Vision, pp. 1457–1464. IEEE (2011)
18. Lucas, S.M., Panaretos, A., Sosa, L., et al.: ICDAR 2003 robust reading competitions. In: Seventh International Conference on Document Analysis and Recognition, 2003. Proceedings, pp. 682–687. IEEE (2003)
19. Karatzas, D., Shafait, F., Uchida, S., et al.: ICDAR 2013 robust reading competition. In: 2013 12th International Conference on Document Analysis and Recognition, pp. 1484–1493 IEEE (2013)
20. Karatzas D, Gomez-Bigorda L, Nicolaou A, et al.: ICDAR 2015 competition on robust reading. In: 2015 13th International Conference on Document Analysis and Recognition (ICDAR), pp. 1156–1160. IEEE (2015)
21. Phan, T.Q., Shivakumara, P., Tian, S., et al.: Recognizing text with perspective distortion in natural scenes. In: Proceedings of the IEEE International Conference on Computer Vision, pp. 569–576 (2013)
22. Risnumawan, A., Shivakumara, P., Chan, C.S., et al.: A robust arbitrary text detection system for natural scene images. Exp. Syst. Appl. **41**(18), 8027–8048 (2014)
23. Wang, Z., Bovik, A.C., Sheikh, H.R., et al.: Image quality assessment: from error visibility to structural similarity. IEEE Trans. Image Process. **13**(4), 600–612 (2004)

An Overlapping Community Detection with Subspaces on Double-Views

Yang Chang[1], Huifang Ma[1,2,3(✉)], Yun Su[1], and Zhixin Li[2]

[1] College of Computer Science and Engineering, Northwest Normal University,
Lanzhou 730070, Gansu, China
`mahuifang@yeah.net`
[2] Guangxi Key Lab of Multi-source Information Mining and Security,
Guangxi Normal University, Guilin 541004, Guangxi, China
[3] Guangxi Key Laboratory of Trusted Software, Guilin University
of Electronic Technology, Guilin 541004, Guangxi, China

Abstract. Community detection algorithms are the basic tools for discovering the internal structure and organizational principles of a community. Ranging from model-based and optimization-based methods, existing efforts typically consider two sources of information, i.e. network structure and node attributes, to obtain communities with both denser network structure and similar attribute information. We argue that an inherent drawback of such methods is that, different impacts of different sources, is ignored during the clustering process. Besides, some existing community detection algorithms typically consider two sources of information but they cannot automatically determine the relative importance between them to reveal subspaces. As such, the detected communities may be unsatisfactory.

In this work, we propose to integrate subspace into a new overlapping community detection framework, an Overlapping Community Detection with Subspaces on Double-Views (CDDV), which exploits the relative importance between structures and attributes. This leads to a better detection result, effectively injecting subspaces to show the diversity of communities into the detection process in an explicit manner. We conduct extensive experiments on four public benchmarks, demonstrating significant improvements over several state-of-the-art models. Further analysis verifies the importance of subspace finding for capturing better communities, justifying the rationality and effectiveness of CDDV.

Keywords: Double views · Overlapping community detection · Subspace · Clustering

1 Introduction

The community detection algorithm is the fundamental tool for discovering the internal structure and organizational principles of the community. It plays an important role in many areas, such as metabolic network analysis in biological

© Springer Nature Switzerland AG 2020
H. Yang et al. (Eds.): ICONIP 2020, LNCS 12533, pp. 164–176, 2020.
https://doi.org/10.1007/978-3-030-63833-7_14

networks and community division in social networks. However, many real-world networks not only contain structural information, but also attached with rich attributes on vertices, obviously it is inadequate that only considering one type of information to determine the community structure. The reasons are twofold. On the one hand, the structure is usually sparse and noisy, if we only use structural information to perform clustering usually leads to poor partitions. On the other hand, if we only use attribute information for clustering [1], irrelevant attribute information may also result in non-optimal clustering results.

In this paper, we propose a community detection method named CDDV (Overlapping Community Detection with Subspaces on Double-Views) to detect communities from attributed graph. Our framework is designed to discover the internal structure and organizational principles of the community on an attributed graph based on both its attributes and topological information. It consists of three phases: (i) the graph is embedded into a low-dimensional space, which can reflect the network structure and maintain the network attributes simultaneously, then structure and attribute view is obtained, (ii) based on the structure and attribute view, it is possible to acquire the weight of two levels and exact subspaces and (iii) an additional step is added in the clustering process which using weight values to automatically calculate the relative importance of dimensions in different communities. What is more, the parameter that controlling the degree of overlap is added. The main contributions of this paper are as follows:

- We propose a new framework to automatically calculate the structural connections and attribute information of each node in an attributed graph. It is a general model that can be used to adapt any two types of heterogeneous information associated with object such as graph clustering.
- We provide an iterative update algorithm to solve this model and analyze its optimization method and experiment. The model simultaneously learns weights of nodes and reveal the latent subspaces automatically.
- We give an analysis about the experiment result and the effectiveness on synthetic and real-world data sets, also evaluated the parameters.

The remainder of this paper is organized as follows. Section 2 gives a brief review of related works on community detection and weighted K-means algorithms. Section 3 gives some preliminaries. Besides the CDDV model and an iterative algorithm are presented in Sect. 4. To verify the effectiveness of CDDV, several experiments on the synthetic and real-world networks are carried out in Sect. 5. The conclusions are drawn in Sect. 6.

2   Related Work

We review three categories of community detection algorithms: structural based, attribute based, and community detection using both types of information. Traditional community detection algorithms typically only focused on one of the structure and the attribute, or linearly superimposes them and then mine the community, so the information source cannot be effectively merged. Some classical works, such as Coscia [2] proposed Demon, which can find complex structures in real networks, but ignored attributes; Approaches in [3] demonstrated the BIGCLAM that can detect densely overlapping hierarchically nested as well as non-overlapping communities in massive networks, but node attributes are not considered at all; However, due to high computational cost, those algorithms cannot handle large scale networks. The algorithm is scalable, but recent research shows it is not robust when running on many common tasks [4].

Naturally, some attribute based community detection method like Whang [5] proposed NEO-K-Means and Jing [6] proposed EWKM, which is an extension of K-Means simultaneously [7], K-Means is attribute based clustering that is widely used in industries and research for its effectiveness and simplicity, so there are a huge number of extensions. The former considered the overlap between clusters and outliers, but neglected the structure and attribute information, the latter considered the subspace and node attributed, however it did not consider the structure; MAC proposed by Frank [8] also ignored the network structure.

Different from the above, a few algorithms are proposed to detect communities based on both structural and attribute information. For example, Li [9] proposed that the CDE algorithm, which utilizes a generative model to predict community assignments and considers the both, but the relative importance cannot be automatically calculated; TW-K-Means proposed by Chen [10] did not consider overlapping cases, Ruan [11] through many experiments to adjust the importance of each information source by domain knowledge and Cohn [12], simply fix the weights of all nodes to specific values. However, not all the structure and attribute information when used to be clustered have the same importance in guide a true partition. Given a large scale network with attributes, existing algorithms are not (at least not directly) applicable. Besides that, how those algorithms adjust the weights between attributes and structural information in an unsupervised manner are not clearly discussed. In addition, as an extension of the traditional clustering algorithm in high-dimensional space, the subspace-based clustering algorithm considers that each cluster is a set of data identified by a subset of attributes, and different clusters can be represented by different subsets of attributes. Therefore, this paper designs a calculation method for cluster subspaces, and updates the attribute subspaces of various clusters during each iteration of the algorithm. The traditional k-means clustering algorithm is modified by defining reasonable constraint conditions of objective function, so as to calculate the weight of each dimension in each cluster, and the weight value was used to identify the relative importance of the dimensions in different clusters. In this paper, the proposed method, CDDV, considers the structure and attribute information in the network simultaneously, which can automatically calculate the relative importance of them and extract the subspaces.

3   Preliminaries

In this section, we introduce some necessary notations to describe the essential basic concepts and the construction process of double-views.

3.1   Problem Definition

An attributed graph can be defined as G = (V, E, F), where $V = \{v_1, v_2, ..., v_n\}$ is the set of n nodes; $E = \{(v_i, v_j)\}$ is the set of edges between nodes and $F = \{f_1, f_2, ..., f_r\}$ is the set of r attributes. The adjacency matrix of G is defined as $\mathbf{A}$. Entry A_{ij} is 1 if nodes v_i and v_j are connected, otherwise, A_{ij} is 0. Supposed the graph is divided into k communities, each community satisfies the internal consistency and external separability, then mining subspaces with its corresponding community sets $C = \{c_1, c_2, ..., c_k\}$.

3.2   The Construction of Double-Views

The attributed graph G is represented as attribute view G_1 and structure view G_2 respectively, where G_1 is denoted by the attribute matrix $\mathbf{A}_F = [f_{ij}] \in R^{n \times r}$ and f_{ij} is the attribute value of node i in the $j - th$ dimension. G_2 is denoted by the structure matrix $\mathbf{A}_s = [S_{ij}] \in R^{n \times d}$, and S_{ij} is the value of node i in the $j - th$ dimension.

Construction of the Attribute View. For every node $v_i \in V$ in the attributed graph, it is related with the attributes that represented by r-dimensional vectors. The elements in the vector are the attribute values of each nodes. The attribute values can be single words, tags, and so on, which depends on the context of the given network. Let $X = \{x_1, x_2, ..., x_n\}$ be a set of data points, each with its own attribute $\{f_1, f_2, ..., f_r\}$ attached, then we obtain the attributed matrix.

Construction of the Structure View. Graph embedding is a common method for representing the structures. LINE [13] is a classical method for graph embedding, which combines first-order and second-order proximity to embed large-scale networks into low-dimensional vector spaces. Recently Chen proposed differentially privacy adjacency spectrum embedding method ASE [14] for Stochastic Block models, which can estimate the potential location close to the Frobenius norm by adjacency spectrum embedding, the accuracy is quite high with the expected parameters in both simulated network and the real network to effectively perform the graph embedding.

Definition 1. (*Structural similarity*) The structural similarity of (v_i, v_j) is the similarity between their neighborhood network structures, where $\mathbf{u_i}$ is the representation of v_i when they are treated as vertex while $\mathbf{u'_i}$ is the representation of v_i when they are treated as specific "contexts" of other vertices. The structural similarity of term v_i and v_j can be described as:

$$p\left(v_j | v_i\right) = \frac{\exp\left(\mathbf{u_j}' \cdot \mathbf{u_i}\right)}{\sum_{k=1}^{n} \exp\left(\mathbf{u_k}' \cdot \mathbf{u_i}\right)} \tag{1}$$

Similar with LINE, minimize the following objective functions:

$$O = \sum_{i \in n} \lambda_i d[\hat{p}(\bullet|v_i), p(\bullet|v_i)] \tag{2}$$

Where λ_i in the objective function to represent the prestige of v_i in the network, which can be measured by the degree or estimated through algorithms such as PageRank [15], n is the number of contexts. The empirical distribution is defined as $\hat{p}(\cdot|v_i) = \frac{S_{ij}}{d_i}$, where S_{ij} is the weight of the edge (v_i, v_j), i.e. if there exists an edge between v_i and v_j, $S_{ij} = N_{(i)} \cap N_{(j)} / N_{(i)} \cup N_{(j)}$, otherwise, $S_{ij} = 0$. Then we have:

$$O = -\sum_{(v_i, v_j) \in E} S_{ij} \log p(v_j|v_i) \tag{3}$$

By learning $\{\mathbf{u}_i\}_{i=1 \cdots n}$ and $\{\mathbf{u}_i\}'_{i=1 \cdots n}$ that minimize this Eq. (3), we are able to represent every vertex v_i with a d-dimensional vector $\mathbf{u_i}$.

4   Methodology

In this section, we introduce the Community Detection algorithm based on Double-Views for overlapping subspaces (CDDV).

4.1   The Objective Function of CDDV

Existing view-based dimension-weighted clustering algorithms such as TW-K-Means [10] can calculate the weights of views and dimensions simultaneously and assign weights to each dimension in the view. In this paper, CDDV combines the attributes and structure information between nodes, which not only can automatically calculate the relative importance of two views but also mine overlapping communities and subspaces. The clustering process of dividing the data matrix $\mathbf{X}$ into k clusters with two views and a single dimension weight is modeled as a minimization of the following objective functions:

$$P(\mathbf{U}, \mathbf{Z}, \mathbf{V}, \mathbf{W}) = \sum_{l=1}^{k} \sum_{i=1}^{n} \sum_{t=1}^{2} \sum_{j \in \mathbf{Gt}} \mathbf{u_{i,l}} \mathbf{w_t} \mathbf{v_j} \mathbf{d}(\mathbf{x_{i,j}}, \mathbf{z_{l,j}})$$
$$+ \sum_{t=1}^{2} \eta \sum_{j \in G_t} v_j \log(v_j) + \lambda \sum_{t=1}^{2} w_t \log(w_t) \tag{4}$$

Where $\mathbf{U}$ is a partition matrix whose elements are binary, where the entry equals 1 indicates the object is allocated to the correspond cluster; $\mathbf{Z}$ is the cluster center matrix; $\mathbf{W}$ is a column vector with elements $\frac{1}{2}$ and representing the relative importance of the view, and $\mathbf{V}$ is a column vector, represents the relative importance of each dimension attribute under each view. The first item on the right side is the sum of the degree of dispersion within the cluster, l represents the cluster, i indicates the node, t represents the view, j represents the dimension of each view, G_1 and G_2 is the structure view dimension and the attribute view dimension respectively; The second entry and the third entry are two negative entropy weights, and λ and η are two positive parameters.

Subject to:

$$\begin{cases} trace(\mathbf{U}^T\mathbf{U}) = (1 + \alpha)n \\ \sum_{t=1}^{2} w_t = 1, 0 \le w_t \le 1 \\ \sum_{j \in G_t} v_j = 1, 0 \le v_j \le 1, 1 \le t \le 2 \end{cases}$$

where α controls the degree of overlap between clusters: $0 \le \alpha \le (k-1)$

4.2   Model Optimization

We can minimize (4) by iteratively solving the following four minimization steps:

Step 1: Fix $\mathbf{Z} = \overset{\wedge}{\mathbf{Z}}$, $\mathbf{V} = \overset{\wedge}{\mathbf{V}}$, and $\mathbf{W} = \overset{\wedge}{\mathbf{W}}$, and solve $P = \left(\mathbf{U}, \overset{\wedge}{\mathbf{Z}}, \overset{\wedge}{\mathbf{V}}, \overset{\wedge}{\mathbf{W}}\right)$;

Step 2: Fix $\mathbf{U} = \overset{\wedge}{\mathbf{U}}$, $\mathbf{V} = \overset{\wedge}{\mathbf{V}}$, and $\mathbf{W} = \overset{\wedge}{\mathbf{W}}$, and solve $P = \left(\overset{\wedge}{\mathbf{U}}, \mathbf{Z}, \overset{\wedge}{\mathbf{V}}, \overset{\wedge}{\mathbf{W}}\right)$;

Step 3: Fix $\mathbf{W} = \overset{\wedge}{\mathbf{W}}$, $\mathbf{Z} = \overset{\wedge}{\mathbf{Z}}$, and $\mathbf{U} = \overset{\wedge}{\mathbf{U}}$, and solve $P = \left(\overset{\wedge}{\mathbf{U}}, \overset{\wedge}{\mathbf{Z}}, \mathbf{V}, \overset{\wedge}{\mathbf{W}}\right)$;

Step 4: Fix $\mathbf{V} = \overset{\wedge}{\mathbf{V}}$, $\mathbf{Z} = \overset{\wedge}{\mathbf{Z}}$, and $\mathbf{U} = \overset{\wedge}{\mathbf{U}}$, and solve $P = \left(\overset{\wedge}{\mathbf{U}}, \overset{\wedge}{\mathbf{Z}}, \overset{\wedge}{\mathbf{V}}, \mathbf{W}\right)$;

The way to optimize the objective function is to partially optimize $\mathbf{U}, \mathbf{Z}, \mathbf{W}$ and $\mathbf{V}$. By the iterative steps, the objective function tends to be local minimum, and each step of the optimization is strictly decremented, so the algorithm converges to the local minimum. Fixed $\mathbf{U}, \mathbf{Z}$, and $\mathbf{V}$, when the objective function is minimized according to $\mathbf{W}$, the following function is used to update the objective function similarly to the literature [10]. The formula for w_t and v_j is as follows:

If and only if $\mathbf{U}, \mathbf{Z}$, and $\mathbf{V}$ are given, the following formula holds:

$$w_t = \frac{exp(-P_t/\lambda)}{\sum_{j=1}^{2} exp(-P_j/\lambda)} \tag{5}$$

where $P_t = \sum_{l=1}^{k} \sum_{i=1}^{n} \sum_{j \in G_t} u_{i,l} v_j (x_{i,j} - z_{l,j})^2$ If and only if $\mathbf{U}, \mathbf{Z}$, and $\mathbf{W}$ are given, the following formula holds:

$$v_j = \exp\left(\frac{-Q_j}{h}\right) \Big/ \sum_{m=1}^{Gt} \exp\left(-\frac{Q_m}{h}\right) \tag{6}$$

where $Q_j = \sum_{l=1}^{k} \sum_{i=1}^{n} u_{i,l} w_j (x_{i,j} - z_{l,j})^2$

4.3   The Algorithm of CDDV

Algorithm 1 illustrates the community detection process as above, in Algorithm 1, x is a set of n data points before preprocessing; k is the number of input

---

Algorithm 1. CDDV

Input: data points x, clusters k, overlap parameter α, positive parameter η, λ
Output: U,Z,V,W
1:
2: Randomly choose k cluster centers **Z**
3: **for** $t = 1$ **to** 2 **do**
4:     $W_t = \frac{1}{2}$
5:     **for all** $j \in G_t$ **do**
6:         $V_j = \frac{1}{G_t}$
7:     **end for**
8: **end for**
9: $r \leftarrow 0$
10: **repeat**
11:     Compute the distance matrix d
12:     Initialize $\mathbf{U} \leftarrow \mathbf{0}$
13:     Initialize $T = \emptyset, p = 0$
14:     **while** $p < (n + \alpha n)$ **do**
15:         $U_{j*,l*} = 1$,where $(j^*, l^*) = \underset{j,l}{argmin}(d_{jl})$ and $\{(j,l)\} \notin T$
16:     **end while**
17:     Update the cluster centers matrix **Z**, Update **W** by (5), Update **V** by (6);
18:     $r \leftarrow r + 1$
19: **until** the objective function obtains its local minimum value

---

clusters; α is a parameter that controls the degree of overlap, and η, λ are two positive parameters. In the initialization process (Lines 1–7), initializing the cluster center matrix **Z**, the view weight vector **W**, and the dimension weight vector V under the view. Then CDDV calculates the weighted distance matrix (Line 10). After that, CDDV initializes some parameters α, η and λ(Line 11–12). CDDV next determines whether the overlap achieves the requirements (Lines 13–15). If not, continuing to allocate data, otherwise stopped. Line 13 performs the assignment to ensure that the objective function satisfies the first constraint. The next step of CCDV updates the cluster center matrix **Z**, the view weight vector **W**, and the weight vector **V** of the dimension under the view (Line 16). Finally, it determines whether the objective function converges (Line 18).

The CDDV algorithm involves three steps in the main calculation steps, we summarize the time complexity of our overall algorithm as follows:

(i) **Division:** we first classified the data into k overlapping clusters, and calculated the weighted distance matrix, the complexity is $O(nk)$;

(ii) **Update cluster center:** Given **U**, updating the cluster center is to find the mean of the data objects in the same cluster. Therefore, for k clusters, the computational complexity of this step is $O\left(nk(|G_1| + |G_2|)\right)$. (iii) **Update the view weight and the view dimension weight:** Given **U**, **Z** and **V**, updated **W** according to formula (5), it is only necessary to traverse the entire data set once to update **W**, so the complexity of this step is $O\left(nk(|G_1| + |G_2|)\right)$; Given **U**, **Z** and **W**, and update **V** according to formula (6). Similarly, it is only necessary to traverse the entire data set once to update $\mathbf{V}$, so the complexity of this step is $O\left(nk(|G_1| + |G_2|)\right)$. If the clustering process requires t iterations to converge, the total computational complexity of the algorithm is $\max(O(tnk(|G_1|+|G_2|)), O(tn^2k))$. The COCD algorithm monotonically reduces the objective function value until it converges to a local minimum.

5   Experimental Evaluation

We perform experiments on five synthetic networks and three real-world datasets to evaluate our proposed method, especially all the synthetic networks and the real-world networks come with ground truth for validation. We aim to answer the following research questions:

RQ1: How do different parameter settings (e.g., overlap parameter α positive parameter η, λ) affect CDDV?

RQ2: How does the scalability of this method perform?

RQ3: How does CDDV benefit from adaptively calculate the importance of double-views?

5.1   Experimental Settings

Synthetic Datasets: The synthetic network with the baseline community is generated based on the LFR BENCHMARK [16], which has similar features to the real-world networks. By setting some important parameters of the synthetic network, five synthetic networks (Syn1–5) with ground truth community is generated and as shown in Table 1.

Table 1. The synthetic network datasets.

Datasets	Nodes	Edges	Attributes
Syn1	1300	34765	$\{f1, f2, f3, f5\}$
Syn2	2100	32987	$\{f1, f5, f7, f9, f12, f15\}$
Syn3	896	12860	$\{f4, f8, f13\}$
Syn4	925	16715	$\{f6, f12, f17\}$
Syn5	1016	87500	—

Real-World Datasets: The network data sets widely used in the existing literature are collected and organized. The Flickr dataset [17] is a picture sharing network; The Amazon data set is a product co-purchase network, which can be obtained from the Stanford large network data set. In this paper, the original Cora data set is simplified, and the words which less than 10 words frequency statistics in the paper are removed. The three real-world network data sets are summarized in Table 2.

Table 2. The real-world network datasets.

Datasets	Number of nodes	Number of edges	Nodes	Edges	Attributes
Flickr	6,710	16,063	User	Friendship	Users' labels
Amazon	5,120	48,406	Product	Co-purchase relationship	Products' features
Cora	2708	5429	Paper	Citation relationship	Words appearing in the paper

Evaluation Metrics: As an overlapping community detection method, the same average F1 score and average NMI score for the improved evaluation index of the classic F1 score and NMI score in the literature [2] and [5] were used for evaluation.A good community detection method should obtain both high average F1 score and high average NMI score.

5.2   Experimental Results and Analysis

Parameter Sensitivity (RQ1): CDDV consists of three important parameters: α, η and λ. In this paper, the parameter α is intuitive, allowing the overlap of specified clusters. η and λ can be verified by experiments to their optimal values. Users can use the domain knowledge to estimate α, otherwise, α can also be estimated by using the heuristics discussed below.

Different data sets may have different levels of overlap between clusters. You can notice $0 \leq \alpha \ll (k-1)$ in the second part. Since α is used to control the overlap, the choice of α should depend on the degree of overlap desired by the users, it can control the number of communities to which a node belongs. Naturally, in larger datasets, the overlap between communities is relatively large, a smaller α will reduce the performance of the method, and a larger α will bring a significant performance improvement. On the contrary, cause the number of communities and the overlap between communities is relatively small, it is not appropriate to set a large α in smaller datasets. Therefore, for different data sets, the degree of overlap between clusters is different, and the requirements for the value of α are also different. For a few clusters(such as $k \approx 10$), it is recommended to use $\alpha = 0.1$, $\alpha = 1$ and $\alpha = \sqrt{k} - 1$ to control the degree of overlap. If the number of cluster centers is relatively huge (such as $k \geq 100$), $\alpha = \frac{1}{\sqrt{k}-1}$, $\alpha = \frac{1}{\log k - 1}$, $\alpha = 1$, $\alpha = \log k - 1$, $\alpha = \sqrt{k} - 1$ would be recommend.

Figures 1 and 2 show the effect of the CDDV algorithm measured by F1 score on the clustering results for different η and λ on five synthetic datasets, respectively. Due to space limitations, and the quantified results using the NMI score are consistent with the F1 score, so only the results quantified using the F1 score are represented. Figures 1 and 2 show that when η and λ changes from 0.5 to 6, the fluctuation of the F1 score is not large, that is, the clustering accuracy is not sensitive to the two parameters. The results show that the clustering results of the CDDV algorithm are robust to the parameters.

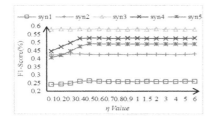

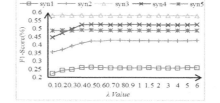

Fig. 1. The effects of η on clustering   **Fig. 2.** The effects of λ on clustering

Evaluating Scalability (RQ2): We evaluate the scalability of CDDV by measuring the running time on an increasing scale of synthetic networks. For evaluating, we consider four types of baseline community detection methods: (1) methods that use only network structures:BIGCLAM; (2) methods that use only node attributes:MAC [7]; and (3) methods that combine the two:CDE [9]; (4) node attributes and subspaces is considered simultaneously:EWKM [6]; (5) multiview clustering method:TW-K-Means [10].

Figure 3 shows the relationship between the running time of the algorithm and the size of the network. In general, CCDV is the fastest algorithm. It can process about 300,000 nodes in an hour or so; MAC is the slowest, and BIG-CLAM is faster than CDE because it uses an optimization process, but node attributes are not considered.

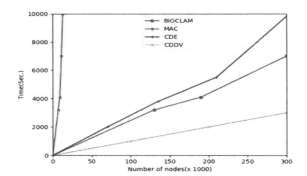

Fig. 3. Algorithm runtime comparison

The Result Analysis on Synthetic Networks and Real-World Networks (RQ3): Table 3 shows the experimental results on the synthetic network dataset. From Tables 3, we notice that CDDV obviously almost outperforms other methods on all synthetic data sets with all metrics, Average F1 score and. Average

Table 3. Average F1 score and average NMI score on five synthetic networks.

Method	Average F1 score					Average NMI score				
	Syn1	Syn2	Syn3	Syn4	Syn5	Syn1	Syn2	Syn3	Syn4	Syn5
BIGCLAM	0.252	0.380	0.552	0.325	0.396	**0.813**	0.552	0.457	0.528	0.532
MAC	0.189	0.189	0.329	0.303	0.258	0.564	0.405	0.492	0.462	0.405
CDE	0.223	0.312	0.498	0.461	0.392	0.505	0.383	0.496	0.623	0.612
EWKM	0.197	0.349	0.414	0.422	0.483	0.608	0.463	0.587	0.525	0.599
TW-K-Means	0.205	0.289	0.395	0.406	0.397	0.576	0.435	0.500	0.497	0.562
CDDV	**0.262**	**0.398**	**0.582**	**0.527**	0.491	0.724	**0.602**	**0.62**	**0.738**	**0.648**

Table 4. Average F1 score and average NMI score on real-world networks.

Method	Average F1 score			Average NMI score		
	Flickr	Amazon	Cora	Flickr	Amazon	Cora
BIGCLAM	0.166	0.276	0.356	0.437	**0.702**	0.426
MAC	0.172	0.252	0.198	0.397	0.476	0.308
CDE	0.179	0.402	**0.432**	0.565	0.415	0.508
EWKM	0.179	0.373	0.436	0.405	0.523	0.506
TW-K-Means	0.176	0.325	0.298	0.388	0.390	0.423
CDDV	**0.211**	**0.458**	0.428	**0.653**	0.626	**0.545**

NMI score. Compared with the other probabilistic models, the performance of CDDV increases by 10% or more on all metrics. Though CDDV does not perform as well as BIGCLAM on syn1 with Average NMI score metric, it is competitive compared against other algorithms. Besides, CDDV outperforms BIGCLAM on other data sets with all metrics. This demonstrates that CDDV has a strong ability to detect community structure.

The experimental results on a real-world network dataset shown in Table 4. Comparing CDDV with the method which ignored node attributes (BIGCLAM), it is noted that CDDV achieves better performance because it combines information from node attributes and networks, also, the most important reason is that CDDV could adaptively calculated the importance of double-views, it significantly reduced runtime complexity. Similarly, CDDV is also better than MAC because the latter only focuses on node attributes. Naturally, CDDV will never perform worse than the most advanced methods using only a single source of information. In addition, the baseline method in this paper does not perform as well as CDE and CDDV on synthetic datasets and real-world datasets. This is because TW-K-Means does not consider the overlap degree in the process of the community detection, but the communities are overlap naturally. So the performance is not good when there are overlapping communities. When comparing the performance of CDDV with the method of considering network structure and node attributes (CDE) and considering the performance of multiview clustering method (TW-K-Means). The powerful performance of CDDV can also be observed again.

The NMI and F1 scores of CDDV on the synthetic data sets are better than the values on the real-world data sets, which is understandable. In addition, for real-world network datasets, CDDV offers a greater advantage in the content sharing network (such as Flickr) than the social network's baseline performance. The possible explanation is that in the content sharing network, the attributes of the nodes plays a bigger role in connection generation. Overall, CDDV produced the best performance in 13 of the 16 cases. In summary, (RQ3) can be understood clearly based on the above analysis.

6 Conclusion and Future Work

This paper proposes a practical framework of community detection algorithm, CDDV, which combines the two information sources i.e. two views: structure information and attribute information, that used to be clustered in the attribute graph. On the one hand, the algorithm automatically calculates the relative importance of the two views. On the other hand, it also assigns weights to each dimension in the corresponding view and extract the subspaces. This is an extensible approach for overlapping community detection in large, complex networks. Experiments illustrate that the CDDV method proposed in this paper shows better performance and improves the effectiveness and efficiency of community detection on the both synthetic network datasets and the real-world network datasets compared with the previous classic community detection algorithms.

In the future, we tend to further improve the research work from two directions. On the one hand, there may be more than one type of nodes and edges in the real network, so community detection on heterogeneous networks will be considered. On the other hand, deep learning frameworks are worth learning and integrating into our method.

Acknowledgment. This work is supported by the National Natural Science Foundation of China (61762078, 61363058, 61966004, 61862058), Research Fund of Guangxi Key Lab of Multi-source Information Mining and Security (MIMS18-08), Northwest Normal University young teachers research capacity promotion plan (NWNU-LKQN2019-2) and Research Fund of Guangxi Key Laboratory of Trusted Software (kx202003).

References

1. Rezvani, M., Liang, W., Liu, C., et al.: Efficient detection of overlapping communities using asymmetric triangle cuts. IEEE Trans. Knowl. Data Eng. **30**(11), 2093–2105 (2018)
2. Coscia, M., Rossetti, G., Giannotti, F., et al.: DEMON: a local-first discovery method for overlapping communities. In: KDD, pp. 615–623 (2012)
3. Yang, J., Leskovec, J.: Overlapping community detection at scale: a nonnegative matrix factorization approach. In: ACM International Conference on Web Search & Data Mining, pp. 587–596. ACM (2013)

4. Yamaguchi, Y., Hayashi, K.: When does label propagation fail? a view from a network generative model. In: IJCAI, pp. 3224–3230 (2017)

5. Whang, J.J., Dhillon, I.S., Gleich, D.F.: Non-exhaustive, overlapping K-means. In: Proceedings of the 2015 SIAM International Conference on Data Mining, pp. 936–944 (2015)

6. Jing, L., Ng, M.K., Huang, J.Z.: An entropy weighting k-means algorithm for subspace clustering of high-dimensional sparse data. IEEE Trans. Knowl. Data Eng. **19**(8), 1026–1041 (2007)

7. Whang, J.J., Dhillon, I.S., Gleich, D.F.: Non-exhaustive, overlapping K-means. In: Proceedings of the 2015 SIAM International Conference on Data Mining, pp. 936–944 (2015)

8. Frank, M., Streich, A.P., Basin, D., et al.: Multi-assignment clustering for boolean data. J. Mach. Learn. Res. **13**, 459–489 (2012)

9. Li, Y., Sha, C., Huang, X., et al.: Community detection in attributed graphs: an embedding approach. In: Thirty-Second AAAI Conference on Artificial Intelligence (2018)

10. Chen, X., Xu, X., Huang, J.Z., et al.: TW-K-means: automated two-level variable weighting clustering algorithm for multiview data. IEEE Trans. Knowl. Data Eng. **25**(4), 932–944 (2013)

11. Ruan, Y., Fuhry, D., Parthasarathy, S.: Efficient community detection in large networks using content and links. In: WWW, pp. 1089–1098. ACM (2013)

12. Cohn, D., Hofmann, T.: The missing link-a probabilistic model of document content and hypertext connectivity. In: International Conference on Neural Information Processing Systems. MIT Press, pp. 430–436 (2000)

13. Tang, J., Qu, M., Wang, M., et al.: LINE: large-scale information network embedding. In: International Conference on World Wide Web. International World Wide Web Conferences Steering Committee, pp. 1067–1077 (2015)

14. Chen, L.: Privacy preserving adjacency spectral embedding on stochastic blockmodels (2019)

15. Berberidis, D., Nikolakopoulos, A.N., Giannakis, G.B.: Adaptive diffusions for scalable learning over graphs. IEEE Trans. Signal Process. **67**(5), 1307–1321 (2019)

16. Lancichinetti, A., Fortunato, S.: Benchmarks for testing community detection algorithms on directed and weighted graphs with overlapping communities. Phys. Rev. E **80**(1), 016118 (2009)

17. Yang, Z., Li, Q., Liu, W., et al.: Dual graph regularized NMF model for social event detection from Flickr data. World Wide Web **20**(5), 995–1015 (2017)

API Based Discrimination of Ransomware and Benign Cryptographic Programs

Paul Black[1(✉)], Ammar Sohail[1], Iqbal Gondal[1], Joarder Kamruzzaman[1],
Peter Vamplew[1], and Paul Watters[2]

[1] Internet Commerce Security Lab, Federation University, Ballarat, Australia
{p.black,iqbal.gondal,joarder.kamruzzaman,p.vamplew}@federation.edu.au,
ammarsohail@gmail.com
[2] Cybersecurity and Networking Group, Latrobe University, Melbourne, Australia
p.watters@latrobe.edu.au

Abstract. Ransomware is a widespread class of malware that encrypts files in a victim's computer and extorts victims into paying a fee to regain access to their data. Previous research has proposed methods for ransomware detection using machine learning techniques. However, this research has not examined the precision of ransomware detection. While existing techniques show an overall high accuracy in detecting novel ransomware samples, previous research does not investigate the discrimination of novel ransomware from benign cryptographic programs. This is a critical, practical limitation of current research; machine learning based techniques would be limited in their practical benefit if they generated too many false positives (at best) or deleted/quarantined critical data (at worst). We examine the ability of machine learning techniques based on Application Programming Interface (API) profile features to discriminate novel ransomware from benign-cryptographic programs. This research provides a ransomware detection technique that provides improved detection accuracy and precision compared to other API profile based ransomware detection techniques while using significantly simpler features than previous dynamic ransomware detection research.

Keywords: Ransomware · Machine learning · Internet security and privacy · Dynamic analysis

1 Introduction

Ransomware seeks to encrypt user data or lock the victim's computer and then extort money to regain access. Although early ransomware programs were detected in 2006, the frequency of ransomware attacks have recently accelerated and have become a significant information security problem [1]. The UK National Health Service (NHS) suffered a ransomware attack in 2017, 80 out of 236 NHS trusts were infected by ransomware; the full cost of this attack is not known [2].

© Springer Nature Switzerland AG 2020
H. Yang et al. (Eds.): ICONIP 2020, LNCS 12533, pp. 177–188, 2020.
https://doi.org/10.1007/978-3-030-63833-7_15

Tangible and non-tangible losses due to ransomware extend the impact beyond reported cash losses. These costs include investigation, new anti-ransomware strategies, data recovery, forensic costs, legal costs, crisis communication, fines, revenue loss, and reputational damage [3].

Prior research [4–8] has shown that novel ransomware families may be detected through the use of machine learning techniques trained on the features of existing ransomware samples. However, this research does not consider that some common programs perform similar cryptographic operations to ransomware. While existing techniques show an overall high accuracy in detecting novel ransomware samples, to the best of our knowledge, none of the previous research investigates the discrimination of ransomware from common programs that share some of the cryptographic characteristics of ransomware. This is a critical, practical limitation of current research; machine learning based techniques would be limited in their practical benefit if they generated too many false positives (at best) or deleted/quarantined critical data (at worst).

The research in this paper provides significantly improved accuracy and precision in ransomware detection when compared with existing Application Programming Interface (API) profile based techniques [7] and does so using simplified features. The experiments in this paper were performed using dynamic analysis of ransomware and benign programs in a Cuckoo sandbox. An SVM model was trained using the API profile extracted from the Cuckoo results for each program. Feature selection was used to select the highest performing API features.

1.1 Command and Control Server Emulation

Historical ransomware samples were used in this research; a Command and Control (C2) server is used by a ransomware sample to obtain a public encryption key and to report successful infections. A common technique to disable a ransomware attack is to remove the Domain Name Server (DNS) entry for the C2 server. A consequence of this is historical ransomware samples may be executed in a virtual machine (VM), but network access to the C2 server will not be available. A C2 emulator is a program that is written to emulate the operation of the absent C2 server. This use of a C2 emulator allows a historical malware sample to exchange the required information with the emulated C2 server and to proceed to searching for and encrypting user files. Running historical malware samples with a C2 emulator provides accurate API profiles containing API calls related to initialization, network communication, file searching, and file encryption.

1.2 Contribution

The contributions of this paper are as follows:

– This research is the first to consider whether machine learning API profile based ransomware detection techniques could distinguish common cryptographic programs from ransomware. In neglecting this question, previous research has focused on detection accuracy but has failed to consider precision as a performance metric.

– Development of techniques to improve the discrimination of ransomware from benign-cryptographic programs. Experimentation showed that API based machine learning has difficulties in discriminating ransomware from benign-cryptographic programs.

– C2 emulators for Cryptowall and CryptoLocker were developed to allow samples of these ransomware families to be run in a simulated environment and to perform the full range of operations that were possible when the malware attack was active.

The remainder of the paper is organized as follows: related work is reviewed in Sect. 2, Sect. 3 presents our research methodology. Section 4 covers feature selection and machine learning. We perform a detailed evaluation of our methodology in Sect. 5, and Sect. 6 concludes the paper.

2   Related Work

Malware analysis may be performed using either static or dynamic analysis. Both static and dynamic analysis techniques have been used to extract specifications of malware behaviour [9,10]. The methodology proposed in this paper uses dynamic analysis.

2.1   Ransomware Detection

Existing research deals with detecting ransomware using machine learning [4–8,11]. *EldeRan* [4] demonstrates the importance of feature selection to reduce the overall complexity of the problem and to improve the performance of machine learning. *EldeRan* uses features from the following classes: API calls, registry key operations, file system operations, file operations per file extension, directory operations, dropped files, and strings. The dataset used in this research consisted of 582 samples of ransomware belonging to 11 malware families and 942 benign programs. The benign programs consisted of generic utilities for Windows, drivers, browsers, file utilities, multimedia tools, developer's tools, network utilities, paint utilities, databases, emulator and virtual machine monitors, office tools. While this is a comprehensive dataset, it does not specifically target programs with cryptographic features that could be misclassified as ransomware. Experiments were performed to test the ability of *EldeRan* to detect known ransomware, and to detect novel ransomware. Testing with known ransomware provided an average accuracy of 0.963, and testing with novel ransomware samples gave a detection rate of 0.933 with 100 features and a detection rate of 0.871 with 400 features [11].

The research in [7] uses Windows API call data from the Cuckoo sandbox to generate a vector model of API calls to train an SVM machine learning model for ransomware detection. This research uses a vector representation that encoded the API call logs using a q-gram frequency and a standardized vector representation. The research uses 312 samples of benign software, further details of these programs are not provided, 276 ransomware programs targeting the Windows Operating System are used in this research. This dataset includes WannaCry, Cerber, Petya, and CryptoLocker, but further details are not provided. The accuracy of this research using the proposed vector format was 0.9352, and 0.9748 using an extension to vector encoding technique. The published results do not include true positive or false positive values. It is noted that the malware samples were not divided into malware families before splitting for training/testing, this allows the possibility that samples of malware families present in the training data were also present in the test data, raising the apparent average detection accuracy.

RansHunt is a hybrid analysis system that used static and dynamic analysis for ransomware detection. RansHunt uses the following feature classes: function length frequency, strings, API calls, registry operations, process operations, and network operations [11]. The dataset used in this research consisted of 360 samples of ransomware from 21 families, 532 different types of malware, and 460 benign software. Details of the types of benign software in the dataset were not provided. This paper uses a 10 fold cross-validation approach. Performing coss-validation selection at the ransomware family level would give a better understanding of research performance. This would avoid the possibility of having samples of the same malware families in both the train and test datasets. Feature selection was performed using Mutual Information criteria. The accuracy and precision values for static analysis were 0.935/0.951, for dynamic analysis was 0.961/0.960 and were 0.971/0.970 using the hybrid approach.

An analysis of the API calls made by malware samples from 14 malware families concluded that it may be feasible to identify ransomware behaviour on the basis of API call profile data alone [6].

GURLS [8] uses API call frequency features and machine learning based on Regularized Least Squares (RLS) for ransomware detection. The highest average binary detection rate of 0.886 was achieved using a radial basis (RBF) kernel. Multiclass classification was used to identify each ransomware family with an average accuracy of 0.867.

3   Research Methodology

The user mode programming interface for the Windows operating system is provided by the Windows API [12]. API calls in malware are readily identified by dynamic analysis techniques. This allows the creation of API profiles for detection and classification [13]. The proposed method is based on the observation that malware samples execute a unique sequence of API calls that can be used to distinguish them from other programs. Our approach uses API call profiles as features and uses feature selection to determine the most significant features.

An SVM machine learning model is used as a classifier to distinguish ransomware from benign programs and benign-cryptographic programs. API calls traces are taken from three datasets (ransomware, benign programs, and benign-cryptographic programs) using the Cuckoo sandbox. The Windows API calls, and the native functions calls are extracted from the API calls trace and are represented as a vector of API call frequency values that are labelled as ransomware/non-ransomware. Mutual Information Criteria (statistical models) is used to extract the most significant features. SVM machine learning is performed on the labelled API call frequency data. The trained model is used to predict whether a sample is ransomware or non-ransomware. The motivation for the use of SVM for learning and classifying ransomware is that, for binary classification, SVM has a high generalization rate and is designed to process large datasets with large feature spaces [7,14]. In our setting, the number of features is relatively high, so linear classifiers are a better choice [15,16].

3.1   Cross-Validation Approach

A cross-validation approach was used where the ransomware samples from one malware family and an equal number of benign programs were used for testing. The remaining ransomware samples and benign programs were used for training. This process was repeated for each ransomware family, and 10 experiments were performed for each ransomware family.

3.2   C2 Emulators

These emulators allow experimentation with historical ransomware samples by providing an emulation of the C2 server used by the malware family. Simulated DNS responses were provided by the Apate DNS simulator [17]; this permits the ransomware process to perform the necessary communications with the emulated C2 server and then continue and encrypt the user files in the test environment. This emulation exercises more of the ransomware capabilities and allows a complete API profile to be collected. C2 emulators were developed for the CryptoWall and CryptoLocker ransomware families.

4   Feature Selection and Classification

The machine learning method consists of two parts: feature selection and classification. Feature selection is performed using statistical and model-based techniques. For the statistical technique, we used Mutual Information Criteria [18] using Python's Scikit-learn library and Information Gain using the Weka machine learning tool [19]. For the model-based technique, we utilized decision trees (Random Forest) [20]. These techniques enable us to choose the most significant features API features.

4.1   Feature Engineering

The detection of ransomware activities may be performed by analysing their API, calls. API call frequency profiles are employed to identify ransomware behaviour in a controlled environment. Feature selection is used to identify the most significant features, allowing the generation of simpler machine learning models, reducing the training and prediction time, and helping to counteract the problem of overfitting. These techniques are not always used in machine learning malware detection approaches [15, 21]. The most significant API calls are selected based on the required level of significance using Mutual Information Criteria. After carrying out several experiments, we found that the highest accuracy could be achieved by utilizing 60% of the most significant selected features.

An API call frequency profile is required for our experiments. This API call profile can be represented by vectors, where each entry is a frequency of a given API function. Let $S = \{a_1, a_2, a_3..., a_n\}$ be a set of all selected features (API calls). A log of an application execution can be recorded as a sequence of API calls of length l, denoted as $A = \{a_1, a_2, ...a_l\}$ where $a_i \in S$ and $l \leq n$.

Let φ be the frequency of a Windows API function, we define a function Ψ that maps A to S and transforms each program's API calls profile to a vector of dimension $|S|$ as shown in Eq. 1.

$$v_{(A)} = \Psi(a)_{a \in S} \tag{1}$$

where

$$\Psi(a) = \begin{cases} \varphi, & \text{Frequency of an API call if present} \\ 0, & \text{otherwise} \end{cases} \tag{2}$$

5   Experiments

In this section, we implement our SVM model and test its ability to discriminate novel ransomware from benign programs and benign-cryptographic programs. We collected 162 benign programs and 14 benign-cryptographic programs. We collected 101 ransomware samples from 15 ransomware families targeting the Windows operating system from Malpedia [22]. The ransomware families used in this research are summarized in Table 1. We collected a dataset of 162 benign programs and 14 benign-cryptographic applications; these include Winzip, SHA256, Crc32, Putty, and John the Ripper. The benign-cryptographic programs that were used in this research are summarized in Table 2.

5.1   Comparison with Existing Research

To evaluate the effectiveness of our SVM model, we selected a comparison with Takeushi's SVM based ransomware detection work [7]. This research was selected for comparison due to its relatively simple ransomware detection techniques, that are still representative of current ransomware detection research. To perform

Table 1. Summary of ransomware families

Family	Year	# Samples
TorrentLocker	2014	4
CryptoFortress	2015	2
TeslaCrypt	2015	9
Locky	2016	20
CryptXXXX	2016	6
CryptoMix	2016	4
CryptoLocker	2013	4
DirCrypt	2014	5
Petya	2016	4
Cerber	2016	10
WannaCrypto	2017	5
CryptoShield	2017	2
CryptoWall	2013	21
Cryptorium	2016	2
PadCrypt	2016	3

Table 2. Summary of Benign-cryptographic programs

Family	# Samples
Cryptographic hashing	03
Error detection	02
File compression	03
Secure data removal	02
Password cracking	02
Secure network file sharing	02

this comparison, we replicated Takeuchi's Extension To Standardized Vectors encoding technique [7]. For the remainder of this paper, we refer to this encoding technique as the *Takeuchi* technique. Although we replicated the vector encoding techniques, we continued to use our cross-validation approach. We were not able to obtain a copy of the dataset used in the *Takeuchi* research. This dataset contained 276 ransomware samples. The ransomware families of these samples are not specified in the paper, and it is assumed that the families of the individual samples are not known. The result of splitting this dataset for machine learning is that training may be performed on ransomware samples that are also present in the test dataset, and this may overstate the accuracy of the technique. The ransomware dataset used in our research contained 101 ransomware samples from

15 ransomware families. We performed cross-validation using one ransomware family at a time for testing.

5.2 Experimental Setup

A Ubuntu 16.04 LTS host operating system and a virtual machine (VM) using host-only networking were used to ensure the containment and isolation of the malware experiments. This research used a TensorFlow version 1 linear SVM model, and an Adam optimizer with a learning rate of 0.001, and 20,000 training iterations. The samples were executed in a Cuckoo sandbox [23] using a Windows XP VM. A second Windows XP VM was used to run the emulated C2 server and an emulated DNS service.

Evaluation Metrics. Each of the programs in our dataset was submitted to the Cuckoo sandbox and an API profile were extracted from the sandbox analysis report. API frequency statistics were calculated, and supervised machine learning was used to predict whether the program was ransomware. The detection was recorded as successful when a ransomware sample was identified correctly (true positive) or benign/benign-cryptographic program was detected as a "not-ransomware" (true negative). The detection fails if ransomware was identified as a "not-ransomware" (false negative) or a benign/benign-cryptographic program was identified as ransomware (false positive). We evaluated the performance using 4 metrics: accuracy, precision, recall, and F1-Score. These metrics are summarized in Table 3.

Table 3. Evaluation metrics

Metrics	Expression	Description
Accuracy	$\dfrac{No.\ of\ correct\ predictions}{Total\ no.\ of\ predictions}$	Correct fraction of predictions
Precision	$\dfrac{TP}{(TP + FP)}$	Rate of relevant results (Trues)
Recall	$\dfrac{TP}{(TP + FN)}$	Sensitivity for the most relevant results
F1-Score	$2 \times \dfrac{Recall \times Precision}{Recall + Precision}$	Estimate of entire system performance

5.3 Feature Selection

Over several experiments, we found that Mutual Information and Random Forest techniques are comparable and outperformed Information Gain for feature selection. Use of the Mutual Information Criteria from *Python's Scikit-learn* machine learning library provided an 11% increase in accuracy compared with the Information Gain algorithm from Weka. We performed several experiments to determine the most significant features and evaluated our SVM model. We observed that our machine learning model performed best when the top 60% of the features were selected after being ranked by the mutual information criteria. While this machine learning approach is able to identify the highest performing set of API features, this approach does not identify the highest performing individual API features. Table 4 summarizes the number of most significant features selected in our experiments.

Table 4. Number of most significant featuress

Experiment type	# Features selected
Ransomware/Benign Programs	118
Ransomware/Benign Cryptographic Programs	90

5.4   Ransomware Against Benign Programs

In this experiment, we test the ability of our technique to distinguish ransomware from benign programs and compare the results with the replicated *Takeuchi* vector encoding technique. Table 5 provides the accuracy, precision, recall, and F1-Score measures of both the methods. Our model substantially outperformed our replication of the *Takeuchi* technique with an improvement of 6.2% in accuracy, 6.2% improvement in precision, and an improvement of 11.1% in recall.

Table 5. Average cross-validation evaluation results

	Ransomware/Benign		Ransomware/Cryptographic	
Metric	Our method	Takeuchi method [7]	Our method	Takeuchi method [7]
Accuracy	93.3%	87.1%	67.3%	57%
Precision	96.2%	90%	67.1%	60%
Recall	90.1%	79%	71.2%	60%
F1-Score	92.2%	82.1%	67.4%	60%

5.5   Ransomware Against Benign-Cryptographic Programs

In this experiment, we test the ability of our technique to distinguish ransomware from benign-cryptographic programs and compared this to the results from the replicated *Takeuchi* vector encoding technique. The summary of these results is shown in Table 5. In this experiment, our model substantially outperformed the

Takeuchi technique with an improvement of 10.3% in accuracy, 7.1% improvement in precision, and an improvement of 11.2% in recall. Our research indicates an accuracy rate of 67.3% in distinguishing ransomware from benign cryptographic programs. Two factors that may account for this relatively low result are, firstly our dataset contained a low number (11) of benign-cryptographic programs, and secondly, our cross-validation approach of testing against program features that were excluded from training, emphasises the need for the machine learning to generalise. This gives a conservative estimate of model accuracy. We acknowledge these limitations but note that our model outperforms existing research.

6   Conclusion

In this research, we developed a technique that detects ransomware with substantially simpler features than existing research. The *Takeuchi* vector encoding technique [7] was replicated, and our model was evaluated against it. The evaluation results demonstrate that our research improves prediction accuracy and is better able to discriminate ransomware from benign-cryptographic programs.

Future research could investigate why the API profiles from some ransomware families and the use of C2 emulators resulted in lower detection rates.

Based on our research, we conclude that machine learning trained on API profile features is limited in its ability to discriminate between ransomware and benign cryptographic programs due to the significant overlap of API calls profiles.

Acknowledgement. This research was funded in part through the Internet Commerce Security Laboratory (ICSL), a joint venture between Westpac, IBM, and Federation University Australia. Paul Black is supported by an Australian Government Research Training Program (RTP) Fee-Offset Scholarship through Federation University Australia. This research was partially supported by funding from the Oceania Cyber Security Centre (OCSC).

References

1. Kharraz, A., Robertson, W., Balzarotti, D., Bilge, L., Kirda, E.: Cutting the gordian knot: a look under the hood of ransomware attacks. In: Almgren, M., Gulisano, V., Maggi, F. (eds.) Detection of Intrusions and Malware, and Vulnerability Assessment. DIMVA 2015. Lecture Notes in Computer Science, vol. 9148, pp. 3–24. Springer, Cham (2015). https://doi.org/10.1007/978-3-319-20550-2_1
2. Morse, A.: Investigation: Wannacry Cyber Attack and the NHS. National Audit Office, London **31**, 2017 (2017)
3. Layton, R., Watters, P.A.: A methodology for estimating the tangible cost of data breaches. J. Inf. Secur. Appl. **19**(6), 321–330 (2014)
4. Sgandurra, D., Muñoz-González, L., Mohsen, R., Lupu, E.C.: Automated dynamic analysis of ransomware: benefits, limitations and use for detection. arXiv preprint arXiv:1609.03020 (2016)

5. Al-rimy, B.A.S., Maarof, M.A., Shaid, S.Z.M.: A 0-day aware crypto-ransomware early behavioral detection framework. In: Saeed, F., Gazem, N., Patnaik, S., Saed Balaid, A., Mohammed, F. (eds.) Recent Trends in Information and Communication Technology. IRICT 2017. Lecture Notes on Data Engineering and Communications Technologies, vol. 5, pp. 758–766. Springer, Cham (2017). https://doi.org/10.1007/978-3-319-59427-9_78

6. Hampton, N., Baig, Z., Zeadally, S.: Ransomware behavioural analysis on windows platforms. J. Inf. Secur. Appl. **40**, 44–51 (2018)

7. Takeuchi, Y., Sakai, K., Fukumoto, S.: Detecting ransomware using support vector machines. In: Proceedings of the 47th International Conference on Parallel Processing Companion, p. 1. ACM (2018)

8. Harikrishnan, N., Soman, K.: Detecting ransomware using gurls. In: 2018 Second International Conference on Advances in Electronics, Computers and Communications (ICAECC), pp. 1–6. IEEE (2018)

9. Christodorescu, M., Jha, S., Seshia, S.A., Song, D., Bryant, R.E.: Semantics-aware malware detection. In: 2005 IEEE Symposium on Security and Privacy (S&P'05), pp. 32–46. IEEE (2005)

10. Black, P., Gondal, I., Layton, R.: A survey of similarities in banking malware behaviours. Comput. Secur. **77**, 756–772 (2018)

11. Hasan, M.M., Rahman, M.M.: RansHunt: a support vector machines based ransomware analysis framework with integrated feature set. In: 2017 20th International Conference of Computer and Information Technology (ICCIT), pp. 1–7. IEEE (2017)

12. Russinovich, M.E., Solomon, D.A., Ionescu, A.: Windows internals. Pearson Education (2012)

13. Qiao, Y., Yang, Y., He, J., Tang, C., Liu, Z.: CBM: free, automatic malware analysis framework using api call sequences. In: Sun, F., Li, T., Li, H. (eds.) Knowledge Engineering and Management. Advances in Intelligent Systems and Computing, vol. 214, pp. 225–236. Springer, Berlin, Heidelberg (2014). https://doi.org/10.1007/978-3-642-37832-4_21

14. Islam, R., Tian, R., Batten, L.M., Versteeg, S.: Classification of malware based on integrated static and dynamic features. J. Netw. Comput. Appl. **36**(2), 646–656 (2013)

15. Kolter, J.Z., Maloof, M.A.: Learning to detect and classify malicious executables in the wild. J. Mach. Learn. Res. **7**, 2721–2744 (2006)

16. Shafiq, mz., Tabish, S.M., Mirza, F., Farooq, M.: PE-miner: mining structural information to detect malicious executables in realtime. In: Kirda, E., Jha, S., Balzarotti, D. (eds.) Recent Advances in Intrusion Detection. RAID 2009. Lecture Notes in Computer Science, vol. 5758, pp. 121–141. Springer, Berlin, Heidelberg (2009). https://doi.org/10.1007/978-3-642-04342-0_7

17. Apatedns:control your responses. https://www.fireeye.com/services/freeware/apatedns.html

18. Cover, T.M., Thomas, J.A.: Elements of Information Theory. Wiley, New York (2012)

19. Weka 3 data mining with open source machine learning software in java. https://www.cs.waikato.ac.nz/ml/weka/

20. Feature selection using random forest. https://towardsdatascience.com/feature-selection-using-random-forest-26d7b747597f

21. Rieck, K., Trinius, P., Willems, C., Holz, T.: Automatic analysis of malware behavior using machine learning. J. Comput. Secur. **19**(4), 639–668 (2011)

22. Plohmann, D., Clauss, M., Enders, S., Padilla, E.: Malpedia: a collaborative effort to inventorize the malware landscape. J. Cybercrime Digit. Invest. **3**(1) (2018). https://journal.cecyf.fr/ojs/index.php/cybin/article/view/17

23. Cuckoo foundation: Cuckoo sandbox - automated malware analysis. https://cuckoosandbox.org/

AutoGraph: Automated Graph Neural Network

Yaoman Li[1,2(✉)] and Irwin King[1]

[1] Department of Computer Science and Engineering,
The Chinese University of Hong Kong, Shatin, NT, Hong Kong
{ymli,king}@cse.cuhk.edu.hk
[2] Lenovo Machine Intelligence Center, Cyberport, Hong Kong

Abstract. Graphs play an important role in many applications. Recently, Graph Neural Networks (GNNs) have achieved promising results in graph analysis tasks. Some state-of-the-art GNN models have been proposed, e.g., Graph Convolutional Networks (GCNs), Graph Attention Networks (GATs), etc. Despite these successes, most of the GNNs only have shallow structure. This causes the low expressive power of the GNNs. To fully utilize the power of the deep neural network, some deep GNNs have been proposed recently. However, the design of deep GNNs requires significant architecture engineering. In this work, we propose a method to automate the deep GNNs design. In our proposed method, we add a new type of skip connection to the GNNs search space to encourage feature reuse and alleviate the vanishing gradient problem. We also allow our evolutionary algorithm to increase the layers of GNNs during the evolution to generate deeper networks. We evaluate our method in the graph node classification task. The experiments show that the GNNs generated by our method can obtain state-of-the-art results in Cora, Citeseer, Pubmed and PPI datasets.

Keywords: Graph Neural Networks (GNNs) · AutoML · Neural Architecture Search (NAS) · Evolutionary Algorithm (EA) · AutoGraph

1 Introduction

Graph Neural Networks (GNNs) are deep learning-based methods that have been successfully applied in graph analysis. It is one of the most important machine learning tools for solving graph problems. Unlike other machine learning data, graphs are non-Euclidean data. Many real-world problems can be modeled as graphs, such as knowledge graphs, protein-protein interaction networks, social networks, etc. The neural networks like Recurrent Neural Networks (RNNs) or Convolutional Neural Networks (CNNs) cannot directly apply to graph data. Hence, GNNs have received more and more attention. Some GNN models have been proposed and obtain promising results on some graph tasks, such as node classification [7,8,14,20], link prediction [23] and clustering [22].

© Springer Nature Switzerland AG 2020
H. Yang et al. (Eds.): ICONIP 2020, LNCS 12533, pp. 189–201, 2020.
https://doi.org/10.1007/978-3-030-63833-7_16

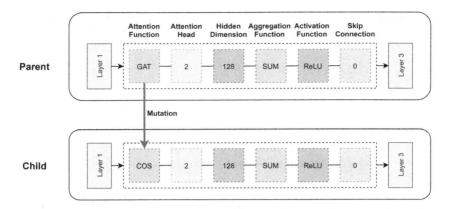

Fig. 1. Graph neural network architecture evolution example. The GNN architecture can be encoded by six states, i.e., Attention Function, Attention Head, Hidden Dimension, Aggregation Function, Activation Function and Skip Connection.

However, most of the GNNs suffer the low expressive power problem due to their shallow architectures. Some works [13,19] have been proposed to solve this problem. The design of deep GNNs requires a huge amount of human effort for neural architecture tuning. GNN models are usually very sensitive to the hyperparameters, for different tasks, we might also need to adjust the hyperparameters to obtain the optimal result. For example, the activation function needs to be carefully selected to avoid features degradation [13], the number of attention heads of GAT [20] needs to be carefully selected for different data, etc. The variants of GNNs may have a better performance in some specific problems. It is impossible to explore all possibilities manually.

We notice that the Neural Architecture Search (NAS) has archived great success in designing the CNNs and RNNs for many computer vision and language modeling tasks [11,17,26]. Many NAS methods for CNNs and RNNs have been proposed recently. For example, Zoph et al. [26] apply reinforcement learning to design CNNs for image classification problems. They use a recurrent network controller to generate CNN models and use the validation result of the CNN models as a reward to update the controller. Real et al. [17] design an evolutionary algorithm to evolve the CNN models from scratch and obtain state-of-the-art results. However, these works cannot be applied to GNNs directly.

Inspired by the success of NAS in designing CNNs and RNNs, recent works [6,25] are tried to apply NAS methods to design GNN models for citation networks. They propose to use reinforcement learning to design the GNN models. However, their proposed method can only generate fixed-length GNN models, and the generated GNN models only have shallow architectures. The deep GNNs generated by their methods will suffer the over-smoothing problem.

To overcome the above-mentioned problem, we propose a new AutoGraph method that applies an evolutionary algorithm to automatically generate deep GNNs. We first design a new search space and schema for the GNN model, which allows GNN with various layers and covers most of the state-of-the-art models. Then we apply evolutionary algorithm and mutation operations to evolve the initial GNN models. Next, we demonstrate a method to search for the best hyperparameters for the new GNN models which allow us to fairly compare the generated models and improve the robustness of our method. Finally, we conduct experiments on both transductive and inductive learning tasks and compare our method with baseline GNNs and the models generated by other reinforcement learning and random search strategies. The results show that we can generate state-of-the-art models for all test data efficiently. In summary, our contributions are:

- To the best of our knowledge, we are the first to study deep GNNs by using NAS. Our method can automate the architecture engineering process for deep GNNs, which can save many human efforts.
- Experiment results show that our proposed method can search for deep GNN models for different tasks efficiently.
- The GNN models generated by our method can outperform the handcrafted state-of-the-art GNN models.

2   Related Work

Inspired by CNNs [9,10] and graph embedding [2,4], GNNs are proposed to collectively aggregate information from graph structure. It is first proposed in [18]. GNNs have been widely applied for graph analysis [21,24] recently. The target of GNNs is to learn a representation of each node $\mathbf{h}_v \in \mathbb{R}^s$ which contains information for its neighborhood. The $\mathbf{h}_v$ also called a state embedding of a node. It can be used to produce an output $\mathbf{o}_v$, e.g., the node labels. They can defined as follows [24]:

$$\mathbf{h}_v = f(\mathbf{x}_v, \mathbf{x}_{co[v]}, \mathbf{h}_{ne[v]}, \mathbf{x}_{ne[v]}), \tag{1}$$

$$\mathbf{o}_v = g(\mathbf{h}_v, \mathbf{x}_v), \tag{2}$$

where f is the transition function that updates the node state according to the neighborhood, g is the output function that generates output from the node state and features. $\mathbf{x}_v$, $\mathbf{x}_{co[v]}$, $\mathbf{x}_{ne[v]}$, $\mathbf{h}_{ne[v]}$ are the features of v, the features of its edges, the features and the states of its neighborhood, respectively.

Let $\mathbf{H}$, $\mathbf{O}$, $\mathbf{X}$ and $\mathbf{X}_N$ be the stacked vectors of $\mathbf{h}_v$, $\mathbf{o}_v$, all features (node features, edge features, neighborhood features, etc.) and all the node features. Then the state embedding and output can be defined as:

$$\mathbf{H} = F(\mathbf{H}, \mathbf{X}), \tag{3}$$

$$\mathbf{O} = G(\mathbf{H}, \mathbf{X}_N). \tag{4}$$

Due to the shallow learning mechanisms of most GNNs, one major problem of GNNs is the low expressive power limit. The main challenge of this problem is that most of the deep GNNs would suffer from the over-smoothing issue, i.e., the deep model would aggregate more and more node and edge information from neighbors which would lead to the representation of node and edge indistinguishable. Some works have been proposed to solve this problem recently. For example, in the work of [13], the authors show that the Tanh activation function may be more suitable for deep GNNs and they also propose a DenseNet like architecture to alleviate the vanish-gradient problem.

To automate neural network exploration, some NAS methods have been proposed. Due to the substantial effort of human experts for discovering the state-of-the-art neural network architectures, there has been a growing interest in developing an automatic algorithm to design the neural network architecture automatically. Recently, the architectures generated by NAS have achieved state-of-the-art results in tasks like image classification, object detection or semantic segmentation. Most of the NAS methods are based on Reinforcement Learning (RL) [15, 26, 27] and Evolutionary Algorithm (EA) [16, 17].

Although the aforementioned NAS methods have successfully designed CNN or RNN architectures for image and language modeling tasks, the GNN is very different from CNN or RNN. Thus they cannot be directly applied to the GNN architecture search. Gao et al. [6] and Zhou et al. [25] propose a new schema to encode the GNN architecture and apply reinforcement learning to search for GNN models, but their methods cannot generate deep GNNs and their methods are not efficient and robust enough.

3   Method

In this section, we first define the AutoGraph problem. Then we describe our search space and schema to represent GNN architectures. Next, we show our evolutionary algorithm for the AutoGraph. Finally, we show a method to improve the robustness of the search process.

3.1   Problem Statement

The AutoGraph problem can be formally defined as follows. Given search space $\mathcal{A}$, the target of our algorithm is to search the optimal GNN architecture $\alpha \in \mathcal{A}$ which minimizes the validation loss $\mathcal{L}_{val}$. It can be written as follows:

$$\min_\alpha \quad \mathcal{L}_{val}(w^*(\alpha), \alpha), \tag{5}$$

$$\text{s.t.} \quad w^*(\alpha) = \text{argmin}_w \mathcal{L}_{train}(w, \alpha), \tag{6}$$

where w^* denotes the optimal parameters learned for the architecture in the training set. This is a bilevel optimization problem [6].

We propose an efficient method to solve this problem based on the evolutionary algorithm. Each generated architecture is trained and obtains the optimal weight of w^* in the training set, then it is evaluated in the validation set. At last, the best architecture in the validation set is reported. The following sections explain the process in more detail.

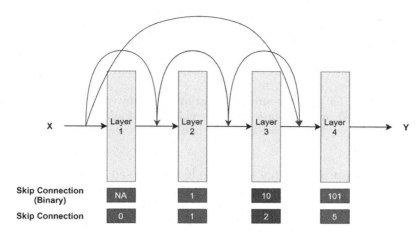

Fig. 2. Graph skip connection example. Binary connection status can be encoded to $[0, 2^{k-1})$, k is the current layer number.

3.2 Search Space

Many state-of-the-art GNNs would suffer from the over-smoothing problem which makes the representation of even distant nodes indistinguishable [24]. The recent work [13] shows that Tanh is better than ReLU for keeping linear independence among column features for GNNs. They propose a densely-connected graph network which is similar to DenseNet as follows:

$$\mathbf{H_0} = \mathbf{X}, \quad \mathbf{H_{l+1}} = f(L[\mathbf{H_0}, \mathbf{H_1}, ..., \mathbf{H_l}]W_l), \quad l = 0, 1, .., n-1, \qquad (7)$$

$$\mathbf{C} = g([\mathbf{H_0}, \mathbf{H_1}, .., \mathbf{H_n}]W_n), \qquad (8)$$

$$\text{output} = \text{softmax}(L^p \mathbf{C} W_C), \qquad (9)$$

where f and g are activation functions; $W_l \in \mathbb{R}^{(\sum_{i=0}^{l} F_i)*F_{l+1}}$, $W_n \in \mathbb{R}^{(\sum_{i=0}^{n} F_i)*F_C}$ and $W_C \in \mathbb{R}^{F_C*F_O}$ are learnable parameters, F_i is the number of input channels in layer i. This architecture stacks all the outputs of previous layers as the input of current layers. It can increase the variety of features for each layer, encourage the feature reuse, alleviate the vanishing gradient problem. However, concatenating all the outputs of previous layers will cause the parameters of the GNNs to increase exponentially.

Inspired by this, we allow each layer of our generated GNN models to connect to a various number of previous layers. To generate deep GNNs, we also allow our method to add a new layer to the GNN model during the searching process. So we define the search space and schema of our method as follows: We first apply the same setting of **Attention Function**, **Attention Head**, **Hidden Dimension**, **Aggregation Function** and **Activation Function** in [6]. Then we introduce two new states:

– **Skip Connection.** It has been observed that most GNN models deeper than two layers could not perform well because of the noisy information from expanding neighbors. This problem usually can be addressed by skip connection. Inspired by Luan et al. [13], we allow skip connections between any previous layers to the current layer. For each previous layer, 0 represents no skip connection, 1 represents there is a skip connection between that layer to the current layer, e.g., Fig. 2.

- **Layer Add**[1]. This state is only used during the mutation process. When this state is selected, we duplicate the current layer and add the new layer after the current layer. This state allows our method to extend the depth of GNNs automatically.

Noted that most of the GNN layers can be represented by the above first six states, as shown in Fig. 1. The above search space can cover a wide variety of state-of-the-art GNN models. If the skip connections are applied then the input dimension of the current layer would be the sum of all the output dimensions of the connected layers.

3.3 Evolutionary Algorithm

Inspired by Real et al. [16], we apply the Aging Evolution Algorithm to search for the deep GNNs. Similar to most of the evolutionary algorithms, our algorithm can be divided into three stages, i.e., initialization, mutation and updating. In the initialization stage, we randomly generate P GNN models with two layers. P is the size of the population. The initial P models are trained and evaluated. Then they are added to the population.

In the mutation stage, we sample S candidates from the population. The candidate with the highest score in the sample set is selected to apply mutation. We randomly select one state in the search space and change it to a new value in the state set. Then the newly generated candidate is trained and evaluated. Next, the new candidate needs to be added to the population. Since we need to keep the population size unchanged, we would select the oldest candidate in the population and remove it before we add the new candidate to the population. This is the main difference between the Aging Evolution Algorithm and other evolutionary algorithms.

We allow multiple skip connections for each layer. The skip connection between the previous layer i to the current layer k can be represented by binary $c_{i,k}$. Since there is always a connection between layer $k-1$ to layer k, we only need to consider $i \in 0, 1, .., k-2$ (0 represents the input of the network). Thus, the skip connections state of layer k can be represented as

$$S_k = \sum_{i=0}^{k-2} c_{i,k} \cdot 2^i, \quad c_{i,k} \in 0, 1, \quad k \geq 2. \tag{10}$$

Then the possible state of S_k is $[0, 2^{k-1})$. When $k = 1$, i.e., the current layer is the first layer, the skip connection state would be always 0. Figure 2 shows an

---

[1] "Layer Add" state is only used in the evolutionary process.

example of skip connection representation. To avoid a significant change of the GNN model, each mutation operation will only change one state of the model. During the search process, every evaluated GNN model is added to the history list. After the whole search process is finished, the model with the highest score in the history list will be reported.

3.4 GNNs Evaluation

We notice that the GNN model is sensitive to change in hyperparameters, such as the learning rate and weight decay. The best performance of a GNN architecture can be achieved at different learning rates, weight decay and iteration number. If we use the same hyperparameters to train and evaluate different GNN architectures, we may miss the best GNN model because the hyperparameters are not set properly. To fairly compare the architecture, we apply the hyperparameters tuning for each generated GNN model.

The work of Bergstra et al. [1] shows that the Tree-structured Parzen Estimator Approach (TPE) performs well on the hyperparameter search. We use the TPE algorithm to search the hyperparameters for each GNN model. To avoid overfitting and speed up the search process. We allow early stops during the training process. For each GNN architecture, we will use the best performance reported by the TPE algorithm as the performance of the architecture. The comparison between different GNN models is based on the performance of their best hyperparameter settings.

4   Experiments

We conduct experiments in both transductive and inductive learning tasks. For the transductive learning task, we test our method on the Cora, Citeseer and Pubmed datasets. For the inductive learning task, we test on the protein-protein interaction (PPI) dataset. Our method is evaluated in the following aspects:

- **Performance.** We evaluate the performance of our AutoGraph method by comparing the generated GNN model with the handcrafted state-of-the-art GNN models.
- **Efficiency.** We analyze the efficiency of our method by comparing it with other search strategies, i.e., GraphNAS (a reinforcement learning-based method) and random search.
- **Scalability.** We analyze the scalability of our method by comparing the performance of GNN models with different layers.

4.1   Experimental Setup

The configuration of our method in the experiments is set as follows. The population size is 100. The max evaluation architecture is 2,000. The maximum training iterations is 1,000. As described in the Methods, the mutation probabilities are

Table 1. Dataset statistic

	Cora	Citeseer	Pubmed	PPI
Task	*Transductive*	*Transductive*	*Transductive*	*Inductive*
# Nodes	2,708 (1 graph)	3,327 (1 graph)	19,717 (1 graph)	56,944 (24 graphs)
# Edges	5,429	4,732	44,338	818,716
# Features/Node	1,433	3,703	500	50
# Classes	7	6	3	121 (multi-label)
# Training nodes	140	120	60	44,906 (20 graphs)
# Validation nodes	500	500	500	6,514 (2 graphs)
# Test nodes	1,000	1,000	1,000	5,524 (2 graphs)

uniform. The generated GNN architecture is trained with the ADAM optimizer. The maximum hyperparameters search number for the TPE algorithm is 50. We run the search algorithm in four RTX 2080 Ti GPU cards. For each task, the best model which has the lowest validation loss is selected as our GNN model to compare with other baseline models.

4.2 Datasets

Transductive Learning. In transductive learning tasks, the same graphs are observed during training and testing. The experiment datasets for the transductive learning are Cora, Citeseer and Pubmed. In these datasets, the nodes represent the documents and the edges (undirected) represent citations. The features of the nodes are got by the bag-of-words representation of the documents. The Cora dataset contains 2,708 nodes and 5,429 edges. We will use 140 nodes for training, 500 nodes for validation and 1,000 nodes for testing. The Citeseer dataset contains 3,327 nodes and 4,732 edges. The training, validation and test set separations are the same as the setup of [20].

Inductive Learning. In inductive learning tasks, the graphs in training and testing are different. The experiment dataset for inductive learning is the protein-protein interaction (PPI). The graphs in this dataset represent different human tissues. There are 20 graphs in the training set, two in the validation set and two in the test set. The data in the test set is completely unobserved during training.

The statistical detail of transductive learning and inductive learning datasets is shown in Table 1. The Cora, Citeseer and Pubmed datasets are classification problems. The PPI dataset is a multi-label problem.

4.3 Baseline Methods

We compare the GNN model generated by our approach with the following state-of-the-arts methods:

- Chebyshev [3]. This method removes the need to compute the eigenvectors of the Laplacian by using K-localized convolution to define a graph convolutional neural network.

Table 2. Experiment results on Cora, Citeseer and Pubmed

Models	Cora	Citeseer	Pubmed
Chebyshev	81.2%	69.8%	74.4%
GCN	81.5%	70.3%	79.0%
GAT	$83.0 \pm 0.7\%$	$72.5 \pm 0.7\%$	$79.0 \pm 0.3\%$
LGCN	$83.3 \pm 0.5\%$	$73.0 \pm 0.6\%$	$79.5 \pm 0.2\%$
GraphNAS	$83.3 \pm 0.6\%$	$73.5 \pm 1.0\%$	$78.8 \pm 0.5\%$
AutoGraph	$\mathbf{83.5 \pm 0.4\%}$	$\mathbf{74.4 \pm 0.4\%}$	$\mathbf{80.3 \pm 0.3\%}$

Table 3. Experiment results on PPI

Models	micro-F1
GraphSAGE (lstm)	0.612
GeniePath	0.979
GAT	0.973 ± 0.002
LGCN	0.772 ± 0.002
GraphNAS	0.985 ± 0.004
AutoGraph	$\mathbf{0.987 \pm 0.003}$

- GCN [8]. This method alleviates the problem of overfitting by limiting the layer-wise convolution operation to $K = 1$.
- GAT [20]. This method introduces the attention mechanism to GNN. It obtains good results in many graph tasks.
- LGCN [5]. It introduces regular convolutional operations to GNN.
- GraphSAGE [7]. This method can be applied to inductive tasks. It samples and aggregates features from a node's neighborhood.
- GeniePath [12]. It uses an adaptive path layer which consists of two complementary functions.

We use the public released implementations of these methods to do the comparisons. The evaluation metric for transductive learning tasks is accuracy. For the inductive learning task, we use the micro-F1 score.

To evaluate the efficiency of our method, we also compare our method with GraphNAS and random search. GraphNAS applies a reinforcement learning controller to generate GNN models. For the random search baseline, we randomly sample GNN models from the same search space in our approach.

4.4 Results

After our algorithm generates 2,000 GNN models, the model which has the lowest loss in the validation set is selected and tested on the test set. The experiment results of transductive learning datasets are summarized in Table 2. The results of the inductive learning dataset are summarized in Table 3.

Performance. For the transductive learning tasks, we compare the classification accuracy with the above-mentioned GNN model and GraphNAS. From Table 2 we can see that our generated model can get the state-of-the-art result in all transductive datasets.

For the inductive task, we compare the micro-F1 score with the popular GNN models and GraphNAS. The result shows that our method also performs well in the inductive dataset.

Table 4. Search strategies comparison

Method	Accuracy	Time (GPU hours)	Best GNN layers
Random search	$81.8 \pm 0.5\%$	10	2
GraphNAS	$83.3 \pm 0.6\%$	10	2
AutoGraph	$\mathbf{83.5 \pm 0.4\%}$	**3**	**4**

Efficiency. To evaluate the effectiveness of our search method, we compare our method with different search strategies, i.e., random search and reinforcement learning-based search method—GraphNAS [6]. Since GraphNAS does not do the hyperparameters tuning when evaluating the GNNs, we also disable our hyperparameters tuning during the search process. During the training process, we record the generated architectures and their performance. From the Table 4, we can see that our method can search for a better GNN model with less time and our method can generate deeper GNNs.

Scalability. We know that most of the handcraft GNNs would suffer from the over-smoothing problem. We compare the performance of the GNNs generated by our method with different layers. Figure 3 shows the best performance of the

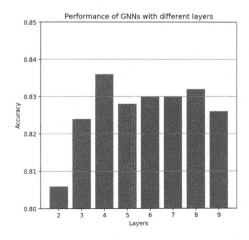

Fig. 3. Comparison of the GNN models with different layers on Cora

GNNs generated by our method from two layers to nine layers. We can see that our generated GNN models have good performance in deep architectures.

5   Discussion and Conclusion

In this work, we study the problem of AutoGraph. We present an efficient evolutionary algorithm to search for GNN models. We can see that our method can generate deep GNNs which alleviate the over-smoothing problem. The experiments show that the generated models can outperform current handcraft state-of-the-art models. In summary, we can see our proposed method has the following advantages:

- It can save substantial efforts to explore good GNN models for different graph tasks.
- Our generated GNN models can get state-of-the-art results.
- Our approach can generate deep GNN models which can alleviate the over-smoothing problem.

Although our proposed method can design state-of-the-art GNNs for graph tasks, it is remarkable that there are still many improvements that can be made. The first problem is that the search process is time-consuming. We notice that some approaches to reduce the search time have been proposed in NAS for CNNs. However, most of them cannot be directly applied to GNNs, we need to design a proper improvement method for GNNs. The second problem is that the search space in our method is still limited, we can try to design a better search space to explore more novel GNNs. We will focus on these two problems in our future works.

Acknowledgments. The work described in this paper was partially supported by the Research Grants Council of the Hong Kong Special Administrative Region, China (CUHK 2410021, Research Impact Fund, No. R5034-18).

References

1. Bergstra, J., Bardenet, R., Bengio, Y., Kégl, B.: Algorithms for hyper-parameter optimization. In: Advances in Neural Information Processing Systems 24: 25th Annual Conference on Neural Information Processing Systems 2011. Proceedings of a meeting held 12–14 December 2011, Granada, Spain, pp. 2546–2554 (2011)
2. Cui, P., Wang, X., Pei, J., Zhu, W.: A survey on network embedding. IEEE Trans. Knowl. Data Eng. **31**(5), 833–852 (2019)
3. Defferrard, M., Bresson, X., Vandergheynst, P.: Convolutional neural networks on graphs with fast localized spectral filtering. In: Advances in Neural Information Processing Systems 29: Annual Conference on Neural Information Processing Systems 2016, December 5–10, 2016, Barcelona, Spain, pp. 3837–3845 (2016)
4. Fu, X., Zhang, J., Meng, Z., King, I.: MAGNN: metapath aggregated graph neural network for heterogeneous graph embedding. In: WWW 2020: The Web Conference 2020, Taipei, Taiwan, April 20–24 2020, pp. 2331–2341 (2020)

5. Gao, H., Wang, Z., Ji, S.: Large-scale learnable graph convolutional networks. In: Proceedings of the 24th ACM SIGKDD International Conference on Knowledge Discovery & Data Mining, KDD 2018, London, UK, 19–23 August 2018, pp. 1416–1424 (2018)
6. Gao, Y., Yang, H., Zhang, P., Zhou, C., Hu, Y.: GraphNAS: graph neural architecture search with reinforcement learning. CoRR abs/1904.09981 (2019)
7. Hamilton, W.L., Ying, Z., Leskovec, J.: Inductive representation learning on large graphs. In: Advances in Neural Information Processing Systems 30: Annual Conference on Neural Information Processing Systems 2017, Long Beach, CA, USA, 4–9 December 2017, pp. 1024–1034 (2017)
8. Kipf, T.N., Welling, M.: Semi-supervised classification with graph convolutional networks. In: 5th International Conference on Learning Representations, ICLR 2017, Toulon, France, 24–26 April 2017, Conference Track Proceedings (2017)
9. Krizhevsky, A., Sutskever, I., Hinton, G.E.: ImageNet classification with deep convolutional neural networks. In: Advances in Neural Information Processing Systems 25: 26th Annual Conference on Neural Information Processing Systems 2012. Proceedings of a meeting held December 3–6 2012, Lake Tahoe, Nevada, United States, pp. 1106–1114 (2012)
10. LeCun, Y., Bottou, L., Bengio, Y., Haffner, P., et al.: Gradient-based learning applied to document recognition. Proc. IEEE **86**(11), 2278–2324 (1998)
11. Li, Y., King, I.: Architecture search for image inpainting. In: Lu, H., Tang, H., Wang, Z. (eds.) ISNN 2019. LNCS, vol. 11554, pp. 106–115. Springer, Cham (2019). https://doi.org/10.1007/978-3-030-22796-8_12
12. Liu, Z., et al.: GeniePath: graph neural networks with adaptive receptive paths. In: The Thirty-Third AAAI Conference on Artificial Intelligence, AAAI 2019, The Thirty-First Innovative Applications of Artificial Intelligence Conference, IAAI 2019, The Ninth AAAI Symposium on Educational Advances in Artificial Intelligence, EAAI 2019, Honolulu, Hawaii, USA, January 27–February 1 2019, pp. 4424–4431 (2019)
13. Luan, S., Zhao, M., Chang, X.W., Precup, D.: Break the ceiling: stronger multi-scale deep graph convolutional networks. In: Advances in Neural Information Processing Systems 32: Annual Conference on Neural Information Processing Systems 2019, NeurIPS 2019, Vancouver, BC, Canada, 8–14 December 2019, pp. 10943–10953 (2019)
14. Manessi, F., Rozza, A., Manzo, M.: Dynamic graph convolutional networks. Pattern Recognit. **97** (2020)
15. Pham, H., Guan, M.Y., Zoph, B., Le, Q.V., Dean, J.: Efficient neural architecture search via parameter sharing. In: Proceedings of the 35th International Conference on Machine Learning, ICML 2018, Stockholmsmässan, Stockholm, Sweden, 10–15 July 2018, pp. 4092–4101 (2018)
16. Real, E., Aggarwal, A., Huang, Y., Le, Q.V.: Regularized evolution for image classifier architecture search. In: The Thirty-Third AAAI Conference on Artificial Intelligence, AAAI 2019, The Thirty-First Innovative Applications of Artificial Intelligence Conference, IAAI 2019, The Ninth AAAI Symposium on Educational Advances in Artificial Intelligence, EAAI 2019, Honolulu, Hawaii, USA, January 27 – February 1 2019, pp. 4780–4789 (2019)
17. Real, E., et al.: Large-scale evolution of image classifiers. In: Proceedings of the 34th International Conference on Machine Learning, ICML 2017, Sydney, NSW, Australia, 6–11 August 2017, pp. 2902–2911 (2017)
18. Scarselli, F., Gori, M., Tsoi, A.C., Hagenbuchner, M., Monfardini, G.: The graph neural network model. IEEE Trans. Neural Netw. **20**(1), 61–80 (2009)

19. Sun, K., Lin, Z., Zhu, Z.: AdaGCN: adaboosting graph convolutional networks into deep models. CoRR abs/1908.05081 (2019)
20. Velickovic, P., Cucurull, G., Casanova, A., Romero, A., Liò, P., Bengio, Y.: Graph attention networks. In: 6th International Conference on Learning Representations, ICLR 2018, Vancouver, BC, Canada, April 30 – May 3 2018, Conference Track Proceedings (2018)
21. Wu, Z., Pan, S., Chen, F., Long, G., Zhang, C., Philip, S.Y.: A comprehensive survey on graph neural networks. IEEE Trans. Neural Netw. Learn. Syst. (2020)
22. Ying, Z., You, J., Morris, C., Ren, X., Hamilton, W.L., Leskovec, J.: Hierarchical graph representation learning with differentiable pooling. In: Advances in Neural Information Processing Systems 31: Annual Conference on Neural Information Processing Systems 2018, NeurIPS 2018, Montréal, Canada, 3–8 December 2018, pp. 4805–4815 (2018)
23. Zhang, J., Shi, X., Zhao, S., King, I.: STAR-GCN: stacked and reconstructed graph convolutional networks for recommender systems. In: Proceedings of the Twenty-Eighth International Joint Conference on Artificial Intelligence, IJCAI 2019, Macao, China, 10–16 August 2019, pp. 4264–4270 (2019)
24. Zhou, J., Cui, G., Zhang, Z., Yang, C., Liu, Z., Sun, M.: Graph neural networks: a review of methods and applications. CoRR abs/1812.08434 (2018)
25. Zhou, K., Song, Q., Huang, X., Hu, X.: Auto-GNN: neural architecture search of graph neural networks. CoRR abs/1909.03184 (2019)
26. Zoph, B., Le, Q.V.: Neural architecture search with reinforcement learning. In: 5th International Conference on Learning Representations, ICLR 2017, Toulon, France, 24–26 April 2017, Conference Track Proceedings (2017)
27. Zoph, B., Vasudevan, V., Shlens, J., Le, Q.V.: Learning transferable architectures for scalable image recognition. In: 2018 IEEE Conference on Computer Vision and Pattern Recognition, CVPR 2018, Salt Lake City, UT, USA, 18–22 June 2018, pp. 8697–8710 (2018)

Automatic Curriculum Generation by Hierarchical Reinforcement Learning

Zhenghua He[ID], Chaochen Gu[(✉)][ID], Rui Xu[ID], and Kaijie Wu[ID]

Key Laboratory of System Control and Information Processing, MOE of China,
Shanghai Jiao Tong University, Shanghai, China
{heeeee,jacygu,xuruihaha,kaijiewu}@sjtu.edu.cn

Abstract. Curriculum learning has the potential to solve the problem of sparse rewards, a long-standing challenge in reinforcement learning, with greater sample efficiency than traditional reinforcement learning algorithms because curriculum learning enables agents to learn tasks in a meaningful order: from simple tasks to difficult ones. However, most curriculum learning in RL still relies on fixed hand-designed sequences of tasks. We present a novel scheme of automatic curriculum learning for reinforcement learning agents. A two-level hierarchical reinforcement learning framework, with a high-level policy called the curriculum generator and a low-level policy called the action policy, is proposed. During training, the curriculum generator automatically proposes curricula for the action policy to learn. Our training methods guarantee that the proposed curricula are always moderately difficult for the action policy. Both levels of policies are trained simultaneously and independently. After training, the low-level policy will be able to finish all tasks without the instructions given by the curriculum generator. Experiment results on a wide range of benchmark robotics environments demonstrate that our method accelerates convergence considerably and improves the training quality compared with the method without the curriculum generator.

Keywords: Deep reinforcement learning · Curriculum learning · Hierarchical reinforcement learning

1 Introduction

Recently, deep reinforcement learning (DRL) has achieved notable progress in solving sequential decision-making problems, including continuous robot control [10,14,17], Go game [24], video games [9,18,25] and automatic driving systems [21]. However reinforcement learning (RL) could be very challenging in tasks with sparse rewards since the agent can hardly get the reward to update the policy. Curriculum learning [4,7] is considered as an effective way to solve the problem of

This work is supported by Shanghai Science and Technology Innovation Action Plan NO. 19511105900, and in part by the National Key Research and Development Project NO. 2018YFB1703201.

sparse rewards. Curriculum learning imitates the mechanism of human learning, starting with simple tasks and then gradually increasing the difficulty of tasks. Take the environment of Reach (moving the end-effector of a robot to a goal point; see Fig. 1) for example, the goal can be firstly set close to the initial position of the end-effector so that the randomly initialized policy can achieve the goal with less effort. Along with the training process, the policy will become more sophisticated, and goals farther from the end-effector can be set to the agent. As the policy can always learn from the current trajectory, curriculum learning greatly improves the efficiency of sampling. However, currently, most applications of curriculum learning on RL [12,22] still rely on hand-designed curricula by domain expertise. This paper explores an automatic way of curriculum learning.

In this paper, a new scheme of curriculum learning is proposed. The main contribution of this paper is a novel way of automatic curriculum generation (ACG). To automatically generate curriculum, a two-level hierarchical reinforcement learning (HRL) architecture is proposed. As shown in Fig. 1, the high-level policy called curriculum generator takes the current state and the final goal g as input and outputs an intermediate goal g_i for the low-level policy called action policy to achieve. Then the action policy uses as input the current state and the intermediate goal g_i and outputs the actions of the robot. In our approach, techniques are designed to guarantee that the goals generated by the curriculum generator are of intermediate difficulty to the action policy. Along with the training, as the action policy become more capable, the curriculum generator will propose more difficult intermediate goals for the action policy. Therefore, the main function of the high-level policy is to generate increasingly difficult curriculum for the low-level policy along with the training. It is worth noting that, different with any other HRL scheme, after training, our low-level policy can achieve goals in the goal space independently, without the instructions given by the high-level policy. Therefore, our approach is a scheme of automatic curriculum learning.

Sufficient experiments are conducted to prove the efficiency of the proposed method. We evaluate our method in a variety set of robotic control and manipulation environments. The experiments show that our approach can accelerate convergence considerably and increase the success rate in some tasks compared with baseline RL algorithms. It is confirmed by our experiments that the intermediate goals generated by the curriculum generator in each episode are increasingly difficult along with the training, which is of critical importance to curriculum learning.

2   Related Works

Many works [2,13,20] research methods to build agents that can learn hierarchical policies. However, these methods cannot learn about multiple levels in the hierarchy simultaneously. Instead, these methods learn each level of policy separately in a bottom-up fashion. Work [15] proposes a framework that successfully learns three-level hierarchies in parallel in tasks with continuous state and

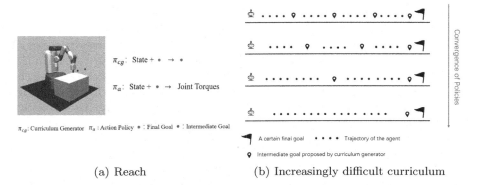

(a) Reach   (b) Increasingly difficult curriculum

Fig. 1. (a): The task is to move the end-effector to a goal point (the red circle). The curriculum generator generates intermediate goals (the blue circle) for the action policy. (b): Alone with the training process, the curriculum generator will generate fewer and fewer intermediate goals in each episode since the low-level policy becomes more and more capable. Therefore, the proposed curriculum becomes increasingly difficult. (Color figure online)

action spaces by integrating hindsight experience replay (HER) [1] and HRL. Our method also learns two-level hierarchies simultaneously. However, our work and [15] have a completely different purpose. Work [15] aims to learn hierarchical policies that work jointly to accomplish tasks even after training, while our method uses the higher level policy to speed up the convergence of the lower level policy. After training, our low-level policy can complete tasks alone.

Many previous works [3,4,26] explore the applications of curriculum learning in supervised tasks using hand-designed curricula. Work [8] proposes a method to build curriculum automatically; however, it is mainly applied in supervised tasks. Most curriculum learning in RL [12] still requires a fixed pre-defined sequence of tasks. Work [11] uses a generative adversarial network to automatically generate goals of intermediate difficulty, but it needs a hand-designed label to label goals. Our work regards the curriculum generator as the high-level policy in hierarchical reinforcement learning and uses the same sparse reward function as the original problem so that no expertise is required.

3 Background

3.1 Multi-goal Reinforcement Learning

In this paper, we consider a specific class of RL called multi-goal reinforcement learning (multi-goal RL) [1], where agents learn to achieve every goal in a goal space. Compared with standard RL, the policy function and value function in multi-goal RL take as input not only the current state $s \in \mathcal{S}$ but also a goal $g \in \mathcal{G}$, where $\mathcal{G}$ denotes the goal space. The correspondence of states and goals can be measured by a function $f_g : \mathcal{S} \to \{0, 1\}$. Whenever agents achieve any states s that satisfy $f_g(s) = 1$, it is considered as agents reaching its goal. It is assumed that for every state $s \in \mathcal{S}$, a corresponding goal g_a satisfying $f_g(s) = 1$ can be easily found by a mapping $m : \mathcal{S} \rightarrow \mathcal{G}$. This goal g_a is called the achieved goal. Besides, the rewards in this setting are sparse and binary given by the reward function:

$$R_g(s_t, a_t, s_{t+1}) = \begin{cases} 0, & f_g(s_{t+1}) = 1 \\ -1, & \text{otherwise} \end{cases} \tag{1}$$

3.2   Deep Deterministic Policy Gradients

We adopt Deep Deterministic Policy Gradients (DDPG) [16], an off-policy RL algorithm for continuous action spaces, to train our policies. DDPG is an actor-critic framework, maintaining two neural networks: a policy network (called the actor) $\pi : \mathcal{S} \rightarrow \mathcal{A}$ and an action-value function approximator (called the critic) $Q : \mathcal{S} \times \mathcal{A} \rightarrow \mathbb{R}$. Because of the multi-goal RL setting in this paper, Universal Value Function Approximators (UVFA) [19] is extended to DDPG. The policy network gets the current state and the current goal as input and output a deterministic action $\pi : \mathcal{S} \times \mathcal{G} \rightarrow \mathcal{A}$. Similarly, the action-value function approximator gets as input not only the state-action pair but also the current goal $Q : \mathcal{S} \times \mathcal{A} \times \mathcal{G} \rightarrow \mathbb{R}$. For more training details about DDPG, see [16].

3.3   Hindsight Experience Replay

Hindsight Experience Replay (HER) [1] is a method to construct experience buffers, which augments data considerably. Unlike the general experience buffer, HER stores not only transition tuples $(s_t, a_t, s_{t+1}, r_t, g)$ but also tuples transformed $\left(s_t, a_t, s_{t+1}, r_t', g'\right)$ where the original goal g is replaced with a goal g', which is achieved during the episode, and reward r_t' is recomputed. Then the buffer constructed by HER is fed to off-policy RL algorithms, such as DDPG, to update policies. As transitions in HER must contain sparse rewards, HER accelerates learning remarkably in tasks with sparse rewards. For more details about HER, see [1].

4   Method

Our method is an approach to automatic curriculum learning. We propose a new scheme of automatic curriculum generation (ACG) that makes action policy π_a explore the whole goal space from simple goals to difficult ones. The overall framework in our method is a two-level hierarchical reinforcement learning, where a high-level policy called curriculum generator π_{cg} generates curricula for the action policy π_a. A method to train the curriculum generator and action policy simultaneously and independently is proposed. Moreover, the training method in our approach guarantees that curricula generated by the curriculum generator π_{cg} are moderately difficult for the action policy π_a.

In every episode, a final goal g is set to the agent to achieve and it is fixed during the whole episode. The curriculum generator π_{cg} first takes the initial state s_0 and the final goal g as input then outputs an intermediate goal g_i for the action policy to achieve: $g_i = \pi_{cg}(s_0, g)$. Each intermediate goal g_i remains unchanged within a maximum number of steps T for the action policy. Then the action policy π_a outputs actions on the agent based on the current state and the intermediate goal $a_t = \pi_a(s_t, g_i)$. After actions are executed, the current state s_t transits to the next state s_{t+1}. After the maximum number of steps T is reached, the state of curriculum generator s_i transits to s_T as the next state s_{i+1}. Then another intermediate goal g_{i+1} is generated based on the new state s_{i+1}. The process above is repeated until the final goal g is reached or the maximum horizon H of intermediate goals have been generated.

Note that, with the training process, the curriculum generator will propose increasingly difficult targets. After training, the action policy can complete tasks individually. In an ideal case, the training process is shown in Fig. 1(b). Initially, to achieve the final goal g, the generator will propose sufficient simple intermediate goals for the action policy. Along with training, the action policy will become more capable so that given a certain final goal, the intermediate goals that need to be generated will become fewer. Fewer intermediate goals means that each intermediate goal is more difficult for the action policy. When both the generator and the action policy converge to the optimal, given any final goal $g \in \mathcal{G}$, the generator will directly guide the action policy to it. This means that the action policy can achieve any goal $g \in \mathcal{G}$ individually. Our experiments confirm this process.

In this section, we will introduce the curriculum generator, the action policy and the algorithm testing in detail.

4.1 Curriculum Generator

The curriculum generator is the high-level policy in our two-level hierarchical reinforcement learning framework. Its input is the current state s_i and the final goal g, and its output is an intermediate goal g_i. Regarding the output intermediate goal as an action, we formulate our curriculum generator as a goal-oriented MDP, which satisfies multi-goal RL setting, described by tuples $\mathcal{M} = \{\mathcal{S}, \mathcal{G}, \mathcal{A}, \mathcal{R}, \mathcal{P}, \gamma\}$, where $\mathcal{S}$ is the state space; $\mathcal{G}$ represents the goal space of the problem; $\mathcal{A}$ denotes the action space of the curriculum generator; $\mathcal{R}$ is the reward; $\mathcal{P}$ is the state transition function; γ is the discount factor. Compared with general goal-oriented MDP, our curriculum generator has some unique features. State transition functions $\mathcal{P}$ in our method are non-stationary. This is because the action policy updates overtime. Given a certain intermediate goal, the state that the agent can achieve changes along with action policy. However, with the convergence of the action policy network, the state transition functions become more stable.

We mainly design the following two techniques to guarantee the convergence of the curriculum generator and ensure that the generated targets are moderately difficult.

Intermediate Goals Between the Achieved Goal and the Final Goal.
The actions of the curriculum generator are intermediate goals of the action
policy. In theory, the action space $\mathcal{A}$ is the same as the goal space of the problem.
However, in our approach, we limit the generated goals g_i between the achieved
goal g_a and the final goal g: $g_i \in (g_a, g]$, where the achieved goal g_a is the goal
currently achieved, obtained by the mapping $m : \mathcal{S} \to \mathcal{G}$ described in Sect. 3.1.
This way of goal generation has three advantages. All these advantages improves
the convergency of the curriculum generator. Firstly, as the intermediate goals
are on the path to the final goal, it is ensured that the intermediate goals guide
the agent to achieve the final goal so that the sparse rewards (described in the
following section) can be obtained with less effort. Secondly, the valid action
space for every state-goal pair (s, g) is much smaller compared with the whole
goal space $\mathcal{G}$; therefore, our method reduces the dimension of action space. Lower
dimensional RL problems are easier to converge. Thirdly, the generator tends to
generate simple goals for the action policy in the initial stage of training, because
generated goals will always be simpler or just as difficult as the original final goal.
A trick to generate a goal that meets not only the constraints of the goal space
but also the restriction described above is not to generate a goal but output
an interpolation coefficient $a_c \in (0, 1]$. The intermediate goal g_i is obtained
by interpolating between the achieved goal g_a and the final goal g using this
interpolation coefficient $g_i = g_a + a_c(g - g_a)$.

Sparse Reward Function Determined by the Final Goal. The reward
of the curriculum generator is obtained by the next state and the final goal:
$r_i = r(s_{i+1}, g)$. Every time an intermediate goal g_i is generated, the action
policy π_a has a certain number T of attempts to achieve it. If the action policy
achieves the final goal g successfully after T time steps, the curriculum generator
π_{cg} receives a reward of 0. Otherwise, it receives a penalty of -1. This is to
ensure that the generated goals are always moderately difficult. Initially, the
capability of the action policy is weak. The generator has to generate multiple
simple intermediate goals, decomposing tasks into simpler tasks, so that the
action policy could reach the final goal g, and the generator could get the sparse
reward. In the meantime, the curriculum generator tends to generate as few
intermediate goals as possible, because every time an intermediate goal g_i is
generated, if the action policy fails to achieve the final goal, the generator will
get a negative penalty of -1. Therefore, as the action policy becomes more
capable, the generator generates fewer intermediate goals, that is, more difficult
subtasks, to maximize rewards.

It is worth noting that HER is adopted to the update of the action policy,
but not to the update of the curriculum generator. We update the curriculum
generator with the method of DDPG. The trajectories are stored in a replay
buffer in the form of tuples $(s_i, g, g_i, s_{i+1}, r_i)$, where g_i denotes the action of the
curriculum generator. If HER is used to construct the replay buffer, another
copy of transition tuples $(s_i, g', g_i, s_{i+1}, r_i')$ will be added to the replay buffer,
but in every tuple, the original goal g is replaced with another goal g', that

the agent later achieved, and the reward r_i' is recomputed based on the new goal g'. As is claimed by work [15], hindsight goal transitions encourage the generator to generate the shortest path of intermediate goals that has been found, ignoring the current capability of the action policy. This does not match the idea of curriculum learning that goals must be slightly beyond the capability of the action policy. Therefore, in our method, the replay buffer only contains real trajectories, where the rewards are obtained based on the capability of the action policy.

4.2 Action Policy

Action policy is the low-level policy in the hierarchy that directly controls the agent. Receiving an intermediate goal given by the curriculum generator, the action policy has a certain number T of attempts to achieve it. Action policy interacting with the environment is another goal-oriented MDP described by tuples $\mathcal{M} = \{\mathcal{S}, \mathcal{G}, \mathcal{A}, \mathcal{R}, \mathcal{P}, \gamma\}$. The goals of the action policy are intermediate goals generated by the curriculum generator. If the agent reaches the intermediate goal, which is simpler than the final goal, it can receive the sparse reward.

The action policy is also updated with the method of DDPG. The difference is that HER is adopted to augment the data and improve sample efficiency. Given an intermediate goal, the action policy controls the agent to interact with the environment T times, thus getting a trajectory. In each episode, the generator generates up to H intermediate goals so that the action policy can collect up to H trajectories. The trajectories are stored in a replay buffer as the form of tuples $(s_t, g_i, a_t, s_{t+1}, r_t)$. With the method of HER, another copy of tuples $(s_t, g', a_t, s_{t+1}, r')$ containing the sparse reward is added to the replay buffer. The original intermediate goal g_i is replaced with the goal g' that the agent later achieved, and the reward r' is recomputed based on the new goal. The tuples are then sampled to update policies using the algorithm of DDPG [16].

It is worth emphasizing that, with the convergence of the action policy, the action policy will have the ability to achieve goals in the goal space independently. This is an important reason why our method is a scheme of curriculum learning.

4.3 Algorithm Testing

The overall training process has been described above. In our method, the curriculum generator and the action policy are trained simultaneously. In training, given a goal $g \in \mathcal{G}$, the generator π_{cg} and action policy π_a work together to complete the task. It should be noted that in the early stage of training, the action policy can only complete some simple tasks independently. Along with the training, the action policy can achieve more and more difficult targets.

After training, it is expected that the action policy can finish tasks alone. Therefore, during testing, the action policy is tested independently. The final goal g sampled from the goal space $\mathcal{G}$ is directly set to the action policy and remains unchanged throughout the whole episode.

5   Experiment

We conduct sufficient experiments on the standard robotic manipulation environments in the OpenAI Gym [5] to confirm the applicability and effectiveness of our method. We mainly compare the training results of our method and the results without the curriculum generator. Without the curriculum generator, the policies are trained with the same method as our action policy: DDPG and HER. All experiments are conducted under the same conditions. In addition, ablation experiments about the horizon H of the curriculum generator are performed to explore its influence on the experiment results. Experiment details and results are clearly explained in this section.

5.1   RL Environments

Our experiments are conducted on the Robotics environments in the OpenAI Gym [5], including several robotic control and manipulation tasks. The tasks are to control the end effector of a robot arm to accomplish some tasks or to control a dexterous hand to complete some manipulation tasks. In these environments, rewards are all sparse and binary. Only when the agent reaches the goal can it get a reward of 0, otherwise, the reward is -1. In addition to the standard settings, we modify some environments as follows. In our method, every episode is divided into H parts at most. For the reason that the action policy to has enough steps to achieve the intermediate goal, the maximum step number of the environments of HandManipulate is modified to 500. It should be noted that all comparative experiments are performed in the same environments.

5.2   Experiment Details

We compare our training results (ACG + HER + DDPG) with the results without the automatic curriculum generator (HER + DDPG). In comparison experiments, we use the implementation of HER in OpenAI Baselines [6]. Almost all hyperparameters in OpenAI Baselines [6] remain unchanged, except that the number of MPI workers is set to 4 and the buffer size is set to 10^5 transitions. For the sake of fairness, our action policy uses the same hyper-parameters (including the number of MPI works and the buffer size) and the network architectures as the comparison experiments. Implementation details can be found in [6].

In our curriculum generator, the policy network and the state-value function network have the same network architecture: three layers generic multilayer perceptron (MLP) with 256 units and ReLU activation. The policies are optimized by Adam optimizer with a learning rate of 10^{-3}. The maximum horizon H of the curriculum generator is set to five. In Sect. 5.5, ablation studies are performed to explore the influence of this hyperparameter.

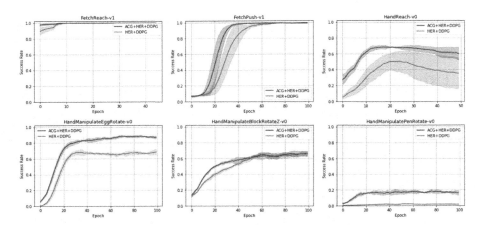

Fig. 2. Training curves of our method (ACG + HER + DDPG) and the baseline algorithm (HER + DDPG) on a variety of robotic manipulation environments. Each epoch contains 50 episodes and the parameters are updated 40 times after each episode. All the experiments use the same set of random seeds (from 1 to 5). The results are averaged across 5 random seeds and shaded areas represent one standard deviation.

As is stated in Sect. 4.1, in our implementation, we do not generate values of intermediate goals but output interpolation coefficient $a_c \in (0, 1]$. The desired intermediate goal g_i is obtained by interpolating between the achieved goal g_a and the final goal g using this interpolation coefficient $g_i = g_a + a_c(g - g_a)$. In Hand environments, the target pose of the manipulated object is given in the form of quaternions. The intermediate goals in Hand environments are computed by spherical linear quaternion interpolation (Slerp) [23].

5.3   Results and Comparison

The results of our experiments are shown in Fig. 2. An episode is considered successful if the agent finishes the task at the end of the episode, which means the distance between the object and the goal is within an acceptable threshold. The success rate is calculated by testing the policies learned 100 times. In each test, the goal g is randomly sampled from the goal space $\mathcal{G}$. It should be noted that when testing our method, only the action policy is used.

From Fig. 2, we can see that our method (ACG + HER + DDPG) shows a clear advantage in comparison with the method without our automatic curriculum generator (HER + DDPG). Our method converges faster on all of the six environments, and after training, our method has higher success rates on three of the six environments. Especially in the experiments of Handreach and HandManipulateEggRotate, our approach has improved the success rate by more than 20%.

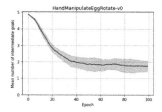

Fig. 3. The average number of the proposed intermediate goals in each episode decreases along with the training, which means each intermediate goal becomes more difficult.

5.4 Increasingly Difficult Curricula

We measure the difficulty of the curricula by the average number of the generated intermediate goals in each episode during training. When the number of intermediate goals is large, it indicates that the task is divided into many small subtasks, and each subtask is simple. When there are fewer intermediate targets, each subtask becomes more difficult. Therefore, the fewer intermediate goals, the more difficult the curricula is.

We only depict the change in the average number of intermediate goals on the environment of HandManipulateEggRotate in Fig. 3 because other environments have very similar results. It is clear that the average number of intermediate goals in each episode decreases considerably alone with the training process, so the curricula proposed by the curriculum generator are increasingly difficult.

5.5 Ablation Studies

In this section, we perform experiments to explore the influence of the hyperparameter H, the maximum horizon of the curriculum generator. Larger H means that the original tasks can be divided into more small tasks and the maximum number of steps T to complete each small task becomes shorter because of the maximum step number limit of the original task T_{ori}. The relation between T_{ori}, T and H can be formulated by: $T = T_{ori}/H$.

The training curves of two environments with different H are shown in Fig. 4. Take the environment of HandManipulateEggRote for example, as H increases, the convergence speed and the final success rate increase at first, reach a peak at $H = 5$, and then decrease. Considering the final success rate, the optimal H in the experiment of HandManipulateEggRotate and HandReach are $H = 5$ and $H = 4$, respectively. Our experiments show that the optimal selection of H is environment dependent. However, the overall trends of the training results, with the increase of H, are similar in different environments. Our explanation is that the increase of H makes the subtasks easier, so the training results become better at the beginning. But when H is further increased, the maximum number of steps T to complete each subtask decreases, limiting the capability of the action policy, so the training results becomes worse.

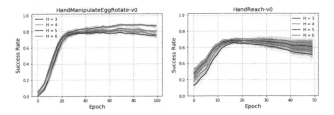

Fig. 4. Training curves with different hyperparameters H.

6 Conclusion

In this paper, a novel method of automatic curriculum learning is proposed to improve the sample efficiency of RL algorithms on problems with sparse rewards. A two-level hierarchical reinforcement learning architecture is proposed, with a high-level policy automatically generating curricula for the low-level action policy. A method to train the curriculum generator and action policy simultaneously and independently is proposed. Our training method guarantees that the proposed curricula are always moderately difficult for the action policy. After training, the action policy is tested alone to complete tasks. Sufficient experiments on the standard robotic manipulation environments are conducted to confirm the applicability and effectiveness of our approach. Experiment results show that our method accelerates the convergence of baseline RL algorithms considerably and improves the training quality in some environments.

References

1. Andrychowicz, M., et al.: Hindsight experience replay. In: Advances in Neural Information Processing Systems, pp. 5048–5058 (2017)
2. Bacon, P.L., Harb, J., Precup, D.: The option-critic architecture. In: Thirty-First AAAI Conference on Artificial Intelligence (2017)
3. Bengio, S., Vinyals, O., Jaitly, N., Shazeer, N.: Scheduled sampling for sequence prediction with recurrent neural networks. In: Advances in Neural Information Processing Systems, pp. 1171–1179 (2015)
4. Bengio, Y., Louradour, J., Collobert, R., Weston, J.: Curriculum learning. In: Proceedings of the 26th Annual International Conference on Machine Learning, pp. 41–48. ACM (2009)
5. Brockman, G., et al.: OpenAI gym (2016)
6. Dhariwal, P., et al.: OpenAI baselines (2017). https://github.com/openai/baselines
7. Elman, J.L.: Learning and development in neural networks: the importance of starting small. Cognition **48**(1), 71–99 (1993)
8. Graves, A., Bellemare, M.G., Menick, J., Munos, R., Kavukcuoglu, K.: Automated curriculum learning for neural networks. In: Proceedings of the 34th International Conference on Machine Learning-Volume 70, pp. 1311–1320. JMLR.org (2017)
9. Guo, X., Singh, S., Lewis, R., Lee, H.: Deep learning for reward design to improve Monte Carlo tree search in ATARI games. arXiv preprint arXiv:1604.07095 (2016)

10. Haarnoja, T., Pong, V., Zhou, A., Dalal, M., Abbeel, P., Levine, S.: Composable deep reinforcement learning for robotic manipulation. In: 2018 IEEE International Conference on Robotics and Automation (ICRA), pp. 6244–6251. IEEE (2018)
11. Held, D., Geng, X., Florensa, C., Abbeel, P.: Automatic goal generation for reinforcement learning agents (2018)
12. Karpathy, A., van de Panne, M.: Curriculum learning for motor skills. In: Kosseim, L., Inkpen, D. (eds.) AI 2012. LNCS (LNAI), vol. 7310, pp. 325–330. Springer, Heidelberg (2012). https://doi.org/10.1007/978-3-642-30353-1_31
13. Konidaris, G., Barto, A.G.: Skill discovery in continuous reinforcement learning domains using skill chaining. In: Advances in Neural Information Processing Systems, pp. 1015–1023 (2009)
14. Levine, S., Finn, C., Darrell, T., Abbeel, P.: End-to-end training of deep visuomotor policies. J. Machine Learn. Res. **17**(1), 1334–1373 (2016)
15. Levy, A., Konidaris, G., Platt, R., Saenko, K.: Learning multi-level hierarchies with hindsight (2018)
16. Lillicrap, T.P., et al.: Continuous control with deep reinforcement learning. arXiv preprint arXiv:1509.02971 (2015)
17. Mahmood, A.R., Korenkevych, D., Komer, B.J., Bergstra, J.: Setting up a reinforcement learning task with a real-world robot. In: 2018 IEEE/RSJ International Conference on Intelligent Robots and Systems (IROS), pp. 4635–4640. IEEE (2018)
18. Mnih, V., et al.: Human-level control through deep reinforcement learning. Nature **518**(7540), 529 (2015)
19. Schaul, T., Horgan, D., Gregor, K., Silver, D.: Universal value function approximators. In: International Conference on Machine Learning, pp. 1312–1320 (2015)
20. Schmidhuber, J.: Learning to generate sub-goals for action sequences. In: Artificial Neural Networks, pp. 967–972 (1991)
21. Shalev-Shwartz, S., Ben-Zrihem, N., Cohen, A., Shashua, A.: Long-term planning by short-term prediction. arXiv preprint arXiv:1602.01580 (2016)
22. Sharma, S., Ravindran, B.: Online multi-task learning using active sampling (2017)
23. Shoemake, K.: Animating rotation with quaternion curves. In: ACM SIGGRAPH Computer Graphics, vol. 19, pp. 245–254. ACM (1985)
24. Silver, D., et al.: Mastering the game of go with deep neural networks and tree search. Nature **529**(7587), 484 (2016)
25. Vinyals, O., et al.: Grandmaster level in StarCraft II using multi-agent reinforcement learning. Nature **575**, 1–5 (2019)
26. Zaremba, W., Sutskever, I.: Learning to execute. arXiv preprint arXiv:1410.4615 (2014)

Boltzmann Exploration for Deterministic Policy Optimization

Shaochen Wang, Yuan Pu, Shangtong Yang, Xin Yao, and Bin Li[✉]

University of Science and Technology of China, Hefei, China
{samwang,puyuan,mxjyst,yaoxin}@mail.ustc.edu.cn, binli@ustc.edu.cn

Abstract. Gradient-based reinforcement learning has gained more and more attention. As one of the most important methods, Deep Deterministic Policy Gradient (DDPG) has achieved remarkable success and has been applied to many challenging continuous scenarios. However, it still suffers from instable training on off-policy data and premature convergence to a local optimum. To deal with these problems, in this paper, we combine Boltzmann exploration with deterministic policy gradient. The candidate policy is represented by a Boltzmann distribution, and updated by Kullback-Leibler (KL) projection. By introducing the Boltzmann policy, the exploration is encouraged to effectively prevent the policy to collapse quickly. Experimental results show that the proposed algorithm outperforms DDPG on most tasks in MuJoCo continuous benchmark.

Keywords: Reinforcement learning · Policy optimization · Boltzmann exploration

1 Introduction

Reinforcement learning (RL) aims to learn an effective behavior through trial and error by interacting with the world. The goal is to optimize the agent's policy, in terms of the cumulative expected reward [16]. With powerful function approximators such as neural networks, reinforcement learning can handle more complex problems. Recently, deep reinforcement learning has been applied to a lot of challenging tasks [11,12,14].

In a supervised learning task, a deterministic gradient-based optimizer can find a pretty good solution. However, in RL, it does not exist explicit label information, and the feedback from the environment is delayed. Without an effective exploration strategy, the agent is prone to get stuck in a local optimum and fails to discover useful policies. Many popular algorithms like Deep Q Network (DQN) [12], Proximal policy optimization (PPO) [14] still rely on naive heuristic exploration strategies. These simple methods may consume large interaction times with the environment and fail on the complex and challenging tasks. In continuous control tasks, deterministic policy optimization [10] plays an important role. The deterministic policy is updated by following the gradient of a

© Springer Nature Switzerland AG 2020
H. Yang et al. (Eds.): ICONIP 2020, LNCS 12533, pp. 214–222, 2020.
https://doi.org/10.1007/978-3-030-63833-7_18

parametric Q function. The generated policy is unimodal and is at high risk of presenting a sub-optimal behavior. It results in unstable training and intensive hyperparameters tuning.

In this paper, we combine Boltzmann exploration with the deterministic policy and extend it to continuous control. Boltzmann exploration attracted a lot of attention in reinforcement learning [1,4,8]. Differently from DDPG which greedily maximizes the Q function, we formulate a Boltzmann optimal policy and minimize the KL divergence between the sampling policy and the Boltzmann optimal policy. The KL projection can be efficiently implemented by stochastic optimization. The benefit of introducing Boltzmann exploration into deterministic policy gradient can be summarized as the following: it assigns positive probability mass to suboptimal actions and provides alternatives to unexpected situations when suggested deterministic policy is not available in the environment. (2) Boltzmann distribution builds up a probability matching framework and draws appealing connections with stochastic policy gradient methods.

2   Related Work

Policy optimization consists of a wide spectrum of algorithms and has a long history in reinforcement learning. The earliest policy gradient method can be traced back to REINFORCE [17] which uses the score function trick to estimate the gradient of the policy. Subsequently, Trust Region Policy Optimization (TRPO) [13] monotonically increases the performance of the policy by limiting update sizes within the trust region. Proximal Policy Optimization (PPO) [14] can be considered as an improvement on TRPO using a heuristic approach to implement KL constraint. Conservative policy updating is helpful to restrict oversized policy update and it also limits the scope of exploration.

Deterministic policy gradient (DPG) [15] as an off-policy algorithm occupies a significant position in policy optimization. DDPG generalizes DPG to high dimensional tasks using neural networks as function approximation. Twin Delayed Deep Deterministic Policy Gradient (TD3) [5] uses the minimum of two Q functions as the Q target to alleviate the overestimation caused by the max operator during policy evaluation. Distributed Distributional DDPG (D4PG) [2] extends DDPG to a distributional fashion that the return is parameterized by a distribution $Z_\theta(s, a)$ and employs several prevalent techniques: parallelizing actors [11], prioritized experience replay and n-step temporal difference update [16].

3   Background

3.1   Markov Decision Process

Sequential decision making problems are often modeled as a Markov Decision Process (MDP). When interacting with the environment, at each step the agent observes a state s_t and chooses an action according to the policy $\pi(a_t|s_t)$. The agent receives a reward signal r_t, and the environment transitions to a next state $s_{t+1} \sim p(s_{t+1}|s_t, a_t)$. The goal of the agent is to maximize the expected total reward in Eq. (1). Hence, it is significant to define the Q function, denoted as $Q^\pi(s, a) = \mathbb{E}\left[\sum_{t=0}^\infty \gamma^t r\left(s_t, a_t\right) | s, \pi\right].$

$$\max_\pi \mathbb{E}\left[\sum_{t=0}^\infty \gamma^t r(s_t, a_t)|\pi, \mathbb{P}_0\right]; \tag{1}$$

where $s_0 \sim \mathbb{P}_0, a_t \sim \mathbb{P}(\cdot|s_t, a_t)$, $\gamma \sim [0, 1)$ is the discounted factor.

3.2 Value Based Methods

Value based methods represent its policy implicitly via the value function or action-value function. In Q-learning, the optimal policy is represented by the optimal $Q^*(s, a)$. In each iteration, the Q function is estimated and the policy is improved by Eq. (2). Thus a higher expected return is generated than previous iterations.

$$Q^\pi(s, a) = \mathbb{E}_{a \sim \pi(\cdot|s)}\left[r(s_t, a_t) + \gamma \max_{a'} Q^*\left(s', a'\right)\right]. \tag{2}$$

DQN [12] combines function approximation with Q-learning using neural networks to represent Q function. Innovatively, it introduces experience replay and target network to handle drastically oscillation during the training procedure. Transition samples (s_t, a_t, r_t, s_{t+1}) are added into the replay pool and this breaks the correlation of the sequential samples and approximately makes samples subject to the i.i.d. assumption. Target network copies the current network parameters periodically and offers a stable supervision signal.

3.3 Deep Deterministic Policy Gradient

DDPG can be considered as a variant of DQN in the continuous domain. DDPG utilizes an actor-critic framework and models the policy as a deterministic decision process. The actor is updated by following the chain rule from the critic in Eq. (3). θ represents the parameters of the policy network and ϕ represents the parameters of the Q value network.

$$\begin{aligned} \nabla_\theta J(\theta) &= \nabla_\theta \mathbb{E}\left[Q(s, a|\phi)|_{a=\mu(s|\theta)}\right] \\ &\approx \frac{1}{n} \sum_i \nabla_a Q(s, a)|a = \mu(s)\nabla_\theta \mu(s|\theta). \end{aligned} \tag{3}$$

Similar to DQN, the value function is optimized with Bellman-error through Eq. (4). To stabilize the training process, it also utilizes a target network and replay buffer to reduce the correlation between samples. ϕ^- is the parameters of the target network. While DDPG also inherits blemishes of DQN and DPG, it has overestimation bias for Q function [5] and may lack of sufficient exploration. These shortcomings make it sensitive to hyperparameters and suffer from fragile convergence.

$$L(\phi) = \mathbb{E}_{s,a \sim \rho(\cdot)}\left[(y - Q\left(s, a; \phi\right))^2\right], \tag{4}$$

where $y = r + \gamma Q\left(s', \mu\left(s'; \theta^-\right); \phi^-\right).$

4   Method

In a classic deterministic policy, the actor is optimized by following the gradient of the critic. Exploration is separated from the optimization and implemented by just adding noise into the action space. However, just by greedily maximizing the Q function, the policy is prone to get stuck in a local optimum. In this paper, a new form of policy called Boltzmann policy, denoted in Eq. (5), is adopted.

$$\pi(\mathbf{a}|\mathbf{s}) \propto \exp\left(Q^{\pi}(\mathbf{s}, \mathbf{a})\right). \tag{5}$$

The probability of selecting each action is proportional to the exponential of the Q function. Actions with a larger Q value estimate are more likely to be sampled, and actions that have small estimations still have a chance. Thus, the policy remains explorative and increases the probability of jumping out of the local optima. When the policy $\pi(a|s) \propto \exp(Q(s,a))$, it is guaranteed to improve and can eventually converge to an optimal policy [7].

An optimal policy, denoted as π_{τ}^{*}, is conducted. We want to establish an actual sampling policy as close as possible with the optimal policy π^{*}. To better measure the distance between two policies, a projection operator Π is defined in Eq. (6). When the $KL(\pi\|\pi^{*})$ equals zero, it means the sampling policy matches the optimal policy.

$$\Pi\pi = \arg\min_{\pi} KL\left(\pi\|\pi_{\tau}^{*}\right)$$
$$\text{where} \quad \pi_{\tau}^{*} = \tfrac{1}{Z}\exp\left(Q^{*}(s,a)\right). \tag{6}$$

However, Eq. (6) is difficult to optimize in practice. In each step, the optimal Q function is not accessible. To make it tractable, we conduct an iterative optimization process and substitute the current Q value into the optimal Boltzmann policy. The denominator of the optimal policy, denoted as $Z = \exp V_{s}^{\pi}$, serves as a normalization factor. It makes sure that the policy subjects to a probability distribution. We utilize an actor-critic framework. The policy network is parameterized by θ and the value network is parameterized by ϕ. Gradient descent is used to minimize the KL divergence between the sampling policy and the optimal policy and the derivative of the KL divergence is showed in Eq. (7). The normalization factor Z doesn't contain the policy term θ and it can be considered as a constant during gradient updates, and thus can be dropped. Empirically, the gradient of the expectation is estimated by the Monte-Carlo return. $\mathcal{D}$ represents the experience replay buffer. Samples from the replay buffer are used to estimate the gradients.

$$\nabla_{\theta} KL\left(\pi_{\theta}\|\pi_{\tau}^{*}\right) = \nabla_{\theta}\mathbb{E}_{\pi}\left[\log\left(\frac{\pi_{\theta}}{\pi_{\tau}^{*}}\right)\right]$$
$$= \nabla_{\theta}\mathbb{E}_{\pi}\left[\log\pi_{\theta} - Q_{\phi}\left(s, \pi_{\theta}(a|s)\right) + \log Z\right] \tag{7}$$
$$= \nabla_{\theta} \sum_{(s,a,r,s')\sim\mathcal{D}} \left(\log\pi_{\theta} - Q_{\phi}\left(s, \pi_{\theta}(a|s)\right)\right).$$

The parameters of the Q value network is updated through fitting the mean-squared-error in Eq. (8). ϕ' is the soft delay target network and does not participate in gradient backpropagation. The optimization procedure alternates between value evaluation and policy improvement. In each iteration, the agent selects the action according to the latest policy and yields a higher cumulative return. In summary, we substitute the original deterministic policy with a more explorative Boltzmann policy and conduct a two-step optimizing procedure. First, we optimize the Q value using the standard Bellman equation in Eq. (8). Second, we project the policy to Boltzmann optimal policy by Eq. (6). Run the above two steps until a desired policy is found (Fig. 1).

$$\min_{\phi} \left(r_t + \gamma Q_{\phi'}^{\pi}(s_{t+1}, a_{t+1}) - Q_{\phi}^{\pi}(s_t, a_t) \right)^2. \tag{8}$$

Fig. 1. Visualization of MuJoCo environments. From left to right: Ant-v2, HalfCheeta-v2, Walker2d-v2, HumanoidStandup-v2.

5   Experiments

We consider the high dimensional action space tasks and evaluate our method on a series of challenging control tasks. Empirical results show that the Boltzmann form policy can lead a better direction to search.

5.1   Environments

Continuous Tasks. We evaluate the performance of our algorithm on MuJoCo in OpenAI Gym [3]. It is the benchmark of continuous control tasks and is widely used in [5,6,14]. Agents are encouraged to go forward as far as possible. Each episode, the agents are allowed to interact with the environment up to 1000 steps. The observation states are the raw sensory inputs, including the locomotion positions and velocities. The action is the corresponding torques applied to the joint. For example, in the ant task, a four-legged creature robot is motivated to move forward as fast as possible and the rewards are proportional to the forward progress.

Sparse Reward Task. We modify the reward function of MountainCar in Gym. In the MountainCar task, a car is driving between two mountains. The goal is to drive to the peak on the right. In a sparse setting, a reward is obtained when the car achieves the top of the mountain.

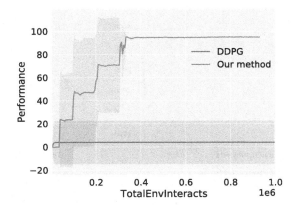

Fig. 2. The result on the sparse environment of MountainCar

5.2 Baseline Methods

For each task, both on-policy and off-policy methods are compared. We compare our algorithm with PPO (clipped version) which uses a parallel setting to explore the environment; Trust Region Policy Optimization (TRPO), an effective on-policy policy gradient method; DDPG, a sample efficient off-policy algorithm.

5.3 Setup

We present the overall training curves in Fig. 3. Each experiment is averaged over 7 different seeds. The solid lines represent the mean of averaged rewards and the shaded region shows the variance during training. To ensure the consistency and comparison with the previous work, we keep the same network architecture and hyperparameters across all the tasks. On most tested tasks, our proposed method outperforms DDPG.

In the implementation of our algorithm, both the actor and critic are parameterized by two layer feedforward networks which use 400 and 300 units respectively. The actor network outputs the mean and variance of the policy. Then an action is sampled according to Gaussian distribution given the state dependent mean and variance. We use Adam [9] to optimize both network parameters with a learning rate of 3×10^{-4} and utilize a replay buffer with size 10^6. Each task runs for 2.5 million steps and the performance is evaluated every 5000 steps. In the first 10000 steps, we use a random policy to explore the environment (Table 1).

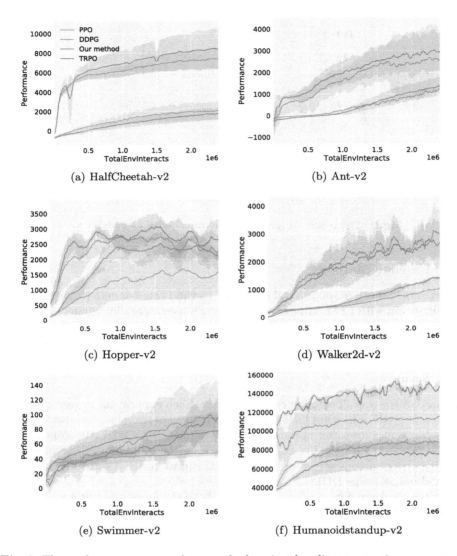

Fig. 3. The performance curves of our method against baselines on continuous control task.

Table 1. Average return over 10 trial

Environment	Halfcheetah	Hopper	Walker2d	Ant	Swimmer	Humanoidstandup
DDPG	7567.3	2010.7	**2403.2**	2749.3	**150.7**	120000.8
Our method	**8276.2**	2303.4	2400.2	**3556.9**	98.4	**145673.2**
TRPO	1892.3	2614.2	1386.3	1152.2	85.9	75424.2
PPO	2282.2	1576.4	1329.7	992.1	52.3	82312.1

5.4 Results and Analysis

Across all the experiments, our method beats the performance on the tasks: HaclfCheetah, Humanoidstandup, and Ant and matches the performance Swimmer and walker. Boltzmann exploration provides a connection with probability matching. The agent selects actions proportional to the expected returns. In Fig. 2, the agent interacts with the environment through sparse rewards. With no clear reward signal, DDPG fails to solve the problem. The car always swings back and forth at the foot of the mountain. Boltzmann policy guides the car to drive up to the mountain and Boltzmann distribution provides more possibilities for other sub-optimal actions. Each action is likely to be sampled.

6 Conclusion

This paper presents a softened deterministic policy gradient which combines Boltzmann distribution to encourage exploration. Extensive empirical evaluations show that our algorithm surpassed the original deterministic policy gradient and express better exploration on the sparse-reward and high dimensional control tasks. In future work, we will consider a more reasonable distance measure as an alternative, filtering out actions that are poor and retain near-optimal actions for exploration.

Acknowledgement. The work is partially supported by the National Natural Science Foundation of China under grand No. U19B2044 and No. 61836011.

References

1. Achbany, Y., Fouss, F., Yen, L., Pirotte, A., Saerens, M.: Tuning continual exploration in reinforcement learning: an optimality property of the Boltzmann strategy. Neurocomputing **71**(13–15), 2507–2520 (2008)
2. Barth-Maron, G., et al.: Distributed distributional deterministic policy gradients. arXiv preprint arXiv:1804.08617 (2018)
3. Brockman, G., et al.: OpenAI gym. arXiv preprint arXiv:1606.01540 (2016)
4. Cesa-Bianchi, N., Gentile, C., Lugosi, G., Neu, G.: Boltzmann exploration done right. In: Advances in Neural Information Processing Systems, pp. 6284–6293 (2017)
5. Fujimoto, S., Hoof, H., Meger, D.: Addressing function approximation error in actor-critic methods. In: International Conference on Machine Learning, pp. 1582–1591 (2018)
6. Haarnoja, T., Zhou, A., Abbeel, P., Levine, S.: Soft actor-critic: off-policy maximum entropy deep reinforcement learning with a stochastic actor. arXiv preprint arXiv:1801.01290 (2018)
7. Hunt, J., Barreto, A., Lillicrap, T., Heess, N.: Composing entropic policies using divergence correction. In: International Conference on Machine Learning, pp. 2911–2920 (2019)
8. Kianercy, A., Galstyan, A.: Dynamics of Boltzmann Q learning in two-player two-action games. Phys. Rev. E **85**(4), 041145 (2012)

9. Kingma, D.P., Ba, J.: Adam: a method for stochastic optimization. arXiv preprint arXiv:1412.6980 (2014)
10. Lillicrap, T.P., et al.: Continuous control with deep reinforcement learning. arXiv preprint arXiv:1509.02971 (2015)
11. Mnih, V., et al.: Asynchronous methods for deep reinforcement learning. In: International Conference on Machine Learning, pp. 1928–1937 (2016)
12. Mnih, V., et al.: Human-level control through deep reinforcement learning. Nature **518**(7540), 529 (2015)
13. Schulman, J., Levine, S., Abbeel, P., Jordan, M., Moritz, P.: Trust region policy optimization. In: International Conference on Machine Learning, pp. 1889–1897 (2015)
14. Schulman, J., Wolski, F., Dhariwal, P., Radford, A., Klimov, O.: Proximal policy optimization algorithms. arXiv preprint arXiv:1707.06347 (2017)
15. Silver, D., Lever, G., Heess, N., Degris, T., Wierstra, D., Riedmiller, M.: Deterministic policy gradient algorithms. In: International Conference on Machine Learning, pp. 387–395 (2014)
16. Sutton, R.S., Barto, A.G.: Reinforcement Learning: An Introduction. MIT Press, Cambridge (2018)
17. Williams, R.J.: Simple statistical gradient-following algorithms for connectionist reinforcement learning. Mach. Learn. **8**(3–4), 229–256 (1992)

Causal Inference for Mixed-Type Data in Additive Noise Models

Xin Liu[1], Zenglin Xu[2(✉)], and Ping Guo[3]

[1] School of Computer Science and Engineering, University of Electronic Science and Technology of China, Chengdu, China
[2] School of Computer Science and Technology, Harbin Institute of Technology Shenzhen, Shenzhen, Guangdong, China
xuzenglin@hit.edu.cn
[3] School of System Science, Beijing Normal University, Haidian, Beijing, China

Abstract. Causal inference between two observed variables has received a widespread attention in science. Generally, most existing approaches are focusing on inferring the casual direction based on data of the same type. However, in practice, it is very common that the observations obtained from different measurements can have different data types. This issue has not been much explored by the causal inference community. In this paper, we generalize the Additive Noise Model (ANM) to mixed-type data where one variable is discrete and the other is continuous, and take an information theoretic approach to find an unequal relationship between the forward and the backward. To conduct model estimation, we propose Discrete Regression model and Continuous Classification model to learn the residual entropy. In addition to the theoretical results, empirical results on synthetic and real data have also demonstrated the effectiveness of our proposed model.

Keywords: Causal inference · Classification · Mixed type data

1 Introduction

Causality has received a widespread attention in the scientific field in recent years [1]. Telling causes from effects is a difficult, expensive, or even impossible task through controlled randomized experiments, so researches on the causality community are mostly based on observational data known as causal discovery [2]. Conditional independence based methods [3] have been proposed to recover the casual structure in causal graphs, typical constraint-based algorithms include PC (named after its authors, Peter and Clark) [4] and Fast Causal Inference (FCI) [2]. PC assumes that there are no confounders, while FCI can return results even in the presence of confounders. Despite their merits, these methods cannot distinguish the two graphs that satisfy the same conditional independence; in

This work was partially supported by the National Key Research and Development Program of China (No. 2018AAA0100204).

H. Yang et al. (Eds.): ICONIP 2020, LNCS 12533, pp. 223–234, 2020.
https://doi.org/10.1007/978-3-030-63833-7_19

other words, they do not provide complete information for causal discovery and thus generate *Markov equivalent classes* only.

To distinguish the causal direction for variables in the same equivalent class, algorithms based on Functional Causal Models (FCMs) have been proposed [5]. Given the joint distribution of two variables X and Y, FCMs assume that the effect Y is a function of the direct cause X and some noise item N, i.e., $Y = f(X, N)$. After proper derivations, certain asymmetric property, which only holds in the true causal direction, can be derived to conduct inference. For example, nonlinear causal discovery with additive noise model (ANM) assumes that the effect Y is a function of the cause X with an additive noise N, i.e., $Y = f(X)+N$, where N is independent of X [6]. It has been shown that the forward model (i.e., $X \to Y$) and the backward model (i.e., $Y \to X$) cannot exist simultaneously for generic choices of f, $p(X)$ and $p(N)$. And the causal direction can be inferred through the p-values after careful model estimation (as a nonlinear regression) and independence tests. Different from ANM which is designed for continuous-valued data, has been extended for discrete data [7]. Other FCM extensions with proper restrictions include LiNGAM [8], and PNL [9], while all of them assume that the variables share the same data types. However, in practice, it could be very common that the two variables X and Y have different data types. For example, in physics, temperature determines the states of water being solid, liquid or gas. Here temperature is a continuous variable, while the states of water is a discrete variable. Under such a condition of mixed data types, regression models on continuous or discrete data are not applicable, and neither are the proposed ANM models.

In this paper, we focus on analyzing observations come from a joint distribution of one variable is continuous and the other is discrete and try to answer the question: can we infer the causal direction of the two variables which have different data types? Our approach is based on information theoretic, due to the independence which between the cause and the noise only holds in the true causal direction, according to the joint entropy and the form of ANM, we can find an unequal relationship between the forward and the backward, and we further propose the estimations for causal inference on mixed-type data and evaluate its performance on both simulated and real data.

The rest of the paper is organized as follows: in Sect. 2, we formalize the model of the mixed-type data based on ANM; in Sect. 3, we use information theoretic approaches to identify the cause and effect; in Sect. 4, model estimation methods are proposed; and we present experiments on synthetic and real world data in Sect. 5 followed by the conclusion in Sect. 6.

2   Model Definition

In this section, we introduce the ANM on mixed-type data where one variable is continuous and the other is discrete. We assume that the observed data are generated by the following model:

$$Y = f(X) + N, \ N \perp\!\!\!\perp X,$$

where $N \perp\!\!\!\perp X$ means that N is independent to X. Without loss of generality, we assume that X is a continuous variable and Y is a discrete variable, and we want to identify which one is the cause and which one is the effect.

The two directions model can be given by

$$Y = f(X) + N_Y, \ X = g(Y) + N_X.$$

Here we can know immediately that $\hat{Y} = f(X)$, N_Y and $\hat{X} = g(Y)$ are discrete variables and N_X is a continuous variable.

In this work, we adapt an information theoretic approach to identify the cause and the effect. Note that the differential entropy has been studied in the statistical consistency of ANM [10]. And the Shannon entropy on ANMs has been studied for discrete data [11]. Followed by model definition and information theories, we have the following theorem:

Theorem 1. *Given samples drawn from the joint distribution $p(X, Y)$ with X being continuous and Y being discrete, in an additive noise model, if $X \to Y$ structure, then it holds that*

$$H(X) + H(N_Y) < H(Y) + H(N_X),$$

where $N_Y \perp\!\!\!\perp X$ while $N_X \not\!\perp\!\!\!\perp Y$ under the ground truth $X \to Y$.

Proof. The joint entropy between X and Y can be given as follows: $H(X, Y) = H(X) + H(Y|X) = H(Y) + H(X|Y)$. Considering the graphical structure $X \to Y$, we have

$$H(X) + H(Y|X) = H(X) + H(N_Y|X) = H(X) + H(N_Y)$$

where $N_Y \perp\!\!\!\perp X$.

In the other direction, we have $N_X \not\!\perp\!\!\!\perp Y$, thus we can derive that

$$H(Y) + H(X|Y) = H(Y) + H(N_X|Y) < H(Y) + H(N_X).$$

Based on Theorem 1, we can have the guideline of causal inference between continuous and discrete data:

- if $H(X) + H(N_Y) < H(Y) + H(N_X)$, we infer that "X causes Y",
- if $H(X) + H(N_Y) > H(Y) + H(N_X)$, we infer that "Y causes X",
- if $H(X) + H(N_Y) = H(Y) + H(N_X)$, we infer that the mixed causal model does not fit the data, or X and Y are not cause-effect pairs.

Therefore, we can discriminate the causal direction by comparing the two sums of entropy in both sides. In addition, a threshold τ can be set to measure the difference between sides of the inequality operators as a criterion of whether we accept this result.

Algorithm 1. Discrete Regression

Input: discrete variable Y and continuous variable X
Model: $X = g(Y) + N_X$
Output: $H(Y) + H(N_X)$

  1: approximate precision of data dis
  2: $supp\, Y \leftarrow domain(Y)$
  3: **for** each $y_i \in supp\, Y$ **do**
  4:    $g_0(y_i) \leftarrow arg\, max_{x \in X} P(Y = y_i, X = x)$
  5: **end for**
  6: $res \leftarrow H(X - g_0(Y))$
  7: **for** each $y_i \in supp\, Y$ **do**
  8:    search $(minx, maxx)$
  9:    **while** $minx < maxx$ **do**
10:     $new \leftarrow H(X - g_i^{minx \rightarrow y_i}(Y))$
11:     **if** $new < res$ **then**
12:      $res \leftarrow new$
13:      $g(y_i) \leftarrow arg\, min_x H(X - g_i^{minx \rightarrow y_i}(Y))$
14:     **end if**
15:     $minx \leftarrow minx + dis$
16:    **end while**
17: **end for**
18: **return** $H(Y) + res$

---

3   Model Estimation

According to the theorem in the previous section, we can know that the difference between the two sums of entropy can give us the information about the true causal direction. We now consider practical estimation methods to infer the entropy of noise variables in the forward and the backward processes. To avoid confusion, we assume that X represents continuous variable and Y represents discrete variable. Thus the inference methods for both directions are (1) "discrete regression" (i.e., learning a function from discrete variables to continuous output) to learn $\hat{X} = g(Y)$ to minimize $H(N_X)$, and (2) "continuous classification" (i.e., learning a function from continuous variables to discrete output) to study $\hat{Y} = f(X)$ to minimize $H(N_Y)$.

3.1   Discrete Regression

Unlike continuous regression, in the case of discrete regression, through functions, each y_i in discrete variable Y will just have one corresponding value. Then we can simply consider all possible (y_i, x_i) mapping relations and get the one which can minimize the value of the loss function, as shown in Algorithm 1.

In order to apply heuristic learning, firstly we learn the approximate precision dis of X which is the minimum difference of data or we can set a small threshold as dis, and learn the support domain $supp\, Y$ of Y (line 1–2). Mapping each discrete value y_i to the most common co-occurring x_i as initial function and

Algorithm 2. Continuous Classification

Input: continuous variable X and discrete variable Y
Model: $Y = f(X) + N_Y$
Output: $H(X) + H(N_Y)$

1: $supp\,Y \leftarrow domain(Y)$
2: $min\,X, max\,X \leftarrow domain(X)$
3: $l \leftarrow [min\,X, x_0], \cdots, [x_{supp\,Y-2}, max\,X]$
4: $p \leftarrow supp\,Y!$
5: $res \leftarrow min\,H(Y - f_0(X))\ where\ f_0 : l \rightarrow p$
6: **for** i in $len(l)$ **do**
7:    $start \leftarrow l[i][0], end \leftarrow l[i][1]$
8:    **while** $start < end$ **do**
9:       $mid \leftarrow (start + end)/2.0$
10:      $new \leftarrow min\,H(Y - f(X))\ where\ f : l_{mid} \rightarrow p$
11:      **if** $new < res$ **then**
12:         $res \leftarrow new$
13:         $end \leftarrow mid$
14:      **else if** $new > res$ **then**
15:         $start \leftarrow mid$
16:      **else**
17:         break
18:      **end if**
19:   **end while**
20: **end for**
21: **return** $H(X) + res$

---

calculate the initial entropy of the residual (line 3–6). Then we iteratively update the mapping relationships for each y_i: find the boundary $minx$ and $maxx$ of x which corresponded to y_i, in the step of dis, search from $minx$ to $maxx$, find the best x^* which minimize the residual entropy under the condition that keeping the mapping relations of the other $\bar{y}$ ($\bar{y} = supp\,Y - y_i$) constant. After the loops, we can get the minimal result (line 7–17). In the end, return the sum of minimal residual entropy and the entropy of Y.

3.2   Continuous Classification

In the case of continuous classification, we need to find the correspondence between continuous intervals $[x_i, x_{i+1}] \subset [min\,X, max\,X]$ and the output $y_i \in supp\,Y$. We first assume that the interval corresponding to y_i is continuous, in other words, there will be no gaps, and all intervals have their own classification, the situations do not exist that one interval has multiple or zero value. These assumptions are in most cases in line with reality. Based on the idea of the Bisection method, we propose the following method for continuous classification and the pseudo code is given in Algorithm 2.

The bisection method is a root-finding method that repeatedly divides the interval into two and then selects a sub-interval in which a root must lie for fur-

---

Algorithm 3. Mixed Causal Inference

Input: continuous variable X and discrete variable Y, threshold τ
Output: causal direction
1: $R_{Y \to X} \leftarrow DiscreteRegression(X, Y)$
2: $R_{X \to Y} \leftarrow ContinuousClassification(X, Y)$
3: **if** $R_{Y \to X} - R_{X \to Y} > \tau$ **then**
4:     **return** $X \to Y$
5: **else if** $R_{X \to Y} - R_{Y \to X} > \tau$ **then**
6:     **return** $Y \to X$
7: **else**
8:     direction undecided
9: **end if**

---

ther processing. We use the idea of sub-interval to divide x into $supp\,Y$ intervals corresponded to each y_i to make the entropy of residual minimal.

We first get the classified size $supp\,Y$ and the boundary $min\,X, max\,X$ of interval of X (line 1–2). Evenly divide the interval into $supp\,Y$ parts to get equal length sub-interval set l. As for the label of these parts remains unknown, so we get the full permutation p of $supp\,Y$, then under every arrangement of labels, calculate the residual entropy and get the minimum value as initial result (line 3–5). Next, iteratively update the interval in l with the extension of bisection method. For every parts in l, we do the following manipulations: set two pointers $start, end$ to the beginning and the end of the interval, at each step divides the interval in two by computing the midpoint mid, update the interval which begins from $start$ and ends with mid, one step further update l and keep others constant, get new correspondences with label set p, calculate the minimum value of loss function as new result new, if new is less than res, it means the interval which is being iterated has not found its minimum boundary value, so we continue to search forward, update the value of res and set the end of the interval as mid. If new is greater than res, it means the interval is divided less than its actual length, so set the beginning of the interval as $start$ to search backward. If new equals to res, it explains the result has fallen into the local optimal solution, so break and iterate next interval. Finally return the sum of minimal residual entropy and the entropy of X.

In Algorithm 2, the full permutation of $supp\,Y$ need to be obtained, so $supp\,Y$ should be in a smaller range that the algorithm efficiency can be guaranteed.

3.3   Mixed Causal Inference

Based on the algorithms mentioned above, we can naturally form the approach for causal inference on mixed-type data, and the pseudo code is given in Algorithm 3. We use $R_{Y \to X}$ represents the result of discrete regression and $R_{X \to Y}$ represents the result of continuous classification. Comparing the difference which between $R_{Y \to X}$ and $R_{X \to Y}$ with τ we can infer the causal direction.

4   Related Work

In recent years, many causal inference methods based on FCM have been proposed. Most of them specify the type of data they are targeting, for continuous variables, or discrete variables, or mixed-type variables.

The causal inference for discrete data, for example, [7] extends the notion of additive noise models to discrete data, they prove the identifiability of the model, propose regression method and use Person's χ^2 test [12] for independence test; [13] proposes to learn the causal direction via comparing the distance correlation between the distribution of the cause $P(X)$ and the conditional distribution $P(Y|X)$, and infer the true causal direction with smaller dependence coefficient; In [14] they propose another method for discrete causal detection based on ANM by analyzing the supports of the conditional distributions $|supp\,P(Y|x_i)|$ and further explore causal discovery with mixture model where the situation is that the function f is changing across the observations.

More causal explorations are based on continuous case. In addition to the aforementioned ANM, LiNGAM, PNL, [15] propose CURE which compare the accuracy of the estimations of $P(Y|X)$ and $P(X|Y)$ by using unsupervised inverse Gaussian process regression, but the method only available for up to 200 data points; [16] propose SLOPE based on the Kolmogorov complexity which believes that the joint distribution $P(X,Y)$ has a simpler description in the causal direction than in the anti-causal direction, it does local and global regression using the method based on the Minimum Description Length (MDL); [17] propose IGCI also following the idea of the Kolmogorov complexity to choose the shortest description as the causal direction, they estimate Shannon entropy or mean of log Jacobi determinant to learn the scores.

Causal methods for mixed-type data are been proposed in recent five years. [18] extends LiNGAM for estimating causal structure in directed acyclic graph (DAG) consisting of both continuous and discrete variables, they use the Bayesian information criterion (BIC) scoring function and logistic regression model to specify the causal model; [19] proposes method based on MDL, uses classification and regression trees to model the dependencies and a greedy algorithm called CRACK to learn the optimal score to determine the causal direction.

5   Experiments

In this section, to show the ability of our approach to distinguish the true causal direction when all the assumptions hold approximately, we apply our method on both synthetic and real data sets.

5.1   Evaluation on Synthetic Data

The aim of our experiments on synthetic data is to show our approach can distinguish the true causal direction no matter the true relation is from discrete

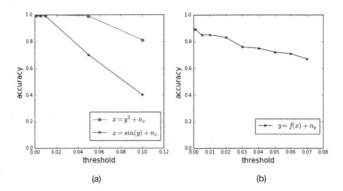

Fig. 1. Accuracy and threshold. (a) the true causal direction is from discrete variable to continuous variable. (b) Contrary to the case of (a). Accuracy decreased with the increase of the threshold.

variable to continuous variable or from continuous variable to discrete variable, and we also set X represents continuous variable and Y for discrete variable.

$Y \rightarrow X$. In this part, we generate synthetic data with assumed ground truth $Y \rightarrow X$. We simulate data using the model $x = y^2 + n_x$ and $x = \sin(y) + n_x$, the random discrete variables y are sampled randomly from $[1, n]$ and the noise variables n_x are sampled from a Gaussian distribution. We use Kozachenko-Leonenko (KL) estimator [20], which is a non-parametric estimator based on k-nearest neighbors of a sample set, to estimate the differential entropy.

Accuracy. We generate 100 different models from each model class and each take 1000 samples, and we set $\tau = [0.001, 0.005, 0.01, 0.05, 0.1], n = 3$. The results are shown in Fig. 1(a). We can see that the greater the threshold, the lower the accuracy. So, it is very important to set the appropriate threshold in practical problems. Compare the two functions $x = y^2 + n_x$ and $x = sin(y) + n_x$ we can find that the greater the interval between data, the better the model can identify the causal relationship.

Sample Size. Next we explore the relationship between sample size and inference accuracy. We take different numbers $size = [500, 1000, 2000, 5000, 10000]$ of samples each from 100 different models which following the function $x = y^2 + n_x$ and we set $\tau = 0.05, n = 3$. The results are shown in Fig. 2, we observe that the accuracy $acc = [0.98, 1.0, 1.0, 1.0, 1.0]$ which means the sample size has little effect on the accuracy of the model but too small sample size will affect the accuracy.

Class Size. In this part we observe the relation between class size and inference accuracy. We set the numbers of the types of y as $supp\,Y = [2, 3, 4, 5, 6]$, and under each $supp\,Y$ we generate 100 different models which following $x = y^2 + n_x$, $\tau = 0.05$ and $size = 1000$. After mixed causal model, we observe that the model achieves 98% to 100% accuracy in all cases. Because the full permutation of

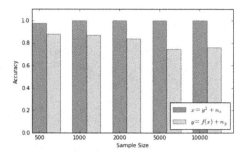

Fig. 2. Accuracy and sample size.

$supp\,Y$ need to be obtained in Continuous Classification, efficiency decreased with the increase of the $supp\,Y$.

$X \rightarrow Y$. In this paragraph we generate synthetic data with assumed the ground truth $X \rightarrow Y$ to argument the effectiveness of our model on continuous-discrete causal relationship. The continuous variables x are sampled from a Gaussian distribution and the noise variables n_y are discrete sampled randomly from $[-t, t]$. We use the piecewise function to classify x with random boundary value

$$f(x) = \begin{cases} y_1 & \text{if } min \leq x < x_0 \\ \cdots & \text{if } x_i \leq x < x_j \\ y_n & \text{if } x_{n-1} \leq x < x_n \end{cases}$$

and take y from $y = f(x) + n_y$.

Accuracy. We generate 100 different models from the model and each take 1000 samples and set $t = 1$, in Fig. 1(b), we show how the accuracy vary if we change the value of τ. One can see that the accuracy for the correct direction decrease with the increase of the threshold, even when $\tau = 0$ we can get the greatest accuracy is 89%. We think the reason for this problem is that in the model, we assume the interval $[x_{i-1}, x_i]$ corresponding to y_i has no gaps which is in line with reality, but in our synthetic data, we can't choose the value of the noise based on the cause because of the independence, so in the process of the stochastic model, some models may have the situation which a y_i corresponding to multiple intervals of x leads to wrong results.

Sample Size. In this part we study the effect of sample size on inference accuracy. We take samples following the $size = [500, 1000, 2000, 5000, 10000]$ each from 100 different models, due to the reason mentioned above, we set $\tau = 0$, $t = 1$. After the mixed causal model, the results shown in Fig. 2 and we obtain the accuracy $acc = [0.88, 0.87, 0.84, 0.75, 0.76]$ which means with the increase of the sample, taking random interference into account, the ability of the model to discriminate causal directions is gradually reduced.

To validate how well the proposed model performs on real-world data, we evaluate the proposed model on three real-world data sets.

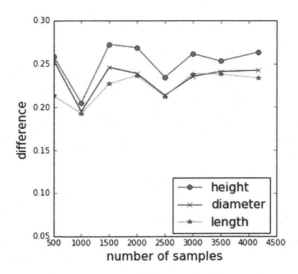

Fig. 3. Abalone data set, the relationship between data size and score difference of the two directions.

Abalone. The data set is available from the UCI machine learning repository[1]. It contains measurements of 4177 abalones (a group of shellfish) and we consider the sex Y of the abalone (male, female or infant) against length (X_1), diameter (X_2) and height (X_3), where sex is discrete which has three choices and the others are continuous data. Since the sex is probably causing the size of abalone, but not vice versa, we regard $Y \rightarrow X_1$, $Y \rightarrow X_2$ and $Y \rightarrow X_3$ as being the ground truth. We set $\tau = 0.05$ and conduct the experiment on whole data and report the result at Table 1. The algorithm identifies the true causal direction in all three cases with a fairish score difference between two directions. Further we include the first n data points of the abalone data set to observe the relationship between the data size and score difference of two directions, the result shown in Fig. 3, and we can see that the difference between the forward and the backward does not depend on the data size, even there has fluctuation, it also remains in a certain range.

Table 1. Results on Abalone data set. The algorithm identifies the true causal direction in all three cases.

Truth	$R_{Y \rightarrow X}$	$R_{X \rightarrow Y}$	diff
sex → *diameter*	7.578982	7.821310	0.24
sex → *height*	6.284361	6.547466	0.26
sex → *length*	7.871185	8.104672	0.23

---

[1] http://archive.ics.uci.edu/ml/.

Table 2. Results on Iris data set. The algorithm identifies the true causal direction in the first three cases and remains unknown in the last case.

Truth	$R_{Y \to X}$	$R_{X \to Y}$	diff
$class \to sepallength$	5.789224	5.926420	0.137
$class \to sepalwidth$	5.278778	5.534991	0.256
$class \to petallength$	5.583935	5.713906	0.129
$class \to petalwidth$	4.490743	4.452572	−0.038

Traffic. The data set is the 47th pair in the well-known Tuebingen cause-effect benchmark pairs[2]. The X variable is denotes the number of cars per 24th at different counting stations in Oberschwaben, Germany; and the Y variable is categorical (denoting whether it is a working day or a holiday). Here the ground truth is that Y is the cause and X is the effect. We apply our algorithm with $\tau = 0.05$ and get the result with $R_{X \to Y} = 10.880608$, $R_{Y \to X} = 10.704337$, $diff = 0.176$. It shows that our algorithm can recover the true causal direction very well.

Iris. The data set is also available from the UCI machine learning repository. The data set contains three classes Y of 50 instances each, where each class refers to a type of iris plant, and has four features sepal length (X_1), sepal width (X_2), petal length (X_3) and petal width (X_4) about each iris plant. We regard $Y \to X_1$, $Y \to X_2$, $Y \to X_3$ and $Y \to X_4$ as being the ground truth, since the class is probably causing the size of the iris plant, but not vice versa, and this is in accordance with our intuition. We conduct the experiments with $\tau = 0.05$ and show the results in Table 2 which we can see that our approach has good performance in the first three cause-effect pairs but remains unknown in the last pair. After analysis, we infer the reason of the failure may caused by that number of samples is too small with 150 and the relation between class and petal width may do not meet our assumptions.

6   Conclusion

We proposed a method based on ANM that is able to infer the cause-effect relationship between two mixed-typed variables (where one is discrete and the other is continuous). Based on information theoretic, we find unequal relationship between the forward and the backward processes, and further propose discrete regression and continuous classification for estimations. We have illustrated our method on both simulated and real-world data sets.

---

[2] http://webdav.tuebingen.mpg.de/cause-effect/.

References

1. Pearl, J., Mackenzie, D.: The Book of Why: The New Science of Cause and Effect, 1st edn. Basic Books Inc., New York (2018)
2. Spirtes, P., et al.: Causation, Prediction, and Search. MIT Press, Cambridge (2000)
3. Pearl, J.: Causality: Models, Reasoning, and Inference. Cambridge University Press, Cambridge (2000)
4. Spirtes, P., Glymour, C., Scheines, R.: Causation, Prediction, and Search, 2nd edn. MIT Press, Cambridge (2000)
5. Pearl, J.: Causal inference in statistics: an overview. Stat. Surv. **3**, 96–146 (2009)
6. Hoyer, P.O., Janzing, D., Mooij, J.M., Peters, J., Schölkopf, B.: Nonlinear causal discovery with additive noise models. In: Advances in Neural Information Processing Systems, pp. 689–696 (2009)
7. Peters, J., Janzing, D., Scholkopf, B.: Causal inference on discrete data using additive noise models. IEEE Trans. Pattern Anal. Mach. Intell. **33**(12), 2436–2450 (2011)
8. Shimizu, S., Hoyer, P.O., Hyvärinen, A., Kerminen, A.: A linear non-Gaussian acyclic model for causal discovery. J. Mach. Learn. Res. **7**(Oct), 2003–2030 (2006)
9. Zhang, K., Hyvärinen, A.: On the identifiability of the post-nonlinear causal model. In: Proceedings of the Twenty-Fifth Conference on Uncertainty in Artificial Intelligence, pp. 647–655. AUAI Press (2009)
10. Kpotufe, S., Sgouritsa, E., Janzing, D., Schölkopf, B.: Consistency of causal inference under the additive noise model. In: International Conference on Machine Learning, pp. 478–486 (2014)
11. Budhathoki, K., Vreeken, J.: Accurate causal inference on discrete data (2017)
12. Lehmann, E.L., Romano, J.P.: Testing Statistical Hypotheses. STS. Springer, New York (2005). https://doi.org/10.1007/0-387-27605-X
13. Liu, F., Chan, L.: Causal inference on discrete data via estimating distance correlations. Neural Comput. **28**(5), 801–814 (2016)
14. Liu, F., Chan, L.: Causal discovery on discrete data with extensions to mixture model. ACM Trans. Intell. Syst. Technol. (TIST) **7**(2), 21 (2016)
15. Sgouritsa, E., Janzing, D., Hennig, P., Schölkopf, B.: Inference of cause and effect with unsupervised inverse regression. In: Artificial Intelligence and Statistics, pp. 847–855 (2015)
16. Marx, A., Vreeken, J.: Telling cause from effect using mdl-based local and global regression. In: 2017 IEEE International Conference on Data Mining (ICDM), pp. 307–316. IEEE (2017)
17. Janzing, D., et al.: Information-geometric approach to inferring causal directions. Artif. Intell. **182**, 1–31 (2012)
18. Li, C., Shimizu, S.: Combining linear non-Gaussian acyclic model with logistic regression model for estimating causal structure from mixed continuous and discrete data. arXiv preprint arXiv:1802.05889 (2018)
19. Marx, A., Vreeken, J.: Causal inference on multivariate and mixed-type data. In: Berlingerio, M., Bonchi, F., Gärtner, T., Hurley, N., Ifrim, G. (eds.) ECML PKDD 2018. LNCS (LNAI), vol. 11052, pp. 655–671. Springer, Cham (2019). https://doi.org/10.1007/978-3-030-10928-8_39
20. Kozachenko, L.F., Leonenko, N.N.: A statistical estimate for the entropy of a random vector. Probl. Inf. Transm. **23**, 9–16 (1987)

CDMC'19—The 10th International Cybersecurity Data Mining Competition

Shaoning Pang[1]([⊠]), Tao Ban[2], Youki Kadobayashi[3], Jungsuk Song[4],
Kaizhu Huang[5], Geongsen Poh[6], Iqbal Gondal[1], Kitsuchart Pasupa[7],
and Fadi Aloul[8]

[1] School of Science Engineering and Information Technology,
Federation University Australia, Ballarat, Australia
p.pang@federation.edu.au
[2] The National Institute of Information and Communications Technology,
Tokyo, Japan
[3] Nara Institute of Science and Technology, Ikoma, Japan
[4] Korea Institute of Science and Technology Information,
Daejeon, Republic of Korea
[5] Xi'an Jiaotong-Liverpool University, Suzhou, China
[6] Malaysian Institute of Microelectronic Systems, Kuala Lumpur, Malaysia
[7] King Mongkut's Institute of Technology Ladkrabang, Bangkok, Thailand
[8] American University of Sharjah, Sharjah, United Arab Emirates

Abstract. CDMC-International Cybersecurity Data Mining Competition (http://www.csmining.org) is a world unique data-analytic competition sitting in the trans-disciplinary area of artificial intelligence and cybersecurity. In this paper, we summarize CDMC'19—the 10th cybersecurity data mining competition, which was held in Sydney Australia—together with a coupled workshop event, the Artificial Intelligence and Cyber Security (AICS) workshop 2019. We introduce the scope and background of the CDMC competition, the competition organizer, International Cyber Security Data-mining Society (ICSDS), and the rules that we followed to manage the competition. We reveal details of CDMC'19 regarding the competition tasks, participating teams, and the results the participants have achieved. Moreover, we publish the collection of CDMC's 10-year competition datasets as the CDMC Cybersecurity Dataset Repository via http://archive.csmining.org. Finally, we conclude the paper with an outlook on the future activities of CDMC.

1 Introduction

Cybersecurity has become more and more a data-driven industry—data becomes both the goals and means of the all the activities in the cyber space. On the one hand, data in the form of digital assets and intellectual properties have grown into the most valuable resources for business and life and thus became the major targets of the cyber attacks. This not only gives rise to the ever-evolving attacks and record-breaking number of attack campaigns toward these data, but also

H. Yang et al. (Eds.): ICONIP 2020, LNCS 12533, pp. 235–245, 2020.
https://doi.org/10.1007/978-3-030-63833-7_20

brings chance to cyber security innovation for the digital economy, e.g., Data Loss Prevention (DLP) and Cloud Security. On the other hand, with the advance of digitization technology, e.g., Industry4.0, AI, 5G, software-defined networking, we are more than ever capable of recording the footprint of any cyber-attacks to enable corresponding defence operations. Hence, the solution to cybersecurity highly relies on systematic collection, management, analysis, interpretation and application of data.

Up to now, Artificial intelligence (AI) has been widely used in analyzing massive cyberspace data, creating so called cyber-threat intelligence to help security operations analysts to identify potential threats and thus stay a step ahead of big incidents. AI also helps to dramatically reduce the incident response time at security operation centers, by instantly extracting insights from the noise of thousands of daily threat alerts [1]. In this intelligence production process, various forms of advanced computational algorithms including statistical analyses, machine-learning algorithms, and deep-learning networks are engaged.

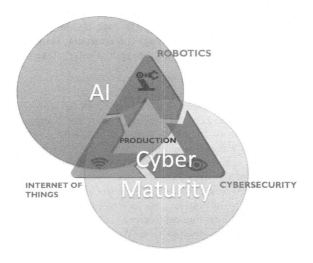

Fig. 1. An illustration of balanced AI and cybersecurity development based on [6].

AI has been attracting investors, inventors, as well as academic researchers worldwide. In the past decade, the publication of AI-focused academic papers, has outpaced the amount of published researches on computer science; and the number of AI-focused startups backed by venture capital was more than doubled, outpacing the increase of the overall pool of startups. Yet as we employ more and more AI and automation technologies in our life and business, e.g. Internet of Thing (IoT) and robotics, they may introduce a potential new opening for electronic intruders. Cybersecurity must be carefully refactored before further application to more major but critical services. Towards a sustainable world through a smart digital transformation, we foster a balanced development scheme between AI and Cybersecurity, as shown in Fig. 1, or namely, cyber maturity of AI. The interdisciplinary field, "AI × Cyber Security" focus on researches to develop AI-enabled defense against increasingly sophisticated cyber attacks.

2 The Activities of ICSDS

In 2008, a remarkable collaboration between National Institute of Information and Communications Technology (NICT) Japan and Auckland University of Technology (AUT) New Zealand led to the formation of a dedicated international academic society in the trans-disciplinary area of computational intelligence and information security, i.e., the International Cyber Security Data-mining Society (ICSDS). ICSDS was founded in 2008, after ICONIP'08 in Auckland, under the leadership of Professor Paul S. Pang of AUT. Tao Ban of NICT, and Prof Youki Kadobayashi of Nara Institute of Science and Technology, Japan. So far, it has grown into a full-fledged research association with 15 governing board members in 10 regions: New Zealand, Japan, Korea, China, Malaysia, Thailand, Australia, United Arab Emirates, Singapore, and Canada. ICSDS seeks more active participation from researchers and professionals specially in the Asia Pacific region.

Aiming at promoting more active interactions of researchers, scientists, and industry professionals, ICSDS engaged in a variety of international research activities soon after it has started.

Since 2008, ICSDS has been hosting the International Workshop on Data Mining and Cybersecurity, which is reforged as the International Workshop on AI and Cybersecurity to incorporate most recent progresses from AI field after 2017. The purpose of AICS is to raise the awareness of cybersecurity, promote the potential of industrial applications, and give young researchers exposure to the key issues related to the topic and to ongoing works in this area. AICS provides a forum for researchers, security experts, engineers, and students to present latest research, share ideas, and discuss future directions in the fields of data mining, artificial intelligence, and cybersecurity. During the past years, we had AICS2010 in Sydney Australia, AICS2011 in Hangzhou China, AICS2012 in Doha Qatar, AICS2013 in Daegu South Korea, AICS2014 in Kuala Lumpur Malaysia, AICS2015 in Istanbul Turkey, AICS2016 in Kyoto Japan, AICS2017 in Guangzhou China, AICS2018 in Siem Reap, Cambodia, and AICS2019 in Sydney Australia respectively. This year's AICS2020 will be held on November 18–22 in Bangkok Thailand.

ICSDS started the CDMC competition since 2010, which turns out to be a popular competition on cybersecurity which is attractive for young researchers. Refer to more detail information, e.g., tasks, evaluation, and yearly statistics, in the next section.

In 2017, ICSDS newly launched the first AI x Cyber Security Summit (ACSS) as an ICSDS-leading international high-tech forum, which is featured as an engagement of academia, industry and venture capital. ACSS'17 was hosted by Xi'an Jiaotong-Liverpool University, in Suzhou, China, and ACSS'18 was hosted by Chongqing University of Science and Technology, in Chongqing, China.

ICSDS collaborates with other organizations and conferences to promote academic and technical activities within its scope of interests. These collaborators includes New Zealand Embassy Beijing, New Zealand Consulate ChengDu, ICONIP conferences, Asia Pacific Neural Network Society (APNNA), IEEE New Zealand Section, Europe Neural Network Society (ENNS), International Neural Network Society (INNS), etc. Over the years, the events hosted by ICSDS have gathered hundreds of researchers, scientists, and professionals who are working in the field of artificial intelligence and/or cybersecurity from more than 68 countries and regions.

This initiative provided funding and infrastructure to foster coordination of a society of international experts, and launch the first ever international Cybersecurity Data Mining Competition (CDMC) in 2010, which not only gathered hundreds of young researchers, but also brought together experts from around the world to facilitate collaboration and accelerate research progress. Since then, great progress has been made. In the following we review the 10-year activities of ICSDS society.

3   CDMC Annual Competition

The CDMC is a challenging, research and practice competition, focusing on application of knowledge discovery and computational intelligence techniques to address cyber security challenges in real world applications. The competition is open to worldwide research teams or individuals, particularly welcomes university students, undergraduate or postgraduate, in the field of data science, network engineering, cyber security, and artificial intelligence.

3.1   Past Statistics

Across the history of this competition series, a wide range of cyber security problems covering 10 different categories of challenges has been studied. In addition to that, a couple of pattern recognition tasks have also been included in the competition series. In total, over 30 original datasets acquired from industry or research experiments were used in the CDMC history. Over 1276 teams/individuals from 68 different countries have registered and/or participated the CDMC over the last 10 years. See Fig. 2 for a record of yearly participation for CDMC.

Figure 3 illustrates the distribution of historical participants by country. As seen, more than half of participants came from the Asia Pacific countries. USA and Canada contributed to 9% of the participants. Europe also contributed to 10%, from which UK is the most active participation country. The remaining 28% participants scattered sparsely in the other part of the world. The participants included commercial ICT companies, universities, and research institutes. It's worth noting that some participants have participated more than one time. Table 1 gives a list of the institution of the winning teams for the annual competition.

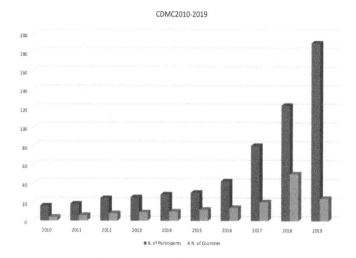

Fig. 2. A summary of 10-year CDMC participation.

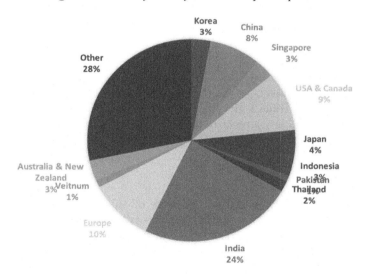

Fig. 3. Distribution of participants by country.

3.2   CDMC Cybersecurity Repository

CDMC has been addressing a number of specific cyber security challenges in the area of network security, IoT security, mobile security, social engineering, and hardware security, as well as a list of pattern recognition tasks towards leveraging AI techniques to meet industry needs. The acquired data range from numerical data, text, image, video/image series, structured & unstructured data, binary & multiple class data, as well as single & multi-task labelled data.

Table 1. A list of institution of the winning teams for the annual CDMC competition.

Institution	Country	Year	Rank
Shandong Uni.	China	2010	1
Tongji Uni.	China	2011	1
Fujitsu R& D Center Co., Ltd.	Japan	2012	1
Uni. of Ottawa	Canada	2013	1
Inst. for Infocomm Research	Singapore	2014	1
Kyoto Women's University	Japan	2015	1
Austral University	Argentina	2016	1
Uni. of Edinburgh	UK	2016	2
Kyoto Women's University	Japan	2017	1
Uni. of Queensland	Australia	2017	2
National Uni. of Defense Tech.	China	2018	1
Washington Uni.	USA	2018	2
Nara Inst. of Sci. & Tech.	Japan	2019	1
Kle Technological University	India	2019	2

As the 10-years anniversary of CDMC, we publish in this paper the whole collection of CDMC datasets as the CDMC Cybersecurity Repository at http://archive.csmining.org. Table 2 gives the list of datasets in the repository. Note that a few tasks which are subject to specific restriction of publication after the competition are excluded from the repository.

3.3   Citation

If you publish your work based on datasets of this repository, you have to acknowledge the contributors of this repository. This will encourage other researchers to conduct a comparison study of different approaches on the same datasets and thus benefit your research as well. We suggest the following pseudo-APA reference format to cite this repository:

Pang, S. Ban, T. Kadobayashi, J. Song, Y. Huang, K. Gondal A. Poh, G. Pasupa, K. and Fadi, A. (2020). CDMC Cybersecurity Dataset Repository http://archive.csmining.org, International Cybersecurity Data-mining Society (ICSDS), hosted by the School of Engineering Information Technology and Physical Sciences, Federation University, Australia.

Note, a few datasets have additional citation requirements which can be found at the bottom of the dataset's web page.

4   CDMC 2019

Taking CDMC 2019 as an example, this section provides more detail information about the events hosted by ICSDS.

4.1   Competition Process

The steps of CDMC process are summarized as:

(a) **Obtaining the Tasks:** The competition tasks will be made available at the CDMC website wwww.csmining.org on the starting date of the competition. To enter the competition, all participants must register and download the assigned tasks at the official website.

(b) **Result Submission:** Submission of the results can be done using the submission form at the competition website. A valid submission should include the predicted results for the testing samples in plain .txt files, where the predictions need to be in the same definitions as the training datasets. In addition to result submission, participants are encouraged to submit a 2–8 page short paper to the AICS workshop.

(c) **Evaluation and Ranking:** The performance evaluation criteria include $precision$, $recall$, F-measure, and $Accuracy$, as defined in Sect. 4.3. Note that while multiple submissions are allowed for one participant/team, only the last valid entry of result submissions will be used for performance evaluation and ranking. Here, a valid entry must include the results for all competition tasks.

(d) **Method Verification:** To prevent cheating, the top ranking teams will be required to fill out a fact sheet to describe their methods used for the competition. The ICSDS governing board will review the method and confirm the ranking. Note that participants might be required to provide their compiled software and/or source code for validation purpose.

(e) **Awarding and Prize Giving:** CDMC will announce 1st-place winner at the AICS workshop, which is collocated with the International Conference on Neural Information Processing (ICONIP). The awarding of cash prize and winner certificate will be at the banquet of the ICONIP conference. The CDMC cash prize is normally set as $3000NZD, the amount of which may be subject to the yearly sponsorship grant received by CDMC.

In tradition, CDMC awards only the 1st-place winner of the competition. The ICSDS governing board reserves the right to also present winner certificates to a short list of top ranking teams.

4.2   Competition Tasks

CDMC'19 came with three competition tasks which includes Task 1&2: SADAVS-sensor array data for autonomous vehicle safety, and Task 3: IoT malware classification [4].

1. SADAVS-Sensor Array Data for Autonomous Vehicle Safety [5]
   Vehicle-based accident detection systems monitor a network of sensors to determine if an accident has occurred. Instances of high acceleration/deceleration are due to a large change in velocity over a very short period of time. In the context of autonomous vehicles, the speeds are hard to attain since a vehicle is not controlled by a human driver. The presented data captured originally in New Zealand gives a collection of a sensor array (160×144) values in monitoring the status of moving vehicle. The objectives of these competition tasks are for early detection of any potential road accidents in two different scenarios.
2. IoT Malware Classification
   Based on the sequence of system calls as discriminant features and the malware families of the programs as training labels, the participants are required to perform a classification task to predict the malware families of the test samples. The dataset consists of 8442 samples generated following the procedure below: First, a collection of potentially malicious Linux programs in CEF format are collected from various sources. Then, each of these programs is executed in a sandbox environment hosted by an emulator that provides the required runtime environment for it. During the runtime, the *strace* command is used to monitor and record the interactions between the processes initialized by the program and the Linux kernel. This process yields a log file that contains lines of system calls. On each line, *strace* records the time stamp, the invoked system call, as long as parameters and results of the calls.

4.3 Performance Evaluation

The results of classification are first represented in a confusion matrix composed of *true positives*, *false positives*, *true negatives*, and *false negatives*. And then, to embrace competition tasks which have imbalanced class distribution, we perform performance evaluation using multiple criteria including *precision*, *recall*, *F-measure*, and *accuracy*.

Precision is a metric that calculates the accuracy for the minority class [3]. For an imbalanced binary classification problem, precision is calculated as the number of true positives divided by the total number of true positives and false positives:

$$Precision = \frac{TP}{(P + FP)}. \tag{1}$$

In an imbalanced classification problem with multiple classes, precision is calculated as the sum of true positives across all classes divided by the sum of true positives and false positives across all classes:

$$Precision = \sum_{c \in C} \frac{TP_c}{\sum_{c \in C}(TP_c + FP_c)}. \tag{2}$$

Table 2. The list of datasets in CDMC Cybersecurity Repository

Category	Dataset Name/CDMC Year
Social Attacks	SNSSR/CDMC'19
	e-News2016/CDMC'16
	PSI/CDMC'14
	e-News2013/CDMC'13
	LingSparm/CDMC'10
	e-News2015/CDMC'15
Sentiment Analysis	Trademe Sentiment/CDMC'15
DDoS Attacks	DDoS-ADENS/CDMC'18
IoT Malware	IoT-Malware/CDMC'19
Network Security	Packet Identification/CDMC'12
Intrusion Detection	IDS-Korea2014/CDMC'14
	IDS-Korea2013/CDMC'13
Mobile Security	Android-API/CDMC'17
	Android-Malware/CDMC'16
Cloud Security	UniteCloud-UTM/CDMC'17
	UniteCloud-Log/CDMC'16
Physical System Security	SADAVS/CDMC'19
Financial Fraud	FDFT/CDMC'17
Pattern Recognition	AAP/CDMC'18
	ESMC/CDMC'14
	DMLI-MTPR/CDMC'13
	5-Disease Diagnosis/CDMC'15

Recall is calculated for binary classification as the number of true positives divided by the total number of true positives and false negatives:

$$Precision = \frac{TP}{TP + FN}. \tag{3}$$

For multiple classes problem, recall is calculated as the sum of true positives across all classes divided by the sum of true positives and false negatives across all classes:

$$Precision = \sum_{c \in C} \frac{TP_c}{\sum_{c \in C}(TP_c + FN_c)}. \tag{4}$$

F-Measure combines precision and recall into a single measure that captures both properties, i.e.,

$$F\text{-}Measure = \frac{2Precision \times Recall}{Precision + Recall}. \tag{5}$$

Accuracy is calculated as the number of correctly classified instances divided by the total number of instances:

$$Accuracy = \frac{1}{\|C\|} \sum_{c \in C} \frac{TP_c + TN_c}{TP_c + TN_c + FP_c + FN_c}, \tag{6}$$

where $\|C\|$ is the total number of classes, $TP_c + TN_c$ and $TP_c + TN_c + FP_c + FN_c$ are the correctly classified instances and the total number of instances of the ith class, respectively.

4.4 Results

CDMC'19 had total 190 teams/participants from 24 different countries. Compared to CDMC'18, the number of participants increased by 67, while the number of countries dropped from 50 to 24. Table 3 gives the top 10 teams and the results they had achieved.

Table 3. Top 10 teams of CDMC'19 and their results

Rank	Name	Institution	Country	Accuracy
1	Masataka Kawai	Nara Institute of Science and Technology	Japan	75.22%
2	Aditya Pandey	Kle Technological University	India	73.42%
3	Shivam Ralli	Kle Technological University	India	73.33%
4	Inzamam Sayyed	Nil	India	72.12%
5	Teoh John	University of Glasgow	United Kingdom	70.63%
6	Qianguang Lin	Hainan University	China	70.41%
7	Syukron Abu Ishaq Alfarozi	King Mongkut's Institute of Technology Ladkrabang	Thailand	69.52%
8	Vadim Borisov	University Tuebingen	Germany	68.93%
9	Binh Nguyen	University of Science, Ho Chi Minh City	Vietnam	66.94%
10	Yoshino Ozawa	Kyoto Women's University	Japan	64.07%

5 Conclusion

Data is driving force for both AI and cybersecurity. It offers opportunities for researchers to discover rules from the practices, for cyber security analysts to find ways to adopt new policies to fortify cyber resilience, and for AI practitioners to explore solutions to transform learned models to businesses. CDMC takes advantage of such interdisciplinary insights, and has contributed to the society in providing a premium forum for discussion and exchange of these experiences.

In 2020, despite of the COVID-19 situation, the 11th CDMC, and the 13th AICS workshop have been set up. You are cordially invited to submit papers to the 13th International Workshop on Artificial Intelligence and Cybersecurity

(AICS2020) and participate in the 11th International Cybersecurity Data Mining Competition (CDMC 2020). The two events are associated with the 27th International Conference on Neural Information Processing (ICONIP 2020) as a special session. ICONIP will be organized in Bangkok, Thailand, November 18–22, 2020. For more information about the CDMC2020, please refer to the competition website at http://wwww.csmining.org. We look forward to meeting you in Bangkok, Thailand.

Acoknowledgement. The authors would like to acknowledge all the participants who had ever take part in the competitions over the last 10 years. We would like to express our great appreciation to Auckland University of Technology, New Zealand, Unitec Institute of Science and Technology, New Zealand, and National Institute of Information and Communications Technology, Japan for their financial sponsorship to CDMC in the past 10 years, and to the Asia Pacific Neural Network Society (APNNS) for 10 years partnership in making CDMC a world known competition in the area of AI × Cybersecurity.

References

1. Aktayeva, A., Niyazova, R., Muradilova, G., Makatov, Y., Kusainova, U.: Cognitive computing cybersecurity: social network analysis. In: Sukhomlin, V., Zubareva, E. (eds.) Convergent 2018. CCIS, vol. 1140, pp. 28–43. Springer, Cham (2020). https://doi.org/10.1007/978-3-030-37436-5_3

2. Sokolova, M., Lapalme, G.: A systematic analysis of performance measures for classification tasks. Inf. Process. Manag. **45**(4), 427–437 (2009)

3. Fernández, A., García, S., Galar, M., Prati, R.C., Krawczyk, B., Herrera, F.: Learning from Imbalanced Data Sets. Springer, Cham (2018). https://doi.org/10.1007/978-3-319-98074-4

4. Pang, S., et al.: CDMC Cybersecurity Dataset Repository. International Cyber Security Data-mining Society (ICSDS), hosted by the Federation University Australia, School of Engineering Information Technology and Physical Sciences (2020). http://archive.csmining.org/

5. Pang, S., Huang, Y.: Sensor array data for autonomous vehicle incident detection. In: The 10th International Cyber Security Data Mining Competition (CDMC 2019). Unitec Institute of Technology, New Zealand (2019)

6. https://projects.tuni.fi/uploads/2019/03/8506038d-erf2019_trinity_cybersecurity_robotics_workshop_slides.pdf

Class-Balanced Loss for Scene Text Detection

Randong Huang[1,2] and Bo Xu[1(✉)]

[1] Institute of Automation, Chinese Academy of Sciences, Beijing, China
{huangrandong2015,xubo}@ia.ac.cn
[2] University of Chinese Academy of Sciences, Beijing, China

Abstract. To address class imbalance issue in scene text detection, we propose two novel loss functions, namely Class-Balanced Self Adaption Loss (CBSAL) and Class-Balanced First Power Loss (CBFPL). Specifically, CBSAL reshapes Cross Entropy (CE) loss to down-weight easy negatives and up-weight positives. However, CBSAL ignores gradient imbalance that CE gives positives and negatives different gradients. Since text detectors need to identify text and background simultaneously, positives and negatives have same importance and should possess equivalent gradients. Thus CBFPL provides equal but opposite gradients for positives and negatives to eliminate this gradient imbalance. Then, CBFPL abandons easy negatives and makes their gradients zero to handle class imbalance. Both CBSAL and CBFPL can focus training on positives and hard negatives. Experimental results show that on the basis of CBSAL and CBFPL, the efficient and accurate scene text detector (EAST) can achieve higher F-score on ICDAR2015, MSRA-TD500 and CASIA-10K datasets.

Keywords: Scene text detection · Class imbalance · Gradient imbalance.

1 Introduction

Scene text detection, which aims to detect text regions in natural scene images, has become increasingly popular, as a result of its great value in real-world applications such as image retrieval, product search, scene understanding and automatic driving.

Due to the fast progress of object detection in recent years, some state-of-the-art object detection frameworks, such as SSD [1] and Faster R-CNN [2], have been employed to localize horizontal and multi-oriented scene text. These text detection methods, which treat words or text lines as general objects and are modified from object detection approaches, can be divided into two categories: (1) Indirect regression based text detection techniques [3–8], which adopt object detectors like Faster R-CNN and SSD to first create region proposals and then regress offset values from these proposals to precise text boxes. (2) Direct

© Springer Nature Switzerland AG 2020
H. Yang et al. (Eds.): ICONIP 2020, LNCS 12533, pp. 246–256, 2020.
https://doi.org/10.1007/978-3-030-63833-7_21

regression based text detection methods [9,10], which adapt DenseBox [11] for extracting scene text. These methods generate a score map and the corresponding offsets. The score map is used to distinguish between text and background. The offsets represent regression distances from one reference pixel to its ground truth. Besides foregoing regression based text detection works, there are also instance segmentation based text detection approaches [12–14], which employ FCIS [15] and Mask R-CNN [16] for text detection task. These approaches first detect individual text instances from images. Afterwards, a minimal area rectangle algorithm is applied to gaining the oriented bounding boxes of text instances as the final detection results.

Just as object detection, class imbalance exists in scene text detection as well. The essential reason of class imbalance is that the quantity of background pixels on the image is far more than that of text pixels. To solve this imbalance, many effective techniques such as Class-Balanced Cross Entropy (CBCE) loss [10], Focal Loss [17] and OHEM [18] have been put forward. Such balancing sample approaches can improve text detectors' performance.

But the aforementioned techniques also own their shortcomings. OHEM chooses N hard samples to train the network. It is boresome that researchers need to spend plenty of time in finding the optimal value of N. Focal Loss downweights the loss of easy samples, while the loss of positives is also restrained at the same time. Focal loss does not differentiate between positive/negative samples. A balancing factor between positives and negatives is offered by CBCE to down-weight the loss assigned to negatives. But CBCE does not consider the distinction between easy and hard negative samples. Hence, CBCE-based text detectors only can obtain sub-optimal performance.

In this paper, we propose two novel loss functions called CBSAL and CBFPL, both of which can act as more valid substitutes to previous methods for coping with class imbalance. The proposed CBSAL considers the distinction between easy/hard samples as well as between positive/negative samples. CBSAL designs a scaling factor as negatives weight to restrain the loss of easy negatives. Meanwhile, the loss of positives is up-weighted by CBSAL to prevent the vast number of negatives from overwhelming the training of positives.

However, the gradient imbalance from CE is not considered by CBSAL which is modified from CE. The gradient imbalance of CE is illustrated in Fig. 3. When sample probability is not equal to 0.5, the gradient of CE w.r.t. positive sample probability has unequal absolute value with that of CE w.r.t. negative sample probability. As scene text detectors not only need to recall text but also remove background regions, positive and negative samples own equal significance for text detectors and should be given equivalent gradients. Therefore, we propose CBFPL to generate equal but opposite gradients for positive and negative samples, which elegantly cancel the gradient imbalance. And then, CBFPL excludes easy negative samples and offers them zero gradients to address class imbalance during training of text detectors. As a result, the proposed CBSAL and CBFPL can make the training of text detectors concentrated on positives and hard negatives. For demonstrating the effectiveness of our proposed loss functions, we

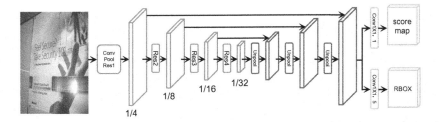

Fig. 1. Reimplemented version of EAST Pipeline. Resnet50 [19] is used as feature extractor. EAST adopts the idea from U-shape [20] to produce the merged feature maps, which are connected to two $conv_{1 \times 1}$ layers to output score map and RBOX.

replace CBCE in EAST [10] with CBSAL and CBFPL, respectively. Experimental results show that two proposed loss functions enable EAST to obtain higher performance on ICDAR2015 [21], MSRA-TD500 [22] and CASIA-10K [23] datasets (Fig. 1).

The contributions of this work are three-fold:

(1) To deal with class imbalance in scene text detection, we first propose one newfangled loss function, named CBSAL. CBSAL can effectively recall challenging text and suppress false positives.
(2) We propose another loss, namely CBFPL, to simultaneously overcome gradient imbalance of CE and class imbalance. CBFPL is a simpler and more effective approach than CBSAL.
(3) We experimentally prove that our CBSAL and CBFPL can boost the performance of text detectors. For testifying the superiority of CBSAL and CBFPL, Focal Loss and OHEM are also adopted for score map [10] in EAST, respectively. The proposed CBSAL and CBFPL significantly outperform previous techniques for addressing class imbalance.

2   Methods

CBSAL and CBFPL are proposed to address foreground-background class imbalance problem encountered during training of scene text detectors. This work describes CBSAL and CBFPL starting from CE for binary classification which corresponds to text/background classification:

$$CE(\hat{Y}, Y^*) = -Y^* \log \hat{Y} - (1 - Y^*) \log(1 - \hat{Y}) \qquad (1)$$

Where $Y^* \in \{1, 0\}$ represents the ground-truth, and $\hat{Y} \in [0, 1]$ indicates the text detector's prediction probability for the class with label $Y^* = 1$.

2.1   Class-Balanced Self Adaption Loss

Due to the drawbacks of CBCE and Focal Loss, we propose a novel CBSAL which balances the importance between easy and hard samples and takes into account the distinction between positive and negative samples. CBSAL down-weights the loss of easy negative samples and up-weights the loss of positive samples to make training focused on positives and hard negatives.

More formally, we propose to add a scaling factor $s(1 - \beta)^{[a(1-\hat{Y})]}$ to CE loss as negatives weight, with two hyperparameters $s > 0$ and $a > 0$. Then, an exponential factor $ge^{1-\hat{Y}}$ is added to CE loss as positives weight, with one tunable parameter $g > 0$. CBSAL is defined as:

$$CBSAL(\hat{Y}, Y^*) = -PY^* \log \hat{Y} - N(1 - Y^*) \log(1 - \hat{Y}) \tag{2}$$

$$P = ge^{1-\hat{Y}} \qquad N = s(1 - \beta)^{[a(1-\hat{Y})]} \tag{3}$$

$$\beta = 1 - \frac{\sum_{y^* \in Y^*} y^*}{|Y^*|} \tag{4}$$

Where β is the same as that of CBCE.

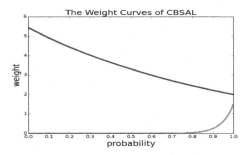

Fig. 2. The weight curves of CBSAL. Red curve indicates the weight of negatives and green curve represents the weight of positives. (Color figure online)

The weight curves of CBSAL is illustrated in Fig. 2. Red curve is the weight of negatives and green curve represents the weight of positives. As the negative sample probability increases, the weight of negatives is increased gradually. Extremely tiny weight values are assigned to easy negative samples. So, these negatives own very small gradients. However, hard negative samples obtain much larger weight values, which can focus training on hard negatives. Besides, the weight values of positives are up-weighted to further alleviate class imbalance problem, because the number of positives is far less than that of negatives. To recall more challenging text in natural scene, the weight values of easy positives are smaller than that of hard positives, which makes training pay more attention on hard positives.

2.2 Class-Balanced First Power Loss

Based on CE, aforementioned loss functions such as CBCE, Focal Loss and CBSAL, design different sample balancing measures to down-weight or up-weight the loss and gradients of samples. These loss functions do not consider the gradient imbalance from CE. The gradient of CE w.r.t. positive sample probability is calculated as:

$$\frac{\partial CE}{\partial \hat{Y}_{pos}} = -\frac{1}{\hat{Y}} \tag{5}$$

The gradient of CE w.r.t. negative sample probability is calculated as:

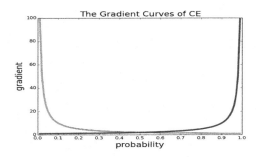

Fig. 3. The gradient curves of CE. Red curve indicates the gradient of CE w.r.t. positive sample probability and green curve respresents gradient of CE w.r.t. negative sample probability. (Color figure online)

$$\frac{\partial CE}{\partial \hat{Y}_{neg}} = \frac{1}{1 - \hat{Y}} \tag{6}$$

We ignore their plus-minus sign and the curves of two gradients can be illustrated in Fig. 3. If the probability is less than 0.5, the gradient of CE w.r.t. positive sample probability is greater than that of CE w.r.t. negative sample probability. However, when the probability exceed 0.5, the gradient of CE w.r.t. positive sample probability is less than that of CE w.r.t. negative sample probability. Such a difference leads to gradient imbalance.

Because scene text detectors not only need to identify text but also recognize background, positives and negatives have equal importance for text detectors. Therefore, positives and negatives should have equal but opposite gradients. As to the class imbalance problem, we propose to get rid of easy negative samples and make their gradients zero to address it. As a result, we define CBFPL as:

$$CBFPL(\hat{Y}, Y^*) = Y^*(1 - \hat{Y}) + H(1 - Y^*)\hat{Y} \tag{7}$$

$$H = \begin{cases} 1, & neg\_pred > hard\_neg\_prob \\ 0, & otherwise \end{cases} \tag{8}$$

Where $neg\_pred = (1 - Y^*)\hat{Y}$ is the prediction for negatives. $hard\_neg\_prob$ is a hyperparameter, whose value is in the range of $[0, 1]$.

If the prediction of negatives neg_pred is greater than $hard\_neg\_prob$, these negatives are hard and other negatives are easy. The gradient of CBFPL w.r.t.

easy negative sample probability is zero. The gradient of CBFPL w.r.t. positive sample probability is calculated as:

$$\frac{\partial CBFPL}{\partial \hat{Y}_{pos}} = -1 \qquad (9)$$

The gradient of CBFPL w.r.t. hard negative sample probability is calculated as:

$$\frac{\partial CBFPL}{\partial \hat{Y}_{neg}} = 1 \qquad (10)$$

The two gradients have same absolute value, which endows positives and negatives with equal significance. In consequence, CBFPL can focus training on positives and hard negatives.

Table 1. Results on ICDAR2015

Algorithm	Recall	Precision	F-score
StradVision1 [21]	0.4627	0.5339	0.4957
StradVision2 [21]	0.3674	0.7746	0.4984
Zhang et al. [24]	0.4309	0.7081	0.5358
Tian et al. [25]	0.5156	0.7422	0.6085
Yao et al. [26]	0.5869	0.7226	0.6477
SegLink [5]	0.768	0.731	0.75
RRPN [27]	0.732	0.822	0.774
EAST [10]	0.7347	0.8357	0.7820
DDR [9]	**0.800**	0.820	0.810
EAST+CBCE	0.7665	0.8032	0.7844
EAST+OHEM	0.7256	**0.8834**	0.7967
EAST+Focal Loss	0.7675	0.8425	0.8032
EAST+CE	0.7612	0.8621	0.8085
EAST+CBSAL	0.7805	0.8743	0.8247
EAST+CBFPL	0.7935	0.8678	**0.829**

3 Experiments

To compare our CBSAL and CBFPL with CBCE, we conduct quantitative and qualitative experiments on three public datasets: ICDAR2015, MSRA-TD500 and CASIA-10K.

3.1 Datasets

ICDAR2015 dataset appears in Challenge 4 of ICDAR 2015 Robust Reading Competition. It includes 500 testing images and 1000 training images. The 229 training images from ICDAR2013 [28] are also used as the training data.

Table 2. Results on MSRA-TD500

Algorithm	Recall	Precision	F-score
TD-ICDAR [22]	0.52	0.53	0.50
TD-Mixture [22]	0.63	0.63	0.60
Yin et al. [29]	0.63	0.81	0.71
Zhang et al. [24]	0.67	0.83	0.74
DDR [9]	0.700	0.770	0.74
Yao et al. [26]	**0.7531**	0.7651	0.7591
EAST [10]	0.6743	0.8728	0.7608
SegLink [5]	0.700	0.860	0.770
EAST+CE	0.7096	0.8343	0.7669
EAST+OHEM	0.6873	**0.8734**	0.7692
EAST+Focal Loss	0.6976	0.8602	0.7704
EAST+CBCE	0.7062	0.8527	0.7726
EAST+CBSAL	0.7354	0.8717	0.7978
EAST+CBFPL	0.7457	0.8697	**0.803**

MSRA-TD500 dataset consists of 300 training images and 200 testing images. Besides text in English, it also contains text in Chinese. The 400 images from HUST-TR400 dataset [30] are also included as the training data.

CASIA-10K is a Chinese scene text dataset provided by Institute of Automation of Chinese Academy of Sciences. This benchmark comprises 7000 training images and 3000 testing images.

Table 3. Results on CASIA-10. * means multi-scale testing

Algorithm	Recall	Precision	F-score
EAST [23]	0.5327	0.7771	0.6321
SegLink [23]	0.6967	0.7275	0.7118
MOML* [23]	**0.7048**	0.8128	**0.7550**
EAST+OHEM	0.5792	**0.8632**	0.6933
EAST+CBCE	0.6115	0.816	0.6991
EAST+Focal Loss	0.6553	0.7691	0.7077
EAST+CE	0.6249	0.8384	0.7161
EAST+CBSAL	0.655	0.8218	0.729
EAST+CBFPL	0.6674	0.8237	0.7374

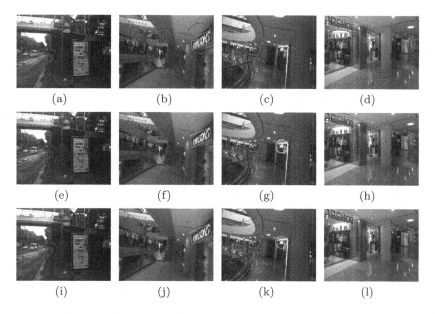

Fig. 4. Detection results from **EAST+CBCE**, **EAST+CBSAL** and **EAST+CBFPL**: (a)–(d) from **EAST+CBCE (Baseline)**, (e)–(h) from **EAST+CBSAL** and (i)–(l) from **EAST+CBFPL**. The boxes in red ellipses represents wrong results while other boxes are correct. (Color figure online)

3.2   Quantitative Results

As shown in Table 1, Table 2 and Table 3, Our reimplemented version of EAST is **EAST+CBCE**, which applies CBCE to score map [10]. **EAST+OHEM** replaces CBCE with CE and OHEM is only applied to negatives. For every image, we choose N hard negatives and all positives to train the score map. Focal Loss is used to replace CBCE to become **EAST+Focal Loss**. Note that our all experimental results are based on single-scale testing.

EAST+CBSAL takes advantage of CBSAL as a more effective alternative to CBCE of **EAST+CBCE**. Our **EAST+CBSAL** reaches a F-score of 0.8247 on ICDAR2015, which is 4.03% higher than **EAST+CBCE**. Meanwhile, CBSAL can also improve EAST by 2.52% on MSRA-TD500 (0.7978 vs. 0.7726) and 2.99% on CASIA-10K (0.729 vs. 0.6991).

Compared with OHEM and Focal Loss, the proposed **EAST+CBSAL** transcends **EAST+OHEM** by 2.8% and **EAST+Focal Loss** by 2.15% on ICDAR2015 benchmark. **EAST+CBSAL** is in excess of **EAST+OHEM** by 2.86% and **EAST+Focal Loss** by 2.74% on MSRA-TD500. In CASIA-10K dataset, **EAST+CBSAL** surpasses **EAST+OHEM** by 3.57% and **EAST+Focal Loss** by 2.13%.

Finally, we also compare proposed CBFPL with CBSAL. On ICDAR2015, **EAST+CBFPL** can achieve a F-score of 0.829, outperforming **EAST+**

CBSAL by 0.43%. Compared with **EAST+CBSAL**, **EAST+CBFPL** brings in 0.52% F-score of revenue on MSRA-TD500 and obtains an increasement of 0.84% in F-score on CASIA-10K. Both CBSAL and CBFPL obtain competitive results, which are slightly lower than the highest result on CASIA-10K. Our CBSAL and CBFPL outperform previous methods by a large margin on ICDAR2015 and MSRA-TD500 datasets. And CBFPL slightly surpasses CBSAL. These experiments validate the efficacy of CBSAL and CBFPL, which can promote the F-score of EAST prominently.

3.3 Qualitative Results

Some detection examples from **EAST+CBCE**, **EAST+CBSAL** and **EAST+CBFPL** are presented in Fig. 4. Compared with CBCE loss, CBSAL and CBFPL possess four advantages: (1) **Eliminate text-like patterns**: A text-like pattern, which is a background region very similar to text, is wrongly recognized as text by Fig. 4(a). However, Fig. 4(e) and Fig. 4(i) can get rid of this text-like pattern, because CBSAL and CBFPL make training focused on hard negatives. (2) **Localize text better**: Fig. 4(f) and Fig. 4(j) can obtain better text localization than Fig. 4(b). Therefore, we believe that training classification branch better can facilitate the learning of localization branch since they share features in EAST. (3) **Recall more challenging text**: Fig. 4(g) and Fig. 4(k) capture more challenging text regions than Fig. 4(c) for the reason that CBSAL and CBFPL also focus training on positives. (4) **Suppress background**: Fig. 4(d) contains a obvious background region which is removed in Fig. 4(h) and Fig. 4(l). This proves that CBSAL and CBFPL have better discriminant ability between text and background.

4 Conclusion

In this paper, we conduct research on class imbalance in scene text detection and analyze the deficiencies of previous methods such as OHEM, CBCE and Focal Loss. To address this problem, we propose two class-balanced loss functions: CBSAL and CBFPL. CBSAL applies a scaling factor and an exponential factor to CE in order to down-weight easy negatives and up-weight positives. CBFPL provides positives and negatives with equal but opposite gradients to solve gradient imbalance from CE. CBFPL gets rid of easy negatives and offer them zero gradients to handle class imbalance. In consequence, the proposed CBSAL and CBFPL can focus training of text detectors on positives and hard negatives. The experiments on three standard benchmarks demonstrate the effectiveness of CBSAL and CBFPL.

Acknowledgement. This work was supported by the Major Project for New Generation of AI (Grant No. 2018AAA0100400).

References

1. Liu, W., et al.: SSD: single shot multibox detector. In: Leibe, B., Matas, J., Sebe, N., Welling, M. (eds.) ECCV 2016. LNCS, vol. 9905, pp. 21–37. Springer, Cham (2016). https://doi.org/10.1007/978-3-319-46448-0_2

2. Ren, S., He, K., Girshick, R., Sun, J.: Faster R-CNN: towards real-time object detection with region proposal networks. In: Advances in Neural Information Processing Systems, pp. 91–99 (2015)

3. Jiang, Y., et al.: R2CNN: rotational region CNN for orientation robust scene text detection. arXiv preprint arXiv:1706.09579 (2017)

4. Liao, M., Shi, B., Bai, X.: Textboxes++: a single-shot oriented scene text detector. IEEE Trans. Image Process. **27**(8), 3676–3690 (2018)

5. Shi, B., Bai, X., Belongie, S.: Detecting oriented text in natural images by linking segments. arXiv preprint arXiv:1703.06520 (2017)

6. Lyu, P., Yao, C., Wu, W., Yan, S., Bai, X.: Multi-oriented scene text detection via corner localization and region segmentation. In: Proceedings of the IEEE Conference on Computer Vision and Pattern Recognition, pp. 7553–7563 (2018)

7. Liao, M., Zhu, Z., Shi, B., Xia, G., Bai, X.: Rotation-sensitive regression for oriented scene text detection. In: Proceedings of the IEEE Conference on Computer Vision and Pattern Recognition, pp. 5909–5918 (2018)

8. Liu, Y., Jin, L.: Deep matching prior network: toward tighter multi-oriented text detection. In: Proceedings of the CVPR, pp. 3454–3461 (2017)

9. He, W., Zhang, X.-Y., Yin, F., Liu, C.-L.: Deep direct regression for multi-oriented scene text detection. arXiv preprint arXiv:1703.08289 (2017)

10. Zhou, X., et al.: EAST: an efficient and accurate scene text detector. In: Proceedings of the CVPR, pp. 2642–2651 (2017)

11. Huang, L., Yang, Y., Deng, Y., Yu, Y.: Densebox: unifying landmark localization with end to end object detection. arXiv preprint arXiv:1509.04874 (2015)

12. Yang, Q., Cheng, M., Zhou, W., Chen, Y., Qiu, M., Lin, W.: Inceptext: a new inception-text module with deformable PSROI pooling for multi-oriented scene text detection. arXiv preprint arXiv:1805.01167 (2018)

13. Liu, J., Liu, X., Sheng, J., Liang, D., Li, X., Liu, Q.: Pyramid mask text detector. arXiv preprint arXiv:1903.11800 (2019)

14. Xie, E., Zang, Y., Shao, S., Gang, Y., Yao, C., Li, G.: Scene text detection with supervised pyramid context network. In: Proceedings of the AAAI Conference on Artificial Intelligence, vol. 33, pp. 9038–9045 (2019)

15. Li, Y., Qi, H., Dai, J., Ji, X., Wei, Y.: Fully convolutional instance-aware semantic segmentation. arXiv preprint arXiv:1611.07709 (2016)

16. He, K., Gkioxari, G., Dollár, P., Girshick, R.: Mask R-CNN. In: 2017 IEEE International Conference on Computer Vision (ICCV), pp. 2980–2988. IEEE (2017)

17. Lin, T.-Y., Goyal, P., Girshick, R., He, K., Dollár, P.: Focal loss for dense object detection. arXiv preprint arXiv:1708.02002 (2017)

18. Shrivastava, A., Gupta, A., Girshick, R.: Training region-based object detectors with online hard example mining. In: Proceedings of the IEEE Conference on Computer Vision and Pattern Recognition, pp. 761–769 (2016)

19. He, K., Zhang, X., Ren, S., Sun, J.: Deep residual learning for image recognition. In: Proceedings of the IEEE Conference on Computer Vision and Pattern Recognition, pp. 770–778 (2016)

20. Ronneberger, O., Fischer, P., Brox, T.: U-Net: convolutional networks for biomedical image segmentation. In: Navab, N., Hornegger, J., Wells, W.M., Frangi, A.F. (eds.) MICCAI 2015. LNCS, vol. 9351, pp. 234–241. Springer, Cham (2015). https://doi.org/10.1007/978-3-319-24574-4_28
21. Karatzas, D., et al.: ICDAR 2015 competition on robust reading. In: 2015 13th International Conference on Document Analysis and Recognition (ICDAR), pp. 1156–1160. IEEE (2015)
22. Tu, Z., Ma, Y., Liu, W., Bai, X., Yao, C.: Detecting texts of arbitrary orientations in natural images. In: 2012 IEEE Conference on Computer Vision and Pattern Recognition, pp. 1083–1090. IEEE (2012)
23. He, W., Zhang, X.-Y., Yin, F., Liu, C.-L.: Multi-oriented and multi-lingual scene text detection with direct regression. IEEE Trans. Image Process. 27(11), 5406–5419 (2018)
24. Zhang, Z., Zhang, C., Shen, W., Yao, C., Liu, W., Bai, X.: Multi-oriented text detection with fully convolutional networks. In: Proceedings of the IEEE Conference on Computer Vision and Pattern Recognition, pp. 4159–4167 (2016)
25. Tian, Z., Huang, W., He, T., He, P., Qiao, Y.: Detecting text in natural image with connectionist text proposal network. In: Leibe, B., Matas, J., Sebe, N., Welling, M. (eds.) ECCV 2016. LNCS, vol. 9912, pp. 56–72. Springer, Cham (2016). https://doi.org/10.1007/978-3-319-46484-8_4
26. Yao, C., Bai, X., Sang, N., Zhou, X., Zhou, S., Cao, Z.: Scene text detection via holistic, multi-channel prediction. arXiv preprint arXiv:1606.09002 (2016)
27. Ma, J., et al.: Arbitrary-oriented scene text detection via rotation proposals. IEEE Trans. Multimed. 20, 3111–3122 (2018)
28. Karatzas, D., et al.: ICDAR 2013 robust reading competition. In: 2013 12th International Conference on Document Analysis and Recognition (ICDAR), pp. 1484–1493. IEEE (2013)
29. Yin, X.-C., Pei, W.-Y., Zhang, J., Hao, H.-W.: Multi-orientation scene text detection with adaptive clustering. IEEE Trans. Pattern Anal. Mach. Intell. 9, 1930–1937 (2015)
30. Yao, C., Bai, X., Liu, W.: A unified framework for multioriented text detection and recognition. IEEE Trans. Image Process. 23(11), 4737–4749 (2014)

Coordinated Behavior for Sequential Cooperative Task Using Two-Stage Reward Assignment with Decay

Yuki Miyashita[1,2]([⊠]) [iD] and Toshiharu Sugawara[1] [iD]

[1] Computer Science and Engineering, Waseda University, Tokyo 1698555, Japan
{y.miyashita,sugawara}@isl.cs.waseda.ac.jp
[2] Shimizu Corporation, Tokyo 1040031, Japan

Abstract. Recently, multi-agent deep reinforcement learning (MADRL) has been studied to learn actions to achieve complicated tasks and generate their coordination structure. The reward assignment in MADRL is a crucial factor to guide and produce both their behaviors for their own tasks and coordinated behaviors by agents' individual learning. However, it has not been sufficiently clarified the reward assignment in MADRL's effect on learned coordinated behavior. To address this issue, using the sequential tasks, *coordinated delivery and execution problem with expiration time*, we analyze the effect of various ratios of the reward given for the task that agent is responsible for to the reward given for the whole task. Then, we propose a two-stage reward assignment with decay to learn the actions for tasks that the agent is responsible for and coordinated actions for facilitating other agents' tasks. We experimentally showed that the proposed method enabled agents to learn both actions in a balanced manner, so they could realize effective coordination, by reducing the number of tasks that were ignored by other agents. We also analyzed the mechanism behind the emergence of different coordinated behaviors.

Keywords: Control and decision theory · Multi-agent deep reinforcement learning · Coordination · Cooperation

1   Introduction

Although one central issue in the study of multi-agent systems is to achieve autonomous coordinated/cooperative behaviors to perform tasks that cannot be solved by a single agent, it is quite difficult to describe these behaviors because their coordination results are subtly affected by many factors such as other agents' actions, environmental characteristics, and task structures. Reinforcement learning (RL) is one method that is expected to identify better coordination for multi-agent systems, but this is challenging due to the sophistication of cooperation and uncertainty in mutual learning. Actually, because all agents individually learn their behaviors simultaneously, the learning results through the training process become meaningless due to the behavioral changes of other

© Springer Nature Switzerland AG 2020
H. Yang et al. (Eds.): ICONIP 2020, LNCS 12533, pp. 257–269, 2020.
https://doi.org/10.1007/978-3-030-63833-7_22

agents by learning. To realize the learning for coordination, a sort of mechanism has to guide the agents to learn the consistent behaviors. Particularly, appropriate design of reward assignment is a crucial issue [1] to encourage agents to learn not only how to do the tasks they are responsible for but also when and where to do them to improve the entire performance.

Recent and ongoing techniques using deep reinforcement learning (DRL) for a single agent have produced several successful results in many application/research fields such as robotics [4,8] and video games [5,6]. Even in multi-agent systems, various approaches to learn coordinated behavior using multi-agent DRL (MADRL) in which each agent has its own deep neural network have been proposed to address non-stationary problems [2,7]. For example, Palmer *et al.* [7] proposed an extension of MADRL called *lenient learning* in which agents possess the temperature value for each state-action pair and decay the value without using the outdated pairs. Foerster *et al.* [3] also proposed an extension of MADRL called COMA in which a centralized critic estimates Q-values for all agents, and the actors for individual agents optimize the agent's policy by using its local observation. In general, DRL requires careful engineering for the reward functions to explore high dimensional state-action spaces. The reward functions for the MADRL are more complicated and difficult to design because the task completion is the result of cooperation and coordination of individual subtasks, and the functions must be able to guide so that agents can learn such coordinated activities. However, the reward assignment method, i.e., how to give the rewards for each subtask completion affected the coordination behaviors generated by MADRL.

Therefore, we first examine how the reward assignment in a MADRL affected the learning of agents' cooperative behaviors using the example task called, which is an abstraction of our target application in a construction site. In this problem, each task consists of two subtasks that should be executed sequentially by two types of agents within a limited time, so these agents have to learn how to coordinate with each other. We experimentally found that different behaviors are generated depending on how rewards are assigned to agents for their responsible subtasks; for example, (1) some agents could not learn behaviors, so their contributions are unbalanced, (2) the learning was very slow while they coordinated to not waste others' efforts, or (3) agents could do their responsible subtasks but ignored coordination with others. We think that these is a trade-off between learning speed, learnability, and the quality of behaviors. Thus, we propose a reward scheme called *two-stage reward assignment with decay* to guide the learning of such coordinated behaviors of agents to ease these problems. We also modified the extended experience replay for the proposed reward assignment method. Our experimental results indicate that the proposed reward scheme enabled agents to produce balanced behaviors for both doing their own tasks and facilitating the subsequent subtasks done by other agents.

2   Model and Problem

2.1   Models of Agents and Environment

In this work, we consider the multi-agent cooperation problem, called *coordinated delivery and execution problem with expiration time*. In this problem, there are two types of agents in an environment, and a task is done by the sequential collaborations of two subtasks. The first type of agent, which is called a *delivery agent*, has the role of picking up a material at the storage area, delivering it, and placing it in a location in the installation area. The second type of agent, called an *execution agent*, has the role of moving to the location where the material has been placed, and executing the finishing task using the delivered material as soon as possible. Delivered materials have an expiration time and become unusable after that time (and then are removed). Thus, an execution agent has to execute the finishing task before the material's expiration after it has been placed; otherwise, the effort of the delivery agents is wasted. The agents in an environment repeat this sequential collaboration of joint tasks until all tasks have been completed in all required areas.

An example of our problem environment is shown in Fig. 1. The environment is a lattice consisting of $N \times N$ cells, where a hollow square is a delivery agent, a hollow pentagon is an execution agent, the group of green cells are the storage areas where a delivery agent can pick up a material, the gray cells are the installation areas where a delivery agent holding a material can put it down, and the yellow cells are the executable cells where one of the delivery agents placed a material, but the execution agents have not completed the finishing task. We represent the remaining time to the expiration of the delivered materials by gradually changing the color of executable cells from yellow to gray as the expiration time approaches.

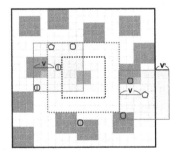

Fig. 1. Environmental state and agent's views (Color figure online)

2.2   Problem Formulation

We introduce discrete time $t \geq 0$. Our problem is described by tuple $\langle I, N, m, E, \{S_i\}_{i \in I}, \{A_i\}_{i \in I} \rangle$, where $I = \{1, \cdots, n\}$ is the set of n agents including two types of agent, N is the side length of the lattice environment, and m is the number of installation cells. $E(\ni e)$ is the set of all possible entire states of the environment including the states of cells and all agents. S_i is the set of the states that can be observed by agent $i \in I$; each agent can observe a limited local area whose center is itself. Thus, its element $s_{i,t} \in S_i$ can be considered as a subset of the entire state e_t at time t $(s_{i,t} \subset e_t \in E)$. Instead, we assume that agents can observe their local areas accurately, so $s_{i,t}$ at t is correct. We also define the set of joint actions $\mathcal{A} = A_1 \times \cdots \times A_n \ni a_t = (a_{1,t}, \ldots, a_{n,t})$, where A_i ($= A = \{up, right, down, left, work\}$) is the set of all possible actions of $\forall i \in I$ regardless of the type of agent, and agents can take one action each time. If more than two agents attempt to move onto the same cell, one of them can successfully move onto the cell, and the other remains in its original position. Agent i cannot communicate with other agents and, of course, cannot know others' next actions before they are performed. The content of action $work$ depends on the agent's type and its location and is described in the next section.

Every time agents take joint actions $a_t \in \mathcal{A}$ in e_t, they may receive $reward$ $r_i(e_t, a_t)$, and then, the environmental state transits to e_{t+1}. The rewards given to delivery agents and execution agents are different due to their different roles, and this will be explained in Sect. 3.2. Because we focus on applying MADRL to our problem, the agents individually learn their own action-value pairs and the associated $policies$ $\pi_i : S_i \to A_i$ by using their own DQNs to increase the rewards that are the results of their own responsible subtasks and coordinated behaviors for the tasks of our problems. Therefore, we are curious if the delivery agents can learn cooperative activities for increasing the success rate of the execution agents' tasks.

2.3   Agents' Behaviors

The storage area consisting of four cells is at the center of the environment, as shown in Fig. 1. Initially (at time $t = 0$), agents $I = \{1, \cdots, n\}$ are scattered within the area surrounded by black dotted lines, and the K installation areas consisting of 3×3 cells are also scattered randomly outside the area specified by the purple dotted lines. Then, all agents take the following steps concurrently.

First, agent $\forall i \in I$ chooses action $a_{i,t}$ in environment e_t on the basis of i's policy π_i at $\forall t$ (so $a_{i,t} = \pi_i(s_{i,t}) \in A$) and performs it in a concurrent manner. Thus, i moves to one of the neighbor cells if the action is not $work$. However, agent's action $work$ is different depending on its types.

Action $work$ by delivery agent: The role of delivery agents is to deliver the materials in the storage area to an installation area for the next subtasks done by execution agents. When i arrives at the storage area, i picks up one material ψ_i automatically and heads toward one of the installation cells (a delivery agent with a material is represented as a hollow square whose inside is green in Fig. 1). Then, if i takes action $work$ on an installation cell at t_d, i puts down ψ_i on the current installation cell. This material is denoted by ψ_{i,t_d}. The installation cell changes to an executable cell, which is represented as a yellow cell in Fig. 1. Agent i receives the reward for delivery r_{d,t_d}; we will describe this reward in detail in Sect. 3.2. Let positive integer p be the fixed expiration time of the delivered material on a cell. Thus, one of the execution agents has to do the finishing task for this material before $t_d + p$; otherwise, it will be removed.

Action $work$ by execution agent: The role of execution agents is to execute the finishing task using the materials delivered by delivery agents before the expiration time. Thus, execution agent j performs one action in $A_j \setminus \{work\}$ to find the executable cell where ψ_i has been placed and, by taking action *work*, executes the finishing task immediately (at least, before the expiration). If it is done before the expiration time, the execution cell changes to an executed cell, which is represented as a white cell in Fig. 1. Then, j receives reward r_{e,t_e} when the finishing task is done at t_e, and an additional reward is given to delivery agent i that delivered ψ_{i,t_d}; the details on the reward scheme are described in Sect. 3.2.

These joint actions of all agents repeat until all the installation areas are filled with executed cells or time t exceeds *epoch length H* (where H is a positive integer); then the epoch ends. Then, another epoch will start after the environment is initialized until F_e epochs have been done, where $F_e > 0$ is also an integer. Note that we assume the materials do not run out at any green cell.

3 Learning Methods

3.1 Deep Q-Network with Local Belief

DQN is a reinforcement learning method in which the action-value function (i.e., Q-function) and the associated policy π are learned using a deep neural network. We use the decentralized concurrent learning approach, meaning that individual agents have their own networks for learning their Q-values. Each deep neural network is specified by its associated parameters θ whose values are updated through the agent's experience. At time t, to obtain the approximated values of the optimal Q-values from the network, parameters $\theta_{i,t}$ of the network of agent i at t are updated to reduce the mean squared loss function $L_{i,t}(\theta_{i,t})$, which is defined as

$$L_{i,t}(\theta_{i,t}) = \mathbb{E}_{(s_i,a_i,r_i,s_i')}[(r_i + \gamma \max_{a_i'} Q_i(s_i', \arg\min_{a_i'} Q_i(s_i', a_i'; \theta_{i,t}); \theta_{i,t}^-)$$
$$- Q_i(s_i, a_i; \theta_{i,t}))^2],$$

where $\gamma \in [0, 1)$ is the discount factor for expected future reward. Note that we use the double DQN (DDQN) [10], i.e., the target network parameters $\theta_{i,t}^-$ are periodically copied from the main Q-network parameters $\theta_{i,t}$ every T_c epochs. Then, actions are selected using the main Q-network, but its parameters $\theta_{i,t}$ are updated using the Q-value based on the target network parameters $\theta_{i,t}^-$.

We propose a policy based on the combination of observations and additional information fed to DQNs. Because we apply MADRL to this problem, each agent decides its next action by using the policy associated with the learned Q-values from its own DQN. Usually, the policy is decided only on the basis of the observation $s_{i,t}$. By defining the *view* of i at time t, $v_{i,t}$, as the aggregation of the observed state $s_{i,t}$ and part of i's local belief, we extend the domain of Q_i and π_i by

$$Q_i : \mathcal{V}_i \times A_i \longrightarrow \mathbb{R}, \text{ and } \pi_i : \mathcal{V}_i \longrightarrow A_i,$$

where $\mathcal{V}$ is the set of observed states with the local belief in i. The details of $v_{i,t}$ are described in Sect. 3.3.

3.2   Two-Stage Reward Assignment and Experience Replay

The design of the reward assignment is an important part of reinforcement learning because in multi-agent systems, agents' rewards depend on not only their individual actions but also others' actions. In particular, in our problem, the desired results come from the sequential subtasks, so the efforts of the first agent (i.e., delivery agent) may become wasted due to the inappropriate behavior of the second agent (i.e., execution agent). Therefore, we should consider how to assign rewards to individual agents and when to do this. We then would like to clarify the learned cooperative behaviors and their differences in cooperation/coordination structure depending on the reward assignment.

For this purpose, we introduce a reward scheme, *two-stage reward assignment (with decay)* into our problem. In this reward scheme, the first agent i receives a reward twice depending on the completion of the subsequent subtask. It receives the first reward $r_1(t_d)$ (which is usually small and can be reduced to 0 if we use the delay mechanism) when it completes its own subtask. Then, i receives the second reward $r_2(t_e)$ when the subsequent subtask has been done by another cooperative agent at time $t_e = t_d + \alpha$. Hence, the required whole task has been completed at t_e, and i receives $r_1(t_d) + r_2(t_e)$ finally. However, i receives only $r_1(t_d)$ if the subsequent subtask cannot be done successfully. It is obvious to extend this reward assignment to a multi-stage reward assignment scheme for tasks consisting of multiple sequential subtasks.

We also modify the *experience reply* to adapt our two-stage reward assignment and reflect sequential collaboration by two types of agent. Experience replay is used in our DQN learning to break the correlation between subsequent experience. In a multi-agent system, agents learn their actions independently and simultaneously, and their selections of actions during learning may lead to DQN learning instability. Experience replay is used to avoid this negative effect. We extended the experience replay by combining it with our multi-agent reward scheme in which the reward may vary due to other agents' subsequent actions within a certain period.

To prevent storing the experience data of delivery agent i whose reward has not been determined yet, i temporally stores the experience data $c_{i,t} = (s_{i,t}, a_{i,t}, r_{i,t}, s_{i,t+1})$ at t in its own *temporal queue* $P_{i,t}$ whose maximal length is $M_p(\geq p > 0)$. Reward $r_{i,t}$ for every action is usually zero, but when i completes its own subtask at t_d, $r_{i,t_d} = r_1(t_d) \geq 0$. When the size of $|P_{i,t}| > M_p$, the top element in $P_{i,t}$ is moved to i's *replay memory* $D_{i,t}$. Then, if i receives reward r_2 at t_e (where $t_d + p \geq t_e > t_d$) as a result of one execution agent at t_e performing the finishing task, the reward of the replay data c_{i,t_d} in P_{i,t_e} is modified as $r_{i,t_d} = r_1(t_d) + r_2(t_e)$. By setting $M_p = p$, where p is the expiration time of the delivered material, when the delivered material could not be executed before its expiration time, the replay data c_{i,t_d} is stored in D_{i,t_d+p+1} without modifying its reward r_{i,t_d} $(= r_1(t_d))$. Note that none of the execution agents have a temporal queue (or its maximal size is set to $M_p = 0$).

We can denote the replay memory at t as $D_{i,t} = \{c_{i,t-p-M_d}, \cdots, c_{i,t-p-1}\}$, where M_d (> 0) is the memory capacity. Then, i updates parameters $\theta_{i,t}$ at every η steps to minimize loss $L_{i,t}(\theta_{i,t})$, which is denoted by

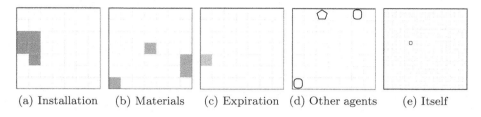

(a) Installation   (b) Materials   (c) Expiration   (d) Other agents   (e) Itself

Fig. 2. Input structure for delivery agent (Color figure online)

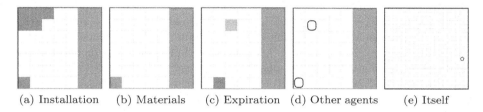

(a) Installation   (b) Materials   (c) Expiration   (d) Other agents   (e) Itself

Fig. 3. Input structure for execution agent (Color figure online)

$$L_{i,t}(\theta_{i,t}) = \mathbb{E}_{(s_i,a_i,r_i,s_i')\sim U(D_{i,t})}[(r_i + \gamma \max_{a_i'} Q_i(s_i', \arg\min_{a_i'} Q_i(s_i', a_i'; \theta_{i,t}); \theta_{i,t}^-)$$
$$- Q_i(s_i, a_i; \theta_{i,t}))^2],$$

where $U(D_{i,t})$ indicates the minibatch, i.e., random sampling from experience memory $D_{i,t}$. To reduce the value of loss function $L_{i,t}(\theta_{i,t})$, we calculated the gradient of loss function $\nabla L_{i,t}(\theta_{i,t})$ and adopted RMSprop [9] this might be useful because independent multi-agent learning is likely to be unsteady.

The goal of the agents is to increase the rewards they receive. Thus, the delivery agents attempt to put down a material on an installation cell so that the material is likely to be executed as soon as possible, and the execution agents have to seek a material on an installation cell or more aggressively seek a delivery agent holding a material to perform the finishing task before the expiration. Our concern is what kind of coordinated behaviors are learned by deep reinforcement learning by defining different ratios between $r_1(t_d)$ and $r_2(t_e)$ in the reward assignment method.

3.3   View Representation

Agent i has a limited observable area specified by the *observable range size* $V_i \geq 0$, where V_i is an integer, and i's observed data at t, $s_{i,t} \in S_i$, is the $(2V_i + 1) \times (2V_i + 1)$ lattice. Two examples of observable areas whose sizes are $V_i = V$ and centers are agents themselves are shown by translucent blue squares in Fig. 1. Regardless of the type of agent, i generates its view $v_{i,t}$ for input to the local DQN by composing its observed state $s_{i,t}$ and i's beliefs about the current environment.

The data in $v_{i,t}$ is divided into five input channels consisting of a number of lattices. These input channels are shown in Figs. 2 and 3. First, we assume that the abstract map of the environment and i's current location are part of the agent's belief; this map is expressed as the fifth channel in Figs. 2e and 3e. The four remaining channels express the observed state $s_{i,t}$, and all input channels except the second one are identical in both types of agent.

The first channel represents the installation cells (using the digit 1 and shown as gray cells in Figs. 2a and 3a). Note that the light blue regions in Fig. 3 mean outside the environment and are unobservable, so their cells are represented by -1. The third channel represents the expiration time of materials put down by delivery agents (Figs. 2c and 3c). The color of a material of ψ_{i,t_d} indicates the ratio of the remaining time to the expiration time, and its cell in the lattice is expressed using $z_t = max((p - t_d)/p, 0)$. The fourth channel includes other agents' IDs, although they cannot observe the types of other agents (Figs. 2d and 3d). The agents' IDs are uniquely expressed by three digits (b_1, b_2, b_3) (where $b_k \in \{0, 1, -1\}$), each of which is expressed by one of three lattices. Note that other agents' IDs are determined in each agent so that no different agents have the same ID. We also note that $(0, 0, 0)$ is not assigned to any agent since it means there is no agent at the cells. Finally, the second input channel (Figs. 2b and 3b) represents the locations of materials that delivery agents are holding (using digit 1). Thus, when a delivery agent holds a material, its material is also shown at the center of this input. In addition, this channel of delivery agents includes the locations of the storage area (Fig. 2b), but that of execution agents do not (Fig. 3b).

We also often add part of the trajectory of itself to the fifth input channels of both types of agent by assuming that any type of agent i can memorize its history of locations. In this representation of the fifth channel, i's current location is represented as 1, and its location k unit time ago is represented as $1 \times \beta^k$ ($= \beta^k$) if $\beta^k > \delta_t$, where $0 < \beta < 1$ is the decay rate, and δ_t is the threshold to decide the length of the trajectory. Note that if the agent visited a certain cell more than twice in this trajectory, only the maximal value is used for the trajectory.

3.4   Architecture of Neural Network

In MADRL, each agent autonomously decides its next action using the policy derived by the local DQN. The specifications of the neural network for deep Q-learning used in our experiments are listed in Table 1; the Q-network consists of some *convolutional layers* (Conv), *max pooling layers*, and three *fully connected network layers* (FCN layers). The sizes of the inputs fed to the Q-network are specified by using M and N, where $M = (2V + 1)$. Note that as described in the previous section, the inputs are six $M \times M$ lattices (Conv-1.1 in Table 1) and one $N \times N$ lattice (Conv-2 in Table 1).

Table 1. Network architecture

Layer	Input	Filter size	Stride	Activation	Next Layer
Conv-1.1	$M \times M \times 6$	2×2	1		Conv-1.2
Conv-1.2	$M \times M \times 32$	2×2	1		Max pooling-1
Max pooling-1	$M \times M \times 32$	2×2	2		FCN-1
Conv-2	$N \times N \times 1$	2×2	1		Max pooling-2
Max pooling-2	$N \times N \times 16$	2×2	2		FCN-1
FCN-1	$M/2 \times M/2 \times 32 +$ $N/2 \times N/2 \times 16$			ReLu	512
FCN-2	512			ReLu	FCN-3
FCN-3	256			Linear	5

We use the ε-greedy learning strategy with decay. Thus, agents choose their actions on the basis of the learning results so far with probability $1 - \varepsilon_{i,t}$; otherwise, choose randomly with probability $\varepsilon_{i,t}$, where $\varepsilon_{i,t}$ gradually decay as $\varepsilon_{i,t} = \max\{\varepsilon_{i,t-1} * \gamma_\varepsilon, \varepsilon_l\}$, and ε_l is the lower limit $\varepsilon_l \geq 0$. By starting with a large value of ε, agents can take various actions in the earlier stages of learning and gradually become conservative.

Table 2. Learning parameters

Parameter	Value
Discount factor γ_q	0.95
Initial value $\varepsilon_i = \varepsilon_{i,0}$	0.99999
Decay rate γ_ε	0.999999
Lower limit ε_l	0.002
Interval of parameter update η	8
Learning rate for RMSprop	0.00001
Momentum for RMSprop	0.90
ε for RMSprop	1e−07
Replay memory capacity M_d	2000
Minibatch size $|U(D_{i,t})|$	32
Copy interval for DDQN T_c	1

Table 3. Experimental parameters

Parameter	Value
Size of environment N	20
No. of agents $n_d + n_e$	12
Observable range size V	3
No. of installation areas K	12
Expiration time p	6
Reward for delivery agents $r_1 + r_2$	1
Reward for execution agents r_e	1
Epoch length H	600
Sum of epochs F_e	13,000
Trajectory decay rate β	0.9
Lower threshold for trajectory δ_t	0.05

4 Experiments and Discussion

4.1 Experimental Setting

We experimentally compare the performance (i.e., the total number of finishing task and the ratio of executable cells to executed cells) and analyze agents'

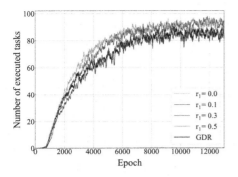
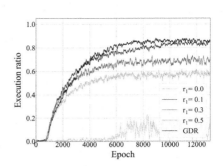

Fig. 4. Executed tasks per epoch

Fig. 5. Execution rate in delivered materials

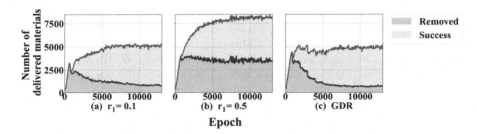

Fig. 6. Number of delivered materials (Color figure online)

behaviors when we used the various values of reward $r_1(t) = r_1$. Note that $r_1 + r_2$ is assumed to be constant. In the first experiment, we assume that r_1 was fixed every epoch, but in the second experiment, we introduced the decay of the first reward, so r_1 was gradually reduced every F_r epochs; thus, r_1 in the h-th epoch was defined as

$$r_1 = r_a - \delta_r \cdot \lfloor h/F_r \rfloor$$

where r_a is the initial reward for r_1, and the δ_r is the decay delta. These values were set as $r_a = 0.5$, $\delta_r = 0.1$, and $F_r = 1000$, so r_1 was zero after 5000 epochs. Therefore, delivery agents could receive high reward in the earlier stage to learn their own behaviors and then gradually learn coordinated behaviors to make sequential subtasks successful with a higher probability. The reward assignment with gradual decay for r_1 is called *gradually decayed reward* (GDR). The other parameters in our experiments are listed in Tables 2 and 3. The number of delivery agents n_d is eight, and the number of execution agents n_e is four in all experiments.

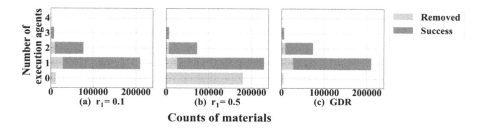

Fig. 7. Number of execution agents (Color figure online)

4.2 Performance Comparison

We first examined the improvement in the performance over time with the various reward assignment methods by setting r_1 to $0, 0.1, 0.3$, and 0.5 and using the GDR. The number of executed finished tasks per epoch from 1 to 13000 epochs is plotted in Fig. 4, where each plot is the moving average of every 50 epochs. This figure indicates that the agents could improve these performances except when $r_1 = 0$; in this situation no agents could learn any meaningful behaviors (the number of executed finishing tasks was almost zero). The learning speed was highest when $r_1 = 0.5$, but its converged value was almost identical to that when $r_1 = 0.3$. When the GDR was adopted, the performance curve was between those when $r_1 = 0.3$ and 0.1.

The ratio of executable cells to executed cells over time is plotted in Fig. 5, in which each plot is also the moving average rate of the ratio every 50 epochs. The high ratio is desirable because it means that the execution agents could utilize the efforts of delivery agents without wasting tasks. The result seemed the opposite of the previous one, i.e., the ratio when $r_1 = 0.1$ is the highest with the fixed r_1 value and gradually decrease with the increase of r_1. When the GDR was used, Fig. 5 indicates that its ratio is slightly higher than that when $r_1 = 0.1$, and we found that the agents try to take into account the behaviors of other agents. The analysis of such coordinated behaviors is shown in the next section. When $r_1 = 0$, the delivery agents could not learn their subtasks, so the execution agents could not learn theirs either because they have few chances to learn their behavior for their subtasks.

4.3 Analyzing Learned Coordinated Behavior

We try to understand the reason for the difference of performance described in the previous section by analyzing the characteristics of the behavior learned by MADRL with the two-stage reward assignment with decay. Figure 6 consist of the stack line graphs that plot the numbers of the delivered materials that were used for the finishing task (green) and those that were not (blue) every 50 epochs.

First, we can see that when $r_1 = 0.5$ (Fig. 6(b)), the number of materials delivered by delivery agents was the largest, but many of them were removed, although the reward of execution agents had many chances to learn their behaviors. Because the delivery agents could get enough rewards by their sole subtasks, they did not need to care the movements/locations of the execution agents; thus, the execution agents could not effectively find the executable cells. In contrast, when r_1 was small, the delivery agents could receive higher rewards after the effort of the execution agents, so their coordination is crucial. Therefore, as shown in Fig. 6(a), the number of removed (so wasted) executable cells was much smaller than that when $r_1 = 0.5$.

When the GDR was used, Fig. 6(c) indicates that the number of removed executable cells was smaller and the number of the executed cells was larger than those when $r_1 = 0.1$. We think that this improvement was caused by the sufficient learning of their own subtasks in the earlier stage and the coordinated behavior after that, so agents could learn the balanced behaviors with less wasted efforts of other agents.

To make sure of the discussion above, we investigated that when the delivery agents put down the materials, how much did they care about the nearby execution agents. We counted the number of execution agents within a delivery agent's observable range when it put down a material in the last 3000 epochs; the results are shown in Fig. 7, where the orange bars indicate the number of removed cells, and the gray ones indicate that of successfully completed cells.

Figure 7(b) shows that when $r_1 = 0.5$, the agents put down many materials even though no execution agent was within its observable range, and almost all delivered materials were removed in vain; these activities are a waste of efforts so are undesirable. When an execution agent happened to be within the observable range, many of the materials were successfully used.

When $r_1 = 0, 1$ or the GDR was used, Figs. 7(a) and 7(c) indicate that the delivery agents did not put down materials without confirming that one or a few execution agents were nearby; actually, we could see the activities of delivery agents. Even if they arrived at one of the installation cells, they wait for an execution agent before putting down the material. These coordinated behaviors made it easier for the execution agents to do the finishing tasks. In particular, when the GDR was introduced, this tendency was more pronounced. As mentioned above, the GDR enabled agents to produce the effective coordinated behaviors for both doing their own tasks and facilitating subsequent tasks done by other agents.

5 Conclusion

This paper proposed a reward scheme called two-stage reward assignment with decay to produce effective coordinated behaviors for sequential coordinated tasks using MADRL, in which each agent has its own DQN to learn actions concurrently. We examined the reward scheme based on completion of a sequential task called *coordinated delivery and execution problem*. Our experiments indicate that when agents could get enough reward just by doing the own subtasks, agents quickly learned the improved behavior for them but did not care about other agents' behaviors that are needed to complete the tasks. On the other hand, when agents could get only a small reward by doing their own subtasks, they could behave to facilitate other agents' subtasks. However, the learning for their own subtasks was not enough, and instead, agents became too cautious about other agents' tasks, so they reduced their efficiency. With learning using our proposed reward assignment, agents could learn both their own responsible subtasks and coordinated behaviors while caring about subsequent subtasks. This learned balanced behavior could reduce the wasted efforts and improved the entire performance of our problem.

We would like to extend our environment and the reward assignment method for more complex tasks in the future. For example, we will explore the problems in which coordination among several types of agent is required.

Acknowledgements. This work was partly supported by JSPS KAKENHI Grant Number 17KT0044, 20H04245.

References

1. Chang, Y.H., Ho, T., Kaelbling, L.P.: All learning is local: multi-agent learning in global reward games. In: Proceedings of the 16th International Conference on Neural Information Processing Systems, NIPS 2003, pp. 807–814. MIT Press, Cambridge (2003)
2. Foerster, J., Nardelli, N., Farquhar, G., Torr, P., Kohli, P., Whiteson, S., et al.: Stabilising experience replay for deep multi-agent reinforcement learning. In: Proceedings of the 34th International Conference on Machine Learning, vol. 70, pp. 1146–1155 (2017)
3. Foerster, J.N., Farquhar, G., Afouras, T., Nardelli, N., Whiteson, S.: Counterfactual multi-agent policy gradients. In: Thirty-Second AAAI Conference on Artificial Intelligence (2018)
4. Gu, S., Holly, E., Lillicrap, T., Levine, S.: Deep reinforcement learning for robotic manipulation with asynchronous off-policy updates. In: 2017 IEEE International Conference on Robotics and Automation (ICRA), pp. 3389–3396. IEEE (2017)
5. Lample, G., Chaplot, D.S.: Playing fps games with deep reinforcement learning. In: AAAI, pp. 2140–2146 (2017)
6. Mnih, V., et al.: Playing Atari with deep reinforcement learning. arXiv preprint arXiv:1312.5602 (2013)
7. Palmer, G., Tuyls, K., Bloembergen, D., Savani, R.: Lenient multi-agent deep reinforcement learning. In: Proceedings of the 17th International Conference on Autonomous Agents and MultiAgent Systems, pp. 443–451. International Foundation for Autonomous Agents and Multiagent Systems (2018)
8. Peng, X.B., Andrychowicz, M., Zaremba, W., Abbeel, P.: Sim-to-real transfer of robotic control with dynamics randomization. In: 2018 IEEE International Conference on Robotics and Automation (ICRA), pp. 1–8. IEEE (2018)
9. Tieleman, T., Hinton, G.: Lecture 6.5-rmsprop: divide the gradient by a running average of its recent magnitude. COURSERA: Neural Netw. Mach. Learn. **4**(2), 26–31 (2012)
10. Van Hasselt, H., Guez, A., Silver, D.: Deep reinforcement learning with double q-learning. In: AAAI, Phoenix, AZ, vol. 2, p. 5 (2016)

Deep Hierarchical Non-negative Matrix Factorization for Clustering Short Text

Wathsala Anupama Mohotti$^{(\boxtimes)}$ and Richi Nayak

School of Computer Science, Queensland University of Technology,
Brisbane, Australia
wathsalaanupama.mohotti@hdr.qut.edu.au, r.nayak@qut.edu.au

Abstract. This paper proposes a deep hierarchical Non-negative Matrix Factorization (NMF) method with Skip-Gram with Negative sampling (SGNS) to learn semantic relationships in short text data. The proposed unsupervised method learns a dense lower-order text presentation by minimizing the encoding and decoding error of factor matrices. Semantically-enriched dense text representation is constructed using the factor matrices where clusters are identified. We empirically evaluate the effectiveness of the method against the state-of-the-art short text clustering methods and deep neural embedding based methods.

Keywords: Deep learning · NMF · SGNS

1 Introduction

Social media platforms are a popular networking mechanism that allow users to disseminate information and assemble social views based on short-text communication [6]. A short text data faces sparsity and low word co-occurrences that create challenges for unsupervised text mining to identify groups or topics or concepts within the data. Recently, supervised deep learning methods based on shallow auto-encoder to deep auto-encoders using Recurrent Neural Networks (RNN) [10] and Convolutional Neural Network (CNN) [10] have been used in learning deep feature representation [24]. However, discovering a dense representation for short text in a fully unsupervised manner is essential in many applications to identify the clusters or concepts or topics.

Non-negative Matrix factorization (NMF) [12], which maps the high dimensional text representation to a lower-dimensional representation, has become popular in text clustering due to its capability to learn part-based lower-order representation where groups can be identified accurately [1,14]. Though the decomposed factor matrices are considerably dense in traditional text data and can be used to identify clusters, extreme-sparseness in short text challenges them in identifying dense factor matrices for the short text data.

In this paper, we present a novel method of deep Hierarchical NMF in which input data undergoes a special normalization which results in an effect similar

© Springer Nature Switzerland AG 2020
H. Yang et al. (Eds.): ICONIP 2020, LNCS 12533, pp. 270–282, 2020.
https://doi.org/10.1007/978-3-030-63833-7_23

to word embedding. This allows NMF to identify dense factor matrices incorporating contexts in the short text data. This technique is similar to Skip-Gram modelling with negative sampling (SGNS) when used with NMF. To the best of our knowledge, the proposed method, named as SG-DHNMF, is the first method that aims to (1) capture the semantic relationship in the short text data by analysing the pairwise documents similarity aligning with SGNS modeling and (2) progressively identify lower-rank factor matrices with each layer of hierarchical NMF that encodes the information of sparse document representation in each iteration. We conjecture that hierarchical factorization of document×document matrix into lower order can embed the geometrical structure of sparse data by combining the nearest neighbor(NN) information in each projection step. Empirical analysis with several Twitter datasets reveals that SG-DHNMF can handle sparsity in short text and outperforms the state-of-the-art short text clustering methods in finding accurate clusters.

2   Related Work

Document Expansion. These methods typically expand the feature vectors by adding relevant terms to deal with the sparsity in data [2,7,9]. A common approach is to use external knowledge sources for document expansion such as Wikipedia [2], WordNet [7], and ontologies [9]. Currently, word embedding-based pre-trained models such as word2vec [15], doc2vec [11], Glove [19] and Skip-Gram [17] have been used by exploiting semantic relationships in the data. However, short text in social media enriched using these static external sources provide inadequate information due to semantic incoherence [18]. Social media data includes unstructured phrases that result in a huge variance to traditional text vocabulary. Self corpus-based expansion is proposed as an alternative semantically aligned method in which concepts are identified in the collection for augmentation using clustering [8] or topics based on term frequency probabilities [18]. However, all these methods face challenges in dealing with the fewer word co-occurrences and the unstructured nature of the micro-blogging data [11].

Supervised Deep Feature Learning. Deep neural networks have been successfully used in feature learning. Deep auto-encoder [4] was one of the first models used in the text representation. Recent research uses advanced versions of NNs to reconstruct text representation [10]. Recurrent and recursive NNs [16,22] have been used with word-embedding to improve the representation learning process by including information of previous nodes for better semantic analysis in the dataset. Recent research applies convolution filters to capture the features similarity and achieve dense feature representation with CNNs [10,24]. CNN-based methods produce promising results in linguistic tasks among all supervised methods due to their ability to detect patterns in the data [3]. However, these methods rely on the label data (i.e. the ground-truths) for feature learning and cannot be applied to tasks where finding labelled data is scarce.

Unsupervised Deep Feature Learning. This emerging research area includes two families of methods, word-embedding-based and matrix factorization. The word embedding methods include language modeling and feature learning techniques where words are mapped to vectors of real numbers using a vocabulary [20]. Word2vec [15], a feed-forward NN based method that efficiently estimates word representations in vector space, is a popular model. Doc2Vec [11] is an extension of Word2vec producing document level embedding using a word vector generated for each word and a document vector generated for each document. Glove [19] is a non-neural network based vector space representation model. It considers a global word×word co-occurrence count matrix and uses the statistics in representing documents. The applicability of these pre-trained word-embedding models to short text data is limited due to vocabulary mismatch that shows a huge variance to the general text, a fewer number of word co-occurrence in short text data and, the noisy nature of micro-blogging data.

Matrix factorization methods are the leading unsupervised text representation methods. NMF has been used in clustering multi-view data by learning latent features embedded in multiple views [14]. It uses assistance provided by many views in identifying the final set of features. In comparison to the one-step dimensional reduction in traditional NMF, the use of progressive dimensional reduction with multiple iterations [5,25] is a recent approach. This is used in [5] with deep learning similar to autoencoder network considering encoding error, trained by a non-negativity constraint algorithm to learn features that show a part-based representation of data with matrix factorization. A handful of methods has been existed that use this type of hierarchical feature learning with NMF in step-by-step fashion [23] for document data to discover feature hierarchies in concepts. However, this stacking of NMF in leaning feature hierarchies considers geometric relationships between features within each iteration and encodes data to factor matrices that could approximate the input matrix more accurately. This method captures latent features precisely that could be neglected by one step dimensional reduction process. In [25], encoding as well as decoding of a factor matrix is considered in the optimization process. It shows that consideration of both the information as in deep auto-encoders is successful in identifying communities through user×user matrix in network data.

The proposed SG-DHNMF performs a progressive factorization on a symmetric document co-relation matrix that encodes normalized neighbourhood information via overlapping terms that results in SGNS through factorization. It considers encoding as well as decoding errors in each layer of the deep NMF process to accurately capture the geometric information in the data in resulting factor matrices. There exist only a very few similar works. In [21], each document in a short text corpus is considered as a window and a term-correlation matrix is factorized to boost the performance of short text-based topic modelling. It models words based on their context through word-correlations to overcome the sparsity. It has been shown that applying factorization on a normalized word-correlation matrix is similar to SGNS that encodes the relationship between the word and its context [13].

3 Deep Hierarchical NMF with SGNS-Based Embedding

Figure 1 illustrates the overall process of SG-DHNMF for identifying clusters in the short text data. Let $D = \{d_1, d_2, \ldots d_n\}$ be the dataset that contain a set of m unique terms after standard prepossessing steps such as lemmatazing and stop word removal. Let $A_1 \in R^{n \times n}$ represent the document×document matrix where a cell models the number of common terms between a document pair. We propose to model A_1 with SGNS that becomes input A to NMF.

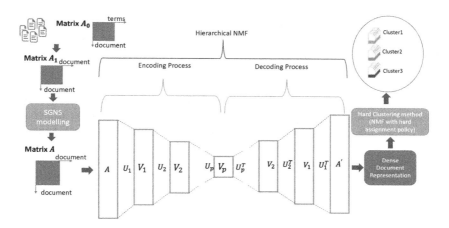

Fig. 1. Overview of SG-DHNMF

SG-DHNMF progressively decomposes A into factor matrices $U_g \in R^{n \times k}$ and $V_g \in R^{k \times n}$ where $g \in p$ and p is the number of layers (or the level of depth) in hierarchical decomposition. It does so by reducing encoding error $\|A - U_1 U_2 \ldots U_p V_p\|$ and decoding error $\|V_p - U_p^T U_{p-1}^T \ldots U_1^T A\|$ in each iteration and optimizes the process to converge for a given lower rank k. The sequential dimensional reduction process allows encoding geometric relationship with nearest neighbour documents to obtain dense representation in each iteration. Finally, SG-DHNMF reconstructs A' (document×document) by multiplying dense factor matrices. A clustering method can be applied to this dense representation to identify cluster assignments. In SG-DHNMF we apply one step NMF to the reconstructed A' with hard assignment policy by setting k as the cluster number.

3.1 Semantic Document Representation Learning with SGNS

SG-DHNMF aims to capture the closeness within documents in A with the SGNS modeling. SGNS has been used to highlight the word embedding in text data when representing them in NNs [17]. It can capture the context of a word in a corpus that the simple bag-of-word model fails [13]. The concept of negative (word, context) sampling is used with the Skip-Gram model to maximize the probability of an observed pair while minimizing the probability of unobserved pairs in distributed word representation [21]. SGNS has been proved to be equivalent to factorizing a word correlation matrix whose cells are the point-wise mutual information of the respective word and context pairs [13]. Specifically, factorizing a word correlation matrix with SGNS can model the closely related words with higher coefficients near to 1 while producing lower coefficients for loosely related words.

Distinct from the previous work, we use SGNS in SG-DHNMF to effectively encode the sparse text data by capturing the documents similarity and represent the input matrix to the NMF process. It maximizes the weight for the document pairs that show closer semantic similarity (i.e., share more common terms) in comparison to the others while minimizing the weight of document pairs that show fewer similarity. We conjecture that representing the input matrix with neighborhood information will capture the geometric structure inherent in the collection and utilise the relatedness while projecting the high-order dimensional data to low-rank data. The low-rank data will exhibit similar documents in closer space and non-similar documents in distant space. Hence, the low-rank representation obtained will improve the accuracy of a clustering solution.

Let d_i, d_j be a document pair in D. we model the closeness between them based on their shared terms with respect to rest of the documents in the collection as in Eq. (1).

$$A_{(d_i,d_j)} = log \left[\frac{c_{(d_i,d_j)} \times T}{\sum_{d_a \in D} c_{(d_a,d_i)} \times \sum_{d_a \in D} c_{(d_a,d_j)}} \right] \; where \; c_{(d_i,d_j)} > 0 \qquad (1)$$

where T is the total number of terms shared by the all the document pairs in D. Equation (1) calculates a ratio of the number of terms shared between a document pair with the number of terms that are shared by each of these documents with others in the collection. Let $c_{(d_i,d_j)}$ be the original cell value that represents the terms shared between d_i and d_j. It is divided by the sum of the values in the d_i row and d_j column. Document pairs that do not share any terms, the SGNS value for them is set to 0. The cell values A_{d_i,d_j} whose arguments of log are less than 0 are converted to 0 to minimize the probability of document pairs that show less similarity [13]. This step ensures that the input to NMF remains positive and will improve group identification.

3.2   Feature Learning with Hierarchical NMF

The matrix A modelled with SGNS becomes input to the deep factorization process. The progressive factorization in NMF enables SG-DHNMF to learn document context relationship accurately as it captures higher-abstract features with every iteration of projection in comparison to learning latent features based on one-step lower dimensional projection. The hierarchical representation of NMF enforces lower to higher level feature learning with each progression. Generally, NMF factorizes a given symmetric input matrix $A \in R^{n \times n}$ into two factor matrices $U \in R^{n \times k}$ and $V \in R^{k \times n}$ as in Eq. (2), where k is the lower rank that generally be the required cluster number.

$$min_{U,V \geq 0}\|A - UV\|_F \qquad s.t \ U \geq 0, V \geq 0 \qquad (2)$$

The SG-DHNMF factorization process with p layers starts factorizing the input matrix A into two non-negative factor matrix pairs (U_1, V_1) and proceeds with factorizing each V_g at each layer $g+1 \in p$ hierarchically. p is an empirical parameter that represents the number of layers on which NMF is applied progressively. In the short text data, the hierarchical representation learning through factorizing a matrix model with SGNS concept allows the data to promote document co-occurrences to deal with sparsity.

$$A \approx U_1 U_2 \ldots U_p V_p \qquad (3)$$

where $V_p \in R^{k \times n}$, $U_g \in R^{r_{g-1} \times r_g}$ where we set $n = r_0 \geq r_1 \geq \ldots \geq r_p = k$ and $1 \leq g < p$.

In the short text data where a fewer number of co-occurrences in terms exist, it becomes difficult to accurately identify the factors only considering encoding of input information. It is important to validate identified factors with decoding that inversely track factors and approximate them through input data. SG-DHNMF attempts to minimise the total approximation error by using both encoding and decoding to obtain an optimum lower-order dense representation. The decoding component, given in Eq. (4), is included in the objective function of SG-DHNMF as follows.

$$V_p \approx U_p^T U_{p-1}^T \ldots U_1^T A \qquad (4)$$

$$min_{U_g,V_g \geq 0}\|A - U_1 U_2 \ldots U_p V_p\|_F^2 + \ min_{U_g,V_g \geq 0}\|V_p - U_p^T U_{p-1}^T \ldots U_1^T A\|_F^2 \\ s.t \ U_g \geq 0, V_g \geq 0 \qquad (5)$$

The objective function of SG-DHNMF as in Eq. (5) calculates the total encoding error within p layer with the first component (Eq. (3)) and the total decoding error within p layer with the second component (Eq. (4)) in each iteration. The total error is attempted to minimize over the iterations in obtaining accurate dense factors for short text. This use of reconstruction loss ensures to capture the geometric information accurately in factor matrices. Figure 2 illustrates the impact of hierarchical NMF in comparison to a single NMF using a toy dataset. The multi-layer progressive factorization ($k = 20, 10, 5$) can achieve the denser and accurate lower-order representation in comparison to the one-step projection ($k = 5$).

Update Rules for SG-DHNMF. We initially pre-train each layer to have initial approximation of factor matrices U_g and V_g by simply decomposing the input matrix for each layer as follows. This pre-training process starts by decomposing A as $A \approx U_1 V_1$ by minimizing $|A - U_1 V_1\|_F^2 + |V_1 - U_1^T A\|_F^2$ where $U_1 \in R^{n \times r_1}$

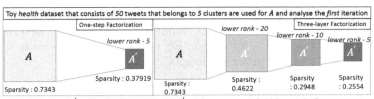

Fig. 2. Impact of hierarchical NMF

and $V_1 \in R^{r_1 \times n}$. Matrix V_1 is then decomposed as $V_1 \approx U_2 V_2$ by minimizing $|V_1 - U_2 V_2\|_F^2 + |V_2 - U_2^T V_1\|_F^2$ where $U_2 \in R^{r_1 \times r_2}$ and $V_2 \in R^{r_2 \times n}$.

This process is continued until all the p layers are pre-trained. This type of pre-training has been found effective and efficient [25]. It has greatly reduced the training time of the model as it gives better initialization for the model. In each iteration of the optimization process of Eq. (5), entries of the factor matrices for each layer are updated sequentially following multiplicative update rule principles. Following update rules for U_g and V_g have been derived based on the objective function using a derivative process similar to [25] for minimizing the total error.

$$U_g \leftarrow U_g \odot \frac{2\Psi_{g-1}^T A V_p^T \Phi_{g+1}^T}{\Psi_{g-1}^T \Psi_{g-1} U_i \Phi_{g+1} V_p V_p^T \Phi_{g+1}^T + \Psi_{g-1}^T A A^T \Psi_{g-1} U_g \Phi_{g+1} \Phi_{g+1}^T} \quad (6)$$

where $\Psi_{g-1} = U_1 U_2 \ldots U_{g-1}$ and $\Phi_{g+1} = U_{g+1} \ldots U_{p-1} U_p$. When $g = 1$ and $g = p$, we set $\Psi_0 = I$ and $\Phi_{p+1} = I$ respectively.

We start updating each matrix U_g as in Eq. (6), and then update V_g for lower rank k within each iteration as follows:

$$V_g \leftarrow V_g \odot \frac{2\Psi_g^T A}{\Psi_g^T \Psi_g V_g + V_g} \quad (7)$$

This process used in SG-DHNMF is illustrated in Algorithm 1.

Algorithm 1. The SG-DHNMF algorithm

Input   : The Document-Document matrix model with SGNS weighting A, Number of layers p with the configuration, Number of Clusters k
Output: The final Document-Cluster matrix C

while *Convergence of Eq. (5) where number of iterations <= 100* **do**
 foreach *g=1: p* **do**
  Compute U_g using Eq. (6)
  Compute V_g using Eq. (7)
 end
end
$A' = V_p U_p$
$C \leftarrow$ Apply NMF with hard clustering policy on A' to assign to k clusters

---

4   Empirical Analysis

Datasets: Two publicly available tweets [26] and stack overflow [24] with their ground-truth labels, used in prior short text clustering research, have also been used. Two additional twitter datasets from Trisma (https://trisma.org/) spanning across discussions on Cancer types and University Education have been used. The DS1:cancer dataset consists of 8 cancer types and the DS2:Edu dataset consists of 7 subject streams. These subgroups are considered as clusters. Stop words were removed. Terms with >90% frequency and <3 were removed. Table 1 reports the details of the datasets as well as the depth of the hierarchical-NMF model and the layer configuration at each depth. The layer configurations have been set based on experiments to systematically reduce the input data matrix.

Baselines: SG-DHNMF is compared with traditional unsupervised clustering methods including traditional NMF, Latent Dirichlet allocation (LDA) and k-means [1]. SG-DHNMF is also compared with the state-of-the-art unsupervised methods proposed to address the sparseness in short text, (1) Gibbs Sampling algorithm for the Dirichlet Multinomial Mixture model for short text clustering (GSDMM) [26], (2) Short-text topic modeling via NMF [21] (SeaNMF) that combines SGNS and NMF, (3) Deep autoencoder-like NMF [25] for community detection with sparse data that combines deep learning and NMF, and (4) k-means clustering with the document-level embedding, Doc2Vec [11] that can be used as an alternative way to obtain dense representation. Doc2Vec which creates a numeric representation that includes positive and negative numbers for a document limits us to use Doc2Vec with NMF.

Additionally, we compare unsupervised SG-DHNMF against the commonly used dense representation learning supervised methods to show how well SG-DHNM can lean deep features without the guidance. State-of-the-art single-layer shallow auto-encoder [10], RNN with Gated Recurrent Unit (RNN-GRU) [10] and CNN [10], that rely on ground-truth labels, are used. Finally, we compare the use of SGNS concept in learning the dense text representation accurately.

Evaluation Metrics: Two measures used to evaluate the accuracy of the short text clustering are standard pairwise F1-score (F1) which calculates the harmonic average of precision and recall, and Normalized Mutual Information (NMI) which measures the purity against the number of clusters [18].

Table 1. Dataset description

Dataset	# Docs	# Terms	#Clusters	Sparsity	Layers configuration
DS1:Cancer	20568	8851	8	0.9913	800-160-32-8
DS2:Uq	7504	5522	7	0.9974	700-140-28-7
DS3:tweet	2472	5077	89	0.9422	356-89
DS4:Stackoverflow	16407	2302	20	0.9413	2000-400-80-20

Comparison with Traditional Clustering Methods: Figure 3 shows the comparative results of SG-DHNMF with traditional shallow clustering methods when the input matrix is document×document representation encoded without SGNS and reconstruction loss. It also shows the results on traditional methods when the input matrix is used in classical way of term×document matrix. Results show that the document×document matrix representation is able to produce better results for traditional clustering methods. Additionally, it depicts that SG-DHNMF is superior to traditional shallow clustering methods as it uses step-wise deep learning to identify dense representation for text. This learning process shows a significant impact on large sparse datasets such as DS1.

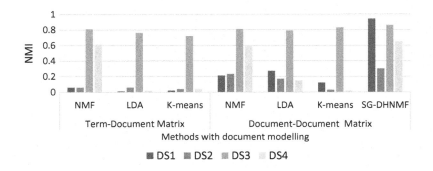

Fig. 3. Comparison with traditional methods and different document modelling

Comparison with Unsupervised Short Text Mining Methods: Table 2 details the performance of SG-DHNMF with state-of-the-art unsupervised methods that have been designed to handle sparsity in the data. The comparison with SeaNMF shows the superiority of using the SGNS concept in SG-DHNMF and with DANMF shows the superiority of using hierarchical learning in SG-DHNMF. The Doc2Vec encoding with k-means that extends the word-embedding concept to document embedding shows the least performance as there exists low word co-occurrences among short documents that this method can be benefited on. GSDMM which uses probability calculation face challenges due to less word co-occurrences and is unable to capture probabilities. DANMF uses reduced representation of network data for community detection balancing encoding and decoding, using a very large user to user network. However, DANMF does not perform well with extremely sparse short text data. SeaNMF uses the term×term relationship to learn word-context relationship with NMF and is able to achieve the best results among the baselines. Datasets DS3 and DS4 show generally good results with other methods such as SeaNMF, DANMF and Doc2Vec in comparison to DS1 and DS2 which are more denser. In highly sparse datasets SG-DHNMF produce superior results.

Comparison with Supervised Deep Learning Methods: Deep learning methods have been commonly developed to learning dense representation with

Table 2. Performance comparison with state-of-the-art unsupervised methods

Dataset	SG-DHNMF		GSDMM		SeaNMF		DANMF		Doc2Vec	
	NMI	F1	NMI	F1	NMI	F1	NMI	F1	NMI	F1
DS1	**0.94**	**0.93**	0.06	0.17	0.11	0.32	0.03	0.2	0.01	0.15
DS2	**0.31**	**0.45**	0.01	0.22	0.08	0.81	0.17	0.39	0.12	0.34
DS3	**0.86**	**0.74**	0.8	0.57	0.87	0.75	0.75	0.54	0.4	0.17
DS4	**0.65**	**0.63**	0.39	0.31	0.6	0.55	0.56	0.57	0.51	0.45
Avg	**0.69**	**0.69**	0.32	032	0.42	0.61	0.38	0.43	0.26	0.28

using ground-truth labels. They have been rarely used in unsupervised setting, except only a handful such as DANMF [25], benchmarked in Table 2. Results in Table 3 show the performance of state-of-the art supervised methods in categorising documents to respective clusters as classes. RNN and CNN based methods trained with ground-truth data only perform 4.35% and 7.25% better than SG-DHNMF in NMI that is not trained with ground-truth data. In spite of training, the shallow auto-encoder method shows inferior performance. It highlights the importance of deep learning embedded with hierarchical NMF in SG-DHNMF without supervision and shows that it even outperforms supervised shallow methods.

Table 3. Comparison of SG-DNMF with Supervised deep learning methods

Dataset	SG-DHNMF		Autoencoder-shallow		RNN-GRU		CNN	
	NMI	F1-score	NMI	F1-score	NMI	F1-score	NMI	F1-score
DS1	0.94	0.93	0.01	0.25	0.86	0.68	0.81	0.81
DS2	0.31	0.45	0.01	0.84	0.92	0.99	0.84	0.99
DS3	0.86	0.74	0.05	0.13	0.77	0.63	0.79	0.71
DS4	0.65	0.63	0.01	0.1	0.65	0.59	0.50	0.50
Avg	0.69	0.69	0.02	0.33	0.8	0.72	0.74	0.75

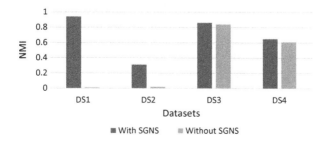

Fig. 4. With and without skip-gram with negative sampling

Impact of SGNS: The SGNS concept has been used in learning the word-context relationship in text mining [13]. In SG-DHNMF, we have proposed to use SGNS in learning the document-context relationship. Figure 4 shows that the document-context relationship can be learnt accurately with the SGNS modelling and factorizing the document correlation matrix. DS1 and DS2 which are extremely sparse show higher boost with SGNS modelling by capturing the document-correlations in modelling.

Sensitivity Analysis. We evaluate the depth of the layers used in NMF for deep learning the low-order features. Results in Table 4 show that this parameter depends on the nature and size of the dataset. This is similar to hyper-parameters tuning in neural network-based methods. The best coefficients for factor matrices are identified in iterative fashion reducing the encoding and decoding error. We measure the total error for 100 iterations and reported the normalized total error as in Fig. 5(a). It depicts that SG-DHNMF converges within this 100 iterations for all the datasets. Figure 5(b) shows how performance varies with sparsity in datasets. We have chosen a subset of cancer dataset (DS1) with different cluster numbers to form varying sparse datasets. Results show that performance increases with the sparsity in datasets. Figure 5(c) shows time taken for hierarchical NMF-based model training against the data size considering the subsets of Cancer dataset. SG-DHNMF shows a trend close to quadratic efficiency when double the sample size, similar to a NMF based method.

Table 4. Performance comparison using different number of layers

Dataset	1 layer		2 layer		3 layer		4 layer		5 layer	
	NMI	F1	NMI	F1	NMI	F1	NMI	F1	NMI	F1
DS1	0.66	0.65	0.81	0.82	0.74	0.74	**0.94**	**0.93**	0.65	0.64
DS2	0.19	0.38	0.3	0.45	0.22	0.39	**0.31**	**0.45**	0.25	0.39
DS3	0.83	0.69	**0.86**	**0.74**	0.83	0.7	-	-	-	-
DS4	0.62	0.59	0.64	0.62	0.64	0.62	**0.65**	**0.63**	-	-

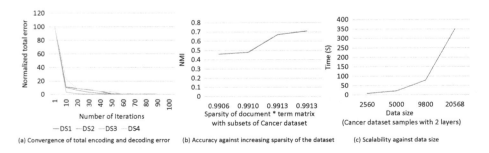

(a) Convergence of total encoding and decoding error   (b) Accuracy against increasing sparsity of the dataset   (c) Scalability against data size

Fig. 5. Convergence and performance with sparsity and scalability

Complexity Analysis. The computational complexity of SG-DHNMF is higher than any methods that use term × document matrix as an input. It includes an additional step calculating the document × document matrix by measuring pair-wise similarity in the document set. Excluding this additional pair-wise comparison, computational complexity of SG-DHNMF is $O(p(n^2r + nr^2))$ where n is the number of documents, p is the number of layers in HNMF with $(n >> p)$, and r is the maximum layer size in layer configuration out of all layers. This is similar to complexity of DANMF. However, seaNMF has $O(n^2)$ complexity [21] while GSDMM [26] and Doc2vec [11] have $O(nlog(v))$ and $O(knl)$ complexity respectively where v is the size of the vocabulary, l is the average document length and k is the number of groups/clusters. In contrast, deep NN models have higher computation complexity to SG-DHNMF.

5   Conclusion

We present a novel unsupervised method for short text clustering by learning a feature representation with deep NMF. The short text data shows extreme sparseness and fewer co-occurrences and creates additional challenges for a clustering algorithm to learn categories. This paper develops a feature learning method with the progressive use of NMF similar to deep NNs to explore the document-context relationships and encoding neighbour information within each step. Empirical analysis shows the superiority of SG-DHNMF.

References

1. Aggarwal, C.C., Zhai, C.: Mining Text Data. Springer, New York (2012). https://doi.org/10.1007/978-1-4614-3223-4
2. Banerjee, S., Ramanathan, K., Gupta, A.: Clustering short texts using Wikipedia. In: SIGIR, pp. 787–788. ACM (2007)
3. He, T., Huang, W., Qiao, Y., Yao, J.: Text-attentional convolutional neural network for scene text detection. IEEE Tran. Image Process. **25**(6), 2529–2541 (2016)
4. Hinton, G.E., Salakhutdinov, R.R.: Reducing the dimensionality of data with neural networks. Science **313**(5786), 504–507 (2006)
5. Hosseini-Asl, E., Zurada, J.M., Nasraoui, O.: Deep learning of part-based representation of data using sparse autoencoders with nonnegativity constraints. IEEE Trans. Neural Netw. Learn. Syst. **27**(12), 2486–2498 (2015)
6. Hu, X., Liu, H.: Text analytics in social media. In: Aggarwal, C., Zhai, C. (eds.) Mining Text Data, pp. 385–414. Springer, Boston (2012). https://doi.org/10.1007/978-1-4614-3223-4_12
7. Hu, X., Sun, N., Zhang, C., Chua, T.S.: Exploiting internal and external semantics for the clustering of short texts using world knowledge. In: CIKM, pp. 919–928. ACM (2009)
8. Jia, C., Carson, M.B., Wang, X., Yu, J.: Concept decompositions for short text clustering by identifying word communities. Pattern Recogn. **76**, 691–703 (2018)
9. Jin, O., Liu, N.N., Zhao, K., Yu, Y., Yang, Q.: Transferring topical knowledge from auxiliary long texts for short text clustering. In: CIKM, pp. 775–784. ACM (2011)

10. Kowsari, K., Jafari Meimandi, K., Heidarysafa, M., Mendu, S., Barnes, L., Brown, D.: Text classification algorithms: a survey. Information **10**(4), 150 (2019)
11. Lau, J.H., Baldwin, T.: An empirical evaluation of doc2vec with practical insights into document embedding generation. arXiv preprint arXiv:1607.05368 (2016)
12. Lee, D.D., Seung, H.S.: Learning the parts of objects by non-negative matrix factorization. Nature **401**(6755), 788–791 (1999)
13. Levy, O., Goldberg, Y.: Neural word embedding as implicit matrix factorization. In: NIPS, pp. 2177–2185 (2014)
14. Luong, K., Balasubramaniam, T., Nayak, R.: A novel technique of using coupled matrix and greedy coordinate descent for multi-view data representation. In: Hacid, H., Cellary, W., Wang, H., Paik, H.-Y., Zhou, R. (eds.) WISE 2018. LNCS, vol. 11234, pp. 285–300. Springer, Cham (2018). https://doi.org/10.1007/978-3-030-02925-8_20
15. Mikolov, T., Chen, K., Corrado, G., Dean, J.: Efficient estimation of word representations in vector space. arXiv preprint arXiv:1301.3781 (2013)
16. Mikolov, T., Kombrink, S., Burget, L., Černockỳ, J., Khudanpur, S.: Extensions of recurrent neural network language model. In: ICASSP, pp. 5528–5531. IEEE (2011)
17. Mikolov, T., Sutskever, I., Chen, K., Corrado, G.S., Dean, J.: Distributed representations of words and phrases and their compositionality. In: NIPS, pp. 3111–3119 (2013)
18. Mohotti, W.A., Nayak, R.: Corpus-based augmented media posts with density-based clustering for community detection. In: ICTAI, pp. 379–386. IEEE (2018)
19. Pennington, J., Socher, R., Manning, C.D.: GloVe: global vectors for word representation. In: 2014 EMNLP, pp. 1532–1543 (2014)
20. Roy, A., Park, Y., Pan, S.: Learning domain-specific word embeddings from sparse cybersecurity texts. arXiv preprint arXiv:1709.07470 (2017)
21. Shi, T., Kang, K., Choo, J., Reddy, C.K.: Short-text topic modeling via non-negative matrix factorization enriched with local word-context correlations. In: 2018 WWW, pp. 1105–1114 (2018)
22. Socher, R., et al.: Recursive deep models for semantic compositionality over a sentiment treebank. In: 2013 EMNLP, pp. 1631–1642 (2013)
23. Song, H.A., Lee, S.-Y.: Hierarchical representation using NMF. In: Lee, M., Hirose, A., Hou, Z.-G., Kil, R.M. (eds.) ICONIP 2013. LNCS, vol. 8226, pp. 466–473. Springer, Heidelberg (2013). https://doi.org/10.1007/978-3-642-42054-2_58
24. Xu, J., Xu, B., Wang, P., Zheng, S., Tian, G., Zhao, J.: Self-taught convolutional neural networks for short text clustering. Neural Netw. **88**, 22–31 (2017)
25. Ye, F., Chen, C., Zheng, Z.: Deep autoencoder-like nonnegative matrix factorization for community detection. In: CIKM, pp. 1393–1402 (2018)
26. Yin, J., Wang, J.: A dirichlet multinomial mixture model-based approach for short text clustering. In: SIGKDD, pp. 233–242 (2014)

Deep Reinforcement Learning
with Temporal-Awareness Network

Ze-yu Liu[1], Jian-wei Liu[1(✉)], Weimin Li[2], and Xin Zuo[1]

[1] Department of Automation, College of Information Science and Engineering,
China University of Petroleum, Beijing Campus (CUP), Beijing 102249, China
`liujw@cup.edu.cn`
[2] School of Computer Engineering and Technology, Shanghai University,
Shanghai, China

Abstract. Advances in deep reinforcement learning have allowed autonomous agents to perform well on video games, often outperforming humans, using only raw pixels to make their decisions. However, timely context awareness is not fully integrated. In this paper, we extend Deep Q-network (DQN) with spatio-temporal architecture - a novel framework that handles the temporal limitation problem. To incorporate spatio-temporal information, we construct variants of architectures by feeding spatial and temporal representations into Deep Q-networks in different ways, which are DQN with convolutional neural network (DQN-Conv), DQN with LSTM recurrent neural network (DQN-LSTM), DQN with 3D convolutional neural network (DQN-3DConv), and DQN with spatial and temporal fusion (DQN-Fusion), to explore the mutual but also fuzzy relationship between them. Extensive experiments are conducted on popular mobile game Flappy Bird and our framework achieves superior results when compared to baseline models.

Keywords: Deep reinforcement learning · Spatio-temporal architecture · Flappy bird

1 Introduction

Learning to play games has been one among of the popular topics researched in AI today. Solving such problems using game theory and search algorithms require careful domain specific feature definitions, making them averse to scalability. Reinforcement learning (RL) algorithms based on neural network hold the promise of allowing autonomous agents, such as robots, to do in a wide variety of tasks such as neural program synthesis [1], high-dimensional robot control [2] and solving autonomous-driving problems [3]. In particular, deep Q-networks (DQN) are shown to be effective in playing Atari video games [4] and more recently, in car racing game [5]. Recent advances in deep learning have made it possible to extract high-level features from raw pixel data, leading to breakthroughs in computer vision [6–8]. These methods utilize a range of neural network architectures,

© Springer Nature Switzerland AG 2020
H. Yang et al. (Eds.): ICONIP 2020, LNCS 12533, pp. 283–294, 2020.
https://doi.org/10.1007/978-3-030-63833-7_24

including convolutional networks, multilayer perceptrons, and recurrent neural networks. Inspired by recent advances, reinforcement learning uses several layers of convolutional neural network to capture spatio information and predict agent's action over time step.

While encouraging performances are reported, a lot of methods predict action from spatio information, without explicitly taking more high-level temporal information from images into account. Furthermore, temporal contexts are properties observed in images over time with rich motion cues and have been proved to be effective in action recognition [9]. Lack of temporal context can lead to partially observable states. In the case of partially observable states, the learning agent needs to remember previous states in order to select optimal actions. A valid question is how to incorporate high-level temporal information into deep neural network as complementary knowledge in addition to spatio image representations. We investigate particularly in this paper the architectures by exploiting the mutual relationship between spatio and temporal image representations for enhancing decision process. More importantly, to better demonstrate the impact of simultaneously utilizing the two kinds of representations, we devise variants of architectures by feeding them into different network, e.g. leveraging spatio and temporal information with two CNN, and combine them in the late fusion step. The main contributions of our work are as follows:

1. We conduct deep reinforcement learning, which incorporates spatial and temporal information. More specifically, we establish four Deep Q-networks: DQN with convolutional neural network (DQN-Conv), DQN with LSTM recurrent neural net- work (DQN-LSTM), DQN with 3D convolutional neural network (DQN-3DConv), and DQN with spatial and temporal fusion (DQN-Fusion).
2. We provide experimental evaluation of multiple approaches for extending reinforcement learning with spatio-temporal architectures, and report significant gains in performance over strong baseline models [20,21].
3. We highlight an architecture (DQN-Fusion) that processes input with two separate network - a spatio single frame network and a temporal multi-frame network - as a promising way to provide complementary signals for deep RL agents.

Our paper is organized as follows. In Sect. 2, we introduce some related works on reinforcement learning; In Sect. 3, we give a detailed description on our proposed frameworks; In Sect. 4, the performance is evaluated and compared to the state of the art models in the environment of Flappy Bird; In Sect. 5, we will make conclusion for this paper.

2 Related Work

The research on deep reinforcement learning has proceeded along three different dimensions: model-based methods [10–12], value-based approaches [13,14] and policy-based models [15,16].

The first direction, model-based methods, learn a transition model that allows for simulation of the environment without interacting with the environment directly. many works learn how the game works and predict which actions will lead to desirable outcomes. Obviously, most of them highly depend on the environment of the game, which may be limited in practice. For example, [10] and [11] employ deep dynamical model to predict high-dimensional observations based on autoencoder. Similarly, [12] utilize auto-encoders to learn a low-dimensional embedding of images jointly with a predictive model in this low-dimensional feature space.

Value-based approaches use value function to select action. The learning process is to optimize the value function. For instance, in [13], deep Q-network agent, receiving only the pixels and the game score as inputs, was able to achieve human-level control over Atari 2600 games. [14] addressed the fundamental instability problem of using function approximation in RL by the use of two techniques: experience replay and target networks.

Different from value-based model, policy-based models do not need to maintain a value function model, but directly search for an optimal policy. In this direction, TRPO [15], directly optimize the quantity of interest while remaining stable under function approximation. In [16], Mnih et al. propose asynchronous DRL, which is an efficient framework for DRL that uses asynchronous gradient descent to optimize the policy.

Although some DRL approaches learn to play games from raw pixel, they suffer from temporal limitations. For this problem, Mnih et al. [17] deploy input with four greyscale frames of the game, concatenated over time, which are initially processed by several convolutional layers in order to extract spatio-temporal features, such as the movement of the ball in "Pong" or "Breakout." Hausknecht et al. [18] extends DQN into an RNN, which allows the network to better deal with POMDPs by integrating information over long time periods. Like recursive filters, recurrent connections provide an efficient means of acting conditionally on temporally distant prior observations.

In short, our work in this paper belongs to the value-based models. Different from most of the aforementioned models which uses layers of CNN or recurrent connections [18], our work contributes by studying not only jointly exploiting image representations and temporal information for deep reinforcement learning, but also how the architecture can be better devised by exploring mutual relationship in between.

3   Deep Reinforcement Learning with Temporal-Awareness Network

In this section, we devise our deep Q-network under the umbrella of additionally incorporating the detected high-level temporal features. Specifically, we begin this section by presenting the problem formulation and followed by four variants of our spatio-temporal frameworks.

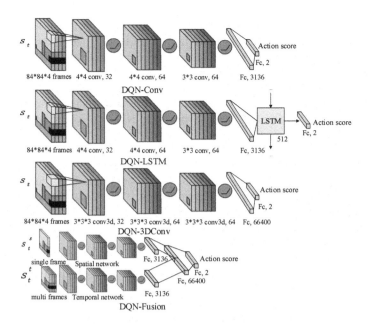

Fig. 1. The illustration of four variants of our spatio-temporal framework.

3.1 Problem Formulation

Suppose we have an agent interacts with an environment $\mathcal{E}$, in a sequence of actions, observations and rewards. At each time step, the agent observes the current state s_t of the environment, and selects an legal action a_t from a set of possible game actions, $A = \{1, \cdots, K\}$. After every interaction with the environment, the agent receives a reward r_t representing the change in game score. The learning process of the game is to find the policy that maximizes the expected cumulated discounted rewards R_t,

$$R_t = \sum_{t'=t}^{T} \gamma^{t'-t} r_{t'} \tag{1}$$

where T is the time at which the game terminates, and $\gamma \in [0, 1]$ is a discount factor that determines the importance of future rewards. The Q value of a given policy π is determined by the expected return from interacting with the environment E by action a,

$$Q^{\pi}(s, a) = \mathbb{E}[R_t | s_t = s, a_t = a] \tag{2}$$

The basic idea behind value-based model is to estimate the action value function. Instead of using an accurate estimate, we follow Bellman equation to obtain an estimate of the Q-function. Specifically, we use a neural network parameterized by θ to estimate reward function Q as follows:

$$Q^*(s,a) = \max_{\pi}\{\mathbb{E}[R_t|s_t = s, a_t = a]\} \tag{3}$$

It is common to use observed image which representing the current screen as state s_t. Since the agent only observes images of the current screen, which is partially observed. It is possible to divide the game states as spatio representations and temporal features to tackle this problem. We therefore consider sequences of actions and observations as temporal features to infer velocity or direction of the objects. Besides, spatio representations provide visual clue for object location and recognition respectively.

The goal of learning is to minimize difference between Q_θ and Q^* which leads to the following square loss function:

$$L_t(\theta_t) = \mathbb{E}_{s,a,r,s'}[(y_t - Q_{\theta_t}(s,a))^2] \tag{4}$$

where $y_t = \mathbb{E}_{s'\sim\epsilon}[r + \gamma \max_{a'} Q(s',a';\theta_{t-1})|s,a]$. Differentiating the loss function with respect to the weights, we arrive the following gradient,

$$\nabla_{\theta_t} L_t(\theta_t) = \mathbb{E}_{s,a,r,s'}[(y_t - Q_{\theta_t}(s,a))\nabla_{\theta_t} Q_{\theta_t}(s,a)] \tag{5}$$

Instead of performing full expectations in the above gradient, it is popular to use experience replay to break correlation between successive samples. Then we arrive at the famous Q-learning algorithm.

Note that, at each time step, the action distribution is selected by an ϵ-greedy strategy that follows the strategy with probability $1 - \epsilon$ selects a random action, and with probability ϵ follow the action with maximum rewards.

3.2   DQN Without Temporal-Awareness

DQN without temporal-awareness considers task in which an agent interacts with an environment, in this case the game engine. The agent observes an image $x_t \in \mathbb{R}^d$ from the engine, which is a vector of raw pixel values representing the current screen. For simplicity the agent utilizes several layers of convolutional neural network for state representation. Since the agent only observes images of the current screen, the task is partially observed.

3.3   DQN with Temporal-Awareness

For DQN with temporal-awareness, the temporal components of game provide an additional and important clue for environment state as velocity or direction can be reliably recognised based on temporal feature. An overview of our proposed architectures is depicted in Fig. 1. We devise four variants of DQN with temporal-awareness network for involvement of two design purposes. The first purpose is about which network is better to capture temporal feature in three architectures, i.e., DQN-Conv (capturing temporal feature with convolutional neural networks), DQN-LSTM (leveraging temporal feature with recurrent neural network) and DQN-3DConv (deriving temporal feature from 3D convolutional neural network). The second purpose is about how to simultaneously utilize spatio and temporal features and we design DQN-Fusion (deploying spatial and temporal networks separately, and combining at late fusion step).

DQN-Conv (DQN with Convolutional Neural Network). Given the current screen of game, one natural way of incorporating temporal features is to consider sequences of actions and observations, $x1, a_1, x_2, \cdots, a_{t-1}, x_t$ and learn game strategies that depend upon these sequences. We consider observations of length n as $s_t = x_{t-n+1}, \cdots, x_t$, where $x_i \in \mathbb{R}^d, s_t \in \mathbb{R}^{n \times d}$. The n frames observations are processed by layers of convolution network and rectified linear unit to obtain high level state representation. In each time step, we use dense layer to predict action. Here max value action ($a_t = arg \max_{a'} Q(s_t, a')$) is selected to interact with the environment $\mathcal{E}$. This kind of CNN architecture with stacked frames input is named as DQN Conv.

DQN-LSTM (DQN with Recurrent Neural Network). To further leverage both image representations and high-level temporal features in the encoding process, we design the second architecture DQN-LSTM. Instead of estimate $Q(s_t, a_t)$, we estimate $Q(s_t, h_{t-1}, a_t)$, where h_{t-1} is an extra input returned by the network at the previous step s_{t-1}. The LSTM updating procedure in DQN-LSTM is designed as

$$x_t = T_x x_0 \text{ and } h_t = f(x_t) \tag{6}$$

where D_e is the dimensionality of LSTM input, $T_x \in \mathbb{R}^{D_e \times D_x}$ is the transformation matrix for screen image representation, and f is the updating function within LSTM unit. In each time step, we use the output h_t of LSTM hidden state to predict the next action.

DQN-3DConv (DQN with 3D Convolutional Neural Network). The third design DQN-3Donv is similar to DQN-Conv as both designs utilize convolutional neural network and high-level temporal features, except that DQN-3DConv use 3-dimensional convolutional neural network. DQN-3DConv model extracts features from both spatial and temporal dimensions by performing 3D convolutions [19], thereby capturing the motion information encoded in multiple adjacent frames. The developed model generates multiple channels of information from the input frames, and the final feature representation is obtained by combining information from all channels. Suppose the current screen of environment as grayscale image $s_t \in R$, multiple adjacent frames add additional temporal dimension of the image which represents motion cues of object. Formally, the value at position (x, y, z) on $j - th$ feature map in the $i - th$ layer is given by

$$v_{ij}^{xyz} = tanh(b_{ij} + \sum_m \sum_{p=0}^{p_i-1} \sum_{q=0}^{Q_i-1} \sum_{r=0}^{R_i-1} w_{ijm}^{pqr} v_{(i-1)m}^{(x+p)(y+q)(z+r)}) \tag{7}$$

where R_i is the size of the 3D kernel along the temporal dimension, w_{ijm}^{pqr} is the $(p, q, r) - th$ value of the kernel connected to the $m - th$ feature map in the previous layer.

DQN-Fusion (DQN with Spatial and Temporal Fusion). Different from the former three designed architectures which mainly explore different network to capture temporal feature, we next deploy spatial network with additional temporal networks for motion features and combine two networks output in the late fusion step. As video game can naturally be decomposed into spatial and temporal components. The spatial part, in the form of individual screen frame, carries information about game scenes and objects depicted in the screen. The temporal part, in the form of motion across the frames, conveys the movement of the observer (the game view) and the objects. In DQN-Fusion, we deploy two convolutional neural network, one for each individual screen, and the other for n adjacent frames, and state can be denoted as $s_t^s = [x_{t-n+1}, \cdots, x_t], s_t^t = x_t$, The ConvNet output is then constructed as follows:

$$x_s = CNN_s(s_t^s) \; and \; x_t = CNN_t(s_t^t) \tag{8}$$

$$u = [T_s x_s, T_t x_t] \tag{9}$$

$$o_t = T_u u \tag{10}$$

where u is the concatenated vector of $T_s x_s$ and $T_t x_t$, $T_s \in \mathbb{R}^{u \times s}$, $T_t \in \mathbb{R}u \times t$, $T_u \in \mathbb{R}^{o \times u}$ are transformation matrices. Next action is selected from max value of o_t.

4   Experiments and Discussion

In this section, we present the Flappy Bird environment with variants of architectures used in our experiments and discuss the effectiveness of spatio-temporal model. The experimental results are listed in the Table 1. Then, it is necessary to restore the above experimental settings and details to facilitate the fair comparison experiments. We conduct the experiments to illustrate the impact of different network involved in deep reinforcement learning and compare with baseline methods.

4.1   Experimental Setting

Environment. The environment, Flappy Bird, is a popular mobile game, in which the player guides the bird, which is the "hero" of the game through the space between pairs of pipes. At each instant there are two actions that the player can take: to press the "up" key, which makes the bird to jump upward or not pressing any key, which makes it descend at a constant rate.

Pre-processing. Since the environment uses a very high dimensional state, we need to perform pre-processing to reduce the dimensionality of state space. The background image does not add important information, so we found a way to erase the background and keep only the birds, pipes and the ground. The original screen size is 512×288 in three channels, we convert the image to grayscale, crop it to 340×288 pixels, rescaled to 84×84 pixels and normalized to the range of $[0, 1]$.

Parameter Settings. The Flappy Bird game was run at 30 frames per second and the number n of adjacent frames was set to 4. All networks are trained using the Adam algorithm with learning rate $1e^{-6}$, $\beta_1 = 0.9$, $\beta_2 = 0.999$ and the minibatch size is 32. Agent experiences (s_t, a_t, r_t, s_{t+1}) are stored in replay memory of size one million, and Q-learning updates are done on batches of experiences randomly sampled from the memory. The discount factor γ was set to 0.99. We use an ϵ-greedy policy during the training process, where ϵ is linearly decreased from 1 to 0.1 over the first million steps, and then fixed to 0.1. The convolution weights are initialized to have a normal distribution with mean 0 and variance $1e^{-2}$.

Evaluation Metrics. We report the performance of our proposed methods using game average score and max score. In addition to the learned agents, we also report the average scores for human game player. The human performance is the median reward achieved after around two hours of playing game.

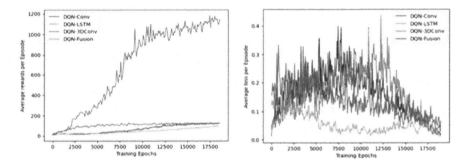

Fig. 2. The two plots of average rewards and losses per episode.

4.2   Comparison Experiments and Discussion

We pick up three methods based on reinforcement learning for Flappy Bird environment as baseline models, i.e., FlapAI SARSA, FlapAI Q-Learning [20] and Mixed-Integer Control Model [21]. The three techniques are totally different. In particular, SARSA and Q-Learning are deep neural network model with discretization and backward updates. Mixed-Integer Control Model was implemented with three different controllers, including flap controller, path plan controller and model-based predictive controller. For these learned methods, we follow the evaluation strategy used in Bellemare et al. [22,23] and report the average score, standard deviation and max score obtained by running an ϵ-greedy policy for a fixed number of steps. The curves in Fig. 2 show how the average rewards and losses evolve during training on the game Flappy Bird. The rewards plots continue improving during training. Besides, the loss plot tends to be very noisy because small changes to the weights of a policy can lead to large changes in the distribution of states the policy visits. In addition to seeing steady improvement of rewards during training our method is able to train deep Q-network with stochastic gradient descent in a stable manner.

Results in Table 1 demonstrate the effectiveness of the spatio-temporal network in improving the performance of our agent. DQN-Spatial is the model without temporal feature which achieves average score of 5.1. DQN-Conv, DQN-LSTM, DQN-3DConv and DQN-Fusion explicitly utilize both spatio and temporal features, which show better average score and max score over DQN-Spatial. In our proposed methods, DQN-Fusion achieves the best performances with average score of 1397. The average score of DQN-Conv is better than DQN-3DConv, but DQN-3DConv has higher max score, which is benefited from the mechanism of convolutional kernel across multiple frames. Besides, our proposed methods achieve much better performance than a human player. Our method boosts reinforcement learning with 3.3 million parameters compared to origin DQN with 1.7 million parameters.

Table 1. Performance of the agents for various learning methods by running ϵ-greedy policy.

Method	Mean	Standard deviation	Max
FlapAI SARSA [20]	117.317	112.998	811
FlapAI Q-Learning [20]	209.298	162.553	1224
Mixed Control [21]	418.6	-	500
DQN-Spatial	5.1	3.534	11
DQN-Conv	111	86.951	249
DQN-LSTM	102	76.422	214
DQN-3DConv	106.2	82.821	264
DQN-Fusion	**1397**	1754.076	**5819**
Human	4.25	10.614	21

To understand more about the working mechanism of the trained CNN model, the screen frame was visualized after the last spatial convolutional layers to highlight regions of interest. OpenCV was used to interpolate the heatmap of activation function and we project it onto the original image. In Fig. 3, we demonstrate weakly-supervised localization in the context of CNN, which capture objects of interest in the environment. Figure 3.a and c are original image, Fig. 3.b and d are corresponding heatmaps of activation function. We see for Flappy Bird, the CNN heatmap focuses on the bird and nearest pipes. This kind of activation function is helpful, likely because the position of the pipe relative to the bird is a very important spatial signals.

In summarize, the DQN-Fusion we proposed achieves superior results when com- pared to baseline deep models. The results basically indicate the advantage of simultaneously utilizing both spatio and temporal features for reinforcement learning. Furthermore, though DQN-Conv, DQN-LSTM and DQN-3DConv all

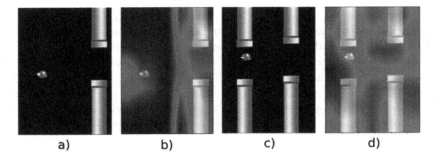

Fig. 3. The screenshots of the game Flappy Bird with heatmaps of activation function.

run involve the utilization of temporal features, DQN-Conv achieves better scores, indicating 2-D convolutional networks is better to deal with temporal features. Compare with other methods such as FlapAI and Mixed Control, our methods are more suitable for game environment with abundant motion information. For game without abundant motion information, our method plays comparable and achieves similar results.

5   Conclusions

In this paper, we construct the four spatio-temporal Deep Q-networks for reinforcement learning: DQN with convolutional neural network (DQN-Conv), DQN with LSTM recurrent neural network (DQN-LSTM), DQN with 3D convolutional neural network (DQN-3DConv), and DQN with spatial and temporal fusion (DQN-Fusion). In addition to spatio signal, the temporal component of the game provides an additional and important clue for environment state as velocity or direction of the bird can be reliably recognized based on the temporal feature. To verify our claim, we have devised variants of architectures for dynamic temporal feature extraction. Experiments conducted on popular game Flappy Bird validate our proposal and analysis. Performance improvements are clearly observed when comparing to previous methods and more remarkably, the performance of our DQN-Fusion achieves average score of 1397 and max score of 5819, which utilizing both spatio and temporal features. Besides, by explicitly visualize heatmaps of activation function of CNN feature map, we found that CNN architectures are capable of learning weakly-supervised localization, which is important for decision process. In order to combine spatio and temporal features, we investigate different architectures based on two separate networks, which are then combined by late fusion. The spatial network performs analysis from single frame, while the temporal network is trained to add motion clues. Taking account two kinds of information leads to significantly improvement over other baseline methods. In fact, we achieved the state-of-the-art results on Flappy Bird environment with two kinds of information.

Here we take a step back from evaluating performance, there are some empirical findings in our results.

1. Temporal and spatial features are complementary, as combination of two kinds of features significantly outperforms baseline models [20,21].
2. 2D-ConvNet trained on multi-frame is better to capture temporal features among our devised three different architectures. The reward signal successfully guides CNN to localize the bird and pipes in the game environments.
3. For game environments with abundant motion clues, DQN-Fusion is capable to utilize meaningful features of the environment, as DQN-Fusion significantly improves average score and max score.

Acknowledgement. This work was supported by the National Key R&D Program of China (No. 2017YFE0117500).

References

1. Bunel, R., Hausknecht, M., Devlin, J., Singh, R., Kohli, P.: Leveraging grammar and reinforcement learning for neural program synthesis. arXiv preprint arXiv:1805.04276 (2018)
2. Zhang, T., Kahn, G., Levine, S., Abbeel, P.: Learning deep control policies for autonomous aerial vehicles with MPC-guided policy search. In: 2016 IEEE International Conference on Robotics and Automation (ICRA), pp. 528–535. IEEE (2016)
3. Kiran, B.R., et al.: Deep Reinforcement Learning for Autonomous Driving: A Survey. arXiv preprint arXiv:2002.00444 (2020)
4. Lipovetzky, N., Ramirez, M., Geffner, H.: Classical planning with simulators: results on the Atari video games. In 24th International Joint Conference on Artificial Intelligence (2015)
5. Aldape, P., Sowell, S.: Reinforcement Learning for a Simple Racing Game (2018)
6. Krizhevsky, A., Sutskever, I., Hinton, G.: ImageNet classification with deep convolutional neural networks. Adv. Neural Inf. Process. Syst. **25**, 1106–1114 (2012)
7. Sermanet, P., Kavukcuoglu, K., Chintala, S., LeCun, Y.: Pedestrian detection with unsupervised multi-stage feature learning. In: Proceedings of the International Conference on Computer Vision and Pattern Recognition (2013)
8. Mnih, V.: Machine Learning for Aerial Image Labeling. Ph.D. thesis, University of Toronto (2013)
9. Karpathy, A., Toderici, G., Shetty, S., Leung, T., Sukthankar, R., Fei-Fei, L.: Large-scale video classification with convolutional neural networks. In: Proceedings of the IEEE Conference on Computer Vision and Pattern Recognition, pp. 1725–1732 (2014)
10. Oh, J., Guo, X., Lee, H., Lewis, R.L., Singh, S.: Action conditional video prediction using deep networks in Atari games. In: Advances in Neural Information Processing Systems, vol. 2863–2871 (2015)
11. Watter, M., Springenberg, J., Boedecker, J., Riedmiller, M.: Embed to control: a locally linear latent dynamics model for control from raw images. In: Advances in Neural Information Processing Systems, pp. 2746–2754 (2015)
12. Wahlström, N., Schön, T.B., Deisenroth, M.P.: Learning deep dynamical models from image pixels. IFAC-PapersOnLine **48**(28), 1059–1064 (2015)

13. Mnih, V., et al.: Human-level control through deep reinforcement learning. Nature **518**(7540), 529–533 (2015)

14. Lin, L.-J.: Self-improving reactive agents based on reinforcement learning, planning and teaching. Mach. Learn. **8**(3–4), 293–321 (1992). https://doi.org/10.1007/BF00992699

15. Schulman, J., Levine, S., Abbeel, P., Jordan, M., Moritz, P.: Trust region policy optimization. In: International Conference on Machine Learning, pp. 1889–1897 (2015)

16. Mnih, V., et al.: Asynchronous methods for deep reinforcement learning. In: International Conference on Machine Learning, pp. 1928–1937 (2016)

17. Mnih, V., et al.: Playing Atari with deep reinforcement learning. arXiv preprint arXiv:1312.5602 (2013)

18. Hausknecht, M., Stone, P.: Deep recurrent Q-learning for partially observable MDPs. In AAAI Fall Symposium Series (2015)

19. Ji, S., Xu, W., Yang, M., Yu, K.: 3D convolutional neural networks for human action recognition. IEEE Trans. Pattern Anal. Mach. Intell. **35**(1), 221–231 (2012)

20. Vu, T., Tran, L.: FlapAI Bird: Training an Agent to Play Flappy Bird Using Reinforcement Learning Techniques. arXiv preprint arXiv:2003.09579 (2020)

21. Piper, M., Bhounsule, P., Castillo-Villar, K.K.: How to beat flappy bird: a mixed-integer model predictive control approach. In: ASME Dynamic Systems and Control Conference. American Society of Mechanical Engineers Digital Collection (2017)

22. Bellemare, M.G., Naddaf, Y., Veness, J., Bowling, M.: The arcade learning environment: an evaluation platform for general agents. J. Artif. Intell. Res. **47**, 253–279 (2013)

23. Bellemare, M., Veness, J., Bowling, M.: Bayesian learning of recursively factored environments. In: Proceedings of the 13th International Conference on Machine Learning, pp. 1211–1219 (2013)

Double Replay Buffers with Restricted Gradient

Linjing Zhang[1] and Zongzhang Zhang[2($\boxtimes$)]

[1] School of Computer Science and Technology, Soochow University,
Suzhou 215006, China
20184227002@stu.suda.edu.cn
[2] National Key Laboratory for Novel Software Technology, Nanjing University,
Nanjing 210023, China
zzzhang@nju.edu.cn

Abstract. In this paper we consider the problem of how to balance exploration and exploitation in deep reinforcement learning (DRL). We propose a generative method called double replay buffers with restricted gradient (DRBRG). DRBRG divides the replay buffer in experience replay into two parts: the exploration buffer and the exploitation buffer. The two replay buffers with different retention policies can increase sample diversity to prevent over-fitting caused by exploiting. In order to avoid the deviation of the current policy from the past behaviors by exploring, we introduce a gradient penalty to limit the policy change into a trust region. We compare our method with other methods using experience replay on continuous-action environments. Empirical results show that our method outperforms existing methods both in training performance and generalization performance.

Keywords: Deep reinforcement learning · Experience replay · Exploration and exploitation

1 Introduction

Deep reinforcement learning (DRL) has achieved great success in many fields by combining deep learning and reinforcement learning [10,15]. From addressing simple physical control problems [2,24] to learning to play Atari games [14] and control robots [12], to defeating expert human players in Go game [19,20], there has been a variety of work to solve sequential decision-making problems. However, one of the biggest challenges in these DRL methods is the exploration-exploitation dilemma [21]. An agent exploits the learned knowledge to take actions for instant reward. Only exploiting will cause the agent to over-fit the current knowledge and fall into a sub-optimal point. Therefore, the agent needs

This work is in part supported by the Natural Science Foundation of China (61876119), the Natural Science Foundation of Jiangsu (BK20181432) and a project funded by the Priority Academic Program Development of Jiangsu Higher Education Institutions.

H. Yang et al. (Eds.): ICONIP 2020, LNCS 12533, pp. 295–306, 2020.
https://doi.org/10.1007/978-3-030-63833-7_25

to explore the environment to find a global optimal point for obtaining more rewards in the future. However, random exploration without guided information sometimes leads the current policy to diverge from the past policy, and cannot even find the optimal point.

Experience replay (ER) [13] is a fundamental technique to improve sample utility and training stabilization for recent DRL methods. Deep Q-network (DQN) [14] uses ER to store experiences generated during interactions between the agent and the environment in a buffer, and samples uniformly from the buffer to train the agent to get the maximum cumulative reward. Prioritized experience replay (PER) [16] is an improvement of DQN. PER distinguishes importance of each experiences and allows the important experiences to be sampled at a higher frequency, thus allowing the agent to learn faster and more effectively. The above ER methods use first-in-first-out (FIFO) to replace experiences in the buffer. Since some experiences which are rare and hard to obtain are flush out the buffer, FIFO causes the problem of catastrophic forgetting [11]. That is, with the training progressing and exploration reducing, experiences in the buffer tend to a small part of the state space, leading the policy to over-fit the current experiences and forget the knowledge acquired from the experiences replaced.

In the paper, we propose a generative method called double replay buffers with restricted gradient (DRBRG) to address the problems mentioned above. DRBRG classifies experiences based on an exploration rate to store in different replay buffers, i.e., the exploration buffer and the exploitation buffer. The two buffers use different retention methods to maintain the state distribution of all experiences. To avoid the current policy diverging from past behaviors, we penalty the policy change during the updating of policy gradient. By calculating the KL divergence between the current policy and the previous policy, we can estimate the change between policies, and add a penalty term when the policy is updated. In the experimental part, we compare DRBRG with other methods in the environments with continuous action spaces, including Pendulum and three MuJoCo environments. The results show that DRBRG can achieve better.

2   Related Work

There are a lot of methods to the problem of exploring and exploiting the interactive environment effectively. Some algorithms improve the two functions of ER (i.e., sampling and retention functions) to obtain more guided information to improve experience utility. PER computers the temporal difference error (TD-error) to assign priority of each experience. PER can allow important experiences with high priority to be replayed at a higher frequency, which allows agent to learn faster and more effectively. de Bruin et al. [5,6,8] store experiences into two replay buffers with FIFO and distance-based retention methods. These methods synthesize new experiences by sampling from the two buffers to train an agent. de Bruin et al. [7] also investigate some proxies to guide the retention and sampling of replay buffer via prior knowledge on control problems. Hindsight experience replay (HER) [1] uses additional goals to guide the agent to deal the problem with sparse and binary rewards. The key idea is to learn from failure by storing the achieved goals to the replay buffer to facilitate learning. However, these ER algorithms usually use FIFO as a retention method. The state distribution in their replay buffers will focus on a small region of the state space, leading catastrophic forgetting.

There are some methods on trust region to improve sample efficiency and training stabilization. Trust region policy optimization (TRPO) [17] optimizes a certain surrogate objective function by controlling the step size, improving training stabilization. Proximal policy optimization (PPO) [18] improves TRPO and is simple to implement. PPO clips a novel surrogate objective function to limit policy change and enables multiple epochs of minibatch updates. It improves sample complexity and training stabilization. Both TRPO and PPO use a KL divergence between the new policy and the old policy to be a trust region constraint. Our method combines ER with the trust region method to balance exploration and exploitation.

3   Background

Reinforcement learning (RL) problems usually can be formalized as Markov decision processes (MDPs). An MDP contains a state set S, an action set A, a transition function $T : S \times A \times S \rightarrow [0, 1]$, a reward function $r : S \times A \rightarrow \mathbb{R}$ and a discount factor $\gamma \in [0, 1]$. When interacting with the environment, an agent takes an action $a_t \in A$ according to a policy π at each step $t \in \{0, 1, 2, \cdots\}$, and moves from the current state $s_t \in S$ to the next state $s_{t+1} \in S$ with the reward $r_{t+1} \in \mathbb{R}$. Meanwhile, the experiences (s_t, a_t, r, s_{t+1}) are stored in a replay buffer $\mathcal{D}$, which are used by off-policy DRL algorithms to train the policy $\pi(a_t|s_t)$. The goal of the agent is to find an optimal policy π^* that maximizes the value function defined as:

$$V_\pi(s) = \left[\sum_{t=0}^{\infty} \gamma^t r(s_t) | s_0 = s, a_t \sim \pi(\cdot|s_t) \right]. \tag{1}$$

The Q-function under the policy π is defined as

$$Q_\pi(s, a) = r(s, a) + \gamma \sum_{s' \in S} T(s, a, s') V_\pi(s'). \tag{2}$$

The Q-function under the optimal policy π^*, denoted Q^*, satisfies the Bellman optimality equation [3]:

$$Q^*(s, a) = r(s, a) + \gamma \sum_{s' \in S} T(s, a, s') \max_{a' \in A} Q^*(s', a'). \tag{3}$$

In order to verify our method, we combine it with deep deterministic policy gradient (DDPG) [12] as an instance to deal with problems with continuous-action spaces. DDPG uses the actor-critic architecture. Thus there are two parts:

an actor and a critic. The actor has two neural networks: target network μ and actor network π. And in the critic, there are also two networks: target network Q' and critic network Q. The critic network is updated to evaluate the action value $Q(s, a)$ to guide the actor to take the action a with the maximum Q-value.

4 Method

Algorithm 1 describes DRBRG. We use two replay buffers to improve experience utility and shift the state distribution towards the entire state space. To avoid the policy diverging from past behaviors, we limit policy change by restricted gradient.

Algorithm 1. Double Replay Buffers with Restricted Gradient

Input: buffer size k, sampling ratio τ, exploration threshold η, interval time C for updating target network

1 Initialize two replay buffers $\mathcal{D}_r$ and $\mathcal{D}_g$ with size $k/2$;
2 Initialize an RL algorithm $\mathbb{A}$;
3 **for** $episode = 1 : M$ **do**
4   **for** $t = 1 : T$ **do**
5   Sample an action a_t using current policy from $\mathbb{A}$;
6   **if** *exploration rate* $> \eta$ **then**
7   Store transition (s_t, a_t, r_t, s_{t+1}) into $\mathcal{D}_r$;
8   **else**
9   Store transition (s_t, a_t, r_t, s_{t+1}) into $\mathcal{D}_g$;
10   **end**
11   Generate the training batch $\mathcal{D}_s$ from $\mathcal{D}_r$ and $\mathcal{D}_g$ according to τ;
12   **for** $i = 1 : \mathcal{D}_s$ **do**
13   Obtain state information s_{t_i};
14   Update $\rho_i = \pi(a_t|s_t)/\mu(a_t|s_t)$;
15   Compute the police gradient with penalty: $\hat{g}_{t_i} = \beta\hat{g}_{t_i} - (1 - \beta)\hat{g}_{t_i}^{\mathrm{KL}}$;
16   **end**
17   Perform optimization update using $\mathbb{A}$ and batch $\mathcal{D}_s$;
18   After C updates, update the target network and τ;
19   **end**
20 **end**

4.1 Double Replay Buffers

The two replay buffers are exploration buffer $\mathcal{D}_r$, and exploitation buffer $\mathcal{D}_g$. The entire buffer $\mathcal{D}$ is the combination of $\mathcal{D}_g$ and $\mathcal{D}_r$. We classify the experiences according the attributions of their actions to store the corresponding buffer. In different problems, the classification methods are different. In discrete action problems, an ϵ-greedy policy can control the magnitude of the exploration. While in continuous action problems, the noise $\mathcal{N}$ is often used to drive exploration. Therefore, we set a threshold η, to determine whether the action belongs to exploration action a_r or exploitation action a_g.

In early period of training process, the agent randomly explores the entire environment, and the experiences are distributed throughout almost the entire state space. When the policy improves, the agent reduces exploration to the environment and exploits learned knowledge to make decisions. The experiences generated by the current policy will be narrowed to a small region of the state space. We use reservoir sampling (RS) [22] in $\mathcal{D}_r$ to preserve earlier experiences to alleviate catastrophic forgetting. RS guarantees that all stored experiences are equally sampled or removed to better maintain the early experiences. $\mathcal{D}_g$ still focuses on the current policy with FIFO.

We introduce an sampling ratio τ which is adaptive to the policy update rate to sample a training batch from the above buffers. τ is a parameter obtained by comparing two networks in off-policy algorithms: the current network π and the target network μ. There are two methods to calculate τ. The first one is to count the number of the same actions n_b. The actions are obtained by inputting the states in training batches into the two networks. The specific expression is as follows:

$$\tau = \frac{n_b}{N_b} \times \mathcal{T}_{\max}, \tag{4}$$

where $\mathcal{T}_{\max} \in [0, 1]$ is a hyper-parameter to control the upper bound of τ and the update of τ is the same as the frequency of the target network. In each training batch with size N_b, τN_b experiences are sampled from exploration buffer $\mathcal{D}_r$ and the rest ones are sampled from exploitation buffer $\mathcal{D}_g$.

The second one is to compute the importance weight (IW) $\rho = \pi(a_t|s_t)/\mu(a_t|s_t)$ of each experience, and to count the number of the experiences with IW below threshold n_{below}. Here, we use the average of the value of all IW as a threshold. The Eq. 5 describes the method:

$$\tau = \frac{n_{below}}{N_b} \times \mathcal{T}_{\max}. \tag{5}$$

4.2   Restricted Gradient

ER uses the samples generated in the training to calculate the gradient estimate for the current network, which improves the experience utility. However, when the difference between the two networks is too large, the accuracy of the estimate and the performance of the algorithm are reduced. In order to address the above problems, we use a gradient penalty term to restrict policy change. This way is combined with DRB, which not only ensures the diversity of samples to alleviate the problem of catastrophic forgetting, but also improves the accuracy of gradient update. When gradient updates, we penalize gradient updates according to the state distribution in the training batch:

$$\hat{g} = \beta\hat{g} - (1 - \beta)\,\hat{g}^{\mathrm{KL}}. \tag{6}$$

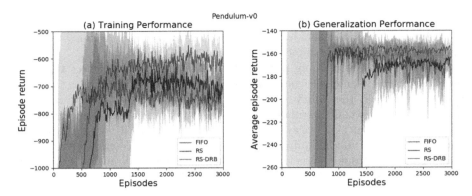

Fig. 1. The performance of RS-DRB vs. other methods on Pendulum

$\beta \in [0,1]$ is a penalty coefficient to penalize the policy change. The update of the coefficient β is related to the learning rate $\alpha \in [0,1]$ of the neural network, and β updates at each step:

$$\beta = (1 - \alpha)\beta + \alpha. \tag{7}$$

In Eq. 6 $\hat{g}$ is the policy gradient function, which is specifically expressed according to different algorithms. $\hat{g}^{\mathrm{KL}}$ is a gradient penalty term that penalizes the experiences where the current policy diverges from the past behaviors. The gradient penalty is related to the KL divergence between the current policy and the previous policy:

$$\hat{g}^{\mathrm{KL}} = \mathrm{E}_{s_i \sim \mathcal{D}}[\nabla_\theta \mathrm{D}_{\mathrm{KL}}(\mu(\cdot|s_i)||\pi(\cdot|s_i)], \tag{8}$$

where θ is the parameter in the neural network to approximate the policy.

The penalty term and related parameter adjustment are related to the current experience distribution. By introducing a penalty term to the gradient update method, the agent can find an optimal policy faster in a right direction.

5   Experiment

In this section we compare our method with DDPG to implement experiments. The comparing methods are two variants of DDPG: DDPG with FIFO and RS. We evaluate the performance of each combination on Pendulum and three MuJoCo environments. The training performance is plotted by the cumulative reward. The average return is used to reflect the generalization performance, which can be used to demonstrate the mitigation of catastrophic forgetting. In order to compare the effect of the two methods of computing τ, we distinguish between RS-DRB, which counts the number of n_b, and RS-DRB with IW counting the number of n_{below}. By comparison, we use the latter in DRBRG. The final result is an average of five trials with different seeds.

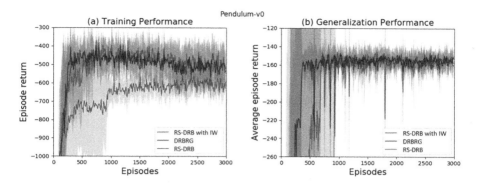

Fig. 2. The performance of DRBRG vs. RS-DRB on Pendulum

5.1   Pendulum

In Pendulum [4], we set the threshold $\eta = 0.01$ to determine the action whether belongs to exploration or exploitation. The evaluation of generalization performs per 10 episodes. The size of replay is 10000 in both FIFO and RS methods, while RS-DRB has two buffers with 5000 capacity. The hyper-parameter $\mathcal{T}_{max}$ in Pendulum is set to 0.4.

We firstly compare our method without gradient penalty, RS-DRB, with FIFO and RS. This was studied in our previous work [23]. The FIFO retention method cannot hold its best performance as time evolves in Fig. 1. It can be explained that the early exploratory experiences begin to reduce in the buffer, while the new generated experiences are concentrated in the current preference direction. The sample diversity of the buffer begins to decrease, which will cause the problem of catastrophic forgetting. From Fig. 1, we can find that both of the generalization and training performances of FIFO decrease after 1000 episodes. In the RS method, the probability that a new experience generated by the learned policy is the same as the probability of an early exploration experience. Therefore the agent cannot concentrate on the learned policy. This causes that the RS method achieves relatively good performance around 1500 episodes, much slower than FIFO and RS-DRB. It can be seen that RS-DRB learned a good policy around 1000 episodes which is the same as the FIFO method. Meanwhile, the training performance of RS-DRB still improves.

Then we compare DRBRG with RS-DRB which is combined with the two methods of computing the sampling ratio τ. From Fig. 2, we can find that the training performance of DRBRG performs significantly better than RS-DRB, but almost the same as RS-DRB with IW. Among the three methods, DRBRG converges and covers the entire state space the fastest around 100 episodes. Meanwhile, the generalization performance of DRBRG converges the fastest among the compared methods and maintains steady. It shows that DRBRG can quickly coverage the entire state space to find an optimal policy and control the policy change in a trust region. The results show that DRBRG can effectively balance

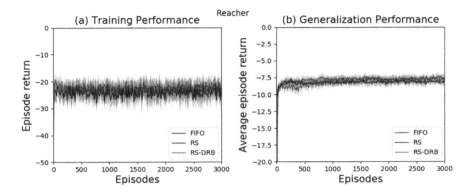

Fig. 3. The performance of RS-DRB vs. other methods on Reacher

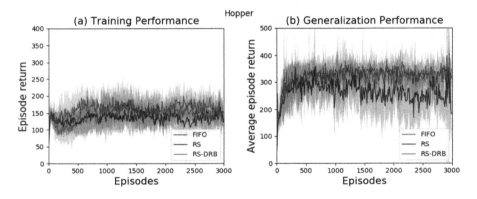

Fig. 4. The performance of RS-DRB vs. other methods on Hopper

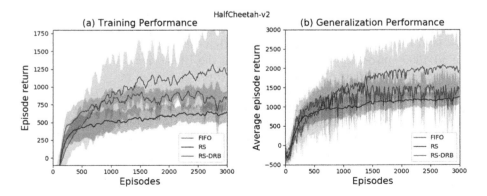

Fig. 5. The performance of RS-DRB vs. other methods on HalfCheetah

exploration and exploitation in training, achieving the best performance in terms of episode return and generalization.

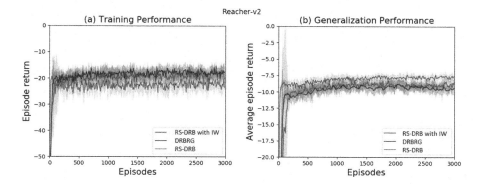

Fig. 6. The performance of DRBRG vs. RS-DRB on Reacher

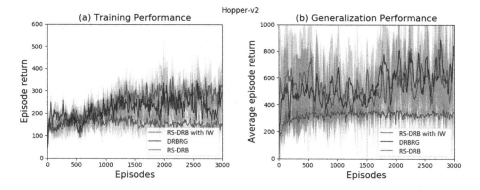

Fig. 7. The performance of DRBRG vs. RS-DRB on Hopper

5.2   MuJoCo

To further verify the effectiveness of our method, we conduct our experiments on three MuJoCo environments: Reacher, Hopper and HalfCheetah. In the three environments, the experimental settings are the same with the one in Pendulum.

From Figs. 3 and 4, we can find that the performance of RS-DRB is not significantly better than the compared methods in Reacher. The generalization performance of RS-DRB and FIFO are better than RS in Hopper. It further demonstrates that RS cannot concentrate on the current policy and leads the policy changes frequently. While RS-DRB significantly outperforms than other methods on HalfCheetah in terms of training and generalization performance. RS-DRB achieves the same performance with FIFO and RS around 500 episodes.

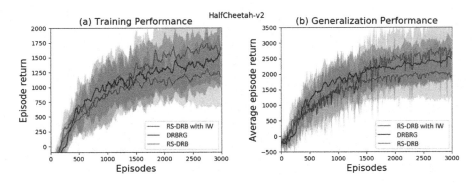

Fig. 8. The performance of DRBRG vs. RS-DRB on HalfCheetah

RS obtains the lowest return among the three methods. Figs. 3, 4 and 5 can explain that the environment setting is not suitable to each environment. The complex controlling problems need efficient implement action and more elaborate adjustments on the parameters [9].

Then we compare the two methods of computing the sampling ratio τ and DRBRG. Figs. 6, 7 and 8 show that RS-DRB with IW is better than the method based on the number of same actions. In Fig. 6, we can find that DRBRG is better than RS-DRB on training performance. In the three environments, the performance of DRBRG is almost the same with the method of RS-DRB with IW. The gradient penalty is not very effective due to the difficulty in handling the complex controlling problem. It may require the more precise estimate to policy change. In the future, we will make further improvement to this situation.

6   Conclusion

In this paper, we proposed DRBRG to deal with the dilemma of exploration and exploitation in reinforcement learning. In our method, we classify the experiences based on the nature of the actions, and store them in two replay buffers with different retention methods. By using an adaptive sampling ratio, our method can control the proportion of sampling from the two buffers. DRBRG efficiently makes the experience distribution cover the entire state space and can explore the potential states to maximize the cumulative reward. In order to perceive the differences between policies, we use a KL divergence to measure the changes and use it as a gradient penalty term. The introduction of gradient penalty limits the policy changes and improves the accuracy of the estimate for the current policy. DRBRG is a generative method that can combine with other DRL algorithms. In order to explain our method in detail, we implement DRBRG by combining it with DDPG to solve the problems with continuous-action spaces. Empirically, we demonstrate that DRBRG outperforms existing methods in both Pendulum and three MuJoCo environments. In the future, we will adjust the algorithm to handle the discrete action space problems and train some challenging environments in the Atari games. We would like to combine our method with other DRL methods that use experience replay to solve the sparse reward problems, and apply it into the training of physical robots.

References

1. Andrychowicz, M., et al.: Hindsight experience replay. In: Advances in Neural Information Processing Systems (NIPS), pp. 5055–5065 (2017)
2. Babuška, R., Groen, F.C.: Interactive Collaborative Information Systems. SCI. Springer, Heidelberg (2010). https://doi.org/10.1007/978-3-642-11688-9
3. Bellman, R.: Dynamic programming and stochastic control processes. Inf. Control **1**(3), 228–239 (1958)
4. Brockman, G., et al.: OpenAI Gym. CoRR abs/1606.01540 (2016). http://arxiv.org/abs/1606.01540
5. de Bruin, T., Kober, J., Tuyls, K., Babuška, R.: Improved deep reinforcement learning for robotics through distribution-based experience retention. In: International Conference on Intelligent Robots and Systems (IROS), pp. 3947–3952. IEEE (2016)
6. de Bruin, T., Kober, J., Tuyls, K., Babuška, R.: Off-policy experience retention for deep actor-critic learning. In: Deep Reinforcement Learning Workshop, Advances in Neural Information Processing Systems (NIPS) (2016)
7. de Bruin, T., Kober, J., Tuyls, K., Babuška, R.: Experience selection in deep reinforcement learning for control. J. Mach. Learn. Res. **19**(1), 347–402 (2018)
8. De Bruin, T., Kober, J., Tuyls, K., Babuška, R.: The importance of experience replay database composition in deep reinforcement learning. In: Deep Reinforcement Learning Workshop, Advances in Neural Information Processing Systems (NIPS) (2015)
9. Dietterich, T.G.: Robust artificial intelligence and robust human organizations. Front. Comput. Sci. **13**(1), 1–3 (2019). https://doi.org/10.1007/s11704-018-8900-4
10. François-Lavet, V., Henderson, P., Islam, R., Bellemare, M.G., Pineau, J.: An introduction to deep reinforcement learning. Found. Trends Mach. Learn. **11**(3–4), 219–354 (2018)
11. Goodfellow, I.J., Mirza, M., Da, X., Courville, A.C., Bengio, Y.: An empirical investigation of catastrophic forgetting in gradient-based neural networks. CoRR abs/1312.6211 (2013). http://arxiv.org/abs/1312.6211
12. Lillicrap, T.P., et al.: Continuous control with deep reinforcement learning. In: International Conference on Learning Representations (ICLR) (2016)
13. Lin, L.J.: Self-improving reactive agents based on reinforcement learning, planning and teaching. Mach. Learn. **8**(3–4), 293–321 (1992)
14. Mnih, V., et al.: Human-level control through deep reinforcement learning. Nature **518**(7540), 529–533 (2015)
15. Liu, Q., et al.: A survey on deep reinforcement learning. Chin. J. Comput. **41**(1), 1–27 (2018)
16. Schaul, T., Quan, J., Antonoglou, I., Silver, D.: Prioritized experience replay. In: International Conference on Learning Representations (ICLR) (2016)
17. Schulman, J., Levine, S., Abbeel, P., Jordan, M.I., Moritz, P.: Trust region policy optimization. In: Proceedings of the 32nd International Conference on Machine Learning (ICML) (2015)

18. Schulman, J., Wolski, F., Dhariwal, P., Radford, A., Klimov, O.: Proximal policy optimization algorithms. CoRR abs/1707.06347 (2017). http://arxiv.org/abs/1707.06347
19. Silver, D., et al.: Mastering the game of Go with deep neural networks and tree search. Nature **529**(7587), 484–489 (2016)
20. Silver, D., et al.: Mastering the game of Go without human knowledge. Nature **550**(7676), 354–359 (2017)
21. Sutton, R.S., Barto, A.G.: Reinforcement Learning - An Introduction, 2nd edn. MIT Press, Cambridge (2018)
22. Vitter, J.S.: Random sampling with a reservoir. ACM Trans. Math. Softw. (TOMS) **11**(1), 37–57 (1985)
23. Zhang, L., et al.: A framework of dual replay buffer: balancing forgetting and generalization in reinforcement learning. In: Workshop on Scaling Up Reinforcement Learning (SURL), International Joint Conference on Artificial Intelligence (IJCAI) (2019)
24. Zhong, S., Liu, Q., Zhang, Z., Fu, Q.: Efficient reinforcement learning in continuous state and action spaces with Dyna and policy approximation. Front. Comput. Sci. **13**(1), 106–126 (2019)

Entropy Repulsion for Semi-supervised Learning Against Class Mismatch

Xuanke You[1], Lan Zhang[1(✉)], Linzhuo Yang[1], Xiaojing Yu[1], and Kebin Liu[2]

[1] University of Science and Technology of China, Hefei, China
yxkyong@mail.ustc.edu.cn, zhanglan@ustc.edu.cn
[2] TNLIST, Tsinghua University, Beijing, China

Abstract. A series of semi-supervised learning (SSL) algorithms have been proposed to alleviate the need for labeled data by leveraging large amounts of unlabeled data. Those algorithms have achieved good performance on standard benchmark datasets, however, their performance can degrade drastically when there exists a class mismatch between the labeled and unlabeled data, which is common in practice. In this work, we propose a new technique, entropy repulsion for mismatch (ERCM), to improve SSL against a class mismatch situation. Specifically, we design an entropy repulsion loss and a batch annealing and reloading mechanism, which work together to prevent potentially mismatched unlabeled data from participating in the early training stages as well as facilitate the minimization of the unsupervised loss term of traditional SSL algorithms. ERCM can be adopted to enhance existing SSL algorithms with minor extra computation cost and no change to their network structures. Our extensive experiments demonstrate that ERCM can significantly improve the performance of state-of-the-art SSL algorithms, namely Mean Teacher, Virtual Adversarial Training (VAT) and Mixmatch in various class-mismatch cases.

Keywords: Semi-supervised learning · Class mismatch

1 Introduction

Deep learning models have achieved remarkable performance on many supervised learning problems by leveraging large labeled datasets [12]. Creating large datasets with high-quality labels, however, is usually very labor-intensive and time-consuming [21,24]. Semi-supervised learning [3] (SSL) provides an attractive way to improve the performance of deep learning models by also utilizing easily obtainable unlabeled data, so as to mitigate the reliance on large labeled datasets. Algorithms for SSL mainly include the following core ideas: consistency regularization [11,14,19], entropy minimization [7,13], and traditional regularization [23]. Recent holistic approaches, Mixmatch [2] and UDA [20] achieve the state-of-the-art performance by combining these ideas above.

Existing SSL algorithms usually demonstrate their successes using fully-labeled classification datasets (e.g., CIFAR-10 [10], SVHN [15] and Imagenet

© Springer Nature Switzerland AG 2020
H. Yang et al. (Eds.): ICONIP 2020, LNCS 12533, pp. 307–319, 2020.
https://doi.org/10.1007/978-3-030-63833-7_26

[5]) by treating most samples of each dataset as unlabeled. Therefore, those evaluation results are based on an implicit assumption that all unlabeled samples come from the same classes as labeled samples. In real world, however, it is very likely that a large portion of the unlabeled samples do not belong to any classes of the labeled data, i.e., there exist a mismatch between class distributions of labeled and unlabeled data. As an example, if you intend to train a model to distinguish between ten classes of animals with only a small amount of labeled images at hand, you may want to employ a large collection of unlabeled animal images to improve the model performance. The unlabeled dataset may contain many images of other animal classes than the ten target classes. Most existing SSL algorithms use a combined loss of a supervised term and an auxiliary (unsupervised) term to achieve high test accuracy as well as generalize better to unseen data. As reported in some recent work, the *class mismatch* issue can make it difficult to minimize the auxiliary loss term [22], furthermore, drastically degrade the performances of SSL algorithms compared to not using any unlabeled data at all [16]. Though class mismatch can actually hurt the applicability of SSL algorithms, it has not received much attention until recently. [11] and [22] consider to evaluate SSL algorithms in class-mismatch cases. Two techniques, Split Batch normalization (Split-BN) [22] and ROI regularization, have been proposed to improve the robustness of existing SSL methods against class mismatch.

In this work, we focus on reducing the performance degradation caused by *class mismatch* problems so as to improve the applicability of existing SSL algorithms. We propose a novel entropy repulsion technique for mismatch (ERCM) to restrict potentially mismatched unlabeled samples from participating in the training process. Specifically, we introduce a new entropy repulsion loss term, which is gradually relaxed to prevent the model from premature overfitting on mismatched unlabelled data. We also design a batch annealing and reloading mechanism to work together with the loss, which dump samples with low-confidence pseudo labels and reload samples with highest-confidence pseudo labels from a temporal pool to make the training more stable. Our contributions are summarized as follows:

- We propose a novel technique ERCM, including an entropy repulsion loss together with a batch annealing and reloading mechanism, which can empower existing SSL algorithms to achieve a significant performance improvement over the state of the art even when there is a significant class mismatch between labeled and unlabeled data. For example, with 250 labeled data and 20000 unlabeled data (mismatched data accounts for 20%) on CIFAR-10, as shown in Table 1, our method achieved 11.3% test error, which is 5.9% lower compared to 17.2% test error of the next-best method (Mix*). Specially, our analysis and ablation experiments show that ERCM can effectively alleviate the difficulty to minimize the auxiliary loss term in class-mismatch cases, which is a challenging issue reported by previous work [22].
- Our design is orthogonal to traditional SSL algorithms and can be effectively adopted by existing SSL methods to improve their performance in

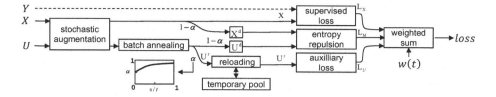

Fig. 1. Workflow of our proposed ERCM technique (details in Sect. 3).

class-mismatch cases. Our ERCM technique is highly portable, requiring no change to network structures and only introducing minor extra computational overhead.

2 Related Work

In this section, we mainly review state-of-the-art SSL techniques and recent efforts to address the class mismatch issue. A more comprehensive survey of SSL is provided in [3]. A common underlying assumption of SSL algorithms is that the decision boundary should pass through the low-density regions of data. One core idea to enforce this is entropy minimization. EntMin [7] makes low-entropy predictions for all unlabeled samples by adding an explicit loss term. Pseudo Label [13] gives pseudo labels for unlabeled data with high-confidence outputs for entropy minimization. Another core idea is consistency regularization that encourages the model to output the same class distribution for various augmentations of an unlabeled sample. Π-Model [11] and Temporal Ensembling [17] generalize ensemble predictions of unlabeled samples by networks with dropout regularization [18]. Mean Teacher [19] averages model weights instead of label predictions in which teacher model is an average of consecutive student models. VAT [14] involves consistency by applying a perturbation to the input. Recently, holistic methods Mixmatch [2] and UDA [20] achieve state-of-the-art performance on benchmark datasets by incorporating several recent advanced techniques. When it comes to a more realistic setting where class mismatch exists, those methods, however, may suffer a significant performance degradation.

The class-mismatch problem has not drawn much attention from traditional SSL methods. It is first considered in [11], which only appears in partial experiments and has not been discussed in depth. Recently, class distribution mismatch is formally discussed in [16], which shows clear performance degradation of various SSL methods in class-mismatch cases. Moreover, class mismatch shares some characteristics with domain adaptation [1,6] in which there are differences between distributions of training data and test data. [9] designs ROI regularization to help VAT perform better against class mismatch. Split-BN [22] uses split batch normalization to improve the performance of Mean Teacher and VAT. And a SSL method named UASD [4] is proposed to mitigate the impact of class mismatch. In this paper, we aim to further enhance existing SSL methods by restricting potentially mismatched unlabeled samples from participating in the training process. Moreover, ERCM can also effectively improve the performance of the holistic method, Mixmatch.

3   Our Method

3.1   Problem Formulation

In SSL, we are given a labeled dataset $\mathcal{D}_L$ and an unlabeled dataset $\mathcal{D}_U$. Let $\mathcal{D}_{\mathcal{Y}} = \{0, 1 .. K - 1\}$ be the set of labels. For each labeled sample $x \in \mathcal{D}_L$, we have $label(x) \in \mathcal{D}_{\mathcal{Y}}$. SSL algorithms aim to leverage unlabeled samples from $\mathcal{D}_U$ to train a model with better performance than what would have been obtained by using $\mathcal{D}_L$ alone. In this work, we consider a situation that is very common in real-world settings, named *class mismatch*. $\mathcal{D}_U$ is very likely to have extra "dirty" data called mismatched samples that do not belong to any of these K classes. As reported in [16], class mismatch can actually hurt the performance of SSL methods. Our goal is to improve the performance of SSL in class-mismatch cases by mitigating the negative impact of mismatched unlabeled samples during the training process.

3.2   Design Overview

In a typical training process of SSL, a minibatch is composed of a labeled batch $\mathcal{X}$ (a set of size $\mathcal{C}$ randomly sampled from $\mathcal{D}_L$), an unlabeled batch $\mathcal{U}$ (a set of size $\mathcal{C}$ randomly sampled from $\mathcal{D}_U$), and corresponding labels $\mathcal{Y}$ of $\mathcal{X}$. Many recent SSL approaches use a combined loss function $\mathcal{L}$ consisting of a supervised part and an auxiliary part:

$$\mathcal{L} = \lambda_{\mathcal{X}} \mathcal{L}_{\mathcal{X}} + \lambda_{\mathcal{U}} \mathcal{L}_{\mathcal{U}}, \tag{1}$$

where $\lambda_{\mathcal{X}}$ and $\lambda_{\mathcal{U}}$ are weights of loss terms. The supervised part $\mathcal{L}_{\mathcal{X}}$ is a loss function of labeled samples like cross-entropy:

$$\mathcal{L}_{\mathcal{X}} = \frac{1}{|\mathcal{X}|} \sum_{x \in \mathcal{X}, \hat{y} \in \mathcal{Y}} \hat{y} \, log(\frac{1}{p(\,y|x, \theta)}). \tag{2}$$

The auxiliary loss $\mathcal{L}_{\mathcal{U}}$ is designed to explore the decision boundary by unlabeled data. For example, in Mixmatch, $\mathcal{L}_{\mathcal{U}}$ is a consistency regularization loss term defined as $|| \hat{g} - p(\,y|u, \theta) ||_2^2, u \in \mathcal{U}$, where $\hat{g}$ represents "guessing label" of unlabeled samples after sharpening.

Entropy Repulsion Loss. In traditional SSL algorithms, combining a cross-entropy loss and a consistency regularization loss leads to a decrease of the entropy of labeled and unlabeled samples, which achieves good performance on standard datasets. In class-mismatch cases, however, blindly reducing the entropy of unlabeled data is not always beneficial and can even hurt the performance. Once the model is over-trained or over-fitted on mismatched unlabeled samples, it will introduce great errors to the model. To address this problem, we propose an entropy repulsion loss term $\mathcal{L}_\mathcal{M}$ (shown in Eq. (3)), which encourages output entropy of labeled samples relatively smaller than that of unlabeled ones during the training process. $\mathcal{L}_\mathcal{M}$ encourages the entropy of $p(y|x', \theta), x' \in \mathcal{X}^d$ to be relatively smaller than the entropy of $p(y|u', \theta), u' \in \mathcal{U}^d$, where $\mathcal{X}^d$ and $\mathcal{U}^d$ are randomly sampled from batch $\mathcal{X}$ and $\mathcal{U}$.

$$\mathcal{L}_\mathcal{M} = E[\mathcal{H}(p(y|x', \theta))] - E[\mathcal{H}(p(y|u', \theta))]$$
$$= \frac{1}{\alpha |\mathcal{U}|} \left(\sum_{x' \in \mathcal{X}^d} \mathcal{H}(p(y|x', \theta)) - \sum_{u' \in \mathcal{U}^d} \mathcal{H}(p(y|u', \theta)) \right) \tag{3}$$

Here the conditional entropy $\mathcal{H}(\mathcal{Y}|\mathcal{X})$ is defined as

$$\mathcal{H}(p(y|x, \theta)) = - \sum_{i=1}^{n} p(y|x, \theta)^i \log p(y|x, \theta)^i \tag{4}$$

The conditional entropy is a measure of class overlap, which is invariant to the parameterization of the model. It is related to the usefulness of unlabeled samples where labeling is indeed ambiguous [7,8].

Batch Annealing and Reloading with Temporal Pool. To further reduce the negative impact of mismatched unlabeled samples, we design a batch annealing mechanism to discard those high-entropy unlabeled samples from batch $\mathcal{U}$ and reserve only low-entropy unlabeled samples in batch $\mathcal{U}^r$ for training. The standard for reserved samples is strict in the early stages and is gradually relaxed as the model gets more accurate. Inspired by [11] and [19] which utilize the temporal information of training process, we propose a reloading mechanism with a temporal pool to refill $\mathcal{U}^r$ with low-entropy unlabeled samples. The temporal pool is a size limited buffer to store the temporal samples with lowest entropy in the training process. The reloading mechanism increases the degree of fitting on low-entropy unlabeled samples as well as enhances training stability. The details of batch annealing and reloading will be presented in Sect. 3.3 and Sect. 3.4.

Based on our batch annealing and reloading mechanism, we redefine the consistency regularization loss term $\mathcal{L}_\mathcal{U}$ in a class mismatch case as

$$\mathcal{L}_\mathcal{U} = \frac{1}{(1-\alpha) |\mathcal{U}|} \| \hat{g} - p(y|u, \theta) \|_2^2 \quad u \in \mathcal{U}^r \tag{5}$$

where $\mathcal{U}^r$ represents unlabeled samples after batch annealing and reloading.

Loss Function in ERCM. By adding our proposed entropy repulsion loss term to supervised loss and consistency regularization loss, the loss function in our method is presented in Eq. (6), which is a weighted combination of $\mathcal{L}_\mathcal{X}$, $\mathcal{L}_\mathcal{U}$, and $\mathcal{L}_\mathcal{M}$. Here, $\lambda_\mathcal{X}$, $\lambda_\mathcal{U}$ and $\lambda_\mathcal{M}$ are weights of loss terms.

$$\mathcal{L} = \lambda_\mathcal{X} \mathcal{L}_\mathcal{X} + \lambda_\mathcal{U} \mathcal{L}_\mathcal{U} + \lambda_\mathcal{M} \mathcal{L}_\mathcal{M} \tag{6}$$

Algorithm 1. Entropy Repulsion for Class Mismatch (ERCM)

Require: the labeled batch $\mathcal{X} = sample\{(x_i)\}_{i=1}^{\mathcal{C}} \sim \mathcal{D}_L$
Require: the corresponding labels $\mathcal{Y}$ of $\mathcal{X}$
Require: the unlabeled batch $\mathcal{U} = sample\{(u_i)\}_{i=1}^{\mathcal{C}} \sim \mathcal{D}_U$
Require: the training step t;
Require: $allocate(\mathcal{T}, \mathcal{M})$, $\mathcal{T}$ is an initialized temporal pool, $\mathcal{M}$ is the pool size;
Require: β and γ are annealing parameters;
Require: k is weights warming step;
Require: $\lambda_{\mathcal{X}}$, $\lambda_{\mathcal{U}}$, $\lambda_{\mathcal{M}}$ are weights of loss term

1:   $\mathcal{X}, \mathcal{U} = augmentation(\mathcal{X}, \mathcal{U})$;
2: **for** s in training steps $\lfloor 1, t \rfloor$ **do**
3:      $\lambda_{\mathcal{U}}, \lambda_{\mathcal{M}} = \begin{cases} \lambda \frac{s}{k} & s < k \\ \lambda & s \geq k \end{cases}$
4:      $\alpha = max(1, update(\beta, \gamma, s, t))$;
5:      $\mathcal{U}^d, \mathcal{U}^r, \mathcal{X}^d = batch\_annealing(\mathcal{U}, \mathcal{X}, \alpha)$
6:      $\mathcal{U}^r, \mathcal{T}' = reloading(\mathcal{U}^r, \mathcal{T}, \alpha)$
7:      $\mathcal{T} = \mathcal{T}'$;   //update temporal pool
8:      $\mathcal{L}_{\mathcal{X}} = cross\_entropy(\mathcal{X}, \mathcal{Y})$;   //supervised loss, e.g., Eq.(1)
9:      $\mathcal{L}_{\mathcal{U}} = consistency\_loss(\mathcal{U}^r)$;   //auxiliary loss, e.g., Eq.(5)
10:     $\mathcal{L}_{\mathcal{M}} = erm\_loss(\mathcal{X}^d, \mathcal{U}^d)$;   //entropy repulsion loss in Eq.(3)
11:     $\mathcal{L} = sum(\lambda_{\mathcal{X}}\mathcal{L}_{\mathcal{X}}, \lambda_{\mathcal{U}}\mathcal{L}_{\mathcal{U}}, \lambda_{\mathcal{M}}\mathcal{L}_{\mathcal{M}})$
12:     $\theta = update(\theta, \nabla_\theta \mathcal{L})$;   //e.g. SGD, Adam
13: **end for**
14: **return** θ

Workflow of ERCM. We illustrate the workflow of ERCM in Fig. 1 and give the detailed algorithm in Algorithm 1. First, we conduct stochastic augmentation (line.1, like random horizontal flips or crops) on the input batch $\mathcal{X}$ and $\mathcal{U}$. At the beginning of training, there will be a warming up process of weights for stability as usually done in traditional SSL approaches (line.3). During training the batch s, batch annealing discards high-entropy parts of $\mathcal{U}$ and reserves $\mathcal{U}^r$ (line.5). We uniformly sample $\mathcal{X}^d$ and $\mathcal{U}^d$ from $\mathcal{X}$ and $\mathcal{U}$. Then, we refill $\mathcal{U}^r$ by reloading low-entropy samples from the temporal pool $\mathcal{T}$ (line.6). Finally, we calculate the supervised loss term $\mathcal{L}_{\mathcal{X}}$ by labeled batch $\mathcal{X}$ and corresponding labels $\mathcal{Y}$, auxiliary loss term $\mathcal{L}_{\mathcal{U}}$ by U^r, and entropy repulsion loss term $\mathcal{L}_{\mathcal{M}}$ by $\mathcal{U}^d$ and $\mathcal{X}^d$ (line.8–10). We update the model by minimizing the total loss $\mathcal{L}$ (line.11).

3.3 Batch Annealing

As shown in Algorithm 2, we first calculate the conditional entropy $\mathcal{H}(p(y|u, \theta))$ of unlabeled samples in $\mathcal{U}$. Then, we reserve the first $\alpha \times \mathcal{C}$ lowest-entropy (most confident) samples from $\mathcal{U}$ to compose $\mathcal{U}^r$ for training. Here, the α is the annealing rate, which is obtained by the following increment function:

$$\alpha = \beta + log(\gamma \frac{s}{t} + 1). \tag{7}$$

t is the total training step number and s is the current training step. β and γ are hyperparameters. With steps of training, the model becomes more accurate and robust, meanwhile α increases so as to gradually relax the standard for selecting reserved samples. In this way, our mechanism improves the model training by restricting potential mismatched unlabeled samples from participating in the training.

For each round of training, to calculate $\mathcal{L}_\mathcal{M}$, we uniformly select $(1 - \alpha) \times \mathcal{C}$ samples from $\mathcal{U}$ to compose $\mathcal{U}^d$ and uniformly select $(1 - \alpha) \times \mathcal{C}$ samples from $\mathcal{X}$ to compose $\mathcal{X}^d$. We note that the limitation of $\mathcal{L}_\mathcal{M}$ will gradually decrease due to the increase of α. The batch annealing mechanism anneals both the loss term $\mathcal{L}_\mathcal{M}$ and unlabeled samples $\mathcal{U}^r$ which will participate in the calculation of $\mathcal{L}_\mathcal{U}$.

Algorithm 2. Batch Annealing

Input: the unlabeled batch $\mathcal{U}$;
        the labeled batch $\mathcal{X}$;
        the annealing rate α;
$\mathcal{H} = cal\_entropy\,(\,p(\mathcal{U},\,\theta)\,)$;
$\mathcal{U}^d = uniform\_sample\,(\,\mathcal{U},\,\lfloor(1 - \alpha) \times \mathcal{C}\rfloor)$
$\mathcal{X}^d = uniform\_sample\,(\,\mathcal{X},\,\lfloor(1 - \alpha) \times \mathcal{C}\rfloor\,)$;
$\mathcal{U}^r = lowest\_k\,(\,\mathcal{H},\,\mathcal{U},\,\lceil\alpha \times \mathcal{C}\rceil)$;
return $\mathcal{U}^d, \overline{\mathcal{U}}^r, \mathcal{X}^d$;

3.4 Reloading with Temporal Pool

Before training, we initialize a temporal pool of size $\mathcal{M}$ to store "very likely matched" unlabeled samples in $\mathcal{D}_U$. We first get the union set $\mathcal{B}$ of current $\mathcal{U}^r$ (output of the batch annealing) and the temporal pool $\mathcal{T}$. Then, top $(1 - \alpha) \times \mathcal{C}$ samples with lowest entropy in $\mathcal{B}$ will be reloaded into $\mathcal{U}^r$ to calculate of the auxiliary loss. The top $\mathcal{M}$ samples with lowest entropy in $\mathcal{B}$ will compose the updated temporal pool. A sample will be reloaded if it keeps high confident pseudo label in several continuous temporal training models. The reloading mechanism improves the model to achieve better fitting on high-confidence unlabeled samples as well as more stable training process.

4 Evaluation

4.1 Experiment Configuration

We use Wide ResNet-28 [16] for all models in experiments. Because traditional SSL methods will be badly hurt by class-mismatch problems in the late training period, for fair comparison, we run 3×2^{23} training steps and report the test error rate of a model with highest valid accuracy.

4.2 Supervised with Mixup

Mixup [23] is a widely adopted data augmentation method. In our experiments, we obtain the performance of supervised learning with Mixup using only labeled data, which is denoted as **Supervised-only**.

4.3 ERCM-SSL Implementations

We combine our design with three state-of-the-art SSL approaches MeanTeacher, VAT, and Mixmatch to obtain ERCM-MT, ERCM-VAT, ERCM-Mix. $\lambda_{\mathcal{X}}$ and $\lambda_{\mathcal{U}}$ in SSL methods refer to the implementation in [2] which achieve good performance. Unless otherwise noted, we use constant ERCM hyperparameters with $k = 100\mathbf{k}$, $\mathcal{M} = 64$, and $\gamma = 0.5$ in our experiments.

ERCM-MT & ERCM-VAT: We use consistency regularization in [19] as the auxiliary loss function. Before feeding the unlabeled data into the model, we add a "guessing label" operation to obtain $p(y|u, \theta)$. In our experiments, we set hyperparameters for all class-mismatch cases, where $\lambda_{\mathcal{X}} = 1$, $\lambda_{\mathcal{U}} = 50$, $\lambda_{\mathcal{M}} = 0.001$, and $\beta = 0.65$. We adopt the loss function of VAT to implement ERCM-VAT with the same $p(y|u, \theta)$ as ERCM-MT. In our experiments, we set hyperparameters for all class-mismatch cases, where $\lambda_{\mathcal{X}} = 1$, $\lambda_{\mathcal{U}} = 0.3$, $\lambda_{\mathcal{M}} = 0.05$, and $\beta = 0.75$.

ERCM-Mix: We adopt square difference between guessing label and output for $\mathcal{L}_{\mathcal{U}}$ as shown in Eq. (5). Moreover, original Mixmatch mixes labeled data with unlabeled data by Mixup for better performance with no mismatched samples. However, in class-mismatch cases, we find that it makes the supervised loss hurt by mismatched samples, especially when the quantity of labeled samples is small as shown in Fig. 2 and Table 1. We adjust Mixmatch to **Mix*** by mixing labeled data and unlabeled data separately. In ERCM-Mix, we set hyperparameters for all class-mismatch cases, where $\lambda_{\mathcal{C}} = 1$, $\lambda_{\mathcal{U}} = 100$, $\lambda_{\mathcal{M}} = 0.5$, and $\beta = 0.75$.

4.4 Results

Table 1. Test error (%) ± standard deviation of methods against different class mismatch rate on CIFAR-10 with 250 label samples and 20k unlabeled samples on different random splits.

	0%	20%	40%	60%	80%	100%
MT	28.4 ± 0.5	28.5 ± 2.6	29.9 ± 0.5	30.0 ± 1.5	29.8 ± 0.4	30.1 ± 0.8
Mix	14.1 ± 0.8	18.0 ± 3.4	17.9 ± 1.1	20.7 ± 1.2	24.6 ± 1.4	28.2 ± 1.0
Mix*	13.4 ± 0.5	17.2 ± 1.2	17.1 ± 1.5	19.0 ± 1.6	21.2 ± 1.8	25.5 ± 1.9
Supervised-only	28.4 ± 0.2					
ERCM-MT	26.4 ± 2.7	26.6 ± 0.7	26.7 ± 2.2	28.3 ± 0.8	28.6 ± 0.4	28.6 ± 1.7
ERCM-Mix	**9.7 ± 1.3**	**11.3 ± 1.3**	**14.3 ± 0.8**	**15.6 ± 0.6**	**18.2 ± 1.5**	**23.6 ± 0.7**

Table 2. Test error (%) ± standard deviation of methods against different class mismatch rate on SVHN with 250 label samples and 20k unlabeled samples on different random splits.

	0%	20%	40%	60%	80%	100%
VAT	4.6 ± 0.3	5.1 ± 0.1	6.1 ± 0.5	7.1 ± 0.7	7.7 ± 0.6	10.5 ± 0.3
Mix	**3.4 ± 0.2**	3.8 ± 0.2	5.2 ± 0.8	6.1 ± 0.7	8.6 ± 0.6	13.8 ± 1.6
Mix*	3.4 ± 0.1	4.0 ± 0.1	4.9 ± 0.2	5.3 ± 0.2	7.2 ± 0.4	14.6 ± 1.3
Supervised-only	21.7 ± 0.2					
ERCM-VAT	4.9 ± 0.5	4.9 ± 0.4	5.8 ± 0.3	6.4 ± 0.3	6.8 ± 0.3	**9.6 ± 0.3**
ERCM-Mix	3.5 ± 0.1	**3.6 ± 0.2**	**4.5 ± 0.3**	**5.0 ± 0.6**	**6.3 ± 0.6**	11.2 ± 1.3

Table 3. Ablation study results on CIFAR-10 with 250 labeled samples and 20k unlabeled samples when mismatch rate is 60%. Average test error ± standard deviation with different entropy repulsion loss weights ($\lambda_{\mathcal{M}} = 0.1, 0.25, 0.5$).

Method	250 labels	2000 labels
ERCM-Mix	**17.1 ± 0.6**	7.8 ± 0.1
ERCM-Mix (mix labeled with unlabeled samples)	18.4 ± 0.8	**7.5 ± 0.1**
ERCM-Mix (without entropy repulsion loss term, $\lambda_{\mathcal{M}} = 0$)	18.4 ± 0.4	8.2 ± 0.1
ERCM-Mix ($\alpha = 1$ and $\lambda_{\mathcal{M}} = 0$, equal to Mix*)	20.8 ± 1.4	8.5 ± 0.2
ERCM-Mix (removing temporal pool, $\mathcal{M} = 0$)	18.1 ± 0.7	7.9 ± 0.2

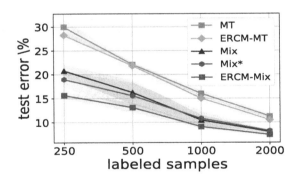

Fig. 2. Test error on various numbers of labeled samples with mismatch rate 60% on splits of CIFAR-10 (6 classes, 400 labels each class). Shaded regions indicate standard deviation over five trials.

In this section, we compare the performances of various methods in class-mismatch cases on different datasets. Mismatch rate represents the proportion of mismatched data among unlabeled data. For example, given 20000 unlabeled samples, 60% mismatch rate means 12000 unlabeled samples are mismatched (Table 3).

CIFAR-10: We first discuss the situation with only a small number of labeled samples. We selected 250 labeled samples, 20k unlabeled samples and 5000 valid samples from CIFAR-10 [10] to train a 5-classes classifier with random splits. We report the average test errors and standard deviations in Table 1. The performances of all three SSL methods decrease gradually as the mismatch rate rises. With the help of our design, ERCM-MT clearly outperforms traditional MT and Supervised-only. ERCM-Mix performs best among all algorithms on CIFAR-10. Compared to the standard Mixmatch, ERCM-Mix achieves up to **6.7%** improvement when the mismatch rate is 20%. Compared to Mix*, ERCM-Mix reduces the error rate by **5.9%** when the mismatch rate is 20%. The results prove that ERCM significantly improves the performance of SSL methods in class-mismatch cases.

We vary the number of labeled samples (250–2000) when the mismatch rate is 60%. The test errors of different methods are presented in Fig. 2. ERCM-Mix still outperforms other methods. We note that the performance of Mix gradually approaches and slightly exceeds Mix* as the number of labeled samples increases. Imbalance between the quantities of labeled and unlabeled samples will introduce uncertainty to training. With smaller quantity of labeled samples, the improvement introduced by ERCM is more significant. Compared to Mix*, the improvement of ERCM-Mix decreases from 3.4% to 0.8% as the number of labeled samples rises.

Table 4. Test error (%) ± standard deviation comparison of 6 classes (400 per class) on CIFAR-10 with mismatch rate of 25% and 75%.

Method	25%	75%
Split-BN+MT	22.4 ± 0.2	22.9 ± 0.4
Split-BN+VAT	23.4 ± 0.3	23.9 ± 0.0
VAT+ROIreg	–	22.3 ± 1.2
ERCM-MT	14.1 ± 0.2	15.6 ± 0.2
ERCM-VAT	16.5 ± 0.4	17.4 ± 0.2
ERCM-Mix	**9.8± 0.1**	**11.8± 0.1**

Table 5. Test error (%) ± standard deviation comparison on 8A8O-Imagenet with mismatch rate of 25% and 75%. Details of 8A8O-Imagenet are described in [22].

Method	25%	75%
Split-BN+MT	44.4 ± 0.5	47.9 ± 0.8
Split-BN+VAT	47.3 ± 0.0	49.3 ± 0.0
ERCM-MT	**32.1 ± 0.5**	**32.7 ± 0.2**
ERCM-VAT	32.5 ± 0.4	33.0 ± 0.6
ERCM-Mix	32.3 ± 0.6	33.4 ± 0.4

To compare with the recent work Split-BN [22] and ROIreg [9], which aims to address the class mismatch issue, we conduct experiments on 6 classes (400 per class) of CIFAR-10 according to [16] and [22]. As shown in Table 4, ERCM-MT and ERCM-VAT significantly outperform Split-BN+MT, Split-BN+VAT and ROIreg+VAT when mismatch rates are 25% and 75%.[1]. Moreover, ERCM-Mix performs best among these methods and achieves 11.8% test error when mismatch rate is 75%.

---

[1] Performances of Split-BN+MT, Split-BN+VAT and ROIreg+VAT are reported in [22] and [9].

SVHN: On SVHN [15], we evaluate traditional VAT and Mixmatch in various class-mismatch cases (0% –100%). We implement ERCM-SSL methods with $\gamma = 0.2$. Table 2 reports the average test error on 250 labeled samples and 20k unlabeled samples over random splits. With no class-mismatch problems, ERCM-SSL methods perform slightly worse than traditional SSL methods. ERCM-SSL methods, however, achieve better performance in all class-mismatch cases. For example, when the mismatch rate is 100%, ERCM-Mix achieves 11.2% test error which is 3.4% lower than Mix*.

8A8O-Imagenet: We conduct evaluations on 8A8O-Imagenet (8 animals and 8 others), a subset of Imagenet [5] described in [22]. We select 600 labeled samples per class for an 8-animals classifier. As shown in Table 5, the performances of ERCM-MT, ERCM-VAT and ERCM-Mixmatch are better than Split-BN+MT and Split-BN+VAT.

4.5 Auxiliary Loss

We explore the impact of our design on auxiliary loss (unsupervised loss). We use 250 labeled samples and 20k unlabeled samples on CIFAR-10 when the mismatch rates is 60%. As shown in Fig. 3, we select uniform batches to observe the auxiliary loss term produced by the unlabeled samples of MT, Mix*, ERCM-MT and ERCM-Mix every 2^{16} steps during training. However, auxiliary loss terms of ERCM-SSL methods are becoming lower than those of traditional SSL methods. ERCM mitigates the harm caused by mismatched data and makes it easier for auxiliary terms to be minimized.

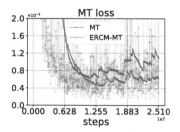 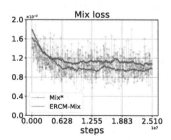

Fig. 3. Auxiliary loss term of SSL methods with and without ERCM when the mismatch rate is 60%. The smoothing rate is 0.95.

4.6 Ablation Study

We conduct ablation study on ERCM-Mix to figure out the importance of each part by removing each part of ERCM separately. We carry out our experiments on CIFAR-10 with 250 labeled and 20k unlabeled samples mentioned in Sect. 4.4

when the mismatch rate is 60% ($\lambda_{\mathcal{M}} = 0.1, 0.25, 0.5$). We measure the impact of using original mixup mode, removing entropy repulsion loss, removing batch annealing operation (i.e. setting $\alpha = 1$ and $\mathcal{L}_{\mathcal{M}} = 0$, equal to Mix*), and removing temporal pool.

5   Conclusion

In this work, we propose ERCM, a new technique that involves a novel entropy repulsion loss together with a batch annealing and reloading mechanism to empower traditional SSL approaches against class-mismatch problems. Compared with the original SSL methods, ERCM-SSL methods can reduce the performance degradation caused by class mismatch samples. Extensive experiments demonstrate a clear performance improvement and strong portability of ERCM. We believe that ERCM has the potential to be combined with more advanced SSL approaches in the future.

Acknowledgments. This research is supported by the National Key R&D Program of China 2018YFB0803400, NSF China under Grants No. 61822209, 61932016, 61751211, China National Funds for Distinguished Young Scientists with No.61625205, the Fundamental Research Funds for the Central Universities.

References

1. Ben-David, S., Blitzer, J., Crammer, K., Kulesza, A., Pereira, F., Vaughan, J.W.: A theory of learning from different domains. Mach. Learn. **79**, 151–175 (2009). https://doi.org/10.1007/s10994-009-5152-4
2. Berthelot, D., Carlini, N., Goodfellow, I., Papernot, N., Oliver, A., Raffel, C.A.: Mixmatch: a holistic approach to semi-supervised learning. In: Advances in Neural Information Processing Systems, pp. 5050–5060 (2019)
3. Chapelle, O., Scholkopf, B., Zien, A.: Semi-supervised learning. IEEE Trans. Neural Netw. **20**(3), 542 (2009). (chapelle, o. et al., eds.; 2006)
4. Chen, Y., Zhu, X., Li, W., Gong, S.: Semi-supervised learning under class distribution mismatch
5. Deng, J., Dong, W., Socher, R., Li, L.J., Li, K., Fei-Fei, L.: ImageNet: a Large-scale hierarchical image database. In: CVPR 2009 (2009)
6. Ganin, Y., et al.: Domain-adversarial training of neural networks. J. Mach. Learn. Res. **17**(1), 2096-2030 (2016)
7. Grandvalet, Y., Bengio, Y.: Semi-supervised learning by entropy minimization. In: Advances in Neural Information Processing Systems, pp. 529–536 (2005)
8. Grandvalet, Y., Bengio, Y.: Entropy regularization. In: Semi-Supervised Learning, pp. 151–168 (2006)
9. Kaizuka, H., Nagasaki, Y., Sako, R.: Roi regularization for semi-supervised and supervised learning. arXiv preprint arXiv:1905.08615 (2019)
10. Krizhevsky, A., Hinton, G., et al.: Learning multiple layers of features from tiny images. Technical report, Citeseer (2009)
11. Laine, S., Aila, T.: Temporal ensembling for semi-supervised learning. arXiv preprint arXiv:1610.02242 (2016)

12. LeCun, Y., Bengio, Y., Hinton, G.: Deep learning. Nature **521**(7553), 436 (2015)
13. Lee, D.H.: Pseudo-label: the simple and efficient semi-supervised learning method for deep neural networks. In: Workshop on Challenges in Representation Learning, ICML, vol. 3, p. 2 (2013)
14. Miyato, T., Maeda, S., Koyama, M., Ishii, S.: Virtual adversarial training: a regularization method for supervised and semi-supervised learning. IEEE Trans. Pattern Anal. Mach. Intell. **41**(8), 1979–1993 (2018)
15. Netzer, Y., Wang, T., Coates, A., Bissacco, A., Wu, B., Ng, A.Y.: Reading digits in natural images with unsupervised feature learning (2011)
16. Oliver, A., Odena, A., Raffel, C.A., Cubuk, E.D., Goodfellow, I.: Realistic evaluation of deep semi-supervised learning algorithms. In: Advances in Neural Information Processing Systems, pp. 3235–3246 (2018)
17. Sajjadi, M., Javanmardi, M., Tasdizen, T.: Mutual exclusivity loss for semi-supervised deep learning. In: 2016 IEEE International Conference on Image Processing (ICIP), pp. 1908–1912. IEEE (2016)
18. Srivastava, N., Hinton, G., Krizhevsky, A., Sutskever, I., Salakhutdinov, R.: Dropout: a simple way to prevent neural networks from overfitting. J. Mach. Learn. Res. **15**(1), 1929–1958 (2014)
19. Tarvainen, A., Valpola, H.: Mean teachers are better role models: weight-averaged consistency targets improve semi-supervised deep learning results. In: Advances in Neural Information Processing Systems, pp. 1195–1204 (2017)
20. Xie, Q., Dai, Z., Hovy, E., Luong, M.T., Le, Q.V.: Unsupervised data augmentation. arXiv preprint arXiv:1904.12848 (2019)
21. Yuan, M., Zhang, L., Li, X.Y., Xiong, H.: Comprehensive and efficient data labeling via adaptive model scheduling. In: 2020 IEEE 36th International Conference on Data Engineering (ICDE), pp. 1858–1861. IEEE (2020)
22. Zając, M., Żołna, K., Jastrzębski, S.: Split batch normalization: Improving semi-supervised learning under domain shift. arXiv preprint arXiv:1904.03515 (2019)
23. Zhang, H., Cisse, M., Dauphin, Y.N., Lopez-Paz, D.: mixup: beyond empirical risk minimization. arXiv preprint arXiv:1710.09412 (2017)
24. Zhang, L., et al.: Crowdbuy: privacy-friendly image dataset purchasing via crowdsourcing. In: IEEE INFOCOM 2018-IEEE Conference on Computer Communications, pp. 2735–2743. IEEE (2018)

Estimating the Performance Indicators of Promotion Efficiency in FMCG Retail

Marcin Blachnik[1]([✉]) [iD] and Joanna Henzel[2] [iD]

[1] Department of Applied Informatics, Silesian University of Technology,
ul. Krasińskiego 8, Katowice, Poland
`marcin.blachnik@polsl.pl`
[2] Department of Computer Networks and System, Faculty of Automatic Control,
Electronics and Computer Science, Silesian University of Technology,
ul. Akademicka 16, 44-100 Gliwice, Poland
`joanna.henzel@polsl.pl`

Abstract. Forecasting promotion efficiency is an important issue in the fast-moving consumer goods sector. The objective of this paper is an analysis of the forecasting performance of two key performance indicators (KPI) used for the assessment of the sales process using machine learning methods. The authors present results of the experiments which were performed for 17 different products on real-life data from a large grocery company. In the paper feature extraction and construction methods are discussed also five different prediction algorithms are compared as well as the feature importance analyses are also provided. Out of the compared algorithms random forest leads and the feature importance are strongly related with the KPI.

Keywords: Applications · Forecasting · Promotions · FMCG

1 Introduction

Promotions play an important role in the modern retail sector. Companies very often are spending a large amount of money on this purpose and sales from promotions make a significant part of a total sale [7].

Over the years many methods and methodologies have been proposed in order to forecast the effect of promotions and optimise them. Very often companies use *judgmental forecasting*. They try to forecast the promotion effect and based on it plan future promotions. Other companies use a simple statistical forecast with judgmental adjustment [8]. Even though many companies still use these simple strategies, a study from 1986 pointed out that using only these kinds of forecasting methods may bring bias [14].

In recent years, with the growing importance of a Data Science field, more research connected with using Machine Learning (ML) methods or Deep Learning (DL) methods were conducted regarding sales forecasting. Different approaches to this problem were taken into consideration, for example, decision

© Springer Nature Switzerland AG 2020
H. Yang et al. (Eds.): ICONIP 2020, LNCS 12533, pp. 320–332, 2020.
https://doi.org/10.1007/978-3-030-63833-7_27

tree based method were considered in [18], extreme learning machine algorithm was proposed in [20], neural networks were used in [4] and [2]. A comparison of different methods regarding this problem can be found in [6] and [13]. However, not so many studies were focused specifically on promotion forecasting using ML or DL methods. Authors of [1] proposed regression trees for a problem of a demand forecasting in the presence of promotions. Method using Principal Component Analysis (PCA) was shown in the paper [19] in order to tackle this problem.

A special kind of sales and promotion forecasting in retail sector is forecasting for fast-moving consumer goods (FMCG). These are products that are consumed quickly e.g. groceries. Forecasting frameworks for this sector was proposed for example in the paper [16] and [10]. The exploratory research presenting the benefits of Machine Learning in sales forecasting for FMCG products can be found in [17].

The objective of this paper is an analysis of forecasting performance of two key performance indicators (KPI) used for assessment of the sales process (a larger group of KPI's were described in [9]). These KPIs are later used for determining the final price of the products, but this aspect is out of the scope of this paper. In the paper, both theoretical aspects and experiments on real data are conducted. The experimental part consists of the data preparation process indicating feature construction from the raw data. Additionally, some methods for preprocessing attributes connected with time are described which are used to increase performance. Finally, five prediction models are evaluated in order to achieve the most accurate prediction.

The paper is organised as follows: the next section describes the problem statement and the data preparation process. In Sect. 3 the explanation of the experiments is provided. The paper ends with some conclusions and discussion of the results.

2   Problem Statement and Data Preparation

In most grocery shops, which are a part of a bigger shop chain, promotions are happening almost non-stop. There are multiple promotions at the same time and they are changing rapidly. When creating a promotion, multiple goals should be taken into consideration. This kind of event should not only make the consumer buy a certain product or buy more of it, but also bring more clients to the shop or encourage them to buy many different products.

In order to capture the effectiveness of a promotion, in [9] we proposed six different indicators which also define the key performance indicators (KPI):

- AVERAGE NUMBER OF SOLD UNITS OR KILOGRAMS EACH DAY (shortcut: AVG. AMOUNT) – This indicator shows how many units or kilograms of the promoted product, on average, were sold during the promotion each day.
- AVERAGE VALUE OF A BASKET CONTAINING THE PROMOTED PRODUCT (shortcut: AVG. BASKET) – This indicator says what an average value of a

basket was where the promoted product appeared. Assuming that customers went for shopping with the will to buy the specific product in promotion, the indicator says how much money they spent in total. The higher the indicator, the more products were bought or the more expensive products were chosen.

- AVERAGE VALUE OF A BASKET CONTAINING THE PROMOTED PRODUCT BUT DISREGARDING THE VALUE OF THE PROMOTED PRODUCT (shortcut: AVG. BASKET WITHOUT ITEM) – This indicator is very similar to the previous one. It shows what an average value of a basket was where the promoted product appeared but the value of the promoted product was not taken into account.
- AVERAGE NUMBER OF RECEIPTS WITH THE PROMOTED PRODUCT (shortcut: AVG. NB. RECEIPTS) – The indicator explains in how many baskets the promoted product appeared, on average, each day during the promotion. It can be treated as an indicator of how many customers bought the product each day.
- AVERAGE NUMBER OF UNIQUE PRODUCTS IN THE BASKET (shortcut: AVG. NB. UNIQUE ITEMS) – It says how varied the basket is. The higher the value of the indicator, the better – it means that the customer not only bought a specific product but also many others.
- AVERAGE NUMBER OF THE BASKETS (shortcut: AVG. NB. CLIENTS) – The indicator shows how many, on average, transactions were performed each day during the promotion. It does not matter if the customer bought a promoted product or not.

The values of indicators are calculated per promotion. It means that each promotion can be described by the 6 proposed indicators. Due to the limited size of the article and in order to present a more in-depth results, we focus only on the first two of them: AVG. AMOUNT and AVG. BASKET. However, a similar analyses can also be conducted for the rest of the indicators.

The forecasting of the promotion effect can be done for every product separately or for a group of products which have a similar response to the promotion. The first aspect is investigated in this paper, while the subsequent was investigated in [9]. In general, having the history of the promotions and their effects, we can model the characteristics of the promotion for the specific product and it is possible to predict what the effect in the future will be. It is important to define proper attributes which will be used to describe each promotion.

2.1 Feature Space of the Input Data

One of the key issues of forecasting the effect of the promotion is the selection of a proper set of attributes. These attributes should cover various aspects such as: being related to price, related to the time of the promotion, describing the advertisement media (promotion channels), describing the store and its surroundings and describing the impact of other promotions.

In the first category, 4 attributes were included: the regular and special price of a product (special is a new price of the product), a change of the price, where

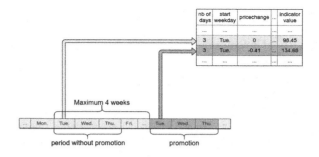

Fig. 1. Finding matching record without promotion.

the change was represented as a relative value, and the reference value of the KPI preceding the promotional period.

The next category represents time attributes, in particular the start day of the promotion, from which we extracted: day of the week (DOW), day of the month (DOM), day of the year (DOY), year.

Considering information about promotion channels, binary attributes were added. They described if the promotion was advertised on TV or the radio, as well as information whether it was specially designed (promotion special design).

The next type of promotion descriptors are attributes representing the store and its surrounding including meta descriptors of the residents and their economic situation. This set of attributes includes: population, the number of city residents, the number of tourists per 1000 residents (in a case when the store is located in the tourists' area), population density, house prices, own parking, the number of parking spaces, average turnover, the number of residents 500 m range, the number of residents 1000 m range, the number of residents 10 min car drive, the number of residents 5 min car drive, the number of cars per 1000 residents, the number of house transactions, unemployment rate, purchasing power resident spendings, average gross salary, competitors rate, the number of competitors, distance to competitors. The information about store ID, product ID, product name, start date and end date of promotion were not included in the data used to train the models, although they were used for data preprocessing.

2.2   Data Preparation

In order to properly forecast the promotion effect some of the attributes required appropriate adaptation and transformation. As mentioned in the previous section one of the attributes which requires to be extracted and adapted from the raw data is the reference value of the KPI preceding the promotional period. This value requires to define the matching period preceding the promotional, in our calculation, it needed to meet the following conditions:

– It considered the same store.
– It had to last as many days as the considered promotion.

- It had to start on the same weekday as the promotion.
- The considered product was not in promotion on any given day.
- The period without promotion could occur maximum 4 weeks and minimum 1 week before the promotion.

The illustration of finding the matching periods is shown in Fig. 1. The matching period was not found for all promotions because of the lack of meeting the requirements. These samples were removed from the data. Similarly, promotions preceding holiday periods have been omitted because they have different characteristics, which will constitute outliers affecting the training process.

One of the important limitations associated with the data representation for the prediction systems is representing cyclical attributes such as weekday, day of the month or day of the year. In this case, typical solutions consist either of a direct numerical representation where, for example, weekday is represented as a number between 1 and 7, or in symbolic form, where weekday is coded using 7 binary attributes, one for each day respectively. Unfortunately, such solutions result in the loss of valuable information. For example, it is not possible to determine the similarity between neighbouring days – for binary representation, the distance between Monday and Friday is identical compared to the distance between Friday and Saturday, and similarly, for numerical representation, the last day of the week is the most dissimilar to the first day of the week (maximum distance = 7). Therefore, the so-called Fourier transformation consisting in replacing a single attribute with a pair of $sin(\frac{2\Pi time}{period})$ and $cos(\frac{2\Pi time}{period})$ attributes was considered. Where $time$ is the value which should be represented in a cyclic form, and $period$ is the value of the period. For example for a week $period = 7$. As a consequence of such transformation, the time values are distributed on a circle as shown in Fig. 2.

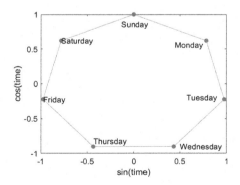

Fig. 2. The effect of day of week transformation.

The final transformation which was applied to the data is within store normalisation. Here the z-score standardisation was used for the attributes representing the KPIs, but for each store separately. The reason for using standardisation for those indicators was that they were referring to the specific values connected with the sale characteristics of a considered product in a given store. For example, the location of the store or other parameters may influence its value.

3   Experiments

The experiments were conducted on a single group of products namely fruits. This group consists of 17 independent products like bananas, apples (3 different apple species, each as a separate product), lemons etc. In order to keep privacy in the remaining part of the article numerical indicators will be used to represent particular products. All of the experiments were conducted for each product independently, so for each product we created a separate prediction model. Initially, we decided to use algorithms that belong to 4 popular and well-known families: tree-based (*Random Forest*), distance-based (*kNN*), neural networks and linear model. We followed the principle popular in meta-learning, where instead of making experiments for a wide range of methods we picked some basic ones to assess them and pick the most promising one. Then the most promising family is evaluated deeper. In our experiments the best results (see Tables 2 and 3) were obtained for the *Random Forest*. As a consequence, we extended the evaluation by an additional tree-based method namely *Gradient Boosted Trees*. In total we evaluated:

- *Random Forest* (RF) [3] - decision tree based ensemble, where the ensemble members collectively vote for the final prediction
- *Gradient Boosted Trees* (GBT) [5] - boosting based approach where the decision trees are constructed one by one in order to minimize some cost function
- *kNN* [12]- k-nearest neighbours
- *Generalized Linear Models* (GLM) [15] - linear model with automatic parameters running model
- Neural networks (*MLP*) [11] - an MLP neural network

The size of the evaluated data varies from 5000 samples up to 20000 samples. Each dataset consists of all of the attributes discussed in Sect. 2.2.

3.1   Prediction Model Pipeline

The experiments were carried out according to the scheme shown in Fig. 3. It starts from loading the raw data and extracting the attributes, then the input attributes are normalised. Here the Z-score was used where the variables after the transformation had a mean value equal to 0, and a standard deviation value equal to 1. Next, the output variables, so the KPIs, were normalised for each shop separately as described in Sect. 2.2. After this, the date variables were transformed with the Fourier transform. This applies to *day of the week*, *day of the month*, and *day of the year*, the date representation also included a year which

Table 1. Parameters used for model optimization within the grid-search procedure.

Model	Parameter	Values
Random Forest	#trees	{50, 60,..., 100 }
Gradient Boosted Trees	#trees	{50, 60,..., 100 }
kNN	k	{1, 3, 5,..., 30}
Generalized Linear Models	Auto	Internally optimized (H_2O package)
MLP	Learning rate	0.01
	Momentum	0.1
	Epochs	2000
	Architecture	{[10], [10, 5], [15], [15, 5]}

could be useful for long term trend prediction, although this variable consists of only 5 unique values (2015, 2016, 2017, 2018 and 2019). After this stage, the parameter optimisation procedure was applied. Here the *grid search* was used where the parameters of the evaluated models are presented in Table 1. For the assessment of the quality of models, the *grid search* procedure included internal 5-fold cross-validation. All of the already described stages were wrapped within the outer cross-validation test which was used to determine the performance of each of the evaluated models. The performance of models was measured using two different metrics namely:

- Root mean square error $RMSE = \sqrt{\frac{1}{n} \sum_{i=1}^{N} (y - \bar{y})^2}$

- Correlation $R = \frac{\sum_{i=1}^{N} (y - mean(y))(\bar{y} - mean(\bar{y}))}{std(y)\, std(\bar{y})}$

where N is the number of samples in the data, y is the true output and $\bar{y}$ is the predicted value. These two measures complement each other and help to understand the value of error. In these experiments different products have different characteristics. Some are sold by the piece and others are measured in kilograms, so for example the RMSE = 5 can have a different meaning for different products, while correlation measure (or the coefficient of determination when squared) has fixed range which can be easily interpreted, but it ignores the bias. Thus in the model optimisation stage, the RMSE was optimised.

3.2   Performance Prediction of the KPI

Obtained results are presented in Tables 2 and 3. They show the results obtained for each product along with its performance, separately for each KPI.

The obtained results indicate that out of the evaluated models *Random Forest* usually leads. In most of the cases, it outperforms other models and competes

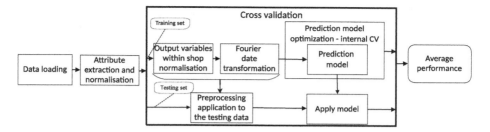

Fig. 3. Scheme of the process of carrying out the experiments.

with the MLP network which takes the second place. We can also observe that for the best models the correlation coefficient is relatively high. For AVG. AMOUNT on average $R \simeq 0.93$, and for AVG. BASKET $\bar{R} \simeq 0.8$, and only for some products it drops to 0.64, while for other it is over 0.96. The issue of the large variance of results for different products has been analysed in more details, and it points out that one of the factors which influence prediction performance is the size of the data. As shown in Fig. 4 the correlation R correlates with the size of the data, and the correlation coefficient is $corr(datasetSize, R) = 0.62$.

Table 2. Prediction results for AVG. AMOUNT for 17 products and 5 prediction models.

Product Id	GBT		kNN		RF		GLM		MLP	
	RMSE	R	RMSE	R	RMSE	R	RMSE	R	RMSE	R
101	53.08	0.919	54.77	0.909	44.79	0.940	61.15	0.885	**44.15**	0.942
102	3.25	0.753	3.29	0.745	**3.12**	0.774	3.30	0.743	3.19	0.765
103	11.42	0.928	9.56	0.935	**7.95**	0.956	10.93	0.914	8.19	0.953
104	9.79	0.876	9.46	0.880	**8.46**	0.905	9.88	0.871	8.53	0.904
105	53.82	0.950	47.60	0.933	33.63	0.966	54.45	0.908	**33.25**	0.967
106	17.14	0.928	15.07	0.935	12.30	0.957	15.93	0.928	**12.05**	0.960
107	4.81	0.888	4.49	0.892	**3.88**	0.920	5.80	0.813	3.98	0.917
108	41.70	0.894	44.37	0.864	35.43	0.915	54.31	0.787	**35.40**	0.917
109	31.53	0.908	30.97	0.895	**24.22**	0.936	33.23	0.876	24.53	0.935
110	16.44	0.917	16.89	0.898	**12.88**	0.942	20.53	0.844	12.96	0.941
111	20.41	0.917	21.36	0.878	**15.04**	0.940	24.81	0.825	15.42	0.936
112	12.59	0.896	12.40	0.883	9.92	0.925	13.42	0.859	**9.88**	0.927
113	12.75	0.926	12.94	0.900	**8.88**	0.952	13.01	0.895	9.11	0.950
114	14.31	0.898	15.02	0.873	**12.11**	0.922	15.77	0.861	12.33	0.921
115	107.74	0.928	103.52	0.908	**73.33**	0.955	108.69	0.899	75.34	0.953
116	11.79	0.895	11.35	0.896	**9.64**	0.926	12.61	0.869	10.20	0.919
117	15.36	0.938	15.11	0.907	**10.55**	0.954	19.87	0.826	10.95	0.951

Table 3. Prediction results for AVG. BASKET for 17 products and 5 prediction models.

Product Id	GBT		kNN		RF		GLM		MLP	
	RMSE	R	RMSE	R	RMSE	R	RMSE	R	RMSE	R
101	4.40	0.953	4.11	0.959	**3.83**	0.964	4.54	0.950	**3.83**	0.965
102	21.48	0.625	21.14	0.640	**21.10**	0.641	21.13	0.640	21.47	0.628
103	8.78	0.867	8.54	0.874	**8.18**	0.885	8.71	0.869	8.26	0.883
104	15.40	0.751	15.23	0.757	**14.89**	0.769	15.27	0.756	15.12	0.763
105	8.96	0.886	8.61	0.893	8.09	0.906	9.21	0.877	**8.08**	0.907
106	5.17	0.946	4.85	0.952	4.55	0.958	5.41	0.941	**4.48**	0.960
107	15.36	0.697	15.31	0.700	**15.03**	0.712	15.26	0.702	15.14	0.708
108	10.29	0.815	9.82	0.834	**9.91**	0.832	10.34	0.816	10.27	0.821
109	8.79	0.828	8.85	0.825	**8.48**	0.840	9.09	0.814	8.59	0.837
110	7.18	0.891	7.12	0.893	**6.85**	0.901	7.14	0.892	6.89	0.900
111	15.49	0.779	15.29	0.783	**14.86**	0.798	15.22	0.786	15.08	0.793
112	13.46	0.781	13.55	0.779	**13.12**	0.795	13.70	0.774	13.17	0.795
113	17.99	0.678	17.98	0.677	**17.41**	0.702	17.85	0.683	17.59	0.695
114	15.13	0.731	15.01	0.737	**14.61**	0.752	15.15	0.730	14.75	0.748
115	7.44	0.895	7.43	0.895	**6.90**	0.910	7.97	0.879	6.94	0.909
116	15.35	0.764	15.11	0.772	**14.81**	0.783	15.05	0.776	15.11	0.777
107	15.44	0.788	15.66	0.779	**14.62**	0.811	15.36	0.789	14.72	0.809

Moreover, obtained results show that the performance obtained by the *Generalized Linear Models* in comparison to the best models is lower. It indicates that the relation between the input variables and the outputs is nonlinear.

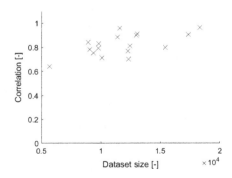

Fig. 4. Relation between dataset size and R performance measure.

3.3   Feature Importance

Next to the prediction quality, one of the most significant factors from the economical point of view is feature importance. It is especially valuable because it indicates which elements of the dataset influence the performance the most. For that reason, we took the *Random Forest* model and extracted feature weights. This procedure was performed independently for each output variable and each product. Then feature importance weights were averaged over products to obtain aggregated indicators separately for each KPI. The obtained results are presented in Fig. 5 and Fig. 6.

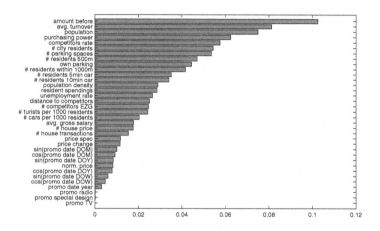

Fig. 5. Feature importance for AVG. AMOUNT.

In both figures, the most significant is the variable indicating the value before promotion. This is reasonable because it defines the reference value. According to these figures, the following most significant features are the meta attributes indicating the welfare and size of the population along with the *competitors rate*. Surprisingly, for AVG. AMOUNT, the *date* variables are not very important, while for AVG. BASKET the *day of month* and *day of year* are among the upper half of features. Moreover, for the AVG. BASKET *special price* and *price change* are the fourth and fifth of the most valuable variables. This is an important factor which indicates the significance of the promotion. In both figures, the least important are variables describing how the advertisement was published like TV or radio. These are binary variables which do not tell how intensive was the advertisement or which media was used, whether it has a global or local range, or was it a prime time or not etc. We believe that this is the reason why their usability is so limited.

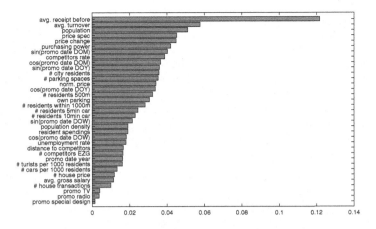

Fig. 6. Feature importance for AVG. BASKET.

4   Conclusion and Discussion

Forecasting plays an important part in the retail sector and promotions forecasting has a big impact on a total sale in a company. Good understanding of the promotions and being able to optimise them is very crucial for maximizing profit in the FMCG sector.

This study has attempted to compare different algorithms for forecasting the efficiency of the promotions. The paper takes into consideration modelling promotions for every product separately. Authors focused on two KPI in order to describe promotions: AVG. AMOUNT and AVG. BASKET. Five different algorithms were investigated and the experiments were conducted for 17 different products (fruits). In the result, the authors got a comparison of the efficiency of different models for these products and for two indicators. The results show that Random Forest in most of the cases outperforms other models. This may be because some attributes were binary and tree-based algorithms handle them very well. Methods based on distances, e.g. kNN, would require appropriate weighting of the attributes and this may be the reason for obtaining worse results for these algorithms. However, these are the authors' assumptions and would require a more in-depth analysis to confirm them.

The paper precisely describes the process of data set preparation. It shows how the records were created and what set of attributes was used. The paper explains also an important process of representing time in order to keep the information about a cyclical characteristic of time attributes. A meaningful part of this paper is a feature importance description. The plots similar to these showed on Figs. 5 and 6 can be used by practitioners and business people to gain knowledge on promotions in their company and to find features that influence the performance the most. This kind of information can be used to optimise future promotions because the user knows which parameters are the most important for the final result.

In conclusion, this paper provides analyses of the forecasting performance of two KPIs that describe the efficiency of the promotions. The experiments were conducted on real-life data. The comparison of different algorithms was presented and the practical aspect of this study was also provided.

Acknowledgments. This work was partially supported by the European Union through the European Social Fund (grant POWR.03.05.00-00-Z305). The work was carried out in part within the project co-financed by European Funds entitled "Decision Support and Knowledge Management System for the Retail Trade Industry (SensAI)" (POIR.01.01.01-00-0871/17-00).

References

1. Ali, Ö.G., Sayin, S., van Woensel, T., Fransoo, J.: SKU demand forecasting in the presence of promotions. Expert Syst. Appl. **36**(10), 12340–12348 (2009). https://doi.org/10.1016/j.eswa.2009.04.052
2. Au, K.F., Choi, T.M., Yu, Y.: Fashion retail forecasting by evolutionary neural networks. Int. J. Prod. Econ. **114**(2), 615–630 (2008). https://doi.org/10.1016/j.ijpe.2007.06.013
3. Breiman, L.: Random forests. Mach. Learn. **45**(1), 5–32 (2001). https://doi.org/10.1023/A:1010933404324
4. Chen, C.Y., Lee, W.I., Kuo, H.M., Chen, C.W., Chen, K.H.: The study of a forecasting sales model for fresh food. Expert Syst. Appl. **37**(12), 7696–7702 (2010). https://doi.org/10.1016/j.eswa.2010.04.072
5. Chen, T., Guestrin, C.: XGBoost: a scalable tree boosting system. In: Proceedings of the 22nd ACM SIGKDD International Conference on Knowledge Discovery and Data Mining, KDD 2016, pp. 785–794. ACM (2016). https://doi.org/10.1145/2939672.2939785
6. Chu, C.W., Zhang, G.P.: A comparative study of linear and nonlinear models for aggregate retail sales forecasting. Int. J. Prod. Econ. **86**(3), 217–231 (2003). https://doi.org/10.1016/S0925-5273(03)00068-9
7. Cohen, M.C., Leung, N.H.Z., Panchamgam, K., Perakis, G., Smith, A.: The impact of linear optimization on promotion planning. Oper. Res. **65**(2), 446–468 (2017). https://doi.org/10.1287/opre.2016.1573
8. Fildes, R., Goodwin, P., Önkal, D.: Use and misuse of information in supply chain forecasting of promotion effects. Int. J. Forecast. **35**(1), 144–156 (2019). https://doi.org/10.1016/j.ijforecast.2017.12.006
9. Henzel, J., Sikora, M.: Gradient boosting application in forecasting of performance indicators values for measuring the efficiency of promotions in FMCG retail. In: Pre Proceedings of the 2020 Federated Conference on Computer Science and Information Systems, pp. 59–68 (2020)
10. Huang, T., Fildes, R., Soopramanien, D.: The value of competitive information in forecasting FMCG retail product sales and the variable selection problem. Eur. J. Oper. Res. **237**(2), 738–748 (2014). https://doi.org/10.1016/j.ejor.2014.02.022
11. Kordos, M., Blachnik, M.: Instance selection with neural networks for regression problems. In: Villa, A.E.P., Duch, W., Érdi, P., Masulli, F., Palm, G. (eds.) ICANN 2012. LNCS, vol. 7553, pp. 263–270. Springer, Heidelberg (2012). https://doi.org/10.1007/978-3-642-33266-1_33

12. Kordos, M., Blachnik, M., Strzempa, D.: Do we need whatever more than k-NN? In: Rutkowski, L., Scherer, R., Tadeusiewicz, R., Zadeh, L.A., Zurada, J.M. (eds.) ICAISC 2010. LNCS (LNAI), vol. 6113, pp. 414–421. Springer, Heidelberg (2010). https://doi.org/10.1007/978-3-642-13208-7_52

13. Krishna, A., Akhilesh, V., Aich, A., Hegde, C.: Sales-forecasting of retail stores using machine learning techniques. In: Sales-Forecasting of Retail Stores using Machine Learning Techniques, pp. 160–166. IEEE (2018). https://doi.org/10.1109/CSITSS.2018.8768765

14. Makridakis, S.: The art and science of forecasting an assessment and future directions. Int. J. Forecast. **2**(1), 15–39 (1986). https://doi.org/10.1016/0169-2070(86)90028-2

15. McCullagh, P., Nelder, J.A.: Generalized Linear Models, vol. 37. CRC Press, Boca Raton (1989)

16. Sen, J., Chaudhuri, T.D.: A predictive analysis of the Indian FMCG sector using time series decomposition-based approach. SSRN Electron. J. (2017). https://doi.org/10.2139/ssrn.2992051

17. Tarallo, E., Akabane, G.K., Shimabukuro, C.I., Mello, J., Amancio, D.: Machine learning in predicting demand for fast-moving consumer goods: an exploratory research. IFAC-PapersOnLine **52**(13), 737–742 (2019). https://doi.org/10.1016/j.ifacol.2019.11.203

18. Thomassey, S., Fiordaliso, A.: A hybrid sales forecasting system based on clustering and decision trees. Decis. Supp. Syst. **42**(1), 408–421 (2006). https://doi.org/10.1016/j.dss.2005.01.008

19. Trapero, J.R., Kourentzes, N., Fildes, R.: On the identification of sales forecasting models in the presence of promotions. J. Oper. Res. Soc. **66**(2), 299–307 (2015). https://doi.org/10.1057/jors.2013.174

20. Xia, M., Zhang, Y., Weng, L., Ye, X.: Fashion retailing forecasting based on extreme learning machine with adaptive metrics of inputs. Knowl.-Based Syst. **36**, 253–259 (2012). https://doi.org/10.1016/j.knosys.2012.07.002

Exploring User Trust and Reliability for Recommendation: A Hypergraph Ranking Approach

Yanbin Jiang[1], Huifang Ma[1,2,3($\boxtimes$)], Yuhang Liu[1], and Zhixin Li[2]

[1] College of Computer Science and Engineering, Northwest Normal University,
Lanzhou 730070, Gansu, China
mahuifang@yeah.net
[2] Guangxi Key Lab of Multi-source Information Mining and Security,
Guangxi Normal University, Guilin 541004, Guangxi, China
[3] Guangxi Key Laboratory of Trusted Software,
Guilin University of Electronic Technology, Guilin 541004, Guangxi, China

Abstract. Recent recommendation strategies attempt to explore relations among both users and items, applying techniques of graph learning and reasoning for solving the so-called information isolated island limitations. However, the graph-based ranking algorithms model the interactions between the user and item either as a user-user (item-item) graph or a bipartite graph that capture pairwise relations. Such modeling cannot capture the complex relationship shared among multiple interactions that can be useful for item ranking.

In this paper, we propose to leverage hypergraph random walk into the ranking process. We develop a new recommendation framework Hypergraph Rank (HGRank), which exploits the weighting methods for hypergraph on both hyperedges and vertices. This leads to the expressive modeling of high-order interactions instead of pairwise relations. Specifically, we take social trust and reliability into the hypergraph weighting process to improve the accuracy of the algorithm. Extensive experimental results demonstrate the effectiveness of our proposed approach.

Keywords: Social trust · Reliability · Hypergraph ranking · Random walk

1 Introduction

Recommendation has been widely applied in various online services, including E-commerce, content sharing, social networking, forum etc. Over the past decade, a vast number of algorithms have been proposed to tackle the top-N recommendation task, aiming at identifying a ranked list of N items users may likely be interested in based on the historical interactions like purchases, view, like etc.

The methods dealt with the top-N recommendation task broadly fall into two classes: the latent space ones and the graph ranking based (which focus

© Springer Nature Switzerland AG 2020
H. Yang et al. (Eds.): ICONIP 2020, LNCS 12533, pp. 333–344, 2020.
https://doi.org/10.1007/978-3-030-63833-7_28

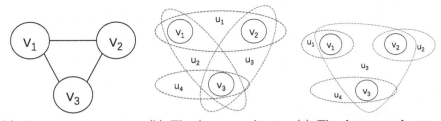

(a) Items co-occurrence graph (b) The hypergraph constructed in case 1 (c) The hypergraph constructed in case 2

Fig. 1. An example illustrates the advantages of hypergraphs over normal graphs in capturing user-item complex relationships

either on users or items or both). The latent space methods [3,13] take users' scoring information in the form of a matrix, and compute a low-rank factorization thus the user and the item are represented by a latent vector through the decomposition of the matrix. The inner product is adopted to describe the pairwise relationship and prediction the score. The graph-based methods [5,7] assume that user-user connections can be established via social relations or similar attributes while user-item interactions involve either implicit feedback or explicit feedback. These investigated objects are usually endowed with pairwise relationships, which can be illustrated as graphs, based on which graph learning and reasoning approaches are applied.

In many real-world problems, however, relationships among the objects of our interest are more complex than pairwise. Simply squeezing the complex relationships into pairwise ones will inevitably lead to loss of information which can be expected valuable for recommendation tasks however. Put it another way, representing a set of complex relational objects as graph is incomplete. Given an item, assuming that the only information we have is which user has interacted with this item. An undirected graph can be constructed in which two vertices are joined together by an edge if there is at least one common user have interacted with these items, and then a graph based ranking approach can applied. Despite their effectiveness, we argue that these methods are not sufficient to yield satisfactory recommendation results because such graph representation obviously misses the information on whether the same users joined interacting with three or more items or not. In the recommendation system, an interaction $v_1 : u_1$ in the history indicates that user u_1 has interacted with item v_1(click, purchase, etc). And we built two interaction sets to illustrate this problem. For different interactions$\{v_1 : u_1, u_2; v_2 : u_1, u_3; v_3 : u_2, u_3, u_4\}$, $\{v_1 : u_1, u_3; v_2 : u_2, u_3; v_3 : u_3, u_4\}$, they have the same structure on the item co-occurrence graph. However, a hypergraph can capture the differences in this complex interaction, as shown in Fig. 1. Such information loss is unexpected and hence utilizes the information is useful for our ranking task.

On the other hand, consideration of user ratings alone may not be sufficient to accurately characterize user similarities. In other words, users may initially be biased toward certain users. Existing studies address the above problems to some extent by integrating additional information into a recommendation system [8]. And they obtained good recommended results. At present, the rapid development of social networking platforms provides information sources for the construction of social relationships among users. In social networks, social relations between users often seem as whether users trust other, which to some extent provides users' preference information. Therefore, on the basis of hypergraph modeling, we use the user's trust relationship to give the explicit coding of user preference. That is, we believe that a higher similarity weight should be assgined for users trusted by the user to be recommended. In addition, in order to make full use of social information and rating data, we model the hyperedge weight using three parts: authority, reliability and the similarity. In addition, algorithm performance can be improved by adjusting the relationship among these three through hyperparameters.

In this paper, we aim to build a recommendation system based on Hypergraph Ranking (HGRank), which can explicitly encode high-order connectivity between items. The theory of hypergraph ranking generalizes the traditional notion of graph ranking, whereby the interaction is now defined over more than a pair of vertices. The proposed recommendation framework have elaborately designed the weighting strategies for both hyperedge and hypervertexes. In particular, social trust and user reliability are considered for hyperedge weighing. Social relations between users often serve as whether users trust each other, which to some extent provides users' preference information whereas the reliability of a user refers to the accuracy of its recommendation, i.e., to what extent a user's recommendations to another user are accurate.

Our major contributions are summarized as follows:

1. We propose a novel hypergraph ranking approach, which can model user-item interaction on the hypergraph instead to completely represent complex relationships among the items of our interest;
2. We provide a principled approach to jointly capture both social trust and reliability for hyperedge weighting;
3. We demonstrate the effectiveness of the proposed framework on various real-world datasets.

The rest of our paper is organised as follows. We summarize the related work in Sect. 2. Then, shortly recapitulate the details of hypergraph and make a brief statement about our task in Sect. 3. Based on this, we give a detailed description about our method in Sect. 4 and Sect. 5. Then, we introduce the datasets, experimental settings and discuss the effectiveness of proposing algorithm in Sect. 6. Finally, we conclude this work in Sect. 7.

2   Related Work

We review the existing approaches that work on trust information and high-order connectivity, which are most relevant with our method. Then we summarize these methods and briefly explain the differences from our method.

Social recommendation, which is capable of conquering the data sparsity problem of traditional recommended systems by considering social information, has attracted a broad range of interest of researchers. With the rapid development of social media data, a large amount of valuable social information can be utilized to effectively solve these problems by modeling the interactions between users. SoRec is considered to be a heuristic work in the field of social recommendation, in which users' rating information and trust information are represented by a vector. Based on SoRec, SocialMF [9] re-formalizes the contributions of trust user to active users, rather than direct predictions of the items. In other words, user preferences are built on the basis of the preferences of friends they trust. TrustSVD [4] is the first work to extend SVD++ with social information. And the method considers that the explicit(true value of ratings and trust) and implicit (who rates what and who trusts whom) effects of user-item score will influence the generation prediction. DSR [10] learns binary codes in a unified framework for users and items, considering social information. It improves the performance of recommendations by learning the binary codes for users and items in a single-stage method.

Due to the prevalence and success of graph neural network technology, many recent studies in the recommendation fields have turned their attention to modeling high-order connectivity. NGCF [14] first constructs the message propagation mechanism and then explicitly integrates the collaborative signal of high-order connectivity into the embedding layer by stacking multiple embedding propagation layers. In addition, due to the advantages of the random walk method in capturing the high-order connectivity of the graph. Researchers also model the high-order connectivity through a random walk. Hop-rec [15] extends the approximate representation of indirect high-order connections to the training set through the random walk, and then applies the matrix decomposition method to recommend.

Although our approach also leverages trust information and makes recommendations by capturing high-order connectivity. We are different from the above methods, they focus more on the unilateral factors. Our approach takes into account both. In this paper, we capture high-order connectivity by representing a user-item bipartite graph as a hypergraph and applying a hypergraph random walk by incorporating the authority and reliability derived from the trust social relationships and rating information into the hypergraph weighting.

3   Preliminaries and Problem Statement

We first review the formalize definition of hypergraph and its related concepts, followed by the definition of the task we study and some commonly used notations.

A simple graph is a representation of a set of vertices that each edge connects two vertices, while hyperedges in a hypergraph are generalizations of edges in a simple graph, a special type of edges that can connect any number of hypervertices [11]. And we follow the nomenclature of [1]. A hypergraph $HG = (V, E)$

contains vertices set V and hyperedges set E. The hyperedge $e \in E$ is essentially a subset of the hypervertices set(i.e. $\bigcup_{e \in E} = V$). For the convenience of description, we use vertices to represent hypervertices in the following. And then, a hypergraph HG is represented by an indicator matrix $\mathbf{H}$ with entries $h(v, e) = 1$ if $v \in e$ and 0 otherwise. For a weighted hypergraph WHG, a weight $w(e)$ is associated with each hyperedge e. And the weights of the vertices on different hyperedges are different. $w(v_e)$ indicates the weights of the vertices v on specific hyperedges e. Similarly, we use the weighted indicator matrix $\mathbf{H}_w$ to represent the weighted hypergraph and the weights $h_w(v, e) = w(v_e)$. In addition, the degree $d(e)$ of a hyperedge e is defined as the sum of all the weights of vertices on the hyperedge e, i.e., $d(e) = \sum_{v \in V} w(v_e)h(v, e)$. The degree of a vertex $d(v)$ is defined as the sum of all the weights of the hyperedges containing v, i.e., $d(v) = \sum_{e \in E} w(e)h(v, e)$.

In the recommendation system, since users are rating in the same item space and the rating for a item is discriminative for different users, it is intuitive to view a user u as a hyperedge and the vertices on this hyperedge represent the item rated by u. By convention, we assume that this recommendation system contains a set of n users $U = \{u_1, u_2, \cdots, u_n\}$ and a set of m items $V = \{v_1, v_2, \cdots, v_m\}$. And $\mathbf{A} = [A_{ij}]_{n \times m}$ is the user-item rating matrix, containing the items that the users have rated. Similarly, we define a trust matrix $\mathbf{T} = [T_{ij}]_{n \times n}$ with $T_{ij} = 1$ if u_i trust u_j and 0 otherwise.

We first generate the hypergraph indicator matrix based on the rating data, and then integrate the trust relationship into the hypergraph weighting process by calculating user authority and reliability. After this, a random walker is used to navigate the generated hypergraph, resulting in a list of length m, each element of which represents the preference of user u. And we can get the top-N recommendation for user u by truncate ordered list.

4   Weighting Strategy

As described, the rules used to weight the hyperedges and the vertex are the essential to a hypergraph. To address this key point, we design the following principle to weight a hypergraph.

4.1   Hyperedges Weighting

When we incorporate social data such as trust information into the recommendation system, for a recommending hyperedge(user) u_i the weight of another users u_j can be considered to be determined by the authority of u_j, the trust relationship and similarity between u_i and u_j. In this paper, we define the above three indicators as $R_{auth}(u_j)$, $R_{reli}^{(u_i)}(u_j)$ and $R_{sim}^{(u_i)}(u_j)$, respectively. Generally speaking, the user trusted by most other users are more authoritative. Trust data is usually represented by a directed graph, so we can do a random walk on this directed matrix $\mathbf{T}$ to get an sorted list $\mathbf{r}$ about authority of all the users to use in the next step of the algorithm. Therefore, the authority of u_j can be defined as the u_j-th element of $\mathbf{r}$,

$$R_{auth}(u_j) = \mathbf{r}_{u_j} \tag{1}$$

Since the authority of u_j is certain for all of users, we omit the superscript u_l. Then, we can construct a function to describe the reliability of user u_j for user u_l as,

$$R_{reli}^{(u_l)}(u_j) = \mathbb{I}(u_l, u_j)e^{-\frac{1}{1+|\mathcal{N}_{u_j}|}} \tag{2}$$

Where $\mathbb{I}(u_l, u_j)$ is an indicator function that implies whether user u_l trusts user u_j(i.e.$\mathbb{I}(u_l, u_j) = 1$ when u_l trusts u_j). If u_l indeed trusts u_j, we can calculate the reliability through the above equation. And the $|\mathcal{N}_{u_j}|$ denotes the number of items rated by u_j. That is, we assume that the more users make ratings, the more reliability they are.

The similarity between the u_l and u_j can be intuitively defined according to the cosine similarity of the user's rating vector for all items as,

$$R_{sim}^{(u_l)}(u_j) = \frac{1 + cos(\mathbf{A}_{u_l}, \mathbf{A}_{u_j})}{2} \tag{3}$$

Finally, the hyperedge weights are calculated as follows,

$$w^{u_l}(u_j) = w_1 R_{auth}(u_j) + w_2 R_{reli}^{(u_l)}(u_j) + w_3 R_{sim}^{(u_l)}(u_j) \tag{4}$$

Where w_1, w_2 and w_3 denote the smoothing factor of the three indicators, which are constrained of: $w_1 + w_2 + w_3 = 1$. Adjust the hyperedge weights according to the different parameters values of w_1, w_2, w_3, and we will explore the optimal selection of them in Sect. 6.2.

4.2   Vertices Weighting

Inspired by the text-word weight strategy, we weight the vertices(items) by the co-occurrence, correlation, and the co-occurrence distance(i.e. similarity of rating) on a particular hyperedge. Let u_l as the user to be recommended. Since the following metrics are different for different users, for the sake of clarity, we omit the superscript u_l that represent the current user to be recommended for all non-critical locations. Given items v_i, v_j and $v_i, v_i \in u_s$, their co-occurrence $co_{u_s}(v_i, v_j)$ for user u_s can be calculated as,

$$co_{u_s}(v_i, v_j) = w^{u_l}(u_s) \times e^{-dist_{u_s}(v_i, v_j)} \tag{5}$$

where $dist_{u_s}(v_i, v_j) = (A_{si} - A_{sj})^2$ denotes the square of the difference in user ratings between the two items, which the larger rating gap between the two items indicate the lower similarity between them, and then multiplied by the user weight to imply the co-occurrence of items for u_s. The co-occurrence between any two items v_i and v_j is the sum of $co_{u_s}(v_i, v_j)$ for all users:

$$co(v_i, v_j) = \sum_{s=1}^{n} co_{u_s}(v_i, v_j). \tag{6}$$

Furthermore, we define the unilateral correlation degree $ucor(v_i, v_j)$ to represent the probability of associating v_j by observing v_i as,

$$ucor(v_i, v_j) = \frac{co(v_i, v_j)}{\sum_{k=1}^{m} co(v_i, v_k)} \times \log_2 \frac{m}{NEI(v_j)} \tag{7}$$

Where the right-hand side of the formula is used to penalize the v_j that are common to a lot of items, and $NEI(v_j)$ represents the number of items that have co-occurred with v_j.

In the same way, the unilateral correlation of $ucor(v_j, v_i)$ can be calculated. Hence, the correlation between v_i and v_j is defined as the mean of the unilateral correlation between $ucor(v_i, v_j)$ and $ucor(v_j, v_i)$ as,

$$cor(v_i, v_j) = \frac{1}{2}(ucor(v_i, v_j) + ucor(v_j, v_i)) \tag{8}$$

The associative weight $cow(v_i, u_l)$ is reflected in the representativeness of vertex v_i in a specific hyperedge u_l. the higher the associative weight of v_i means that when v_i appears, the higher the probability of other vertices v_j appearing in hyperedge u_s. Further, calculate the correlation weight of an item in a particular user as,

$$cow(v_i, u_s) = \frac{\sum_{j=1}^{m} cor(v_i, v_j) * H(v_j, u_s)}{|\mathcal{N}_{u_s}|} \tag{9}$$

The correlation weight is combined with the global statistical weight, that is, inverse frequency statistics (if) considers that items with fewer global statistics have higher weights for the users interacting with them. Based on this, the weight calculation formula of vertex v_i on hyperedge u_s is as follows,

$$w(v_i, u_s) = cow(v_i, u_s) \times log_2 \frac{1+m}{1+|\mathcal{N}_{v_i}|} \tag{10}$$

5   Hypergraph Random Walk

Unlike simple graphs, a hypergraph usually contains more than two vertices on a single hyperedge. Therefore, a more general random walk method is needed for hypergraphs. Bellaachia et al. generalized the random walk method on hypergraphs [1]. Its random walk process is as follows: First, select the starting vertex u, and select a specific hyperedge e containing the current vertex u in proportion to the probability of the size of the hyperedge weight $w(e)$; Then, in the determined hyperedge, the transfer is carried out according to the calculated probability of the weight of vertices. Let $\mathbf{P}$ be the probability matrix of the random walk of the hypergraph. And the calculation method is as follows,

$$P(v_i, v_j) = \sum_{u \in U} w(u) \frac{h(v_i, u)}{\sum_{\hat{u} \in \mathcal{N}_{v_i}} w(\hat{u})} \frac{h_w(v_j, u)}{\sum_{\hat{v} \in u} h_w(\hat{v}, u)} \tag{11}$$

For the convenience of calculation, [1] gave the matrix form calculation method as,

$$\mathbf{P} = \mathbf{D}_v^{-1}\mathbf{H}\mathbf{W}_e\mathbf{D}_e^{-1}\mathbf{H}_w^T \tag{12}$$

where $\mathbf{D}_v, \mathbf{D}_e$ is the degree diagonal matrix of vertices and hyperedges, $\mathbf{H}$ is the hypergraph indicator matrix, $\mathbf{W}_e$ is the diagonal matrix of the hyperedge weights and $\mathbf{H}_w$ is the weighting hypergraph indicator matrix.

Once the transition matrix is constructed, the obstacles that prevent the random walk process are all removed. First set the initial distribution vector $\mathbf{v}^0$ equally probabilistically. Then, after iterating a number of steps according to Eq. 13, $\mathbf{v}$ will no longer change significantly, that is, the convergence is completed. β is the smoothing factor.

$$\mathbf{v}^{i+1} = \beta\mathbf{P}^T\mathbf{v}^i + (1-\beta)\mathbf{v}^0 \tag{13}$$

After the iteration finally stops, we can sort the convergent vector $\mathbf{v}$ to get the first N items that the recommended user might be interested in.

6   Experiments

We implement our method in Python, and all the experiments are implemented on a computer with a 4.0 GHz CPU and 32 GB memory. In this section, we perform experiments on two real-world datasets to evaluate our proposed method and answer the following research questions:

- **RQ1** How does our proposed method as compared with state-of-the-art methods?
- **RQ2** How can we benefit from modeling authority and reliability?

6.1   Datasets

For evaluate the effectiveness of our method, we conduct series of experiments on two datasets:Filetrust and CiaoDVD, which are publicly accessible and we summarize the characteristics of the two datasets in Table 1.

Table 1. Statistics of the datasets

Dataset	User#	Item#	Rating#	Trust#	Density (Rating/Trust)
Filmtrust	1,508	2,071	35,497	1,853	0.011400/0.000815
CiaoDVD	17,615	16,121	72,665	40,133	0.000256/0.000129

The two datasets are widely used in previous studies [2,12]. We adopt the leave-one-out protocol, for each dataset we randomly select one of historical positive ratings of each user to constitute the training set.

6.2   Experimental Settings

Evaluation Metrics. Following the neural collaborative filtering [6], we use Hit Rate(HR@k) and Normalized Discount Cumulative Gain (NDCG@k) to measure the performance of different recommendation algorithms.

Baselines. To demonstrate the effectiveness our proposed method, we compare it with the following methods:

- MF: This is the most classical personalized recommendation algorithm based on latent factor factorization.
- SocialMF [9]: This method assumes that the user's preference information is largely dependent on the preference information of the trusted friend
- Hop-rec [15]: This is a graph-based model, where high-order neighborhood generated by random walk is used to enrich user-item interaction data.
- NGCF [14]: This is a new recommendation framework based on graph neural network, which model the collaborative signal in the form of high-order connectivities by performing embedding propagation.

Parameter Settings. The latent factor size is fixed to 32 for all MF-based models. For NGCF, we use the proposed epoch of 10 and set the number of embedded propagation layers to 3. For our method, there are three non-negative parameters in the hyperedge weighting process:w_1, w_2, w_3, considering $w_1 + w_2 + w_3 = 1$, we only need to consider two independent parameters. To this end, we use grid search to find the optimal parameters of w_1 and w_3. The results are visualized in Fig. 2. We find that the best Settings for parameters were [0.2, 0.1, 0.7] on the Filmtrust dataset. And the best parameter in CiaoDVD is similar. In addition, we use the recommended value of 0.85 for the random walk parameter β.

6.3   Performance Comparison

We first compare the performance of all the baselines. To be fair, we experiment three times and reported the average results of the performance comparison in the Table 2. From the result, We summarize several important observations.

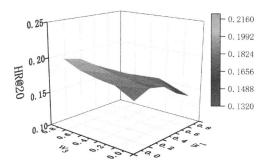

Fig. 2. Impact of parameters$[w_1, w_2, w_3]$ on HGRank performance

1) *Social information can help improve recommendation performance.* In general, adding more attributes to a recommendation system can improve recommendation performance. Social recommendation methods capture more relationships between users and users than normal recommendation methods that only consider rating information. As the results show, social recommendation is superior to the MF approach. Experimental results from both datasets demonstrate that incorporating social information into recommendations improves accuracy.

2) *Capturing the high-order connectivity is more efficient.* Hop-rec and NGCF are two of the latest recommendation models to capture high-order connectivity, and the difference between them is how to model the high-order connectivity. Specifically, Hop-rec performs a random walk on a bipartite graph to obtain the user's interaction with multi-hop neighbors. NGCF use a GNN to propagate the embedding representation to capture the information of high-order connectivity. Thus, high-order propagation is explicitly coded in model training. In all cases, Hop-rec and NGCF have achieved performance improvements that demonstrate the importance of capturing high-order connectivity rather than local information on interactive bipartite graph.

3) *Hypergraphs have advantages over simply bipartite graphs.* Our method performed best on most of the results, demonstrating the effectiveness of using the side information of the trust relationship and building bipartite graph of user-item interactions as hypergraphs. Experimental results show that by representing bipartite graphs as hypergraphs, we can capture more complex relationships between user-items to improve recommendation performance.

However, in CiaoDVD, HGRank underperformance NGCF *w.r.t.* ndcg. The reason might be that our random-walk approach, compared to NGCF, which captures high-order connectivity through a multi-layer messaging mechanism, transfers the weight correlation of all nodes at each iteration. And without considering the connectivity of each layer, this can lead to suboptimal results.

6.4   Effect of Social Information

We now investigate the second question, i.e., how can we benefit from modeling authority and reliability? In our proposed hypergraph design, the trust and reliability factors are incorporated into recommendation systems. Three variations of HGRank are used for this ablation study. Their performance on Filmtrust is included in Table 3, Results on CiaoDVD show the same observations.

- HGRank-*1*. HGRank without using authority factor in hyperedges weighting process.
- HGRank-*2*. hyperedges weighting strategy of HGRank with reliability factor removed.
- HGRank-*3*. An extremely simplified version of HGRank, which combine the above two variants means removing both the authority and reliability factors.

From the results in Table 3, we have the following observations:

Table 2. The performance of all methods on all the two datasets

	Filmtrust				CiaoDVD			
	hr@20	ndcg@20	hr@50	ndcg@50	hr@20	ndcg@20	hr@50	ndcg@50
MF	0.1406	0.0567	0.2248	0.0762	0.0079	0.0039	0.0125	0.0048
SocialMF	0.1916	0.0745	0.2865	0.1110	0.0230	0.0075	0.0354	0.0109
Hop-rec	0.1942	0.0824	0.3627	0.1242	0.0246	0.0089	0.0468	0.0143
NGCF	0.2056	0.0918	0.3804	0.1304	0.0298	**0.0128**	0.0499	0.0168
HGRank	**0.2141**	**0.1003**	**0.4083**	**0.1353**	**0.0315**	0.0119	**0.0511**	**0.0183**
impr.%	4.13%	9.26%	7.33%	3.76%	5.70%	–	2.40%	8.93%
p-value	2.14e−3	7.01e−4	4.52e−5	2.34e−3	4.26e−2	–	4.35e−3	2.10e−2

Table 3. The performance of all variants on filmtrust in terms of $k = 20$

	HGRank-*1*	HGRank-*2*	HGRank-*3*	HGRank
HR@20	0.1923	0.1856	0.1710	0.2141
NDCG@20	0.0938	0.0874	0.0811	0.1021

1) HGRank is consistently superior to all variants. We attribute this improvement to the use of trust relationships, authority and reliability measures. Thus, we verify the rationality and effectiveness of introducing trust relationship and measuring reliability in the hyperedge weighting.

2) In all cases, HGRank-2 underperforms HGRank-1. This shows that considering authority alone is not enough to obtain better hyperedge weights to improve recommendation performance without introducing the direct trust relationships that help capture information about preferences between users.

7   Conclusion

The development of social network provides a new opportunity to improve the recommendation algorithm. In this paper, in order to make full use of social information, we design a new framework, which converts the user-item bipartite graph of the traditional recommendation system into a hypergraph, and models the user's trust social relationship into a hyperedge weighting through authority and reliability indicators. The constructed probability transfer matrix is applied to the hypergraph random walk to capture the high-order connectivity. Through these designs, our method can make the recommended results achieve higher precision. The experimental results on the real datasets confirm the rationality and effectiveness of our method.

Acknowledgments. This work is supported by the National Natural Science Foundation of China (61762078, 61363058, 61966004), Research Fund of Guangxi Key Lab of

Multi-source Information Mining and Security (MIMS18-08), Northwest Normal University young teachers research capacity promotion plan (NWNU-LKQN2019-2) and Research Fund of Guangxi Key Laboratory of Trusted Software (kx202003).

References

1. Bellaachia, A., Al-Dhelaan, M.: Hg-rank: a hypergraph-based keyphrase extraction for short documents in dynamic genre. In: # MSM, pp. 42–49 (2014)
2. Chen, C., Zhang, M., Liu, Y., Ma, S.: Social attentional memory network: modeling aspect-and friend-level differences in recommendation. In: The 12th ACM International Conference on Web Search and Data Mining, pp. 177–185 (2019)
3. Ding, J., Yu, G., He, X., Feng, F., Li, Y., Jin, D.: Sampler design for Bayesian personalized ranking by leveraging view data. IEEE Trans. Knowl. Data Eng. (2019)
4. Guo, G., Jie, Z., Yorkesmith, N.: TrustSVD: collaborative filtering with both the explicit and implicit influence of user trust and of item ratings (2015)
5. He, X., Gao, M., Kan, M.Y., Wang, D.: Birank: towards ranking on bipartite graphs. IEEE Trans. Knowl. Data Eng. $29(1)$, 57–71 (2016)
6. He, X., Liao, L., Zhang, H., Nie, L., Hu, X., Chua, T.S.: Neural collaborative filtering. In: The 26th International Conference on World Wide Web, pp. 173–182 (2017)
7. Hu, X., Mai, Z., Zhang, H., Xue, Y., Zhou, W., Chen, X.: A hybrid recommendation model based on weighted bipartite graph and collaborative filtering. In: The International Conference on Web Intelligence Workshops, pp. 119–122 (2016)
8. Indra, R., Thangaraj, M.: An integrated recommender system using semantic web with social tagging system. Int. J. Semant. Web Inf. Syst. $15(2)$, 47–67 (2019)
9. Jamali, M., Ester, M.: A matrix factorization technique with trust propagation for recommendation in social networks. In: The 4th Conference on Recommender Systems (2010)
10. Liu, C., Wang, X., Lu, T., Zhu, W., Sun, J., Hoi, S.C.: Discrete social recommendation. In: The 33rd AAAI Conference on Artificial Intelligence (2019)
11. Mao, M., Lu, J., Han, J., Zhang, G.: Multiobjective e-commerce recommendations based on hypergraph ranking. Inf. Sci. 471, 269–287 (2019)
12. Sun, P., Wu, L., Wang, M.: Attentive recurrent social recommendation. In: The 41st International ACM SIGIR Conference on Research and Development in Information Retrieval, pp. 185–194 (2018)
13. Tran, T., Lee, K., Liao, Y., Lee, D.: Regularizing matrix factorization with user and item embeddings for recommendation. In: The 27th ACM International Conference on Information and Knowledge Management, pp. 687–696 (2018)
14. Wang, X., He, X., Wang, M., Feng, F., Chua, T.: Neural graph collaborative filtering. In: The 42nd International ACM SIGIR Conference on Research and Development in Information Retrieval, pp. 165–174 (2019)
15. Yang, J.H., Chen, C.M., Wang, C.J., Tsai, M.F.: HoP-rec: high-order proximity for implicit recommendation. In: The 12th ACM Conference on Recommender Systems, pp. 140–144 (2018)

Facial Action Units Intensity Estimation via Graph Relation Network

Ce Wang, Fei Jiang$^{(\boxtimes)}$, and Ruimin Shen

Department of Computer Science and Engineering, Shanghai Jiao Tong University,
Shanghai, China
{dirtyface,jiangf,rmshen}@sjtu.edu.cn

Abstract. Facial action units (AUs) intensity estimation is a fundamental task for facial expression analysis, emotion recognition and affective computing. Since AUs commonly appear in specific combinations and are highly related to each other, modeling relations among multiple AUs is expected to improve the estimation performance. In this paper, we propose a novel end-to-end Graph Relation Networks approach to efficiently capture hidden relations among AUs. Firstly, we model AU intensity estimation tasks in a weighted directed graph. Secondly, we design an attention-based graph relation framework to capture dynamic relations and perform information sharing between tasks. To the best of our knowledge, we are the first to introduce graph neural networks into the AU intensity estimation. Experimental results on two public benchmark databases, BP4D and DISFA, show that our method achieves the state-of-the-art performance.

Keywords: Facial expression analysis · Action units intensity · Graph neural networks

1 Introduction

Facial expression recognition enjoys increasing attention due to their potential applications in emotion recognition, affective computing and human-computer interaction. Facial action units describe the facial expression locally, representing movements of one or more muscles in the face. Facial Action Coding System (FACS) [3] was designed to systematically depict those facial muscle movements. It defines AUs as movements of one or a group of muscles, and a group of AUs can code nearly any possible facial expression. FACS also divides AUs intensity into six intensity levels: Neutral < Trace(A) < Slight (B) < Pronounced (C) < Extreme (D) < Maximum (E). Given an input facial image, the goal of AU intensity estimation is to predict the intensity level of each AU.

The work was supported by National Nature Science of Science and Technology (No. 61671290), China Postdoctoral Science Foundation (No. 2018M642019), Shanghai Municipal Commission of Economy and Information (No. 2018-RGZN-02052).

© Springer Nature Switzerland AG 2020
H. Yang et al. (Eds.): ICONIP 2020, LNCS 12533, pp. 345–356, 2020.
https://doi.org/10.1007/978-3-030-63833-7_29

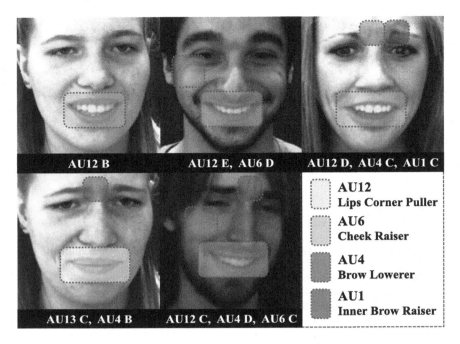

Fig. 1. Facial action units are highly related to each other. For example, AU6 is highly related to the intensity of AU12, and AU4 appears differently depending on whether it occurs alone or in combination with AU1.

Since AUs are temporal actions and each AU is only related to different region features, current works of AU detection and AU intensity estimation mostly focus on temporal [6,21] or regional [1,8,17,23] feature extraction. Most of those works consider each AU independently and ignore relations among AUs. Actually, facial muscle interactions are controlled by certain anatomical mechanisms [3]. AUs are highly related to each other and significantly affect each other's appearance.

For example (Fig. 1), high intensity of AU12 (lip corner puller) always results in AU6 (chin raiser). AU1, AU4 (eyebrow lower) and AU9 (nose wrinkler) are associated with the group of muscles near the glabellae. And AU17 (chin raiser), AU14 (dimpler), AU15 (lip corner depressor) are all related to the group of muscles near the mouse and chin. Those AUs always occur in a group and appear differently with or without others. Therefore, modeling relations among multiple AUs is expected to achieve a more accurate estimation of the target AU intensity.

Researchers have recently begun to consider AU relations in AU recognition and intensity estimation. Some multi-task learning approaches [19], additional constraints approaches [4,22] and probabilistic graphical approaches [7,18] have been investigated. The multi-task learning approaches simultaneously deal with multiple AUs, while lacking a specific mechanism to capture relation representation within each AU. The additional constraints can only capture local or fixed AU relations but are unable to model the variations in AU relations. The probabilistic graphical models can capture complex, and global AU relations but cannot be trained end-to-end. In this paper, we adopt the end-to-end trainable graph neural network structure to extract dynamic relations between AU intensity estimation tasks.

Graph is a kind of data structure that models a set of objects (nodes) and their relations (edges). Due to the great expressive power of graphs, analyzing graphs with machine learning has been receiving more and more attention. Graph neural networks (GNNs) are connectionist models that capture the dependence of graphs via messages passing between the nodes of graphs. Unlike standard neural networks, GNNs retain a state that can represent information from its neighborhoods with arbitrary depth. Recent progress in network architectures, optimization techniques, and parallel computation have enabled many groundbreaking applications of GNNs [2,10,16].

The attention mechanism has been successfully used in many sequence-based tasks such as natural language processing, machine translation, and so on. Driven by the task goal, an attention mechanism can model dependency between the elements without knowing their locations and feature distributions. Graph attention network (GAT) [16] incorporates the self-attention mechanism into the propagation step, which simultaneously computes the hidden states of each node by attending over its neighbors, following a multi-head attention strategy.

Inspired by the GAT, we propose a novel graph relation network structure, which uses a self-attention mechanism to efficiently generate nodes relations in graph structure. Then, the message propagates through the graph. Using this graph relation network structure, we can easily model the relations and information sharing between AU intensity estimation tasks.

Our contributions can be summarized as follows:

- We propose a novel graph relation network model to extract dynamic relations better and achieve information sharing between AU intensity estimation tasks. To the best of our knowledge, it is the first time to introduce GNN into AU intensity estimation.
- Our Graph Relation Network model is end-to-end trainable, and can be easily plugged into most backbone networks.
- We show that our model achieves the state-of-the-art performance on two public AU intensity estimation benchmark databases.

2   Related Work

2.1   Facial Action Units Analyses

Facial action units analysis works can be divided into two categories: AU recognition and AU intensity estimation. AU recognition task only predicts the appearance of each AU, while AU intensity estimation also estimates AU intensity level. Current deep learning works in AU detection and AU intensity estimation mainly focus on temporal or regional feature extraction. Zhao et al. [23] designed a region layer to model regional feature of AUs. Li et al. [8] improved VGG with an attention map for regional feature extracting of AUs. Hu et al. [6] designed a cross-concat and temporal neural network to consider physical features and the distribution differences of AUs. However, all above approaches ignored relations between AUs.

2.2   Researchs on AU Relations

Recently, some researchers have begun to examine how AU relations could improve AU recognition and AU intensity estimation. Zhao et al. [22] leverages group sparsity by setting constraints to select a sparse subset of facial patches for multiple AU recognition. Eleftheriadis et al. [4] proposed a multi-conditional approach with Bayesian learning strategy based on Monte Carlo sampling. Robert Walecki et al. [18] combined conditional random field (CRF) with deep learning to encode AU pairs dependencies. Rudovic et al. [12] proposed a conditional ordinal random field model for context-sensitive modeling of AU intensity. Kaltwang et al. [7] proposed a generative latent tree model to represent the joint distribution of AU intensities and facial features. Although current works explored AU relations for facial action units analyses works, few end-to-end trainable deep learning approaches were proposed to capture dynamic and global relations between AUs efficiently.

2.3   Graph Neural Networks

Graph neural networks (GNNs) are deep learning based methods that operate on graph domain, to capture the dependence of graphs via message passing between the nodes of graphs. Due to its convincing performance and high inductive ability, GNNs have been widely applied in graph analysis tasks.

GNNs were first introduced in Gori et al. [5] and Scarselli et al. [13] as a generalization of recursive neural networks that can directly deal with a more general class of graphs. Recent researches mainly focus on spectral [2] and spatial [10,16] Graph Convolutional Neural Networks(GCNs), which aim to generalize convolutional neural networks to graph-structured data. CNNs intrinsically exploit the regular grid-like structure of data defined on the euclidean domain (e.g. images), GCNs extend this concept to non-regularly structured data (e.g. social/brain networks). Graph attention network(GAT) [16] is a GCN structure method which incorporates the self-attention mechanism into the propagation step, simultaneously computes the hidden states of each node by attending over its neighbors, following a self-attention strategy. Due to GNN's interpreting ability and convincing performance on dependence capturing, it is natural to introduce graph neural network structure to model relations between AUs.

3   Proposed Methods

Figure 2 summarizes the overall structure of our graph relation networks. Firstly, given an input face image, we apply a pre-defined convolutional network (VGG19

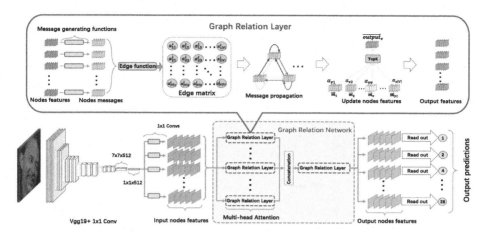

Fig. 2. The overall architecture of the proposed graph relation network. The vgg19 backbone extracts deep features from input face images, then the features are sent through our two layer structure graph relation network to generate global representation of each task, finally we use those representations to predict each AU's intensities.

[14] in our experiments) to extract high semantic image features. Secondly, we adopt 1×1 Convs on those image features to generate input nodes features for each AU, since each AU relies on different semantic features from different regions. Thirdly, we define the graph relation layer, which adopts relation capturing and information sharing within graph structure. Then we stack graph relation layers to construct the graph relation network. Fourthly, we send nodes features through the graph relation network to get the final high representative global features for each task. Finally, the readout functions use those global features to predict the intensities for each AU intensity estimation task directly.

3.1 Graph Relation Layer

Our graph relation layer takes nodes features as input, then builds a graph structure to update those features. Formally, let $\mathcal{G} = (\mathcal{V}, \mathcal{E})$ denotes the graph structure we need to build. Nodes $v \in \mathcal{V}$ take unique values from $\{1, \cdots, |\mathcal{V}|\}$ ($|\mathcal{V}|$ equals to the number of AUs to estimate). Then we define edges as a matrix $\mathcal{E}$ ($\mathcal{E} \in \mathbb{R}^{|\mathcal{V}| \times |\mathcal{V}|}$), and each element in matrix denotes relations between nodes.

Firstly, we adopt message generating functions $\mathcal{M}_v : \mathbb{R}^{|f^l|} \to \mathbb{R}^{|m^l|}$ for each node v to compute incoming messages respectively. At each graph relation layer l, the messages are computed as:

$$m_v^l = \mathcal{M}_v \left(f_v^l \right) \tag{1}$$

m_v^l denotes messages of node v in the $l-th$ layer, f_v^l denotes input features of node v in the $l-th$ layer. In our experiments, we defined the message generate functions $\mathcal{M}_v$ as convolutional layers with 1×1 kernels.

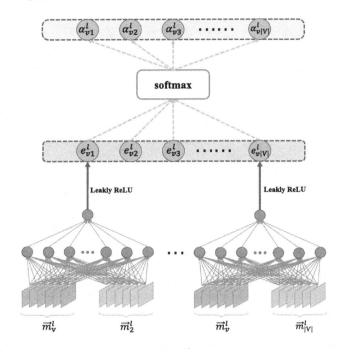

Fig. 3. Self-attention based edge function E^l takes message tuples (v, u) as input to calculate the importance (α_{vu}^l) of the message of node u^l to node v^l

Secondly, with messages generated for each node, we adopt a self-attention based edge function to generate directed edges for each node tuples to represent relations between them. For each layer l, we define a shared self-attention based edge function $E^l : \mathbb{R}^{2|m^l|} \to \mathbb{R}$ to take messages from each nodes tuples and compute attention coefficients:

$$e_{vu}^l = E^l \left(m_v^l, m_u^l \right) \tag{2}$$

Figure 3 summarizes the structure of the edge function. This e_{vu}^l indicates the importance of node u's message to node v. Then we normalize them across all nodes using the softmax function. This step can be seen as generating the graph topology structure. We define edge function E^l as a single-layer fully connected neural network, parameterized by a weight vector $\boldsymbol{a}$ ($\boldsymbol{a} \in \mathbb{R}^{2|m^l|}$), then apply an activation function σ (LeakyReLU in our experiments). The inputs are concatenated node message vectors. So the edge function can be written as:

$$\alpha_{vu}^l = \mathrm{softmax}_u \left(\sigma \left(\overrightarrow{a}^T \left[m_v^l \| m_u^l \right] \right) \right) \tag{3}$$

where $\|$ represents the concatenation operation and α_{vu}^l is normalized e_{vu}^l.

Once obtained, the normalized attention coefficients α_{ij}^l are used to aggregate messages from nodes set $\mathcal{V}$, to generate output features for every node.

Until now, our model allows every node to focus on every other node. In order to learn a sparse graph structure focusing only on the most relevant nodes, we adopt a top-k strategy to retain only k most important messages of each node. In our experiments, we use a linear combination with activation function σ (LeakyReLU) to implement this operation:

$$\boldsymbol{output}_v^l = \sigma \left(\sum_{u \in \mathcal{V}_v^k} \alpha_{vu}^l \boldsymbol{m}_u^l \right) \tag{4}$$

$\boldsymbol{output}_v^l$ denotes the output features of node v in $l - th$ layer, $\mathcal{V}_v^k$ denotes a set consists of nodes with k largest attention coefficients to node v.

This step can be seen as message propagation through the graph. With the topology structure generating step and message propagation step, the graph relation layer obtains the ability of setting up node relations and share information through each task. With the above definition, we can build graph relation network by stacking graph relation layers.

3.2   Multi-head Attention Strategy

In our experiments, we build a two-layer graph relation network to generate high representative global features for each AU. We employ the multi-head attention strategy on the first graph relation layer to stabilize the learning process of self-attention, since multi-head attention strategy can capture variable AU relations parallelly [15,16]. The multi-head attention strategy allows the model to jointly focus on information from different representation subspaces at different positions. To achieve this, we ensemble K independent graph relation layer, then calculate K group independent feature vectors to generate the final output of the first layer. We combine K groups of feature vectors by concatenation:

$$\boldsymbol{output}_v = \|_{k=1}^{K} \sigma \left(\sum_{u \in \mathcal{V}_v^k} \alpha_{vu}^{lk} \boldsymbol{m}_u^{lk} \right) \tag{5}$$

where $\|_{k=1}^{K}$ denotes concatenating K feature vectors and σ denotes activation function (LeakyReLU).

3.3   Intensity Regression

The second layer in graph relation network takes concatenated multi-head outputs as input, then calculates the final hidden representations of each AU intensity estimation task. Finally, we use those hidden representations to generate the final estimation results. Since we treat AU intensity estimation as a regression task, we define readout functions R_v to directly output the predictions y_v for each AU intensity:

$$y_v = R_v \left(\hat{f}_v \right) \tag{6}$$

Where $\hat{f}_v$ denotes the final representation of AU v. We implement readout functions by several fully connected layers(parameterized by $\mathbf{W}^R$) followed by an activation function σ (LeakyReLU):

$$y_v = R_v\left(\hat{f}_v\right) = \sigma\left(\mathbf{W}^R\hat{f}_v\right) \tag{7}$$

3.4 Loss Function

We choose the Smooth L1 Losses to evaluate each estimation task, and sum them up with equal weights to obtain the total training loss:

$$\text{Loss} = \sum_{v \in \mathcal{V}} \text{Smooth L}_1(y_v, \widehat{y}_v) \tag{8}$$

where $\mathcal{V}$ is the group of AUs to estimate, y_v and $\widehat{y}_v$ denote the prediction and the ground truth of AU v. The Smooth L1 Losses can be writen as:

$$\text{Smooth L}_1(x) = \begin{cases} 0.5x^2 & \text{if } |x| < 1 \\ |x| - 0.5 & \text{otherwise} \end{cases} \tag{9}$$

Since our graph relation network structure enables the consistent, efficient propagation of gradients in the complete pipeline, we can adopt any gradient-based stochastic optimizer to train our model end-to-end.

4 Experiments

4.1 Settings

Datasets. We evaluate the proposed graph relation network on two large spontaneous benchmark datasets – Denver Intensity of Spontaneous Facial Action (DISFA) [9] and BinghamtonPittsburgh 4D database (BP4D) [20]. BP4D consists of 328 sequences from 41 subjects. Around 140, 000 frames are annotated with AU intensity for 5 AUs. In our experiments, we use 27 subjects for training and 13 for testing. DISFA consists of 27 sequences from 27 subjects. Around 130, 000 frames are annotated with AU intensity for 12 AUs. In our experiments, we use 18 subjects for training and 9 for testing.

Pre-processing. Our data prepossessing includes face image cropping, image normalization, and data augmentation. Firstly we register face images by using facial landmarks provided in each database. Then, we crop the face image and resize it into 224 × 224. For data augmentation, we randomly adopt rotation, crop and horizontal flip to face images and perform contrast normalization to alleviate the influence of illumination changes.

Implementation Metails. In our experiments, we train our model with mini-batches of 64, initial learning rate of 0.001 and weights decay of 5e−4. All

Table 1. Comparing with VGG19

	Datasets	BP4D						DISFA												
	AU	6	10	12	14	17	Avg	1	2	4	5	6	9	12	15	17	20	25	26	Avg
MAE	VGG19	1.24	1.39	1.14	1.80	1.19	1.35	.68	.52	1.31	.16	.76	.59	.67	.43	.59	.47	1.33	.76	.69
	Ours	.65	.77	.53	.92	.63	.70	.18	.10	.29	.04	.25	.09	.23	.09	.31	.18	.26	.42	.20
	Datasets	BP4D						DISFA												
	AU	6	10	12	14	17	Avg	1	2	4	5	6	9	12	15	17	20	25	26	Avg
ICC(3,1)	VGG19	.63	.61	.73	.25	.31	.51	.19	.14	.19	.02	.39	.33	.68	.14	.27	.03	.59	.38	.28
	Ours	.66	.72	.84	.28	.54	.60	.43	.57	.78	.40	.48	.59	.84	.47	.37	.02	.93	.48	.53

experiments are based on the pytorch toolbox, and are performed on NVIDIA Tesla K40c GPU.

Evaluation Metrics. We use the Mean Absolute Error (MAE) and Intra-class Correlation ICC(3,1) as the measures. MAE is widely used to evaluate regression and ordinal classification performances. ICC(3,1) is widely used in behavioral sciences to measure agreement between annotators (AU intensity labels and output predictions).

Baseline Models. We compare our approach with the original VGG19 [14] network with fully connected layers and other 6 state-of-the-art approaches: CNN [1], OR-CNN [11], CCNN-IT [17], EAC [8], DRML [23] and KJRE [21]. All baselines compared enjoy the same pre-processing as we mentioned above.

4.2   Results

Our Model vs VGG19. Our model use convolutional layers of VGG19 as image features extractor, then use the proposed graph relation network to predict the intensities. So we compare our model with the original VGG19 to verify the validity of the proposed graph relation network. As showen in Table 1, our model achieve better performance on BP4D and DISFA datasets.

Multi-head Attention Strategy. To verify the validity of the Multi-head attention strategy in our model, we evaluate our model with different number of heads. As shown in Table 2, results on BP4D and DISTF datasets show that mulit-heads attention strategy can achieve better performance than single-head attention strategy.

Comparing with the State-of-the-art Models. As shown in Table 3, we compare our method with the state-of-the-art ones on BP4D and DISFA datasets. For the BP4D dataset, our method achieve the best performance on average in MAE and achieves the best performance of AU10 and AU17 in ICC(3,1). For the DISFA dataset, our method achieves the best performance of most AUs and average both in MAE and ICC(3,1). Furthermore, compared with baseline models, our method achieves better improvements on AU1, 2, 4, 9, which are highly related since they are all occurred by the group of muscles near glabellae. In summary, our method uses graph relation structures to model

Table 2. Comparison between graph relation networks with different number of heads. (K denotes the number of heads)

	Datasets	BP4D						DISFA												
	AU	6	10	12	14	17	Avg	1	2	4	5	6	9	12	15	17	20	25	26	Avg
MAE	K = 1	.80	.91	.71	1.14	.75	.86	.21	.34	.51	.08	.37	.23	.35	.22	.38	.41	.31	.52	.33
	K = 4	.77	.79	.65	1.02	.68	.78	.19	.12	.23	.03	.28	.11	.30	.10	.33	.23	.23	.39	.21
	K = 8	.65	.77	.53	.92	.63	.70	.18	.10	.29	.04	.25	.09	.23	.09	.31	.18	.26	.42	.20
	Datasets	BP4D						DISFA												
	AU	6	10	12	14	17	Avg	1	2	4	5	6	9	12	15	17	20	25	26	Avg
ICC(3,1)	K = 1	.62	.60	.82	.22	.41	.52	.20	.23	.65	.34	.22	.38	.60	.31	.43	.03	.76	.46	.38
	K = 4	.65	.64	.83	.30	.43	.57	.24	.32	.70	.42	.30	.48	.80	.45	.46	.03	.88	.52	.46
	K = 8	.66	.72	.84	.28	.54	.60	.43	.57	.78	.40	.48	.59	.84	.47	.37	.02	.93	.48	.53

Table 3. Comparison with the state-of-the-art methods on the DISFA & BP4D datasets. The best results are shown in bold and in brackets. The second best results are shown in bold only.

	Datasets	BP4D						DISFA												
	AU	6	10	12	14	17	avg	1	2	4	5	6	9	12	15	17	20	25	26	avg
	CNN [1]	1.30	1.35	1.28	1.80	1.14	1.37	1.62	1.09	1.44	.23	.86	.71	.83	.50	.63	.47	1.71	.84	.91
	OR-CNN [11]	1.37	1.39	1.37	1.80	1.19	1.42	1.05	.87	1.47	.17	.79	.70	.69	.44	.59	.50	1.33	.86	.79
	CCNN-IT [17]	1.14	1.30	.99	1.65	1.08	1.23	.87	.63	.86	.26	.73	.57	.55	.38	.57	.45	.81	.64	.61
MAE	EAC [8]	.76	**.86**	**.61**	**1.06**	.72	**.80**	.48	.46	.85	.09	.40	.41	.44	.24	.37	.23	**.50**	.51	.42
	DRML [23]	**.73**	.86	.67	1.27	**.71**	.85	**.44**	**.38**	**.80**	**.07**	.35	.29	.36	.16	[.28]	[.14]	.53	[.38]	**.35**
	KJRE [21]	.82	.95	.64	1.08	.85	.87	1.02	.92	1.86	.70	.79	.87	.77	.60	.80	.72	.96	.94	.91
	Ours	[.65]	[.77]	[.53]	[.92]	[.63]	[0.70]	[.18]	[.10]	[.29]	[.04]	[.25]	[.09]	[.23]	[.09]	.31	.18	[.26]	.42	[.20]
	Datasets	BP4D						DISFA												
	AU	6	10	12	14	17	avg	1	2	4	5	6	9	12	15	17	20	25	26	avg
	CNN	.67	**.69**	.77	.35	.33	.56	.05	.04	.36	.02	.44	.27	.67	.25	.08	.03	.46	.22	.23
	OR-CNN	.60	.61	.59	.25	.31	.47	.03	.07	.01	.00	.29	.08	.67	.13	**.27**	.00	.59	.33	.20
	CCNN-IT	[.75]	**.69**	**.86**	[.40]	.45	[.63]	.18	.15	**.61**	.07	[.65]	**.55**	**.82**	.44	[.37]	[.28]	.77	[.54]	**.45**
ICC(3,1)	EAC	.70	.64	**.84**	.32	.45	.59	.08	.07	.30	.14	.46	.15	.70	.09	**.27**	.14	**.82**	.36	.30
	DRML	**.73**	.67	.81	.36	**.47**	.61	.09	.05	.41	.15	.40	.26	.71	.17	.18	.11	.80	[.54]	.32
	KJRE	.71	.61	[.87]	**.39**	.42	.60	**.27**	**.35**	.25	**.33**	**.51**	.31	.67	.14	.17	**.20**	.74	.25	.35
	Ours	.66	[.72]	.84	.28	[.54]	.60	[.43]	[.57]	[.78]	[.40]	.48	[.59]	[.84]	[.47]	[.37]	.02	[.93]	.48	[.53]

relations among multiple AUs, and achieves the state-of-the-art performance on BP4D and DISFA datasets.

5   Conclusion

In this paper, we introduce graph structure neural network into AU intensity estimation tasks. The proposed graph relation network models AU intensity estimation tasks in graph structure to adopt relation capturing and information sharing through graph. Our method can extract highly related global features for each AU, and then leads to a more accurate estimation of the target AU intensities. Evaluations on two benchmark AU datasets: BP4D and DISFA demonstrate that our method can achieve the state-of-the-art performance on AU intensity estimation tasks.

References

1. Amogh, G., Tasli, H.E., Den Uyl, T.M., Maroulis, A.: Deep learning based facs action unit occurrence and intensity estimation. In: 2015 11th IEEE International Conference and Workshops on Automatic Face and Gesture Recognition (FG), vol. 6, pp. 1–5. IEEE (2015)
2. Defferrard, M., Bresson, X., Vandergheynst, P.: Convolutional neural networks on graphs with fast localized spectral filtering. In: Advances in Neural Information Processing Systems, pp. 3844–3852 (2016)
3. Ekman, P., Rosenberg, E.: What the Face Reveals: Basic and Applied Studies of Spontaneous Expression Using the Facial Action Coding System (FACS). Oxford University Press, New York (1997)
4. Eleftheriadis, S., Rudovic, O., Pantic, M.: Multi-conditional latent variable model for joint facial action unit detection. In: Proceedings of the IEEE International Conference on Computer Vision, pp. 3792–3800 (2015)
5. Gori, M., Monfardini, G., Scarselli, F.: A new model for learning in graph domains. In: Proceedings, 2005 IEEE International Joint Conference on Neural Networks, vol. 2, pp. 729–734. IEEE (2005)
6. Hu, Q., Jiang, F., Mei, C., Shen, R.: CCT: a cross-concat and temporal neural network for multi-label action unit detection. In: 2018 IEEE International Conference on Multimedia and Expo (ICME), pp. 1–6. IEEE (2018)
7. Kaltwang, S., Todorovic, S., Pantic, M.: Latent trees for estimating intensity of facial action units. In: Proceedings of the IEEE Conference on Computer Vision and Pattern Recognition, pp. 296–304 (2015)
8. Li, W., Abtahi, F., Zhu, Z., Yin, L.: EAC-Net: a region-based deep enhancing and cropping approach for facial action unit detection. In: 2017 12th IEEE International Conference on Automatic Face & Gesture Recognition (FG 2017), pp. 103–110. IEEE (2017)
9. Mavadati, S.M., Mahoor, M.H., Bartlett, K., Trinh, P., Cohn, J.F.: DISFA: a spontaneous facial action intensity database. IEEE Trans. Affect. Comput. 4(2), 151–160 (2013)
10. Niepert, M., Ahmed, M., Kutzkov, K.: Learning convolutional neural networks for graphs. In: International Conference on Machine Learning, pp. 2014–2023 (2016)
11. Niu, Z., Zhou, M., Wang, L., Gao, X., Hua, G.: Ordinal regression with multiple output CNN for age estimation. In: Proceedings of the IEEE Conference on Computer Vision and Pattern Recognition, pp. 4920–4928 (2016)
12. Rudovic, O., Pavlovic, V., Pantic, M.: Context-sensitive dynamic ordinal regression for intensity estimation of facial action units. IEEE Trans. Pattern Anal. Mach. Intell. 37(5), 944–958 (2014)
13. Scarselli, F., Gori, M., Tsoi, A.C., Hagenbuchner, M., Monfardini, G.: The graph neural network model. IEEE Trans. Neural Netw. 20(1), 61–80 (2008)
14. Simonyan, K., Zisserman, A.: Very deep convolutional networks for large-scale image recognition. arXiv preprint arXiv:1409.1556 (2014)
15. Vaswani, A., et al.: Attention is all you need. In: Advances in Neural Information Processing Systems, pp. 5998–6008 (2017)
16. Veličković, P., Cucurull, G., Casanova, A., Romero, A., Lio, P., Bengio, Y.: Graph attention networks. arXiv preprint arXiv:1710.10903 (2017)
17. Walecki, R., Pavlovic, V., Schuller, B., Pantic, M., et al.: Deep structured learning for facial action unit intensity estimation. In: Proceedings of the IEEE Conference on Computer Vision and Pattern Recognition, pp. 3405–3414 (2017)

18. Walecki, R., Rudovic, O., Pavlovic, V., Pantic, M.: Variable-state latent conditional random fields for facial expression recognition and action unit detection. In: 2015 11th IEEE International Conference and Workshops on Automatic Face and Gesture Recognition (FG), vol. 1, pp. 1–8. IEEE (2015)
19. Wang, S., Yang, J., Gao, Z., Ji, Q.: Feature and label relation modeling for multiple-facial action unit classification and intensity estimation. Pattern Recogn. **65**, 71–81 (2017)
20. Zhang, X., Yin, L., Cohn, J.F., Canavan, S., Reale, M., Horowitz, A.: BP4D-spontaneous: a high-resolution spontaneous 3D dynamic facial expression database. Image Vis. Comput. **32**(10), 692–706 (2014)
21. Zhang, Y., et al.: Joint representation and estimator learning for facial action unit intensity estimation. In: The IEEE Conference on Computer Vision and Pattern Recognition (CVPR) (2019)
22. Zhao, K., Chu, W.S., De la Torre, F., Cohn, J.F., Zhang, H.: Joint patch and multi-label learning for facial action unit detection. In: Proceedings of the IEEE Conference on Computer Vision and Pattern Recognition, pp. 2207–2216 (2015)
23. Zhao, K., Chu, W.S., Zhang, H.: Deep region and multi-label learning for facial action unit detection. In: Proceedings of the IEEE Conference on Computer Vision and Pattern Recognition, pp. 3391–3399 (2016)

Feature Selection Using Sparse Twin Bounded Support Vector Machine

Xiaohan Zheng[1] , Li Zhang[1,2(✉)] , and Leilei Yan[1]

[1] School of Computer Science and Technology and Joint International Research Laboratory of Machine Learning and Neuromorphic Computing, Soochow University, Suzhou 215006, Jiangsu, China
{20184227056,20184227032}@stu.suda.edu.cn,
zhangliml@suda.edu.cn
[2] Provincial Key Laboratory for Computer Information Processing Technology, Soochow University, Suzhou 215006, Jiangsu, China

Abstract. Although twin bounded support machine (TBSVM) has a lower time complexity than support vector machine (SVM), TBSVM has a poor ability to select features. To overcome the shortcoming of TBSVM, we propose a sparse twin bounded support machine (STBSVM) inspired by the sparsity of the ℓ_1-norm. The objective function of STBSVM contains the hinge loss and the ℓ_1-norm terms, both which can induce sparsity. We find solutions in the primal space instead of the dual space and avoid the operation of matrix inversion. All of these can assure the sparsity of STBSVM, or the ability to select features. Experiments carried out on synthetic and UCI datasets show that STBSVM has a good ability to select features and simultaneously enhances the classification performance.

Keywords: Supervised learning · Twin support vector machine · ℓ_1-norm regularization · Sparsity · Feature selection

1 Introduction

With the development of big data technology, the amount of data has being growing rapidly, and the dimensionality of data is getting higher and higher. High-dimensional data may contain some redundant features, which would increase the complexity of processing data. Fortunately, feature selection techniques can help us eliminate redundant features and retain valuable features, which can

This work was supported in part by the Natural Science Foundation of the Jiangsu Higher Education Institutions of China under Grant No. 19KJA550002, by the Six Talent Peak Project of Jiangsu Province of China under Grant No. XYDXX-054, by the Priority Academic Program Development of Jiangsu Higher Education Institutions, and by the Collaborative Innovation Center of Novel Software Technology and Industrialization.

H. Yang et al. (Eds.): ICONIP 2020, LNCS 12533, pp. 357–369, 2020.
https://doi.org/10.1007/978-3-030-63833-7_30

improve the efficiency of processing data [5]. In addition, feature selection has been widely used in practical applications [6, 7].

According to the relationship between feature selection algorithms and subsequent learners, feature selection algorithms can be divided into three types: filter, wrapper and embedded [1–3]. Filter methods are independent of subsequent learners and directly utilize the statistical performance of all training data to evaluate features. Wrapper methods utilize subsequent learners to identify the pros and cons of selected feature subsets. Each measurement of feature subsets requires a training and test process. Embedded methods embed feature selection as a component into the learning algorithms, the most representative one is support vector machine (SVM).

As a famous learner, SVM was proposed based on the statistical learning theory [8, 9]. However, the sparse performance of the traditional SVM seldom achieves feature selection in practice, which is due to the model representation with kernel functions. To enhance the ability of SVM to select features, ℓ_1-norm SVMs were proposed [4, 10, 11]. Moreover, SVM has an issue of computational complexity. For a given learning task, SVM needs to solve a large quadratic programming problem (QPP) if the number of training samples is huge. To avoid solving a large QPP, twin support vector machine (TSVM) solves two smaller QPPs, which results in a lower computational complexity than SVM [12]. Further, in order to improve the performance of TSVM, many variants of TSVM have been proposed, such as twin bounded support vector machine (TBSVM) [13] and LS-TSVM [14]. TBSVM adds the term of minimizing the ℓ_2-norm of model coefficients into TSVM, and then achieves the principle of structural risk minimization [13]. LS-TSVM uses the square loss function instead of the hinge loss function in TSVM [14]. To deal with multi-class tasks, Dou and Zhang [15] developed a decision tree twin support vector machine (DTTSVM) that uses the kernel K-means clustering to generate a decision tree and trains a binary TSVM on each non-leaf node. Further, Ju and Jing [16] presented an improved fuzzy multi-class twin support vector machine (IF-MTSVM) that is insensitive to outliers.

To make TSVM-like methods having the ability to select features, some methods have been proposed, such as ℓ_p-norm least square twin support vector machine (ℓ_p-LSTSVM) [17], and new linear programming twin support vector machines (NLPTSVM) [18]. ℓ_p-LSTSVM was proposed by using an adaptive learning procedure with the ℓ_p-norm ($0 < p < 1$) [17]. However, the computational complexity of ℓ_p-LSTSVM is high owing to determining the optimal p. NLPTSVM improves sparsity by using the ℓ_1-norm [18]. However, NLPSVM is too sparse to get a bad performance.

To provide a better alternative method for embedded feature selection, this paper proposes a sparse twin bounded support vector machine (STBSVM) by introducing the ℓ_1-norm into TBSVM. STBSVM consists of three terms: the average distance between positive (negative) samples to the positive (negative) hyperplane, the ℓ_1-norm of model coefficients and the hinge loss. Both the ℓ_1-norm and hinge loss can make models sparse. In addition, STSVM solves a pair of

QPPs in the primal space instead of the dual space to prevent the disappearance of sparsity. On the basis of those ways, STBSVM could has a sparse decision model and implement feature selection.

The rest of this paper is organized as follows. Section 2 proposes STBSVM. In Sect. 3, numerical experiments are given to demonstrate the ability of STBSVM to select features. Finally, we conclude this paper in Sect. 4.

2   STBSVM

2.1   Notations

At first, we describe the framework of learning task and introduce main notations. Consider a binary classification task with a set X of n training samples.

Let $\mathbf{X}$ be the total sample matrix, $\mathbf{X}_1$ and $\mathbf{X}_2$ be the positive and negative sample matrices, respectively, where $\mathbf{X}_1 = [\mathbf{x}_{11}, ..., \mathbf{x}_{1n_1}]^T \in \mathbb{R}^{n_1 \times m}$, $\mathbf{X}_2 = [\mathbf{x}_{21}, ..., \mathbf{x}_{2n_2}]^T \in \mathbb{R}^{n_2 \times m}$, $\mathbf{x}_{ji} \in \mathbb{R}^m$, m is the number of features, n_1 and n_2 are the number of positive and negative samples, respectively, and $n = n_1 + n_2$. Without loss of generality, let y_{ji} denote the label of $\mathbf{x}_{ji}$, where $y_{1i} = 1$ and $y_{2i} = -1$.

Let $\mathbf{e}_{n_j}$ be a vector of all zeros with length n_j, and $\mathbf{0}_{n_j}$ be to a vector of all zeros with length n_j. Function $\| \cdot \|_1$ is the ℓ_1-norm and $\| \cdot \|_2$ is the ℓ_2-norm. $\mathbf{O}_{n' \times n''}$ denotes the matrix of all zeros with the size of $n' \times n''$ and $\mathbf{I}_{n' \times n'}$ is the identify matrix with size of $n' \times n'$.

2.2   Objective Functions

The main idea behind TSVM-like algorithms is to find two hypothesis functions:

$$f_1(\mathbf{x}) = \mathbf{w}_1^T \mathbf{x} + b_1 \tag{1}$$

and

$$f_2(\mathbf{x}) = \mathbf{w}_2^T \mathbf{x} + b_2 \tag{2}$$

where $\mathbf{w}_1 \in \mathbb{R}^m$ and $\mathbf{w}_2 \in \mathbb{R}^m$ are the weight vectors for positive-class and negative-class hypothesis functions, respectively, b_1 and b_2 are the thresholds for these functions. To obtain these hypothesis functions, TBSVM tries to solve the following optimization problems [13]:

$$\min_{\mathbf{w}_1, b_1, \boldsymbol{\xi}_{n_2}} \quad \frac{1}{2} \|\mathbf{X}_1 \mathbf{w}_1 + b_1\|_2^2 + \frac{1}{2} C_1 (\|\mathbf{w}_1\|_2^2 + b_1^2) + C_2 \mathbf{e}_{n_2}^T \boldsymbol{\xi}_{n_2}$$
$$s.t. \quad -(\mathbf{X}_2 \mathbf{w}_1 + \mathbf{e}_{n_2} b_1) + \boldsymbol{\xi}_{n_2} \geq \mathbf{e}_{n_2}, \quad \boldsymbol{\xi}_{n_2} \geq \mathbf{0}_{n_2} \tag{3}$$

and

$$\min_{\mathbf{w}_2, b_2, \boldsymbol{\xi}_{n_1}} \quad \frac{1}{2} \|\mathbf{X}_2 \mathbf{w}_2 + b_2\|_2^2 + \frac{1}{2} C_3 (\|\mathbf{w}_2\|_2^2 + b_2^2) + C_4 \mathbf{e}_{n_1}^T \boldsymbol{\xi}_{n_1}$$
$$s.t. \quad (\mathbf{X}_1 \mathbf{w}_2 + \mathbf{e}_{n_1} b_2) + \boldsymbol{\xi}_{n_1} \geq \mathbf{e}_{n_1}, \quad \boldsymbol{\xi}_{n_1} \geq \mathbf{0}_{n_1} \tag{4}$$

where $C_i > 0, i = 1, 2, 3, 4$ are regularization parameters, and $\boldsymbol{\xi}_{n_i}, i = 1, 2$ are slack variables.

By introducing the ℓ_1-norm into TBSVM, we propose STBSVM and have the following optimization problems:

$$
\min_{\boldsymbol{\beta}_+^*, \boldsymbol{\beta}_+, \gamma_+^*, \gamma_+, \boldsymbol{\xi}_{n_2}} \frac{1}{2} \|\mathbf{X}_1 \left(\boldsymbol{\beta}_+^* - \boldsymbol{\beta}_+ \right) + \mathbf{e}_{n_1} (\gamma_+^* - \gamma_+)\|_2^2
$$

$$
+ C_1 \left(\|\boldsymbol{\beta}_+^*\|_1 + \|\boldsymbol{\beta}_+\|_1 + \gamma_+^* + \gamma_+ \right) + C_2 \mathbf{e}_{n_2}^T \boldsymbol{\xi}_{n_2}
$$

$$
s.t. \quad - \left(\mathbf{X}_2 \left(\boldsymbol{\beta}_+^* - \boldsymbol{\beta}_+ \right) + \mathbf{e}_{n_2} \left(\gamma_+^* - \gamma_+ \right) \right) + \boldsymbol{\xi}_{n_2} \geq \mathbf{e}_{n_2}
$$

$$
\boldsymbol{\xi}_{n_2} \geq \mathbf{0}_{n_2}, \quad \boldsymbol{\beta}_+^* \geq \mathbf{0}_m, \quad \boldsymbol{\beta}_+ \geq \mathbf{0}_m, \gamma_+^* \geq 0, \gamma_+ \geq 0 \quad (5)
$$

and

$$
\min_{\boldsymbol{\beta}_-^*, \boldsymbol{\beta}_-, \gamma_-^*, \gamma_-, \boldsymbol{\xi}_{n_1}} \frac{1}{2} \|\mathbf{X}_2 \left(\boldsymbol{\beta}_-^* - \boldsymbol{\beta}_- \right) + \mathbf{e}_{n_2} (\gamma_-^* - \gamma_-)\|_2^2
$$

$$
+ C_3 \left(\|\boldsymbol{\beta}_-^*\|_1 + \|\boldsymbol{\beta}_-\|_1 + \gamma_-^* + \gamma_- \right) + C_4 \mathbf{e}_{n_1}^T \boldsymbol{\xi}_{n_1}
$$

$$
s.t. \quad \left(\mathbf{X}_1 \left(\boldsymbol{\beta}_-^* - \boldsymbol{\beta}_- \right) + \mathbf{e}_{n_1} \left(\gamma_-^* - \gamma_- \right) \right) + \boldsymbol{\xi}_{n_1} \geq \mathbf{e}_{n_1}
$$

$$
\boldsymbol{\xi}_{n_1} \geq \mathbf{0}_{n_1}, \quad \boldsymbol{\beta}_-^* \geq \mathbf{0}_m, \quad \boldsymbol{\beta}_- \geq \mathbf{0}_m, \gamma_-^* \geq 0, \gamma_- \geq 0 \quad (6)
$$

where $\boldsymbol{\beta}_+^* - \boldsymbol{\beta}_+ = \mathbf{w}_1$, $\boldsymbol{\beta}_-^* - \boldsymbol{\beta}_- = \mathbf{w}_2$, $\gamma_+^* - \gamma_+ = b_1$, $\gamma_-^* - \gamma_- = b_2$, $\boldsymbol{\beta}_+^*, \boldsymbol{\beta}_+, \boldsymbol{\beta}_-^*, \boldsymbol{\beta}_- \geq \mathbf{0}_m \in \mathbb{R}^m, \gamma_+^*, \gamma_+, \gamma_-^*, \gamma_- \geq 0 \in \mathbb{R}$.

Since the formulation of (6) is similar to that of (5), we mainly discuss (5) in the following for the sake of for simplicity. The first term in (5) is to minimize the distance between the positive samples to the positive-class hyperplane $f_1(\mathbf{x}) = 0$, the second term is to minimize the ℓ_1-norm of model coefficients and the third term is to minimize the hinge loss.

The first term of (5) can be derived as follows:

$$
\frac{1}{2} \|\mathbf{X}_1 (\boldsymbol{\beta}_+^* - \boldsymbol{\beta}_+) + \mathbf{e}_{n_1} (\gamma_+^* - \gamma_+)\|_2^2
$$

$$
= \boldsymbol{\beta}_+^{*T} \mathbf{X}_1^T \mathbf{X}_1 \boldsymbol{\beta}_+^* - \boldsymbol{\beta}_+^T \mathbf{X}_1^T \mathbf{X}_1 \boldsymbol{\beta}_+^* - \boldsymbol{\beta}_+^{*T} \mathbf{X}_1^T \mathbf{X}_1 \boldsymbol{\beta}_+ + \boldsymbol{\beta}_+^T \mathbf{X}_1^T \mathbf{X}_1 \boldsymbol{\beta}_+
$$

$$
+ \boldsymbol{\beta}_+^{*T} \mathbf{X}_1^T \mathbf{e}_{n_1} \gamma_+^* - \boldsymbol{\beta}_+^{*T} \mathbf{X}_1^T \mathbf{e}_{n_1} \gamma_+ - \boldsymbol{\beta}_+^T \mathbf{X}_1^T \mathbf{e}_{n_1} \gamma_+^* + \boldsymbol{\beta}_+^T \mathbf{X}_1^T \mathbf{e}_{n_1} \gamma_+
$$

$$
+ \gamma_+^* \mathbf{e}_{n_1}^T \mathbf{X}_1 \boldsymbol{\beta}_+^* - \gamma_+ \mathbf{e}_{n_1}^T \mathbf{X}_1 \boldsymbol{\beta}_+^* - \gamma_+^{*T} \mathbf{e}_{n_1}^T \mathbf{X}_1 \boldsymbol{\beta}_+ + \gamma_+^T \mathbf{e}_{n_1}^T \mathbf{X}_1 \boldsymbol{\beta}_+ \quad (7)
$$

$$
+ \gamma_+^* \mathbf{e}_{n_1}^T \mathbf{e}_{n_1} \gamma_+^* - \gamma_+^* \mathbf{e}_{n_1}^T \mathbf{e}_{n_1} \gamma_+ - \gamma_+ \mathbf{e}_{n_1}^T \mathbf{e}_{n_1} \gamma_+^* + \gamma_+ \mathbf{e}_{n_1}^T \mathbf{e}_{n_1} \gamma_+
$$

$$
= \frac{1}{2} \boldsymbol{\alpha}'^T \mathbf{Q}_1 \boldsymbol{\alpha}'
$$

where $\boldsymbol{\alpha}' = \left[\boldsymbol{\beta}_+^{*T}, \boldsymbol{\beta}_+^T, \gamma_+^*, \gamma_+ \right]^T$, and

$$
\mathbf{Q}_1 = \begin{bmatrix}
\mathbf{X}_1^T \mathbf{X}_1 & -\mathbf{X}_1^T \mathbf{X}_1 & 0.5\mathbf{X}_1^T \mathbf{e}_{n_1} & -0.5\mathbf{X}_1^T \mathbf{e}_{n_1} \\
-\mathbf{X}_1^T \mathbf{X}_1 & \mathbf{X}_1^T \mathbf{X}_1 & -0.5\mathbf{X}_1^T \mathbf{e}_{n_1} & 0.5\mathbf{X}_1^T \mathbf{e}_{n_1} \\
0.5\mathbf{e}_{n_1}^T \mathbf{X}_1 & -0.5\mathbf{e}_{n_1}^T \mathbf{X}_1 & \mathbf{e}_{n_1}^T \mathbf{e}_{n_1} & -\mathbf{e}_{n_1}^T \mathbf{e}_{n_1} \\
-0.5\mathbf{e}_{n_1}^T \mathbf{X}_1 & 0.5\mathbf{e}_{n_1}^T \mathbf{X}_1 & -\mathbf{e}_{n_1}^T \mathbf{e}_{n_1} & \mathbf{e}_{n_1}^T \mathbf{e}_{n_1}
\end{bmatrix}
$$

Let $\boldsymbol{\alpha} = [\boldsymbol{\alpha}'^T, \boldsymbol{\xi}_{n_2}^T]^T = [\boldsymbol{\beta}_+^{*T}, \boldsymbol{\beta}_+^T, \gamma_+^*, \gamma_+, \boldsymbol{\xi}_{n_2}^T]^T$. The first term of (5) can be further rewritten as:

$$\frac{1}{2}\|\mathbf{X}_1(\boldsymbol{\beta}_+^* - \boldsymbol{\beta}_+) + \mathbf{e}_{n_1}(\gamma_+^* - \gamma_+)\|_2^2 = \frac{1}{2}\boldsymbol{\alpha}^T \begin{bmatrix} \mathbf{Q}_1 & \mathbf{O}_{(2m+2)\times n_2} \\ \mathbf{O}_{n_2\times(2m+2)} & \mathbf{O}_{n_2\times n_2} \end{bmatrix} \boldsymbol{\alpha} = \frac{1}{2}\boldsymbol{\alpha}^T\mathbf{Q}\boldsymbol{\alpha} \quad (8)$$

The second and third terms of (5) can be combined and represented as in matrix form:

$$C_1\left(\|\boldsymbol{\beta}_+^*\|_1 + \|\boldsymbol{\beta}_+\|_1 + \gamma_+^* + \gamma_+\right) + C_2\mathbf{e}_{n_2}^T\boldsymbol{\xi}_{n_2} = \boldsymbol{\zeta}^T\boldsymbol{\alpha} \quad (9)$$

where $\boldsymbol{\zeta} = \left[C_1\mathbf{1}_m^T, C_1\mathbf{1}_m^T, C_1, C_1, C_2\mathbf{e}_{n_2}^T\right]^T$.

The inequality constraints in (5) can be rewritten as:

$$- (\mathbf{X}_2(\boldsymbol{\beta}_+^* - \boldsymbol{\beta}_+) + \mathbf{e}_{n_2}(\gamma_+^* - \gamma_+)) + \boldsymbol{\xi}_{n_2} \geq \mathbf{e}_{n_2}$$
$$\Rightarrow \mathbf{P}\boldsymbol{\alpha} \geq \mathbf{e}_{n_2} \quad (10)$$

where $\mathbf{P} = \left[-\mathbf{X}_2, \mathbf{X}_2, -\mathbf{e}_{n_2}, \mathbf{e}_{n_2}, \mathbf{I}_{n_2\times n_2}\right]$. For the rest of bounded constraints, we have

$$\boldsymbol{\xi}_{n_2} \geq \mathbf{0}_{n_2}, \boldsymbol{\beta}_+^* \geq \mathbf{0}_m, \boldsymbol{\beta}_+ \geq \mathbf{0}_m, \gamma_+^* \geq 0, \gamma_+ \geq 0$$
$$\Rightarrow \boldsymbol{\alpha} \geq \mathbf{0}_{(2m+2+n_2)} \quad (11)$$

Thus, (5) can be rewritten as in matrix form:

$$\min_{\boldsymbol{\alpha}} \quad \frac{1}{2}\boldsymbol{\alpha}^T\mathbf{Q}\boldsymbol{\alpha} + \boldsymbol{\zeta}_1^T\boldsymbol{\alpha} \quad (12)$$
$$s.t. \quad \mathbf{P}\boldsymbol{\alpha} \geq \mathbf{e}_{n_2}, \boldsymbol{\alpha} \geq \mathbf{0}_{(2m+2+n_2)}$$

Then $f_1(\mathbf{x})$ can be derived from $\boldsymbol{\alpha}$ easily by solving (12). Further, using the above way, we can also get $f_2(\mathbf{x})$.

2.3   Solutions and Property Analysis

The optimization problem (12) is a convex programming, that is to say, (12) has a globally optimal solution. The Lagrange multiplier method, the interior point method, and the effective set method [19] are sophisticated ways that can be used to solve the optimization problem (12).

Creatively, we find the solution $\boldsymbol{\alpha}$ to (12) in the primal space instead of the dual space, which can avoid the operation of matrix inversion and the disappearance of sparsity. Moreover, we use the hinge loss and the ℓ_1-norm, which can induce the sparsity of $\boldsymbol{\alpha}$. Both of two aspects can guarantee the sparsity of solution $\boldsymbol{\alpha}$. As a part of solution, the weight vector $\mathbf{w}_1$ is hence no doubt sparse, which implements the task of feature selection. That is, STBSVM can deal with feature selection and classification tasks at the same time.

For a given $\mathbf{x} \in \mathbb{R}^m$, we need to determine its label. Let $\rho(\mathbf{x}) = |f_2(\mathbf{x})| - |f_1(\mathbf{x})|$ be the distance difference of $\mathbf{x}$ from the positive-class hyperplane to the negative-class hyperplane, which can be used to predict the label information for $\mathbf{x}$. Namely,

$$\hat{y} = \begin{cases} +1, & if \;\; \rho(\mathbf{x}) > 0 \\ -1, & otherwise \end{cases} \tag{13}$$

where $\hat{y}$ is the estimated label for $\mathbf{x}$. The separating hyperplane can be obtained where $\rho(\mathbf{x}) = 0$ for $\forall \mathbf{x}$. According to (13), whether a training sample is misclassified is determined by both positive-class and negative-class hypothesis function values.

In fact, the negative (or positive) samples lied between H_1 (or H_2) and $f_1(\mathbf{x}) = -1$ (or $f_2(\mathbf{x}) = 1$) have losses and the negative (or positive) ones with $f_1(\mathbf{x}) < -1$ (or $f_2(\mathbf{x}) > 1$) have no loss. The closer the negative (or positive) sample is to the positive-class (or negative-class) hyperplane, the greater the possible loss is. In conclusion, whether a training sample has a loss can be determined by the opposition-class hypothesis function value. The relationship between losses of negative samples and the positive-class hypothesis function can be described as:

$$\begin{cases} f_1(\mathbf{x}_{2i}) > -1, & if \;\; \xi_{2i} > 0 \\ f_1(\mathbf{x}_{2i}) \le -1, & otherwise \end{cases} \tag{14}$$

Similarly, the relationship between losses of positive samples and the negative-class hypothesis function is

$$\begin{cases} f_2(\mathbf{x}_{1i}) < 1, & if \;\; \xi_{1i} > 0 \\ f_2(\mathbf{x}_{1i}) \ge 1, & otherwise \end{cases} \tag{15}$$

3   Numerical Experiments

To validate the performance of STBSVM, we carry out experiments on one artificial and 8 real-world datasets. All experiments are performed on a personal computer with operation system 3.0 GHZ Intel Core and 8 G bytes of memory. This computer runs Windows 10, with Matlab R2016a.

3.1   Toy Dataset

This section mainly analyzes the ability of STBSVM to perform feature selection and compares it with TBSVM. First, we define the degree of feature selectivity (FS) to measure this ability that can be described as:

$$FS = \left(1 - \frac{\|\mathbf{w}_1\|_0 + \|\mathbf{w}_2\|_0}{length(\mathbf{w}_1) + length(\mathbf{w}_2)} \right) \times 100\% \tag{16}$$

where $\| \cdot \|_0$ is the ℓ_0-norm, and the function $length(\cdot)$ is to find the dimension of vector $\cdot$. Note that the greater FS is, the better the feature selection performance is. In experiments, to eliminate the computational error, let $w_i = 0$ if $|w_i| \le 10^{-8}$.

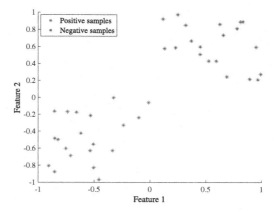

Fig. 1. Data distribution of the toy dataset

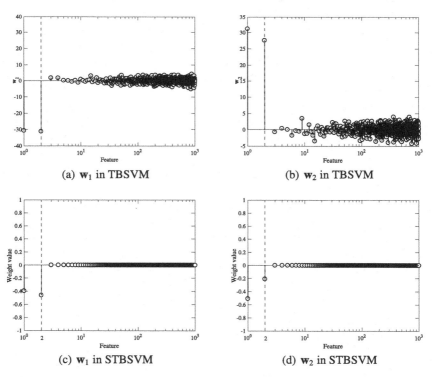

Fig. 2. Values of $\mathbf{w}_1$ and $\mathbf{w}_2$ for TBSVM and STBSVM on the toy dataset.

To observe the ability of TBSVM and STBSVM to select features, we randomly generate a toy dataset. This toy dataset contains 20 positive and 20 negative samples that have 1000 features, where the first two features are valid and the others are noise. The valid features of positive samples are drawn from the uniform distribution with $[-1,0] \times [-1,0]$, and those of negative samples with $[0,1] \times [0,1]$, while each noise feature is drawn from the Gaussian distribution with zero mean and 0.01 variance. Figure 1 plots the first two valid features.

In a very general way, we empirically set $C_i = 1, i = 1, \cdots, 4$ in both TBSVM and STBSVM. The values of $\mathbf{w}_1$ and $\mathbf{w}_2$ obtained by TBSVM and STBSVM on the toy dataset are shown in Fig. 2, where the dotted lines are the division between the valid features and the noise ones. From Fig. 2, we can see that the first two components in $\mathbf{w}_1$ and $\mathbf{w}_2$ of TBSVM and STBSVM have relatively large absolute values. The weights of rest features are relatively small in TBSVM, while the weights of rest features are zero in STBSVM. In addition, STBSVM can obtain $FS = 99.9\%$ on the toy dataset, while TBSVM obtains $FS = 0\%$. The results indicate that the contribution of noise attributes to the STBSVM model is negligible, and also imply that STBSVM has a better feature selection performance than TBSVM.

3.2   UCI Datasets

In this section, we conduct experiments on 8 UCI datasets [20] and compare the performance of STBSVM and the related methods, including SVM [8], TSVM [12], TBSVM [13], ℓ_p-LSTSVM [17] and NLPTSVM [18]. Table 1 shows the details of these datasets.

FS obtained by most methods can be computed by (16). But, the definition of FS for SVM is slightly different. Namely,

$$FS = \left(1 - \frac{|SV|}{n} \right) \times 100\% \tag{17}$$

where SV is the set of support vectors, and n is the number of training samples.

In order to obtain more effective results, we repeat 5 time experiments on the partitioned datasets. Five-fold cross validation is used here to obtain the average accuracy [21]. In each partition, regularization parameters in all algorithms are determined from the set $\{2^{-3}, ..., 2^3\}$ by using the training sets.

Table 1. The details of UCI datasets

Dateset	Australian	Diabetes	German	Heart	Sonar	Tic_tac_toe	Wdbc	Wpbc
#Sample	690	768	1000	270	208	958	569	194
#Feature	14	8	24	13	60	9	30	33
# Class	2	2	2	2	2	2	2	2

The average test accuracy and FS obtained by six methods are shown in Table 2, where the best performance among those algorithms is in bold type (so do tables below). The experimental results in Table 2(a) shows the superiority of STBSVM over other methods in four out of eight datasets. Observation on Table

2(b) indicates that NLPTSVM has the highest FS, followed by STBSVM. But, the classification accuracy of NLPTSVM is almost the worst among the compared methods. It can be seen that the values of FS obtained by SVM, TBSVM, TSVM, ℓ_p-LSTSVM are almost 0, which demonstrates that these methods have a poor feature selection performance. It is concluded that STBSVM has a better classification performance and a better ability to perform feature selection according to Table 2.

Table 2. Mean of test accuracy (%) and FS(%) obtained by six linear methods

(a) Mean and standard deviation of test accuracy (%)					
Datasets	SVM	TBSVM	TSVM	ℓ_p-LSTSVM	NLPTSVM	STBSVM
Australian	86.96 ± 2.98	87.25 ± 2.88	87.25 ± 2.75	87.54 ± 3.10	86.96 ± 2.98	$\mathbf{88.27 \pm 3.40}$
Diabetes	78.00 ± 1.37	$\mathbf{78.26 \pm 2.36}$	77.22 ± 4.12	78.25 ± 2.79	46.84 ± 16.52	77.47 ± 2.72
German	74.90 ± 1.75	77.40 ± 1.43	77.30 ± 1.75	77.40 ± 0.96	54.50 ± 21.23	$\mathbf{78.20 \pm 1.82}$
Heart	$\mathbf{84.44 \pm 1.66}$	82.59 ± 3.36	83.33 ± 2.62	82.59 ± 2.48	46.67 ± 4.97	$\mathbf{84.44 \pm 3.36}$
Sonar	$\mathbf{77.47 \pm 5.13}$	76.48 ± 7.31	75.44 ± 6.28	76.50 ± 8.37	53.37 ± 0.56	75.03 ± 4.43
Tic_tac_toe	65.34 ± 0.15	63.88 ± 7.83	67.75 ± 2.52	68.79 ± 1.48	65.34 ± 0.15	$\mathbf{69.31 \pm 1.92}$
Wdbc	97.54 ± 0.97	95.96 ± 1.71	97.37 ± 1.07	$\mathbf{97.72 \pm 0.99}$	91.91 ± 3.78	97.71 ± 1.19
Wpbc	81.98 ± 4.77	$\mathbf{82.06 \pm 6.21}$	78.41 ± 5.92	77.84 ± 2.91	76.30 ± 0.79	82.02 ± 5.55
(b) FS (%) and standard deviation					
Datasets	SVM	TBSVM	TSVM	ℓ_p-LSTSVM	NLPTSVM	STBSVM
Australian	0.00 ± 0.00	0.00 ± 0.00	0.00 ± 0.00	0.00 ± 0.00	$\mathbf{95.71 \pm 1.60}$	2.14 ± 1.96
Diabetes	0.00 ± 0.00	0.00 ± 0.00	0.00 ± 0.00	0.00 ± 0.00	$\mathbf{100.00 \pm 0.00}$	1.25 ± 2.80
German	0.00 ± 0.00	0.00 ± 0.00	0.00 ± 0.00	0.00 ± 0.00	$\mathbf{98.75 \pm 1.14}$	0.83 ± 1.14
Heart	0.00 ± 0.00	0.00 ± 0.00	0.00 ± 0.00	0.00 ± 0.00	$\mathbf{100.00 \pm 0.00}$	8.46 ± 1.72
Sonar	0.00 ± 0.00	0.00 ± 0.00	0.00 ± 0.00	0.00 ± 0.00	$\mathbf{55.83 \pm 0.59}$	46.50 ± 4.31
Tic_tac_toe	0.00 ± 0.00	61.11 ± 21.15	45.56 ± 31.53	0.00 ± 0.00	$\mathbf{100.00 \pm 0.00}$	50.00 ± 0.00
Wdbc	0.00 ± 0.00	0.00 ± 0.00	0.00 ± 0.00	0.00 ± 0.00	11.33 ± 3.61	$\mathbf{38.65 \pm 9.75}$
Wpbc	0.00 ± 0.00	0.00 ± 0.00	0.00 ± 0.00	0.00 ± 0.00	60.61 ± 1.07	$\mathbf{43.64 \pm 9.49}$

In addition, we add 50 randomly generated noise characteristics to those eight datasets to further test the feature selection performance of these six methods, where each noise feature obeys the Gaussian distribution with zero mean and 0.01 variance. Table 3 shows the results. From Table 3(a), it can be seen that STBSVM performs better on seven out of eight datasets. In theory, FS increases as noise features increase since it requires excluding more features in such a situation. Comparing Table 3(b) with Table 2(b), STBSVM shows the most obvious increase of FS in the case of existing noise features. Although the ability of NLPTSVM to select features is stronger than STBSVM, NLPTSVM is much less sensitive to noise features than STBSVM. In other words, the ability of STBSVM to remove noise features is stronger than that of NLPTSVM.

To graphically illustrate the advantage of STBSVM, we plot the weight vectors obtained by TSVM-like algorithms on the Wpbc dataset in Figs. 3 and 4, where the computational error threshold is still 10^{-8}, and the dashed lines are the boundaries between the original features of datasets and the added noise features. From these figures, we can see that ℓ_p-LSTSVM, TBSVM and TSVM generate irregular weight values for noise features, while NLPTSVM and STB-SVM can assign the value of zero or close to zero to the weights of noise features. That reveals that NLPTSVM and STBSVM have a better feature selection performance and are robust to noise. However, the solution of NLPTSVM is so sparse that $\mathbf{w}_1$ is a vector with all zeros as shown in Fig. 3(b) that is abnormal. Thus, extreme sparsity would lead to a bad classification performance.

Table 3. Mean of test accuracy (%) and FS(%) on UCI datasets with 50 noise features

(a) Mean and standard deviation of test accuracy (%)

Datasets	SVM	TBSVM	TSVM	ℓ_p-LSTSVM	NLPTSVM	STBSVM
Australian	86.96 ± 2.98	85.65 ± 3.12	86.09 ± 3.48	86.96 ± 3.08	86.23 ± 3.72	**88.12 ± 2.30**
Diabetes	76.30 ± 0.55	75.52 ± 1.66	74.87 ± 3.24	76.43 ± 1.39	77.48 ± 2.81	**77.73 ± 2.70**
German	73.80 ± 2.39	75.40 ± 1.47	73.90 ± 2.22	74.90 ± 1.19	74.70 ± 2.28	**76.70 ± 2.93**
Heart	**85.56 ± 2.41**	80.37 ± 5.94	82.59 ± 4.26	80.74 ± 5.49	81.85 ± 2.75	**85.56 ± 3.04**
Sonar	76.51 ± 7.01	61.09 ± 2.34	61.52 ± 3.47	73.48 ± 7.26	74.58 ± 8.59	**78.51 ± 9.25**
Tic_tac_toe	65.34 ± 0.15	59.07 ± 2.61	64.52 ± 2.68	**66.81 ± 3.59**	65.34 ± 0.15	64.51 ± 3.30
Wdbc	97.55 ± 1.28	95.26 ± 1.29	96.66 ± 1.91	97.01 ± 1.83	96.66 ± 1.92	**97.71 ± 1.34**
Wpbc	78.90 ± 3.58	76.30 ± 0.79	75.80 ± 2.60	76.84 ± 2.15	79.38 ± 6.33	**79.90 ± 3.78**

(b) FS (%) and standard deviation

Datasets	SVM	TBSVM	TSVM	ℓ_p-LSTSVM	NLPTSVM	STBSVM
Australian	0.00 ± 0.00	0.00 ± 0.00	0.00 ± 0.00	0.00 ± 0.00	**91.56 ± 7.66**	24.53 ± 17.84
Diabetes	0.00 ± 0.00	0.00 ± 0.00	0.00 ± 0.00	0.00 ± 0.00	43.79 ± 10.42	**68.79 ± 12.73**
German	0.00 ± 0.00	0.00 ± 0.00	0.00 ± 0.00	0.00 ± 0.00	**55.00 ± 23.9**	1.22 ± 0.74
Heart	0.00 ± 0.00	0.00 ± 0.00	0.00 ± 0.00	0.00 ± 0.00	70.16 ± 9.15	**77.14 ± 1.30**
Sonar	0.00 ± 0.00	0.00 ± 0.00	0.00 ± 0.00	0.00 ± 0.00	64.91 ± 9.70	**75.55 ± 4.50**
Tic_tac_toe	0.00 ± 0.00	9.32 ± 20.84	3.90 ± 8.25	0.00 ± 0.00	75.93 ± 22.02	**89.32 ± 23.41**
Wdbc	0.00 ± 0.00	0.00 ± 0.00	0.00 ± 0.00	0.00 ± 0.00	**51.25 ± 14.71**	47.25 ± 16.89
Wpbc	0.00 ± 0.00	0.00 ± 0.00	0.00 ± 0.00	0.00 ± 0.00	**91.20 ± 0.54**	60.48 ± 16.15

In a summary, we observe two situations and have some facts. In the situation of datasets without noise features, STBSVM is superior to other methods on four out of eight datasets, followed by SVM and TBSVM. In the situation of datasets with 50 noise features, STBSVM is superior to other methods on seven out of eight datasets, followed by ℓ_p-LSTSVM and SVM. Unlike NLPTSVM, STBSVM does not one-sidedly focus on sparsity, disregarding classification performance. Compared with other TSVM-like algorithms except for NLPTSVM, the ability of STBSVM to select feature is rather strong.

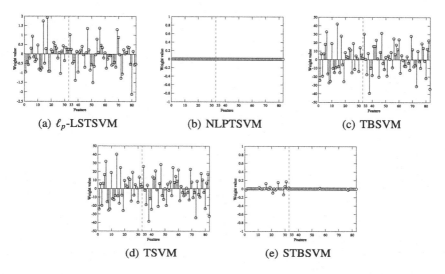

(a) ℓ_p-LSTSVM   (b) NLPTSVM   (c) TBSVM

(d) TSVM   (e) STBSVM

Fig. 3. Weight vector $\mathbf{w}_1$ obtained by TSVM-like methods on Wpbc with 50 noise features

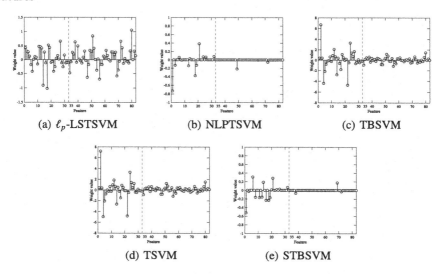

(a) ℓ_p-LSTSVM   (b) NLPTSVM   (c) TBSVM

(d) TSVM   (e) STBSVM

Fig. 4. Weight vector $\mathbf{w}_2$ obtained by TSVM-like methods on Wpbc with 50 noise features

4   Conclusion

This paper proposes a novel algorithm named sparse twin bounded support vector machine for binary classification tasks, which has a good sparse solution to implement feature selection. In STBSVM, sparsity is induced by introducing the ℓ_1-norm and the hinge loss function, and by searching solutions in the primal space. Numerical experiments on synthetic and several UCI datasets show that STBSVM can obtain a good classification performance and generate a proper sparsity, that is STSVM has a favorable performance of feature selection, especially for the noise situation.

Moreover, it is also a promising avenue of research to expand STSVM to its kernelized version to filter samples. In addition, how to improve the training efficiency of STBSVM is an issue that needs to be solved in future.

References

1. Zhang, X., Wu, G., Dong, Z., Crawford, C.: Embedded feature-selection support vector machine for driving pattern recognition. J. Franklin Inst. **352**, 669–685 (2015)
2. Guyon, I., Elisseeff, A.: An introduction to variable and feature selection. J. Mach. Learn. Res. **3**, 1157–1182 (2003)
3. Chandrashekar, G., Sahin, F.: A survey on feature selection methods. Comput. Electr. Eng. **40**, 16–28 (2014)
4. Bi, J., Bennett, K., Embrechts, M., Breneman, C., Song, M.: Dimensionality reduction via sparse support vector machines. J. Mach. Learn. Res. **3**, 1229–1243 (2003)
5. Jain, A.K., Duin, R.P.W., Mao, J.: Statistical pattern recognition: a review. IEEE Trans. Pattern Anal. Mach. Intell. **22**(1), 4–37 (2000)
6. Huffener, F., Niedermeier, R., Wernicke, S.: Techniques for practical fixed-parameter algorithms. Comput. J. **51**(1), 7–25 (2008)
7. Niedermeier, R.: Invitation to Fixed-Parameter Algorithms. Oxford University Press, Oxford (2006)
8. Cortes, C., Vapnik, V.: Support vector networks. Mach. Learn. **20**(3), 273–297 (1995). https://doi.org/10.1007/BF00994018
9. Vladimir N.V.: The Nature of Statistical Learning Theory. Springer, Switzerland (2000). https://doi.org/10.1007/978-1-4757-3264-1
10. Zhou, W., Zhang, L., Jiao, L.: Linear programming support vector machines. Pattern Recognit. **35**(12), 2927–2936 (2002)
11. Zhang, L., Zhou, W.: On the sparseness of 1-norm support vector machines. Neural Netw. **23**(3), 373–385 (2010)
12. Khemchandani, R., Chandra, S.: Twin support vector machine for pattern classification. IEEE Trans. Pattern Anal. Mach. Intell. **29**(5), 905–910 (2007)
13. Shao, Y.H., Zhang, C.H., Wang, X.B., Deng, N.Y.: Improvements on twin support vector machine. IEEE Trans. Neural Netw. **22**(6), 962–968 (2011)
14. Kumar, M.A., Gopal, M.: Least squares twin support vector machine for pattern classification. Exp. Syst. Appl. **36**, 7535–7543 (2009)
15. Dou, Q., Zhang, L.: Decision tree twin support vector machine based on kernel clustering for multi-class classification. In: Cheng, L., Leung, A.C.S., Ozawa, S. (eds.) ICONIP 2018. LNCS, vol. 11304, pp. 293–303. Springer, Cham (2018). https://doi.org/10.1007/978-3-030-04212-7_25
16. Ju, H. and Jing, L.: an improved fuzzy multi-class twin support vector machine. In: 2019 6th International Conference on Systems and Informatics (ICSAI), pp. 393–397 (2019)
17. Zhang, Z., Zhen, L., Deng, N., Tan, J.: Sparse least square twin support vector machine with adaptive norm. Appl. Intell. **4**(41), 1097–1107 (2014). https://doi.org/10.1007/s10489-014-0586-1

18. Tanveer, M.: Robust and sparse linear programming twin support vector machines. Cognit. Comput. **7**(1), 137–149 (2015). https://doi.org/10.1007/s12559-014-9278-8
19. Hertog, D.D.: Interior point approach to linear, quadratic and convex programming: algorithms and complexity. Topics in Engineering Mathematics. Springer, Netherlands (1992)
20. UCI Machine Learning Repository (2017). http://archive.ics.uci.edu/ml
21. Duda, R.O., Hart, P.E., Stork, D.G.: Pattern Classification, 2nd edn. John Wiley, New York (2001)

Few-Shot Classification with Transductive Data Clustering Transformation

Haojie Wang, Jieya Lian, and Shengwu Xiong[✉]

School of Computer Science and Technology, Wuhan University of Technology,
Wuhan, China
{wanghj18,jylian,xiongsw}@whut.edu.cn

Abstract. Few-shot classification aims to recognize unlabeled samples from unseen classes given only a small number of labeled examples. Most methods addressing few-shot problem through meta-learning. They focus on learning a generic classifier across a large number of multiclass classification tasks and generalizing the model to a new task. However, the low-data problem in the novel classification task still remains. In this paper, we propose Transductive Data Clustering Transformation (TDCT), a novel and simple method which can potentially be applied to any metric-based few-shot classification approaches. TDCT exploits the task-specific knowledge and enhances the data representations by using a transformation that incorporates data clustering. This transformation implicitly does transductive inference by leveraging the relationships between all samples within a task, alleviating the low data problem. Extensive experiments show that TDCT is an effective and computationally efficient method which can improve few-shot learning performance by a large margin on two benchmarks.

Keywords: Few-shot classification · Transductive inference · Data clustering transformation

1 Introduction

Deep learning has achieved impressive results in a variety of tasks, such as visual recognition [6], machine translation [1] and speech modeling [11]. However, deep learning methods always require a large number of labeled data. On the contrary, humans can learn new concepts with very little supervision. For example, children can learn the concept of "birds" from only a single picture ("one-shot") in a book.

Few-shot learning is proposed to tackle the learning problem with limited labeled data. It aims to recognize unlabeled samples (the query set) from unseen classes given only a small number of labeled examples (the support set). Most approaches addressing this problem are based on the meta-learning(learning to learn) paradigm which relies on an episodic training framework, we will detail it in Sect. 2.

H. Yang et al. (Eds.): ICONIP 2020, LNCS 12533, pp. 370–380, 2020.
https://doi.org/10.1007/978-3-030-63833-7_31

Methods specifically designed for few-shot learning fall into two categories. The first line of methods are based on optimization learning [3,14,16]. Instead of training a learner on a single task, they train a meta-learner to control how a classifier for the target task should be constructed. Most optimization-based approaches are built on the basis of model-agnostic meta-learning (MAML) [3], which aims to train a model's parameters such that a small number of gradient updates will lead to fast learning on a new task. Afterward, LSTM-based meta-learner model [14] is proposed to learn appropriate parameter updates specifically for the scenario where a set amount of updates will be made. Meta-SGD [7] has a much higher capacity than MAML by learning to learn not just the learner's initialization, but also the learner update direction and learning rate, all in a single meta-learning process. Compared to the popular meta-learner LSTM, Meta-SGD is conceptually simpler, easier to implement, and can be learned more efficiently. Latent Embedding Optimization (LEO) [16], learns a low-dimensional latent embedding of model parameters and performs optimization-based meta-learning in this space.

Another line of approaches are based on metric learning [12,18,22,24]. They aim to learn proper representations, which minimize intra-class distances and maximize inter-class distances. Matching Networks [24] make use of advances in attention and memory to learn to map a small labeled support set and an unlabeled query data to its label. Prototypical networks [18] aim to learn a metric space in which classification can be performed by computing Euclidean distances to every class prototypes, which is computed as the mean of embedded support examples for each class. In contrast to computing softmax over Euclidean distance or cosine similarity, Relation network [22] is proposed to learn a relation module that determines if the query embeddings and support set embeddings are from the same categories or not. Task dependent adaptive metric (TADAM) [12] aims to learn a task-dependent scaled metric via conditional batch normalization. They introduce a task representation computed as the mean of the task class centroids (class prototypes). Both lines of methods are framed in meta-learning paradigm as we mentioned before.

Although meta-learning paradigms are effective for few-shot problem, the fundamental difficulty of learning a novel classifier with scarce data remains. One way to improve performance in the low-data regime is to consider relationships between all samples (unlabeled data and labeled data), which is referred to as transductive inference. We can explore the interrelationships among these data points and make an assessment (perhaps preliminary) of their structure by data clustering [4].

In this paper, we propose a novel method for few-shot classification called TDCT (Transductive Data Clustering Transformation), which makes use of relationships between all samples within a task. By using a transformation that incorporates data clustering technique, our model exploits task-specific knowledge and enhances the data representations. With the enhanced representations and ground truth labels of the query set, we compute the cross-entropy loss and update all parameters in an end to end manner.

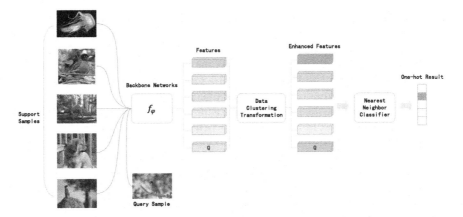

Fig. 1. The overall architecture of our method. The backbone networks are basic network in prototypical network. Our model is built on prototypical network and we use data clustering module to enhance the original feature. The classification result in this picture is $[0, 1, 0, 0, 0]^T$.

Our proposed TDCT is effective and computationally efficient. It can potentially be applied to any metric-based few-shot classification approaches. In this paper, we build our model on top of prototypical network due to its simplicity. Experiments show that it outperforms the baseline by a large margin. Besides, our model can be easily extended to semi-supervised few-shot learning. Experimental results also show that our proposed method achieves higher performance than semi-supervised few-shot learning baselines.

2   Methodology

In this section, we first give the formal definition of the few-shot classification tasks, then we do a small review about the prototypical network. Finally we detail our proposed method.

2.1   Problem Statement

In few-shot classification tasks, we need to predict the classes for some unlabeled query samples (D_{query}) with only a small set of labeled examples ($D_{support}$). Typically, the support set ($D_{support}$) includes $N \times K$ labeled examples where N denotes the number of classes, each class contains K different examples. This few-shot problem setting is always called the N-way K-shot problem, and each few-shot classification task is called an episode.

We follow the episodic paradigm [24] which is an effective training strategy for the few-shot classification. It is commonly employed in various literature [3,12,14,18,21]. Concretely, we have two different meta-sets for meta-training and meta-testing ($D_{meta-training}$, $D_{meta-testing}$). Each meta-set contains multiple regular datasets, i.e., episodes. In each meta-training iteration, we sample episodes from $D_{meta-training}$ as "training samples" to mimic the situation that will be encountered at test time. Then we evaluate its generalization performance on $D_{meta-testing}$.

Episodic training strategy is based on meta-learning and performs well on few-shot classification tasks. However, the fundamental low-data problem in the novel few-shot classification task still remains. This motivates us to assume a transductive setting, in which we utilize the union of support set and query set rather than predict each example independently. For simplicity, we build our TDCT model on prototypical network [18]. And we consider two backbone convolution networks as embedding function.

2.2 Review of Prototypical Network

Prototypical network is a classic metric-based method in few-shot learning literature. It aims to learn an embedding function f which maps examples into space where examples from the same class are close and those from different classes are far. Then the classification of query sample q_t can be completed by finding its nearest prototype. The following equation describes this process:

$$p(y = c | q_t, P) = \frac{exp(-d(f(q_t), p_c))}{\sum_{k=1}^{N} exp(-d(f(q_t), p_k))} \tag{1}$$

where $P = \{p_1, ... p_N\}$ denotes the set of class prototypes. The prototype p_c of class c is the mean of support samples. d denotes the Euclidean distance. The loss function can be cross-entropy loss.

We build our method on prototypical network for its simplicity, but it can be applied to any metric-based few-shot learning approaches.

2.3 Method

Our method follows the transductive inference settings and incorporates the idea of data clustering. Figure 1 illustrates how our method is built on the metric-based methods, after origin feature embedding, we enhance the data representations by Transductive Data Clustering Transformation (TDCT), then utilize these enhanced features to classify the query data.

Clustering is one of the most widely used techniques for exploratory data analysis. We use data clustering as a post processing in our methods to enhance the original feature. Spectral clustering has been used in many recent works [19, 20]. Here we use spectral clustering, but other data clustering methods are also available. Readers can get more details about Spectral clustering in [9].

As described in [10], the spectral clustering process can be seen as a Markov random walk process. So we can define a Markov Chain random walk process in an episode to cluster these data. The concrete method is as follow:

Suppose $X \in R^{n \times d}$ is the embedded data points (all samples in support set and query set) of an episode. n denotes the number of data points, and d denotes the dimensions of the feature vector. W is the weighted adjacency matrix. Each element of W is defined as:

$$w_{ij} = exp(\frac{\boldsymbol{x}_i^T \boldsymbol{x}_j}{\sigma \cdot \|\boldsymbol{x}_i\|_2 \|\boldsymbol{x}_j\|_2}) \tag{2}$$

w_{ij} represents the similarity between data points i and j. And σ is a temperature parameter. We add an exponential operation on cosine similarity as our similarity, cause we want to make the similarity positive. By normalizing the rows of W to 1, we can easily find that this defines a Markov Chain random walk process on a graph which is constructed with data points X. Specifically, we use T denotes the normalized W, then T is a transition probability matrix of a Markov Chain. The normalizing process can be completed by:

$$T = D^{-1}W, \tag{3}$$

where D is a diagonal matrix. The ith diagonal element of D is the degree of data point i, i.e. $d_i = \sum_{j=1}^{n} w_{ij}$. This operation can be implemented by applying softmax function on the pre-exponential weighted adjacency matrix S ($W = exp(S)$, in which $exp()$ is element-wise exponential function).

In this random walk process. We use $P(A \rightarrow B)$ to represent the transition probability from subset A to subset B in one step if the current state is A. For the few-shot classification tasks, a subset A denotes the set of samples belonging to class A. Then the transition probability $P(A \rightarrow \bar{A})$ is essentially the probability that a sample belonging to class A getting misclassified. This probability was called the evading probability in [10]. We follow their definition of this evading probability as follow:

$$P(A \rightarrow B) = \frac{\sum_{i \in A, j \in B} \pi_i T_{ij}}{\sum_{i \in A} \pi_i}. \tag{4}$$

Wherein π_i is defined as:

$$\pi_i = \frac{d_i}{\sum_{j,k \in X} w_{jk}} \tag{5}$$

It is easy to verify that π_i is a stationary distribution of the Markov Chain. And here we assume that the random walk is started in its stationary distribution.

A small evading probability requires strong intra-class connections and weak inter-class connections which is a desirable property in classification problem. From the Markov walks view of point, a step of Markov walks can be seen as a step of clustering. The evading probability in Eq. 4 can be minimized by optimizing the cross-entropy loss after a step of clustering with standard SGD in an end-to-end manner.

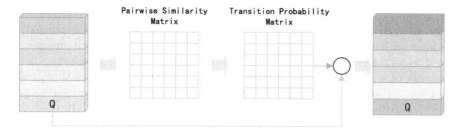

Fig. 2. The illustration of clustering transformation operation. First, calculate the pairwise similarity matrix with respect to the original features (left). Then calculate the transition probability matrix by normalizing the similarity matrix. Finally do the matrix multiply operation. Q can contains multiple samples.

Our Data clustering transformation can be implemented by a step of data cluster, multiplying T with original features X, i.e.

$$X_{enhanced} = TX \qquad (6)$$

The transition probability matrix T contains the relationship information between all samples within the episode. Data clustering operation using such episode-specific information can enhance the original feature and boost the model performance.

Detail of this clustering transformation operation can be found in Fig. 2. We only apply one step of random walk operation. Because experimental results show that one step is enough.

3 Experiment

3.1 Dataset

We validate TDCT on Mini-ImageNet and Caltech-UCSD Birds (CUB) 200-2011 dataset. The Mini-ImageNet dataset is originally proposed by Vinyals et al. [24]. It is a subset of the ImageNet dataset [2] that includes a total number of 100 classes and 600 examples per class. We follow the split provided by Ravi and Larochelle [14]. CUB dataset is initially designed for fine-grained classification. It contains 11788 images of birds over 200 species. We follow the split proposed by Ye et al. [22].

For all images in the CUB dataset, we use the provided bounding box to crop the images as a pre-processing [23]. And we use Euclidean distance as the distance function in the nearest neighbor algorithm.

3.2 Implementation Detail

We consider two backbone convolution networks as embedding function. The first is a four-layer convolution network (ConvNet) which has the same architecture used in several recent works [8,18,21,24]. It contains 4 repeated convolutional blocks. In each block, there are a 64-filter 3×3 convolution, a batch normalization layer, a ReLU, and a max-pooling with size 2. And the input images are resized to $84 \times 84 \times 3$. The second is a wide residual network (ResNet) [12,16]. It consists of a convolutional layer with stride 1 and padding 1 and three residual blocks followed by a global average pool. Three residual blocks have channels 160/320/640, stride 2, and padding 2. And the input images are resized to $80 \times 80 \times 3$.

We use an additional pre-training strategy, which is suggested by [12,16]. At first, the backbone network is followed by a fully-connected layer with SoftMax. It is trained to classify all classes in the meta-training set. The trained weights are then used to initialize the embedding function.

Inspired by Oreshkin et al. [12], we use the metric scale technique which has a great influence on the model performance. we set the metric scale coefficient to 32 and 64 for ConvNet and ResNet respectively. During the training, stochastic gradient descent (SGD) with Adam [5] optimizer is employed, with the initial learning rate set to be 1e−3. We use both the original features and the enhanced features during training to make the training process steady. The original feature and the enhanced feature share the same classifier. The temperature parameter σ of the TDCT layer is set to 0.1.

We batch 15 query images per class in each episode for evaluation. The classification accuracies are computed by averaging over 10,000 randomly generated episodes from the meta-test set.

Again, in this paper, we build our model on prototypical network [18] for simplicity. But TDCT can also be applied on other metric learning based few-shot learning methods.

3.3   Results and Discussion

From Table 1 and Table 2, we can see that our model improves performance on both 1-shot and 5-shot settings on Mini-ImageNet and CUB dataset. Note that some recent works such as TADAM [12] and LEO [16] use the pre-training strategy as we do. Our model which built on prototypical network [18] can still achieve competitive performance compared with them.

We can apply TDCT multiple times, but we can hardly observe improvement when we apply TDCT more than once. So it is enough to apply TDCT only once.

Among the approaches we compared, TPN is the only approach that modeled transductive inference explicitly. Both of us construct a graph on the support set and the entire query set. However, we use TDCT on features directly rather than propagate labels. Even without the example-wise temperature parameter σ, our experimental results have higher accuracy. Moreover, compared to TPN, our method is more computationally efficient cause we do not need to compute the matrix inversion.

Our model can easily extend to the semi-supervised few-shot regime without changing the structure of the model. In the semi-supervised few-shot setting, when we sample support examples from the dataset, some extra unlabeled samples are provided. We can directly make those extra unlabeled samples

Table 1. Few-shot classification results with 95% confidence interval on Mini-ImageNet.

Setups	5-way 1-shot		5-way 5-shot	
Backbone	ConvNet	ResNet	ConvNet	ResNet
MatchNet [24]	$43.40_{\pm 0.78}$	–	$51.09_{\pm 0.71}$	–
MAML [3]	$48.70_{\pm 1.84}$	–	$63.11_{\pm 0.92}$	–
ProtoNet [18]	$49.42_{\pm 0.78}$	–	$68.20_{\pm 0.66}$	–
RelationNet [21]	$51.38_{\pm 0.82}$	–	$67.07_{\pm 0.69}$	–
PFA [13]	$54.53_{\pm 0.40}$	$59.60_{\pm 0.41}$	$67.87_{\pm 0.20}$	$73.74_{\pm 0.19}$
TADAM [12]	–	$58.50_{\pm 0.30}$	–	$76.70_{\pm 0.30}$
LEO [16]	–	$61.76_{\pm 0.08}$	–	$77.59_{\pm 0.12}$
TPN [8]	$53.75_{\pm 0.86}$	59.46	$69.43_{\pm 0.67}$	75.65
TDCT [Our]	$\mathbf{55.35}_{\pm 0.22}$	$\mathbf{62.56}_{\pm 0.22}$	$\mathbf{71.23}_{\pm 0.16}$	$\mathbf{78.53}_{\pm 0.15}$

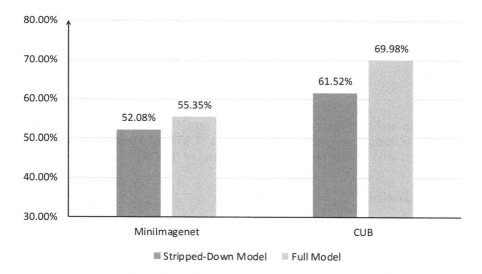

Fig. 3. Comparison of our full model and stripped down model on two datasets.

participate in the construction of the graph, then apply TDCT as usual. We use the same semi-supervised settings and backbone networks(ConvNet) as TPN [8]. We construct the graph with only one query sample each time for fairness. Table 3 shows that with the same setting, TDCT outperforms other semi-supervised methods.

We also compare our full model with a stripped-down version (without the TDCT module) to further analyze the effectiveness of TDCT. The experimental results are shown in Fig. 3. We evaluate these two trained models with **10,000** same tasks in the 5-way 1-shot setting. TDCT module increases the average

Table 2. Few-shot classification results with 95% confidence interval on CUB with ConvNet backbone.

Setups	5-way 1-shot	5-way 5-shot
MatchNet [24]	61.16 ± 0.89	72.86 ± 0.70
MAML [3]	55.92 ± 0.95	72.09 ± 0.76
ProtoNet [18]	51.31 ± 0.91	70.77 ± 0.69
RelationNet [21]	62.45 ± 0.98	76.11 ± 0.69
TDCT [Our]	**69.98 ± 0.23**	**82.24 ± 0.16**

Table 3. Semi-supervised Few-shot classification results with 95% confidence interval on Mini-ImageNet. "w/D" means with distractors. Part of the unlabeled data does not belong to any of the classes in the episode.

Model	1-shot	5-shot	1-shot w/D	5-shot w/D
k-Means [15]	$50.09_{\pm 0.45}$	$64.59_{\pm 0.28}$	$48.70_{\pm 0.32}$	$63.55_{\pm 0.28}$
k+1 Cluster [15]	$49.03_{\pm 0.24}$	$63.08_{\pm 0.18}$	$48.86_{\pm 0.32}$	$61.27_{\pm 0.24}$
k+Masked [15]	$50.41_{\pm 0.31}$	$64.39_{\pm 0.24}$	$49.04_{\pm 0.31}$	$62.96_{\pm 0.14}$
TPN-semi [8]	$52.78_{\pm 0.27}$	$66.42_{\pm 0.21}$	$50.43_{\pm 0.84}$	$64.95_{\pm 0.73}$
TDCT [Our]	$\mathbf{54.63_{\pm 0.22}}$	$\mathbf{69.09_{\pm 0.16}}$	$\mathbf{52.47_{\pm 0.21}}$	$\mathbf{66.93_{\pm 0.15}}$

classification accuracy from 61.52% to 69.98% on CUB dataset and increases the average classification accuracy from 52.08% to 55.35% on MiniImagenet dataset. The experimental results prove the effectiveness of TDCT.

4   Conclusion

In this paper, we introduce a novel and effective method called Transductive Data Clustering Transformation for few-shot classification. Our method follows the transductive inference settings and incorporates the idea of data clustering. It can potentially be applied to any metric-based few-shot classification approaches. By leveraging the relationships between all samples within a task, TDCT exploits the task-specific knowledge and enhances the data representations. Thus it boosts the performance of metric-based few-shot classification methods. We also achieve higher performance in semi-supervised few-shot learning settings compared with some other semi-supervised few-shot classification methods.

Acknowledgments. This work was in part supported by the National Key Research and Development Program of China (Grant No. 2017YFB1402203), the National Natural Science Foundation of China (Grant No. 61702386). the Defense Industrial Technology Development Program (Grant No. JCKY2018110C165), Major Technological Innovation Projects in Hubei Province (Grant No. 2019AAA024).

References

1. Bahdanau, D., Cho, K., Bengio, Y.: Neural machine translation by jointly learning to align and translate. In: 3rd International Conference on Learning Representations, ICLR, Conference Track Proceedings, San Diego, CA, USA, 7–9 May (2015)
2. Deng, J., Dong, W., Socher, R., Li, L., Li, K., Li, F.: Imagenet: a large-scale hierarchical image database. In: IEEE Computer Society Conference on Computer Vision and Pattern Recognition (CVPR), Miami, Florida, USA, 20–25 June, pp. 248–255 (2009). https://doi.org/10.1109/CVPRW.2009.5206848
3. Finn, C., Abbeel, P., Levine, S.: Model-agnostic meta-learning for fast adaptation of deep networks. In: Proceedings of the 34th International Conference on Machine Learning, Sydney, NSW, Australia, 6–11 August, pp. 1126–1135 (2017)
4. Jain, A.K., Murty, M.N., Flynn, P.J.: Data clustering: a review. ACM Comput. Surv. **31**(3), 264–323 (1999). https://doi.org/10.1145/331499.331504
5. Kingma, D.P., Ba, J.: Adam: a method for stochastic optimization. In: 3rd International Conference on Learning Representations, ICLR, Conference Track Proceedings, San Diego, CA, USA, 7–9 May (2015)
6. Krizhevsky, A., Sutskever, I., Hinton, G.E.: Imagenet classification with deep convolutional neural networks. In: Advances in Neural Information Processing Systems, pp. 1106–1114 (2012)
7. Li, Z., Zhou, F., Chen, F., Li, H.: Meta-sgd: learning to learn quickly for few-shot learning. arXiv preprint arXiv:1707.09835 (2017)
8. Liu, Y., et al.: Learning to propagate labels: transductive propagation network for few-shot learning. In: 7th International Conference on Learning Representations, ICLR 2019, New Orleans, LA, USA, 6–9 May 2019 (2019). https://openreview.net/forum?id=SyVuRiC5K7
9. von Luxburg, U.: A tutorial on spectral clustering. Stat. Comput. **17**(4), 395–416 (2007). https://doi.org/10.1007/s11222-007-9033-z
10. Meila, M., Shi, J.: A random walks view of spectral segmentation. In: Proceedings of the Eighth International Workshop on Artificial Intelligence and Statistics, Key West, Florida, USA, 4–7 January (2001)
11. van den Oord, A., et al.: Wavenet: a generative model for raw audio. In: The 9th ISCA Speech Synthesis Workshop, Sunnyvale, CA, USA, 13–15 September, p. 125 (2016)
12. Oreshkin, B.N., López, P.R., Lacoste, A.: TADAM: task dependent adaptive metric for improved few-shot learning. In: Advances in Neural Information Processing Systems, NeurIPS, Montréal, Canada,3–8 December, pp. 719–729 (2018)
13. Qiao, S., Liu, C., Shen, W., Yuille, A.L.: Few-shot image recognition by predicting parameters from activations. In: 2018 IEEE Conference on Computer Vision and Pattern Recognition, CVPR, Salt Lake City, UT, USA, 18–22 June, pp. 7229–7238 (2018). https://doi.org/10.1109/CVPR.2018.00755
14. Ravi, S., Larochelle, H.: Optimization as a model for few-shot learning. In: 5th International Conference on Learning Representations, Conference Track Proceedings, Toulon, France, 24–26 April (2017)
15. Ren, M., et al.: Meta-learning for semi-supervised few-shot classification. In: 6th International Conference on Learning Representations, ICLR 2018, Conference Track Proceedings, Vancouver, BC, Canada, 30 April–3 May 2018 (2018). https://openreview.net/forum?id=HJcSzz-CZ
16. Rusu, A.A., et al.: Meta-learning with latent embedding optimization. In: 7th International Conference on Learning Representations, ICLR (2019)

17. Shi, J., Malik, J.: Normalized cuts and image segmentation. IEEE Trans. Pattern Anal. Mach. Intell. **22**(8), 888–905 (2000). https://doi.org/10.1109/34.868688
18. Snell, J., Swersky, K., Zemel, R.S.: Prototypical networks for few-shot learning. In: Advances in Neural Information Processing Systems, Long Beach, CA, USA, 4–9 December, pp. 4080–4090 (2017)
19. Luo, C., et al.: Spectral feature transformation for person re-identification. In: Proceedings of the IEEE International Conference on Computer Vision, pp. 4976–4985 (2019)
20. Yang, X., et al.: Deep spectral clustering using dual autoencoder network. In: Proceedings of the IEEE Conference on Computer Vision and Pattern Recognition, pp. 4066–4075 (2019)
21. Sung, F., Yang, Y., Zhang, L., Xiang, T., Torr, P.H.S., Hospedales, T.M.: Learning to compare: relation network for few-shot learning. In: IEEE Conference on Computer Vision and Pattern Recognition, CVPR, 18–22 June, Salt Lake City, UT, USA, pp. 1199–1208 (2018). https://doi.org/10.1109/CVPR.2018.00131
22. Sung, F., Yang, Y., Zhang, L., Xiang, T., Torr, P.H., Hospedales, T.M.: Learning to compare: Relation network for few-shot learning. In: Proceedings of the IEEE Conference on Computer Vision and Pattern Recognition, pp. 1199–1208 (2018)
23. Triantafillou, E., Zemel, R.S., Urtasun, R.: Few-shot learning through an information retrieval lens. In: Advances in Neural Information Processing Systems, 4–9 December, Long Beach, CA, USA, pp. 2252–2262 (2017)
24. Vinyals, O., Blundell, C., Lillicrap, T., Kavukcuoglu, K., Wierstra, D.: Matching networks for one shot learning. In: Advances in Neural Information Processing Systems, 5–10 December, Barcelona, Spain, pp. 3630–3638 (2016)

Forward Iterative Feature Selection Based on Laplacian Score

Qing-Qing Pang[1] and Li Zhang[1,2(✉)]

[1] School of Computer Science and Technology and Joint International Research Laboratory of Machine Learning and Neuromorphic Computing, Soochow University, Suzhou 215006, Jiangsu, China
20184227025@stu.suda.edu.cn, zhangliml@suda.edu.cn
[2] Provincial Key Laboratory for Computer Information Processing Technology, Soochow University, Suzhou 215006, Jiangsu, China

Abstract. As a feature selection method, Laplacian score (LS) is widely used for dimensionality reduction in the unsupervised situation. However, LS separately measures the importance of each feature, and does not consider the association of features. To remedy it, this paper proposes an improved version of LS, called forward iterative Laplacian score (FILS). The goal of FILS is to maintain the local manifold structure of original data with the least number of features. The proposed FILS introduces a recursive scheme to pick up features one-by-one, and evaluates the feature importance according to the joint locality preserving ability. Extensive experiments are conducted on UCI and microarray gene datasets. Experimental results confirm that FILS can achieve a good performance.

Keywords: Unsupervised learning · Feature selection · Laplacian score · Manifold

1 Introduction

As a technique of dimensionality reduction, feature selection has attracted a lot of attentions in pattern recognition, machine learning and data mining. Feature selection can eliminate irrelevant and redundant features, which promotes the computational efficiency, and keep the interpretation of reduced description [1,2]. According to the situations of data labels, feature selection methods can be divided into three types: supervised, unsupervised and semi-supervised ones [3].

In supervised methods, the correlation between features and class labels can be used to assess the importance of features [4,5]. Fisher score, for example, seeks features by making the within-class distance of data as small as possible and the

Supported by the Natural Science Foundation of the Jiangsu Higher Education Institutions of China under Grant No. 19KJA550002, the Six Talent Peak Project of Jiangsu Province of China under Grant No. XYDXX-054, and the Priority Academic Program Development of Jiangsu Higher Education Institutions.

H. Yang et al. (Eds.): ICONIP 2020, LNCS 12533, pp. 381–392, 2020.
https://doi.org/10.1007/978-3-030-63833-7_32

between-class distance of data as large as possible [6]. Semi-supervised feature selection methods use a limited label information and the whole distribution of data to evaluate features [7], such as semi-Fisher score [8] that combines the local structure preserving criterion and the variance strategy. These two kinds of feature selection methods, to some extent, depend on the label information to guide the feature evaluation by encoding features' discriminative information in labels [9].

For unsupervised methods, feature correlation is assessed by the ability to maintain specific features of data, such as the variance value [10], and Laplacian score (LS) [11]. LS was proposed based on the spectral graph theory and uses a neighborhood graph to determine optimal features. However, LS separately measures the importance of each feature, and does not consider the association of features. Zhu and Miao et al. [12] proposed an iterative Laplacian score (IterativeLS). This method progressively changes the nearest neighbor graph by discarding the least relevant features in each iteration, and assesses the importance of the feature by its local retention capabilities. In each iteration, IterativeLS would reconstruct a nearest neighbor graph using the rest features. In doing so, the local structure of the original data would be ruined.

To enhance both LS and IterativeLS, this paper presents a Forward iterative feature selection based on Laplacian score (FILS) method for unsupervised feature selection. The goal of FILS is to maintain the local manifold structure of original data with the least number of features. Inspired by IterativeLS, FILS adopts a recursive scheme to select features one-by-one. The criterion of evaluating the feature importance in FILS is different from those in LS and IterativeLS. A feature subset would be an optimal one if and only if this subset has the closest local preserving ability to the whole featureset. In doing so, the selected feature subset could maintain the local structure of data. Compared with IterativeLS, FILS does not need to construct a nearest neighbor graph in each iteration. The validity and stability of FILS is confirmed by experimental results.

The rest of this paper is organized as follows. In Sect. 2, we review two unsupervised feature selection methods. Section 3 proposes the Forward feature selection based on Laplacian score. In Sect. 4, we conduct experiments on UCI and gene datasets to compare the proposed method with LS and IterativeLS. This paper is summarized in Sect. 5.

2   Related Methods

This section briefly reviews two unsupervised feature selection methods: Laplacian score and iterative Laplacian score, which are very related to our work.

Assume that there has a set of unlabeled data $X = \{\mathbf{x}_1, \cdots, \mathbf{x}_u\}$, where $\mathbf{x}_i \in \mathbf{R}^n$, n is the number of features, u is the number of samples. Let $F = \{f_1, \cdots, f_n\}$ be the feature set with features f_k, $k = 1, \cdots, n$ and $\mathbf{Z} \in \mathbf{R}^{u \times n}$ be the sample matrix with column feature vectors $\mathbf{z}_k \in \mathbf{R}^u$, $k = 1, \cdots, n$ and row sample vectors $\mathbf{x}_i$, $i = 1, \cdots, u$.

2.1 Laplacian Score

LS aims to maintain the local structure of data during the feature selection process [11]. A key assumption of the LS algorithm is that samples from the same class are closer to each other than samples of different classes, that is, LS focuses on the local structure of data rather than the global structure.

For the given dataset $\mathbf{X}$, LS first constructs a nearest neighbor graph G with u nodes. The i-th node corresponds to the sample $\mathbf{x}_i$. We put an edge between nodes i and j if $\mathbf{x}_i$ is among K nearest neighbors of $\mathbf{x}_j$ or $\mathbf{x}_j$ is among K nearest neighbors of $\mathbf{x}_i$. The graph G can be represented by the weight matrix $\mathbf{S}$:

$$S_{ij} = \begin{cases} \exp\left\{-\gamma\|\mathbf{x}_i - \mathbf{x}_j\|^2\right\}, & if \ (\mathbf{x}_i \in KNN(\mathbf{x}_j) \ \vee \ \mathbf{x}_j \in KNN(\mathbf{x}_i)) \\ 0, & otherwise \end{cases} \quad (1)$$

where $\gamma > 0$ is a constant to be tuned, and $KNN(\mathbf{x}_i)$ denotes the set of K nearest neighbors of $\mathbf{x}_i$.

In LS, the score for measuring feature f_k can be computed by

$$J_{LS}(f_k) = \frac{\sum_{i,j}(z_{ki} - z_{kj})^2 S_{ij}}{\sum_i (z_{ki} - \mu_k)^2 D_{ii}} \quad (2)$$

where z_{ki} denotes the kth feature of ith samples, $\mu_k = \frac{1}{u}\sum_{i=1}^{u} z_{ki}$ denotes the mean of all samples on feature f_k, and $\mathbf{D}$ is a diagonal matrix with $D_{ii} = \sum_j S_{ij}$.

The smaller $J_{LS}(f_k)$ is, the greater the contribution of the kth feature to the local structural of the retained data, so LS always selects features with smaller scores. The computational complexity of constructing $\mathbf{S}$ is $O(u^2)$, and the computational complexity of calculating scores for n features is $O(nu^2)$. Hence, the overall computational complexity of LS is $O(nu^2)$.

2.2 Iterative Laplacian Score

The iterative Laplacian score algorithm proposed by Zhu et al. [12] introduces the iterative idea into LS. Experimental results in [12] indicated that IterativeLS outperforms LS on both classification and clustering tasks.

The key idea of IterativeLS is to gradually improve the nearest neighbor graph by discarding the least relevant features in each iteration. As with LS, IterativeLS evaluates the importance of an feature by its locality preserving ability. In Algorithm 1, we describe IterativeLS in details. For each iteration, the computational complexity of constructing $\mathbf{S}$ is still $O(u^2)$, and computational complexity of calculating scores for n features is $O(nu^2)$. Hence, the overall computational complexity of the method is $O(n^2u^2)$.

3 Forward Iterative Feature Selection Based on Laplacian Score

This section presents the novel feature selection method: FILS, which is an extension of LS. Both LS and IterativeLS measure the importance of feature separately

---

Algorithm 1: Iterative Laplacian Score

Input: Dataset X with n features, target feature number r and nearest neighbor
number K;

Output: feature subset with r features;

  repeat

    Construct the nearest neighbor graph G using the dataset X and calculate the
corresponding weight matrix $\mathbf{S}$;

    Compute the scores J_{LS} for the features in X using LS (2);

    Rank all features in ascending order according to scores J_{LS};

    Discard the last one feature and update the dataset X consisting of only the rest
features;

  until X contains at most r features

---

and then maintain the local structure of data by the important features. Similar
to LS and IterativeLS, FILS also considers maintaining the local structure of
data. Unlike them, FILS measures the local preserving ability of feature subsets.
Similar to IterativeLS, FILS adopts a recursive scheme. The difference is that
IterativeLS discards the least relevant feature according to their current Laplacian scores in each iteration, and FILS selects the most relevant feature to form
an optimal subset in each iteration. An feature subset would be an optimal one if
and only if this subset has the closest local preserving ability to the whole feature
set. In doing so, the selected feature subset could maintain the local structure
of data. Moreover, FILS does not need to reconstruct a nearest neighbor graph
in each iteration.

Quite simply, FILS requires constructing a nearest neighbor graph, and calculating Laplacian scores of features sets respectively. First, we use the adaptive method [13] to construct the nearest neighbor graph, and have the weight
matrix $\mathbf{S}$:

$$S_{ij} = \exp\left(-\frac{d^2(\mathbf{x}_i, \mathbf{x}_j)}{\sigma_i \sigma_j}\right) \tag{3}$$

where $d(\mathbf{x}_i, \mathbf{x}_j)$ is the Euclidean distance between vectors $\mathbf{x}_i$ and $\mathbf{x}_j$, σ_i is the
local scale and $\sigma_i = d(\mathbf{x}_i, \mathbf{x}_i^K)$, $\mathbf{x}_i^K$ is the Kth nearest neighbor of $\mathbf{x}_i$.

Next, we discuss the Laplacian score of feature sets and give a definition
below.

Definition 1. *Given a dataset X, an feature subset $A \subseteq F$ the Laplacian score
of feature set A is defined as*

$$J(A) = \frac{trace(\mathbf{Z}_A^T \mathbf{L} \mathbf{Z}_A)}{trace(\tilde{\mathbf{Z}}_A^T \mathbf{D} \tilde{\mathbf{Z}}_A})) \tag{4}$$

$$\tilde{\mathbf{Z}}_A^T = \mathbf{Z}_A - \frac{\mathbf{Z}_A^T \mathbf{D} \mathbf{1}}{\mathbf{1}^T \mathbf{D} \mathbf{1}},$$

*where $\mathbf{Z}_A$ is the sample sub-matrix of X with the feature set A, and $\mathbf{L} = \mathbf{D} - \mathbf{S}$
is the Laplacian matrix of the dataset X with all feature, and $\mathbf{1}$ is the vector of
all ones.*

In the following, we describe the criterion of FILS. Assume that we have a target feature subset $A \subseteq F$ in the current iteration, the goal of FILS is to find the most important feature f_k from the complement set of A and add it into A, where $f_k \in \bar{A} = F - A$. FILS uses the incremental search technique to determine the optimal feature subset. In FILS, we define a new criterion to pick up the most important feature:

$$f_k^* = \arg\min_{f_k} J_{FILS}(f_k) = \arg\min_{f_k} \left| 1 - \frac{J(A \cup f_k)}{J(F)} \right| \tag{5}$$

where $J(A \cup f_k)$ and $J(F)$ are the Laplacian scores of feature set $A \cup f_k$ and F, respectively.

In (4), $J(F)$ reflects the information on the local structure of original data, which is the standard conforming to. $J(A \cup f_k)$ is the degree of maintaining local structure of original data if only a part of features are selected. If the Laplacian score of the feature subset $A \cup f_k$ is equal to that of feature set F, then we thought that the feature subset $A \cup f_k$ can represent the whole feature set F. Thus, we except that $J(A \cup f_k)$ is as close to $J(F)$ as possible. In other words, the smaller the score $J_{FILS}(f_k)$ is, the more important the feature f_k is.

The detail algorithm description of FILS is shown in Algorithm 2. First, the target subset A is initialized to be an empty set. $F = \{f_1, \cdots, f_n\}$ is the feature set, A is the target feature subset, and $\bar{A} = F - A$ is the candidate feature subset. The parameter r is used to terminate the main loop in this algorithm, which needs to be set in advance. The parameter K is required for the nearest neighbor graph. Step 2 constructs the nearest neighbor graph G using the dataset X. Step 3 computes the weight matrix $\mathbf{S}$ by (3). Step 5 computes the Laplacian score for feature set F by (4). Steps 6–10 calculate the importance of features in the candidate feature subset $\bar{A}$. Step 11 finds out the most important feature in $\bar{A}$ with the minimum value $J_{FILS}(f_k)$. Steps 12–17 update A and $\bar{A}$ when the size of A is smaller than r or $\bar{A}$ is non-empty. The algorithm jumps out of the loop when the termination conditions are satisfied. Step 19 returns the target feature subset A.

The computational complexity of constructing $\mathbf{S}$ is $O(u^2)$. For each iteration, the computational complexity of calculating the score in (4) for n features is $O(nu^2)$. Hence, the overall computational complexity of FILS is about $O(n^2u^2)$. Therefore, FILS has the same computational complexity as IterativeLS.

4   Experimental Analysis

In order to verify the feasibility and effectiveness of FILS, simulation experiments were carried out on UCI datasets [14] and microarray gene expression datasets [15]. We compared FILS with both LS and IterativeLS and used the nearest neighbor classifier to measure the discriminant ability of selected features. In our experiments, we choose $K = 5$ to construct the nearest neighbor graph G.

Algorithm 2: Forward iterative feature selection based on Laplacian Score (FILS)

Input: Dataset X, target feature number r and nearest neighborhood K;
Output: Target feature subset A;
1: Initialize $A = \varnothing$, $F = \{f_1, \cdots, f_n\}$, $\bar{A} = F - A$ and $start = 1$;
2: Construct the nearest neighbor graph G using the dataset $\mathbf{X}$;
3: Compute the weight matrix $\mathbf{S}$ by (3);
4: Compute the Laplacian score of the feature set F: $J(\mathbf{F})$ by (4);
5: **while** $star = 1$ **do**
6:     **for** each $f_k \in \bar{A}$ **do**
7:         Let $A^k = A \cup \{f_k\}$;
8:         Compute the Laplacian score for feature set A^k: $J(A^k)$ by (4);
9:         Compute $J_{RFR_LS}(f_k)$ by (5);
10:     **end for**
11:     Find f_k with the minimum value $J_{RFR_LS}(f_k)$;
12:     **if** $(|A| < r) \wedge (\bar{A} \mathrel{!=} \varnothing)$ **then**
13:         $A \leftarrow A \cup \{f_k\}$;
14:         $\bar{A} \leftarrow F - A$;
15:     **else**
16:         $star = 0$;
17:     **end if**
18: **end while**
19: **return** A

4.1   UCI Dataset

We considered 8 UCI datasets here and compared FILS with LS and IterativeLS algorithms. The related information of 8 UCI datasets, including Australian, Heart, Pima, Segment, Spambase, Vehicle, Wdbc and Wine, is shown in Table 1. For these UCI datasets, the original features are normalized to the interval $[0, 1]$. In order to obtain more convincing comparison results and eliminate accidental errors, we used 10-fold cross-validation. That is to say, the original dataset is randomly divided into ten equal-sized subsets. Then 9 subsets are used as the training set and the rest one is used as the test set. The 10 subsets are used as test sets in turn, and then the average of 10 times is calculated as the final result of classification. Owing to the small number of UCI dataset features, r is set to the number of features of each dataset. Namely, we perform feature ranking.

Figure 1 shows the classification accuracy vs. feature number on 8 UCI datasets. Observation on Fig. 1 implies that the three methods have a similar curve variation with the increase of feature number. For example, the accuracy increases with increasing feature number on most datasets, such as Pima, Segment, Spambase, Vehicle and Wine. When the feature number grows to a determinate value, the accuracy varies slightly. In this case, less features would result in a fast test when holding the classification performance. Conversely, the

Table 1. Description of UCI data sets

No	Dataset	#Sample	#feature	#Class
1	Australian	690	14	2
2	Heart	301	13	2
3	Pima	768	8	2
4	Segment	2310	19	7
5	Spambase	4601	57	2
6	Vehicle	846	18	4
7	Wdbc	569	30	2
8	Wine	178	13	3

Table 2. Accuracy (%) and standard deviation obtained of different methods on UCI datsets

Dataset	FILS	LS	IterativeLS
Australian	**70.13** ± 3.31 (4)	66.96 ± 3.66 (13)	66.96 ± 3.60 (13)
Heart	**66.71** ± 4.63 (10)	62.03 ± 2.27(10)	62.01 ± 2.62 (3)
Pima	**62.63** ± 2.36 (6)	60.41 ± 2.38 (6)	60.41 ± 2.36 (6)
Segment	**93.38** ± 15.93 (15)	93.20 ± 10.39 (19)	93.20 ± 11.48 (19)
Spambase	**90.59** ± 5.19 (54)	86.57 ± 5.06 (56)	86.57 ± 5.55 (56)
Vehicle	**67.23** ± 8.70 (11)	66.88 ± 5.12 (16)	66.88 ± 5.65 (16)
Wdbc	**93.32** ± 1.72 (24)	90.87 ± 2.28 (17)	90.69 ± 2.28 (17)
Wine	**82.46** ± 6.03 (9)	82.46 ± 6.61(10)	82.46 ± 6.67 (10)

*Numbers in parentheses are optimal feature ones.

accuracy fluctuates irregularly on Australian, Heart and Wdbc datasets, which means that no all features help classification tasks.

Table 2 shows the highest average accuracy with the corresponding standard deviation and optimal feature number of all compared algorithms, where the best values among compared methods are in bold. We can see that FILS is superior to LS and IterativeLS on all datasets. For example, FILS achieves the accuracy 93.32% on the Wdbc dataset, followed by LS 90.87%. In a nutshell, FILS can effectively rank features and make discriminant ones at the top of feature list.

4.2   Microarray Gene Datasets

In this section, FILS was applied to microarray gene datasets, including Leukemia [16], St. Jude Leukemia (SJ-Leukemia) [17], Lungcancer [18] and the central nervous system (CNS) [19]. It is well-known that the number of features is much greater than the number of samples in the gene datasets. The gene

Table 3. Description of microarray gene data sets

No.	Data set	#Sample	#feature	#Class
1	Lungcancer	197	1000	4
2	SJ–Leukemia	248	985	6
3	Leukemia	38	999	3
4	CNS	42	989	5

expression datasets we used have been processed as described in [15]. Further biological details about these datasets can be found in the referenced papers. Most data were processed on the Human Genome U95 Affymetrix ©microarrays. The leukemia dataset was from the previous-generation Human Genome HU6800 Affymetrix ©microarray. The relevant information of these datasets is summaries in Table 3, and the detail description of these gene datasets is given as follows:

- Leukemia: Bone marrow samples were obtained from acute leukemia patients at the time of diagnosis. The dataset includes 11 acute myeloid leukemia (AML) samples, 8 T-lineage acute lymphoblastic leukemia (ALL) samples and 19 B-lineage ALL samples.
- SJ-Leukemia: Diagnostic bone marrow samples were from pediatric acute leukemia patients corresponding to 6 prognostically important Leukemia subtypes. The dataset includes 43 T-lineage ALL, 27 E2A-PBX1, 15 BCR-ABL, 79 TEL-AML1, 20 MLL rearrangements and 64"hperdiploid>50" chromosomes.
- Lung cancer: This dataset includes 4 known classes: 139 adenocarcinomas (AD), 21 squamous cell carcinomas (SQ), 20 carcinoids (COID), and 17 normal lung (NL). The AD class is highly heteroge-neous, and substructure is known to exist, although not well understood.
- CNS: The embryonal tumors of CNS dataset include 10 medulloblastomas (MD), 8 primitive neuroectodermal tumors (PNET), 10 atypical teratoid/rhabdoid tumors (Rhab), 10 malignant gliomas (Glio) and 4 normal cerebellum (Ncer).

Here, we also compared FILS with the related methods: LS and IterativeLS. In order to obtain convincing comparison results and eliminate accidental errors, as in the previous section, we used 3-fold cross-validation. In each trail, we randomly selected 2/3 of the samples as the training set, and the remaining 1/3 of samples as the test set. The experimental results were reported on the well-defined test sets. According to the statement in [20], we can know that we need 400 genes at most to complete the classification task of microarray gene data. Therefore, in order to save time, let $r = 400$ for all compared algorithms.

Figure 2 gives the classification accuracy vs. feature number on four microarray gene datasets. From Fig. 2, we can see that FILS is obviously superior to

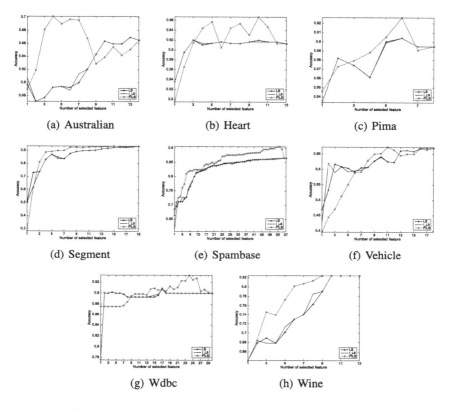

(a) Australian   (b) Heart   (c) Pima

(d) Segment   (e) Spambase   (f) Vehicle

(g) Wdbc   (h) Wine

Fig. 1. Accuracy vs. feature number on 8 UCI datasets

other two methods on both four gene datasets. In addition, FILS can quickly achieve a better classification performance. We summarized the highest accuracy of compared methods in Table 4 according to Fig. 2, where bold numbers are the best results among compared methods. On the SJ-Leukemia datasets, FILS is just 0.03% better than IterativeLS. FILS achieves much better accuracies on the other three gene datasets. On the CNS dataset, for example, the accuracy of FILS is almost 14.88% higher than LS.

4.3 Statistical Comparison on Multiple Datasets

In order to give a comprehensive comparison on UCI and gene datasets, we used the Friedman test [21] and the Bonferroni-Dunn test [22]. The Friedman test with the Bonferroni-Dunn test is used to test whether all the methods are equivalent, which has the null hypothesis that all the methods are equivalent. If the ranks of all methods are equal to each other, the test result accepts the null hypothesis; otherwise the test result rejects the null hypothesis and the Bonferroni-Dunn test is carried out to reveal the significant differences. The critical difference between two methods is defined as:

Table 4. Average accuracy and standard deviation comparison on four microarray gene datasets

Dataset	FILS	LS	IterativeLS
Lungcancer	**93.39** ± 3.41 (331)	86.36 ± 3.59 (368)	75.58 ± 2.65(397)
SJ–Leukemia	**95.14** ± 5.07 (102)	95.11 ± 5.51 (346)	95.11 ± 5.63 (255)
Leukemia	**85.16** ± 4.71 (72)	63.90 ± 7.30 (276)	71.41 ± 5.01 (356)
CNS	**76.41** ± 6.36 (326)	61.53 ± 3.93 (102)	53.08 ± 5.06 (171)

* The numbers in parentheses are optimal feature numbers.

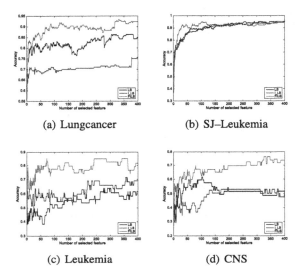

(a) Lungcancer   (b) SJ–Leukemia

(c) Leukemia   (d) CNS

Fig. 2. Accuracy vs. feature number on 4 gene datasets

$$CD = q_\alpha \sqrt{\frac{j(j+1)}{6N}} \tag{6}$$

where j is the number of methods, N is the number of datasets, q_α is the critical value and α is the threshold value. In our experiments, $j = 3$ and $N = 12$. Generally, let $\alpha = 0.1$ [23,24]. Then we have $q_\alpha = 1.96$ [22]. Thus, the critical difference in our experiment is $CD = 0.79$.

Table 5 gives the mean rank of three methods and the rank difference between FILS and other two methods. According to the Friedman test, the results of mean ranks from Table 5 can reject the null hypothesis, which means that the three methods in comparison are not equivalent and there are significant differences among different methods. The rank differences between FILS and both LS and IterativeLS are greater than the critical difference 0.79, which means that FILS is significantly better than these two methods in this current experimental setting.

Table 5. Statistical comparison of three methods

Methods	FILS	LS	IterativeLS
Mean rank	1	2.29	2.71
Rank difference	0.00	1.29	1.71

5 Conclusion

This paper concentrates on an unsupervised feature selection method and proposes an algorithm called FILS. FILS aims to maintain the local manifold structure of original data with the least number of features. Different from existing LS-like methods, FILS evaluates the locality preserving ability of feature subsets instead of single features. In doing so, FILS can pick up the feature subset which maintains the local structure of data as possible. On 8 UCI and 4 microarray gene datasets, a series of experiments were conducted for evaluating the proposed method. FILS retains the highest classification accuracy on most datasets. The final statistical results confirm that FILS is significantly better than other LS-like methods.

Although the classification accuracy of FILS on majority datasets have been enhanced, the running time required by FILS would increase as the feature dimension of data grows. In future, we try to reduce the time complexity of the algorithm.

References

1. Sheikhpour, R., Sarram, M.A., Gharaghani, S., Chahooki, M.A.Z.: A Survey on semi-supervised feature selection methods. Pattern Recognit. **64**, 141–158 (2017)
2. Song, X., Han, Y., Jiang, J.: Semi-supervised feature selection via hierarchical regression for web image classification. Multimed. Syst. **22**(1), 41–49 (2016)
3. Sooksatra, K., Li, R., Li, Y., Guan, X., Li, W.: Fairness-aware auction mechanism for sustainable mobile crowdsensing. In: Biagioni, E.S., Zheng, Y., Cheng, S. (eds.) WASA 2019. LNCS, vol. 11604, pp. 310–321. Springer, Cham (2019). https://doi.org/10.1007/978-3-030-23597-0_25
4. Battiti, R.: Using mutual information for selecting features in supervised neural net learning. IEEE Trans. Neural Netw. **5**(4), 537–550 (1994)
5. Wang, C., Hu, Q., Wang, X., Chen, D., Qian, Y., Dong, Z.: Feature selection based on neighborhood discrimination index. IEEE Trans. Neural Netw. Learn. Syst. **29**(7), 2986–2999 (2018)
6. Amiri, S.H., Jamzad, M.: Automatic image annotation using semi-supervised generative modeling. Pattern Recognit. **48**(1), 174–188 (2015)
7. Han, Y., Yang, Y., Yan, Y., Ma, Z., Sebe, N., Member, S.: Semisupervised feature selection via spline regression for video semantic recognition. IEEE Trans. Neural Netw. Learn. Syst. **26**(2), 252–264 (2015)
8. Yang, M., Chen, Y., Ji, G.: Semi-Fisher score: a semi-supervised method for feature selection. In: Proceedings of the International Conference on Machine Learning and Cybernetics, vol. 1, pp. 527–532 (2010)

 9. Luo, M., Nie, F., Chang, X., Yang, Y., Hauptmann, A.G., Zheng, Q.: Adaptive unsupervised feature selection with structure regularization. IEEE Trans. Neural Netw. Learn. Syst. **29**(4), 944–956 (2018)
10. Bishop, C.M.: Neural networks for pattern recognition. Agric. Eng. Int. CIGR J. Entific Res. Dev. Manuscript PM **12**(5), 1235–1242 (1995)
11. He, X., Cai, D., Niyog, P.: Laplacian score for feature selection. In: International Conference on Neural Information Processing Systems, pp. 507–541 (2005)
12. Zhu, L., Miao, L., Zhang, D.: Iterative laplacian score for feature selection. In: Liu, C.-L., Zhang, C., Wang, L. (eds.) CCPR 2012. CCIS, vol. 321, pp. 80–87. Springer, Heidelberg (2012). https://doi.org/10.1007/978-3-642-33506-8_11
13. Zelnik-manor, L., Perona, P.: Self-tuning spectral clustering. In: Saul, L.K., Weiss, Y., Bottou, L. (eds.) Advances in Neural Information Processing Systems, vol. 17, pp. 1601–1608. MIT Press (2005)
14. Dheeru, D., Karra Taniskidou, E.: UCI machine learning repository (2017). http://archive.ics.uci.edu/ml
15. Monti, S., Tamayo, P., Mesirov, J., Golub, T.: Consensus clustering: a resampling-based method for class discovery and visualization of gene expression microarray data. Mach. Learn. **52**(1–2), 91–118 (2003)
16. Golub, T.R., et al.: Molecular classification of cancer: class discovery and class prediction by gene expression. Science **286**(5439), 531–537 (1999)
17. Yeoh, E.J., et al.: Classification, subtype discovery, and prediction of outcome in pediatric acute lymphoblastic leukemia by gene expression pro ling. Cancer Cell **1**(2), 133–143 (2002)
18. Bhattacharjee, A., et al.: Classification of human lung carcinomas by mRNA expression pro lingreveals distinct adenocarcinomas sub-classes. Proc. Natl. Acad. Sci. **98**(24), 13790–13795 (2001)
19. Pomeroy, S., et al.: Gene expression-based classification and outcome prediction of central nervous system embryonal tumors. Nature **415**(6870), 436–442 (2001)
20. Shieh, M.D., Yang, C.C.: Multiclass SVM-REF for product from feature selection. Exp. Syst. Appl. **35**(1–2), 531–541 (2008)
21. Friedman, M.: The use of ranks to avoid the assumption of normality implicit in the analysis of variance. Publ. Am. Stat. Assoc. **32**(200), 675–701 (1937)
22. Dunn, O.J.: Multiple comparisons among means. J. Am. Stat. Assoc. **56**(293), 46–52 (1961)
23. Huang, X., Zhang, L., Wang, B., Li, F., Zhang, Z.: Feature clustering based support vector machine recursive feature elimination for gene selection. Appl. Intell. **48**(3), 594–607 (2018)
24. Chen, H., Tinňo, P., Yao, X.: Predictive ensemble pruning by expectation propagation. IEEE Trans. Knowl. Data Eng. **21**(7), 999–1013 (2009)

Functional Data Clustering Analysis via the Learning of Gaussian Processes with Wasserstein Distance

Tao Li and Jinwen Ma[✉]

Department of Information Science, School of Mathematical Sciences and LMAM,
Peking University, Beijing, China
li_tao@pku.edu.cn, jwma@math.pku.edu.cn

Abstract. Functional data clustering analysis becomes an urgent and challenging task in the new era of big data. In this paper, we propose a new framework for functional data clustering analysis, which adopts a similar structure as the k-means algorithm for the conventional clustering analysis. Under this framework, we clarify three issues: how to represent functions, how to measure distances between functions, and how to calculate centers of functions. We utilize Gaussian processes to represent the clusters of functions which are actually their sample curves or trajectories on a finite set of sample points. Moreover, we take the Wasserstein distance to measure the similarity between Gaussian distributions. With the choice of Wasserstein distance, the centers of Gaussian processes can be calculated analytically and efficiently. To demonstrate the effectiveness of the proposed method, we compare it with existing competitive clustering methods on synthetic datasets and the obtained results are encouraging. We finally apply the proposed method to three real-world datasets with satisfying results.

1 Introduction

Functional data analysis [18] is a branch of modern statistics that deals with learning and inference problems of functional data. The atoms of functional data are functions, and one sample in a functional dataset consists of random samples of the underlying function. To get a flavor of functional data, let us consider an electrical load analysis problem for example. Suppose we record the electrical load of a city every 15 min for 50 days, and the electrical load records in one day are regarded as a curve (versus time). We may want to analyze whether there are certain patterns underlying this dataset. One possible way is to view each curve as a vector of length $96(= 24$ h $\times$ 4 records/hour). However, this point of view has some severe drawbacks. Firstly, the dimension is relatively high, thus the subsequent processing is challenging. Secondly, there may be missing values or extra measurements due to technical reasons, which results in varying lengths. Besides, there may exist a time-warping problem between curves, since it is difficult to control the measurement time accurately in practice. It would

© Springer Nature Switzerland AG 2020
H. Yang et al. (Eds.): ICONIP 2020, LNCS 12533, pp. 393–403, 2020.
https://doi.org/10.1007/978-3-030-63833-7_33

be better to regard each curve as a function mapping from $\mathbb{R}$ to $\mathbb{R}$, mapping a time variable t to the electrical load measured at t. In this way, we do not need to bother with varying signal lengths, and there exists a variety of alignment algorithms to sidestep the time-warping problem. In fact, there are many such examples in data analysis and signal processing, which face similar problems mentioned above and it is beneficial to view them as functional data to sidestep these problems.

We mainly concern the functional data clustering problem in this work. Clustering [8] is the main task in unsupervised learning, which aims to partition the data into groups such that samples in the same group are similar. Back to the example discussed above, we can discover intrinsic patterns of electrical loads by applying functional data clustering to the electrical load dataset. For example, we can discover in which days the electrical load trends are similar, and use such information to help prediction.

There are various clustering algorithms for finite-dimensional (vector-valued) data [8], but little is known about functional data clustering. The main challenge of functional data analysis is the data is intrinsically infinite-dimensional, thus classical clustering methods for finite-dimensional data are invalid in this circumstance. To tackle the difficulty caused by infinite dimension, one usually find finite-dimensional representations for the functional data, then perform classical clustering algorithms based on such representations. Popular dimension reduction methods include functional basis expansion, functional principal component analysis, and subspace projection.

In this paper, we first formulate the functional data clustering problem in a similar way as k-means. Then we point out that to devise a practical algorithm based this formulation we need to consider three questions: how to find finite-dimensional representations for the functional data, how to measure distance or similarity between functions based on the finite-dimensional representation, and how to solve the optimization problem, or more precisely, how to calculate barycenters based on the representations and distance. We propose to use Gaussian processes [19] to fit each function separately, and use the learned parameters as the finite-dimensional representation. Gaussian processes are the dominant Bayesian non-parametric non-linear model for inference over functions. The key point is that Gaussian processes allow us to learn uncertainty about posterior functions, and the information about posterior functions can be efficiently retrieved from the learned parameters. After obtaining the representations, we propose to calculate approximate Wasserstein distance [11,16] between Gaussian processes by their finite-dimensional distributions. Intuitively, finite-dimensional distributions of the posterior functions provide us good approximations of functions as long as the number of sample points is large enough. There are numerous criteria to measure the distance between Gaussians, such as Kullback-Leibler divergence [6], Jensen-Shannon divergence [6] and so on, but we choose the Wasserstein 2-distance. Wasserstein distances arise naturally in the optimal transport theory [23], which aims to match two measures with the least cost. In addition, Wasserstein 2-distance enables us to calculate the centers of functions easily, and the iteration process can be further accelerated by Anderson acceleration [25] technique.

2   Functional Data Clustering: Problem Formulation

Suppose we have functional data $\mathcal{D} = \{\mathcal{D}_i\}_{i=1}^N$, where $\mathcal{D}_i = \{(\mathbf{x}_{ij}, \mathbf{y}_{ij})\}_{j=1}^{N_i}$, and $\mathbf{x}_{ij} \in \mathbb{R}^{D_x}, \mathbf{y}_{ij} \in \mathbb{R}^{D_y}$. Usually we have $D_x = D_y = 1$ in real applications as discussed in Sect. 1, but we develop the theory here for general dimensions. The philosophy of functional data analysis is to play with functions, or more concretely, every $\mathcal{D}_i$ is viewed as observations of an underlying function $f_i : \mathbb{R}^{D_x} \to \mathbb{R}^{D_y}$ which maps $\mathbf{x}_{ij}$ to $\mathbf{y}_{ij}$. The goal of functional data clustering is to split the data into K groups such that similar functions are in the same group. Formally, we want to find a map $c : \{1, 2, \cdots, N\} \to \{1, 2, \cdots, K\}$ such that $c(i) = c(j)$ if and only if f_i and f_j are "similar". Note that different samples may have different number of data-points, i.e., N_i may vary as i, which makes common clustering algorithms invalid for functional data clustering. Nevertheless, we adapt the idea of k-means, and formulate functional data clustering as the following optimization problem:

$$\min_{c, m_1, m_2, \cdots, m_K} \sum_{k=1}^K \sum_{c(i)=k} d^2(f_i, m_k), \tag{1}$$

where m_k is the center (function) of k-th cluster and $d(\cdot, \cdot)$ is a distance function.

However, there are several vital problems in formulation (1). Firstly, it is usually intractable to store and calculate $\{f_i\}_{i=1}^N$ directly in the computer, and we must find effective finite-dimensional representations of these functions. Secondly, how to measure the distance of two functions based on their finite-dimensional representations? Indeed, there are various metrics for a different type of functions as studied in functional analysis, but most of them are difficult to calculate analytically, especially when only the finite-dimensional representations are given. One possible choice is to use metrics for finite-dimensional vectors directly, such as Euclidean norm, Mahalanobis distance, and so on, but this may loss certain global or temporal information of original functions. Last but not least, problems like k-means and (1) are usually solved by the coordinate descent algorithm, which optimizes the objective function with respect to c and $m_1, \cdots, m_K$ alternately. For problem (1), the sub-problem of finding c is trivial (as long as the previously mentioned two problems are solved), but the optimization with respect to $m_1, m_2, \cdots, m_K$ remains difficult and highly dependent on the choice of representations and distance function.

3   Proposed Method

3.1   Gaussian Process Representation of Functional Data

The key-point in functional data analysis is to find a finite dimensional representation $\mathcal{R}(f_i | \mathcal{D}_i)$ for function f_i based on observations $\{(\mathbf{x}_{ij}, \mathbf{y}_{ij})\}_{j=1}^{N_i}$. Traditional methods include basis function expansion, functional principal component analysis (FPCA) and so on. Once functional data is transformed to finite dimensional representations, we can apply a variety of machine learning techniques to solve functional data problems. We assume $D_y = 1$ temporarily. Functional principal component analysis focus on model the correlation between $\{f_i\}_{i=1}^N$, and use this relationship to find d eigen-functions $\{\phi_l\}_{l=1}^d$ that best explain the data, then representation is given by coefficients as $\mathcal{R}(f_i|\mathcal{D}_i) = [\langle f_i, \phi_1 \rangle, \cdots, \langle f_i, \phi_d \rangle]^T$.

Gaussian processes are a natural and fruitful way of specifying prior and inferencing over functions. Instead of considering the correlation between functions, we propose to specify Gaussian process prior for each sample independently. For each sample $\mathcal{D}_i = \{(\mathbf{x}_{ij}, \mathbf{y}_{ij})\}_{j=1}^{N_i}$, we assume the underlying function f_i is a Gaussian process, i.e., $f_i \sim \mathcal{GP}(\mu(\mathbf{x}), k(\mathbf{x}, \mathbf{x}'))$ where $\mu(\mathbf{x})$ and $k(\mathbf{x}, \mathbf{x}')$ are mean function and covariance function respectively. In this work, we only consider zero mean function and squared exponential covariance function as follows:

$$\mu(\mathbf{x}) = 0 \quad , \quad k(\mathbf{x}, \mathbf{x}') = \theta_0 \exp\left(-\frac{\sum_{k=1}^{d_x} \theta_k (\mathbf{x}_k - \mathbf{x}'_k)^2}{2} + \sigma^2 \mathbb{I}(\mathbf{x} = \mathbf{x}')\right). \quad (2)$$

Let $\mathbf{K_{xx}} \in \mathbb{R}^{N_i \times N_i}$ with $[\mathbf{K_{xx}}]_{mn} = k(\mathbf{x}_{im}, \mathbf{x}_{in})$, then the Gaussian process prior is equivalent to say we assume $[\mathbf{y}_{i1}, \cdots, \mathbf{y}_{iN_i}]^T \sim \mathcal{N}(\mathbf{0}, \mathbf{K_{xx}})$. Therefore, the parameters can be learned by the Type-II maximum likelihood method, which has been implemented effectively in the GPML toolbox [20]. Once these parameters are learned, given any new input $\{\mathbf{z}_k\}_{k=1}^D$, let $[\mathbf{K_{zz}}]_{mn} = k(\mathbf{z}_m, \mathbf{z}_n)$ and $[\mathbf{K_{zx}}]_{mn} = k(\mathbf{z}_m, \mathbf{x}_{in})$, from the conditional property [4] of Gaussian distributions we immediately have

$$[f_i(\mathbf{z}_1), \cdots, f_i(\mathbf{z}_D)]^T \sim \mathcal{N}\left(\mathbf{K_{zx}} \mathbf{K_{xx}^{-1}} [\mathbf{y}_{i1}, \cdots, \mathbf{y}_{iN_i}]^T, \mathbf{K_{zz}} - \mathbf{K_{zx}} \mathbf{K_{xx}^{-1}} \mathbf{K_{zx}^T}\right). \quad (3)$$

Thus, we have access to any finite dimensional joint distribution of $f_i|\mathcal{D}_i$. For abbreviation, we use $\boldsymbol{\theta}_i$ to denote all the parameters of the Gaussian process learned from $\mathcal{D}_i$, then $\mathcal{R}(f_i|\mathcal{D}_i) = \boldsymbol{\theta}_i$ is a reasonable representation of $f_i|\mathcal{D}_i$ since we can restore all the information of $f_i|\mathcal{D}_i$ from $\boldsymbol{\theta}_i$. Therefore, we have transformed the original functional data $\{\mathcal{D}_i\}_{i=1}^N$ to a collection of Gaussian process representations $\{\boldsymbol{\theta}_i\}_{i=1}^N$.

3.2   Approximate Wasserstein Distance of Gaussian Processes

Then, we consider the problem of how to define the distance function $d(\cdot, \cdot)$ in (1). Ideally, we hope the cluster centers $\{m_k\}_{k=1}^K$ are also Gaussian processes. However, it's still difficult to work with Gaussian processes directly. The trick here is to approximate a Gaussian process by its finite-dimensional distributions, which are multivariate Gaussian distributions, and thus we may view $\{m_k\}_{k=1}^K$ as multivariate Gaussian distributions too. Let $\mathbf{Z} = \{\mathbf{z}_k\}_{k=1}^D$ be D equally-spaced grids in the input region, then $[f_i(\mathbf{z}_1), \cdots, f_i(\mathbf{z}_D)]^T$ can be seen as a good estimation of $f_i|\mathcal{D}_i$ as long as $\{\mathbf{z}_k\}_{k=1}^D$ are dense enough. We use $f_i(\mathbf{Z})$ to denote the distribution of $[f_i(\mathbf{z}_1), \cdots, f_i(\mathbf{z}_D)]^T$ as derived in (3), then the problem of defining $d(\cdot, \cdot)$ becomes how to measure the distance between two Gaussian distributions $f_i(\mathbf{Z})$ and m_k. Fortunately, optimal transportation theory [16] provides us a powerful tool. The basic problem of optimal transportation is to investigate how to match two measures with the least cost, and the corresponding cost is called the Wasserstein distance, which is a distance function of the measures. Formally[1], suppose we have two Radon measures $\alpha, \beta \in \mathcal{M}(\mathcal{X})$, $d_{\mathcal{X}}(\cdot, \cdot)$ is a distance function on $\mathcal{X}$, then the Wasserstein 2-distance between α and β is defined as:

$$\mathcal{W}_2(\alpha, \beta) = \left(\inf_{\pi \in \mathcal{U}(\alpha, \beta)} \int_{\mathcal{X} \times \mathcal{Y}} d_{\mathcal{X}}^2(x, y) \mathrm{d}\pi(x, y) \right)^{1/2}, \tag{4}$$

$$\mathcal{U}(\alpha, \beta) = \{\pi \in \mathcal{M}_+^1(\mathcal{X} \times \mathcal{Y}), \pi(A \times \mathcal{Y}) = \alpha(A), \pi(\mathcal{X} \times B) = \beta(B), \forall A \subset \mathcal{X}, B \subset \mathcal{Y}\}.$$

It has been proved [16] that for two Gaussian measures $\alpha = \mathcal{N}(\boldsymbol{\mu}_\alpha, \boldsymbol{\Sigma}_\alpha)$ and $\beta = \mathcal{N}(\boldsymbol{\mu}_\beta, \boldsymbol{\Sigma}_\beta)$, the Wasserstein 2-distance when $d_{\mathcal{X}}$ is Euclidean distance is

$$\mathcal{W}_2(\alpha, \beta) = \left(\|\boldsymbol{\mu}_\alpha - \boldsymbol{\mu}_\beta\|_2^2 + \mathrm{tr}\left(\boldsymbol{\Sigma}_\alpha + \boldsymbol{\Sigma}_\beta - 2(\boldsymbol{\Sigma}_\alpha^{1/2} \boldsymbol{\Sigma}_\beta \boldsymbol{\Sigma}_\alpha^{1/2})^{1/2} \right) \right)^{1/2}. \tag{5}$$

Therefore, we may use the Wasserstein 2-distance between $\mathcal{N}_i$ and m_k as a distance of f_i and m_k, i.e., $d(f_i, m_k) := \mathcal{W}_2(f_i(\mathbf{Z}), m_k)$. With this definition, the functional data clustering formulation (1) becomes:

$$\min_{c, m_1, m_2, \cdots, m_K} \sum_{k=1}^{K} \sum_{c(i)=k} \mathcal{W}_2^2(f_i(\mathbf{Z}), m_k). \tag{6}$$

We point out that similar ideas have been suggested in [14]. However, our goal is to cluster functional data, while the main purpose of [14] is to prove theoretical results about Gaussian processes from the optimal transport perspective and learn uncertain curves.

3.3   Barycenter Calculation

Th last problem is to derive how to optimize (6). Similar to k-means, we perform optimization with respect to cluster labels c and centers $m_1, \cdots, m_k$ alternatively. The update of c is trivial: we only need to calculate pairwise distances between f_i and centers then assign $c(i) = \arg\min_k d(f_i, m_k)$. With a little abuse of notation, we suppose $\{c(i) = k\} = \{1, 2, \cdots, N\}$, then the problem of updating m_k can be written as

$$\min_{m_k} \sum_{i=1}^{N} \mathcal{W}_2^2(f_i(\mathbf{Z}), m_k). \tag{7}$$

In fact, this problem is equivalent to find the Wasserstein barycenter [16] of Gaussian measures. In optimal transport theory, the concept of barycenter can

---

[1] The details here are not so important, and the definition of Wasserstein 2-distance of Gaussian measures is enough for the development of this work. We present the formal definition here for completeness.

be regarded as a natural extension of "mean" to measures. For a collection of S Gaussian distributions $\{\alpha_s = \mathcal{N}(\boldsymbol{\mu}_s, \boldsymbol{\Sigma}_s)\}_{s=1}^{S}$ and positive weights satisfying $\sum_{s=1}^{S} \lambda_s = 1$, the Wasserstein barycenter is defined as

$$\arg \min_{\beta} \sum_{s=1}^{S} \lambda_s \mathcal{W}_2^2(\beta, \alpha_s). \tag{8}$$

Note that β can be any Radon measure by definition, but it has been shown in [1] the Wasserstein barycenter of Gaussian distributions is itself a Gaussian $\mathcal{N}(\boldsymbol{\mu}_*, \boldsymbol{\Sigma}_*)$. Making use of (5), we can easily show that $\boldsymbol{\mu}_* = \sum_{s=1}^{S} \lambda_s \boldsymbol{\mu}_s$, but there is no closed formula for $\boldsymbol{\Sigma}_*$. The first-order optimality condition shows that $\boldsymbol{\Sigma}_*$ satisfies

$$\boldsymbol{\Sigma}_* = \Phi(\boldsymbol{\Sigma}_*) \quad \text{where} \quad \Phi(\boldsymbol{\Sigma}) = \sum_{s=1}^{S} \lambda_s (\boldsymbol{\Sigma}^{1/2} \boldsymbol{\Sigma}_s \boldsymbol{\Sigma}^{1/2})^{1/2}. \tag{9}$$

Based on the optimality condition, we can perform the fixed-point iteration $\boldsymbol{\Sigma}^{(t+1)} = \Phi(\boldsymbol{\Sigma}^{(t)})$, and this iteration process has been proved to converge to $\boldsymbol{\Sigma}_*$. Back to our problem (8), we notice that this is equivalent to find the Wasserstein barycenter of N Gaussian distributions $f_i(\mathbf{Z})$ with equal weights $1/N$.

We need to calculate K barycenters in each iteration, which is very time consuming since every barycenter calculation relies on a fixed-point iteration over the covariance matrix. To reduce the computational burden and speed up iteration, we apply the Anderson acceleration [25] technique, which is a general acceleration scheme for fixed-point iteration. Anderson acceleration has a close relationship with quasi-Newton method [15], and is essentially equivalent to generalized minimal residual method [22] for linear problems [21,25] in numerical linear algebra. However, no global or even local convergence guarantee has been proved for general Anderson accelerated fixed-point iteration. Nevertheless, Anderson acceleration usually has faster convergence speed in practice. Furthermore, this procedure can also be parallelized naturally.

3.4   Clustering Algorithm and Extensions

The framework of the proposed method is summarized in Algorithm 17. We refer to this algorithm as **GPWC**, which means **G**aussian **P**rocess and **W**asserstein distance based **C**lustering. The entire process is very similar to k-means, the major difference is we are clustering Gaussian distributions here instead of vectors. Similar to the k-means algorithm, in the initialization step, we should choose Gaussian distributions that are relatively far from each other. Besides, it is also possible to employ other clustering methods such as fuzzy c-means [3] and hierarchical clustering [8].

Until now, we have developed the theory for the one-dimensional ($D_y = 1$) case. It is natural to extend the method to deal with multivariate functional data, *i.e.*, where $D_y > 1$. Each component of mapping $f_i : \mathbb{R}^{D_x} \rightarrow \mathbb{R}^{D_y}$ is a

---

Algorithm 1. (GPWC) Functional data clustering based on Gaussian processes and Wasserstein distance.

1: **Input:** Dataset $\mathcal{D} = \{\mathcal{D}_i\}_{i=1}^N$, $\mathcal{D}_i = \{(\mathbf{x}_{ij}, \mathbf{y}_{ij})\}_{j=1}^{N_i}$, number of clusters K.
2: **Output:** cluster labels $\{c(i)\}_{i=1}^N$ and cluster centers $\{m_k\}_{k=1}^K$.
3: // *Learn finite-dimensional representations of functions.*
4: **for** each $i = 1, 2, \cdots, N$ **do**
5:     Learn a Gaussian process by fitting $\mathcal{D}_i$ and obtain $\boldsymbol{\theta}_i$.
6:     Approximate the posterior $f_i | \mathcal{D}_i$ by discretizing the Gaussian process on equally-spaced grids $\mathbf{Z}$ to obtain $f_i(\mathbf{Z})$.
7: **end for**
8: // *Update cluster labels and cluster means alternately.*
9: Randomly choose K Gaussian distributions obtained above as K initial centers.
10: **while** Not converged **do**
11:     **for** each $i = 1, 2, \cdots, N$ **do**
12:         Assign $c(i) = \arg\min_{k=1,2,\cdots,K} \mathcal{W}_2^2(f_i(\mathbf{Z}), m_k)$.
13:     **end for**
14:     **for** each $k = 1, 2, \cdots, K$ **do**
15:         Update centers $m_k = \arg\min_m \sum_{c(i)=k} \mathcal{W}_2^2(f_i(\mathbf{Z}), m)$.
16:     **end for**
17: **end while**

---

function $f_i^{(r)} : \mathbb{R}^{D_x} \to \mathbb{R}$, and we can learn Gaussian processes for the components $\{f_i^{(r)}\}_{r=1}^{D_y}$ independently. In this case, the centers $\{m_k\}_{k=1}^K$ are no longer Gaussian distributions, each center m_k is a collection of D_y Gaussian distributions $\{m_k^r\}_{r=1}^{D_y}$ corresponds to D_y dimensions. As for the distance measure, we simply define $d(f_i, m_k) := \left(\sum_{r=1}^{D_y} d^2(f_i^{(r)}, m_k^{(r)}) \right)^{1/2}$. The optimization procedure is similar to the univariate case discussed in Sect. 3.3, since the problem is separable with respect to each dimension.

4   Experimental Results

4.1   On Synthetic Data

We first evaluate the proposed method on synthetic data. We generate three datasets with $K = 3, 5, 7$ respectively, and each cluster contains 20 curves. Without loss of generality, we constrain the input domain to be $[0, 1]$. For each cluster, the mean function is the predictive mean of a Gaussian process with 20 samples, and the covariance function is squared exponential covariance function as in Eq. 2 with random parameters. Then, we generate curves using these Gaussian processes, with each curve containing 25–40 observations. The synthetic dataset is shown in Fig. 1a.

Table 1. Performances of various clustering methods on the synthetic dataset. All the results are averaged over 10 trials. Best results are in bold.

K	Method	RI	ARI	HI	NMI	χ^2	Cramer	MOC
3	CC (lrm)	0.7289	0.4567	0.4579	0.6679	60.3	0.7089	0.5026
	CC (lrm)	**1.0000**	**1.0000**	**1.0000**	**1.0000**	**120.0**	**1.0000**	**1.0000**
	CC (lrm_b)	0.9373	0.8570	0.8746	0.8679	104.3	0.9325	0.8696
	FPCA+k-means	0.9775	0.9495	0.9549	0.9592	114.0	0.9708	0.9501
	FPCA+fcm	0.9436	0.8866	0.8872	0.9305	108.0	0.9414	0.9000
	GPWC	**1.0000**	**1.0000**	**1.0000**	**1.0000**	**120.0**	**1.0000**	**1.0000**
5	CC (lrm)	0.9263	0.7794	0.8526	0.8480	316.8	0.8879	0.7919
	CC (lrm_d)	0.9037	0.7345	0.8073	0.8637	306.5	0.8732	0.7662
	CC (lrm_b)	0.8941	0.6947	0.7882	0.8225	284.1	0.8418	0.7103
	FPCA+k-means	0.9524	0.8578	0.9047	0.8989	345.9	0.9283	0.8646
	FPCA+fcm	0.9345	0.8076	0.8689	0.8835	331.0	0.9080	0.8275
	GPWC	**0.9645**	**0.9024**	**0.9290**	**0.9372**	**359.5**	**0.9439**	**0.8988**
7	CC (lrm)	0.9515	0.8131	0.9029	0.9044	704.2	0.9147	0.8384
	CC (lrm_d)	0.9395	0.7780	0.8789	0.9047	657.5	0.8995	0.7964
	CC (lrm_b)	0.9469	0.7943	0.8939	0.9040	684.9	0.9024	0.8154
	FPCA+k-means	0.9902	0.9624	0.9805	0.9844	811.0	0.9819	0.9655
	FPCA+fcm	0.9789	0.9180	0.9578	0.9614	774.9	0.9592	0.9225
	GPWC	**0.9949**	**0.9803**	**0.9898**	**0.9923**	**826.0**	**0.9913**	**0.9833**

For comparison, we consider five competing methods: CC (lrm), CC (lrm_d), CC (lrm_b), FPCA+k-means and FPCA+fcm. The first three methods were proposed in [10], corresponding to different transformations. The details can be found in [9]. FPCA+k-means and FPCA+fcm first transform the curves to vectors using functional principal component analysis, then perform k-means or fuzzy c-means on the vectors. The dimension of vectors is determined by cross-validation. For the proposed method, we set $D = 30$. The number of components is set to the correct number of clusters in all methods.

To evaluate the clustering performance, we use Rand Index (RI), Adjusted Rand Index (ARI), Hubert Index (HI), Normalized Mutual Information (NMI), χ^2-statistics (χ^2), Cramer's coefficient (Cramer), Measure of Concordance (MOC) as evaluation metrics. Details about these metrics can be found in [2,7,24]. In general, higher values correspond to better performances. We report the results in Table 1. Since there is randomness in all these methods, the reported results are averaged over 10 trials. From Table 1, we can see that when $K = 3$, GPWC and CC (lrm) achieves perfect clustering results. As we increase K to 5 and 7, the proposed method outperforms competing methods significantly. We also show the clustering results of these methods and Gaussian distributions with barycenters when $K = 5$ in Fig. 1.

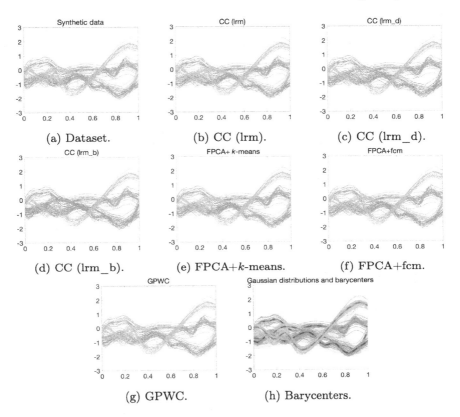

(a) Dataset.   (b) CC (lrm).   (c) CC (lrm_d).

(d) CC (lrm_b).   (e) FPCA+k-means.   (f) FPCA+fcm.

(g) GPWC.   (h) Barycenters.

Fig. 1. Synthetic datasets and clustering results of various methods.

4.2   On Real-World Data

To further validate the performance, we apply the proposed clustering method to three real-world datasets:

- **Weather:** the weather dataset [13] recorded temperatures from 1961 to 1994 in 35 Canadian weather stations. The 35 weather stations correspond to 35 curves, and each curve consists of 73 (365/5) observations, which are mean temperatures of every five days within one year.
- **Electrical load:** the electrical load dataset was issued by the Northwest China Grid Company. This dataset consists of 50 curves, corresponding to 50 days. In each day, the electrical load was recorded every 15 min and therefore each curve contains 96 (24 × 4) samples.
- **Gait:** the gait dataset [17] is collected by the Motion Analysis Laboratory at Children's Hospital, San Diego, CA. This dataset consists of the angles formed by the hip and knee of each of 39 children over each child's gait cycle. There are 78 curves in total, half of them are hip angles and the rest are knee

angles. Time is measured in terms of the individual gait cycle, which has been translated into $[0, 1]$.

We set $D = 30$ and $K = 2$ in this experiment. The clustering results are shown in Fig. 2. For the weather dataset, the two components correspond to the observations of stations in the south and north of Canada respectively. For the gait dataset, the two clusters correspond to hip angles and knee angles respectively. This demonstrates the proposed curve clustering method can reveal information and structure underlying observational data.

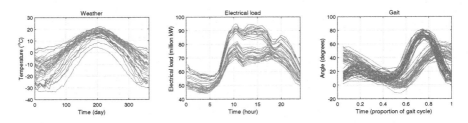

Fig. 2. Clustering results of the proposed method on three real-world datasets.

5   Conclusions and Discussions

We propose a functional data clustering algorithm based on Gaussian processes and Wasserstein distance. Experimental results on both synthetic and real-world datasets demonstrate the effectiveness of the proposed method. There are several promising research directions. For example, we can incorporate the mixture of Gaussian processes [5] to fit non-stationary data, and employ sparse Gaussian processes to accelerate the learning process. We can also consider automated model selection techniques [12] to determine the number of clusters in practice. Finally, it is interesting to use multi-output Gaussian processes to model multivariate functions.

Acknowledgements. This work was supported by the National Key R & D Program of China (2018YFC0808305).

References

1. Agueh, M., Carlier, G.: Barycenters in the wasserstein space. SIAM J. Math. Anal. **43**(2), 904–924 (2011)
2. Amigó, E., Gonzalo, J., Artiles, J., Verdejo, F.: A comparison of extrinsic clustering evaluation metrics based on formal constraints. Inf. Retriev. **12**(4), 461–486 (2009)
3. Bezdek, J.C., Ehrlich, R., Full, W.: FCM: The fuzzy c-means clustering algorithm. Comput. Geosci. **10**(2–3), 191–203 (1984)

4. Bishop, C.M.: Pattern Recognition and Machine Learning. Springer, Heidelberg (2006)
5. Chen, Z., Ma, J., Zhou, Y.: A precise hard-cut EM algorithm for mixtures of gaussian processes. In: Huang, D.-S., Jo, K.-H., Wang, L. (eds.) ICIC 2014. LNCS (LNAI), vol. 8589, pp. 68–75. Springer, Cham (2014). https://doi.org/10.1007/978-3-319-09339-0_7
6. Cover, T.M., Thomas, J.A.: Elements of Information Theory. John Wiley, Hoboken (2012)
7. Desgraupes, B.: Clustering indices. University of Paris Ouest-Lab Modal'X, vol. 1, p. 34 (2013)
8. Friedman, J., Hastie, T., Tibshirani, R.: The Elements of Statistical Learning, Springer series in statistics New York, vol. 1 (2001)
9. Gaffney, S.: Probabilistic curve-aligned clustering and prediction with regression mixture models. Ph.D. thesis, University of California, Irvine (2004)
10. Gaffney, S.J., Smyth, P.: Joint probabilistic curve clustering and alignment. In: Advances in Neural Information Processing Systems, pp. 473–480 (2005)
11. Kolouri, S., Park, S.R., Thorpe, M., Slepcev, D., Rohde, G.K.: Optimal mass transport: signal processing and machine-learning applications. IEEE Signal Process. Mag. **34**(4), 43–59 (2017)
12. Li, T., Ma, J.: Fuzzy clustering with automated model selection: entropy penalty approach. In: 2018 14th IEEE International Conference on Signal Processing (ICSP), pp. 571–576 (2018)
13. López-Pintado, S., Romo, J.: On the concept of depth for functional data. J. Am. Stat. Assoc. **104**(486), 718–734 (2009)
14. Mallasto, A., Feragen, A.: Learning from uncertain curves: the 2-wasserstein metric for gaussian processes. In: Advances in Neural Information Processing Systems, pp. 5660–5670 (2017)
15. Nocedal, J., Wright, S.: Numerical Optimization. Springer, Heidelberg (2006). https://doi.org/10.1007/978-0-387-40065-5
16. Peyré, G., et al.: Computational optimal transport. Found. Trends® Mach. Learn. **11**(5—-6), 355–607 (2019)
17. Ramsay, J.O., Hooker, G., Graves, S.: Functional Data Analysis with R and MATLAB, 1st edn. Springer, Heidelberg (2009). https://doi.org/10.1007/978-0-387-98185-7
18. Ramsay, J.O.: Functional data analysis. Encycl. Stat. Sci. **4** (2004)
19. Rasmussen, C.E.: Gaussian processes in machine learning. In: Bousquet, O., von Luxburg, U., Rätsch, G. (eds.) ML -2003. LNCS (LNAI), vol. 3176, pp. 63–71. Springer, Heidelberg (2004). https://doi.org/10.1007/978-3-540-28650-9_4
20. Rasmussen, C.E., Nickisch, H.: Gaussian processes for machine learning (GPML) toolbox. J. Mach. Learn. Res. **11**(Nov), 3011–3015 (2010)
21. Toth, A., Kelley, C.: Convergence analysis for anderson acceleration. SIAM J. Numeric. Anal. **53**(2), 805–819 (2015)
22. Trefethen, L.N., Bau III, D.: Numerical Linear Algebra, vol. 50. SIAM (1997)
23. Villani, C.: Optimal Transport: Old and New, vol. 338. Springer, Heidelberg (2008). https://doi.org/10.1007/978-3-540-71050-9
24. Wagner, S., Wagner, D.: Comparing clusterings: an overview. Universität Karlsruhe, Fakultät für Informatik Karlsruhe (2007)
25. Walker, H.F., Ni, P.: Anderson acceleration for fixed-point iterations. SIAM J. Numeric. Anal. **49**(4), 1715–1735 (2011)

GPU-Based Self-Organizing Maps for Post-labeled Few-Shot Unsupervised Learning

Lyes Khacef[1]([✉]), Vincent Gripon[1,2], and Benoît Miramond[1]

[1] Université Côte d'Azur, CNRS, LEAT, Sophia Antipolis, France
{lyes.khacef,benoit.miramond}@univ-cotedazur.fr
[2] Electronics Department, IMT Atlantique, Brest, France
vincent.gripon@imt-atlantique.fr

Abstract. Few-shot classification is a challenge in machine learning where the goal is to train a classifier using a very limited number of labeled examples. This scenario is likely to occur frequently in real life, for example when data acquisition or labeling is expensive. In this work, we consider the problem of post-labeled few-shot unsupervised learning, a classification task where representations are learned in an unsupervised fashion, to be later labeled using very few annotated examples. We argue that this problem is very likely to occur on the edge, when the embedded device directly acquires the data, and the expert needed to perform labeling cannot be prompted often. To address this problem, we consider an algorithm consisting of the concatenation of transfer learning with clustering using Self-Organizing Maps (SOMs). We introduce a TensorFlow-based implementation to speed-up the process in multi-core CPUs and GPUs. Finally, we demonstrate the effectiveness of the method using standard off-the-shelf few-shot classification benchmarks.

Keywords: Brain-inspired computing · Self-Organizing Map · Few-shot classification · Post-labeled unsupervised learning · Transfer learning · Feature extraction

1 Introduction

In the last decade, Deep Learning (DL) techniques have achieved state-of-the-art performance in many classification problems. However, DL heavily relies on supervised learning with abundant labeled data. With the fast expansion of Internet of Things (IoT) devices, a huge amount of unlabeled data is gathered everyday, but labeling these data is a very difficult task because of the human annotation cost as well as the scarcity of data in some classes [4]. Finding methods to learn to generalize to new classes with a limited amount of labeled examples for each class is therefore a very active topic of research. This is the main motivation for few-shot learning. Recently, three main approaches have been proposed in the literature:

© Springer Nature Switzerland AG 2020
H. Yang et al. (Eds.): ICONIP 2020, LNCS 12533, pp. 404–416, 2020.
https://doi.org/10.1007/978-3-030-63833-7_34

- **Hallucination methods** where the aim is to augment the training sets by learning a generator that can create novel data using data-augmentation techniques [4]. However, these methods lack precision which results in coarse and low-quality synthesized data that can sometimes lead to very poor gains in performance [29].
- **Meta-learning** where the goal is to train an optimizer that initializes the network parameters using a first generic dataset, so that the model can reach good performance with only a few more steps on the new dataset [26]. This type of solution suffers from the domain shift problem [4] as well as the sensitivity of hyper-parameters.
- **Transfer learning** where a model developed for a given task is reused as the starting point for a model on a different task. In real-world problems, it happens that we have a classification task in one domain of interest, but we only have sufficient training data in another domain of interest. Therefore, knowledge transfer would greatly improve the performance of learning by avoiding much expensive data-gathering and data-labeling efforts [19]. Hence, transfer learning has emerged as the new learning framework for the few-shot classification task.

The problem becomes even harder when facing technical limitations, such as using embedded implementations for real-time processing on the edge. As a matter of fact, in many real-world scenarios, the training data is acquired using the same device that will later be used for training and inference, and labels could be given at any time of the process. To encompass for this added difficulty, we consider in this work the problem of post-labeled few-shot unsupervised learning. In this problem, learning algorithms can be deployed using no annotated data, for example to learn representations using the data acquired by the considered device. These algorithms can later be adjusted using a few labeled samples so that they become able to make predictions, at the condition that this adjustment comes with almost no added complexity to the process, so that it can be performed on the edge.

To address this problem, we propose a solution that combines transfer learning with a recently introduced algorithm [9] using Self-Organizing Maps (SOM). On the one hand, transfer learning is used to exploit a Deep Neural Network (DNN) trained on a large collection of labeled data as a "universal" feature extractor. On the other hand, a post-labeled clustering algorithm is used to leverage the obtained features and make predictions. This algorithm works in two steps: in a first step, clusters prototypes are learned using no annotated data, then the prototypes are named (labeled) using a few available annotated samples.

The motivation for using the SOM, initially proposed in [11], comes from the fact they are known to be a very effective clustering method. Indeed, it has been shown that SOMs perform better in representing overlapping structures compared to classical clustering techniques such as partitive clustering or K-means [2]. In addition, SOMs are well suited to hardware implementation based on cellular neuromorphic architectures [10,21,25]. Thanks to a fully distributed architecture with local connectivity amongst hardware neurons, the energy-efficiency of the SOM is highly improved since there is no communication between a centralized controller and a shared memory unit, as it is the case in classical Von-Neumann architectures. Moreover, the connectivity and time complexities of the SOM become scalable with respect to the number of neurons [21]. SOMs are used in a large range of applications [12] going from high-dimensional data analysis to more recent developments such as identification of social media trends [23], incremental change detection [18] and energy consumption minimization on sensor networks [13].

This work is an extension of [9], where we used the SOM for MNIST [14] classification with unsupervised learning, and compared different training and labeling techniques. Here, we focus on the case of few-shot learning, and demonstrate the ability of the proposed method in reaching top performance with the challenging benchmark of mini-ImageNet classification task. We introduce a TensorFlow (TF) software implementation for the proposed method, and compare execution times when using multi-core CPUs or GPUs.

The outline of the paper is as follows. Section 2 details the SOM training and labeling algorithms and describes the transfer learning methods. Then, Sect. 3 presents the mini-ImageNet few-labels classification problem. Next, Sect. 4 presents the TF-based SOM implementation and shows the multi-core CPU and GPU speed-ups. Afterwards, Sect. 5 presents the experiments and results on transfer learning with few labels using a SOM classifier. Finally, Sect. 6 and Sect. 7 discuss and conclude our work.

2   Proposed Methodology

In this section, we review the proposed methodology. We begin with the transfer learning part, then how to train the SOM, and we finally explain the labeling procedure.

Let us consider that we are given a dataset $X = \{x, x \in X\}$, that we initially consider to be unlabeled. Our first step consists in extracting relevant features from these inputs.

2.1   Transfer Learning

In this work, we follow the approach proposed by [8] and train a supervised feature extractor f_φ that we call a *backbone* on a large annotated dataset. The parameters of the backbone are then fixed and used to obtain *generic* features from any input. In our case, we therefore transform X into $V = f_\varphi(X) = \{f_\varphi(x), x \in X\}$.

2.2   Self-Organizing Maps Learning Procedure

The next step consists in training a SOM using the transformed representations in V, i.e. the extracted features. To this end, we use a two-dimensional array of k neurons, that are randomly initialized and updated thanks to the following algorithm, based on the one in [11]:

Initialize the network as a two-dimensional array of k neurons, where each neuron n with m inputs is defined by a two-dimensional position p_n and a randomly initialized m-dimensional weight vector w_n.

for t from 0 to t_f **do**

  for every input vector v **do**

    for every neuron n in the network **do**

      Compute the afferent activity a_n from the distance d:

$$d = \|v - w_n\| \tag{1}$$

$$a_n = e^{-\frac{d}{\alpha}} \tag{2}$$

    end for

    Compute the winner s such that:

$$a_s = \max_{n=0}^{k-1}(a_n) \tag{3}$$

    for every neuron n in the network **do**

      Compute the neighborhood function $h_\sigma(t, n, s)$:

$$h_\sigma(t, n, s) = e^{-\frac{\|p_n - p_s\|^2}{2\sigma(t)^2}} \tag{4}$$

      Update the weight w_n of the neuron n:

$$w_n = w_n + \epsilon(t) \times h_\sigma(t, n, s) \times (v - w_n) \tag{5}$$

    end for

  end for

  Update the learning rate $\epsilon(t)$:

$$\epsilon(t) = \epsilon_i \left(\frac{\epsilon_f}{\epsilon_i}\right)^{t/t_f} \tag{5}$$

  Update the width of the neighborhood $\sigma(t)$:

$$\sigma(t) = \sigma_i \left(\frac{\sigma_f}{\sigma_i}\right)^{t/t_f} \tag{6}$$

end for

It is to note that t_f is the number of epochs, i.e. the number of times the whole training dataset is presented. The α hyper-parameter is the width of the

Gaussian kernel. Its value in Eq. 2 is fixed to 1 in the SOM training, but it does not have any impact in the training phase since it does not change the neuron with the maximum activity. Its value becomes critical though in the labeling process. The SOM hyper-parameters are reported in Table 1.

At the end of the training process, each neuron of the SOM corresponds to a cluster prototype in the considered problem. At this stage, these prototypes are anonymous and cannot be directly used to perform predictions. The next step explains the neurons labeling process for transforming the SOM into a classifier.

2.3   SOM Labeling

The labeling is the step between training and test where we assign each neuron the class it represents in the training dataset. We proposed in [9] a labeling algorithm based on very few labels. The idea is the following: we randomly considered a labeled subset of the training dataset, and we tried to minimize its size while keeping the best classification accuracy. Our study showed that we only need 1% of randomly taken labeled samples from the training dataset for MNIST classification. The labeling algorithm detailed in [9] can be summarized in five steps:

- First, we calculate the neurons activations based on the labeled input samples from the euclidean distance following Eq. 2, where v is the input vector, w_n and a_n are respectively the weights vector and the activity of the neuron n. The parameter α is the width of the Gaussian kernel that becomes a hyper-parameter for the method.
- Second, the Best Matching Unit (BMU), i.e. the neuron with the maximum activity is elected.
- Third, each neuron accumulates its normalized activation (simple division) with respect to the BMU activity in the corresponding class accumulator, and the three steps are repeated for every sample of the labeling subset.
- Fourth, each class accumulator is normalized over the number of samples per class.
- Fifth and finally, the label of each neuron is chosen according to the class accumulator that has the maximum activity.

The complete GPU-based source code for the SOM training, labeling and test is available in https://github.com/lyes-khacef/GPU-SOM.

3   Datasets and Implementation Details

3.1   mini-ImageNet Few-Shot Learning

In this work, we perform experiments using the mini-ImageNet [28] benchmark. mini-ImageNet is a subset of ImageNet [22] that contains 60,000 images divided into 100 classes of 600 images, each image has 84×84 pixels. Following the standard approach [20], we use 64 base classes with labels to train the backbone and 20 novel classes to draw the novel datasets from. For each run, 5 classes are drawn uniformly at random among these 20 classes, then q unlabeled inputs and s labelled inputs per class are chosen uniformly at random among the 5 drawn classes. The features of the $(q + s) \times 5$ samples are used to train the SOM, then the s labeled samples are used to label the SOM neurons. Finally, the $Q = q \times 5$ unlabeled samples are classified and produce a classification accuracy for each run. We run 10,000 random draws to obtain a mean accuracy score and indicate confidence scores (95%) when relevant.

3.2   WRN Training

The feature extractor we use is the same as in [8]. It is mostly based on Wide Residual Networks (WRN) [30] as a backbone extractor, with 28 convolutional layers and a widening factor of 10. As a result, the output feature size (the dimension of a vector $v \in V$) is 640. Let us insist on the fact the backbone is trained on a completely disjoint dataset with the tasks we consider thereafter.

3.3   Cosine Distance

In transfer learning, the backbone feature extractor is trained with 80 classes that are different from the 20 classes we classify using the SOM. Hence, the features amplitude is not relevant, and the Euclidean distance of the SOM does not provide the best performance. Therefore, we replace the Euclidean distance in Eq. 1 with the Cosine distance in Eq. 1.

$$d = 1 - \cos(v, w_n) = 1 - \frac{v.w_n}{\|v\| \times \|w_n\|} \tag{1}$$

The Cosine distance is also used in the labeling and test phases. The comparison to Euclidean distance is discussed in Sect. 6.

4   SOM Software Implementation

4.1   TensorFlow-Based SOM

The SOM was implemented using TF [1] 2.1, an end-to-end open source platform for machine learning that uses dataflow graphs to represent computation, shared state, and the operations that mutate that state. It maps the nodes of a dataflow graph across multiple computational devices including multi-core CPUs, general-purpose GPUs and custom-designed ASICs known as Tensor Processing Units (TPUs) [1]. TF facilitates the design of many machine learning models providing built-in functionalities such as convolution, pooling and dense (i.e. fully connected) layers. However, TF does not provide computational neuroscience models, and to the best of our knowledge, there is no efficient implementation for SOMs using TF.

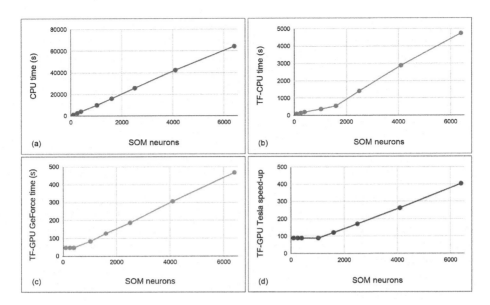

Fig. 1. SOM training speed on MNIST database for 10 epochs (i.e. 600,000 samples of 784 dimensions) VS. number of SOM neurons: (a) CPU (mono-core) implementation; (b) TF-CPU (muti-core) implementation; (c) TF-GPU GeForce implementation; (d) TF-GPU Tesla implementation.

4.2 CPU and GPU Speedups

The SOMs of different sizes were trained for 10 epochs on MNIST database, i.e. 600,000 samples of 784 dimensions. The CPU mono-core implementation is based on NumPy [27] and run on an Intel Core i9-9880H CPU (16 cores), while the GPU implementation is based on TF 2.1 [1] and run on two different GPUs: Nvidia GeForce RTX 2080 and Nvidia Tesla K80 freely available on Google Colab cloud service [3]. Interestingly, the TF-based SOM can also run on the multiple cores of the CPU, providing a speed-up even without access to GPU.

Figures 1-a , 1-b, 1-c and 1-d show that the time complexities of the CPU, TF-CPU and TF-GPU implementations are all linear. It is to note that the time complexity slope of the TF-CPU, TF-GPU GeForce and TF-GPU Tesla implementations changes at 1600 neurons, 400 and 1024 neurons respectively, which is due to their different degrees of parallelism.

As shown in Fig. 2, we achieved a minimum speedup of 12× (22×) and a maximum speedup of 161× (138×) with the TF-GPU Tesla (TF-GPU GeForce) implementation, with an increasing speedup with respect to the number of neurons. Our GPU implementation is therefore scalable in simulation time with respect to the SOM size, which is an important aspect to accelerate the simulations and hyper-parameters exploration. In addition, we achieved a minimum

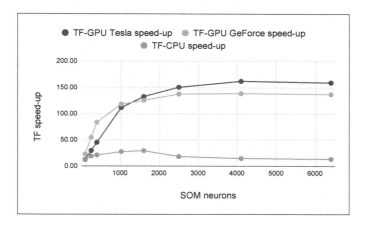

Fig. 2. TF-CPU and TF-GPU speed-ups compared to CPU.

speedup of $11\times$ times and a maximum speedup of $29\times$ times with the TF-CPU implementation, which runs the 16 cores of the CPU. Nevertheless, the gap between the GPU and CPU speed-ups increases with the number of neurons, which is expected due to the highly parallel computation of the GPU hardware.

Recent works have tried an other approach using CUDA acceleration on Nvidia GPUs. They showed relative gains to CPU of $44\times$ [17], $47\times$ [6] and $67\times$ [16]. Our implementation reaches an average gain of $19\times$ in a multi-core Intel Core i9 CPU, $100\times$ in a Nvidia Tesla GPU and $102\times$ in a Nvidia GeForce GPU. A fair comparison is difficult since we do not target the same hardware, but the order of magnitude is comparable and our results are in the state of the art. Another advantage of our TF-based approach is the easy integration of the SOM layer into Keras [5], a high-level neural networks API capable of running on top of TF with a focus on enabling fast experimentation.

5   Experiments and Results

The SOM training hyper-parameters for the different settings were found with a grid search and are reported in Table 1.

Table 1. SOM training hyper-parameters.

Dataset	ϵ_i	ϵ_f	η_i	η_f	Epochs	α
mini-ImageNet	1	0.01	10	0.1	10	1

First, we investigated the impact of the SOM size on the classification accuracy for the commonly used number of unlabeled samples $q = 15$ and labeled

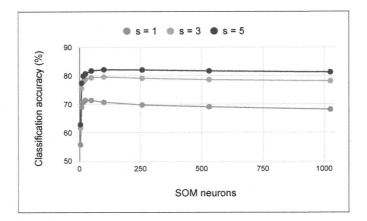

Fig. 3. SOM classification accuracy on mini-ImageNet transfer learning for different numbers of labeled samples s vs. number of SOM neurons.

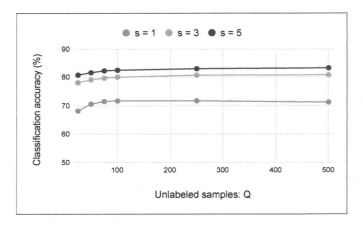

Fig. 4. SOM classification accuracy on mini-ImageNet transfer learning for different numbers of labeled samples s vs. number of unlabeled samples to classify Q.

samples $s = [1, 3, 5]$ [8]. Figure 3 shows that there is an optimal point at 25 neurons for $s = 1$ and 100 neurons for $s = 3$ and $s = 5$. There is a tradeoff between the number of neurons that learn different prototypes and the quality of the learning/labeling of these neurons. The more neurons we have, the more potential to learn different prototypes of the data but the more fuzzy the prototypes become, which makes the labeling part more difficult. For example, a neuron may be assigned a class "A" with respect to the labeled subset, but will be more active for a class "B" with respect to the test set. When we only have one labeled sample per class, i.e. $s = 1$, then a SOM of only 25 neurons achieves the best accuracy because more neurons will not converge as well.

Next, we varied the number of unlabeled data $Q = q \times 5$ with the above mentioned SOM sizes. Figure 4 shows that even though the labels are only used for the neurons class assignment and not in the training process, they still have a large impact on the accuracy. Naturally, the more labeled data we have, the better accuracy we get. A second remark is that the more unlabeled data we have, the better accuracy we get too. This is not intuitive, because the unlabeled data are the queries, i.e. the samples to classify, so the more we have the harder the classification task becomes. However, since the SOM is trained on these data, its adaptation capabilities makes the accuracy increase with the number of unlabeled data for the same number of labels. The only exception is when $s = 1$, where there is a small decrease in accuracy between $Q = 250$ ($71.74\% \pm 0.21$) and $Q = 500$ ($71.27\% \pm 0.21$). A third remark is that the SOM reaches the same accuracy for $[s = 5, Q = 25]$ and $[s = 3, Q = 250]$, which means that the lack of labeled data can be compensated by more unlabeled data. In fact, it is a very interesting property since unlabeled data can be gathered much more easily, and no extra-effort for labeling these data is needed.

6 Discussion

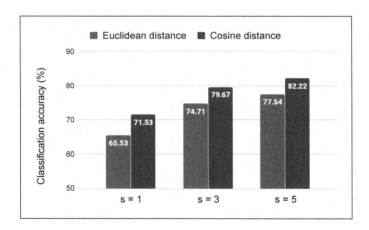

Fig. 5. SOM classification accuracy on mini-ImageNet transfer learning with few labels using Euclidean distance and Cosine distance.

The choice of using the Cosine distance in the SOM computation (training, labeling and test) was inspired from the work of [8]. In fact, Fig. 5 shows that replacing the Euclidean distance by the Cosine distance significantly improves the SOM classification accuracy, with a gain of $+5.9\%$, $+4.96\%$ and $+4.68\%$ for s = 1, s = 3 and s = 5, respectively. It validates our hypothesis about the non-effectiveness of the Euclidean distance when using transfer learning.

Table 2. mini-ImageNet few labels transfer learning with q = 15 (Q = 75): state of the art reported from [8].

Method	Backbone	Classifier	1-shot (%)	5-shot (%)
wDAE-GNN [7]	WRN	Supervised	61.07 ± 0.15	76.75 ± 0.11
ACC+Amphibian [24]	WRN	Supervised	64.21 ± 0.62	**87.75 ± 0.73**
BD-CSPN [15]	WRN	Supervised	70.31 ± 0.93	81.89 ± 0.60
Transfer+SGC [8]	WRN	Supervised	**76.47 ± 0.23**	85.23 ± 0.13
Transfer+SOM [Our work]	WRN	**Unsupervised**	71.53 ± 0.23	82.22 ± 0.15

Finally, Table 2 reports the recent works that proposed solutions to the mini-ImageNet few labels classification problem using transfer learning with the WRN backbone feature extractor. The SOM reaches top-2 accuracy for $s = 1$ and top-3 accuracy for $s = 5$, which is a good result that proves the SOM ability to handle complex datasets. Nevertheless, one has to keep in mind that while the other works use the few labels in the training process, we only use them for neurons labeling phase. Our accuracy performance is therefore obtained with fully unsupervised learning followed by post-labeling, which we believe is the right approach for the few-shot classification problem in the context of embedded systems on the edge.

7   Conclusion and Further Works

We introduced in this work the problem of post-labeled few-shot unsupervised learning and proposed a solution that combines transfer learning and SOMs. Transfer learning was used to exploit a WRN backbone trained on a base dataset as a feature extractor, and the SOM was used to classify the obtained features from the target dataset. The SOM is trained with no label, then labeled with the few available annotated samples. We show that we reach a good performance on the mini-ImageNet few shot classification benchmark with an unsupervised learning method. Furthermore, the SOM is suitable for hardware implementations based on a cellular neuromorphic architecture, which enables its application on the edge. Finally, to speed-up the SOM simulation process, we proposed a novel TF-based GPU implementation which is about 100× faster than the classical CPU implementation.

Acknowledgment. This work has been supported by the French government, through the UCAJEDI Investments in the Future project managed by the National Research Agency (ANR) with the reference number ANR-15-IDEX-01.

References

1. Abadi, M., et al.: TensorFlow: a system for large-scale machine learning. In: Proceedings of the 12th USENIX Conference on Operating Systems Design and Implementation, OSDI 2016, pp. 265–83. USENIX Association, USA (2016)

2. Budayan, C., Dikmen, I., Birgonul, M.T.: Comparing the performance of traditional cluster analysis, self-organizing maps and fuzzy c-means method for strategic grouping. Exp. Syst. Appl. **36**(9), 11772 – 11781 (2009). https://doi.org/10.1016/j.eswa.2009.04.022

3. Carneiro, T., Nobrega, R.V.M.D., Nepomuceno, T., Bian, G.B., de Albuquerque, V.H.C., Filho, P.P.R.: Performance analysis of Google colaboratory as a tool for accelerating deep learning applications. IEEE Access **6**, 61677–61685 (2018)

4. Chen, W.Y., Liu, Y.C., Kira, Z., Wang, Y.C.F., Huang, J.B.: A closer look at few-shot classification (2019)

5. Chollet, F., et al.: Keras (2015). https://github.com/fchollet/keras

6. Gavval, R., Ravi, V., Harshal, K.R., Gangwar, A., Ravi, K.G.: CUDA-self-organizing feature map based visual sentiment analysis of bank customer complaints for analytical CRM. ArXiv abs/1905.09598 (2019)

7. Gidaris, S., Komodakis, N.: Generating classification weights with GNN denoising autoencoders for few-shot learning (2019)

8. Hu, Y., Gripon, V., Pateux, S.: Exploiting unsupervised inputs for accurate few-shot classification (2020)

9. Khacef, L., Miramond, B., Barrientos, D., Upegui, A.: Self-organizing neurons: toward brain-inspired unsupervised learning. In: 2019 International Joint Conference on Neural Networks (IJCNN), pp. 1–9, July 2019. https://doi.org/10.1109/IJCNN.2019.8852098

10. Khacef, L., Girau, B., Rougier, N.P., Upegui, A., Miramond, B.: Neuromorphic hardware as a self-organizing computing system. In: IJCNN 2018 Neuromorphic Hardware in Practice and Use Workshop, Rio de Janeiro, Brazil (2018)

11. Kohonen, T.: The self-organizing map. Proc. IEEE **78**(9), 1464–1480 (1990). https://doi.org/10.1109/5.58325

12. Kohonen, T., Oja, E., Simula, O., Visa, A., Kangas, J.: Engineering applications of the self-organizing map. Proc. IEEE **84**(10), 1358–1384 (1996). https://doi.org/10.1109/5.537105

13. Kromes, R., Russo, A., Miramond, B., Verdier, F.: Energy consumption minimization on LoRaWAN sensor network by using an artificial neural network based application. In: 2019 IEEE Sensors Applications Symposium (SAS), pp. 1–6, March 2019. https://doi.org/10.1109/SAS.2019.8705992

14. LeCun, Y., Cortes, C.: MNIST handwritten digit database (1998). http://yann.lecun.com/exdb/mnist/

15. Liu, J., Song, L., Qin, Y.: Prototype rectification for few-shot learning (2019)

16. McConnell, S., Sturgeon, R., Henry, G., Mayne, A., Hurley, R.: Scalability of self-organizing maps on a GPU cluster using OpenCL and CUDA. J. Phys. Conf. Ser. **341**, 012018 (2012). https://doi.org/10.1088/1742-6596/341/1/012018

17. Moraes, F.C., Botelho, S.C., Filho, N.D., Gaya, J.F.O.: Parallel high dimensional self organizing maps using CUDA. In: 2012 Brazilian Robotics Symposium and Latin American Robotics Symposium, pp. 302–306, October 2012. https://doi.org/10.1109/SBR-LARS.2012.56

18. Nallaperuma, D., Silva, D.D., Alahakoon, D., Yu, X.: Intelligent detection of driver behavior changes for effective coordination between autonomous and human driven vehicles. In: 44th Annual Conference of the IEEE Industrial Electronics Society, IECON 2018, pp. 3120–3125 (2018)

19. Pan, S.J., Yang, Q.: A survey on transfer learning. IEEE Trans. Knowl. Data Eng. **22**(10), 1345–1359 (2010)

20. Ravi, S., Larochelle, H.: Optimization as a model for few-shot learning. In: ICLR (2017)

21. Rodriguez, L., Khacef, L., Miramond, B.: A distributed cellular approach of large scale SOM models for hardware implementation. In: IEEE Image Processing and Signals, Sophia-Antipolis, France (2018)
22. Russakovsky, O., et al.: ImageNet large scale visual recognition challenge. Int. J. Comput. Vis. **115**(3), 211–252 (2015). https://doi.org/10.1007/s11263-015-0816-y
23. Silva, D.D., et al.: Machine learning to support social media empowered patients in cancer care and cancer treatment decisions. PLoS One **13**, e0205855 (2018)
24. Snell, J., Swersky, K., Zemel, R.: Prototypical networks for few-shot learning. In: Guyon, I., Luxburg, U.V., Bengio, S., Wallach, H., Fergus, R., Vishwanathan, S., Garnett, R. (eds.) Advances in Neural Information Processing Systems, vol. 30, pp. 4077–4087. Curran Associates, Inc. (2017). http://papers.nips.cc/paper/6996-prototypical-networks-for-few-shot-learning.pdf
25. de Abreu de Sousa, M.A., Del-Moral-Hernandez, E.: An FPGA distributed implementation model for embedded SOM with on-line learning. In: 2017 International Joint Conference on Neural Networks (2017). https://doi.org/10.1109/IJCNN.2017.7966351
26. Thrun, S., Pratt, L. (eds.): Learning to Learn. Springer, Heidelberg (2012)
27. van der Walt, S., Colbert, S.C., Varoquaux, G.: The numPy array: a structure for efficient numerical computation. Comput. Sci. Eng. **13**(2), 22–30 (2011). https://doi.org/10.1109/MCSE.2011.37
28. Vinyals, O., Blundell, C., Lillicrap, T., kavukcuoglu, k., Wierstra, D.: Matching networks for one shot learning. In: Lee, D.D., Sugiyama, M., Luxburg, U.V., Guyon, I., Garnett, R. (eds.) Advances in Neural Information Processing Systems, vol. 29, pp. 3630–3638. Curran Associates, Inc. (2016). http://papers.nips.cc/paper/6385-matching-networks-for-one-shot-learning.pdf
29. Wang, Y., Yao, Q., Kwok, J., Ni, L.M.: Generalizing from a few examples: a survey on few-shot learning (2019)
30. Zagoruyko, S., Komodakis, N.: Wide residual networks (2016)

Gradient-Based Adversarial Image Forensics

Anjie Peng[1], Kang Deng[1], Jing Zhang[2], Shenghai Luo[1], Hui Zeng[1(✉)], and Wenxin Yu[1]

[1] School of Computer Science and Technology,
Southwest University of Science and Technology, Mianyang 621010, Sichuan, China
zengh5@mail2.sysu.edu.cn
[2] China University of Mining and Technology, Xuzhou 221116, Jiangsu, China

Abstract. Adversarial images which can fool deep neural networks attract researchers' attentions to the security of machine learning. In this paper, we employ a blind forensic method to detect adversarial images which are generated by the gradient-based attacks including FGSM, BIM, RFGSM and PGD. Through analyzing adversarial images, we find out that the gradient-based attacks cause significant statistical changes in the image difference domain. Besides, the gradient-based attacks add different perturbations on R, G, B channels, which inevitably change the dependencies among R, G, B channels. To measure those dependencies, the 3rd-order co-occurrence is employed to construct the feature. Unlike previous works which extract the co-occurrence within each channel, we extract the co-occurrences across from the 1st-order difference of R, G, B channels to capture the inter dependence changes. Due to the shift of difference elements caused by attacks, some co-occurrence elements of the adversarial images have distinct larger values than those of legitimate images. Experimental results demonstrate that the proposed method performs stable for different attack types and different attack strength, and achieves detection accuracy up to 99.9% which exceeding state-of-the-art much.

Keywords: Deep learning · Digital image forensics · Co-occurrence

1 Introduction

Deep Neural Networks (DNNs) have achieved excellent performances in image classification tasks [1], and show great potential on many artificial intelligence applications. However, recent works show that DNNs are vulnerable to adversarial attacks [2–8]. An adversarial image is crafted by adding small imperceptive yet effective perturbations on a legitimate image, and forces DNNs to give an

Thanks Weilin Xu et al. and Cleverhans for providing the codes of attacks. This work was partially supported by NSFC (No. 61702429), Sichuan Science and Technology Program (No. 19yyjc1656).

H. Yang et al. (Eds.): ICONIP 2020, LNCS 12533, pp. 417–428, 2020.
https://doi.org/10.1007/978-3-030-63833-7_35

error output class, i.e. either to a particular class (targeted attack) or to any class rather than the original class of the legitimate image (un-targeted attack). Cleverhans [3] listed some typical attack methods: fast gradient sign method (FGSM) [4], randomized FGSM (RFGSM) [5], basic iterative method (BIM) [6], Deep Fool [7], projected gradient descent (PGD) [21], and Carlini&Wagner method (CW) [8]. FGSM, BIM, RFGSM, and PGD generate adversarial images directly based on the gradient, which will be called as gradient-based attacks in the following. Compared with CW and Deep Fool, the gradient-based attacks are much faster, thus are more likely to be used to launch large-scale attacks. The BIM and PGD approximate the optimal attack under L_∞ constraint and have been empirically identified as the most effective approach for L_∞ attacks [21]. Hence, the detection of the gradient-based attacks is of great interest for the security of DNN model.

The defenses against adversarial attacks aim at developing robust DNN to maintain the performances via classifying adversarial examples correctly [5, 9, 29]. However, constructing a robust DNN is difficult when contesting with a tricky attacker [30]. One of the defense ways is to detect adversarial images and disallow them as inputs to DNNs [10–15, 26–28, 31]. This defense is based on the discrepancy between legitimate image and adversarial image, which does not need to modify the architecture of DNN and is usually time efficient. Some works studied statistically difference between adversarial images and legitimate images, such as principle components analysis (PCA) [11, 12], maximum mean discrepancy (MMD) test [27]. Some works depended on the changes of DNN units (such as ReLU, hidden nodes) caused by the adversarial attack to detect adversarial images. Lu et al. [13] captured the patterns of the last ReLU in the network to detect adversarial images. Metzen et al. [28] employed the values of hidden nodes of DNN as inputs to train a binary classifier to detect adversarial images. Meng et al. [10] first learned a manifold of legitimate images, and then predicted the input which is far from the manifold as an adversarial image. Guo et al. [15] found out that adversarial images have larger prediction differences for various DNN models than legitimate images have, and used those differences called transfer-ability prediction difference to detect adversarial images. Assuming that DNN model is robust against image manipulations, some works depended on the consistency of predicted outputs of manipulated samples to detect adversarial images. Xu et al. [14] proposed a feature squeezing method to modify an input image, and compared predict outputs of original input and its modified version to detect adversarial images. If these two outputs are different, the input image will be considered as adversarial. Similarly, Liang et al. [26] utilized scalar quantization and smooth spatial filter to detect adversarial images. Guo et al. [31] applied bit-depth reduction, JPEG compression, total variance minimization, and image quilting to counter adversarial images.

To further improve detection performance, some steganalysis methods which aim at detecting subtle changes caused by steganography are employed to detect adversarial images. Pascal SchÖttle et al. [16] first highlighted the parallels between steganalysis and adversarial images detection. They proposed a similar version of early steganalysis method to detect PGD [21] adversarial images. Fan et al. [17] proposed an integrated detector to detect attacks. They first used subtractive pixel adjacency matrix (SPAM [18]) to detect FGSM, R-FGSM and BIM adversarial images, and then used Gaussian noise injection detector to detect legitimate images from DeepFool and CW adversarial images. Inspired by spatial rich model (SRM [19]), Liu et al. [20] proposed an enhanced SRM (ESRM) method. They considered the modification probability of each pixel and allocated large weights to probably modified pixels when calculating co-occurrences. The above-mentioned methods achieved great improvements for the gradient-based attacks. However, their performances decrease sharply with reducing attack strength, and thus are still needed to be improved for the attacks using weak attack strength.

In this paper, we focus on detecting the gradient-based attacks with weak attack strength. Such attack leaves behind weak visual traces which can hardly be detected by human eyes (please see Fig. 1(b), (c), (e), (f)), yet can fool DNN model with high probability. To do this, we propose to detect the gradient-based adversarial images from the view of blind forensics which is also use to detect visual fidelity forged images from pristine images. Such method can distinguish the image is adversarial attacked or not without any prior knowledge. Along with the pipeline of forensics, we first analyze the statistic property of adversarial images, and find out that the gradient-based attacks would leave statistical traces in the image difference domain. Based on these traces, we extract the 3rd-order co-occurrences across from R, G, B channel as features, and construct a fast yet effective detector. Experimental results show that the proposed detector can accurately detect adversarial images generated by typical attacks.

2   Background

Given a legitimate image x with label y, an un-targeted adversarial attack seeks to find an image $x' = x + z$ to mislead a DNN model $F(.)$, subjecting to a distortion constraint between x and x'. This can be formulized as an optimization problem in (1), where $d(.)$ is a Lp norm distance metric. To train the model $F(.)$, the gradient descent methods are used to update parameters, which aim at minimizing the loss function value of samples. Intuitively, the attacker could find the perturbations z along the direction of gradient, which probably increases the loss function value of the adversarial image x' and thus misleading the DNN model $F(.)$. Some typical gradient-based attacks are introduced in the sequel.

$$\arg\min_{x'} d(x', x), \quad s.t. F(x') \neq y, d(x', x) < \varepsilon \tag{1}$$

Fast Gradient Sign Method (FGSM). Hypothesizing that DNNs are nearly linear in the high dimensional space, Goodfellow et al. [6] proposed to directly add the sign of gradient of the loss function $J(.)$ onto the benign image x to generate the adversarial image x'. To control the L_∞ norm distortion, they limited the strength of perturbations by a parameter $\varepsilon(0 < \varepsilon \leq 255)$. Formally, the un-targeted adversarial image x' is generated as (2).

$$x' = x + \varepsilon sign(\nabla_x J(F(x), y)) \qquad (2)$$

FGSM is a fast method. However, it introduces large perturbations.

Randomized Fast Gradient Sign Method (RFGSM). To defeat adversarial training defense [4], Tramèr et al. [5] proposed a simple yet effective method called RFGSM. They first added a small random noise controlled by $\alpha(\alpha < \varepsilon)$ into the legitimate image x to generate x^{1st} as (3), then generated x' using FGSM as (4).

$$x^{1st} = x + \alpha sign(rand(m, n)) \qquad (3)$$

$$x' = x^{1st} + (\varepsilon - \alpha)sign(\nabla_{x^{1st}} J(F(x^{1st}), y)) \qquad (4)$$

In (3), $rand(m, n)$ will generate a normally distributed random matrix of size mn which has equal size with the image x.

Basic Iterative Method (BIM). Kurakin et al. [8] proposed an iterative version of FGSM. Rather than adding large perturbation directly, they added a small perturbation many times using FGSM, and clipped the perturbation within ε-neighborhood to control the distortion.

$$x'_0 = x; \quad x'_{i+1} = x'_i + clip_{x,\varepsilon}(\alpha sign(\nabla'_{x_i} J(F(x'_i), y))) \qquad (5)$$

In (5), $clip_{x,\varepsilon}(.)$ tunes the perturbation to ε -neighborhood. BIM improves the attacking success rate of FGSM, and produces smaller distortions than FGSM.

Projected Gradient Descent (PGD). Madry et al. [6] first added a random initialization on x, and then used iterative attack as BIM to generate adversarial image. PGD can be viewed as an iterative version of RFGSM.

3   The Proposed Method

3.1   Statistical Property of Adversarial Images

Inspired by the blind forensics and steganalysis [18,19,22,23], we analyze the statistics of adversarial images in the difference domain. This domain benefits to suppress image contents, thus stresses the perturbations and highlights the statistical differences between legitimate images and adversarial images. The 1st-order differences of horizontal direction $s_{i,j}^{\rightarrow} = x_{i,j+1} - x_{i,j}$ and vertical direction $s_{i,j}^{\uparrow} = x_{i,j+1} - x_{i,j}$ are used, where $x_{i,j}$ is the pixel at the position (i,j).

The gradient-based attacks add some perturbations into a legitimate image which will make some shifts between the legitimate image and its adversarial image in the difference domain, thus change the distributions of the 1st-order difference. Taken the difference $s_{i,j}^{\rightarrow}$ for example, it will be shifted to $s_{i,j}^{\rightarrow'} = x'_{i,j+1} - x'_{i,j} = s_{i,j}^{\rightarrow} + z_{i,j+1} - z_{i,j}$, where $z_{i,j}$ is the added perturbation. In the following, we focus on analyzing the item $z_{i,j+1} - z_{i,j}$. As the sign(.) value

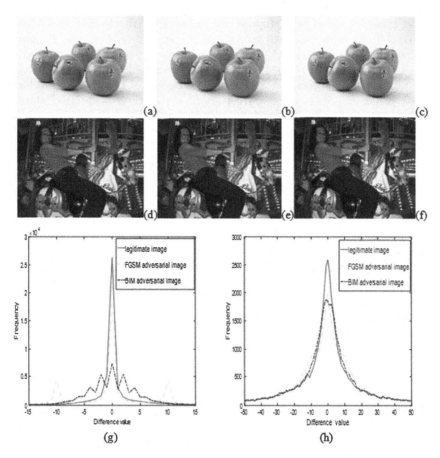

Fig. 1. A smooth(a)/textured(d) legitimate image, its FGSM image (b)/(e), BIM image (c)/(f) and the histograms of the vertical 1st-order image differences of green channel (g)/(h). Both FGSM and BIM use the attack strength $\varepsilon = 5$. (Color figure online)

is 0, -1 or 1, it can be inferred from Eq. (2) that the FGSM attack adds the perturbation $z_{i,j}$ with the value $-\varepsilon$, or ε or 0 into a legitimate image. As a result, $z_{i,j+1} - z_{i,j}$ takes the value from the set $\{-2\varepsilon, -\varepsilon, 0, \varepsilon, 2\varepsilon\}$. For RFGSM, BIM and PGD attacks, as these attacks add the perturbation using values from the set $\{-\varepsilon, -\varepsilon+1, ..., 0, \varepsilon-1, \varepsilon\}$, $z_{i,j+1} - z_{i,j}$ takes the value from the set $\{-2\varepsilon, -2\varepsilon+1, ..., -1, 0, 1, ..., 2\varepsilon-1, 2\varepsilon\}$. Those shifts change the histogram of the 1st-order difference. For a smooth image (Fig. 1(a)) which has dominant 0 elements in the 1st-order difference, it has only one peak at zero bin (red plot in Fig. 1 (g)). After attacks, the zero elements in $s_{i,j}^{\rightarrow}$ will be shifted to non-zero elements (i.e. $-2\varepsilon, -2\varepsilon+1, ..., 2\varepsilon-1, 2\varepsilon$) in the $s_{i,j}^{\rightarrow'}$ of the adversarial image. As a result, the histogram of the 1st-order difference for the adversarial image probably have peaks at these non-zero bins. Figure 1(g) shows that the FGSM ($\varepsilon = 5$) image (green double dot plot) has two peaks at $2\varepsilon = 10$ and $-2\varepsilon = -10$. Figure 1(g) also shows that the BIM ($\varepsilon = 5$) image (blue dash dot plot) has four clear non-zero peaks in the range of $[-10, 10]$. Obviously, one possible method for detecting adversarial images is to check whether the histogram of the 1st-order difference has clear peaks at non-zero bins. We tested such method on MNIST database whose images are smooth (most of pixel values in the black area are equal to 0), and obtained detection accuracy about 95%.

For the textured image (Fig. 1(d)), due to the peaks at non-zero bin is not prominent or may be disappeared as shown in Fig. 1(h), we cannot directly rely on the non-zero peaks to predicted whether it is an adversarial image or not. However, for the legitimate images and their adversarial versions, there are distinct statistical differences in the histogram of the 1st-order difference as shown in Fig. 1(g) and (h). The gradient-based attacks make the histograms become flat. It means that some zero bins shift to non-zero bins. Analogically, the shifts also occur among non-zero bins. Consequently, such shift will cause some unusual co-occurrence elements (nearly 0) become much larger. The co-occurrence will be employed to construct the proposed feature. Considering that most of elements of the 1st-order difference locate in $[-2\varepsilon, 2\varepsilon]$, we will extract the feature from the differences ranging in $[-2\varepsilon, 2\varepsilon]$.

3.2   The Proposed Feature

In this subsection, the co-occurrence which can be viewed as a high order of histogram is used to construct the feature. Unlike the methods proposed in [17,20] which calculates the co-occurrence within each channel to capture the intra dependencies, we form the co-occurrences across from R, G, B channels to capture the inter dependencies. For color images, when calculating the gradient $\nabla_x J(F(x), y)$ to generate the adversarial image, the same loss value $J(F(x), y)$ is used for R, G, B channel, it causes $\nabla_{x^r} J(F(x), y)$, $\nabla_{x^g} J(F(x), y)$, $\nabla_{x^b} J(F(x), y)$ have different values, thus leading different perturbations on R, G, B channels. It can be assumed that the gradient-based attacks will inevitably change the inter dependencies among R, G, B channel in the legitimate image, which is similar with the steganography done on RGB cover image [24]. Besides, since the gradient-based attack treats each channel as an isolate channel, it makes the correlations among the 1st-order differences of R, G, B channel become weak. Such correlations are expected to be beneficial for constructing a stable feature which is less affected by the changes of image content and attack strength.

Let us represent a RGB mn image x by $x=\{R, G, B\}=\{x_{ij}^r, x_{ij}^g, x_{ij}^b\}$, x_{ij}^r, x_{ij}^g, $x_{ij}^b \in \{0, 1, \ldots, 255\}, 1 \leq i \leq m, 1 \leq j \leq n$. We first calculate the 1st-order difference $s_{i,j}^{\rightarrow}, s_{i,j}^{\uparrow}$ from each channel, and truncate these six differences into $[-T, T]$. Then calculate the 3rd-order co-occurrences $C_{d_0 d_1 d_2}^{\rightarrow}$ and $C_{d_0 d_1 d_2}^{\uparrow}$ as (6) respectively. In (6), T is the truncating threshold, d_0, d_1, and d_2 are the integer bins which range in $[-T, T]$, [.] is the Iverson bracket which is 1 if satisfying the conditions in the bracket, and 0 otherwise, $s_{i,j}^{\rightarrow(r)}$ is the 1st-order difference of horizontal direction for R channel, $i.e.$ $s_{i,j}^{\rightarrow(r)} = x_{ij+1}^r - x_{ij}^r$, the meaning of other symbol can be obtained analogically.

$$C_{d_0d_1d_2}^{\rightarrow} = \sum_{i,j=1}^{m,n-1} [s_{i,j}^{\rightarrow(r)} = d_0, s_{i,j}^{\rightarrow(g)} = d_1, s_{i,j}^{\rightarrow(b)} = d_2]/\{M \times (N-1)\}$$

$$C_{d_0d_1d_2}^{\uparrow} = \sum_{i,j=1}^{m,n-1} [s_{i,j}^{\uparrow(r)} = d_0, s_{i,j}^{\uparrow(g)} = d_1, s_{i,j}^{\uparrow(b)} = d_2]/\{(M-1) \times N\} \quad (6)$$

$$F = (h(C^{\rightarrow}) + h(C^{\uparrow}))/2$$

To visually demonstrate differentiated ability of the proposed 3rd-order co-occurrence, we select top three Fisher Criterion Scores (FCS) of $C_{d_0d_1d_2}^{\rightarrow}$ to draw scatter plots. FCS defined in (7) measures the differentiated ability of feature element, where $u_k^1(u_k^0)$ is the mean of the kth element of adversarial (legitimate) class and $\sigma_k^1(\sigma_k^0)$ is the respective standard variance. The feature element with higher FCS usually means it has stronger differentiated ability. We evaluate FCS via 1000 randomly selected images of the validation data-set of ImageNet-1000(ILSVRC-2012) and their corresponding FGSM ($\varepsilon = 1$) adversarial images. With setting $T=3$, $C_{2,3,2}^{\rightarrow}$, $C_{-2,-3,-2}^{\rightarrow}$, and $C_{2,1,2}^{\rightarrow}$ are selected as the top three FCS elements. The plot in Fig. 2(b) demonstrates that most of legitimate images and adversarial images are separated by these three feature elements. Furthermore, the top one element $C_{2,3,2}^{\rightarrow}$ also can detect adversarial images as shown in Fig. 2(a). Due to the shift among the 1st-order difference bins caused by adversarial attack, most of the adversarial images have larger value at $C_{2,3,2}^{\rightarrow}$ than legitimate images do. These two plots empirically indicate that the proposed co-occurrence feature is effective in detecting adversarial images.

$$FCS(k) = (u_k^1 - u_k^0)^2/((\sigma_k^1)^2 + (\sigma_k^0)^2) \quad (7)$$

The 3rd-order co-occurrence matrix $C^{\rightarrow}$ and $C^{\uparrow}$ each have $(2T+1)^3$ elements. Based on the symmetric property of co-occurrence [19], we use a dimension reduction function $h(.)$ to reduce the feature dimension. Please refer to Ref [19] for more details. After applying $h(.)$, the feature F has $\sum_{k=0}^{3}(2T+1)^k/4$ elements. We categorize the image into two parts, and concatenate the feature extracted from these two parts to get the final feature which having $\sum_{k=0}^{3}(2T+1)^k/2$ elements. Specially, an image x of size $m \times n$ is equally divided into non-overlapped blocks of size $m/2 \times n/2$. These four blocks are categorized into two parts. The first part is composed by the block which has the most 0 elements in the 1st-order difference, while the second part is composed by the remaining three blocks. To do this, we expect the first part will have distinctive properties (such as non-zero peaks in Fig. 1(g)) to enhance the differentiated ability of the feature.

As discussed in Subsect. 3.1, the adversarial attacks cause statistical changes mainly within $[-2\varepsilon, 2\varepsilon]$, so we set $T = 2\varepsilon$. Considering that the strong attack will cause perceptible traces that can be detected by human eyes, we only focus on the attack with $\varepsilon \leq 5$. Under $T = 10$, the proposed feature has $(1 + 21 + 212 + 213)/2 = 4862$ dimensions in total. We use the FLD ensemble classifier with default settings [25] to construct the proposed binary detector which detects adversarial images from legitimate images.

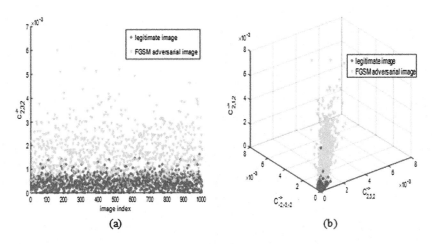

(a)   (b)

Fig. 2. Scatter plots of the top one (a) and top three (b) *FCS* elements estimated from the proposed 3rd-order co-occurrence.

4   Experimental Results

4.1   Experiment Settings

The proposed method is tested against typical gradient-based attacks by FGSM, BIM, RFGSM and PGD. Like the previous work [20], the un-targeted attacks with L_∞ norm are used in experiments. We use the code provide by Weilin Xu et al. [14] from to generate adversarial images. As the attack with strength $\varepsilon > 5$ may leave visually perceptual traces, we only consider $\varepsilon = 1, 3, 5$ for all attacks. For BIM, we set the iteration attack strength with default value $\alpha = 1$.

We randomly select 15000 RGB images of size 224×224 from the validation data-set of ImageNet-1000(ILSVRC-2012) as legitimate images. All attacks take Resnet-50 [1] as the target model. 9000 correctly predicted images are randomly selected for attacking. Only the successfully attacked images and their legitimate counterparts are tested in experiments. The number of adversarial images is listed in Table 2. For all experiments, half of adversarial samples and legitimate samples are used for training, the rest half is used for testing. Under equal priors, we report the detection accuracy (*Acc*) in the test.

The truncated threshold T is an important factor for tuning feature dimension and detection accuracy. We test our method with setting $T = 3, 5, 10$ for

FGSM using $\varepsilon = 5$. After attacks, we get 8644 successfully attacked images. These 8644 adversarial images along with their corresponding 8644 legitimate images are used in this test. The results in Table 1 show that $T = 10$ is the best choice for our method, which verify that FGSM $\varepsilon = 5$ mainly causes statistical changes within $[-10, 10]$. The results in Table 1 also tell us that the proposed methods with $T = 3, 5$ still obtain satisfactory performance, and can be applied in the low dimension feature required situations.

Table 1. The *Acc* (%) of the proposed method using different truncated threshold T. All results are detected FGSM($\varepsilon = 5$) images from legitimate images.

T	Feature dimension	*Acc* (%)
3	200-D	96.94
5	732-D	98.75
10	4862-D	99.92

4.2 Compared with Previous Works

We then test the proposed method for detecting FGSM and BIM with $\varepsilon = 1, 3, 5$. The steganalysis method SPAM [19] and SRM [19,20] are compared in these tests. To our best knowledge, the SRM-based method ESRM [20] is state-of-the-art. However, ESRM is more time-consuming than SRM but gains only about 0.5% higher *Acc* than SRM for detecting attacks with $\varepsilon \leq 5$ as reported in [20], so we do not compare ESRM with our method directly. As done in [17,20], SPAM/SRM is calculated within each R, G, B channel, and averaged 3 SPAM/SRM features to get the final feature with dimension 686-D/34671-D. Feature squeezing (FS) [14] which detects adversarial images by comparing the output consistence of the sample and its squeezed version is also tested. For FS, the best joint detection method is used, where the threshold value is determined by the false positive rate FPR<5%. The results in Table 2 show that the proposed method achieves nearly perfect performance (*Acc* > 99.95%) in all tests, and performs much better than SPAM, SRM and FS. We also test the SPAM using $T = 10$ which has the same dimension with our method for detecting BIM with $\varepsilon = 1$. This SPAM obtains *Acc* = 80.75%, which is also much lower than the result of our method. As the proposed method achieves *Acc* > 99.95%, we can infer that the proposed method also performs better than ESRM. SPAM [17], SRM [19], ESRM [20] capture the intra dependency within each R, G, B channel, wherein the modified degree of such dependence is decreased with decreasing attack strength, so the detection accuracy is reduced when reducing the attack strength. This phenomenon accords with the result that the steganalysis tool

Table 2. The *Acc* (%) of detecting legitimate images from their adversarial images generated by FGSM and BIM. The number in the bracket represents the total number of adversarial images.

	SPAM [17]	SRM [20]	FS [14]	Proposed
FGSM, $\varepsilon = 1(7659)$	82.61	88.90	70.80	**99.97**
FGSM, $\varepsilon = 3(8555)$	92.54	96.62	62.49	**99.98**
FGSM, $\varepsilon = 5(8644)$	95.41	98.33	61.07	**99.92**
BIM, $\varepsilon = 1(8850)$	79.11	87.42	80.13	**99.99**
BIM, $\varepsilon = 3(8992)$	88.17	94.57	56.58	**99.98**
BIM, $\varepsilon = 5(8996)$	91.05	96.50	51.39	**99.98**

[18, 19] obtains lower detection accuracy when lowering embedding rate. Rather than intra dependence, the proposed method captures the inter dependency across R, G, B channel, which is stable for different attack strength as shown in Table 2. The other advantage of the proposed method against SRM and ESRM lies in time efficiency. This is because we only need to calculate two kinds of the 3rd-order co-occurrence matrices, whereas SRM and ESRM need to calculate 78 kinds of the 4th-order co-occurrence matrices. As indicated by FS [14], FS is well suited for the CW attack, but not for the gradient-based attacks. Hence, the performance of FS is inferior to that of SPAM, SRM and the proposed method in the experiments.

In practice, we probably do not know attack type and attack strength. To solve this problem, a mixed detector is trained from samples generated by FGSM and BIM with different attack strengths. Specially, both FGSM($\varepsilon = 1, 3, 5$) and BIM($\varepsilon = 1, 3, 5$) are used for training. We select 8400 legitimate images and 8400 adversarial images (each attack has 1400 images), where the training set and testing set have 4200 legitimate images and 4200 adversarial images respectively. In this test, the proposed method achieves $Acc = 99.90\%$. We also use this mixed detector to detect unseen adversarial images, i.e. different attack type or attack strength not used in the training set. The results in Table 3 show that the proposed detector can accurately ($Acc > 99.9\%$) detect the attack with same type but different attack strength and the different attack types. Notice that the RGSM with $\varepsilon = 1$, $\alpha = 1$ just adds the Gaussian noise to the legitimate image according to (3) and (4), so there are only a few successfully attacked images in this test which is also indicated in [4] that adding random noise is hardly to fool DNN model. The proposed method can also accurately detect the RGSM with $\varepsilon = 1$, $\alpha = 1$. All results indicate that the proposed method owns good generalization ability, and can resist random noise attack to some extent.

Table 3. The *Acc* (%) of the proposed mixed detector for detecting unseen adversarial images which never be used in the training set. The bracket gives the number of adversarial images.

Attack type	*Acc* (%)
BIM, $\varepsilon = 2(8989)$	99.98
PGD, $\varepsilon = 1(8883)$	99.92
BIM, $\varepsilon = 3(8996)$	99.98
RFGSM, $\varepsilon = 1(51)$	100
RFGSM, $\varepsilon = 3(8366)$	99.98

5   Conclusion

This paper proposes a blind forensic method to detect gradient-based attacks applied on RGB color images. Through analyzing the inter dependencies among R, G, B channel which are disturbed by the attacks, we extract the feature from the co-occurrences across from the difference of R, G, B channel. Experimental results show that the proposed detector is fast and effective, and achieves stable detection accuracy up to 99.9% for different attack types and attack strength. It is expected that the proposed detector can enrich the arsenal for defending against adversarial attacks. We will integrate the proposed detector along with some other detectors to detect more attack types in the future.

References

1. He, K., Zhang, X., Ren, S., Sun, J.: Deep residual learning for image recognition. In: Proceedings of the IEEE Conference on Computer Vision and Pattern Recognition, pp. 770–778 (2016)
2. Szegedy, C., Zaremba, W.: Intriguing properties of neural networks. In: Proceedings of International Conference on Learning Representations arxiv: 1312.6199 (2014)
3. Cleverhans. https://github.com/tensorflow/cleverhans
4. Goodfellow, I., Shlens, J., Szegedy, C.: Explaining and harnessing adversarial examples. In: Proceedings of ICML, pp. 1–10 (2015)
5. Tramèr, F., Kurakin, A.: Ensemble adversarial training: attacks and defenses. In: Proceedings of ICLR, pp. 1–20 (2018)
6. Kurakin, A., Goodfellow, I., Bengio, S.: Adversarial examples in the physical world. In: Proceedings of International Conference on Learning Representations arxiv: 1607.02533 (2016)
7. Moosavidezfooli, S., Fawziand, A., Frossard, P.: DeepFool: a simple and accurate method to fool deep neural networks. In: Proceedings of Computer Vision and Pattern Recognition, pp. 2574–2582 (2015)
8. Carlini, N., Wagner, D.: Towards evaluating the robustness of neural networks. In: IEEE Symposium on Security and Privacy, pp. 39–57 (2017)
9. Papernot, N.: Distillation as a defense to adversarial perturbations against deep neural networks. In: IEEE Symposium on Security and Privacy, pp. 582–597 (2016)
10. Meng, D., Chen, H.: MagNet: a two-pronged defense against adversarial examples. In: Proceedings of the 2017 ACM SIGSAC Conference on Computer and Communications Security, pp. 135–147 (2017)

11. Hendrycks, D., Gimpel, K.: Early methods for detecting adversarial images arXiv:1608.00530 (2016)
12. Li, X., Li, F.: Adversarial examples detection in deep networks with convolutional filter statistics. In: Proceedings of IEEE International Conference on Computer Vision, pp. 5775–5783 (2017)
13. Lu, J., Issaranon, T., Forsyth, D.: SafetyNet: detecting and rejecting adversarial examples robustly. In: Proceedings of IEEE International Conference on Computer Vision, pp. 446–454 (2017)
14. Xu, W., David, Y., Yan, J.: Feature squeezing: detecting adversarial examples in deep neural networks. In: Network and Distributed System Security Symposium. arXiv:1704.01155 (2017). https://evadeML.org/zoo
15. Guo, F.: Detecting adversarial examples via prediction difference for deep neural networks. Inf. Sci. **501**, 182–192 (2019)
16. Schöttle, P., Schlögl, A., Pasquini, C.: Detecting adversarial examples-a lesson from multimedia security. In: Proceedings of the 26th IEEE European Signal Processing Conference, pp. 947–951 (2018)
17. Fan, W., Sun, G., Su, Y., Liu, Z., Lu, X.: Integration of statistical detector and Gaussian noise injection detector for adversarial example detection in deep neural networks. Multimedia Tools Appl. **78**(14), 20409–20429 (2019). https://doi.org/10.1007/s11042-019-7353-6
18. Pevny, T., Bas, P., Fridrich, J.: Steganalysis by subtractive pixel adjacency matrix. IEEE Trans. Inf. Forensics Secur. **5**(2), 215–224 (2010)
19. Fridrich, J., Kodovsky, J.: Rich models for steganalysis of digital images. IEEE Trans. Inf. Forensics Secur. **7**(3), 868–882 (2012)
20. Liu, J., Zhang, W., Zhang, Y.: Detecting Adversarial Examples Based on Steganalysis arXiv:1806.09186 (2018)
21. Madry, A.: Towards deep learning models resistant to adversarial attacks. In: Proceedings of International Conference on Learning Representations arXiv:1706.06083 (2017)
22. Chen, J., Kang, X., Liu, Y.: Median filtering forensics based on convolutional neural networks. IEEE Signal Process. Lett. **22**(11), 1849–1853 (2015)
23. Belhassen, B., Stamm, M.C.: Constrained convolutional neural networks: a new approach towards general purpose image manipulation detection. IEEE Trans. Inf. Forensics Secur. **13**(11), 2691–2706 (2018)
24. Goljan, M., Cogranne, R.: Rich model for steganalysis of color images. In: Proceedings of IEEE WIFS, pp. 185–190 (2014)
25. Kodovsky, J., Fridrich, J., Holub, V.: Ensemble classifiers for steganalysis of digital media. IEEE Trans. Inf. Forensics Secur. **7**(2), 432–444 (2012)
26. Liang, B., Li, H., Su, M., Li, X., Shi, W., Wang, X.: Detecting adversarial image examples in deep neural networks with adaptive noise reduction. IEEE Trans. Dependable Secure Comput. https://doi.org/10.1109/TDSC.2018.2874243
27. Grosse, K., Manoharan, P., Papernot, N., Backes, M.: On the (statistical) detection of adversarial examples. arXiv preprint arXiv:1702.06280 (2017)
28. Metzen, J. H., Genewein, T., Fischer, V., Bischoff, B.: On detecting adversarial perturbations arXiv:1702.04267 (2017)
29. Liao, F., Liang, M., Dong, Y., Pang, T., Zhu, J.: Defense against adversarial attacks using high-level representation guided denoiser arXiv:1712.02976 (2017)
30. Athalye, A., Carlini, N., Wagner, D.: Obfuscated gradients give a false sense of security: circumventing defenses to adversarial examples arXiv:1802.00420 (2018)
31. Guo, C., Rana, M., Cisse, M., van der Maaten, L.: Countering adversarial images using input transformations arXiv:1711.00117 (2017)

Hindsight-Combined and Hindsight-Prioritized Experience Replay

Renzo Roel P. Tan[1,2]([✉]) [ID], Kazushi Ikeda[1] [ID], and John Paul C. Vergara[2] [ID]

[1] Division of Information Science, Nara Institute of Science and Technology,
Takayama Town, Ikoma, Nara 6300192, Japan
{tan.renzo_roel_perez.tp7,kazushi}@is.naist.jp
[2] School of Science and Engineering, Ateneo de Manila University,
Katipunan Avenue, National Capital Region, 1108 Quezon City, Philippines
{rrtan,jpvergara}@ateneo.edu

Abstract. Reinforcement learning has proved to be of great utility; execution, however, may be costly due to sampling inefficiency. An efficient method for training is experience replay, which recalls past experiences. Several experience replay techniques, namely, combined experience replay, hindsight experience replay, and prioritized experience replay, have been crafted while their relative merits are unclear. In the study, one proposes hybrid algorithms – hindsight-combined and hindsight-prioritized experience replay – and evaluates their performance against published baselines. Experimental results demonstrate the superior performance of hindsight-combined experience replay on an OpenAI Gym benchmark. Further, insight into the nonconvergence of hindsight-prioritized experience replay is presented towards the improvement of the approach.

Keywords: Experience replay · Deep Q-Network · Reinforcement learning · Sample efficiency · Hybrid algorithm

1 Introduction

Reinforcement learning [20] has been the subject of research. Its uncomplicated formulation is capable of capturing a vast number of problems in artificial intelligence. Fields such as resource management [13], traffic signal control [2], and robotics [8] abound with practical applications.

Generally, the learning problem is to control a system so as to maximize a numerical value representing a long-term objective [7]. One calls the learner the agent and the agent is established to be in an environment. The standard reinforcement learning formalism, therefore, concurs with a decision making framework consisting of an agent that interacts with an environment and improves its performance based on feedback. At each time step, the agent is given a state and

Supported by the Japan Society for the Promotion of Science through the Grants-in-Aid for Scientific Research Program (KAKENHI 18K19821).

H. Yang et al. (Eds.): ICONIP 2020, LNCS 12533, pp. 429–439, 2020.
https://doi.org/10.1007/978-3-030-63833-7_36

it selects an action; the environment then presents a reward and a new state. By and large, the goal is to maximize the cumulative reward.

While reinforcement learning shows promise, implementation in real-world contexts can be costly because of sampling inefficiency. This means that a multitude of runs are needed for the algorithm to achieve success. A way to address such a complication is through the utilization of experience replay [11], where previous experiences are reused. As an aside, there are other methods through which one may grapple with the problem. Recent alternatives include using Gaussian processes [5] and using babbling [9,14] in speeding up learning. The paper, notwithstanding, focuses on experience replay and its variants.

By experience replay, the agent remembers past events and presents them to the algorithm as if to experience what it had before. More concretely, this means maintaining a buffer memory of experiences. An experience or transition is a quadruple (s, a, r, s'), meaning that doing action a given state s results in reinforcement r and the new state s' [11]. Batches of transitions are drawn from the buffer and used for training the agent. With the replay technique, the weights in the mapping function are amended not only once upon the completion of the task. A batch update to the parameters is done via experiences sampled randomly from the replay buffer which stores recent transitions.

The usefulness of experience replay primarily lies in its ability to speed up the process of reward propagation [11]. It encourages sample-efficient training by pulling experience from the memory [18]. On top of that, the agent may get chances to refresh what it has learned before. When training a network, for instance, if an input pattern is not encountered for quite a while, the network usually forgets what it has learned for that pattern and hence would need to relearn the output [11]. Lastly, applying the experience replay buffer breaks up the correlation in data and improves network convergence [18]. Do note, nonetheless, that experience replay implementation is best suited for environments which do not rapidly change over time. Should this be the case, past transitions may become irrelevant or even misleading [11].

To date, three techniques for experience replay have been widely deemed fundamental. These are combined experience replay [23], hindsight experience replay [1], and prioritized experience replay [18]. The standalone methods are well documented; nevertheless, the three have never been run on the same environment and their combinations have yet to be put forward in literature.

Hybrid algorithms are an ongoing trend in reinforcement learning [4,12,21, 22]. A notable study, for example, has concluded that a straightforward combination of the various modifications to the Deep Q-Network [17] resulted in superior performance [6]. The idea of hybridization on the level of experience replay then came to light. Rooted in this motivation, the study has the following objectives:

- Incorporate a comprehensive buffer into reinforcement learning algorithms;
- Juxtapose variations of replay stemming from different combinations; and
- Evaluate the performance of the proposed hybrid algorithms.

The succeeding sections are organized as follows. The second section looks into the existing methods for experience replay in three subsections. In the third section, one details the set-up and the experimentation process. Results on the hybrid replay algorithms from the specified configurations are discussed in the fourth section. To conclude, a synthesis is contained in the fifth section.

2   Literature

2.1   Prioritized Experience Replay

In baseline experience replay, all transitions are sampled uniformly, irrespective of usefulness. Prioritized experience replay [18], as the name suggests, prioritizes stored transitions by assigning greater weights to transitions with high expected learning progress.

A focal aspect in prioritized experience replay is quantifying the importance of a transition. How much one expects to learn from a transition may be estimated by the temporal-difference error δ. For transition i, the weight is

$$w_i = \left(\frac{1}{N} \cdot \frac{1}{P\left(i\right)} \right)^{\beta},$$

where N is the batch size and β is a parameter between 0 and 1 determining importance sampling. In addition, $P\left(i\right)$ is the sample probability

$$\frac{p_i^{\alpha}}{\sum_k p_k^{\alpha}},$$

where $p_i > 0$ is the priority. This is simply the last known temporal-difference error added to a positive constant to avoid disregard should the error be 0. The parameter α regulates how much prioritization is utilized, with $\alpha = 0$ for uniform sampling and $\alpha = 1$ for proportional sampling.

2.2   Combined Experience Replay

Since the conceptualization of experience replay, a new hyperparameter – the buffer size N – has required careful adjustment. As there has been neglect, investigations gave light to the fact that a large replay buffer may crucially influence performance [23]. The stabilization of the training system is extremely sensitive to the size of the replay buffer. Experiments show that learning in experiments hiring larger replay buffers is hindered. A simple $\mathcal{O}\left(1\right)$ experience replay method was then created in response to the $\mathcal{O}\left(\log N\right)$ prioritized experience replay technique, resolving the negative effects of a large buffer size.

Treated inexactly a special case of prioritized experience replay is combined experience replay [23]. While both assigns the larges priority to the latest transition, the distinction lies in the guarantee. Combined experience replay includes the latest transition to the batch invariably and samples the rest uniformly.

2.3   Hindsight Experience Replay

The key idea in hindsight experience replay [1] is that the machine can learn just as much from undesirable outcomes as from desirable ones. Essentially, there are two perspectives in an unsuccessful trajectory – the performed actions result in a failure or would have been a success. Hindsight experience replay enables the software agent to reason from the latter viewpoint. One takes an extended input comprising the current state, action, reward, next state, and goal state. The method is particularly useful in environments where a number of states may be treated as a separate goal. It is also effective when rewards are binary.

For an episode with sequence of states $s_0, s_1, s_2, \ldots, s_T$ which does not hit target g, one probes with a different goal to be assigned g'. Despite the fact that the above transition did not reach original goal g, it holds information on how to get to state s_T or any other state in the sequence. The knowledge may be gathered using an algorithm where one replaces g with $g' = s_i$, $0 \leq i \leq T$. Rewards are modified as a result and learning becomes less complicated.

3   Methodology

3.1   Replay Techniques

Sampling methods include baseline experience replay, combined experience replay, hindsight experience replay, hindsight-combined experience replay, hindsight-prioritized experience replay, and prioritized experience replay. The baseline and independent algorithms – combined, hindsight, and prioritized experience replay – had been made known in the previous section; thus, one elaborates on the proposed hybrid replay approaches.

Hindsight-Combined Experience Replay. For the method of hindsight-combined experience replay, what is done first is to activate hindsight experience replay. As revealed in the literature review, hindsight experience replay adds a new dimension to the transition (s, a, s', r) by storing the additional information of a goal g'. Typically, the transitions are sampled uniformly during training. With the integration of combined experience replay, weights are assigned in accord with combined experience replay after the inclusion of a goal dimension to the transition. One puts sizeable priority on the latest extended transition. Algorithm 1 describes the procedure. In the algorithm, the storing of an extended tuple in line 9 and the generation of multiple goals in lines 10 to 14 are motivated by hindsight experience replay. Combined experience replay is seen in line 16, where one updates using a batch including the latest transition.

Hindsight-Prioritized Experience Replay. Comparable to the first hybrid algorithm is the method of hindsight-prioritized experience replay. Upon the transformation of transitions of four dimensions to five-dimensional ones via the inclusion of g', transitions are given weights following the conventional prioritized

Algorithm 1. Hindsight-Combined Experience Replay

1: Initialize value function Q
2: Initialize ϵ-greedy policy $\mathcal{P}$
3: Initialize replay buffer $\mathcal{M}$
4: **while** *not converged* **do**
5:     Sample goal G and get initial state S
6:     **while** S *is not the terminal state* **do**
7:         Select action A according to $\mathcal{P}$ with respect to Q
8:         Execute A, get reward R, and next state S'
9:         Store transition $t = (S, A, R, S', G)$ into $\mathcal{M}$
10:         Sample a set $\mathcal{G}$ of additional goals for replay
11:         **for** $G' \in \mathcal{G}$ **do**
12:             Compute new reward R'
13:             Store transition (S, A, R', S', G') into $\mathcal{M}$
14:         **end**
15:         Sample batch $\mathcal{B}$ from $\mathcal{M}$
16:         Update Q with $\mathcal{B}$ and t
17:         $S \leftarrow S'$
18:     **end**
19: **end**

experience replay – the greater the error, the less likely it is for the transition to be recalled. One uses prioritization based on temporal-difference error. The pseudocode is provided as Algorithm 2. Similar to Algorithm 1, hindsight learning is activated in lines 10 to 15. The involvement of prioritized experience replay is most apparent in lines 4 and 16, indicating prioritized sampling.

3.2   Learning Algorithm

In the experiments, the learning algorithm initiated is the Deep Q-Network [17]. A four-layer neural network architecture is hired. The replay buffer accommodating all experience replay types is composed with the aid of a segment tree structure. Refer to the accompanying repository for the complete implementation of both the algorithm and the memory.[1]

3.3   Testing Environment

The environment utilized is the Lunar Lander from the OpenAI Gym suite.[2] Lunar Lander is a two-dimensional environment featuring a landing pad at the origin. The goal is to move from the top of the screen to the landing pad without crashing. It loses rewards if it moves away from the landing pad. An episode finishes when the lander crashes or comes to rest. A crash would yield an additional -100 points and coming to rest would yield an additional 100. Each leg ground

---

[1] This is found in: https://github.com/renzopereztan/HyER.
[2] The link is: https://gym.openai.com/envs/LunarLander-v2.

Algorithm 2. Hindsight-Prioritized Experience Replay

1: Initialize value function Q
2: Initialize ϵ-greedy policy $\mathcal{P}$
3: Initialize replay buffer $\mathcal{M}$
4: Initialize weighting scheme ω
5: **while** *not converged* **do**
6:    Sample goal G and get initial state S
7:    **while** S *is not the terminal state* **do**
8:       Select action A according to $\mathcal{P}$ with respect to Q
9:       Execute A, get reward R, and next state S'
10:      Store transition $t = (S, A, R, S', G)$ into $\mathcal{M}$
11:      Sample a set $\mathcal{G}$ of additional goals for replay
12:      **for** $G' \in \mathcal{G}$ **do**
13:         Compute new reward R'
14:         Store transition (S, A, R', S', G') into $\mathcal{M}$
15:      **end**
16:      Sample batch $\mathcal{B}$ from $\mathcal{M}$ following ω
17:      Update Q with $\mathcal{B}$
18:      $S \leftarrow S'$
19:   **end**
20: **end**

---

contact is 10 and firing the main engine is -0.3 points each frame. If the agent solves the game then the reward is 200 points. This has states $s \in \mathbb{R}^8$:

- The x and y positions;
- The x and y velocities;
- The lander angle and angular velocity; and
- The right and left leg ground contact information.

The actions are do nothing, fire left orientation engine, fire right orientation engine, and fire main engine.

3.4 Machine Specifications

Experiments were done in the Python programming language. Concerning the machine, the operating system was the Ubuntu 16.04 LTS, the central processing unit was the Intel® Core™ i7-8565U processor working at 1.80 GHz, the graphics card was the NVIDIA® GeForce® MX250, and the memory was 16 GB.

3.5 Investigation Outline

There are two main factors that may influence the results greatly – the hyperparameters chosen for weight assignment and the size of the buffer memory. Hyperparameters determine the weight distribution for the transitions most evidently during prioritized experience replay; the buffer size, on the other hand, is the number of transitions to be included in the memory at a given time. Suitably,

Table 1. First points of convergence in the initial experiment.

Experience replay	First point of convergence	
	Training	Testing
Baseline	4900	5500
Combined	3600	4000
Hindsight	4800	5000
Hindsight-Combined	2000	2000
Hindsight-Prioritized	—	—
Prioritized	4400	4500

one directs the analysis towards the management of the effects of the selected hyperparameters and buffer size.

Hyperparameters for prioritized experience replay and hindsight-prioritized experience replay are first derived from a set of optimal values [18]. An exhaustive grid search based on the aforementioned environment confirms the sweet spot being $\alpha = 0.60$ and $\beta = 0.40$.

An initial experiment is then conducted. Every variation of experience replay was maintained for 10000 episodes each. The buffer size is set at 100000 transitions. For each training run, one initializes testing mode every 500 episodes. Convergence is checked for every algorithm.

A decrease in the buffer size is then put to effect afterwards. The memory was limited to 10000 experiences. Apart from the reason of ensuring consistency, a smaller buffer size is handled better by some of the chosen replay algorithms [23]. Earlier suppositions may be confirmed and more comments may be made.

The runs are evaluated with a defined metric – the first point of convergence. This is simply the number of the episode in training or testing on which the agent has acquired a reward of greater than or equal to the target score.

4   Results

4.1   The Hybrid Replay Algorithms

For better comparison, the different algorithms are set side by side. The variability between the random initializations for each replay type is not substantial; any such measure is omitted from the graph for visual brevity. Figure 1 shows the average training and testing rewards of all algorithms. Moreover, Table 1 presents the first points of convergence for each experience replay type.

An immediate complication is the nonconvergence of the run with hindsight-prioritized experience replay. For prioritized experience replay, a large buffer is not suitable; this may prove to be a problem during endeavors in empirical algorithmics. Furthermore, when both hindsight and prioritized experience replay are activated, multi-goal learning is initiated and rewards may be noisy. In this scenario, the temporal-difference error can be a poor estimate [18].

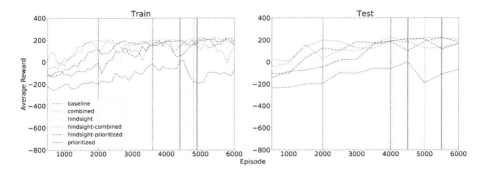

Fig. 1. Training and testing curves with lines emphasizing convergence for the experiment with buffer size $N = 100000$.

Table 2. Convergences for the implementation with reduced buffer.

Experience replay	First point of convergence	
	Training	Testing
Baseline	5100	5500
Combined	4300	4500
Hindsight	4800	5000
Hindsight-Combined	2400	2500
Hindsight-Prioritized	—	—
Prioritized	4200	4500

Setting that result aside, one sees that the outcome is consistent with existing study. The runs with published experience replay techniques fared better than the baseline. Among the standalone algorithms, the run with combined experience replay did best due to its suitability for a large memory. The run with prioritized experience replay came in close second. To reiterate, this is expected because of the buffer size. Hindsight experience replay did relatively good according to what is written in literature [1].

The most interesting result, albeit up to verification in the next subsection, is the extraordinary performance of hindsight-combined experience replay algorithm. It reached convergence at least 2000 episodes ahead of the others.

4.2   The Effect of Buffer Size Reduction

The buffer size is decreased from 100000 to 10000 to validate the outcome in the previous section. The objective of the subsection is to check whether the buffer size significantly affects learning. The experiment has results shown in Table 2. Figure 2 comprising the six learning curves presents an outcome consistent with the previous subsection.

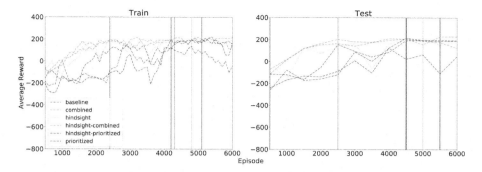

Fig. 2. Training and testing curves with lines emphasizing convergence for the experiment with buffer size $N = 10000$.

Runs with hindsight-prioritized experience replay perform as poorly as before. Regardless the buffer size, the said hybrid algorithm does not yield good results. The possibility of the large buffer size being the problem is eliminated; hence for this observation, one may look into prioritization instead. Prioritization variants are possible for prioritized experience replay [18]. This means to say that the assignment of weights based on the temporal-difference error δ may be altered and variations for prioritization emerge. With inconclusive results are taking the derivative, considering the norm of the weight-change induced by replay, introducing asymmetry to prioritize positive errors, basing prioritization on episodic return, and others.

Another possible alternative could be a measure that would prioritize transitions based on the error if the entire distribution instead of just the expectation. This is tantamount to moving from considering the absolute value of the summation of differences to looking at the summation of the absolute value of each difference. Should such a measure be found, the problem of noise brought about by the combination of hindsight experience replay and prioritized experience replay based on temporal-difference error could be solved. Instead of combining hindsight experience replay with the conventional prioritized experience replay, the former may be integrated with a modified prioritized experience replay that prioritizes based on a different measure. A promising measure is the Wasserstein metric [3], appearing in numerous machine learning papers.

There is improvement with the prioritized experience replay runs. Decreasing the buffer size has helped the run with prioritized experience replay, enabling it to be the best standalone method conforming to literature. Combined experience replay yields the same in accord with its function.

Based on the point of convergence measures throughout the section, the hybrid algorithm hindsight-combined experience replay shows promise. It has once again converged almost twice as fast as the other variations. In summary, the run with hindsight-combined experience replay continues to produce exceptional outcomes, proving to be more efficient and effective than the rest.

4.3   A Change in Environment

As a remark, supplementary experiments were done on some Box2D and classic control tasks from the OpenAI Gym to ensure greater consistency. In brief, it is discovered that hindsight-combined experience replay steadily outperforms the other methods and hindsight-prioritized experience replay fails to converge across the chosen environments.

5   Conclusion

The paper has shown the integration of an extensive buffer into the Deep Q-Network algorithm. The subsequent investigation exhibited progressive findings with respect to both existing methods and proposed techniques. Once consistent with literature, the exceptional performance of hindsight-combined experience replay was confirmed.

While the other hybrid algorithm has not been effective, a direction for its improvement was pointed out. A discussion on prioritization variants contends that the use of another error measure in lieu of the temporal-difference error may resolve the resulting noise when hindsight experience replay is combined with the conventional prioritized experience replay.

For future work, the code base may manage the change in learning algorithm and testing environment. The Advantage Actor-Critic [15,16] and Deep Deterministic Policy Gradient [10,19] are comfortably implemented through the same process as foray. This encourages the validation of both hindsight-combined and hindsight-prioritized experience replay beyond the OpenAI Gym standard.

References

1. Andrychowicz, M., et al.: Hindsight experience replay. In: Advances in Neural Information Processing Systems (2017)
2. Arel, I., Liu, C., Urbanik, T., Kohls, A.: Reinforcement learning-based multi-agent system for network traffic signal control. Institution of Engineering and Technology Intelligent Transport Systems (2010)
3. Dobrushin, R.: The definition of random variables by conditional distributions. Probab. Theory Appl. **15**, 458–486 (1970)
4. Fan, Z., Su, R., Zhang, W., Yu, Y.: Hybrid actor-critic reinforcement learning in parametrized action space. In: Proceedings of the International Joint Conference on Artificial Intelligence (2019)
5. Grande, R., Walsh, T., How, J.: Sample efficient reinforcement learning with Gaussian processes. In: Proceedings of the International Conference on Machine Learning (2014)
6. Hessel, M., et al.: Rainbow: combining improvements in deep reinforcement learning. arXiv Preprint arXiv:1710.02298 (2017)
7. Kapoor, S.: Multi-agent reinforcement learning: a report on challenges and approaches. arXiv Preprint arXiv:1807.09427 (2018)
8. Kober, J., Bagnell, A., Peters, J.: Reinforcement learning in robotics: a survey. Int. J. Robot. Res. **32**, 1238–1274 (2013)

9. Kwiatkowski, R., Lipson, H.: Task-agnostic self-modeling machines. Sci. Robot. **4**(26), 4 (2019)
10. Lillicrap, T., et al.: Continuous control with deep reinforcement learning. arXiv Preprint arXiv:1509.02971 (2015)
11. Lin, L.J.: Self-improving reactive agents based on reinforcement learning, planning, and teaching. Mach. Learn. **8**, 293–321 (1992). https://doi.org/10.1007/BF00992699
12. Ma, C., Li, J., Bai, J., Wang, Y., Liu, B., Sun, J.: A hybrid deep reinforcement learning algorithm for intelligent manipulation. In: Yu, H., Liu, J., Liu, L., Ju, Z., Liu, Y., Zhou, D. (eds.) ICIRA 2019. LNCS (LNAI), vol. 11743, pp. 367–377. Springer, Cham (2019). https://doi.org/10.1007/978-3-030-27538-9_31
13. Mao, H., Alizadeh, M., Menache, I., Kandula, S.: Resource management with deep reinforcement learning. In: Proceedings of the Fifteenth Association for Computing Machinery Workshop on Hot Topics in Networks (2016)
14. Marjaninejad, A., Urbina-Melendez, D., Cohn, B., Valero-Cuevas, F.: Autonomous functional movements in a tendon-driven limb via limited experience. Nat. Mach. Intell. **1**, 144–154 (2019)
15. Mirowski, P., et al.: Learning to navigate in complex environments. arXiv Preprint arXiv:1611.03673 (2016)
16. Mnih, V., et al.: Asynchronous methods for deep reinforcement learning. In: Proceedings of the International Conference on Machine Learning (2016)
17. Mnih, V., et al.: Human-level control through deep reinforcement learning. Nature **518**, 529–533 (2015)
18. Schaul, T., Quan, J., Antonoglou, I., Silver, D.: Prioritized experience replay. arXiv Preprint arXiv:1511.05952 (2015)
19. Silver, D., Lever, G., Heess, N., Degris, T., Wierstra, D., Riedmiller, M.: Deterministic policy gradient algorithms. In: Proceedings of the International Conference on Machine Learning (2014)
20. Sutton, R., Barto, A.: Reinforcement Learning: An Introduction. Massachusetts Institute of Technology Press, Cambridge (1998)
21. Tesauro, G., Jong, N., Das, R., Bennani, M.: A hybrid reinforcement learning approach to autonomic resource allocation. In: Proceedings of the International Conference on Autonomic Computing (2006)
22. Wang, Z., Qiu, X., Wang, T.: A hybrid reinforcement learning algorithm for policy-based autonomic management. In: Proceedings of the International Conference on Services Systems and Services Management (2012)
23. Zhang, S., Sutton, R.: A deeper look at experience replay. arXiv Preprint arXiv:1712.01275 (2017)

HPSGD: Hierarchical Parallel SGD
with Stale Gradients Featuring

Yuhao Zhou, Qing Ye, Hailun Zhang, and Jiancheng Lv$^{(\boxtimes)}$

College of Computer Science, Sichuan University, Chengdu, China
sooptq@gmail.com, fuyeking@stu.scu.edu.cn, tamakokoodaza@gmail.com,
lvjiancheng@scu.edu.cn

Abstract. While distributed training significantly speeds up the training process of the deep neural network (DNN), the utilization of the cluster is relatively low due to the time-consuming data synchronizing between workers. To alleviate this problem, a novel Hierarchical Parallel SGD (HPSGD) strategy is proposed based on the observation that the data synchronization phase can be paralleled with the local training phase (i.e., Feed-forward and back-propagation). Furthermore, an improved model updating method is unitized to remedy the introduced stale gradients problem, which commits updates to the replica (i.e., a temporary model that has the same parameters as the global model) and then merges the average changes to the global model. Extensive experiments are conducted to demonstrate that the proposed HPSGD approach substantially boosts the distributed DNN training, reduces the disturbance of the stale gradients and achieves better accuracy in given fixed wall-time.

Keywords: Distributed training · Parallel SGD · Hierarchical computation · Large scale · Optimization

1 Introduction

While the synchronous stochastic gradient descent (SSGD) remarkably reduces the training time of the large-scale DNN on the complex dataset by allocating the overall workload to multiple workers, it is additionally required to synchronize local gradients of the workers to keep the convergence of the models [2]. Hence, the introduced gradient synchronizing phase in the cluster will consume much time, making the acceleration effect of the distributed training non-linear and deteriorating the cluster's scalability. Thus, the communication

This work is supported in part by the National Key Research and Development Program of China under Contract 2017YFB1002201, in part by the National Natural Science Fund for Distinguished Young Scholar under Grant 61625204, and in part by the State Key Program of the National Science Foundation of China under Grant 61836006.

© Springer Nature Switzerland AG 2020
H. Yang et al. (Eds.): ICONIP 2020, LNCS 12533, pp. 440–451, 2020.
https://doi.org/10.1007/978-3-030-63833-7_37

cost caused by the network I/O and transmission of the synchronization generally becomes the most significant bottleneck of the distributed DNN training with the increasing number of workers and model parameters [14], especially when the communication-to-computation ratio is high (e.g., Gate Recurrent Unit (GRU) [1]).

Flourish developments have been made to overcome this problem, including batch-size enlarging [7], periodically synchronizing [16] and data compressing [8,13,19,21]. However, although these methods considerably reduce the communication load, many side effects are brought by them to the distributed DNN training process as well, respectively be it the generalization ability degradation [5], the added performance-influential hyper-parameter γ (i.e., configuration of the interval between synchronizations) or the introduced time-consuming extra phases during training (e.g., sampling, compressing, decompressing, etc.). Moreover, they are all focused on reducing either the worker-to-worker communication rounds or the data transfer size, which limits the results they can achieve since neither the round nor the size can be reduced to 0.

In this paper, we propose Hierarchical Parallel SGD (HPSGD) algorithm that not only fully overlaps the synchronization phase with the local training phase with hierarchical computation but also mitigates the gradients staleness problem and therefore achieves high performances. The desired timeline of HPSGD is illustrated in Fig. 1, which implies that it also ensures synchronous training progresses between workers (i.e., workers start to feed-forward at the same time). The main challenge of all algorithms that separate the local training phase and synchronization phase, including HPSGD, is the gradients staleness problem, meaning the model is updated using stale gradients, which is a detriment to the model convergence. However, Unlike previous literature that tries to counteract stale gradients' effects [22], HPSGD treats these gradients as the features of unknown global optimization surface and thus uses these features to optimize the global training function. In this scenario, the local training phase that overlaps with the synchronization phase helps the global training function collect valuable gradients information and optimize. As a result, the HPSGD algorithm fully utilizes the computational performance, and also maintains model convergence.

The contributions of this paper are summarised as follow:

– We entirely overlap the synchronization phase with the local training phase by utilizing hierarchical computation, which significantly boosts the distributed training process.
– We utilize an optimized algorithm based on hierarchical computation to address the gradients staleness problem, and therefore improving the training speed, stability, and model accuracy of the distributed DNN training process.
– We demonstrate and verify the reliability and effectiveness of HPSGD by applying it to sufficient experiments with various approaches to extensive models. The source code and parameters of all experiments are open-sourced for reproducibility[1].

---

[1] https://github.com/Soptq/Hierarchical_Local_SGD.

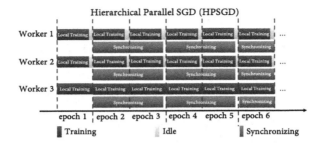

Fig. 1. In HPSGD, every worker has two processes. One of them is the local training process doing continuous model training and the other one is the synchronizing process doing continuous data exchanging. These two processes run in parallel.

The rest of this paper is organized as follows. The literature review is illustrated in Sect. 2, where some background information is introduced. In Sect. 3, the structure and implementation of the proposed HPSGD algorithm are presented. Then the experimental design and result analysis are detailedly documented in Sect. 4. Finally, the conclusions of this paper are drawn in Sect. 5.

2   Literature Review

Synchronous and asynchronous SGD: Synchronous SGD (SSGD) is generally a distributed training's model updating strategy that evenly distributes the workload among multiple workers. Then, it updates the model parameters by utilizing SGD algorithm with global gradients aggregated by averaging all local gradients of the different workers. Particularly, the convergence of the model is unaffected with SSGD since it ensures the synchronized gradients are the latest. SSGD can be employed in both centralized [10,11] and decentralized [12] architectures and the timeline of decentralized SSGD is drawn in Fig. 2a, where it can be noticed that before starting synchronizing, there is a waiting phase where some workers might have already finished local training and wait for the slower workers to catch up, which leads to a wasted resource. Asynchronous SGD (ASGD) overcomes this problem by allowing workers to work independently. Specifically, fast workers instantly *push* the calculated local gradients to the parameter servers once they finished training. The timeline of ASGD is shown in Fig. 2b. Although ASGD eliminates the waiting time before synchronizing, it can be utilized only in the centralized architecture, indicating that the cluster is more likely to incur communication overload.

Moreover, as workers are not aware of other workers' status, the gradients staleness problem can be easily triggered. For example, $worker_i$ uses W_0 to compute local gradients ∇_0 and synchronizes ∇_0 to parameter servers to start a global model updating operation. However, the global model is updated to W_1 during its synchronization phase due to the faster synchronization speed of another worker. Thus $worker_i$ eventually updates the global model W_1 to W_2

using ∇_0 computed by W_0, which will considerably impact the convergence of the model.

Fig. 2. (a) A worker waits for other workers to finish local training before starting data exchanging. (b) A worker instantly exchanges data with the parameter server and then steps into the next epoch.

It is worth noting that in both centralized and decentralized architectures, synchronizations are processed in the worker's main thread, implying the next epoch's training will be prevented until the current epoch's synchronization phase is completed. Consequently, in both SSGD and ASGD, the processing units of the worker (e.g., CPU, GPU) are idle during the synchronization phase, which is generally a much bigger waste of the worker's performance compared to the waiting phase in SSGD, considering synchronizing usually takes much more time than waiting in practice.

Local SGD: Local SGD [16] is a well-known algorithm that utilizes periodically model averaging to reduce the number of synchronization rounds. It is capable of achieving good performance both theoretically and practically. Specifically, it introduces a new hyper-parameter γ that configures the frequency of the model synchronizing. When synchronizing, workers synchronize the model parameters in place of gradients. However, there are several drawbacks of Local SGD and its variations [3,18]. 1) Local SGD delivers a relatively slow convergence rate per epoch, and the introduced hyper-parameter γ is required to be configured manually to achieve the model's best performance. 2) Although Local SGD reduces the number of synchronization rounds, the computing performance is still idle and not been fully utilized during synchronizing. The pseudo-code of the Local SGD is illustrated in Algorithm 1.

3   Methodology

In this section, we present the implementation of the proposed HPSGD detailedly, which includes: 1) Spawning two process P_s and P_t to Perform data synchronizing and local training, respectively. 2) Applying a model updating algorithm to alleviate the gradients staleness problem.

Algorithm 1. Local SGD

1: Initialization: Cluster size n. Learning rate $\mu \geq 0$. Max training *epoch*. Local gradient $\widehat{\nabla}^e$. Synchronous period γ;
2: **for all** $i \in 1, ..., n$ **do in parallel**
3:     **for** $e \in 1, ..., epoch$ **do**
4:         Update local model: $w_i^{e+1} = w_i^e - \mu \widehat{\nabla}^e$;
5:         **if** $e \bmod \gamma == 0$ **then**
6:             Average model: $w_i^{e+1} = AllReduce(w_i^{e+1})$;
7:         **end if**
8:     **end for**
9: **end for**

3.1   Implementation of Hierarchical Computation

Hierarchical computation enables the synchronization phase to be fully overlapped with the local training phase and is achieved by spawning a dedicated process P_s responsible for data synchronizing. P_t and P_s are located in the same worker and they share the same rank in the distributed system. These two processes connect and communicate via shared memory, and there are typically the following variables that need to be shared.

– *status*: The variable that indicates the status of P_s. It has two states: *synchronizing* and *idling*.
– *replica*: The replica of the latest global model, which will be detailedly discussed in Sect. 3.2.
– ∇_i^a: The i-th worker's accumulated gradients when workers are performing local training.
– $\widehat{\nabla}_i^e$: The local gradients calculated by i-th worker at epoch e.
– *counter*: The integer that represents how many times has P_t trained locally.

specifically, when *status* is *synchronizing*, P_t will firstly make a replica of the global model if the *counter* equals 0, then it will accumulated the calculated gradients to the ∇_i^a, updating *replica* and finally increasing the *counter* by 1. On the other hand, P_t will activate P_s to start to synchronize and mark the *status* as *synchronizing* when the *status* is *idling*, and P_s will firstly update the global model using the global gradients which was synchronized in the last time, and then *AllReduce*ing the ∇_i^a, resetting *counter* to 0, and finally marking the *status* as *idling*. The workflow of the HPSGD algorithm is demonstrated as Algorithm 2.

In step 14, *StartSync* is a function to activate P_s to start to synchronize. It is an operation processed in P_s, which does not block the training process in P_t, so that P_t can keep training instead of idling. Furthermore, *StartSync* function also reduces the gradients staleness effects and will be explained in detail below.

Algorithm 2. Hierarchical Parallel SGD Algorithm

1: Initialization: Cluster size n. Learning rate $\mu \geq 0$. Max training $epoch$;
2: **for all** $i \in 0, ..., n-1$ **do in parallel**
3:     **for** $e \in 0, ..., epoch - 1$ **do**
4:         **if** $state == synchronizing$ **then**
5:             **if** $counter == 0$ **then**
6:                 Make replica: $r_i = clone(w_i^e)$;
7:             **end if**
8:             Update replica: $r_i = r_i - \mu\widehat{\nabla}_i^e$;
9:             Accumulate gradients: $\nabla_i^a = \nabla_i^a + \widehat{\nabla}_i^e$;
10:            Increase counter: $counter = counter + 1$;
11:        **else**
12:            Accumulate gradients: $\nabla_i^a = \nabla_i^a + \widehat{\nabla}_i^e$
13:            Mark $status$: $status = synchronizing$;
14:            Instruct P_s: $StartSync(P_s)$;
15:        **end if**
16:    **end for**
17: **end for**

3.2 Gradients Utilization

HPSGD considers stale gradients advanced instead of stale and lets workers make replicas of the current global model when starting to synchronize. When performing local training, calculated $\widehat{\nabla}_i^e$ will be applied to workers' respective replicas and be added to ∇_i^a. Finally, when the synchronization completes, accumulated ∇_i^a during the local training will be committed to the previous global model. The key concept of HPSGD's model updating algorithm is that with the increase of the dataset size, the local optimization surface modeled by a worker with a sub-dataset becomes more similar to the global optimization surface. Therefore, the local gradients of different workers on its own sub-dataset can be utilized to help optimize the global training function, which is illustrated in Fig. 3. The HPSGD's model updating function can be formulated as:

$$W_i^{e_2+1} = W_i^{e_1} - \mu\frac{\sum_{i=0}^{n}\sum_{e=e_1}^{e_2}\widehat{\nabla}_i^e}{n} \tag{1}$$

Where W_e denotes the model parameters at epoch e, n refers to the number of workers in the distributed system and e_1, e_2 denote the epoch range of local training. The equation can be further simplified to:

$$W_i^{e_1} - \mu\frac{\sum_{i=0}^{n}\sum_{e=e_1}^{e_2}\widehat{\nabla}_i^e}{n} = \frac{\sum_{i=0}^{n}W_i^{e_2}}{n} \tag{2}$$

Which suggests that HPSGD is generally a Local SGD's deformation in the form of gradients, thus the convergence of HPSGD can be proved by [16,20]. Although the formula is essentially the same for HPSGD and Local SGD, HPSGD focuses on achieving a lock-free and highly paralleled model updating algorithm with minimal influence from gradients staleness problem while

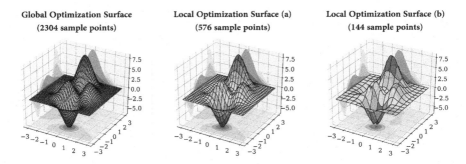

Fig. 3. Intersections on the surface represent sample points. The local optimization surface with massive reduced number of sampling points still retains similar features as the global optimization surface, including the coordinates of the extremums, the variations of partial derivatives over an interval, etc.

Local SGD specializes in synchronization rounds reducing. In addition, since the synchronization phase is overlapped with the local training phase, it is deemed impossible to synchronize and update model parameters at the same time, because the model parameters would then be written simultaneously by P_s and P_t at the next epoch. Thus the simplified model updating function Eq. 2 can not be used in the real scenario and therefore the gradients are synchronized between workers instead of model parameters to update the model with Eq. 1 as gradients are generally intermediate data in the model update process and do not need to be persisted. Furthermore, HPSGD eliminates the hyper-parameter γ. Instead, it continuously performs synchronization in another process (i.e, performs synchronizing whenever possible), making the synchronization phase highly flexible and bringing two benefits to the distributed DNN training process: 1) Improved robustness. Methods like Local SGD fix γ, assuming the synchronization time is stable throughout the training process, which is not practical in real scenarios where exceptions are unpredictable. 2) the maximal number of synchronizations, which improves the convergence rate and stability of the distributed DNN training process.

For example, assume the local training time is t_{train} seconds and synchronization time is t_{sync} seconds where $t_{sync} = k \cdot t_{train}$. In Local SGD, a complete loop that contains γ local trainings and one data synchronizing takes $((\gamma + 1) \cdot t_{train} + t_{sync})$ seconds. For the same period of time, HPSGD can synchronize for averagely $(\frac{\gamma+1}{k} + 1)$ times and train locally for $(\gamma + k + 1)$ times, suggesting that in a given fixed time, HPSGD could sample more features in optimization surfaces and perform data synchronizing more times. Consequently, HPSGD eliminates the hyper-parameter γ, while making the global model iterate faster, sample more features and achieve better accuracy. The pseudo-code of these behaviors is presented in Algorithm 3.

As the pseudo-code has shown, P_s *AllReduces* the gradients and will perform global model updating with these synchronized gradients at the next synchronization's beginning. Thus, step 1 is used for ensuring that the synchronized

Algorithm 3. P_s's behavior

1: **if** $e - counter > 0$ **then**
2:     Update global model: $w_i^{e+1} = w_i^{e-counter} - \mu \widehat{\nabla}_i^{e-counter}$
3: **end if**
4: *AllReduce* gradients: $\widehat{\nabla}_i^e = AllReduce(\nabla_i^a)$
5: Reset the counter: $counter = 0$
6: Mark the *status*: $status = idle$

gradients exist when performing the first global model updating, since the first epoch is almost certainly used for local training instead of synchronizing in practice. Consequently, the global model updating operation is always 1 synchronization delayed compared to the corresponding *AllReduce* operation due to the above updating strategy of P_s.

4 Experiments

4.1 Experimental Setup

Hardwares: An Nvidia DGX-Station is employed to set up the environment of the experiments with 4 Nvidia Tesla V100 32G GPU and Intel(R) Xeon(R) CPU E5-2698 v4 @ 2.20 GHz.

Softwares: All experiments are done in an nvidia-docker environment with CUDA 9.0.176. Pytorch 1.6.0[2] is utilized to simulate the distributed training process of the cluster by spawning multiple processes, whereas each stands for an individual worker.

Methods: SSGD, HPSGD, Local SGD [16], and purely offline training (PSGD, no communication between workers during training). PSGD will only be presented in the convergence rate comparison since it is utilized to only give a reference of fast training speed.

Models: ResNet [4], DenseNet [6], MobileNet [15] and GoogLeNet [17].

Datasets: Cifar-10 [9] dataset, which consists of $60,000$ 32×32 images in total with both RGB channels.

Other Settings: Learning rate: 0.01, batch size: 128, γ of Local SGD: 8, epoch size 100, loss function: cross entropy, optimizer: SGD.

4.2 Experiment Design and Analysis

Convergence and Training Loss: Various models have been trained with different methods in order to verify the convergence of HPSGD. Accuracy curve of 4 workers with different models and of different cluster size with the ResNet-101 model is presented in Figs. 4a and b, respectively.

---

[2] https://pytorch.org.

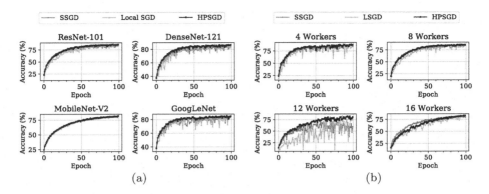

Fig. 4. (a) Accuracy comparisons of various models each with different methods. (b) The accuracy comparison with 4 workers, 8 workers, 12 workers and 16 workers.

As shown in Fig. 4a, the reached accuracy of HPSGD at epoch 100 is generally identical to the SSGD's among all experiments, suggesting HPSGD maintains the convergence of the model by utilizing local gradients to help global training function optimize. Moreover, the accuracy curve of HPSGD is significantly smoother than other methods, especially in GoogLeNet. This phenomenon is caused by the fact that the global model in HPSGD is updated by gradients that are repeatedly sampled on the sub-dataset between synchronizations, which could considerably reduce the instability of mini-batch SGD.

Furthermore, in some cases, the accuracy of HPSGD even outperforms SSGD (e.g., DenseNet-121). We believe it is mainly due to the HPSGD algorithm's characters that it is less likely to be trapped in a local optimum. Notably, as HPSGD generally lets workers independently compute their solutions and lastly applies them to the global model, it can be seen as the global model is simultaneously optimized toward multiple directions in the global optimization surface. Thus, the probability of multiple models simultaneously trapping in their local optimums is significantly reduced compared to a single model updating toward one direction.

Scale Efficiency: The scale efficiency of different methods with the ResNet-101 is shown in Fig 5a. As the figure demonstrated, both HPSGD and Local SGD have a much larger scale efficiency than SSGD when the number of workers is relatively small with significantly reduced network traffic. However, with the cluster size increase, Local SGD's scale efficiency is drastically decreased, indicating the communication jam is triggered, and its synchronization interval γ needs to be larger. However, enlarging the γ could lead to a slower convergence rate. Meanwhile, the impacts on HPSGD are considerably smaller than other methods when the number of workers increases. Specifically, HPSGD obtains the same performance compared to Local SGD and 133% more performances compared to SSGD when four workers participated in the training. When there are 16 workers, HPSGD obtains 75% more performance compared to Local SGD and

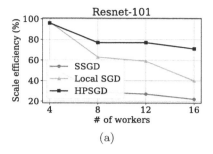
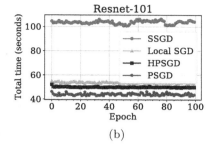

Fig. 5. (a) The scale efficiency of different cluster size with ResNet-101 model. (b) The total time of different methods training Cifar-10 with ResNet-101.

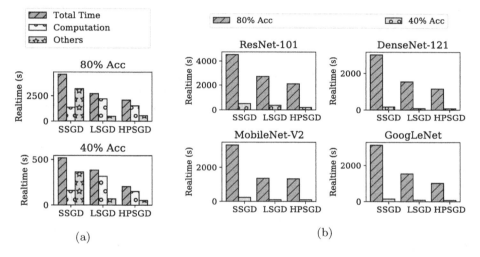

Fig. 6. a) Upper: Time breakdown for reaching 80% accuracy. Lower: Time breakdown for reaching 40% accuracy. b) Total training time used to reach 80% and 40% accuracy on different models with different methods.

250% more performance compared to SSGD. This is mainly because, theoretically, in HPSGD, there is no extra synchronizing time at all during distributed DNN training procedure as it is entirely overlapped with the local training phase. Thus, the main reason for the decreasing scale efficiency and the performance loss for HPSGD is the increasing time spent in the waiting phase, which is caused by different computational performance of workers (gray part in Fig. 1). This will be left as our future optimization direction.

Convergence Rate: Figure 5b illustrates the cost time for each epoch with different methods. Here LSGD refers to Local SGD due to tight space. PSGD serves as the lower bound of the cost time of the distributed DNN training. The time difference between HPSGD and PSGD is mainly due to the impact of limited CPU time and workers' performance difference. The breakdown of the total training time is presented in Fig. 6a. It can be shown that the computation time of SSGD and HPSGD is roughly the same when reaching either 80% accuracy or 40% accuracy, suggesting that while HPSGD shares the same converge rate as SSGD, it drastically reduce the non-computation-related time and thereby boosting the distributed training process. On the other hand, although the total time of Local SGD is shorter than SSGD when reaching either 80% accuracy or 40% accuracy, the computation time is relatively longer, indicating that the converge rate of Local SGD is lower than the SSGD and HPSGD. This phenomenon matches and verifies the explanation in Sect. 3.2. To avoid chance, we performed more experiments on total training time (wall time) of four different models with the same configuration, which is illustrated in Fig. 6b

5   Conclusions and Future Work

In this paper, we propose a novel Hierarchical Parallel SGD (HPSGD) algorithm that firstly overlaps the time-consuming synchronization phase with the local training phase by deploying hierarchical computation across two processes, which significantly boosts the distributed training. Then it alleviates the stale gradients problem by utilizing the sub-gradients calculated by different workers to help global model update. Detailedly, workers perform training on a replica of the global model independently, recording local gradients and lastly committing these gradients to the global model. In such circumstances, the sub-gradients of different workers are not stale but advanced and can be taken advantage of. Extensive experiments and comparisons turn out that the performance of HPSGD surpasses SSGD and Local SGD, which actively verifies its effectiveness and high efficiency. However, although HPSGD drastically drops the synchronization time of the distributed training process, the waiting phase remains, which is caused by workers' imbalanced performances. In future work, we would like to investigate methods capable of reducing such waiting costs and thereby further improving the scalability of the cluster in distributed training.

References

1. Cho, K., et al.: Learning phrase representations using RNN encoder-decoder for statistical machine translation. arXiv preprint arXiv:1406.1078 (2014)
2. Dean, J., Corrado, G.S., Monga, R., Kai, C., Ng, A.Y.: Large scale distributed deep networks. In: Advances in Neural Information Processing Systems (2012)
3. Haddadpour, F., Kamani, M.M., Mahdavi, M., Cadambe, V.: Local SGD with periodic averaging: tighter analysis and adaptive synchronization. In: Advances in Neural Information Processing Systems, pp. 11080–11092 (2019)
4. He, K., Zhang, X., Ren, S., Sun, J.: Deep residual learning for image recognition. In: Proceedings of the IEEE Conference on Computer Vision and Pattern Recognition, pp. 770–778 (2016)
5. Hoffer, E., Hubara, I., Soudry, D.: Train longer, generalize better: closing the generalization gap in large batch training of neural networks. In: Advances in Neural Information Processing Systems, pp. 1731–1741 (2017)

6. Huang, G., Liu, Z., Weinberger, K., van der Maaten, L.: Densely connected convolutional networks. arXiv preprint arXiv:1608.06993 (2017)
7. Jia, X., et al.: Highly scalable deep learning training system with mixed-precision: training imagenet in four minutes. arXiv preprint arXiv:1807.11205 (2018)
8. Karimireddy, S.P., Rebjock, Q., Stich, S.U., Jaggi, M.: Error feedback fixes SignSGD and other gradient compression schemes. arXiv preprint arXiv:1901.09847 (2019)
9. Krizhevsky, A., Hinton, G., et al.: Learning multiple layers of features from tiny images (2009)
10. Li, M., et al.: Scaling distributed machine learning with the parameter server. In: 11th {USENIX} Symposium on Operating Systems Design and Implementation ({OSDI 2014}), pp. 583–598 (2014)
11. Li, M., Andersen, D.G., Smola, A.J., Yu, K.: Communication efficient distributed machine learning with the parameter server. In: Advances in Neural Information Processing Systems, pp. 19–27 (2014)
12. Lian, X., Zhang, C., Zhang, H., Hsieh, C.J., Zhang, W., Liu, J.: Can decentralized algorithms outperform centralized algorithms? A case study for decentralized parallel stochastic gradient descent. In: Advances in Neural Information Processing Systems, pp. 5330–5340 (2017)
13. Lin, Y., Han, S., Mao, H., Wang, Y., Dally, W.J.: Deep gradient compression: reducing the communication bandwidth for distributed training. arXiv preprint arXiv:1712.01887 (2017)
14. Ouyang, S., Dong, D., Xu, Y., Xiao, L.: Communication optimization strategies for distributed deep learning: a survey. arXiv e-prints arXiv:2003.03009 (2020)
15. Sandler, M., Howard, A., Zhu, M., Zhmoginov, A., Chen, L.C.: MobileNetV2: inverted residuals and linear bottlenecks. In: Proceedings of the IEEE Conference on Computer Vision and Pattern Recognition, pp. 4510–4520 (2018)
16. Stich, S.U.: Local SGD converges fast and communicates little. arXiv preprint arXiv:1805.09767 (2018)
17. Szegedy, C., et al.: Going deeper with convolutions. In: Proceedings of the IEEE Conference on Computer Vision and Pattern Recognition, pp. 1–9 (2015)
18. Haddadpour, F., Kamani, M.M., Mahdavi, M., Cadambe, V.: Trading redundancy for communication: speeding up distributed SGD for non-convex optimization. In: International Conference on Machine Learning, pp. 2545–2554 (2019)
19. Xu, H., et al.: Compressed communication for distributed deep learning: survey and quantitative evaluation. Technical report (2020)
20. Yu, H., Yang, S., Zhu, S.: Parallel restarted SGD with faster convergence and less communication: Demystifying why model averaging works for deep learning. In: Proceedings of the AAAI Conference on Artificial Intelligence, vol. 33, pp. 5693–5700 (2019)
21. Yu, M., et al.: GradiVeQ: vector quantization for bandwidth-efficient gradient aggregation in distributed CNN training. In: Advances in Neural Information Processing Systems, pp. 5123–5133 (2018)
22. Zhang, W., Gupta, S., Lian, X., Liu, J.: Staleness-aware Async-SGD for distributed deep learning. arXiv preprint arXiv:1511.05950 (2015)

Implicit Posterior Sampling Reinforcement Learning for Continuous Control

Shaochen Wang and Bin Li[✉]

University of Science and Technology of China, Hefei, China
samwang@mail.ustc.edu.cn, binli@ustc.edu.cn

Abstract. Value function approximation has achieved notable success in reinforcement learning. Many popular algorithms (e.g. Deep Q Network) maintain a point estimation of the parameters in the value network or policy network. However, the frequentist perspective is prone to overfitting and lacks uncertainty representation. In this paper, we perform Bayesian analysis on the value function. Following the principle "optimism in the face of uncertainty", we conduct a posterior sampling of the value or policy network which implicitly captures the posterior distribution via a Bayesian hypernetwork. Experimental results show that the implicit posterior distribution for modeling the structural dependencies between parameters can better balance exploration and exploitation, and it is competitive to state-of-the-art methods on MuJoCo continuous benchmark.

Keywords: Reinforcement learning · Bayesian optimization · Implicit distribution

1 Introduction

Reinforcement learning can obtain flexible and powerful behaviors through trial and error just by simple rewards guidance. Recently, deep reinforcement learning with parameterized neural networks has made remarkable progress across many domains. Without experienced human experts providing training samples, the agents can play video games from raw pixels [15], dexterously manipulate robotics [13], and learn strategies to outperform humans by self-play at board games [21]. Armed with the neural networks, the agents can confront higher dimensional state spaces and tackle more complex action spaces. However, most of the achievements made by deep reinforcement learning are under a frequentist view. The value or policy networks generally keep a point estimate about the cumulative expected reward or the underlying policy, where parameters are treated as unknown constants. As more observations are collected, parameters are prone to convergent to a mode where agents present a sub-optimal behavior. Simultaneously the deterministic output is easy to overfitting and the agent is under-explored.

© Springer Nature Switzerland AG 2020
H. Yang et al. (Eds.): ICONIP 2020, LNCS 12533, pp. 452–460, 2020.
https://doi.org/10.1007/978-3-030-63833-7_38

Over the past decades, artificial intelligence plays a central role in assisting human beings for decision making. For building the safety AI and keeping driving force for exploration, maintaining uncertainty is indispensable in real life. For instance, in healthcare, the computer-aided diagnosis system should keep a certain degree of uncertainty estimate in his diagnosis to prevent over-confident judgment and misdiagnosis of some rare diseases. A principled approach to the aforementioned problems is the probabilistic motivated method. The uncertainty in reinforcement learning primarily [16] originates from two parts: parameter uncertainty and return uncertainty. The parameter uncertainty consists of exerting perturbations into the parameter space [20] or explicitly obtains a posterior distribution of the model parameters through theoretical Bayesian inference. The return uncertainty does not account for parameter uncertainty but formulates a probability distribution over the model outputs, learning a distribution over the action value function, not just the mean [1]. Treating parameters as random variables to explore in reinforcement learning is still an open question. A fully-factorized posterior approximation [3] was employed on contextual Bandits problems. Lipton et al. [14] applied a parameter-independent Bayes Q learning approach to dialogue systems. Yet neural networks exhibit strong correlations between parameters. The dependencies between parameters of the value network and policy network are crucial and complex and seem to have an impact on the final performance and learning speed.

In this paper, we are not limited to account for a simple form of the posterior distribution e.g. a unimodal posterior. We build an implicit posterior sampling method for the Q value network, which focuses on the correlations between parameters. The implicit posterior distribution is approximated by a Bayesian hypernetwork and can better balance exploration and exploitation. We evaluate our approach against state-of-the-art methods on a series of continuous control tasks and the results show that the expressive posterior distribution for the Q value network is effective for challenging tasks.

2   Background

2.1   Markov Decision Process

Classical reinforcement learning is often modeled as a Markov Decision Process (MDP), which is described as a tuple $(\mathcal{S}, \mathcal{A}, \mathcal{P}, \mathcal{R})$. $\mathcal{S}$ is the state space, $\mathcal{A}$ is the action space, $\mathcal{P}$ is the environment transition probability, and $\mathcal{R}$ is the reward function. For a robot locomotion task, the state $s \in \mathcal{S}$ contains the joint angles and velocities of the robot, and the action $a \in \mathcal{A}$ is the corresponding control torques applied to the robot. A policy $\pi(a|s)$ is a distribution over actions, which is a mapping from states to actions. An episode begins at an initial state s_0. At each time step, the agent selects an action $a_t \sim \pi(a|s)$ and acts the action. Then the environment transitions to a new state s_{t+1} according to the dynamics $p(s_{t+1}|s_t, a_t)$ and emits a reward r_t. Thus a trajectory $\tau : (s_0, a_0, r_0, s_1, ..., s_t, a_t, r_t)$ is generated. The cumulative reward function is often decomposed as $G(\tau) = \sum_{t=0}^{T} r(s_t, a_t)$. The goal is to find the optimal policy $\pi^\star$ which maximizes the expected return. A core concept is the value function, denoted as $V^\pi(s) = \mathbb{E}\left[\sum_{t=0}^{\infty} \gamma^t r(s_t, a_t) | s_0 = s, \pi\right]$, which measures the expected return from a state s following the policy π. Similarly, we can define action value function, $Q^\pi(s, a) = \mathbb{E}\left[\sum_{t=0}^{\infty} \gamma^t r(s_t, a_t) | s, a\right]$.

2.2 Deep Q Network

It is intractable for tabular reinforcement learning methods to solve continuous tasks or large-scale MDP. It is necessary to use function approximation to tackle high dimensional problems. Deep Q Network (DQN) [15] incorporates a neural network to approximate action value $Q(s, a)$, minimizing the Bellman residual in Eq. (1). θ represents the parameters of the neural network and DQN seeks to find a point estimate of the optimal action value function, $Q(s, a)$. θ^- is the parameters of the target network and periodically copies from θ. D represents the experience replay that stores the transition samples, $e = (s_t, a_t, r_t, s_{t+1})$. Samples drawn from the replay buffer are used to minimize Bellman residual. Innovatively, experience replay breaks the correlation of the sequential samples and the target network offers a stable supervision signal. The conventional exploration method in DQN is ϵ-greedy where actions with the greatest value estimate is chosen most of the time.

$$L(\theta) = \mathbb{E}_D\left[(y - Q(s, a; \theta))^2\right], \tag{1}$$

where $y = r + \gamma \max_{a'} Q(s', a'; \theta^-)$.

3 Methodology

3.1 Posterior Reinforcement Learning

Some promising actions with poorly point estimates are difficult to be selected in interaction with the environment. This inevitably brings difficulties to exploration, and more samples and interaction times are needed to obtain a relatively accurate estimate of $Q(s, a)$ value. In Vanilla DQN [15], with a naive ϵ-greedy strategy, an underestimated action may be executed after thousands of steps. In fact, DQN obtains a human-like policy after millions of frames interacting with video games. This problem can be mitigated by considering the uncertainty of the value function estimate [18]. The variance brought by the uncertainty will naturally trigger exploration and has a potential to improve policy (Fig. 1).

Posterior sampling methods are major ingredients for modeling uncertainty. In this section, we conduct a fully Bayesian analysis about the $Q(s, a)$ value, considering a probabilistic representation. Bayesian reinforcement learning can better incorporate prior knowledge and provide uncertainty for action selection. Our goal is to fit the cumulative return when an agent takes an action at the state s, and learn the conditional probability $P(Q(s, a)|\mathcal{D}; \mathcal{W})$ where $\mathcal{W}$ is the

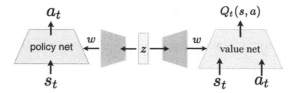

Fig. 1. Visualization of the architecture of implicit inference sampling. A sample z from the latent space is as input to the hypernetwork, and then the hypernetwork transforms it into the parameters of the policy network and value network. Finally the policy network and value network use the parameters generated by the hypernet to predict the action and action value function, $Q(s, a)$.

parameters of the model. Hence, instead of parameterizing the Q function under a specific parameter, we obtain a probabilistic expression of the Q function. We marginalize the Q function over the entire posterior distribution in Eq. (2).

$$p(Q(s,a)|\mathcal{D}) = \int P(Q(s,a)|\boldsymbol{w})P(\boldsymbol{w}|\mathcal{D})\mathrm{d}\boldsymbol{w} \tag{2}$$

The estimation is an ensemble averaging over all model parameters rather than a single point estimation. With the Bayes' rule, we can compute the exact posterior through $p(\mathbf{w}|\mathcal{D}) = p(\mathbf{w})p(\mathcal{D}|\mathbf{w})/p(\mathcal{D})$, where $p(w)$ is the prior over the model, $p(\mathcal{D})$ is the marginal likelihood. It suffers from a high dimensional integral of the denominator. As more data is collected, the beliefs over model parameters are also updated. Due to the expensive computation cost of integral, performing an exact inference is intractable. A common alternative is to use approximate inference methods such as Markov chain Monte Carlo (MCMC) or variational inference. However, MCMC still takes a long time to converge to the posterior distribution, and samples are heavily correlated [2]. In this paper, we adopt variational inference to optimize the approximate posterior distribution of the value network.

3.2   Implicit Inference for Model Parameter

In contrast with minimizing the Bellman residual, we optimize the evidence lower bound (ELBO), also referenced as variational free energy in Eq. (3). $q(\mathbf{w}|\theta)$ is the approximate posterior and $P(\mathbf{w}|\mathcal{D})$ is the true posterior of the model. The optimization objective includes two parts: the first term is to regularize the approximate posterior distribution of the weight not too far from the prior with KL divergence, the second term is a reconstruction error term which increases the log-likelihood of the optimal Q function. Moreover, a better approximate posterior expression can make the ELBO tighter, and improve the performance of the model.

$$\theta^* = \arg\min_\theta KL(q(\mathbf{w}|\theta)\|P(\mathbf{w}|\mathcal{D}))$$

$$= \arg\min_\theta \int q(\mathbf{w} \mid \theta) \log \frac{q(\mathbf{w} \mid \theta)}{P(\mathbf{w})P(Q(s,a) \mid \mathcal{D}; \mathbf{w})} \mathrm{d}\mathbf{w} \tag{3}$$

$$= \arg\min_\theta KL(q(\mathbf{w}|\theta)\|p(\mathbf{w})) - \mathbb{E}_{w \sim q(\mathbf{w}|\theta)}[\log P(Q(s,a)|\mathbf{w})]$$

Inspired by recent advances in Bayesian inference [12, 19], we utilize Bayesian hypernetworks to capture the posterior distributions of the value network and policy network. The hypernetwork [9] is a neural network that generates the model parameters of other networks. Similar to the generator in generative adversarial networks (GANs) [8], the hypernetwork implicitly approximates the posterior distribution via variational inference. Let $w = G_\theta(z)$, where w is the sample of parameters of the policy network or value network, $z \sim P_z$ which is sampled from a latent space, and G_θ is the hypernet which itself expresses the distribution. Compared with the previous Bayesian reinforcement learning framework where the candidate posterior distribution is often limited to a specific parametric family of distributions, we utilize the hypernetwork to implicitly learn the correlations between the parameters without a hand-crafted assumption.

$$- \log (Q(s,a) \mid w) = \frac{\log \sigma^2}{2} + \frac{(y - Q(s,a \mid w))^2}{2\sigma^2} + \text{constant} \tag{4}$$

The parameters of the Q function are usually optimized by minimizing the mean square Bellman error. In this paper, we minimize the negative log-likelihood of the Q function modeled as a Gaussian distribution in Eq. (4). σ is a hyperparameter to tune and $y = r + \gamma Q(s', a'; w^-)$ which is a periodically fixed network.

Since we are using an implicit distribution approximated by neural networks, an exact probability density is not available. It's not tractable to compute the analytical form of KL divergence between the posterior distribution and the prior distribution. We employ a sample-based estimation in Eq. (5) to approximate KL divergence [11] where m is the batch size of samples, d is the dimension of the policy network or value network. For prior distribution, samples are sampled from $N(0, \sigma I)$ distribution.

$$KL(q(w|\theta)\|p(w)) = \frac{d}{n} \sum_{i=1}^n \log \frac{\min_j \left\| w_q^i - w_p^j \right\|}{\min_{j \neq i} \left\| w_q^i - w_q^j \right\|} + \log \frac{m}{n-1} \tag{5}$$

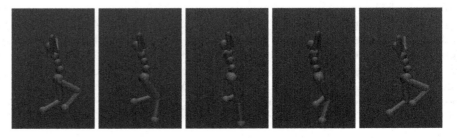

Fig. 2. Visualization after 6M training steps. The agent learns a human-like running posture.

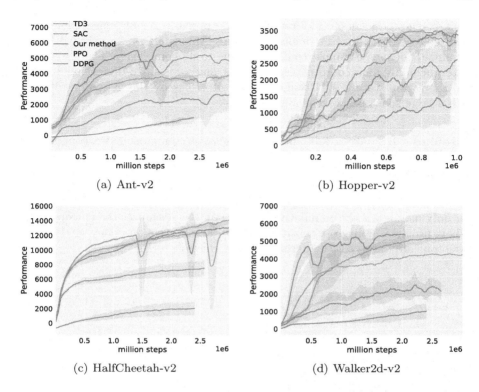

Fig. 3. Plots show the performance curves on the continuous control tasks. The solid line represents the mean and the shaded region corresponds to the variance of the returns.

4   Experiments

In this section, we evaluate our implicit posterior sampling approach against the state-of-the-art reinforcement learning algorithms on continuous control benchmarks from OpenAI Gym [4] simulated in MuJoCo physical simulator [22]. We test the algorithm on HalfCheetah-v2 (a cheetah like robot), Hopper-v2 (a single legged robot), Walker2d-v2 (a bipedal robot), Ant-v2 (a four legged robot), and Humanoid-v2 (a humanoid robot) (Fig. 2).

4.1   Task Descriptions

Locomotion Tasks. Training bipedal robotics to learn to walk has always been a persistently hot topic under automatic control and reinforcement learning [5,6]. The well-trained robots can be capable to provide various reliable assistance across many circumstances such as disaster relief. In OpenAI Gym continuous tasks, the agents learn to manipulate locomotion from raw sensory inputs. In each episode, the agents are allowed to interact with the environment up to 1000 steps.

Tasks encourage agents to go forward as far as possible. A sufficient exploration is required to avoid falling into a local optimum. Among all the legged robotics, the humanoid robot is the most challenging due to its complexity. The observation state (including joint coordinates, velocities, angles) is a 376-dimensional vector, and the action state is a 17-dimensional vector. Discovering a good strategy in such a large space is a very challenging task.

Baselines. To make a fair and comprehensive comparison, we compare our method with the current state-of-the-art methods, including deep deterministic policy gradient (DDPG) [13], a continuous variant of DQN; twin delayed deep deterministic policy gradient algorithm (TD3) [7], an algorithm that tackles the over-estimation of the Q function; soft actor critic (SAC) [10], an algorithm that not only maximizes the cumulative reward but also optimizes the entropy of the policy; proximal policy optimization (PPO), an effective on-policy method that monotonically increases the performance of the policy (Fig. 4).

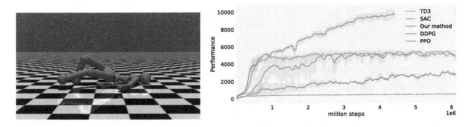

Fig. 4. The left is a humanoid agent without training. The agent does not know how to control its limbs and lies on the ground to move randomly. The right is the learning curves of the humanoid agent. After 3M training steps, under our method, the agent learned to stand and gradually started to run.

Implementation. Our algorithm shares the same architecture and hyperparameter configuration across all the experiments. For baselines, we use the hyperparameters recommended in their papers. The discounted factor γ is 0.99. Our algorithm builds on an off-policy actor critic framework [13]. The hypernet, value, and policy networks are both parameterized by feedforward neural networks. The hypernet has 2 hidden layers with 32 and 16 units. The Q value network has one hidden layer with 256 units. We use a double Q function trick to provide an accurate Q target estimation. For fitting the Q value, we utilize weighted squared Huber loss [17]. We train the BayesHypernet instead of the Q value or policy network and it is optimized by Adam with a learning rate of 3e−4. In the first 10000 time steps, we use a purely random policy to explore the environment.

4.2 Results and Analysis

We present the learning curves in Fig. 3. Each experiment is executed on 4 instances of the environment with different random seeds. The x-axis shows the interaction steps with the environment and the y-axis shows the average cumulative rewards during the training steps. The results illustrate our method is better than DDPG on all tasks and is competitive to state-of-art methods, in terms of convergent performance and learning speed. Our method is superior in the tasks of Ant and Walker2d and matches the performance in Hopper and HalfCheetah. In the humanoid task, DDPG fails to get a good policy. When the agent is hovering at the origin or marching forward slowly, the uncertain estimate brought by Bayesian inference will help Q function improve exploration. The implicit posterior can better capture the stochasticity of the environment and use the structure information of parameters to enhance exploration. At the same time, the uncertainty of each state is also updated and it provides a good trade-off between exploration and exploitation.

5   Conclusion

In this work, we propose an implicit posterior sampling method for reinforcement learning. Our method uses a Bayesian hypernetwork to inference the posterior distribution of the value network and policy network, which learns an expressive distribution. The experimental results confirm that our approach outperforms previous algorithms on the continuous benchmark, especially for solving very complex tasks, such as human gait tasks. More Bayesian inference methods can be incorporated into reinforcement learning and using the model uncertainty to explore is a promising direction for future research.

Acknowledgements. The work is partially supported by the National Natural Science Foundation of China under grand No. U19B2044 and No. 61836011.

References

1. Bellemare, M.G., Dabney, W., Munos, R.: A distributional perspective on reinforcement learning. arXiv preprint arXiv:1707.06887 (2017)
2. Bishop, C.M.: Pattern Recognition and Machine Learning (2006)
3. Blundell, C., Cornebise, J., Kavukcuoglu, K., Wierstra, D.: Weight uncertainty in neural networks. arXiv, Machine Learning (2015)
4. Brockman, G., et al.: Openai gym. arXiv preprint arXiv:1606.01540 (2016)
5. Cao, Z., Lin, C.T.: Reinforcement learning from hierarchical critics. arXiv preprint arXiv:1902.03079 (2019)
6. Cao, Z., Wong, K., Lin, C.T.: Human preference scaling with demonstrations for deep reinforcement learning. arXiv preprint arXiv:2007.12904 (2020)
7. Fujimoto, S., Hoof, H., Meger, D.: Addressing function approximation error in actor-critic methods. In: International Conference on Machine Learning, pp. 1582–1591 (2018)

8. Goodfellow, I., et al.: Generative adversarial nets. In: Advances in Neural Information Processing Systems 27, pp. 2672–2680 (2014)
9. Ha, D., Dai, A.M., Le, Q.V.: Hypernetworks. In: International Conference on Learning Representations 2017, ICLR 2017 (2017)
10. Haarnoja, T., Zhou, A., Abbeel, P., Levine, S.: arXiv preprint arXiv:1801.01290 (2018)
11. Jiang, B.: Approximate Bayesian computation with Kullback-Leibler divergence as data discrepancy. In: International Conference on Artificial Intelligence and Statistics, pp. 1711–1721 (2018)
12. Krueger, D., Huang, C.W., Islam, R., Turner, R., Lacoste, A., Courville, A.: Bayesian hypernetworks. arXiv, Machine Learning (2018)
13. Lillicrap, T.P., et al.: Continuous control with deep reinforcement learning. arXiv preprint arXiv:1509.02971 (2015)
14. Lipton, Z., Li, X., Gao, J., Li, L., Ahmed, F., Deng, L.: BBQ-networks: efficient exploration in deep reinforcement learning for task-oriented dialogue systems. In: AAAI Conference on Artificial Intelligence, AAAI 2018, pp. 5237–5244 (2018)
15. Mnih, V., et al.: Human-level control through deep reinforcement learning. Nature **518**(7540), 529–533 (2015)
16. Moerland, T., Broekens, D., Jonker, C.: Efficient exploration with double uncertain value networks. arXiv preprint arXiv:1711.10789 (2017)
17. Nachum, O., Norouzi, M., Xu, K., Schuurmans, D.: Trust-PCL: an off-policy trust region method for continuous control. In: International Conference on Learning Representations 2018, ICLR 2018 (2018)
18. Osband, I., Blundell, C., Pritzel, A., Roy, B.V.: Deep exploration via bootstrapped DQN. In: Proceedings of the 30th International Conference on Neural Information Processing Systems, NIPS 2016, pp. 4033–4041 (2016)
19. Pawlowski, N., Rajchl, M., Glocker, B.: Implicit weight uncertainty in neural networks. arXiv preprint arXiv:1711.01297 (2017)
20. Plappert, M., et al.: Parameter space noise for exploration. In: International Conference on Learning Representations 2018, ICLR 2018 (2018)
21. Silver, D., et al.: Mastering the game of go without human knowledge. Nature **550**(7676), 354–359 (2017)
22. Todorov, E., Erez, T., Tassa, Y.: Mujoco: a physics engine for model-based control. In: 2012 IEEE/RSJ International Conference on Intelligent Robots and Systems, pp. 5026–5033 (2012)

Improving Multi-view Stereo with Contextual 2D-3D Skip Connection

Liang Yang[1], Xin Wang[1], and Biao Leng[1,2,3(✉)]

[1] School of Computer Science and Engineering, Beihang University,
Beijing 100191, China
{yangliang1526,wangxin_ivy,lengbiao}@buaa.edu.cn
[2] Shenzhen Institute of BeiHang University, Shenzhen 518057, China
[3] Beijing Advanced Innovation Center for Big Data and Brain Computing,
Beihang University, Beijing 100191, China

Abstract. Learning-based methods have shown their strong competitiveness in estimating voxel for multi-view stereo. However, due to the modality gap between 2D and 3D space, the quality of the estimated 3D object is limited by the reconstruction of some detailed structures. To tackle this problem, we regard the 3D voxel reconstruction as a semantic segmentation task where skip connections between the 2D encoder and 2D decoder are usually utilized to incorporate significant contextual, aiming to segment more details. Thus, we propose an approach to improve the multi-view 3D voxel reconstruction via contextual 2D-3D skip connection. In our method, a 2D-3D skip connection branch embedded with feature visual hull is designed and plugged into the standard 2D encoder-3D decoder reconstruction architecture, which enables 2D contextual information to be effectively transmitted into the 3D domain. Then, an attention-guided module is designed to adaptively combine the transmitted features with the original 3D decoded features. Finally, a 3D RNN layer is built at the end of network to aggregate individual 3D features from different views. Extensive results have shown that the contextual information from our 2D-3D skip connections can significantly improve the reconstruction performance, especially for the detailed structures recovering.

Keywords: Deep learning · 3D reconstruction · Skip connection

1 Introduction

Multi-view stereo (MVS) aims to estimate a geometric representation from a set of images with known camera parameters. It is a fundamental issue in computer

This work is supported by Science, Technology and Innovation Commission of Shenzhen Municipality Foundation (No. JCYJ20180307123632627), the Beijing Municipal Natural Science Foundation (No. L182014), the National Natural Science Foundation of China (No. 61972014), the project of the State Key Laboratory of Software Development Environment (No. SKLSDE-2019ZX-19), and the National Key R&D Program of China (No. 2019YFB2102400).

© Springer Nature Switzerland AG 2020
H. Yang et al. (Eds.): ICONIP 2020, LNCS 12533, pp. 461–473, 2020.
https://doi.org/10.1007/978-3-030-63833-7_39

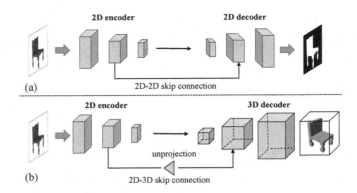

Fig. 1. Motivation. View-based 3D voxel reconstruction can be regarded as a binary segmentation task (foreground/background). Inspired by that semantic segmentation usually utilizes contextual information via feature skip connections in (a), we introduce a 2D-3D skip connection, which embeds with feature unprojection operation, to improve 3D reconstruction quality with significant 2D contextual information in (b).

vision and graphics, which has a wide range of applications in robotics, virtual reality, 3D shape recognition, *etc.* Recently, convolution neural network has been extended in estimating voxel for multi-view stereo reconstruction and get great success compared with the traditional methods. Most deep-learning-based multi-view reconstruction methods use the standard encoder-decoder structure to infer 3D structure [5,9,15]. However the standard encoder-decoder structure cannot fully utilize the shape information in 2D images and the gap between 2D and 3D space is difficult to be alleviated, which limits the reconstruction of shape details.

To resolve the problems mentioned above, we treat the general 3D voxel reconstruction as a segmentation task, segmenting foreground and background. Reviewing on 2D segmentation field [2] in Fig. 1(a), the contextual information of multi-level features is utilized to enhance the final segmentation because of the clear geometric correspondence between 2D encoded features and 2D decoded features. Similarly, for 3D reconstruction, it is believed that the contextual information of 2D encoder can provide significant complementary information for the 3D decoder to recover more subtle structures. However, due to the model difference between 2D encoded features and 3D decoded representations, there is an ill-conditioned geometric relationship between 2D-3D skip connections.

Motivated by recent visual hull based 3D reconstruction methods [8,9,15] providing an explicit photo-consistence between rendered views and 3D shapes, we employ the feature visual hull to achieve the skip connection between 2D features and 3D features, which is computationally efficient and differentiable. Figure 1(b) is a semantic description of our method. Specifically, given multiple views and their corresponding camera poses, we firstly extract certain middle image features from the 2D encoder which contains abundant part information, and such 2D features are unprojected into rough 3D features according to camera poses. Then, a set of CNNs are adopted to further refine these rough 3D features and combine them with the corresponding 3D decoded representations using an attention combination module. Finally, the 3D features from different views are fused by a 3D RNN structure and refined to be a final 3D volume. During 2D-3D skip connection, unprojection operation can offer a photo-consistence correlation between 2D space and 3D world, effectively transmitting abundant contextual information to the 3D domain. Traditional multi-view stereopsis is able to recover both objects and scenes, while our network is only suitable for objects reconstruction and the scenes reconstruction task is left for future work. Compared to the existing visual hull based method following a cascaded pipeline, our method aggregates the learned interpretation from image to shape based on network, and projective geometry based on the visual hull for 3D reconstruction in a unified end-to-end system.

In summary, our main contributions are as follows:

– We propose a novel approach leveraging significant contextual information of 2D images to improve the 3D voxel reconstruction for multi-view stereo.
– A flexible 2D-3D skip connection with unprojection and attention architectures is presented to effectively transmit multi-scale 2D encoded features to the corresponding 3D decoded representations with a feature visual hull.
– Extensive experiments prove the effectiveness of our 2D-3D skip connections and demonstrate that utilizing abundant contextual information can significantly improve the reconstruction results, especially for detailed structures recovering.

2   Related Work

Our method is related to two fields including learning-based 3D reconstruction and visual hull. Next we discuss representative work in these areas.

Learning-Based 3D Reconstruction. Due to the impressive achievements made by deep learning in image understanding [7,16] and 3D vision [11,12], deep learning methods have been widely introduced in the view-based 3D reconstruction task. For example, [6] proposed TL-embedding network to build a predictive and generative shape representation for 3D reconstruction. [19] constructed an encoder-decoder network with a projection loss that minimized the difference between the predicted silhouette and groundtruth silhouette of views. Furthermore, [14] incorporated a differentiable ray consistency term in the reconstruction scheme, leveraging different types of observations for learning 3D prediction such as depth, color images and semantics. [13] firstly applied pose prediction task to single-view reconstruction and unknown-pose natural images with annotated masks can be used for training. [10] proposed a 3D reconstruction framework based on a new 3D geometry representation which directly learn the continuous space occupancy function.

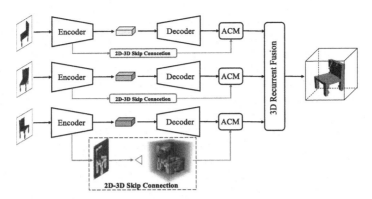

Fig. 2. An overview of the proposed method, which reconstructs 3D volume using one or more views and camera poses. The 2D-3D skip connection can transfer abundant 2D encoded features to 3D decoded representation by feature unprojection operation and Attention Combination Module (ACM).

In the multi-view setting, [5] presented a unified recurrent architecture for single-view and multi-view reconstruction, learning a mapping from observations of the object to underlying 3D shapes via RNN fashion. [3] efficiency fused 3D geometry priors and 2D texture information into a feature-augmented point cloud for multi-view reconstruction. [18] proposed a context-aware structure to adaptively fuse 3D volumes from different views and a refiner to recover more details.

Visual hull for 3D Reconstruction. Visual hull is able to provide the significant regional correspondence between 3D shape structures and projected images using pose information, which is beneficial to recover more 3D details. [9] use the projective geometry of visual hull to differentially map 2D features to 3D feature grids. SurfaceNet [8] learned photo-consistency and geometric context for dense 3D reconstruction via color visual hull. [15] conducted a 'soft' visual-hull embedding strategy for single-view 3D estimation. Compared to the above stage-wise (cascaded) visual hull based 3D reconstruction methods, our method is constructed in a simple and elegant one-stage training and testing manner.

3   Method

Our goal in this paper is to improve multi-view stereo reconstruction via making use of the contextual information of 2D images. To this end, we propose an end-to-end unified 3D reconstruction system to recover 3D shape from one or more images and their corresponding camera poses. As shown in Fig. 2, the proposed method consists of three parts, the *Main Net* which is a standard 2D encoder-3D decoder with a 3D GRU layer, the *2D-3D Skip Connection* embedding with feature visual hull to pass detailed contextual information of multiple images

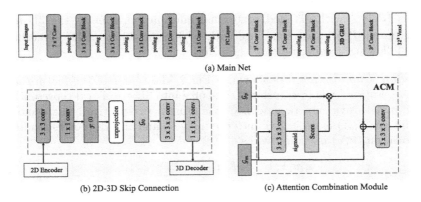

(a) Main Net

(b) 2D-3D Skip Connection

(c) Attention Combination Module

Fig. 3. The architectures of the key components in our method. We introduce 2D-3D skip connection (b) to transmit contextual 2D information to 3D space and combine (b) with (a) using an attention combination module (c).

into 3D middle representations, and the *Attention Combination Module* which aims to select complementary information from skip connections to enhance the 3D middle features from Main Net.

3.1   Main Net

The Main Net is composed of a 2D encoder and a 3D decoder. The architecture of Main Net is adapted from 3D-R2N2 [5]. The main difference is that we replace the middle LSTM layer with a fully connected layer and add a well-designed 3D GRU layer adapted from [9] to the end of 3D decoder for multi-view stereo. Figure 3(a) shows the detailed architecture of the Main Net. First, the 2D encoder encodes the input images $\{I\}_{i=1}^n$ to the latent shape vectors and the decoder decodes them to the 3D feature grids $\{\mathcal{G}_i^m\}_{i=1}^n$. In this procedure, the multi-resolution 2D middle features $\{\mathcal{M}_i\}_{i=1}^n$ and multi-level 3D middle features $\{\mathcal{G}_i^m\}_{i=1}^n$ are extracted for the parallel 2D-3D Skip Connection. Then, the 3D GRU layer fuses the multiple 3D grids $\{\mathcal{G}_i\}_{i=1}^n$ into a single grid $\mathcal{G}_f$. The fused grid $\mathcal{G}_f$ is further refined by a series of CNNs to generate the final voxel prediction $\mathcal{V}$.

3.2   2D-3D Skip Connection

The goal of 2D-3D Skip Connection is to pass the detailed contextual information of the multiple 2D images into the 3D decoder space with the help of authentic geometric correlation. Figure 3(b) shows the detailed architecture of the 2D-3D Skip Connection. Given 2D middle features $\{\mathcal{M}_i\}_{i=1}^n$ and 3D middle grid features $\{\mathcal{G}_i^m\}_{i=1}^n$, $\{\mathcal{M}_i\}_{i=1}^n$ firstly feed into convolutional layers and the learned features $\{\mathcal{F}_i\}_{i=1}^n$ are unprojected to 3D world frame with a feature visual hull method [9], which output rough feature visual hull $\{\mathcal{G}_i^p\}_{i=1}^n$. The rough unprojected 3D features $\{\mathcal{G}_i^p\}_{i=1}^n$ are further refined to be fine-grained 3D features $\{\mathcal{G}_i^{'}\}_{i=1}^n$ with detailed contextual information. Finally, the 3D middle grid features $\{\mathcal{G}_i^m\}_{i=1}^n$ are combined with $\{\mathcal{G}_i^{'}\}_{i=1}^n$ utilizing attention mechanism for the following decoder part of Main Net.

2D Feature Unprojection. As shown in Fig. 2, the target of the feature unprojection operation [9] is to transfer 2D features into 3D voxel grid representations via the perspective camera model. Considering a perspective camera model, a 3D point (X, Y, Z) is projected onto the image plane and the corresponding 2D pixel location (u, v) is computed as

$$Z[u, v, 1]^T = K[R|T] \cdot [X, Y, Z, 1]^T \tag{1}$$

where $K = \begin{bmatrix} f & 0 & u_0 \\ 0 & f & v_0 \\ 0 & 0 & 1 \end{bmatrix}$ is the camera intrinsic matrix including the focal length f and principle point (u_0, v_0). $[R|T]$ is an extrinsic camera matrix. We assume that the camera intrinsic matrix and extrinsic camera matrix are known in our method.

Given a 2D middle feature $\mathcal{M}^{(i)}$, we replicate representation $\mathcal{M}^{(i)}(\mathbf{x})$ of each pixel $\mathbf{x}$ along the viewing ray into the corresponding location in the 3D voxel grid to construct a rough feature visual hull $\mathcal{G}_0$. Specifically, we assign $\mathcal{M}^{(i)}(\mathbf{x}) = \mathcal{G}_0(\mathbf{X})$, where we project each 3D location $\mathbf{X}$ onto the image feature plane of $\mathcal{M}^{(i)}$ by perspective transformation Eq. 1 with the scale ratio of feature and original input, to obtain the corresponding pixel location $\mathbf{x}$ of feature $\mathcal{M}^{(i)}$. The feature unprojection process is differentiable and the gradients could backpropagate to 2D features along the skip connection branch. Similar to [9], we employ the bilinear sampling to sample from discrete 2D features to make the obtained feature visual hull $\mathcal{G}_0$ smooth and gradient stable.

Multi-scale Context Enhancement. In order to recover the full 3D shape, especially for some detailed structures, the network requires more contextual information at different scales. We use three different scales $\{14^2, 28^2, 56^2\}$ of 2D middle features to skip connect to three different scales $\{8^3, 16^3, 32^3\}$ of 3D middle features in our work. Since the unprojected features are very rough, if all unprojected features are connected to the final 32^3 3D voxel representation, such global 3D voxel representation could not effectively capture multi-scale contextual information.

3.3 Attention Combination Module

The 3D features from skip connection $\mathcal{G}_p$ are unprojected along the viewing ray from 2D space, while the 3D grids from Main Net $\mathcal{G}_m$ are produced by a latent vector using reshape operation. Considering that $\mathcal{G}_p$ and $\mathcal{G}_m$ have different spatial distribution, directly add or concatenate them may negatively affect the robustness of information.

Inspired by the reverse attention mechanism [4], we propose an Attention Combination Module(ACM) to select the complementary spatial information from $\mathcal{G}_p$ to enhance $\mathcal{G}_m$ in the combination stage. As shown in Fig. 3 (c), We first convert $\mathcal{G}_m$ to $\mathcal{G}'_m$ using a 3×3 convolution with a sigmoid function and get the attention scores A. It is computed as:

$$A = 1 - sigmoid(\mathcal{G}'_m) \tag{2}$$

Then, the attention features $\mathcal{G}_a$ is generated as:

$$\mathcal{G}_a = dot(A, \mathcal{G}_p) \tag{3}$$

Under the attention step, the unprojected features $\mathcal{G}_a$ and middle features $\mathcal{G}_m$ are complementary. Futhermore, we add $\mathcal{G}_a$ and $\mathcal{G}_m$ and leverage a 3×3 convolution to generate the final fused 3D features $\mathcal{G}_f$.

3.4   Network Training

We employ a one-stage training manner to train the proposed network. The Main Net and 2D-3D skip connections are jointly trained and optimized. With the help of contextual information, the whole network could converge more quickly and stable. Compared to other step-wise visual hull based method [8,15] where a 2D feature extractor is firstly trained and then a global shape visual hull is refined by 3D encoder-decoder network, our method is easy to implement and uses fewer parameters.

Objective Loss Function. Since the voxel grid representation is a binary matrix, we adopt the binary cross-entropy loss to supervise the proposed network. Let the predicted output at each voxel n be Bernoulli distributions $[1 - p(n), p(n)]$ and the corresponding ground truth label be $y(n) \in \{0, 1\}$. The objective loss is defined as

$$L = \frac{1}{N} \sum_n y(n) \cdot \log p(n) + (1 - y(n)) \cdot \log(1 - p(n)) \tag{4}$$

where N is the number of the voxel cells.

4   Experiments

In this section, we evaluate the capability of our method to reconstruct a full 3D shape with extensive experiments using the large-scale 3D shape dataset.

4.1   Dataset and Implementation Details

Dataset. We use the ShapeNet dataset [1] to generate projected images and camera poses. The ShapeNet dataset is a large 3D CAD models repository, which consists of 44k shapes and 13 major categories. We use the ShapeNet object images with clean background and camera poses released by [9] to train

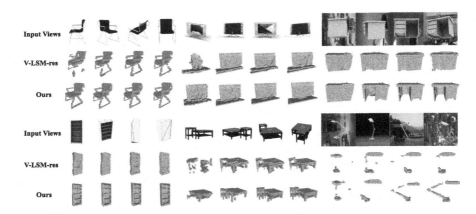

Fig. 4. Reconstruction comparison with the V-LSM-res on the test set of ShapeNet objects. The left two columns show the results with clean background and the right column shows the results with noisy background. It is shown that our method can recognize more details than V-LSM-res in each view.

Table 1. Multi-view reconstruction performance on ShapeNet dataset compared by Intersection-over-Union (IoU). Note that V-LSM-res is trained with a residual structure which is adopted in our method. Ours w/bg indicates that the proposed network was trained and tested on the images with noisy background from natural image crops.

Views	1	2	3	4
3D-R2N2	0.560	0.603	0.617	0.625
Pix2Vox	0.661	0.686	0.693	0.697
V-LSM	0.615	0.721	0.762	0.782
V-LSM-res	0.640	0.722	0.755	0.773
Ours	**0.714**	**0.790**	**0.823**	**0.841**
Ours w/bg	0.708	0.782	0.810	0.830

and test the proposed method. During the rendering process, the objects are resized to place in the unit cube centered at the origin. Each model has 20 images sampled from a viewing sphere with $\theta_{az} \in [0, 360)$ and $\theta_{el} \in [-20, 30]$ degrees and lighting variations. Following [9], we use 70% 3D models for training, 10% for validating and the remaining 20% for testing. In order to evaluate the reconstruction quality, we binarize the probability with threshold 0.4 and use the voxel Intersection-over-Union (IoU) as the metric.

Implementation Details. Our network is trained to reconstruct a full 3D voxel grid using multi-view RGB images and their corresponding camera poses. The rendered image size is $224 \times 224 \times 3$ and the output 3D voxel is $32 \times 32 \times 32$. During the training phase, the rendered images of the object and their corresponding camera pose are fed to the network as input. The whole network is end-to-end one-stage training and the Main Net does not need pre-training.

In our experiment, we employ batch size of 3 and 4 views per object to train the proposed method for 70 epochs using ADAM solver. The initial learning rate is set to $1e-4$ and is dropped by 10 at 40 epochs and 60 epochs.

4.2   Results on ShapeNet Objects Reconstruction

In this section, we evaluate our method on the ShapeNet testing set consisted of 8770 models in 13 major categories.

Baseline Setup. Our method is compared to the following baselines - V-LSM [9], a proposed system which uses the feature visual hull to perform multi-view reconstruction, V-LSM-res which is an extension of V-LSM where we modify the original V-LSM to a residual structure adopted in our method for a fair comparison. In addition, we compare our method with 3D-R2N2 and Pix2Vox which perform multi-view reconstruction but do not use camera poses.

Table 2. The reconstruction performance (per category reconstruction IoU) of 4 views on the test set of the ShapeNet dataset.

Category	V-LSM	V-LSM-res	Main Net	Ours
airplane	0.777	0.766	0.704	**0.847**
bench	0.723	0.709	0.649	**0.811**
cabinet	0.795	0.792	0.774	**0.857**
car	0.849	0.844	0.832	**0.877**
chair	0.757	0.751	0.670	**0.834**
display	0.749	0.750	0.638	**0.815**
lamp	0.698	0.686	0.532	**0.761**
speaker	0.773	0.775	0.687	**0.797**
rifle	0.832	0.821	0.681	**0.888**
sofa	0.810	0.809	0.752	**0.861**
table	0.745	0.739	0.684	**0.833**
telephone	0.859	0.851	0.830	**0.899**
watercraft	0.737	0.736	0.645	**0.798**
mean	0.777	0.771	0.698	**0.837**

Quantitative Results. Table 1 shows the mean voxel IoU (across 13 categories) for sequences of $\{1, 2, 3, 4\}$ views. The IoU increases with the number of views in all methods, while the jump is less for the methods without camera poses such as 3D-R2N2, Pix2Vox. Our method achieves the best performance in all number of views and get a boost of 5.9% IoU in 4 views compared with V-LSM. The performance of V-LSM is limited for the rough geometric correlation of unprojection operation, while our method could combine the learned interpretation from 2D image to 3D grid based on Main Net and projective geometry based on skip connection branches to enhance the final 3D output.

We also train our method using images with random crops of natural images from Pascal 3D+ [17] and the bottom row in Table 1 shows that the performance of our method with noisy background is only slightly decreased indicating the robustness of our method. Table 2 shows the performance of 4-view reconstruction in each category. Our method obtains top performance in each category and increases mean-category IoU over V-LSM by 6.0%.

Qualitative Results. Figure 4 presents some reconstruction results of our method and V-LSM-res on ShapeNet testing set. It is observed that reconstruction quality improves as the number of views increases and our method recovers more detailed structures. A typical example is the cabinet on the bottom left. The cabinet partitions are missed by V-LSM-res, while our method reliably recognizes such local structure in 3D space with the help of abundant contextual information. Moreover, the reconstruction performance is also robust to the noisy background. For example, the V-LSM-res has failed to recover the skeleton of lamp on the bottom right in 4 views, while our method can reconstruct this structure using only 3 views.

Table 3. The reconstruction performance (mean voxel IoU) in different units. Note that SC represents the proposed 2D-3D skip connection. SC w/o pose represents the proposed skip connection without camera poses and the visual hull structure is replaced by reshape operation. AVG and ACM represent two combination modules with average summation and attention mechanism.

Conf	(a)	(b)	(c)	(d)
+SC		√		√
+SC w/o pose			√	
+ACM		√	√	
+AVG				√
	ShapeNet			
1 view	0.642	0.714	0.650	0.698
2 views	0.684	0.790	0.692	0.774
3 views	0.699	0.823	0.708	0.809
4 views	0.708	0.841	0.717	0.828

4.3   Ablation Study

Table 3 compares the influence of different units in our method. We first evaluate the 2D-3D skip connection by method(a) and method(b). Method(a) is the

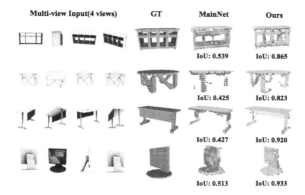

Fig. 5. Reconstruction comparison with Main Net on ShapeNet test set. The color of the voxels denotes the predicted confidence. Note that yellow refers to high confidence and red means low confidence. (Color figure online)

baseline Main Net and method(b) is the proposed method. As shown in Table 3, the baseline method(a) obtains an IoU of 70.8% in 4 views, while the method(b) with 2D-3D skip connections get a boost of 13.3% IoU compared to method(a). The significant improvement is caused by the abundant contextual information transferred by the skip branch which alleviates the gap between 2D space and 3D world. Considering that the proposed 2D-3D skip connections use visual hull to transform the space of features and adopt ACM to combine with Main Net, we further perform two other comparisons method(c) and (d) to verify the effectiveness of the visual hull and ACM. In the first experiment(c), we replace the visual hull architecture with reshape operation and the IoU of 4 views is decreased to 71.7% compared with method(b). This is because the projective geometry of visual hull can establish consistency between 2D and 3D space and help to explore more geometric structures in 3D domain. In the second experiment(d), we replace the ACM with average summation fusion. It can be seen that compared with method(d), the proposed method(b) get a improvement by 1.3% IoU of 4 views demonstrating that the ACM can adaptively select the complementary spatial information from skip connection branch and enhance the 3D decoded features.

In order to visually explore the enhancement of contextual information from 2D-3D skip connections, we visualize the estimated probability of foreground, which is shown in Fig. 5. Note that yellow denotes high confidence and red means low confidence. As we can see, the promotion of our contextual information to Main Net is dedicated to two aspects. On one hand, contextual information makes the network more confident to predict the foreground. For example, the table object in row 3, although the Main Net predicts the shape of the table, it has low confidence for the voxels on the border of the object. Our method performs higher confidence for voxels on this location, demonstrating the overall improvement for the foreground. On the other hand, contextual information helps to recover more detailed structures such as the table leg in row 2.

5　Conclusion

In this paper, we propose a novel approach to improve multi-view stereo for voxel estimation by incorporating significant contextual information of 2D image features. Moreover, a 2D-3D skip connection based on the feature visual hull is designed to effectively convey useful 2D encoded features to 3D decoded representations. Our future work is extending the proposed method to the area of scene reconstruction.

References

1. Chang, A.X., et al.: ShapeNet: an information-rich 3D model repository. arXiv preprint arXiv:1512.03012 (2015)
2. Chen, L.C., Papandreou, G., Schroff, F., Adam, H.: Rethinking atrous convolution for semantic image segmentation. arXiv preprint arXiv:1706.05587 (2017)
3. Chen, R., Han, S., Xu, J., Su, H.: Point-based multi-view stereo network. In: Proceedings of the IEEE International Conference on Computer Vision, pp. 1538–1547 (2019)
4. Chen, S., Tan, X., Wang, B., Hu, X.: Reverse attention for salient object detection. In: Proceedings of the European Conference on Computer Vision (ECCV), pp. 234–250 (2018)
5. Choy, C.B., Xu, D., Gwak, J.Y., Chen, K., Savarese, S.: 3D–R2N2: a unified approach for single and multi-view 3D object reconstruction (2016)
6. Girdhar, R., Fouhey, D.F., Rodriguez, M., Gupta, A.: Learning a predictable and generative vector representation for objects. In: Leibe, B., Matas, J., Sebe, N., Welling, M. (eds.) ECCV 2016. LNCS, vol. 9910, pp. 484–499. Springer, Cham (2016). https://doi.org/10.1007/978-3-319-46466-4_29
7. He, K., Gkioxari, G., Dollár, P., Girshick, R.: Mask R-CNN. In: Proceedings of the IEEE International Conference on Computer Vision, pp. 2961–2969 (2017)
8. Ji, M., Gall, J., Zheng, H., Liu, Y., Fang, L.: SurfaceNet: an end-to-end 3D neural network for multiview stereopsis. In: Proceedings of the IEEE International Conference on Computer Vision, pp. 2307–2315 (2017)
9. Kar, A., Häne, C., Malik, J.: Learning a multi-view stereo machine. In: Advances in Neural Information Processing Systems, pp. 365–376 (2017)
10. Mescheder, L., Oechsle, M., Niemeyer, M., Nowozin, S., Geiger, A.: Occupancy networks: learning 3D reconstruction in function space. In: Proceedings of the IEEE Conference on Computer Vision and Pattern Recognition, pp. 4460–4470 (2019)
11. Mo, K., et al.: PartNet: a large-scale benchmark for fine-grained and hierarchical part-level 3D object understanding. In: Proceedings of the IEEE Conference on Computer Vision and Pattern Recognition, pp. 909–918 (2019)
12. Qi, C.R., Liu, W., Wu, C., Su, H., Guibas, L.J.: Frustum pointnets for 3D object detection from RGB-D data. In: Proceedings of the IEEE Conference on Computer Vision and Pattern Recognition, pp. 918–927 (2018)

13. Rui, Z., Galoogahi, H.K., Wang, C., Lucey, S.: Rethinking reprojection: closing the loop for pose-aware shape reconstruction from a single image. In: IEEE International Conference on Computer Vision (2017)
14. Tulsiani, S., Zhou, T., Efros, A.A., Malik, J.: Multi-view supervision for single-view reconstruction via differentiable ray consistency (2017)
15. Wang, H., Yang, J., Liang, W., Tong, X.: Deep single-view 3D object reconstruction with visual hull embedding. arXiv preprint arXiv:1809.03451 (2018)
16. Wang, X., Girshick, R., Gupta, A., He, K.: Non-local neural networks. In: Proceedings of the IEEE Conference on Computer Vision and Pattern Recognition, pp. 7794–7803 (2018)
17. Xiang, Y., Mottaghi, R., Savarese, S.: Beyond pascal: a benchmark for 3D object detection in the wild. In: IEEE Winter Conference on Applications of Computer Vision, pp. 75–82. IEEE (2014)
18. Xie, H., Yao, H., Sun, X., Zhou, S., Zhang, S., Tong, X.: Pix2vox: context-aware 3D reconstruction from single and multi-view images. arXiv preprint arXiv:1901.11153 (2019)
19. Yan, X., Yang, J., Yumer, E., Guo, Y., Lee, H.: Perspective transformer nets: learning single-view 3D object reconstruction without 3D supervision (2016)

Improving Self-Organizing Maps
with Unsupervised Feature Extraction

Lyes Khacef[✉], Laurent Rodriguez, and Benoît Miramond

Université Côte d'Azur, CNRS, LEAT, Sophia Antipolis, France
{lyes.khacef,laurent.rodriguez,benoit.miramond}@univ-cotedazur.fr

Abstract. The Self-Organizing Map (SOM) is a brain-inspired neural model that is very promising for unsupervised learning, especially in embedded applications. However, it is unable to learn efficient prototypes when dealing with complex datasets. We propose in this work to improve the SOM performance by using extracted features instead of raw data. We conduct a comparative study on the SOM classification accuracy with unsupervised feature extraction using two different approaches: a machine learning approach with Sparse Convolutional Auto-Encoders using gradient-based learning, and a neuroscience approach with Spiking Neural Networks using Spike Timing Dependant Plasticity learning. The SOM is trained on the extracted features, then very few labeled samples are used to label the neurons with their corresponding class. We investigate the impact of the feature maps, the SOM size and the labeled subset size on the classification accuracy using the different feature extraction methods. We improve the SOM classification by +6.09% and reach state-of-the-art performance on unsupervised image classification.

Keywords: Brain-inspired computing · Self-organizing map · Unsupervised learning · Feature extraction · Sparse convolutional auto-encoders · Spiking neural networks

1 Introduction

With the fast expansion of Internet of Things (IoT) devices, a huge amount of unlabeled data is gathered everyday. While it is a big opportunity for Artificial Intelligence (AI) and Machine Learning (ML), the difficult task of labeling these data makes Deep Learning (DL) techniques slowly reaching the limits of supervised learning [5,8]. Hence, unsupervised learning is becoming one of the most important and challenging topics in ML. In this work, we use the Self-Organizing Map (SOM) proposed by Kohonen [20], an Artificial Neural Network (ANN) that is very popular in the unsupervised learning category [22]. Inspired from the cortical synaptic plasticity and its self-organization properties, the SOM is a powerful vector quantization algorithm which models the probability density function of the data into a set of prototype vectors that are represented by the neurons synaptic weights [34]. It has been shown that SOMs perform better in

ⓒ Springer Nature Switzerland AG 2020
H. Yang et al. (Eds.): ICONIP 2020, LNCS 12533, pp. 474–486, 2020.
https://doi.org/10.1007/978-3-030-63833-7_40

representing overlapping structures compared to classical clustering techniques such as partitive clustering or K-means [3].

In addition, SOMs are well suited to hardware implementation based on cellular neuromorphic architectures [15,33,37]. Thanks to a fully distributed architecture with local connectivity amongst hardware neurons, the energy-efficiency of the SOM is highly improved since there is no communication between a centralized controller and a shared memory unit, as it is the case in classical Von-Neumann architectures. Moreover, the connectivity and computational complexities of the SOM become scalable with respect to the number of neurons [33]. SOMs are used in a large range of applications [21] going from high-dimensional data analysis to more recent developments such as identification of social media trends [36], incremental change detection [28] and energy consumption minimization on sensor networks [23].

This work is an extension of the work done in [14], where we introduced the problem of post-labeled unsupervised learning: no label is available during training and representations are learned in an unsupervised fashion, then very few labels are available for assigning each representation the class it represents. The latter is called the labeling phase. In [14], we used the MNIST dataset [24] to demonstrate the potential of this unsupervised learning method on the classification problem and compared different training and labelling techniques. In order to improve the classification accuracy of the SOM and be able to work with more complex datasets, we need to extract useful features from the raw data that will then be classified with the SOM. In the context of unsupervised learning, feature extraction can be done using two different approaches: a classical "machine learning approach" using Sparse Convolutional Auto-Encoders (SCAEs), and a "neuroscience approach" using Spiking Neural Networks (SNNs). The SCAE is trained using gradient back-propagation while the SNN is trained using Spike Timing Dependant Plasticity (STDP). The goal of this work is to compare the performance of both approaches when using a SOM classifier. We also experiment a supervised Convolutional Neural Network (CNN) with the same topology for approximating the best accuracy we can expect from the feature extraction.

Section 2 describes the unsupervised feature extraction methods and details the SOM training and labeling algorithms. Then, Sect. 3 presents the implementation details of each feature extractor. Next, Sect. 4 presents the experiments and results on MNIST unsupervised classification. Finally, Sect. 5 and Sect. 6 discuss and conclude our work.

2   Related Work and Methodology

In this section, we review the related work and present the proposed methodology. We begin with the unsupervised feature extraction learning part, then how to train the SOM, and we finally explain the labeling procedure. Our first step is to extract relevant features from the raw data using unsupervised learning.

2.1 Unsupervised Feature Extraction

Sparse Convolutional Auto-Encoders (SCAEs). Introduced by Rumelhart, Hinton and Williams [35], AEs were designed to address the problem of back propagation without supervisor via taking the input data itself as the supervised label [1]. Today, AEs are typically used for dimensionality reduction or weights initialization in CNNs to improve the classification accuracy [19,26]. In this work, we want to use AEs as feature extractors with unsupervised learning. In such cases, the feature map representation of a Convolutional AE (CAE) is most of the time of a much higher dimensionality than the input image. While this feature representation seems well-suited in a supervised CNN, the overcomplete representation becomes problematic in an AE since it gives the autoencoder the possibility to simply learn the identity function by having only one weight "on" in the convolutional kernels [26]. Without any further constraints, each convolutional layer in the AE could easily learn a simple point filter that copies the input onto a feature map [19]. While this would later simplify a perfect reconstruction of the input, the CAE does not find any more suitable representation for our data. To prevent this problem, some constraints have to be applied in the CAE to increase the sparsity of the features representation.

The concept of sparsity was introduced in computational neuroscience, as sparse representations resemble the behavior of simple cells in the mammalian primary visual cortex, which is believed to have evolved to discover efficient coding strategies [31]. It has been proven that encouraging sparsity when learning the transformed representation can improve the performance of classification tasks [11]. Indeed, the overcomplete architecture of a CAE allows a larger number of hidden units in the code, but this requires that for the given input, most of hidden neurons result in very little activation [30]. In a Sparse CAE (SCAE), activations of the encoding layer need to have low values in average. Units in the hidden layers usually do not fire [4] so that the few non-zero elements represent the most salient features [30].

In order to increase the sparsity of the CAE's feature representation, several methods can be found in the literature. In [26], the authors use max-pooling to enforce the learning of plausible filters, but the filters are then fine-tuned with supervised learning for the classification. Since we do not want to use any label in the training process, we apply additional constraints in the SCAE, namely weights and activity constraints of types L2 and L1, respectively [29].

Spiking Neural Networks (SNNs). Spiking Neural Networks (SNNs) are a brain-inspired family of ANNs used for large-scale simulations in neuroscience [10] and efficient hardware implementations for embedded AI [6]. SNNs are characterized by the spike-based information coding, a computational model of the electrical impulses amongst the biological neurons. The amplitude and duration of all spikes are almost the same, so they are mainly characterized by their emission time [17]. Furthermore, spiking neurons appear to fire a spike only when they have to send an important message, which leads to the fast and extremely energy-efficient neural computation in the brain.

Moreover, SNNs have a great potential for unsupervised learning through STDP [7], a biologically plausible local learning mechanism that uses the spike-timing correlation to update the synaptic weights. Kheradpisheh et al. proposed in [17] a SNN architecture that implements convolutional and pooling layers for spike-based unsupervised feature extraction. The SNN processes image inputs as follow. The first layer of the network uses Difference of Gaussians (DoG) filters to detect contrasts in the input image. It encodes the strength of the edges in the latencies of its output spikes, i.e. the higher the contrast, the shorter the latency. On the one hand, neurons in convolutional layers detect complex features by integrating input spikes from the previous layer, and emit a spike as soon as they detect their "preferred" visual feature. A Winner-Take-All (WTA) mechanism is implemented so that the neurons that fire earlier perform the STDP learning and prevent the others from firing. Hence, more salient and frequent features tend to be learned by the network. On the other hand, neurons in the pooling layers provide translation invariance by using a temporal maximum operation, and help the network to compress the flow of visual data by propagating the first spike received from neighboring neurons in the previous layer which are selective to the same feature. However, in [17], the extracted features were classified using a supervised Support Vector Machine (SVM). In this work, we use the unsupervised SOM classifier to keep the unsupervised training from end to end.

2.2 Unsupervised Classification with Self-Organizing Maps (SOMs)

SOM Learning. The next step consists in training a SOM using the extracted features. We use a two-dimensional array of k neurons, that are randomly initialized and updated thanks to the following algorithm based on [20]:

Initialize the network as a two-dimensional array of k neurons, where each neuron n with m inputs is defined by a two-dimensional position p_n and a randomly initialized m-dimensional weight vector w_n.

for t from 0 to t_f **do**
    for every input vector v **do**
        for every neuron n in the network **do**
        Compute the afferent activity a_n from the distance d:

$$d = \|v - w_n\| \qquad (1)$$

$$a_n = e^{-\frac{d}{\alpha}} \qquad (2)$$

    end for
    Compute the winner s such that:

$$a_s = \max_{n=0}^{k-1}(a_n) \qquad (3)$$

for every neuron n in the network **do**
    Compute the neighborhood function $h_\sigma(t, n, s)$:

$$h_\sigma(t, n, s) = e^{-\frac{\|p_n - p_s\|^2}{2\sigma(t)^2}} \tag{4}$$

Update the weight w_n of the neuron n:

$$w_n = w_n + \epsilon(t) \times h_\sigma(t, n, s) \times (v - w_n) \tag{5}$$

end for
end for
Update the learning rate $\epsilon(t)$:

$$\epsilon(t) = \epsilon_i \left(\frac{\epsilon_f}{\epsilon_i}\right)^{t/t_f} \tag{6}$$

Update the width of the neighborhood $\sigma(t)$:

$$\sigma(t) = \sigma_i \left(\frac{\sigma_f}{\sigma_i}\right)^{t/t_f} \tag{7}$$

end for

It is to note that t_f is the number of epochs, i.e. the number of times the whole training dataset is presented. The α hyper-parameter is the width of the Gaussian kernel. Its value in Eq. 2 is fixed to 1 in the SOM training, but it does not have any impact in the training phase since it does not change the neuron with the maximum activity. Its value becomes critical though in the labeling process. The SOM hyper-parameters are reported in Sect. 4.

SOM Labeling. The labeling is the step between training and test where we assign each neuron the class it represents in the training dataset. We proposed in [14] a labeling algorithm based on very few labels. We randomly took a labeled subset of the training dataset, and we tried to minimize its size while keeping the best classification accuracy. Our study showed that we only need 1% of randomly taken labeled samples from the training dataset for MNIST classification. In this work, we will extend the so-called post-labeled unsupervised learning to SOM classification with features extracted by different means.

The labeling algorithm detailed in [14] can be summarized in five steps. First, we calculate the neurons activations based on the labeled input samples from the euclidean distance following Eq. 2, where v is the input vector, w_n and a_n are respectively the weights vector and the activity of the neuron n. The parameter α is the width of the Gaussian kernel that becomes a hyper-parameter for the method. Second, the Best Matching Unit (BMU), i.e. the neuron with the maximum activity is elected. Third, each neuron accumulates its normalized activation (simple division) with respect to the BMU activity in the corresponding class accumulator, and the three steps are repeated for every sample of the labeling subset. Fourth, each class accumulator is normalized over the number of samples per class. Fifth and finally, the label of each neuron is chosen according to the class accumulator that has the maximum activity. The complete GPU-based source code is available in https://github.com/lyes-khacef/GPU-SOM.

3   Implementation Details

MNIST [24] is a dataset of 70000 handwritten digits (60000 for training and 10000 for test) of 28×28 pixels. In order to compare the feature extraction performance, we use the following topologies for the two approaches: $28 \times 28 \times 1 - 64c5 - Xc5 - p5$ for the SCAE and $28 \times 28 \times 1 - 64c5 - p2 - Xc5 - p2$ for the SNN, i.e. two convolutional layers of 64 maps and X maps respectively. Each uses 5×5 kernels followed by a max-pooling layer. The reason for the different pooling mechanisms is explained in Sect. 3.3. We explore the impact of the number of features X on the classification accuracy.

3.1   CNN Training

The CNN is modeled in TensorFlow/Keras and trained with Adadelta [39] gradient-based algorithm for 100 epochs with a learning rate of 1.0. Since the goal is to estimate the maximum accuracy we can expect from each topology, the CNN is trained with the labeled training set by using 10 neurons with a Softmax activation function on top of the last pooling layer. This network is noted as CNN+MLP in the following.

3.2   SCAE Training

The SCAE is also modeled in TensorFlow/Keras and trained using Adadelta [39] gradient-based algorithm for 100 epochs with a learning rate of 1.0. However, no label is used in the training process, as the goal of the SCAE is to reconstruct the input in the output. The complete SCAE topology is $28 \times 28 \times 1 - 64c5 - Xc5 - p5 - u5 - 64d5 - 1d5$, where u stands for up-sampling and d stands for deconvolution (or transposed convolution) layers. The complete architecture is thus symetric. We add to every convolution and deconvoltion layer a weight constraint of type L2, and we add to the second convolution layer that produces the features an activity constraint of type L1. The weights and activity regularisation rates are set to 10^{-4}. Therefore, the objective function of the SCAE takes in account both the image reconstruction and the sparsity constraints.

3.3 SNN Training

The SNN is modeled in SpykeTorch [27], an open-source simulator of convolutional SNNs based on PyTorch [32]. The SNN is trained with STDP layer by layer, with a different pooling mechanism than the CNN and SCAE. Except for the number of feature maps and kernel sizes, we kept the same hyper-parameters as the original implementation of [17] that can be found on [27]. Hence, we used a pooling layer of 2×2 after each convolutional layer, with a padding of 1 before the second convolutional layer. The threshold of the neurons in the last convolutional layer were set to be infinite so that their final potentials can be measured [17]. Finally, the global pooling neurons compute the maximum potential at their corresponding receptive field and produce the features that will be used as input for the SOM. Our experimental study showed that the added padding and the pooling mechanism proposed in [27] performs better than the one used in the CNN and SCAE (i.e. no pooling and one polling layer), with a gain of 1.43% on the maximum achievable accuracy.

4 Experiments and Results

The SOM training hyper-parameters were found with a grid search: $\epsilon_i = 1.0$, $\epsilon_f = 0.01$, $\eta_i = 10.0$, $\eta_f = 0.01$, $\alpha = 1.0$ and the number of epochs is 10.

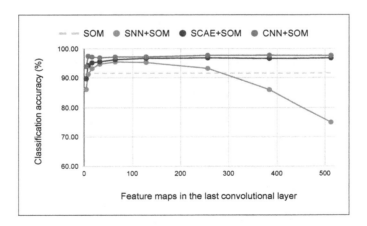

Fig. 1. SOM classification accuracy using CNN, SCAE and SNN feature extraction vs. number of feature maps with 256 SOM neurons and 10% of labels.

First, Fig. 1 shows the impact of the number of feature maps in the second convolutional layer, using 256 neurons in the SOM and 10% of labels. We deliberately use a large number of labels to avoid any bias due to the labeling performance, and focus on the impact of the feature maps. The accuracy of the CNN+SOM and SCAE+SOM is increasing with respect to the number of feature maps, reaching a maximum at 256 maps. Interestingly, the CNN+SOM performs better with 8 maps (97.56%) than with 16 (97.25%), 32 (97.00%), 64 (97.26%) or 128 (97.31%) maps. This is due to the tradeoff between additional information and additional noise induced by more feature maps according to the SOM classification. In fact, the CNN+MLP supervised baseline accuracy is increasing from 98.7% to 99% when the feature maps increase from 8 to 512. This observation is more pronounced when we look at the SNN+SOM that reaches a maximum accuracy for 64 maps then drastically decreases with more feature maps. Following the approach of [17], we used a SNN+SVM supervised baseline and its accuracy increases from 97% to 98% when the feature maps increase from 64 to 512. It means that the increasing number of feature maps for the SNN produces noisy features that do not affect the supervised classification but do decrease the unsupervised classification accuracy, because the SOM prototypes overlap and become less descriminative. Thus, we choose 256 maps for the CNN and SCAE that produce a feature size of 4096, and 64 maps for the SNN that produces feature maps of size 3136. We remark that the SNN features size is different from the CNN/SCAE features size, which is due to the to the added padding and the different pooling mechanism as explained in Sect. 3.3.

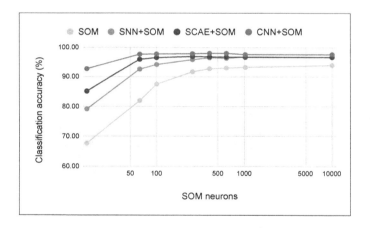

Fig. 2. SOM classification accuracy using CNN, SCAE and SNN feature extraction vs. number of SOM neurons with the optimal topologies and 10% of labels.

Second, with the above mentioned topologies, we investigate the impact of the SOM size with 10% of labels, from 16 to 10000 neurons. We see in Fig. 2 that the accuracy of the four systems is increasing with respect to the number of neurons. We notice that the SNN-SOM reaches the same accuracy as the SCAE+SOM starting from 1024 neurons. Nevertheless, for the next step of the study, it is important to keep the same number of neurons. Hence, we have chosen the number of neurons for which one of the SCAE+SOM or SNN+SOM reaches the maximum accuracy, which is equal to 256 neurons with respect to the SCAE+SOM accuracy.

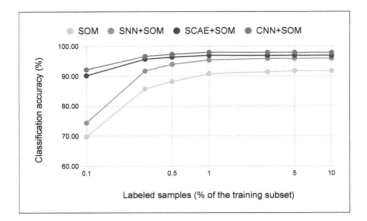

Fig. 3. SOM classification accuracy using CNN, SCAE and SNN feature extraction vs. % of labeled data from the training subset for the neurons labeling with the optimal topologies and 256 SOM neurons.

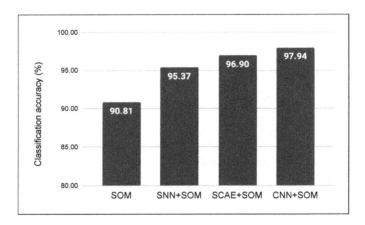

Fig. 4. SOM classification accuracy using CNN, SCAE and SNN feature extraction vs. summary of the comparative study with the optimal topologies, 256 SOM neurons and 1% of labels.

Third, using 256 neurons for the SOM, we investigate the impact of the labeling subset size in terms of % of the training set. Figure 3 shows that the accuracy increases when the labeled subset increases. Interestingly, the CNN+SOM and SCAE+SOM reach their maximum accuracy with only 1% of labeled data, while the SNN+SOM and SOM need approximately 5% of labeled data. Since the SCAE+SOM performs better than the SNN+SOM, we only need 1% of labeled data. It confirms the results obtained in [14].

Finally, the comparative study of the four settings with the best topology of each, using 256 neurons for the SOM and 1% of labeled data for the neu-

Table 1. Comparison of unsupervised feature extraction and classification techniques in terms of accuracy and hardware cost.

Feature extraction		Classification		Performance		
Model	Learning	Model	Learning	Accuracy (%)	Error (%)	Hardware cost
CNN	Supervised	MLP	Supervised	99.00	1.00	High
CNN	Supervised	SOM	Unsupervised	97.94	2.06	Medium
SCAE	Unsupervised	SOM	Unsupervised	96.90	3.10	Medium
SNN	Unsupervised	SOM	Unsupervised	95.37	4.63	Low

Table 2. MNIST unsupervised learning with AE-based feature extraction: state of the art reported from [12].

Method	Accuracy (%)
AE + K-means [2]	81.2
Sparse AE + K-means [30]	82.7
Denoising AE + K-means [38]	83.2
Variational Bayes AE + K-means [18]	83.2
SWWAE + K-means [40]	82.5
Adversarial AE [25]	95.9
Sparse CAE + SOM [Our work]	**96.9**

rons labeling is summarized in Fig. 4. As expected, the SOM without feature extraction has the worst accuracy of $90.91\% \pm 0.15$ and the CNN+SOM with supervised feature extraction reaches the best accuracy of $97.94\% \pm 0.22$. More interestingly, with fully unsupervised learning, the SCAE performs better than the SNN ($+1.53\%$), with $96.9\% \pm 0.24$ and $95.37\% \pm 0.58$ respectively.

5   Discussion

Table 1 shows the gap between supervised and unsupervised methods for feature extraction and classification. Interestingly, we only lose about 1% of accuracy when going from CNN+MLP to CNN+SOM, and another 1% when going from CNN+SOM to SCAE+SOM. The gap is slightly higher when going from SCAE+SOM to SNN+SOM, which is about 1.5%. In return, the hardware cost decreases when using SOMs and SNNs, thanks to the brain-inspired computing paradigm (distributed and local). Indeed, we showed in [13] that the SNN has a gain of approximately 50% in hardware resources and power consumption when implemented in dedicated FPGA and ASIC hardware.

Overall, the SCAE+SOM reaches the best accuracy of $96.9\% \pm 0.24$ on MNIST classification with unsupervised learning. As shown in Table 2, we achieved state of the art accuracy compared to similar works that followed an AE-based approach. The sparsity constraints of the SCAE through the weights and activities regularization significantly improved the SOM classification accuracy. Indeed, without these constraints, the CAE+SOM with the same configuration achieves an accuracy of $94.9\% \pm 0.24$, which means a loss of -2%.

A similar study was conducted in [9], but it was limited to one layer SCAE and SNN, and a supervised SVM was used for classification. The authors concluded that the SCAE reaches a better classification accuracy. Our study extands their finding to multiple convolutional layers by using unsupervised learning for both feature extraction and classification. Nevertheless, the SNN+SOM remains attractive due to the hardware-efficient computation of spiking neurons [13] associated to the cellular neuromorphic architecture of the SOM [33].

6   Conclusion and Further Works

In the context of unsupervised learning, we conducted a comparative study for unsupervised feature extraction, and concluded that the SCAE+SOM achieves a better accuracy thanks to the sparsity constraints that were applied to the SCAE through weights and activities regularization. However, the SNN+SOM remains interesting due to the hardware efficiency of spiking neurons. We achieved state of the art performance on MNIST unsupervised classification, using post-labeled unsupervised learning with the SOM. The future works will focus on using the feature extraction on more complex datasets to improve the accuracy of a multimodal unsupervised learning mechanism [16] based on SOMs.

Acknowledgment. This material is based upon work supported by the French National Research Agency (ANR) and the Swiss National Science Foundation (SNSF) through SOMA project ANR-17-CE24-0036.

References

1. Baldi, P.: Autoencoders, unsupervised learning, and deep architectures. In: Guyon, I., Dror, G., Lemaire, V., Taylor, G., Silver, D. (eds.) Proceedings of ICML Workshop on Unsupervised and Transfer Learning. Proceedings of Machine Learning Research, vol. 27, pp. 37–49. PMLR, Bellevue, 02 July 2012. http://proceedings.mlr.press/v27/baldi12a.html

2. Bengio, Y., Lamblin, P., Popovici, D., Larochelle, H.: Greedy layer-wise training of deep networks. In: Proceedings of the 19th International Conference on Neural Information Processing Systems, NIPS 2006, pp. 153–160. MIT Press, Cambridge (2006)

3. Budayan, C., Dikmen, I., Birgonul, M.T.: Comparing the performance of traditional cluster analysis, self-organizing maps and fuzzy c-means method for strategic grouping. Expert Syst. Appl. **36**(9), 11772–11781 (2009)

4. Charte, D., Charte, F., García, S., del Jesus, M.J., Herrera, F.: A practical tutorial on autoencoders for nonlinear feature fusion: taxonomy, models, software and guidelines. Inf. Fusion **44**, 78–96 (2018). https://doi.org/10.1016/j.inffus.2017.12.007. http://www.sciencedirect.com/science/article/pii/S1566253517307844

5. Chum, L., Subramanian, A., Balasubramanian, V.N., Jawahar, C.V.: Beyond supervised learning: a computer vision perspective. J. Indian Inst. Sci. **99**(2), 177–199 (2019)

6. Davies, M., et al.: Loihi: a neuromorphic manycore processor with on-chip learning. IEEE Micro **38**(1), 82–99 (2018)
7. Diehl, P., Cook, M.: Unsupervised learning of digit recognition using spike-timing-dependent plasticity. Front. Comput. Neurosci. **9**, 99 (2015). https://doi.org/10.3389/fncom.2015.00099
8. Droniou, A., Ivaldi, S., Sigaud, O.: Deep unsupervised network for multimodal perception, representation and classification. Robot. Auton. Syst. **71**, 83–98 (2015). https://doi.org/10.1016/j.robot.2014.11.005. http://www.sciencedirect.com/science/article/pii/S0921889014002474. Emerging Spatial Competences: From Machine Perception to Sensorimotor Intelligence
9. Falez, P., Tirilly, P., Bilasco, I.M., Devienne, P., Boulet, P.: Unsupervised visual feature learning with spike-timing-dependent plasticity: how far are we from traditional feature learning approaches? Pattern Recogn. **93**, 418–429 (2019). https://doi.org/10.1016/j.patcog.2019.04.016.http://www.sciencedirect.com/science/article/pii/S0031320319301621
10. Furber, S.B., Galluppi, F., Temple, S., Plana, L.A.: The spinnaker project. Proc. IEEE **102**(5), 652–665 (2014)
11. Hoyer, P.O.: Non-negative matrix factorization with sparseness constraints. J. Mach. Learn. Res. **5**, 1457–1469 (2004)
12. Ji, X., Vedaldi, A., Henriques, J.F.: Invariant information clustering for unsupervised image classification and segmentation. In: 2019 IEEE/CVF International Conference on Computer Vision (ICCV), pp. 9864–9873 (2018)
13. Khacef, L., Abderrahmane, N., Miramond, B.: Confronting machine-learning with neuroscience for neuromorphic architectures design. In: 2018 International Joint Conference on Neural Networks (IJCNN) (2018). https://doi.org/10.1109/IJCNN.2018.8489241
14. Khacef, L., Miramond, B., Barrientos, D., Upegui, A.: Self-organizing neurons: toward brain-inspired unsupervised learning. In: 2019 International Joint Conference on Neural Networks (IJCNN), pp. 1–9, July 2019. https://doi.org/10.1109/IJCNN.2019.8852098
15. Khacef, L., Girau, B., Rougier, N.P., Upegui, A., Miramond, B.: Neuromorphic hardware as a self-organizing computing system. In: IJCNN 2018 Neuromorphic Hardware in Practice and Use Workshop, Rio de Janeiro, Brazil (2018)
16. Khacef, L., Rodriguez, L., Miramond, B.: Brain-inspired self-organization with cellular neuromorphic computing for multimodal unsupervised learning (2020)
17. Kheradpisheh, S.R., Ganjtabesh, M., Thorpe, S.J., Masquelier, T.: STDP-based spiking deep convolutional neural networks for object recognition. Neural Netw. **99**, 56–67 (2018). https://doi.org/10.1016/j.neunet.2017.12.005. http://www.sciencedirect.com/science/article/pii/S0893608017302903
18. Kingma, D.P., Welling, M.: Auto-encoding variational bayes (2013)
19. Kohlbrenner, M.: Pre-training CNNs using convolutional autoencoders (2017)
20. Kohonen, T.: The self-organizing map. Proc. IEEE **78**(9), 1464–1480 (1990). https://doi.org/10.1109/5.58325
21. Kohonen, T., Oja, E., Simula, O., Visa, A., Kangas, J.: Engineering applications of the self-organizing map. Proc. IEEE **84**(10), 1358–1384 (1996). https://doi.org/10.1109/5.537105
22. Kohonen, T., Schroeder, M.R., Huang, T.S. (eds.): Self-organizing Maps, 3rd edn. Springer, Heidelberg (2001)

23. Kromes, R., Russo, A., Miramond, B., Verdier, F.: Energy consumption minimization on lorawan sensor network by using an artificial neural network based application. In: 2019 IEEE Sensors Applications Symposium (SAS), pp. 1–6, March 2019. https://doi.org/10.1109/SAS.2019.8705992
24. LeCun, Y., Cortes, C.: MNIST handwritten digit database (1998). http://yann.lecun.com/exdb/mnist/
25. Makhzani, A., Shlens, J., Jaitly, N., Goodfellow, I., Frey, B.: Adversarial autoencoders (2015)
26. Masci, J., Meier, U., Cireşan, D., Schmidhuber, J.: Stacked convolutional autoencoders for hierarchical feature extraction. In: Honkela, T., Duch, W., Girolami, M., Kaski, S. (eds.) ICANN 2011. LNCS, vol. 6791, pp. 52–59. Springer, Heidelberg (2011). https://doi.org/10.1007/978-3-642-21735-7_7
27. Mozafari, M., Ganjtabesh, M., Nowzari-Dalini, A., Masquelier, T.: Spyketorch: efficient simulation of convolutional spiking neural networks with at most one spike per neuron. Front. Neurosci. 13, 625 (2019). https://doi.org/10.3389/fnins.2019.00625
28. Nallaperuma, D., Silva, D.D., Alahakoon, D., Yu, X.: Intelligent detection of driver behavior changes for effective coordination between autonomous and human driven vehicles. In: 44th Annual Conference of the IEEE Industrial Electronics Society, IECON 2018, pp. 3120–3125 (2018)
29. Jiang, N., Rong, W., Peng, B., Nie, Y., Xiong, Z.: An empirical analysis of different sparse penalties for autoencoder in unsupervised feature learning. In: 2015 International Joint Conference on Neural Networks (IJCNN), pp. 1–8, July 2015. https://doi.org/10.1109/IJCNN.2015.7280568
30. Ng, A.: Sparse autoencoder. In: Lecture Notes CS294A. Stanford University. Stanford, CA (2011). https://web.stanford.edu/class/cs294a/sparseAutoencoder.pdf
31. Olshausen, B.A., Field, D.J.: Sparse coding with an overcomplete basis set: a strategy employed by v1? Vis. Res. 37(23), 3311–3325 (1997)
32. Paszke, A., et al.: Pytorch: an imperative style, high-performance deep learning library. In: Wallach, H., Larochelle, H., Beygelzimer, A., d' Alché-Buc, F., Fox, E., Garnett, R. (eds.) Advances in Neural Information Processing Systems 32, pp. 8024–8035. Curran Associates, Inc. (2019)
33. Rodriguez, L., Khacef, L., Miramond, B.: A distributed cellular approach of large scale SOM models for hardware implementation. In: IEEE Image Processing and Signals, Sophia-Antipolis, France (2018)
34. Rougier, N., Boniface, Y.: Dynamic self-organising map. Neurocomputing 74(11), 1840–1847 (2011). https://doi.org/10.1016/j.neucom.2010.06.034
35. Rumelhart, D.E., Hinton, G.E., Williams, R.J.: Learning Internal Representations by Error Propagation, pp. 673–695. MIT Press, Cambridge (1988)
36. Silva, D.D., et al.: Machine learning to support social media empowered patients in cancer care and cancer treatment decisions. PloS One (2018)
37. de Abreu de Sousa, M.A., Del-Moral-Hernandez, E.: An FPGA distributed implementation model for embedded SOM with on-line learning. In: 2017 International Joint Conference on Neural Networks (2017). https://doi.org/10.1109/IJCNN.2017.7966351
38. Vincent, P., Larochelle, H., Lajoie, I., Bengio, Y., Manzagol, P.A.: Stacked denoising autoencoders: learning useful representations in a deep network with a local denoising criterion. J. Mach. Learn. Res. 11, 3371–3408 (2010)
39. Zeiler, M.D.: Adadelta: an adaptive learning rate method. CoRR abs/1212.5701 (2012)
40. Zhao, J., Mathieu, M., Goroshin, R., LeCun, Y.: Stacked what-where auto-encoders (2015)

Information Security Implications of Machine-Learning-Based Automation in ITO Service Delivery – An Agency Theory Perspective

Baber Majid Bhatti[1]([⊠]) [iD], Sameera Mubarak[1] [iD], and Sev Nagalingam[2] [iD]

[1] UniSA STEM, University of South Australia, Adelaide, SA, Australia
baber.bhatti@mymail.unisa.edu.au
[2] UniSA Business, University of South Australia, Adelaide, SA, Australia

Abstract. The trend of information technology outsourcing (ITO) to service providers (SPs) is growing. SPs bring improvements through transformation projects and migrate outsourced scopes to their service delivery platforms (SDPs). For realizing economies of scales for themselves, and improving the information security and bringing efficiencies for their clients, the SPs implement machine-learning-based automation (MLA) for ITO service delivery on SDPs. However, MLA is not a silver bullet and exposes the outsourced scopes to new types of information security risks (ISRs). This paper aims at exploring those ISRs and understanding their implications. It applies agency theory to examine differing viewpoints of multiple organizations engaged in an ITO relationship. The study investigates an ITO setup of three organizations in the telecom industry. To gain insights into ISR implications, a qualitative approach was followed using a case study method and data was collected through interviews. Adversarial attack scenarios, ISRs and ISR implications on ITO service delivery are presented. To the best of our knowledge, it is the first study investigating the ISRs of MLA in ITO service delivery.

Keywords: Information technology outsourcing · Service delivery platform · Machine learning-based automation · Information security risk

1 Introduction

Businesses often delegate the delivery of their information technology (IT) services to other organizations [11]. It is a popular practice, generally known as information technology outsourcing (ITO), where client organizations outsource to one or more service providers (SPs) [16]. ITO clients naturally prefer SPs who specialize in delivering outsourced services in terms of their capacity, capability and compliance with the best practices [12]. Consequently, to meet the expectations of clients, large-scale SPs tend to establish their own service delivery platforms (SDPs), which are usually cloud-based [2,24]. It also gives SPs

© Springer Nature Switzerland AG 2020
H. Yang et al. (Eds.): ICONIP 2020, LNCS 12533, pp. 487–498, 2020.
https://doi.org/10.1007/978-3-030-63833-7_41

the advantage of specialized capability, sufficient capacity and process maturity. Since SPs invest significant effort and money in developing their SDPs, they strive to implement industry best practices and service delivery models to attract clients on their SDPs. Therefore, SPs enjoy economies of scale by migrating ITO services of multiple clients to their SDPs [17]. For clients, migrating their ITO services to SDP is an opportunity to transform their IT processes and achieve compliance with the best practices. Hence, migrating the services to SDPs is a win-win proposition for both, clients and SPs engaged in ITO relationship [23].

However, migration to SDP comes with challenges [14]. Soon the interests of both types of parties, i.e., clients and SPs, may not remain aligned due to conflicting priorities [5,9]. To understand the differing priorities that may arise, this study applies agency theory. According to this theory, when a client (principal) delegates work to the SP (agent), two potential issues may arise in this relationship: (i) conflicting priorities or goals of principal and agent, and (ii) difficulty for the principal to verify actual service delivery or performance of the agent [5]. Figure 1 presents a summary of the contradicting scenarios and their solution through MLA in ITO service delivery on SDPs. It illustrates strengths, weaknesses, opportunities and threats (SWOT) analysis for migration of ITO service delivery to SDP from clients' and SPs' perspectives. The automation of ITO service delivery on SDP using machine-learning (ML) is the popular choice and a solution to several concerns of clients and SPs [19,23].

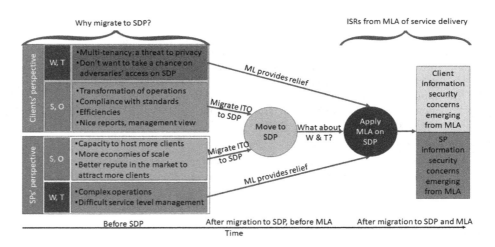

Fig. 1. Using MLA in ITO service delivery on SDPs.

While MLA in service delivery is witnessing a sharp increase, its use comes with challenges [19]. One of the most critical types of these challenges is the information security risks (ISRs) emerging from MLA in ITO service delivery [24,26]. It is contrary to the common perception that MLA is the panacea to solve challenges without adding to issues [20]. MLA is an evolving field itself, and therefore new developments are coming rapidly. Consequently, unprecedented ISR scenarios resulting from new developments in MLA often come to surface [22]. Hence, many businesses discover them as a surprise, and several others struggle to manage the problem [20]. Practitioners often look at the research literature to seek solutions to this problem. But there is insufficient coverage in the literature on information security risk management (ISRM) of MLA in ITO service delivery [18]. Hence, there is a need to identify the ISRs that result from MLA and to learn to manage them.

This research aims at understanding the information security implications of MLA in ITO service delivery and proposes solutions for the management of resulting ISRs. To achieve this aim, the following research questions are investigated in this study:

RQ1: How MLA in service delivery impacts information security in ITO?
RQ2: What are the ISRs of MLA in ITO service delivery?

This study brings visibility to the potential information security challenges of MLA evolving in the industry practising ITO. It contributes new knowledge by investigating ISRs of MLA in ITO service delivery and by proposing their solutions. To the best of our knowledge, it is the first study to apply agency theory to investigate ISRs arising from MLA in ITO service delivery. ITO clients in the industry will be helped to realize their information security risk exposures and strategize appropriate measures to manage those risks. SPs will be helped to discover information security weakness in their SDPs and to improve the reliability of their ITO service delivery.

2   Related Work

Organizations often rely on other business entities to deliver or manage their IT services [11]. Figure 2 shows a typical ITO lifecycle [6]. When the client has taken the business decision to go for ITO, it is reflected in the Business and IT strategies of the organization, and the ITO lifecycle starts with ITO strategic analysis. The clients then engage potential SPs through the procurement process. A transition project is undertaken to handover the outsourced to the SP. Upon completing the transition, SPs are fully accountable for the operations and delivery of ITO services to the clients [1,4].

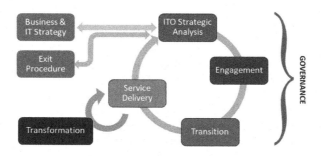

Fig. 2. Typical ITO lifecycle.

2.1 ITO Service Transformation

Migration of ITO service delivery to SDP is an example of ITO service transformation projects. Its objectives are standardization and automation of services, their consolidation, smooth integration, reduced investments, better utilization of competence, geographic independence, increased operational efficiency, and vendor and technology neutrality [4]. It is often easier to achieve these objectives by migrating ITO services to SDP of SPs than to optimize ITO services within client environments [23]. Hence, migrating to SDP of the SPs is a crucial consideration for the clients. Modern SDPs leverage technologies, for example, Artificial Intelligence (AI) through MLA [17, 19].

2.2 ITO Service Delivery Automation Using ML

The collection of ITO services on an SDP depends on its setup, strategy and design choices made by the SP [3, 7, 21]. The examples of candidate ITO services for MLA are as follows:

- **Fault management system** collects, consolidates and correlates events and alarms from clients' information systems, irrespective of ownership [3].
- **Trouble management system** manages trouble tickets from creation (based on input from fault management system), handling, escalation to closure. It enriches information from the inventory management system [3, 19].
- **Inventory management system** records and maintains a consolidated high-quality inventory of the physical and logical resources of clients [17, 21].
- **Workforce management system** creates, schedules and handles work orders for teams including subcontractors. It also issues job notifications [19].
- **Data warehouse and business intelligence systems** visualize, analyze and generate reports for operational efficiencies and contract fulfillments. They help to find bottlenecks in operational efficiencies [17, 21].
- **Performance management system** monitors and assures service performance by proactive surveillance, timely alarms' generation, troubleshooting, and planning and operations support [17].
- **Service level management system** manages service levels based on data from alarms, KPIs and KQIs, trouble ticket and inventory. Service modeling, threshold-based warnings, root-cause analysis, impact analysis and SLA monitoring are supported [3].

2.3 Lifecycle of MLA

The implementation of MLA follows a lifecycle approach [10], as shown in Fig. 3. It starts with the preparation of input data on which the MLA training is intended. In the first phase, the data is processed, cleaned and prepared in the form of (input, output: labels). In the second phase, the model (the MLA instance) is trained using this data. After training, the model is tested. In the third phase, the trained model is deployed into operational use, where it receives operational data and classifies (predicts) the output based on the training provided to it earlier. In view of the results, the lifecycle is repeated if desired [10].

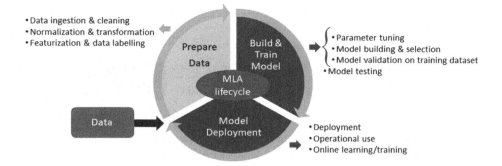

Fig. 3. MLA lifecycle.

3   Research Methodology

This study adopted a qualitative approach [8] for an in-depth investigation of the research questions posed in Sect. 1. A literature review was conducted to grasp the latest information security trends and issues in the field of MLA as they are evolving rapidly. To gain qualitative insights into first-hand experiences faced in the industry, a case study method was used [25]. Three organizations were chosen from the ICT industry to gain industry-focused knowledge, and their relationship is presented in Fig. 4. The first organization is an ITO client organization in the ICT industry, anonymized as Telco. The second organization, named as Contractor, is a large-scale multi-national SP, who undertook ITO contract from Telco. Contractor owns SDP and migrated ITO Service delivery of Telco to their SDP. They implemented MLA in ITO service delivery of their clients, including Telco. The third organization, called Subcontractor, is a medium-scale SP organization with a focused capability on telecom billing systems. After due permissions from Telco, the Contractor practised subcontracting by further outsourcing the operations of billings systems of Teleco to Subcontractor. The Contractor in this scenario had a dual role: SP to Telco, and the client to Subcontractor.

The data were collected through interviews based on a semi-structured, open-ended questionnaire [8]. This design was intended to keep discussions on track while allowing the participants to express their opinions freely. Five participants were interviewed: two from Telco, two from Contractor, and one from Subcontractor. All participants were technical experts having professional experiences in the range of fifteen to twenty years in information security, automation, ITO service delivery and machine learning. The typical duration of each interview was around one-and-a-half hours. The interviews were audio-recorded and were transcribed later. Coding was performed on the interview transcripts in NVivo tool [15]. The codes were then interrelated and aggregated to identify themes from the participants' opinions [13].

Agency theory was used to understand the underlying perspectives of organizations. This theory is relevant because it helps to understand the differing priorities of parties engaged in an ITO setup [5]. As depicted in Fig. 4, each party has its obligations and rights concerning other parties. Moreover, it has

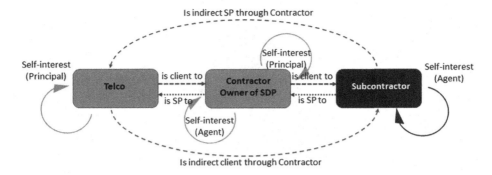

Fig. 4. Organizations in ITO relations and agency theory perspectives.

its own interest as well. The self-interests of one party are often in conflict with that of the other, which gives rise to distinct perspectives of information security risks. For example, Telco does not want to share SDP with its competitors in the market because of a possible compromise on data confidentiality. However, the Contractor wants to migrate all its clients to SDP for economies of scale. Therefore, agency theory was applied in identifying the problem, posing the research questions, formulating the interview questionnaire, interpreting the views of participants and understanding their opinions on the root-causes of ISRs of MLA in ITO service delivery.

4   Findings and Analysis

This section presents the classification of adversarial attacks on MLA in ITO, techniques used by the adversaries and the implications of resulting ISRs.

4.1   Classification of Attack-Based ISR Factors in MLA

The attack-based ISR factors of MLA in ITO can be classified as follows:

- **Data poisoning.** In this class, the training data of MLA instance (model) is incorrectly altered to cause either weak classification or misclassification.
- **Evasion.** In service operations, inputs to MLA instances are altered to mislead it, resulting in weak confidence or misclassification by the MLA instance.
- **Model inversion or extraction.** Using partial access or information about the MLA instance during service operations, the adversary prepares another MLA instance which gives similar results. Later, using new MLA instance, they generate incorrect input data to evade the results of the original MLA instance in operations.
- **Trojaning or backdooring.** The MLA instance is made to learn to misbehave on a targetted "backdoor" pattern. The MLA instance behaves normally until it finds the specific "backdoor" pattern in the input data during service operations.

4.2   Adversarial Techniques on MLA in ITO

Common adversarial techniques to attack MLA instances in SDPs are:

– **Data injection.** No access of adversary to training data and internal architecture of MLA instance, but adds new data to training dataset and corrupts training of MLA instance.
  ITO service delivery attack scenario 1. Wrong training data (input: alarm, output: action) about fault management system added by the adversary for corrupting the training of MLA instance.
– **Data modification.** No access of adversary to the internal architecture of the MLA, but they have access to training data. They poison the training data by tampering it and corrupt the training of MLA instance.
  ITO service delivery attack scenario 2. Training data (input: service level info, output: action) about service level management system incorrectly modified by the adversary for corrupting the training of MLA instance.
– **MLA instance corruption.** The adversary has access to the internal architecture of MLA instance and corrupts its internal architecture or weights.
  ITO service delivery attack scenario 3. MLA instance re-modelled to later generate false service alerts during operational use.
– **White-box attack.** The adversary has total knowledge of the MLA instance, i.e., its internal architecture, weights and training data characteristics. Using this knowledge and by contaminating the input data to the operational MLA instance, the attack during operations is substantiated.
  ITO service delivery attack scenario 4. The adversary replaces true ERP system incident alarm with a normal service event to mislead MLA instance during operations that no action is required.
– **Black-box attack type-1 (using statistical characteristics of input training data).** The adversary has access only to statistical characteristics of input training data. They train another MLA instance, which behaves similar to the original MLA when the input training data with the same statistical characteristics is used. Now, the adversary uses white-box attack strategies to prepare corrupt input samples from this new model which give similar outcomes as expected from the original model. The prepared corrupt input sample is then fed to the operational (original) MLA instance, to lead misclassification, i.e., produce wrong outputs.
  ITO service delivery attack scenario 5. The adversary has access to the statistical characteristics of training data (input: ticket, output: site access authorization), which they use to mislead MLA instance during operations to create a work order with wrong site authorization for site access.
– **Black-box attack type-2 (using the functionality of MLA instance).** This attack type is similar to previous black-box attack with the difference that the adversary has access only to the functionality of MLA instance. Exploiting this access, they generate (input, output) dataset.
  ITO service delivery attack scenario 6. The adversary misleads MLA instance during operations to incorrectly correlate a service alert to the wrong classification in the fault management system.

- **Black-box attack type-3 (restrained).** In this type of attack the adversary has access to only observe information from the operational environment on SDP to compile a factual (input, output) dataset. The remaining details remain the same as that of adaptive black-box attack.
  ITO service delivery attack scenario 7. The adversary misleads MLA instance during operations to send Telco (client)'s dashboard data to Subcontractor (cascaded SP).

The relationship of adversarial techniques with the classification of attacks is presented in Table 1. The table is sorted on adversarial techniques, based on the emphasis placed by the research participants.

Table 1. Adversarial attack techniques on MLA in ITO.

Adversarial technique	Attacking time	Attack classification	Service delivery attack scenario
Data injection	Training	Data poisoning	Attack scenario 1
Data modification	Training	Data poisoning	Attack scenario 2
MLA instance corruption	Training	Trojaning, backdooring	Attack scenario 3
White-box attack	Testing or operations	Evasion	Attack scenario 4
Black-box attack type-1	Testing or operations	Inversion leading to subsequent evasion	Attack scenario 5
Black-box attack type-2	Testing or operations	Inversion leading to subsequent evasion	Attack scenario 6
Black-box attack type-3	Testing or operations	Inversion leading to subsequent evasion	Attack scenario 7

4.3   Implications of ISRs on ITO Service Delivery

The occurrence of ISRs from MLA in ITO service delivery can have the following implications, and their implication scenarios are presented in Table 2, sorted on ISR implications from the viewpoint of research participants:

- **Reduced confidence.** The prediction confidence of MLA instance on SDP can get deteriorated. An example of this implication is the automated classification of an alarm with low confidence in the fault management system.
- **Misclassification.** A worse outcome of occurrence of ISR from MLA in ITO service delivery is when the MLA instance starts classifying incorrectly. For example, system alarms denoting an incident start getting classified as normal system events.
- **Targeted misclassification.** The outputs from MLA in ITO service delivery get converged incorrectly to a targeted classification. For example, all alarms and tickets get classified as incidents during an attack. It is misleading and beyond the handling capacity of the incident management team.

– **Source and target misclassification.** The MLA instance in this type only
impacts specific inputs and results in their misclassifications to specific outputs. For example, the SDP will behave normally, except for the incident
alarms from Customer Relationship Management (CRM) system of Telco. In
such a case, the MLA instance will create a work order for the Subcontractor's
team to conduct corrective action in the billing system.

Table 2. ISR implication scenarios of ITO service delivery systems.

ITO Service Delivery System	ISR implication scenario of adversarial attack
Fault management	Alarms incorrectly classified
Trouble management	Incident alerts from core router depicted normal status events
Inventory	Tickets enriched with incorrect data from inventory
Workforce management	Work orders issued to irrelevant teams, or irrelevant authorization issued to Subcontractor teams
Data warehouse and business intelligence	Business intelligence system predicted need for capacity expansion, but the request does not reach intended user
Performance management	During incident, service alerts from Telco's provisioning system are evaded & normal alerts generated (false positive)
Service level management	SLA breach notifications suppressed & Contractor's operations staff incorrectly assume service levels are met

5 Discussion and Future Directions

5.1 Current Constraints of ISRM in MLA of ITO Service Delivery

The current ISRM strategies face the following challenges [22] when responding
to ISRs of MLA in ITO service delivery:

– **Adversarial attacks on MLA instances.** MLA instances on SDPs are
vulnerable to adversarial attacks, which may be formulated using a variety of
techniques, for example, data poisoning, evasion or model inversion.
– **Unavailability of data sets for training and validation.** Preparing a
response to adversarial ISRs essentially requires data sets from ITO service
delivery on SDPs for the training of MLA instances and the validation of ISR
response strategies. Such data sets are not yet available for research.
– **Resource requirements.** As perceived from a game-theoretic perspective,
the security considerations in MLA pose a competition with adversaries.
Hence, the ISRM of MLA in ITO service delivery requires significant resources

of infrastructure (memory, data, processing capability) as well as skilled human resources of ITO clients and SPs. The skilled human resources in this specialized field are scarce.

- **False alarms.** In the current practice of MLA, a significant number of false alarms are generated. The processing of false alarms wastes processing resources, which may cause the MLA instance to miss an attack altogether. The refinement of false alarms is a compromise between minimizing false alarms versus the level of ISRM.

The practitioners of MLA in ITO service delivery must keep themselves updated with these current constraints of ISRM, and the researchers are encouraged to investigate ways to overcome these limitations.

5.2   Theoretical Perspective

Theories help to discover a complex and comprehensive understanding of a field [8]. They give different perspectives to the researchers to explore and analyze a concept. Game theory is the most popular choice of studies which investigate adversarial attacks or propose their solutions in the field of MLA [20]. This study employs agency theory to investigate ISRM in MLA. Agency theory is relevant in this context because it helped to approach the problem from the perspectives of ITO clients and SPs (contractors and their subcontractors). The application of agency theory helped this research unfolding the information security concerns and risks of multiple parties engaged in ITO service delivery relationship. As shown in Fig. 4, an adversary may exploit conflicting self-interests of any party. Similarly, the use of this theory can guide future studies to explore the inter-party ISRs highlighted by this study.

5.3   Limitations

This study suffers the following limitations. Firstly, this study investigated a single ITO setup comprising three organizations, i.e., Telco (client), Contractor (SP) and Subcontractor (cascaded SP). To inform the body of knowledge with experiences on diverse technologies and ITO practices employing MLA in ITO service delivery, future studies can enrich their findings by examining a variety of ITO setups. Secondly, this study conducted five interviews with expert practitioners, and the results could be limited to their knowledge and experiences. Thirdly, the use of MLA is still evolving as the technologies get mature, the success of information security measures lies in staying ahead of the adversaries. Hence, the knowledge needs to be abreast of the new ISR scenarios in MLA as they emerge. Lastly, the research on ISRM in ITO is a fast-growing research area. Although this study attempted to sufficiently cover the topic, more research is need for comprehensive coverage of information security implications of MLA in ITO service delivery.

6   Conclusion

This paper investigates the perspective of information security when using MLA in ITO service delivery. It uses a qualitative approach using a case study method to investigate the research questions and collects data through interviews based on a semi-structured, open-ended questionnaire. Agency theory is applied to gain insights into the ISR viewpoints of multiple stakeholders involved in ITO relationship. The first research question about the information security impact of MLA in ITO service delivery is answered by exploring the classification of attacks, the adversarial techniques experienced by the practitioners in the telecom industry and identifying ITO service delivery attack scenarios. The second research question is answered by analyzing the ISR impact scenarios and examining the implication scenarios of those ISR of MLA in ITO service delivery. Further research is recommended on exploring more ISR scenarios of MLA in ITO service delivery, understanding their implications and proposing mitigation to those ISRs.

References

1. ISO 37500:2014 Guidance on outsourcing. Standard, International Organization for Standardization, November 2014. https://www.iso.org/standard/56269.html
2. Ahmed Nacer, A., Godart, C., Rosinosky, G., Tari, A., Youcef, S.: Business process outsourcing to the cloud: balancing costs with security risks. Comput. Ind. **104**, 59–74 (2019). https://doi.org/10.1016/j.compind.2018.10.003
3. Al-Hawari, F., Barham, H.: A machine learning based help desk system for it service management. J. King Saud Univ. Comput. Inf. Sci. 17 (2019). https://doi.org/10.1016/j.jksuci.2019.04.001
4. Babin, R., Quayle, A.: ISO 37500 - comparing outsourcing life-cycle models. Strateg. Outsourcing Int. J. **9**(3), 271–286 (2016)
5. Bahli, B., Rivard, S.: The information technology outsourcing risk: a transaction cost and agency theory-based perspective. In: Willcocks, L.P., Lacity, M.C., Sauer, C. (eds.) Outsourcing and Offshoring Business Services, pp. 53–77. Springer, Cham (2017). https://doi.org/10.1007/978-3-319-52651-5_3
6. Bhatti, B.M., Mubarak, S., Nagalingam, S.: A framework for information security risk management in it outsourcing. In: Australasian Conference on Information Systems, December 2017
7. Chelliah, P.R., Kumar, S.A.: A cloud-based service delivery platform for effective homeland security. In: IEEE 4th International Conference on Cyber Security and Cloud Computing, pp. 157–162 (2017)
8. Creswell, J.W., Creswell, J.D.: Research Design: Qualitative, Quantitative, and Mixed Methods Approaches, 5th edn. SAGE Publications Inc., Thousand Oaks (2018)
9. Dhillon, G., Syed, R., de Sá-Soares, F.: Information security concerns in it outsourcing: identifying (in) congruence between clients and vendors. Inf. Manag. **54**(4), 452–464 (2017). https://doi.org/10.1016/j.im.2016.10.002
10. Garcia, R., Sreekanti, V., Yadwadkar, N., Crankshaw, D., Gonzalez, J.E., Hellerstein, J.M.: Context: the missing piece in the machine learning lifecycle. In: KDD CMI Workshop, vol. 114 (2017)

11. Gartner: IT Outsourcing. Report, Gartner Inc. (2017). http://www.gartner.com/it-glossary/it-outsourcing
12. González, R., Gascó, J., Llopis, J.: Information systems outsourcing reasons and risks: review and evolution. J. Glob. Inf. Technol. Manag. **19**(4), 223–249 (2016). https://doi.org/10.1080/1097198x.2016.1246932
13. Harreveld, B., Danaher, M., Lawson, C., Knight, B.A., Busch, G. (eds.): Constructing Methodology for Qualitative Research. Springer, Heidelberg (2016)
14. Hong, J.B., Nhlabatsi, A., Kim, D.S., Hussein, A., Fetais, N., Khan, K.M.: Systematic identification of threats in the cloud: a survey. Comput. Netw. **150**, 46–69 (2019). https://doi.org/10.1016/j.comnet.2018.12.009
15. Jackson, K., Bazeley, P.: Qualitative Data Analysis with NVivo. SAGE Publications Limited (2019)
16. Könning, M., Westner, M., Strahringer, S.: A systematic review of recent developments in it outsourcing research. Inf. Syst. Manag. **36**(1), 78–96 (2019). https://doi.org/10.1080/10580530.2018.1553650
17. Marcilla, J.S., de la Cámara, M., Arcilla-Cobián, M.: Do outsourcing service providers need a methodology for service delivery? Int. J. Softw. Eng. Knowl. Eng. **25**(07), 1153–1169 (2015)
18. Miller, D.J., Xiang, Z., Kesidis, G.: Adversarial learning targeting deep neural network classification: a comprehensive review of defenses against attacks. Proc. IEEE **108**(3), 402–433 (2020)
19. Sailer, A., Mahindru, R., Song, Y., Wei, X.: Using machine learning and probabilistic frameworks to enhance incident and problem management: automated ticket classification and structuring, pp. 2975–3012. IGI Global (2017)
20. National Academies of Sciences, Engineering, and Medicine: Implications of Artificial Intelligence for Cybersecurity: Proceedings of a Workshop. The National Academies Press, Washington, DC (2019). https://doi.org/10.17226/25488
21. Tambo, T., Filtenborg, J.: IT Service Management Architectures, pp. 409–421. IGI global (2019)
22. Truong, T.C., Diep, Q.B., Zelinka, I.: Artificial intelligence in the cyber domain: offense and defense. Symmetry **12**(3), 410 (2020)
23. Willcocks, L., Lacity, M., Craig, A.: Robotic process automation: strategic transformation lever for global business services? J. Inf. Technol. Teach. Cases **7**(1), 17–28 (2017). https://doi.org/10.1057/s41266-016-0016-9
24. Wulf, F., Strahringer, S., Westner, M.: Information security risks, benefits, and mitigation measures in cloud sourcing. In: 21st Conference on Business Informatics, vol. 01, pp. 258–267. IEEE (2019). https://doi.org/10.1109/CBI.2019.00036
25. Yin, R.K.: Case Study Research: Design and Methods, 5th edn. SAGE, Thousand Oaks (2014)
26. Youssef, A.E.: A framework for cloud security risk management based on the business objectives of organizations. Int. J. Adv. Comput. Sci. Appl. **10**(12), 186–194 (2020)

Key Nodes Cluster Augmented Embedding for Heterogeneous Information Networks

Hongyan Xu[1], Wenjun Wang[1,3(✉)], Hongtao Liu[1], Mengxuan Zhang[1], Qiang Tian[1], and Pengfei Jiao[2]

[1] College of Intelligence and Computing, Tianjin University, Tianjin, China
{hongyanxu,wjwang,htliu,mxzhang,tianqiang}@tju.edu.cn
[2] Center for Biosafety Research and Strategy, Law School, Tianjin University, Tianjin, China
pjiao@tju.edu.cn
[3] State Key Laboratory of Communication Content Cognition, Beijing, China

Abstract. Heterogeneous Information Networks (HINs), composed of multiple types of node and relation, usually have more expressive ability for complex relational data. Recently, network embedding aiming to project the network into a low-dimensional vector space has received much attention. Most existing embedding methods for HINs utilize meta-path to capture the proximity of node. However, these methods usually ignore the inequivalence of different types of node and clustering structure of network, which are important characteristics of HINs. Hence, we propose a key node based heterogeneous network embedding method enhanced by the clustering information. In our method, we first utilize a meta-path guided random walk to obtain general node representations in terms of rich heterogeneous semantic features in HINs. To indicate different equivalence of nodes, we define key nodes which are usually in the essential location in HINs, such as *paper* in the bibliographic HINs. Afterwards, we incorporate the clustering structure of the key nodes into network embedding learning via Gaussian Mixture Model to further enhance the representations of nodes. Lastly, we design a unified objective function to mutually learn the two parts effectively. Extensive experiments are conducted and the results validate the effectiveness of our model.

Keywords: Network embedding · HINs · Cluster

1 Introduction

With the growth of network data, network analysis, such as node classification, link prediction and clustering, has become an important research field. In order to improve the efficiency and effectiveness of network analysis task, network embedding, projecting the nodes into the low-dimensional space and preserving the network structure, has attracted more and more attention. Previous

© Springer Nature Switzerland AG 2020
H. Yang et al. (Eds.): ICONIP 2020, LNCS 12533, pp. 499–511, 2020.
https://doi.org/10.1007/978-3-030-63833-7_42

network representation methods can be divided into three categories, matrix decomposition, random walk and deep learning based. These methods usually obtain the node embeddings by preserving the local [14], global proximities or high-order [16] of networks. However, these algorithms are only designed for homogeneous networks, which usually consist of only one type of nodes and relations.

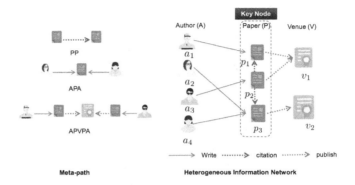

Fig. 1. An illustration of a bibliographic HIN. The left are some meta-paths in the HIN schema. The right is an example of bibliographic HIN, in which "paper" nodes are key nodes.

In contrast to homogeneous networks, heterogeneous information networks (HINs) contain multi-type nodes and relations. For example, as shown in Fig 1, a bibliographic network contains multiple types of node (paper, author, venue, etc.) and multiple types of relationship (citation, co-author, etc.) among these nodes. Similar to homogeneous network embedding, an important step in HINs embedding is to capture the proximities similarity of nodes in the same or different types. Recently, Meta-path [12], a path consisting of a sequence of relations defined between different node types, has been introduced to HINs embedding. For example, in the bibliographic networks of Fig 1, **APVPA** (i.e., "Author-Paper-Venue-Paper-Author") represents two author's papers are published in the same venue, **APA** represents co-author relationship. Some works perform meta-path based random walks to learn node embedding for HINS [2]. Especially, Metapath2vec [2] exploits meta-path based walk to sample heterogeneous neighbors, and leverages the skip-gram to learn node embeddings. In addition, there are some other methods for HINs embedding. PTE [13] decomposes heterogeneous network into three bipartite networks and performs representation learning on these subnetworks via [14].

These methods have been successfully applied to heterogeneous network representation learning. However, most existing methods ignore the inequivalence of different types of node and the inherent clustering structure, which have potential to improve the network representations. On one hand, different types of node are usually inequivalent, and there is a type of node in a relative essential position, i.e., key nodes, which should receive more attention than other types nodes. As shown in Fig 1, node *paper* links with all other types of node in the HINs schema, becoming a "bridge", thus we denote it as the `key nodes` in HINs. Hence, the different equivalence of different nodes should be exploited in network embedding learning. On the other hand, similar to the homogeneous networks, nodes of the same type in HINs usually present a clustering structure, within nodes in each cluster are more similar than nodes in other clusters. For example, the nodes (papers) are usually formed different clusters that correspond to different academic area, papers in the same area are more similar and so are their embedding vectors, although there is no direct links between them. Hence it is necessary to reflect the clustering structure for heterogeneous network embeddings. Nevertheless, most existing methods have ignored the prominent features in HINs.

To this end, we propose a Key Nodes Cluster-Augmented (KNCA) model to learn embeddings for heterogeneous information networks. Firstly, our model exploits meta-path based random walk and heterogeneous skip-gram model to learn the general embeddings for all nodes, which is capable of capturing the heterogeneous features of HINs. Secondly, we define the key nodes which are on critical location in HIN schema to highlight their important roles among different types of nodes. Meanwhile, inspired by the effective ability of Gaussian Mixture Model (GMM) in mining the cluster features of homogeneous networks, in our model we introduce GMM to further integrate the clustering structure of the key nodes into the network embedding learning phase above. Finally, we design a principle and unified objective function which can effectively jointly optimize the two phases in a mutual learning manner (i.e., meta-path random walk model and the GMM). Besides, the key nodes clustering structure can inherently affect the embeddings of other nodes with different types. We conduct extensive experiments including node classification, link prediction and visualization to evaluate our proposed method. And the results show that our KNCA can effectively improve the quality of embeddings in HINs and outperform many competitive baselines.

The main contributions of our works are: (1) we propose a key nodes based and cluster-augmented (KNCA) heterogeneous network embedding model; (2) we define key nodes and use key nodes clustering structure to enhance node representation learning; (3) Extensive experimental results prove the effectiveness of our model in various heterogeneous network mining tasks.

2 Related Work

Network representation learning or network embedding aims to automatically learn low-dimensional representation for nodes or edges in network, which is useful in a variety of applications such as node classification [14], link prediction [18], recommendation systerm [5], text classification [13]. Recently, network embedding has attracted a lot of attention and some methods have been proposed to obtain embeddings for nodes or edges [4,6–8,16,18].

Some prior works focus on homogeneous information network, which are only able to capture network structure by preserving local, global or high-order proximity of the network. For example, inspired by word2vec [10], DeepWalk [11] and node2vec [4] learn feature vector for nodes via random walk over the network and skip-gram model. LINE [14] preserve first-order and second-order neighbor structure of nodes. Besides, SDNE [16] uses Deep Neural Network (DNN) to preserve network properties.

However, different from homogeneous network, heterogeneous information networks have multiple types of nodes and relations. Some existing methods focus on heterogeneous network embedding and have achieved expected performance in a various applications [2,3,13]. Most of methods for HIN embedding are based on meta-path shema [12]. For example, Metapath2vec [2] exploit meta-path based walk to sample heterogeneous neighbors. In addition, there are some other methods for HINs embedding. PTE [13] decomposes heterogeneous network into three bipartite networks and learns representation separately. SHINE [17] uses autoencoder to extract three kinds of embedding, and gets the final embeddings by aggregating these embeddings. However, the above-mentioned methods ignore the inequivalence of different types of nodes and the clustering structure in HINs which can improve embedding model [1]. In this paper, we propose a key nodes cluster-augmented (KNCA) heterogeneous network embedding model, which learns node embeddings by meta-path guided random walk and integrating the clustering structure of key nodes into embedding learning.

3   Proposed Method

In this section, we present proposed model KNCA, a novel key nodes based and cluster-augmented HIN embedding method. As shown in Fig 2, a meta-path guided random walk model is utilized to capture the rich semantic relations in terms of heterogeneous features. Secondly, we learn node embeddings with the constrains of clustering structure of the key nodes via Gaussian Mixture Model (GMM). Finally, we design a principle and unified objective function to jointly optimize the two modules effectively.

3.1   Problem Definition

We first introduce the following definitions in our paper.

Definition 1. **Heterogeneous Information Network** is defined as $G = (V, E, A, R)$, where V and E denote the sets of nodes and edges respectively. Each node v and each edge e is associated with a specific type with mapping functions $\phi(v) \rightarrow A$ and $\varphi(e) \rightarrow R$, where A and R denote the sets of node and edge types respectively. In heterogeneous networks, $|A| + |R| > 2$.

Definition 2. **Meta-Path** $\mathcal{P}$ of length l is defined as a sequence of node types A_i and edge types R_i in the form of $A_1 \xrightarrow{R_1} A_2 \xrightarrow{R_2} \cdots A_t \xrightarrow{R_t} A_{t+1} \cdots \xrightarrow{R_{l-1}} A_l$,

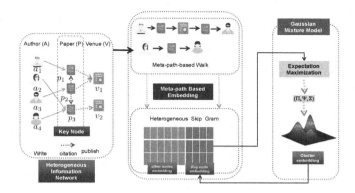

Fig. 2. Overview of the KNCA framework.

where $A_i \in A$, $R_i \in R$, $R_1 \circ R_2 \circ \cdots \circ R_{l-1}$ represents a composite relations between node types A_1, A_l.

Definition 3. **Key Nodes** In a HIN G with multiple node types, key nodes H are defined as a type of node which have maximum $out\_degree(H) + in\_degree(H)$ in heterogeneous network schema. Usually, Key nodes belong to the most critical node type, such as paper nodes, linking all other types of node, are key nodes in heterogeneous citation network schema.

Definition 4. **Cluster Embedding** With the assumption that there are K clusters for key nodes, the embedding of the cluster k in d-dimensional space is a multivariate Gaussian distribution $\mathcal{N}(\psi_k, \sum_k)$, where $k \in 1, \ldots, K$, $\psi_k \in \mathbb{R}^d$ is mean vector and $\sum_k \in \mathbb{R}^{d \times d}$ is covariance matrix.

3.2   Meta-path Based Embedding for HINs

Meta-path-Based Walk. In heterogeneous networks, as demonstrated by [12], random walks are biased to highly visible nodes and concentrated nodes. To avoid this problem, we use meta-paths to guide random walks for HINs since the meta-path can capture semantic and structural correlations between different types of node. Specifically, given a meta-path scheme $A_1 \xrightarrow{R_1} A_2 \xrightarrow{R_2} \cdots A_t \xrightarrow{R_t} A_{t+1} \cdots \xrightarrow{R_{l-1}} A_l$ on the HIN, the transition probability of walk at step i is defined as:

$$p(v^{i+1}|v_t^i, \mathcal{P}) = \begin{cases} \frac{1}{|N_{t+1}(v_t^i)|}, & (v^{i+1}, v_t^i) \in E, \phi(v^{i+1}) = t + 1 , \\ 0, & (v^{i+1}, v_t^i) \in E, \phi(v^{i+1}) \neq t + 1 , \\ 0, & (v^{i+1}, v_t^i) \notin E , \end{cases} \quad (1)$$

where $v_t^i \in A_t$, $v^{i+1} \in A_{t+1}$ and $N_{t+1}(v_t^i)$ denotes v_t^i's neighborhood with the A_{t+1} type of node, which means $v^{i+1} \in A_{t+1}$. We generate nodes sequences guided by meta-path $\mathcal{P}$. For instance, if we get an author node, we need to sample paper nodes in next step under the meta-path schema "Author-Paper-Author". This strategy ensures that we can obtain such node sequences that

contains the underlying semantic features in HINs instead of randomly sampling
arbitrary nodes.

Heterogeneous Skip Gram. Inspired by [11] and [2], we adopt heterogeneous
skip gram model to learn the representation for all nodes. Given a collection of
meta-path-based walks $W_{\mathcal{P}}$, the loss for maximizing the probability of neighbor
nodes is defined as:

$$\mathcal{O}_1(X) = - \sum_{w \in W_{\mathcal{P}}} \sum_{v \in w} \sum_{p_u \in [p_v - \tau, p_v + \tau], p_u \neq p_v} log p(u|v, \mathcal{P}), \qquad (2)$$

where $X = \{X_i\}$, τ is the context window size of node v, p_v/p_u indicates the
position of v/u in walk w. The conditional probability of reaching context node
u given a node v is defined as $p(u|v, \mathcal{P}) = \frac{exp(X_u X_v)}{\sum_{u' \in C_{\mathcal{P}}} X_v X_{u'}}$, where $X_u, X_v \in \mathbb{R}^d$ is
the vector representation of u and v. $C_{\mathcal{P}}$ denotes the set of all nodes in corpus
$W_{\mathcal{P}}$. In real networks, there are millions of nodes in corpus, the skip cost is
too expensive. Following PTE [13], we propose to apply heterogeneous negative
sampling approach, in which the sampling distribution is specified by targeted
node u's type:

$$log p(u|v, \mathcal{P}) \approx log \sigma(X_u X_v) + \sum_{i=1}^{n} \mathbb{E}_{u_i' \sim p(u')} [log \sigma(-X_v X_{u_i'})] , \qquad (3)$$

where u_i' is negative node sampled from a given noise distribution $p(u')$ on $C_{\mathcal{P}}$,
and n negative nodes are sampled for each positive node v.

3.3 Key Nodes Cluster Augmentation

Clustering structure can be used to optimize the node embedding by introducing
a high-order proximity under a mesoscopic perspective. We first exploit the key
nodes clustering structure to enhance the node representation, since the key
nodes with the maximum degree in HIN schema can indirectly influence other
node embeddings. We suppose 1) each key node is assigned with a specific cluster.
2) each key node v_h's embedding X_h are generated by a multivariate Gaussian
distribution. For all key nodes in H, the likelihood is expressed as:

$$\prod_{h=1}^{|H|} \sum_{k=1}^{K} p(z_h = k) p(v_h|z_h = k; X_h, \psi_k, \Sigma_k), \qquad (4)$$

where $p(z_h = k)$ is the probability that node v_h belongs to k-th cluster, repre-
sented as π_{hk} later and $\sum_{k=1}^{K} \pi_{hk} = 1$. In addition, $p(v_h|z_h = k; X_h, \psi_k, \Sigma_k)$ is
a multivariate Gaussian distribution that can be defined as $\mathcal{N}(X_h|\psi_k, \Sigma_k)$.

Suppose that we have got model parameters π_{hk} and the cluster embedding
$\mathcal{N}(\psi_k, \sum_k)$ by Eq. 4. And key node embedding X_v as unknown, we can re-use

Eq. 4 to optimize key node embeddings by setting key nodes embedding X_v as unknown. In this way, we will make nodes within the same cluster closer to the corresponding cluster center ψ_k. In other words, Nodes belonging to the same cluster have similar embedding. Based on Eq. 4, We define the objective function for key nodes cluster embedding as:

$$\mathcal{O}_2(X', \Pi, \Psi, \Sigma) = \sum_{h=1}^{|H|} log \sum_{k=1}^{K} \pi_{hk} \mathcal{N}(X_h | \psi_k, \Sigma_k), \tag{5}$$

where $X' = \{X_h\}$, $\Pi = \{\pi_{hk}\}$, $\Psi = \{\psi_k\}$ and $\Sigma = \{\Sigma_k\}$. When a gaussian component collapsed to a sample point, there exists the singularity issue that $diag(\Sigma_k)$ will become zero and further O_2 will become negative infinity. Inspired by ComE [1] we constrain $diag(\Sigma_k) > 0$.

3.4   A Unified Optimization Method

Finally, we propose to jointly optimize the two modules by minimizing the following combined loss function:

$$\mathcal{L} = O_1(X) + \alpha O_2(X', \Pi, \Psi, \Sigma), \tag{6}$$

where $\alpha \geq 0$ is a trade-off hyper parameter.

We adopt the iterative optimization strategy to train KNCA. In each iteration, we alternate the training between the meta-path based embedding and key nodes cluster-augmented module. Specifically, we first fix node embedding X' and update clustering embedding parameters (Π, Ψ, Σ) by EM algorithm and thus improve key nodes cluster-augmented module. Next, we fix (Π, Ψ, Σ), and optimize key node embedding X' by Stochastic Gradient Descent (SGD) algorithm. We repeat the above process until the maximum iteration.

4   Experiments

In this section, we evaluate the performance of our model in three important heterogeneous network mining tasks: multi-class node classification, link prediction and visualization.

4.1   Experimental Setup

Dataset. To evaluate our model, we use three datasets from different domains. 1) AMiner(CS) [15]: a bibliographic network in computer science, which contains 9,323,739 authors and 3,194,405 papers from 3,883 computer science venues. 2) DBLP: an academic paper citation network extracted from DBLP, including 14,328 papers (P), 4,057 authors (A), 20 conferences (C), and 8,789 keywords (T). In addition, in DBLP, four research fields are used as labels, if the author has a publication at a venue in that field, a label corresponding to that field is assigned to the author for author classification. The meta-path set {APA, APCPA, APTPA} is used for experiments. 3) IMDB: a movie network based on movie reviews. It includes 4,780 films (M), 5,841 actors (A) and 2,269 directors (D). In IMDB, movies are divided into three types: action, comedy and drama. The meta-path set MAM, MDM is used for experiments.

Baselines. We evaluate KNCA against four competitive embedding methods. Among them, DeepWalk and LINE are designed for homogeneous information networks, PTE and Metapath2vec are designed for heterogeneous networks. DeepWalk [11]: To learn node embeddings, DeepWalk gets node sequences by random walk on network and uses skipgram to get d-dimensional node vectors. LINE [14]: To preserve network properties, LINE considers first-order and second-order neighborhood proximity of node. PTE [13]: PTE decomposes heterogeneous network into three bipartite networks (paper-author, paper-venue, author-venue) according to edge types like PTE, and performs representation learning on these subnetworks. Metapath2vec [2]: Based on meta-path guided random walk, Metapath2vec uses skip gram to capture node similarity within a fixed window size.

Parameter Settings. We set the dimension of the node vector X 128 for all models. For the random walk based method, we set the window size to 5, the walk length to 100, and the length of walk 100. For model optimized by negative sampling, we set the number of negative samples n to 5. We use SGD to optimize training and set the initial learning rate 0.025. For the proposed KNCA, set $\alpha = 0.3$, the number of cluster $K = 7$ for all the experiments. We use meta-path "APA" and "$APVPA$" to guide random walk, in which "APA" represent coauthor semantics and "$APVPA$" means that two author publish papers at the same venue. Our experiment results show that node embeddings learned through these meta-paths perform well in a variety of tasks.

4.2   Multi-label Classification

In this section, we evaluate the model on the multi-label classification task. After learning node representations, we assign labels to training nodes and use their embeddings as input to train a logistic regression classifier. We adopt micro-f1 score and macro-f1 score as evaluation metric.

In AMiner (CS) dataset, 8 field labels of venue are extracted from Google Scholar Metrics (Computational Linguistics, Computer Graphics, Computer Networks & Wireless Communication, Computer Vision & Pattern Recognition, Computing Systems, Databases & Information Systems, Human Computer Interaction, and Theoretical Computer Science). After matching, 133 venues are labeled. 282361 papers are labeled from the venue in which paper was submitted. 246,678 authors' labels are assigned to the label with the majority of his / her publications. In DBLP dataset, We extracted a total of 14,328 papers from 20 venues in 4 research fields (database, data mining, machine learning, information retrieval) from DBLP knowledge base. 4057 authors were labeled according to four research fields, if the author has a publication at a conference in that field. In the IMDB dataset, we extracted 4,780 movies based on three movie genres (action, comedy and drama). In the movie dataset, there are 5,841 actors and 2,269 directors. We vary the ratio of the training data from 20% to 80% , and the remaining nodes for the test.

Table 1. Node classification results on DBLP and IMDB

Dataset		DBLP					IMDB				
Metric	train%	LINE	DeepWalk	PTE	Metapath2vec	KNCA	train%	LINE	DeepWalk	PTE	Metapath2vec	KNCA
Macro-F1	20%	0.8305	0.7743	0.8788	0.9016	**0.9110**	20%	0.4510	0.4072	0.400 0	0.4116	**0.5601**
	40%	0.8395	0.8102	0.8936	0.9082	**0.9164**	40%	0.4690	0.4519	0.4247	0.4422	**0.5943**
	60%	0.8467	0.8367	0.9086	0.9132	**0.9216**	60%	0.4822	0.4813	0.4331	0.4511	**0.5802**
	80%	0.8502	0.8481	0.9122	0.9189	**0.9280**	80%	0.4831	0.5035	0.4320	0.4515	**0.5894**
Micro-F1	20%	0.8385	0.7937	0.8848	0.9153	**0.9269**	20%	0.3902	0.4638	0.3387	0.4565	**06252**
	40%	0.8632	0.8273	0.8990	0.9203	**0.9318**	40%	0.4254	0.4999	0.3796	0.4824	**0.6727**
	60%	0.8701	0.8527	0.9137	0.9248	**0.9370**	60%	0.4367	0.5221	0.3857	0.4909	**0.6691**
	80%	0.8712	0.8626	0.9150	0.9280	**0.9327**	80%	0.4404	0.5433	0.3890	0.4881	**0.6864**

Table 2. Author node classification and paper node classification results on AMiner.

Dataset		Author node classification					Paper node classification				
Metric	training%	LINE	DeepWalk	PTE	Metapath2vec	KNCA	training%	LINE	DeepWalk	PTE	Metapath2vec	KNCA
Macro-F1	20%	0.8911	0.7256	0.897	0.9292	**0.9320**	20%	0.9315	0.9428	0.9369	0.9923	**0.9986**
	40%	0.8926	0.7273	0.8987	0.9309	**0.9332**	40%	0.9322	0.9433	0.9461	0.9928	**0.9985**
	60%	0.8934	0.7273	0.8997	0.9315	**0.9338**	60%	0.9327	0.9437	0.9500	0.9931	**0.9987**
	80%	0.8938	0.7275	0.9002	0.9319	**0.9343**	80%	0.9341	0.9451	0.9530	0.9933	**0.9988**
Micro-F1	20%	0.8993	0.7402	0.9051	0.9346	**0.9376**	20%	0.9456	0.9443	0.9550	0.9940	**0.9989**
	40%	0.9007	0.7418	0.9066	0.9361	**0.9385**	40%	0.9461	0.9447	0.9611	0.9944	**0.9989**
	60%	0.9015	0.7419	0.9075	0.9365	**0.9387**	60%	0.9467	0.9450	0.9636	0.9946	**0.9990**
	80%	0.9018	0.7425	0.9079	0.9367	**0.9388**	80%	0.9469	0.9451	0.9658	0.9949	**0.9990**

Table 1 list the node classification results on DBLP and IMDB respectively. Table 2 list the experiment results of author and paper classifications on AMiner dataset. Overall, KNCA outperform all baselines in terms of both macro-f1 and micro-f1 metrics.

Firstly, KNCA performs better than LINE and DeepWalk, because we consider the heterogeneous structure and semantics in network embedding. Applying homogeneous network embedding methods on heterogeneous network leads to loss of node type information and rich semantic relation hidden in the network. For instance, the metapath "APA" represents coauthor semantics, but the authors are not connected directly.

Secondly, we also observe that KNCA outperforms PTE which decomposes network into several bipartite networks. It leads to the loss of structural and semantic information in networks, especially PTE only preserves low-order proximity of network. Besides, KNCA achieves better result than Metapath2vec on paper and author node classification tasks, because KNCA learns node representation with the constrains of clustering structure of paper nodes.

4.3   Link Prediction

In this task, we conduct paper-author link prediction on AMiner. We randomly hide 20%, 40% and 60% paper-author links from the original network in our experiment as the ground truth or test set. Subsequently, we use the residual subnetwork to learn network embedding. We use AUC to evaluate all models on the link prediction problem based on the assumption that the node embeddings similarity linked by real links should be larger than that linked by non-existent links.

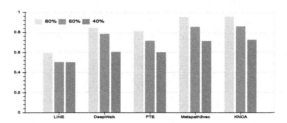

Fig. 3. Link prediction results.

As shown in Fig 3, KNCA significantly outperform other baseline models. This is because that LINE and PTE only capture low-order similarity of neighbors, and random walk-based method, DeepWalk and Metapath2vec, preserve the neighborhood of nodes with more hops. The important clustering features of networks have not been exploited. Our method can capture high-order similarity from clustering structure of key nodes (i.e., paper nodes) to improve the performance of node embeddings on link prediction.

4.4   Visualization

To evaluate the network representations intuitively, we visualize the paper node embeddings in Aminer using the t-SNE [9] algorithm. We sample 600 paper nodes in proportion to the number of papers in each academic area. There are 8 colors corresponded to 8 categories.

From Fig 4, we can see that LINE and DeepWalk cannot effectively identify different academic areas on account of ignoring the heterogeneity. On the other hands, PTE and Metapath2vec have significant improvement, but the boundary is blurry. It is worth noting that Metapath2vec can not separate well purple and orange-red nodes (Computing Systems and Theoretical Computer Science, respectively), because the two academic areas are similar. However, our model clearly clusters the paper nodes, since our model not only uses meta-paths to capture semantic information, but also exploits the clustering structure to regularize the learning of node representations.

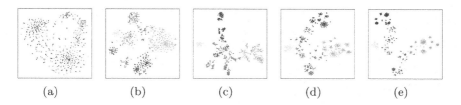

Fig. 4. Visualization embedding on DBIS. Each point indicates one paper and its color indicates the field. (a): LINE, (b): DeepWalk, (c): PTE, (d): Metapath2vec, (e): KNCA.

4.5   Parameter Analysis

In this part, we evaluate KNCA in paper node classification task to investigate the influence of parameter settings to node representation learning. Specifically, we explore the sensitivity of four parameters, including the trade-off factor α, the length of the walk, the number of cluster and the dimension of node embedding.

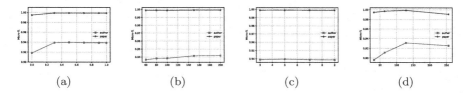

Fig. 5. Parameter sensitivity of KNCA. (a): trade-off factor $alpha$, (b): Length of walk, (c): number of clusters, (d): Dimension of the embeddings.

Figure 5(a) shows that as the trade-off factor α increases, the performance of our model first grows and becomes stable. In Fig. 5(b) and 5(c), it can be seen that our model is not sensitive to length of walk and number of cluster k. As shown in Fig. 5(d), the metric first increases and then decreases when embedding dimension d increases. Thus, we set the dimension of node embedding 128 in our experiment.

5   Conclusion and Future Work

In this paper, we propose a key nodes cluster augmented (KNCA) heterogeneous network embedding model. Our model utilizes random walk with selected meta-paths to learn node embeddings. Meanwhile, we introduce Gaussian Mixture Model (GMM) to further integrate the clustering structure of the key node based on the learned embeddings above. Finally, we design a principle and unified objective function which can effectively join the two phases. Expensive experimental results have verified the effectiveness of KNCA on various tasks.

Acknowledgement. This work was supported by the National Key R&D Program of China (2018YFC0809800), the Science and Technology Key R&D Program of Tianjin (18YFZCSF01370), the National Natural Science Foundation of China (61902278).

References

1. Cavallari, S., Zheng, V.W., Cai, H., Chang, K.C.C., Cambria, E.: Learning community embedding with community detection and node embedding on graphs. In: Proceedings of the 2017 ACM on Conference on Information and Knowledge Management, pp. 377–386. ACM (2017)
2. Dong, Y., Chawla, N.V., Swami, A.: metapath2vec: Scalable representation learning for heterogeneous networks. In: Proceedings of the 23rd ACM SIGKDD International Conference on Knowledge Discovery and Data Mining, pp. 135–144. ACM (2017)
3. Fu, T.Y., Lee, W.C., Lei, Z.: Hin2vec: explore meta-paths in heterogeneous information networks for representation learning. In: Proceedings of the 2017 ACM on Conference on Information and Knowledge Management, pp. 1797–1806. ACM (2017)
4. Grover, A., Leskovec, J.: node2vec: Scalable feature learning for networks. In: Proceedings of the 22nd ACM SIGKDD International Conference on Knowledge Discovery and Data Mining, pp. 855–864. ACM (2016)
5. Hu, B., Shi, C., Zhao, W.X., Yang, T.: Local and global information fusion for top-n recommendation in heterogeneous information network. In: Proceedings of the 27th ACM International Conference on Information and Knowledge Management, pp. 1683–1686. ACM (2018)
6. Jiao, P., Tang, M., Liu, H., Wang, Y., Lu, C., Wu, H.: Variational autoencoder based bipartite network embedding by integrating local and global structure. Inf. Sci. **519**, 9–21 (2020)
7. Jin, D., et al.: Detecting communities with multiplex semantics by distinguishing background, general and specialized topics. IEEE Trans. Knowl. Data Eng. (2019)
8. Lu, C., Jiao, P., Liu, H., Wang, Y., Xu, H., Wang, W.: SSNE: status signed network embedding. In: Yang, Q., Zhou, Z.-H., Gong, Z., Zhang, M.-L., Huang, S.-J. (eds.) PAKDD 2019. LNCS (LNAI), vol. 11441, pp. 81–93. Springer, Cham (2019). https://doi.org/10.1007/978-3-030-16142-2_7
9. van der Maaten, L., Hinton, G.: Visualizing data using t-SNE. J. Mach. Learn. Res. **9**(Nov), 2579–2605 (2008)
10. Mikolov, T., Sutskever, I., Chen, K., Corrado, G.S., Dean, J.: Distributed representations of words and phrases and their compositionality. In: Advances in Neural Information Processing Systems, pp. 3111–3119 (2013)
11. Perozzi, B., Al-Rfou, R., Skiena, S.: Deepwalk: online learning of social representations. In: Proceedings of the 20th ACM SIGKDD International Conference on Knowledge Discovery and Data Mining, pp. 701–710. ACM (2014)
12. Sun, Y., Han, J., Yan, X., Yu, P.S., Wu, T.: Pathsim: meta path-based top-k similarity search in heterogeneous information networks. Proc. VLDB Endow. **4**(11), 992–1003 (2011)
13. Tang, J., Qu, M., Mei, Q.: PTE: predictive text embedding through large-scale heterogeneous text networks. In: Proceedings of the 21th ACM SIGKDD International Conference on Knowledge Discovery and Data Mining, pp. 1165–1174. ACM (2015)

14. Tang, J., Qu, M., Wang, M., Zhang, M., Yan, J., Mei, Q.: Line: large-scale information network embedding. In: Proceedings of the 24th International Conference on World Wide Web, pp. 1067–1077. International World Wide Web Conferences Steering Committee (2015)

15. Tang, J., Zhang, J., Yao, L., Li, J., Zhang, L., Su, Z.: Arnetminer: extraction and mining of academic social networks. In: Proceedings of the 14th ACM SIGKDD International Conference on Knowledge Discovery and Data Mining, pp. 990–998. ACM (2008)

16. Wang, D., Cui, P., Zhu, W.: Structural deep network embedding. In: Proceedings of the 22nd ACM SIGKDD International Conference on Knowledge Discovery and Data Mining, pp. 1225–1234. ACM (2016)

17. Wang, H., Zhang, F., Hou, M., Xie, X., Guo, M., Liu, Q.: Shine: signed heterogeneous information network embedding for sentiment link prediction. In: Proceedings of the Eleventh ACM International Conference on Web Search and Data Mining, pp. 592–600. ACM (2018)

18. Xu, H., Liu, H., Wang, W., Sun, Y., Jiao, P.: NE-FLGC: network embedding based on fusing local (first-order) and global (second-order) network Structure with Node Content. In: Phung, D., Tseng, V.S., Webb, G.I., Ho, B., Ganji, M., Rashidi, L. (eds.) PAKDD 2018. LNCS (LNAI), vol. 10938, pp. 260–271. Springer, Cham (2018). https://doi.org/10.1007/978-3-319-93037-4_21

Processing of Incomplete Images by (Graph) Convolutional Neural Networks

Tomasz Danel[ID], Marek Śmieja[(✉)][ID], Łukasz Struski[ID], Przemysław Spurek[ID], and Łukasz Maziarka[ID]

Faculty of Mathematics and Computer Science, Jagiellonian University,
Łojasiewicza 6, 30-428 Krakow, Poland
{tomasz.danel,marek.smieja}@ii.uj.edu.pl

Abstract. We investigate the problem of training neural networks from incomplete images without replacing missing values. For this purpose, we first represent an image as a graph, in which missing pixels are entirely ignored. The graph image representation is processed using a spatial graph convolutional network (SGCN) – a type of graph convolutional networks, which is a proper generalization of classical CNNs operating on images. On one hand, our approach avoids the problem of missing data imputation while, on the other hand, there is a natural correspondence between CNNs and SGCN. Experiments confirm that our approach performs better than analogical CNNs with the imputation of missing values on typical classification and reconstruction tasks.

Keywords: Graph convolutional networks · Convolutional neural networks · Missing data

1 Introduction

Learning from missing data is one of the basic challenges in machine learning and data analysis [7]. In a typical pipeline, missing data are first replaced by some values (imputation) and next the complete data are used for training a given machine learning model [13]. The above approach depends strictly on the imputation procedure – if we accurately predict missing values, then the other model that operates on completed inputs can obtain good performance. However, it is not obvious how to select imputation method for a given problem, because it is difficult to validate its performance in a real-life scenario. Thus, there appears a natural question: *can we learn from missing data directly without using any imputation at the preprocessing stage?*

While it is difficult to answer this problem in general, a few approaches have already been designed for particular machine learning models [4,5]. In [1] a

---

A preliminary version of this paper appeared as an extended abstract [2] at the ICML Workshop on The Art of Learning with Missing Values.

© Springer Nature Switzerland AG 2020
H. Yang et al. (Eds.): ICONIP 2020, LNCS 12533, pp. 512–523, 2020.
https://doi.org/10.1007/978-3-030-63833-7_43

modified SVM classifier is trained by scaling the margin according to observed features only. In [8], the embedding mapping of feature-value pairs is constructed together with a classification objective function. Pelckmans et al. [15] model the expected risk, which takes into account the uncertainty of the predicted outputs when missing values are involved. In a similar spirit, a random forest classifier is modified to adjust the voting weights of each tree by estimating the influence of missing data on the decision of the tree [19]. The authors of [9] design an algorithm for kernel classification that performs comparably to the classifier which has access to complete data. Goldberg et al. [6] treat class labels as an additional column in the data matrix and fill missing entries by matrix completion. The work [17] shows how to generalize fully connected neural networks to the case of missing data given only an imprecise Gaussian estimate of missing data. In the similar spirit, RBF kernel can be calculated for missing data [16]. Liu et al. [12] introduce partial convolutions, where the convolution is masked and renormalized to be conditioned on only observed pixels.

In this paper we interpret the image as a graph, in which each node coincides with a visible pixel, while edges connect neighboring pixels, see Fig. 1. Since missing values are not mapped to graph nodes, we avoid the problem of missing data imputation. In order to efficiently process such an image representation, we use spatial graph convolutional neural networks (SGCN) [3]. In contrast to typical graph convolutions [10,18], which consider graph as a relational structure invariant to rotations and translations, SGCN introduces a theoretically-justified mechanism to take into account spatial coordinates of nodes. More precisely, it has been proven that any layer of convolutional neural networks (CNNs) can be represented as a spatial graph convolution. This fact allows us to think about SGCN as a generalization of CNNs, which is able, in particular, to process incomplete images without imputation.

To verify the introduced procedure, we consider MNIST [11] and SVHN [14] image datasets. Experimental results show that SGCN performs better than typical CNNs with imputations on the tasks of image classification and reconstruction.

2 Graph-Based Model for Processing Incomplete Images

In this section, we introduce our model for processing incomplete images. First, we describe how to interpret images as graphs. Next, we recall basic idea of graph convolutional networks (GCNs). Finally, we show the construction of SGCN and discuss it from an intuitive point of view.

2.1 General Idea

Images can be interpreted as vectors (tensors) of fixed sizes. If the values of selected pixels are unknown, then the vector structure is destroyed. To recover this structure, we need to replace missing attributes with some values. Substituting unknown inputs carries the risk of introducing unreliable information and noise to initial data. This may have negative consequences on data interpretation as well as can decrease the performance of subsequent machine learning algorithms applied to completed inputs.

Our idea is to interpret an incomplete image as a graph. Graphs represent a relational structure, in which the number of nodes and edges is not fixed. If some pixels in the image are unknown, then the corresponding graph contains less nodes, but the way it is processed does not change. In consequence, graph-based representation of incomplete images is more natural than using imputation.

It is well-known that CNNs are state-of-the-art feature extractors for images. However, as explained above, it is not obvious how to apply CNNs to incomplete data without replacing missing values. In this paper, we use SGCN, which is a type of graph convolutional networks, that takes spatial coordinates of nodes into account. It has been proven that SGCN can mimic any image convolution and, in consequence, SGCN is able to work comparably to CNNs using analogical network architecture (number of layers, size of filters, etc.).

2.2   Graph-Based Representation of Incomplete Images

Formally, the image is represented as a tensor $H = (\mathbf{h}_{ijk}) \in \mathbb{R}^{n \times m \times l}$, where n, m denote height and width of the image, and l is the number of channels. In the case of missing data, we do not have information about pixels values at some coordinates. Thus the incomplete image is denoted by a pair (H, J), where $J \subset \{1, \ldots, n\} \times \{1, \ldots, m\}$ indicates pixels which are unknown. In other words, $\mathbf{h}_{ijk}$ is unknown for every $(i, j) \in J$. Clearly, for a fully-observed image, $J = \emptyset$.

To construct a graph-based image representation, we create a node for every visible pixel of H, i.e.

$$V = \{v_{ij} : (i, j) \in J'\},$$

where J' is the set of indices of the observed components. The edge is defined only for nodes that represent adjacent pixels. Formally,

$$E = \{(v_{ij}, v_{pq}) : (i, j) - (p, q) \in \{-1, 0, 1\}^2\}.$$

Observe that for a complete image, every "non-boundary" pixel (node) has exactly 8 neighbors. In the case of incomplete data, the number of neighbors can be smaller, as the unknown pixels are not converted to nodes and, in consequence, the corresponding edge is not created, see Fig. 1. The information about pixels brightness is supplied with a feature vector $\mathbf{h}_{ij} \in \mathbb{R}^l$ that corresponds to a node:

$$H = \{\mathbf{h}_{ij} : (i, j) \in J'\}.$$

For a gray-scale image, $\mathbf{h}_{ij} \in \mathbb{R}$, while for a color picture $\mathbf{h}_{ij} \in \mathbb{R}^3$.

2.3   Graph Convolutions

Let $G = (V, E, H)$ be a graph (representing the image H) with n nodes. To avoid multiple indexes in the following description, the node and the corresponding

Fig. 1. Graph construction for an incomplete image of the size 4×4 with a missing region of the size 2×2.

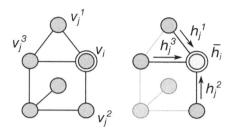

Fig. 2. Basic idea of GCNs. Every filter is responsible for defining a pattern used to aggregate feature vectors from adjacent nodes.

feature vector are denoted by v_i and $\mathbf{h}_i$, respectively, while e_{ij} is the edge between v_i and v_j. To make a natural correspondence between graphs and images, we put $\mathbf{i} = \begin{pmatrix} i_x \\ i_y \end{pmatrix}$ to denote both pixel coordinates and index in graph.

Basic idea of GCNs is to aggregate the information of feature vectors from neighboring nodes over multiple layers, see Fig. 2. To build a diverse set of patterns, GCNs use filters for defining a specific aggregation. Information from higher-level neighborhoods are fused by combining many layers together.

The above goal is realized by combining two operations. For each node v_i, feature vectors of its neighbors are first aggregated:

$$\bar{\mathbf{h}}_i = \sum_{(v_i, v_j) \in E} u_{ij} \mathbf{h}_j. \tag{1}$$

Observe that the aggregation is performed only over neighbor nodes, i.e $(v_i, v_j) \in E$. The weights u_{ij} are either trainable [18] or determined from a graph [10]. Next, a standard MLP is applied to transform the intermediate representation $\bar{\boldsymbol{H}} = [\bar{\mathbf{h}}_1 \ldots, \bar{\mathbf{h}}_n]$ into the final output of a given layer:

$$\mathrm{MLP}(\bar{\boldsymbol{H}}; \boldsymbol{W}) = \mathrm{ReLU}(\boldsymbol{W}^T \bar{\boldsymbol{H}} + \mathbf{b}), \tag{2}$$

where $\boldsymbol{W} \in \mathbb{R}^{I \times O}$ is a trainable weight matrix, $\boldsymbol{b} \in \mathbb{R}^O$ is a trainable bias vector (added column-wise), and I and O is the size of input and output layer respectively. A typical GCN is composed of a sequence of graph convolutional

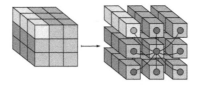

Fig. 3. Convolutional kernel for images (left) can be translated to spatial graph convolutions. Vectors at different positions (see e.g. the orange vector in the figure) are multiplied by different weights. In spatial graph convolutions (right) weights are modified by relative positions of graph neighbors to achieve different weights for each vector. (Color figure online)

layers (described above). Finally, its output is aggregated to the network response using a global pooling or a dense layer depending on a given task, e.g. node or graph classification.

2.4 Spatial Graph Convolutions

In contrast to typical GCNs described above, SGCN uses spatial coordinates of nodes, see Fig. 3. In the case of images, spatial coordinates allows to identify a given pixel in the image grid, which is not possible using only the information about neighborhood. What is more important, the convolution defined by SGCN is constructed so that it is able to reflect any convolutional filter of typical CNNs. In other words, any image convolution can be obtained by a specific parametrization of SGCN. This makes a natural correspondence between SGCN and CNNs. This property cannot be obtained by simply adding spatial coordinates to feature vectors in classical GCNs.

From a formal side, SGCN replaces (1) by:

$$\bar{\mathbf{h}}_i(\boldsymbol{U}, \mathbf{b}) = \sum_{(v_i, v_j) \in E} \mathrm{ReLU}\left(\boldsymbol{U}\left[\begin{pmatrix} j_x \\ j_y \end{pmatrix} - \begin{pmatrix} i_x \\ i_y \end{pmatrix}\right] + \mathbf{b}\right) \odot \mathbf{h}_j, \tag{3}$$

where $\boldsymbol{U} \in \mathbb{R}^{I \times 2}$, $\mathbf{b} \in \mathbb{R}^I$ are trainable, and I is the dimension of the previous layer vectors. The operator $\odot$ is element-wise multiplication. The relative positions in the neighborhood are transformed using a linear operation combined with non-linear ReLU function. This is used to weight the feature vectors $\mathbf{h}_j$ in a neighborhood. By analogy with classical convolution, this transformation can be extended to multiple filters. Let $\boldsymbol{U} = [\boldsymbol{U}_1, \ldots, \boldsymbol{U}_k]$ and $\boldsymbol{B} = [\mathbf{b}_1, \ldots, \mathbf{b}_k]$ define k-filters. The intermediate representation $\bar{\mathbf{h}}_i$ is a vector defined by:

$$\bar{\boldsymbol{H}}_i = \left[\bar{\mathbf{h}}_i(\boldsymbol{U}_1, \mathbf{b}_1), \ldots, \bar{\mathbf{h}}_i(\boldsymbol{U}_k, \mathbf{b}_k)\right].$$

Finally, MLP transformation is applied in the same manner as in (2) to transform these feature vectors.

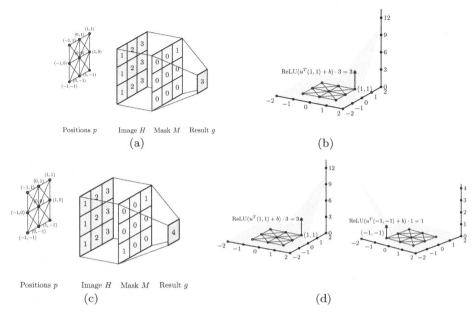

Fig. 4. Replicating two convolution operations by SGCN (top and bottom), see Examples 1 and 2 for details. On the left, the result of applying a convolutional filter **M** to the image **H**. The positions grid p represents spatial coordinates of the pixels; the neighbors are connected with edges. Analogous convolution can be applied to a spatial graph representation, as shown on the right. In the first case (top) ReLU applied to the linear transformation of the spatial features of the image graph (with $\mathbf{u} = (2,2)^T$ and $b = -3$) allows to select (and possibly modify) the top-right neighbor. In the second case (bottom), convolution operation can be obtained by extracting two opposite corner values (with $\mathbf{u}_1 = (2,2)^T, b_1 = -3$ and $\mathbf{u}_2 = (-2,-2)^T, b_2 = -3$) and summing them.

2.5   Intuition Behind Spatial Graph Convolutions

As mentioned, the formulas of SGCN allow to imitate the filters of classical CNNs. While the formal proof of this fact can be found in [3], in this section, we demonstrate this property on toy examples, which help to better understand the reasoning behind SGCN. Let us recall that the classical convolution operation (without pooling) defined by the mask $\mathbf{M} = (m_{i'j'})_{i',j' \in \{-k..k\}}$ applied to the image $\mathbf{H} = (h_{ij})_{i \in \{1..n\}, j \in \{1..m\}}$ is given by

$$\mathbf{M} * \mathbf{H} = \mathbf{G} = (g_{ij})_{i \in \{1..n\}, j \in \{1..m\}},$$

where

$$g_{ij} = \sum_{\substack{i'=-k..k:\, i+i' \in \{1..n\}, \\ j'=-k..k:\, j+j' \in \{1..m\}}} m_{i'j'} h_{i+i',j+j'}.$$

For simplicity, we consider gray-scale images with a single channel.

The following examples illustrate how to construct filters of SGCN that produce identical results to classical convolution operations.

Example 1. First, let us consider a linear convolution given by the mask

$$M = \begin{bmatrix} 0 & 0 & 1 \\ 0 & 0 & 0 \\ 0 & 0 & 0 \end{bmatrix}.$$

Observe that as the result of this convolution on the image H, every pixel is exchanged by its right upper neighbor, see Fig. 4 (top). Now the image is represented as a graph, where the neighborhood N_i of the pixel with coordinates $\mathbf{i} = \begin{pmatrix} i_x \\ i_y \end{pmatrix}$ is given by the pixels with coordinates $\mathbf{j} = \begin{pmatrix} j_x \\ j_y \end{pmatrix}$ such that $\mathbf{j} - \mathbf{i} \in \{-1, 0, 1\}^2$.

Given a vector[1] $\mathbf{u} \in \mathbb{R}^2$ and a bias $b \in \mathbb{R}$ the (intermediate) graph operation is defined by

$$g_i = \bar{h}_i(\mathbf{u}, b) = \sum_{j \in N_i} \text{ReLU}\left(\mathbf{u}^T \left[\begin{pmatrix} j_x \\ j_y \end{pmatrix} - \begin{pmatrix} i_x \\ i_y \end{pmatrix} \right] + b \right) \cdot h_j.$$

Consider now the case when $\mathbf{u} = 2 \cdot \mathbb{1}, b = -3$, where $\mathbb{1} = (1, 1)^T$. One can easily observe that

$$\text{ReLU}(\mathbf{u}^T \mathbf{z} - b) = \begin{cases} 0, & \text{for } \mathbf{z} \neq \mathbb{1} \\ 1, & \text{for } \mathbf{z} = \mathbb{1}, \end{cases}$$

where $\mathbf{z} = \mathbf{j} - \mathbf{i} \in \{-1, 0, 1\}^2$.

Consequently, we obtain that $g_i = h_{i+\mathbb{1}}$, which equals the result of the considered linear convolution.

Example 2. Now, let us consider the mask, see Fig. 4 (bottom):

$$M = \begin{bmatrix} 0 & 0 & 1 \\ 0 & 0 & 0 \\ 1 & 0 & 0 \end{bmatrix}.$$

This convolution cannot be obtained from graph representation using a single transformation as in previous example.

To formulate this convolution, we define two intermediate operations for $k = 1, 2$:

$$\bar{h}_i(\mathbf{u}_k^T, b_k) = \sum_{j \in N_i} \text{ReLU}\left(\mathbf{u}_k \left[\begin{pmatrix} j_x \\ j_y \end{pmatrix} - \begin{pmatrix} i_x \\ i_y \end{pmatrix} \right] + b_k \right) \cdot h_j.$$

where $\mathbf{u}_1 = 2 \cdot \mathbb{1}, b_1 = -3$ and $\mathbf{u}_2 = -2 \cdot \mathbb{1}, b_2 = -3$. The first operation extracts the right upper corner, while the second one extracts the left bottom corner, i.e.

$$\bar{h}_i(\mathbf{u}_1, b_1) = 1 \cdot h_{i+\mathbb{1}} \text{ and } \bar{h}_i(\mathbf{u}_2, b_2) = 1 \cdot h_{i-\mathbb{1}}.$$

---

[1] For simplicity, we consider an image with a single channel.

Finally, we put

$$\bar{H}_i = \begin{bmatrix} \bar{h}_i(\mathbf{u}_1, b_1) \\ \bar{h}_i(\mathbf{u}_2, b_2) \end{bmatrix} = \begin{bmatrix} h_{i+1} \\ h_{i-1} \end{bmatrix}.$$

Making an additional linear transformation defined by (2) with $\mathbf{w}_i = (1,1)^T$, we obtain:

$$g_i = \mathbf{w}_i^T \bar{H}_i = 1 \cdot h_{i+1} + 1 \cdot h_{i-1}.$$

Following the above examples, to obtain the result of applying an arbitrary 3×3 filter, we may need at most 9 operations using (3) (one for replicating each of 9 entries). SGCN is also capable of imitating larger filters, see [3] for details.

3   Experiments

In this section, we evaluate our model on two machine learning tasks and compare it with related approaches.

3.1   Reconstruction

First, as a proof of concept, we take into account MNIST database and consider the problem of restoring corrupted images, in which a part of data is hidden. To prepare this task, for each image of the size 28×28, we remove a square patch of the size 13×13. The location of the patch is uniformly sampled for each image.

The reconstruction models are instantiated using the auto-encoder architecture (AE). In the case of our model, the encoder is implemented as SGCN with 5 spatial graph convolutional layers while the decoder is a typical deconvolutional neural network, which returns the image in the form of tensor. We emphasize that graph neural network is only needed for initial stage of processing (encoder) to avoid replacing missing values. Subsequent stages, e.g. decoder, can be implemented using typical non-graph networks. The result of our model is compared to CNNs combined with typical imputation techniques: (i) **mask**, which is a zero imputation with an additional binary channel indicating unknown pixels (ii) **mean** imputation, where absent attributes are replaced by mean values for a given coordinate (iii) **k-nn** imputation, which substitutes missing features with mean values of those features computed from the k-nearest training samples (we use $k = 5$). For a fair comparison, every architecture (SGCN and CNNs) has the same structure, i.e. the number of layers and filters as well as the type of regularization.

We assume that the complete data are not available in the training phase. Therefore, for all models, the loss is defined as the mean-square error (MSE) calculated outside the missing region. This makes the problem more difficult than a typical inpainting task. To isolate the effect of processing incomplete images and directly compare SGCN with CNNs, we do not introduce additional losses and use only MSE in training.

It can be seen from the Fig. 5 that SGCN gives similar results to CNN (mask). The reconstructions coincide on average with ground-truth and are free

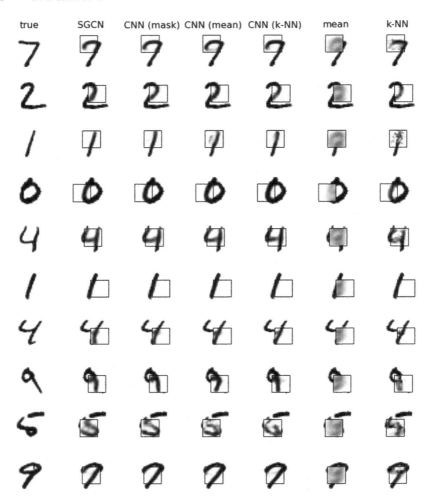

Fig. 5. Reconstructions obtained for MNIST dataset (the first 10 images of test set). To demonstrate the influence of initial imputations on the final reconstructions returned by CNNs, we show the results of applying mean and k-NN imputations (last two columns).

of artifacts. There was a problem in restoring digit "9" (last row), but the same holds for other methods. The results produced by CNN (mean) and CNN (k-NN) are sometimes blurry. To support the visual inspection with quantitative assessment, we calculate MSE inside the missing region, see the first row of Table 1. Surprisingly, CNN (k-NN) gives the highest resemblance with ground-truth in terms of MSE. While the reconstructions look visually less plausible than the ones returned by SGCN and CNN (mask), the pixel-wise agreement with ground-truth is higher. SGCN gives the second best result. It is important to observe the influence of initial imputation on the performance of CNNs. Since

Table 1. Performance on three machine learning tasks (lower is better). The first row shows mean-square error for reconstructing incomplete MNIST images, while the last two rows present test errors for classifying incomplete MNIST and SVHN images.

Dataset	SGCN	GCN	CNN (mask)	CNN (mean)	CNN (k-NN)
MNIST (MSE)	0.0755	–	0.0760	0.0787	**0.0725**
MNIST (Error)	**4.6%**	31.4%	4.9%	5.9%	5.7%
SVHN (Error)	**16.6%**	74.6%	18.6%	19.9%	22.4%

the k-NN imputation is significantly better than the mean imputation[2], the corresponding CNN model is able to restore input image more reliably. However, it is evident that mistakes made by k-NN imputation also negatively affect the performance of CNN (3rd and 9th rows). In contrast, our method is more stable, because it does not depend on imputation strategy. In consequence, it may give worse results than CNN when it is easy to predict missing values, but, at the same time, it should perform better if the imputation problem is more difficult. Another advantage is that SGCN is trained end-to-end (no preprocessing of missing values).

3.2 Classification

In the second experiment, we consider the classification task, in which incomplete data appear in both train and test phase. In addition to gray-scale handwritten digits retrieved from MNIST database, we also use color house-number images of the SVHN dataset. In the case of SVHN images of the size 32×32, we use patches of the size 15×15.

For a comparison, we use CNN models combined with the same imputation techniques as before, but, additionally, we consider "vanilla" **GCN** [10], which is one of the simplest GCNs that ignores spatial coordinates[3]. Classification network is composed of 8 convolutional layers. Each one contains 64 filters of the size 3×3. Batch normalization is used after every convolutional layer. As mentioned, we use analogical architecture for both graph convolutions and typical image convolutions.

It is evident from Table 1 (2nd and 3rd rows) that SGCN performs significantly better than the other version of GCN. It is not surprising because, in contrast to typical GCNs, SGCN introduces information about spatial coordinates to the model. Next observation is that SGCN gives lower errors than CNNs combined with imputation strategies. While the advantage of SGCN over the second best method in the case of MNIST is slight, the difference in accuracy is higher in the case of SVHN, which is significantly harder dataset to classify. As mentioned earlier, it may be difficult to reliably predict missing values for

---

[2] k-NN imputation obtains MSE = 0.0807 while mean imputation gives MSE = 0.1265.

[3] We also experimented with graph attention network [18], but the results did not improve.

hard tasks, which, in consequence, negatively affects CNN models applied to completed data. Moreover, it can be seen that k-NN imputation is not so beneficial in classification problems as in reconstruction task – CNN (k-NN) performs even worse than CNN (mean) on SVHN. As can be seen, the knowledge about missing pixels is more important for the success of classification CNNs than using specific imputation technique[4]. In contrast to CNN (mask), which uses an additional binary channel to pass the information about unknown values to the neural model, SGCN directly ignores missing pixels, which is more natural.

4   Conclusion

We presented an alternative way of learning neural networks from incomplete images, which does not require replacing missing values at the preprocessing stage. While graph representation of incomplete images avoids the problem of imputation, applying SGCN allows us to reflect the action of classical CNNs. Our model is trained end-to-end without any missing data preprocessing (imputation) as in the case of CNNs. Since the proposed model completely ignores the information about missing values, it is especially useful in the case of complex tasks, where imputation strategies introduce noise and unreliable information. The main disadvantage of our approach is the computational cost of using GCNs. In contrast to classical CNNs, the current implementations of GCNs are less efficient and it is difficult to manage large graphs created from high dimensional images.

Acknowledgements. The work of M. Śmieja was supported by the National Science Centre (Poland) grant no. 2017/25/B/ST6/01271. The work of Ł. Struski was supported by the Foundation for Polish Science Grant No. POIR.04.04.00-00-14DE/18-00 co-financed by the European Union under the European Regional Development Fund. The work of P. Spurek was supported by the National Centre of Science (Poland) Grant No. 2019/33/B/ST6/00894. The work of Ł. Maziarka was supported by the National Science Centre (Poland) grant no. 2018/31/B/ST6/00993.

References

1. Chechik, G., Heitz, G., Elidan, G., Abbeel, P., Koller, D.: Max-margin classification of data with absent features. J. Mach. Learn. Res. **9**, 1–21 (2008)
2. Danel, T., Śmieja, M., Struski, Ł., Spurek, P., Maziarka, L.: Processing of incomplete images by (graph) convolutional neural networks. In: ICML Workshop on the Art of Learning with Missing Values (Artemiss), p. 6 (2020)
3. Danel, T., et al.: Spatial graph convolutional networks. arXiv preprint arXiv:1909.05310 (2020)
4. Dekel, O., Shamir, O., Xiao, L.: Learning to classify with missing and corrupted features. Mach. Learn. **81**(2), 149–178 (2010)

---

[4] We verified that combining masking with mean/k-nn imputation does not lead to further improvement of CNNs.

5. Globerson, A., Roweis, S.: Nightmare at test time: robust learning by feature deletion. In: Proceedings of the International Conference on Machine Learning, pp. 353–360. ACM (2006)
6. Goldberg, A., Recht, B., Xu, J., Nowak, R., Zhu, X.: Transduction with matrix completion: three birds with one stone. In: Advances in Neural Information Processing Systems, pp. 757–765 (2010)
7. Goodfellow, I., Bengio, Y., Courville, A.: Deep Learning. MIT Press, Cambridge (2016)
8. Grangier, D., Melvin, I.: Feature set embedding for incomplete data. In: Advances in Neural Information Processing Systems, pp. 793–801 (2010)
9. Hazan, E., Livni, R., Mansour, Y.: Classification with low rank and missing data. In: Proceedings of the 32nd International Conference on Machine Learning, pp. 257–266 (2015)
10. Kipf, T.N., Welling, M.: Semi-supervised classification with graph convolutional networks. arXiv preprint arXiv:1609.02907 (2016)
11. LeCun, Y., Bottou, L., Bengio, Y., Haffner, P.: Gradient-based learning applied to document recognition. Proc. IEEE **86**, 2278–2324 (1998)
12. Liu, G., Reda, F.A., Shih, K.J., Wang, T.-C., Tao, A., Catanzaro, B.: Image inpainting for irregular holes using partial convolutions. In: Ferrari, V., Hebert, M., Sminchisescu, C., Weiss, Y. (eds.) ECCV 2018. LNCS, vol. 11215, pp. 89–105. Springer, Cham (2018). https://doi.org/10.1007/978-3-030-01252-6_6
13. McKnight, P.E., McKnight, K.M., Sidani, S., Figueredo, A.J.: Missing Data: A Gentle Introduction. Guilford Press (2007)
14. Netzer, Y., Wang, T., Coates, A., Bissacco, A., Wu, B., Ng, A.Y.: Reading digits in natural images with unsupervised feature learning. In: NIPS Workshop on Deep Learning and Unsupervised Feature Learning (2011)
15. Pelckmans, K., De Brabanter, J., Suykens, J.A., De Moor, B.: Handling missing values in support vector machine classifiers. Neural Netw. **18**(5), 684–692 (2005)
16. Śmieja, M., Struski, Ł., Tabor, J., Marzec, M.: Generalized RBF kernel for incomplete data. Knowl.-Based Syst. **173**, 150–162 (2019)
17. Śmieja, M., Struski, Ł., Tabor, J., Zieliński, B., Spurek, P.: Processing of missing data by neural networks. In: Advances in Neural Information Processing Systems, pp. 2719–2729 (2018)
18. Veličković, P., Cucurull, G., Casanova, A., Romero, A., Lio, P., Bengio, Y.: Graph attention networks. arXiv preprint arXiv:1710.10903 (2017)
19. Xia, J., et al.: Adjusted weight voting algorithm for random forests in handling missing values. Pattern Recogn. **69**, 52–60 (2017)

Multi-agent Cooperation and Competition with Two-Level Attention Network

Shiguang Wu[1,2], Zhiqiang Pu[1,2(✉)], Jianqiang Yi[1,2], and Huimu Wang[1,2]

[1] School of Artificial Intelligence, University of Chinese Academy of Sciences, Beijing 100049, China
shiguang.wu@outlook.com,
{zhiqiang.pu,jianqiang.yi,wanghuimu2018}@ia.ac.cn
[2] Institute of Automation, Chinese Academy of Sciences, Beijing 100190, China

Abstract. Multi-agent reinforcement learning (MARL) has made significant advances in multi-agent systems. However, it is hard to learn a stable policy in complicated and changeable environment. To address these issues, a two-level attention network is proposed, which is composed of across-group observation attention network (AGONet) and intentional communication network (ICN). AGONet is designed to distinguish the different semantic meanings of observations (including friend group, foe group, and object/entity group) and extract different underlying information of different groups with across-group attention. Based AGONet, the proposed network framework is invariant to the number of agents existing in the system, which can be applied in large-scale multi-agent systems. Furthermore, to enhance the cooperation of the agents in the same group, ICN is used to aggregate the intentions of neighbors in the same group, which are extracted by AGONet. It obtains the understanding and intentions of their neighbors in the same group and enlarges the receptive filed of the agent. The simulation results demonstrate that the agents can learn complicated cooperative and competitive strategies and our method is superiority to existing methods.

Keywords: Graph attention network · Deep reinforcement learning · Multi-agent system

1 Introduction

Cooperation and competition widely exist in nature from bacteria, social animals, and humans, in which they consist of many roles including friend group, foe group, object/entity group and so on. Individuals in the same group need to cooperate to accomplish some tasks or to compete against foe groups. The

Research supported by the National Key Research, Development Program of China under Grant 2018AAA0102404, and Innovation Academy for Light-duty Gas Turbine, Chinese Academy of Sciences, No. CXYJJ19-ZD-02.

H. Yang et al. (Eds.): ICONIP 2020, LNCS 12533, pp. 524–535, 2020.
https://doi.org/10.1007/978-3-030-63833-7_44

research of cooperation and competition in multi-agent systems have promising applications in engineering systems, such as smart grids [12], games [17], resource management [7] and robot soccer, which have attracted a lot of attentions by many researchers.

Recently, deep reinforcement learning (DRL) [14] has shown the human-level or higher performance in sequential decision-making problems. With the advance of DRL, it has been combined with multi-agent reinforcement learning (MARL) to solve complex and large-scale problems [11]. Based on the common paradigm of centralized learning with decentralized execution, some MARL algorithms learn centralized critics for multiple agents and determine the decentralized action solely based on local observation. However, they have some limitations. On the one hand, the algorithms based on attention mechanisms [3,5] can effectively extract valuable information from their neighbors via communication protocol. However, the observation of each agent is directly encoded into a feature vector, where it never distinguishes different groups of observation and ignores different underlying influences brought by different groups of observation. On the other hand, by creating a shared agent-entity graph, the agents learn cooperative behaviors by exchanging messages with each other along the edges of this graph in [1]. It is assumed that the entity can attend messages to other agents, which is unsuitable in real applications that the entity only can be observed.

To address the limitations mentioned above, a two-level attention network is proposed in this paper for promoting the cooperative or competitive behavior in multi-agent systems. The network is divided into two parts including across-group observation attention network (AGONet) and intentional communication network (ICN). To distinguish different semantic meanings of observations including friend group, foe group, and object group, AGONet is designed to extract different underlying information of different groups. It is helpful to learn a stable policy and be applied to some cooperation and competitive environments or more complex systems. With AGONet, the network framework is invariant to the number of agents in the multi-agent system, which can be applied in large-scale multi-agent systems. Furthermore, ICN is used to aggregate the intentions of neighbors in the same group, aiming to enhance the cooperation of the agents in the same group. By means of ICN, It can obtain the understanding and intentions of their neighbors in the same group and enlarge the receptive filed of each agent.

The effectiveness and superiority of the proposed framework is verified in different environments including cooperation navigation, 3 vs. 1, and 5 vs. 2 predator-prey games. The simulation results indicate that our method enables the agents to learn complicated cooperative and competitive strategies and substantially outperforms existing methods.

2   Related Work

Multi-Agent Deep Deterministic Policy Gradient (MADDPG) [9] is extended from the Deep Deterministic Policy Gradient (DDPG) [8] to multi-agent systems for mixed cooperative-competitive environments. Counterfactual Multi-agent (COMA) [2] is proposed to solve the multi-agent credit assignment by isolating the effect of each agent's action. MADDPG and COMA are based on the framework of the centralized learning with decentralized execution, where the observations and actions of all agents are used to compute a centralized critic. They need to train an independent policy for each agent. Therefore, it is hard to be applied in large-scale environments. To solve the limitation, the Mean-Field approach [18] considers the mean action of agents to derive their actions, which can be applied in large-scale environments. However, it ignores different impacts from their neighbor agents.

Recently, due to the similarity of the graph structures and multi-agent systems, the combination of graph neural network and multi agent reinforcement learning has attracted a lot of attention by many researchers, especially in the graph convolutional network (GCN), which has been successfully applied in many domains including social network [6], natural language processing [15], and knowledge graph [19]. GCN is an effective framework to extract the locally connected features from arbitrary graphs. Based on GCN, some achievements have made in MARL. MAAC [3] models a centralized critic by using the attention network and graph neural network and derives the decentralized actors with soft actor-critic. Attentional Communication (ATOC) [5] enable agents to extract information from their neighbors effectively via attention network mechanism. DGN [4] uses the graph convolutional network for communication using a deep Q-network [10] for training. These models prove that communication between the agents is useful for cooperation based on GCN. However, they ignore underlying structure of observation in multi-agent systems and encode the observation of the agents into a feature vector without considering the correlation among different groups of observation. HAMA [13] employs a hierarchical graph neural network to effectively model the inter-agent and inter-group relationships. However, it ignores the communications between the agents in the same group.

To the best of our knowledge, none of existing work in MARL studies the across-group observation attention network and intentional communication network simultaneously, in which we consider the correlation among each group of observation and intentional communication among the agents in the same group.

3   Background

3.1   Partially Observable Markov Game (POMG)

Partially observable markov game is an extension of partially observable markov decision process. POMG for N agents is defined by a global state S, a set of local observations $O_1, \cdots, O_N$, and a set of actions $A_1, \cdots, A_N$. The action for agent i is determined by a learned policy π_i. Giving the current states and actions of the agents, the next states of the agents are determined by the transition model $T : S \times A_1 \times \cdots \times A_N \to S'$. The reward for agent i denoted as r_i is computed after all agents taking actions: $S \times A_1 \times A_2 \cdots \times A_N \to \mathbb{R}$. The agent i aims to maximize its discounted return $R_i = \sum_{t=0}^{T} \gamma^t r_i^t$ with the learned policy, where γ is a discount factor.

3.2 Proximal Policy Optimization (PPO)

PPO [14] holds the stability and reliability of trust-region methods while is easy to implement. It strikes a balance between implementation, sample complexity, and tuning, which tries to compute an update at each step that minimizes the cost function while ensuring the deviation from the previous policy is relatively small. Let

$$l_t(\theta) = \frac{\pi_\theta(a_t|s_t)}{\pi_{\theta_{old}}(a_t|s_t)} \tag{1}$$

is the likelihood ratio. π_θ is the policy with network parameters θ. $\pi_{\theta_{old}}$ is the old policy, so $l_t(\theta_{old}) = 1$. s_t, a_t denote the state and action of agent, respectively. Then, PPO algorithm aims to optimize the following objective:

$$L(\theta) = E[\min(l_t(\theta)\hat{A}_t^{\theta_{old}}(s_t, a_t), clip(l_t(\theta), 1 - \epsilon, 1 + \epsilon)\hat{A}_t^{\theta_{old}}(s_t, a_t))] \tag{2}$$

where $\hat{A}_t^{\theta_{old}}(s_t, a_t)$ is an estimator of the generalized advantage, $clip(l_t(\theta), 1 - \epsilon, 1+\epsilon)$ limits the value of $l_t(\theta)$ to $(1-\epsilon, 1+\epsilon)$ with the parameters ϵ, which makes the deviation between π_θ and $\pi_{\theta_{old}}$ not too big. This ensures the rationality of the importance sampling.

4 Method

In this section, a two-level attention network is proposed, which is composed of across-group observation attention network (AGONet) and intentional communication network (ICN) as shown in Fig. 1. First, AGONet is designed to extract different underlying information of different groups by distinguishing different semantic meanings of observation including friend group, foe group, and object/entity group, which are represented as green, blue, black in Fig. 1, respectively. Furthermore, ICN is used to aggregate the intentions of neighbors in the same group, which are exacted by AGONet, aiming to enhance the cooperation of the agents in the same group. Finally, the captured node-embedding vector for each agent is subsequently used as the input of the critic network and the actor network.

4.1 Across-Group Observation Attention Network (AGONet)

For some complex tasks, the observation of each agent may contain some different properties. In general, the observation of the agent can be divided into friend group, foe group, object/entity group (obstacles or common goal) and so on. Intuitively, each group of observation have different importance for policy at different time. Therefore, to describe the different influences of different groups of observation on the agent, the across-group observation attention network with across-group softmax function is proposed, which is shown in Fig. 2.

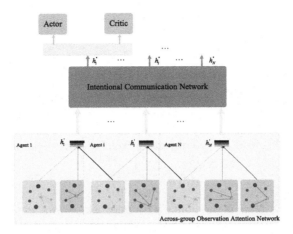

Fig. 1. Two-level attention network.

In multi-agent systems, the agents can be classified to R group with N^R agent or entity according to the property of tasks. Therefore, the observation of agent also can be classified to R group with N^R individuals. Considering a partially observable environment, where at each time-step t, each agent i receives an observation $O_i^t = [o_i^1, o_i^r, \ldots, o_i^R]$, where $o_i^r = [s_1^r, s_i^r \cdots, s_{N^r}^r]$ is the states set of the agent in the r-th group. Then, the relationship between the agent i and the agents of observation o_i^r in the r-th group can be represented as a graph $G^r = (N^r + 1, E^r)$ called observation graph, consisting of the set of $N^r + 1$ nodes (the number of the node in the r-th group and agent i) and the set of E^r edges.

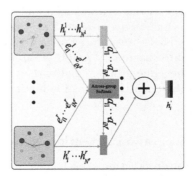

Fig. 2. Across-group observation graph attention network.

The features vector of agent i is encoded by a full connected layer denoted as $h_i = f(s_i)$, where s_i is the state of agent i, including the position and the velocity. The features vector of the agent j of the observation o_i^r is also encoded by a full connected layer denoted as $h_j^r = f^r(s_j^r), j \in N_r$. Therefore, the attention coefficient between agent i and the agent j of observations o_i^r is denoted as e_{ij}^r, which is computed as

$$e_{ij}^r = a_G^r(W_r^T h_i, W_r^T h_j^r), \qquad (3)$$

where W_r is a linear learnable weight matrix, a_G^r is a single-layer feed forward neural network.

To effectively express attention coefficients across different nodes, common approach is to use the softmax function to normalize the attention coefficients across different nodes for each group. However, it only represents the influence from each node in the same group to the agent without from a global perspective, which ignores the influence from the agents of other groups to the agent. Therefore, an across-group softmax is designed to normalize the attention coefficients across every nodes in the whole observation. The influence from each node of observation to the agent is described, which is computed as following:

$$a_{ij}^r = \frac{exp(LeakyReLU(e_{ij}^r))}{\sum_{r \in R} \sum_{j \in N_r} exp(LeakyReLU(e_{ij}^r))}, \sum_{r \in R} \sum_{j \in N_r} a_{ij}^r = 1, \qquad (4)$$

where the LeakyReLU is a nonlinearity activate function with negative input slope. Combining Eqs. 3 and 4, the whole observation embedding aggregated from each node is given as

$$h_i' = \sigma(\sum_{r \in R} \sum_{j \in N_r} \alpha_{ij}^r W_r^T h_j^r), \qquad (5)$$

where σ represents an optional nonlinearity. It concatenates the different groups of observation with different weight matrix W_r. It is worth noted that to avoid the state repeatedly, only the friend group consider the self-attention mechanism while other groups not adopt the self-attention mechanism.

Giving the credit to AGONet, it describes different attentions from each node of observation to the agent, which represents effectively the intentions and understanding of current agent. It is noteworthy that AGONet can handle any number of agent or entity, owing to distinguishing different semantic meanings of observations, which enables the method to learn more complex task and be applied in the large-scale multi-agent systems.

4.2   Intentional Communication Network (ICN)

The policy of the agent is influenced by the behaviors of neighbors in the same group, which should be treated differently through some pattern for cooperation. Based on AGONet, the understanding of the environments and intentions of agent can be acquired. If other agents can know the understanding and intentions of the neighbors, it can promote the cooperation behavior in multi-agent systems. Hence, as shown in Fig. 3. ICN is used to extract the different understanding

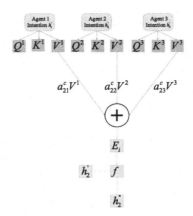

Fig. 3. Intentional communication network.

and intentions attended by neighbors, which quantifies the different importance of the incoming messages.

For each agent that has the communication capacity in the same group, the communication structure can also be represented as an undirected graph $G_c = (V_c, E_c)$ called communication graph, where $V_c = \{1, \cdots, N\}$ denotes the set of nodes, which is composed of the agents in the same group, $E_c \subseteq V_c \times V_c$ denotes the edge set. It is worthy highlight that the communication graph is different from the above mentioned observation graph. The neighbors set of the agent i is determined as $\mathcal{N}_i^c = \{j | (j, i) \in E_c\}$. By using linear weight matrices W^Q, W^K, W^V, the understanding and intentions h_i' of the agent i that is the output of AGONet are transformed to a different space denoted as $Q_i = W^Q h_i'$, $K_i = W^K h_i'$, $V_i = W^V h_i'$ to passing the information. After receiving the query-value pair from the neighbors $k \in \mathcal{N}_i^c$, the attention coefficients a_{ik}^c for agent k to agent i is computed and the messages of the neighbors is aggregated according to attention coefficients a_{ik}^c as following:

$$e_{ik}^c = \frac{(W^Q h_k')(W^K h_i')}{d_K}, \ a_{ik}^c = \frac{exp(e_{ik}^c)}{\sum_{k \in \mathcal{N}_i^c} exp(e_{ik}^c)}, \tag{6}$$

$$E_i = \sigma\left(\sum_{k \in \mathcal{N}_i^c} a_{ik}^c V_k\right) \tag{7}$$

where d_K is the dimensionality of keys, E_i is the aggregated messages from other agents with nonlinear activation function σ. Ref [16] shows that multi-head attention is beneficial to learn the attention. Therefore, we aggregate messages attended from other agents with multi-head as following:

$$E_i = ||_{m=1}^n \sigma\left(\sum_{k \in \mathcal{N}_i^c} a_{ik}^{c,m} V_k^m\right) \tag{8}$$

where $\|$ represents the concatenation operation, n is the number of the attention heads, $a_{ij}^{c,m}$ is the attention coefficient of the m-th attention head and V_k^m is the features vector of the m-th attention head. Then, the agent i updates the embedding information $h_i^{''}$ by a non-linear transformation of its current embedding information $h_i^{'}$ concatenated with E_i by using a neural network f. Furthermore, K-hop communication is used to enlarge the receptive field of agent i, which is expressed as $h_i^{'}(1) \to$ ICN $\to h_i^{''}(1) \to$ ICN $\to h_i^{''}(2) \to \cdots \to h_i^{''}(K)$. With ICN, the agent can know the intentions of the neighbors and enlarge the receptive field, which is beneficial to the cooperation with other agents in the same group.

4.3   Training

Based on the two-level attention network, the embedding information $h_i^{''}$ can be obtained to compute the actor network and critic network. The PPO algorithm in the Actor-Critic framework is adopted to train the proposed method, which is trained and executed both in a completely decentralized manner. What's more, the parameters sharing method is applied to train all the agents. It enables the proposed method applied to the large-scale multi-agent systems.

5   Simulations

In this section, the effectiveness of the proposed framework is verified by several simulations, which include three scenarios as shown in Fig. 4: cooperation navigation, 3 vs. 1 predator-prey, and 5 vs. 2 predator-prey. These simulations are implemented based on Multi-Agent Particle Environment [9], which has been widely used in existing studies. For comparing the performance of our method denoted (AGONet-ICN), the baseline algorithms TRANSFER [1] and a variant of our method that only adopt the across-group observation attention network (AGONet) are taken into consideration.

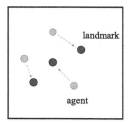 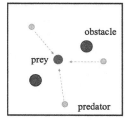 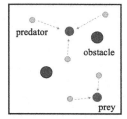

(a) Cooperation navigation   (b) 3 vs. 1 Predator-prey   (c) 5 vs. 2 Predator-prey

Fig. 4. Illustration of the simulation environments.

5.1 Cooperation Navigation

In this scenario, all the agents are required to reach the landmarks without colliding with each other, where only cooperation needs to be considered. It is noted that the landmarks can not transmit any message about their state to agent. Therefore, the agent can only observe state information of other agents and landmarks. The observation of agent is composed of our group and entity group, which means that the number of observation group R is 2.

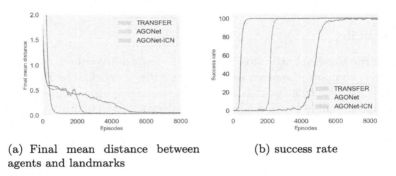

(a) Final mean distance between agents and landmarks

(b) success rate

Fig. 5. Simulation results in the cooperation navigation with three agents.

Figure 5 shows simulation results in the cooperation navigation with three agents and three landmarks. As shown in Fig. 5, AGONet-ICN and AGONet has a higher convergence rate and higher performance compared to TRANS-FER. This is possible because AGONet-ICN and AGONet represent effectively the features of each agent by considering each group of observation with across-group softmax. Compared to AGONet-ICN, AGONet needs more time to learn an effective policy, which ignores the intentions of neighbors of the agent in the same group. It also indicates the effectiveness of AGONet and the superiority of AGONet-ICN. Table 1 presents the evaluation results during tested with 1000 episodes with the trained models, which further shows the superior performance of AGONet-ICN. Therefore, the performance difference among AGONet-ICN, AGONet and algorithm demonstrates that the cooperative strategies trained by AGONet-ICN can promote the cooperation and learn an effective strategy quickly. To further verify the generalization of the proposed method, the policy trained for five agents and five landmarks denoted $m = 5$ without any fine-tuning is evaluated directly on different number of agents and landmarks. The generalization results are shown in Table 2 with evaluation index. These results indicate that AGONet-ICN have satisfactory generalization success rate and steps. Furthermore, the results manifest that the proposed method can handle any number agents and easily transfer to more agents to complete more complicated tasks, which is practical in large-scale multi-agent systems.

Table 1. The evaluation results of cooperation navigation with three agents.

Algorithm	Success rate(%)	Steps	Collisions(%)
TRANSFER	99.6	11.13	0.6
AGONet	100	10.34	0.5
AGONet-ICN	**100**	**9.88**	**0.5**

Table 2. Generalization results: policy trained for five agents is evaluated directly for m agents and m landmarks without any fine-tuning.

Evaluation index	$m=1$	$m=2$	$m=3$	$m=4$	$m=5$	$m=6$	$m=7$	$m=8$	$m=9$
Success rate(%)	100	100	100	100	100	100	100	100	100
Steps	10.25	10.5	10.36	10.24	10.35	10.68	11.1	11.51	12.6
Collisions(%)	0	0.09	0.14	0.13	0.26	0.37	0.41	0.46	0.49

5.2 Predator-Prey

In this scenario, there are two groups of agents competing with each other and one group of obstacles are hindering the agents. The predator group is aimed to capture the prey and avoid the obstacle collisions, while the prey group needs to escape from the predators and avoid the obstacle collisions. In the environments, the predator move slower than the preys. For the predators, they need to learn how to cooperate with each other. Each agent aims to maximize their accumulated rewards, which results in the competition between predators and preys and the cooperation among the predators. Two scenarios including the 3 vs. 1 and 5 vs. 2 predator-prey are designed to evaluate the performance of AGONet-ICN. The preys are trained with TRANSFER. The predators are trained with AGONet-ICN, AGONet, and TRANSFER, respectively.

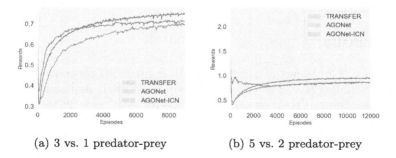

(a) 3 vs. 1 predator-prey   (b) 5 vs. 2 predator-prey

Fig. 6. The training reward in predator-prey.

Figure 6 shows the training reward in predator-prey. It is indicated that AGONet-ICN is slower to learn a policy compared to TRANSFER in the early

Table 3. Evaluating results of 3 vs. 1 predator-prey

Algorithm	Success rate(%)	Steps	Rewards	Collisions(%)
TRANSFER	99.5	13.05	0.72	2.7
AGONet	99.7	12.99	0.71	1.3
AGONet-ICN	**100**	**12.91**	**0.75**	**0.67**

Table 4. Evaluating results of 5 vs. 2 predator-prey

Algorithm	Success rate(%)	Steps	Rewards	Collisions(%)
TRANSFER	95.9	18.94	0.90	4.16
AGONet	99.7	16.10	0.90	1.7
AGONet-ICN	**99.9**	**15.31**	**0.95**	**1.65**

stage. We think that is because the framework of our method is more complex, which needs to learn more network parameters while learn an effective policy. The final results demonstrate that AGONet-ICN is superiority to TRANSFER. Although AGONet also can learn an effective policy, it is weaker than two-level attention network. This is because the features of neighbors of the agent in the same group is ignored in AGONet, which is useful to cooperation among the agents in the same group. To further show the effectiveness of AGONet-ICN, some evaluation results are conducted in Tables 3 and 4. The evaluation results show that AGONet-ICN can perform better than TRANSFER and AGONet for complex tasks. The performance difference among AGONet-ICN and TRANS-FER in the evaluation index is salient in 5 vs. 2 predator-prey compared to 3 vs. 1 predator-prey. It further represents the power of AGONet-ICN when dealing with the complex tasks. In addition, the performance of AGONet is better than TRANSFER, which is owing to distinguishing each group of their observation with across-group softmax.

6   Conclusion

In this paper, a two-level attention network is proposed for multi-agent cooperation and competitive. The two-level attention network distinguishes different semantic meanings of observation and aggregates features from their neighbors in the same group to promote the cooperation and competition. In addition, the proposed framework exhibits the superiority performance and is invariant to the numbers of agents existing in the system. In the future, large-scale multi-agent system will be taken into consideration and be conducted in unmanned ground vehicles.

References

1. Agarwal, A., Kumar, S., Sycara, K.: Learning transferable cooperative behavior in multi-agent teams. arXiv preprint arXiv:1906.01202 (2019)
2. Foerster, J.N., Farquhar, G., Afouras, T., Nardelli, N., Whiteson, S.: Counterfactual multi-agent policy gradients. In: Thirty-Second AAAI Conference on Artificial Intelligence (2018)
3. Iqbal, S., Sha, F.: Actor-attention-critic for multi-agent reinforcement learning. In: International Conference on Machine Learning, pp. 2961–2970 (2019)
4. Jiang, J., Dun, C., Huang, T., Lu, Z.: Graph convolutional reinforcement learning. arXiv preprint arXiv:1810.09202 (2018)
5. Jiang, J., Lu, Z.: Learning attentional communication for multi-agent cooperation. In: Advances in Neural Information Processing Systems, pp. 7254–7264 (2018)
6. Kipf, T.N., Welling, M.: Semi-supervised classification with graph convolutional networks. arXiv preprint arXiv:1609.02907 (2016)
7. Li, X., Zhang, J., Bian, J., Tong, Y., Liu, T.Y.: A cooperative multi-agent reinforcement learning framework for resource balancing in complex logistics network. In: Proceedings of the 18th International Conference on Autonomous Agents and MultiAgent Systems, pp. 980–988 (2019)
8. Lillicrap, T.P., et al.: Continuous control with deep reinforcement learning. arXiv preprint arXiv:1509.02971 (2015)
9. Lowe, R., Wu, Y.I., Tamar, A., Harb, J., Abbeel, O.P., Mordatch, I.: Multi-agent actor-critic for mixed cooperative-competitive environments. In: Advances in Neural Information Processing Systems, pp. 6379–6390 (2017)
10. Mnih, V., et al.: Human-level control through deep reinforcement learning. Nature **518**(7540), 529–533 (2015)
11. Nguyen, H.T., et al.: A deep hierarchical reinforcement learner for aerial shepherding of ground swarms. In: Gedeon, T., Wong, K.W., Lee, M. (eds.) ICONIP 2019. LNCS, vol. 11953, pp. 658–669. Springer, Cham (2019). https://doi.org/10.1007/978-3-030-36708-4_54
12. Radhakrishnan, B.M., Srinivasan, D.: A multi-agent based distributed energy management scheme for smart grid applications. Energy **103**, 192–204 (2016)
13. Ryu, H., Shin, H., Park, J.: Multi-agent actor-critic with hierarchical graph attention network. arXiv preprint arXiv:1909.12557 (2019)
14. Schulman, J., Wolski, F., Dhariwal, P., Radford, A., Klimov, O.: Proximal policy optimization algorithms. arXiv preprint arXiv:1707.06347 (2017)
15. Vashishth, S., Yadati, N., Talukdar, P.: Graph-based deep learning in natural language processing. In: Proceedings of the 7th ACM IKDD CoDS and 25th COMAD, pp. 371–372 (2020)
16. Veličković, P., Cucurull, G., Casanova, A., Romero, A., Lio, P., Bengio, Y.: Graph attention networks. arXiv preprint arXiv:1710.10903 (2017)
17. Vinyals, O., et al.: Grandmaster level in starcraft ii using multi-agent reinforcement learning. Nature **575**(7782), 350–354 (2019)
18. Yang, Y., Luo, R., Li, M., Zhou, M., Zhang, W., Wang, J.: Mean field multi-agent reinforcement learning. In: 35th International Conference on Machine Learning, ICML 2018, vol. 80, pp. 5571–5580. PMLR (2018)
19. Zhang, Y., Dai, H., Kozareva, Z., Smola, A.J., Song, L.: Variational reasoning for question answering with knowledge graph. In: Thirty-Second AAAI Conference on Artificial Intelligence (2018)

Multi-view Subspace Adaptive Learning via Autoencoder and Attention

Jian-wei Liu$^{(\boxtimes)}$, Hao-jie Xie, Run-kun Lu, and Xiong-lin Luo

Department of Automation, College of Information Science and Engineering, China University of Petroleum, Beijing Campus (CUP), Beijing 102249, China
liujw@cup.edu.cn

Abstract. Multi-view learning can cover all features of data samples more comprehensively, so multi-view learning has attracted widespread attention. Traditional subspace clustering methods, such as sparse subspace clustering (SSC) and low-ranking subspace clustering (LRSC), cluster the affinity matrix for a single view, thus ignoring the problem of fusion between views. In our article, we propose a new Multi-view Subspace Adaptive Learning based on Attention and Autoencoder (MSALAA). This method combines a deep autoencoder and a method for aligning the self-representations of various views in Multi-view Low-Rank Sparse Subspace Clustering (MLRSSC), which can not only increase the capability to non-linearity fitting, but also can meets the principles of consistency and complementarity of multi-view learning. We empirically observe significant improvement over existing baseline methods on six real-life datasets.

Keywords: Multi-view learning · Subspace self-representation · Autoencoder · Attention · Spectral clustering

1 Introduction

In real-world machine learning problems, the same data consists of several different representations or views. For example, a traditional web page contains a lot of information. We can use pictures as one view and text features as another view. Therefore, each view reflects some properties of object. Although we can utilize a single view for learning tasks, integrating supplementary information from different views can reduce the complexity for a given task [1]. In recent years, multi-view learning has attracted more attention. Multi-view learning methods can learn each view with the help of consistency and complementarity on multiple views. Therefore, Multi-view learning not only can effectively use the special information of each view, but also take advantage of the common information of multiple views.

In recent years, many multi-view learning algorithms have been developed. For example, non-negative matrix factorization (NMF) [2] based multi-view

© Springer Nature Switzerland AG 2020
H. Yang et al. (Eds.): ICONIP 2020, LNCS 12533, pp. 536–545, 2020.
https://doi.org/10.1007/978-3-030-63833-7_45

learning algorithms [3–7]. These multi-view learning methods all consider consistency and complementarity in multi-view. However, the subspace clustering algorithm often ignores this information.

The subspace clustering algorithm is commonly divided into two stages. First, we need to construct an affinity matrix for each pair of data points. Second, we use this affinity matrix to implement spectral clustering. Of course, up to now self-representation subspace clustering algorithm is the most representative one. Self-representation-based methods represent data points as a linear combination of other points in the same subspace. These approaches are becoming increasing popular due to their excellent performance. Most recent work on self-representation-based method has focused on incorporating regularization term to make the self-representation matrix more robust, which is better than the factorization method, and can make full use of all data points to obtain a better representation. In this decade, several attempts have been made in this direction. The typical variants are Sparse Subspace Clustering (SSC) [8], Low Rank Subspace Clustering (LRSC) [9], Consistent and Specific Multi-view Subspace Clustering (CSMSC) [10], Multi-view Low-Rank Sparse Subspace Clustering (MLRSSC) [11]. Among them, CSMSC realizes the principle of consistency and complementarity by decomposing the self-representing coefficient matrix, and MLRSSC explores the alignment of the self-representing coefficient matrix. In addition, there are many methods for subspace self-representation learning using deep neural networks, such as Generalized Latent Multi-view Subspace Clustering (gLMSC) [12], Deep Subspace Clustering Networks (DSCN) [13].

In this paper, inspired by the autoencoder [16], attention mechanism [17] and MLRSSC, we develop a new Multi-view Subspace Adaptive Learning based on Attention and Autoencoder (MSALAA). First, we map different views to the same dimension, fuse each view with other views through the attention mechanism, and then construct the self-representation. Finally the self-representation output of each view is input to the decoder to reconstruct the original data, which are trained according to our designed loss function. Since the traditional subspace clustering algorithm needs to use ADMM algorithm to iteratively update the training parameters, and our algorithm is implemented by neural network, we only need to choose the optimization strategy, such as SGD, Adam, RMSProp, etc., and the deep learning optimizer will can automatically help us to update parameters. Our contributions in this article are as follows:

(1) We incorporate autoencoder with attention mechanism. The entire deep network is divided into four consecutive parts: encoder layer, multi-view attention layer, self-representation layer, and decoder Layer. The off-the-shelf optimizer of deep learning framework is used to automatically derive and update network parameters.

(2) Inspired by MLRSSC, we fuse the encoders' output with the same dimensions, so that learning hidden representation for each view incorporates the characteristics of hidden representation of other views. We improve the effect of the self-representation matrix of each view by adapting each view with

other views, and explicitly carry out the consistency and complementarity in multi-view learning.

(3) We have conducted extensive experiments on six real-life datasets that have different properties and scales to demonstrate the effectiveness and efficiency of our proposed formulation.

2   Based Method

Before introducing our method, we will introduce some basic concepts and representative methods of multi-view subspace clustering.

2.1   Self-representation of Data

We briefly introduce the self-representation method for training data. The meaning of data self-representation is that each data point in a union set of subspace can be effectively reconstructed by combining other points in the data set. More precisely, each data point x_i can be expressed as:

$$x_i = \mathbf{X}c_i, c_{ii} = 0 \tag{1}$$

where $c_i = [c_{i1}, c_{i2}, ..., c_{iN}]$ and N represents the number of samples. In addition, $c_{ii} = 0$ represents a simple way of eliminating the linear combination of writing points as themselves. In this way, we can represent each data point in the data point matrix X as a linear combination of other data points. Since the above problem has solution vectors of infinite number, incorporating constraint, the problem (1) is transformed into the following minimizing problem:

$$\min \|c_i\|_q \quad s.t. \quad x_i = \mathbf{X}c_i, \quad c_{ii} = 0 \tag{2}$$

where different choices for q have different effects in the obtained solutions, such as L_1 norm, kernel norm, and F-norm, etc. $q = 1$ is used in the SSC algorithm. We can also rewrite the above problem in matrix form:

$$\min \|\mathbf{C}\|_q \quad s.t. \quad \mathbf{X} = \mathbf{X}\mathbf{C}, \quad diag(\mathbf{C}) = 0 \tag{3}$$

where $\mathbf{C} = [c_1, c_2, \cdots, c_N] \in R^{N \times N}$ is a matrix of self-representation coefficients.

2.2   MLRSSC

Our work are inspired by MLRSSC, before we dive into the details of our proposed framework, let's briefly introduce MLRSSC based on pairwise similarity. Because this one is intendedly designed for multi-view data, it has good reference value. MLRSSC mainly concerns the similarity between matrix pairs represented by self-representation matrices. MLRSSC solves the following joint optimization problems with n_v views:

$$\min_{C^{(1)}, C^{(2)}, \cdots, C^{(n_v)}} \sum_{v=1}^{n_v} (\beta_1 \|C^{(v)}\|_* + \beta_2 \|C^{(v)}\|_1) + \sum_{1 \leq v, w \leq n_v, v \neq w} \lambda_{(v)} \|C^{(v)} - C^{(w)}\|_F^2 \tag{4}$$

$$s.t. X^{(v)} = X^{(v)}C^{(v)}, diag(C^{(v)}), v = 1, \cdots, n_v \tag{5}$$

where $C^{(v)} \in R^{N \times N}$ is the self-representation matrix of the view v. β_1, β_2, and $\lambda^{(v)}$ indicate the trade-off parameters between low-rank, sparse, and consistency constraints between views, respectively. In order to solve the convex optimization problem, MLRSSC used the alternating direction multiplier method (ADMM) [14].

3   Our Proposed Framework

In this section we exposure our proposed multi-view subspace adaptive learning paradigm in which we incorporate autoencoder with attention mechanism. The network structure in MSALAA is shown in Fig. 1. In Fig. 1, for simplicity, we only take the v-th sample with 3 views as a demonstration. The entire devised framework is divided into four consecutive stages: encoder layer, multiview attention layer, self-representation layer, and decoder Layer. First, let us briefly explain some denotation. Suppose that training data set contains multiple views, the v-th view is expressed as follows:

$$X^v = [x_1^v, x_2^v, \cdots, x_n^v, \cdots, x_N^v], X^v \in R^{F^v \times N}, v \in \{1, \cdots, M\}, n \in \{1, \cdots, N\} \tag{6}$$

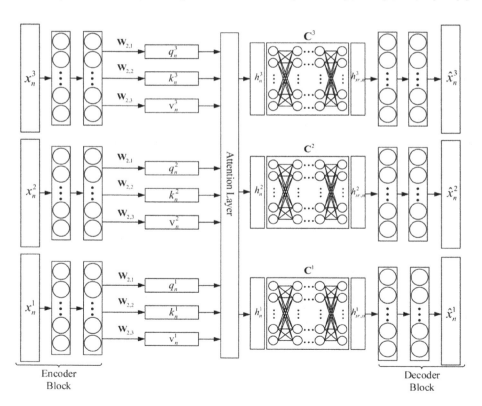

Fig. 1. Network structure of MSALAA.

where v is the number of v-th view, and N is the number of samples. F_v represents the number of features for the v-th view.

3.1 Encoder Layer

In the training data set, the feature dimensions in different views are different. hence we need to map examples in different views into the same dimension, so that the examples for each view has the same dimension$R$:

$$z_n^v = f(W_{1,l}^v \cdots f(W_{1,2}^v f(W_{1,1}^v x_n^v + b_{1,1}) + b_{1,2}) \cdots + b_{1,l}) \tag{7}$$

where $Z^v = [z_1^v, z_2^v, \cdots, z_n^v, \cdots, z_N^v] \in R^{R \times N}$. $W_{1,1}^v \in R^{R \times F^v}, \cdots, W_{1,l}^v \in R^{R \times F^v}$ and $b_{1,1} \in R^R, \cdots, b_{1,l} \in R^R$ are the weight matrices and biases of the fully connected layer, l is the number of network layers, and $f(\cdot)$ is the activation function of the fully connected layer. Here we use the ReLu as activation function.

3.2 Multi-view Attention Layer

In this layer, we need to use attention mechanism to process each view in order to achieve the fusion of the content of multiple views for each view. First, we construct a query matrix Q, a key matrix K, and a value matrix V by following identities:

$$Q^v = W_{2,1} Z^v, K^v = W_{2,2} Z^v, V^v = W_{2,3} Z^v \tag{8}$$

where $Q^v = [q_1^v, \cdots, q_n^v, \cdots, q_N^v], K^v = [k_1^v, \cdots, k_n^v, \cdots, k_N^v], V_v = [v_1^v, \cdots, v_n^v, \cdots, v_N^v], q_n^v, k_n^v, v_n^v \in R^{R \times 1}$, and $W_{2,1}, W_{2,2}, W_{2,3} \in R^{R \times R}$ are the linear transform matrices. Secondly, we need to calculate alignment weight a_i^v for the context vector h_i^v of i-th sample in the v-th view. We first define the score function:

$$score(q, k) = q \cdot k \tag{9}$$

The alignment weight is defined as follows:

$$a_i^v = \frac{exp(score(q_i^v, k_i^v))}{\sum_{j=1}^M exp(q_i^v, k_i^j)} \tag{10}$$

By getting the alignment weight a_i^v we can derive the context vector h_i^v:

$$h_i^v = a_i^v v_i^v \tag{11}$$

3.3 Self-representation Layer

For the subspace clustering, we need to utilize self-representation coefficient matrix $C^v \in R^{N \times N}$ and $H^v = h_1^v, \cdots, h_N^v \in R^{R \times N}$ to obtain representation recombination matrix $H_{sr}^v = [h_{sr,1}^v, h_{sr,2}^v, \cdots, h_{sr,n}^v, \cdots, h_{sr,N}^v] \in R^{R \times N}$, the formula is as follows:

$$H_{sr}^v = H^v C^v \tag{12}$$

What we need to note is that in subspace clustering, the self-representation coefficient matrix C^v needs to satisfy the constraint $diag(C^v) = 0$.

3.4   Decoder Layer

In this layer, we utilize the self-representation $h_{sr,n}^v$ as input of the fully connected layer to reconstruct the original data x_n^v of each view:

$$\hat{x}_n^v = f(W_{3,l}^v \cdots f(W_{3,2}^v f(W_{3,1}^v h_{sr,n}^v + b_{2,1}) + b_{2,2}) \cdots + b_{2,l}) \qquad (13)$$

where $\hat{x}_n^v$ is reconstruction of x_n^v. In addition, $W_{3,1}^v \in R^{F^v \times R}, \cdots, W_{3,1}^v \in R^{F^v \times R}$ and $b_{2,1} \in R^{F^v}, \cdots, b_{2,l} \in R^{F^v}$ are the weight matrices and bias vectors of the fully connected layer, and $f(\cdot)$ is the activation function of the fully connected layer, which uses the ReLu as activation function. We concatenate $\hat{x}_n^v$, $n \in \{1, \cdots, N\}$ to form the matrix $\hat{X}^v = [\hat{x}_1^v, \hat{x}_2^v, \cdots, \hat{x}_n^v, \cdots, \hat{x}_N^v] \in R^{F^v \times N}$.

3.5   Loss Function

This loss function can be divided into two parts. The first part is related to the encoder and decoder:

$$\sum_{v=1}^{M} \frac{1}{2NM} \left\| X^v - \hat{X}^v \right\|_F^2 + \beta_2 \Omega(W_{1,1}^v, \cdots, W_{1,l}^v, W_{3,1}^v, \cdots, W_{3,l}^v) \qquad (14)$$

where $\Omega(\cdot)$ is a regularization term, and its role is to constrain the parameters in the encoder and decoder. In this paper, $\Omega(\cdot)$ mainly have two forms, one is L_1 norm regularization term and the other is L_2 norm regularization term, and β_2 is a trade-off parameter. The other part is related to the self-representation of the subspace.

$$\sum_{v=1}^{M} \frac{1}{2N} \|H_{sr}^v - H^v\|_F^2 + \|C^v\|_F^2 + \beta_1 \sum_{1 \leq v,w \leq M, v \neq w} \frac{1}{2N} \|C^v - C^w\|_F^2 \qquad (15)$$

Here we first impose the alignment constraint for the self-representation matrices of multiple views in the form of $\sum_{1 \leq v,w \leq M, v \neq w} \frac{1}{2N} \|C^v - C^w\|_F^2$, by imposing the alignment constraint, the information of each view and other views can be fused with each other, so that the multi-views are complementary. We introduce the F norm for the self-representation coefficient matrix C^v. In the subspace self-representation learning task, we can apply different regularization constraints on the self-representation coefficient matrix C^v, for example, L_1 norm regularization term, kernel norm Constraints, and the F-norm constraints witch we use. In addition, β_1 is a trade-off parameter. Due to the use of deep autoencoder architecture, an additional benefit is that we do not need to resort complex optimization approaches, such as the ADMM algorithm for iterative updates, which is used in SSC and LRSC. We can directly update the self-representation coefficient matrix C^v by the off-the-shelf Stochastic Gradient Descent (SGD) approaches in Tensorflow.

Table 1. Characteristics of six datasets

Datasets	M	C	N	F^v
ORL	3	40	400	[4096, 3304, 6750]
Reuters	5	6	600	[21526, 24892, 34121, 15487, 11539]
3-sources	3	6	169	[3560, 3631, 3068]
Yale	3	15	165	[4096, 3304, 6750]
UCI digit	3	10	2000	[216, 76, 64]
Prokaryotic	3	4	551	[438, 3, 393]

4   Experimental Results

In this section, we design a series of experiments to demonstrate the effectiveness of MSALAA on real-world data sets. We first introduce the data set we use. Secondly, we describe the baseline methods, and finally we elaborate the configuration of hyper-parameters, and perform spectral clustering with MSALAA and six baseline methods on six data sets.

4.1   Datasets

To explore the performance of our proposed network, we have performed comparative experiments on six datasets, which are ORL, Reuters, 3-sources, Yale, UCI-digit, and Prokaryotic dataset. The characteristics of six datasets are summarized in Table 1. Where M, C, N and F^v represents the number of views, the sample category, the number of samples, and the feature dimensions of each view, respectively.

4.2   Baseline Methods

To demonstrate the efficiency of our proposed MSALAA, some state-of-the-art subspace clustering methods are chosen as the baseline methods: SSC, LRSC, LMSC [15], CSMSC, and MLRSSC and its three improved variants (MLRSSC-Centroid, KMLRSSC, KMLRSSC-Centroid).

4.3   The Configuration of Hyper-parameters

In our experiments, we only need to set the number of layers of the network and the number of neurons in each layer, and determine whether to perform batch normalization. In this article, all our datasets are initialized with the LeCun normal distribution. We choose Adam optimization algorithm as optimizer, the learning rate is 0.001, and the decay value of learning rate is 0.99. In addition, we set the trade-off parameters β_1 and β_2 to 0.1. In the experiment, we will get the self-representation matrix of multiple views. Here we will use the self-representation matrix C^v of the best experimental view to construct the affinity matrix $A^v = |C^v| + |C^v|^T$ for spectral clustering. In addition, we performed residual processing on the network to prevent the gradient from disappearing.

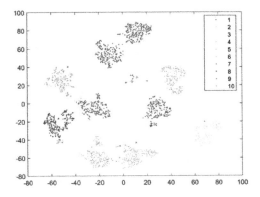

Fig. 2. Visualization of H^v on UCI-digit.

Table 2. Experimental comparison results on three datasets

Method for ORL	ACC	NMI	ARI	Precision	Recall	F-score
SSC	72.13%(0.017%)	87.33%(0.001%)	63.26%(0.023%)	74.25%(0.023%)	72.13%(0.017%)	70.84%(0.022%)
LRSC	72.23%(0.027%)	84.06%(0.005%)	59.82%(0.032%)	74.78%(0.041%)	72.23%(0.027%)	71.79%(0.033%)
LMSC	82.18%(0.136%)	92.74%(0.018%)	77.39%(0.188%)	81.25%(0.148%)	82.18%(0.136%)	80.21%(0.168%)
CSMSC	84.55%(0.004%)	92.74%(0.001%)	79.78%(0.004%)	85.30%(0.006%)	84.55%(0.004%)	83.78%(0.005%)
MLRSSC	63.68%(0.116%)	81.a32%(0.029%)	52.41%(0.136%)	63.97%(0.108%)	63.68%(0.116%)	62.27%(0.113%)
MLRSSC-C	78.03%(0.075%)	91.65%(0.008%)	72.88%(0.085%)	78.46%(0.106%)	78.03%(0.075%)	76.11%(0.101%)
KMLRSSC	78.55%(0.166%)	90.27%(0.026%)	72.05%(0.175%)	79.32%(0.163%)	78.55%(0.166%)	77.43%(0.177%)
KMLRSSC-C	78.25%(0.094%)	90.71%(0.007%)	72.06%(0.069%)	79.34%(0.153%)	78.25%(0.094%)	77.16%(0.123%)
MSALAA	86.40%(0.009%)	93.16%(0.009%)	80.91%(0.032%)	86.43%(0.023%)	86.40%(0.009%)	85.38%(0.015%)
Method for Reuters	ACC	NMI	ARI	Precision	Recall	F-score
SSC	50.85%(0.000%)	35.05%(0.002%)	24.69%(0.000%)	51.93%(0.001%)	50.85%(0.000%)	44.89%(0.000%)
LRSC	31.12%(0.000%)	14.18%(0.000%)	3.257%(0.000%)	58.52%(0.000%)	31.12%(0.000%)	28.13%(0.000%)
LMSC	41.53%(0.010%)	33.04%(0.024%)	17.34%(0.039%)	40.56%(0.014%)	41.53%(0.010%)	33.46%(0.005%)
CSMSC	42.42%(0.000%)	32.63%(0.000%)	18.28%(0.000%)	46.20%(0.000%)	42.42%(0.000%)	34.49%(0.000%)
MLRSSC	52.95%(0.108%)	38.22%(0.024%)	28.18%(0.070%)	50.09%(0.308%)	52.95%(0.108%)	48.30%(0.259%)
MLRSSC-C	51.35%(0.124%)	36.96%(0.013%)	26.76%(0.070%)	48.36%(0.160%)	51.35%(0.124%)	45.89%(0.216%)
KMLRSSC	57.12%(0.054%)	37.38%(0.037%)	30.41%(0.040%)	61.79%(0.135%)	57.12%(0.054%)	56.67%(0.088%)
KMLRSSC-C	55.12%(0.057%)	35.69%(0.026%)	29.38%(0.027%)	59.05%(0.208%)	55.12%(0.057%)	53.97%(0.071%)
MSALAA	57.88%(0.034%)	40.69%(0.020%)	31.02%(0.013%)	65.14%(0.025%)	57.88%(0.034%)	57.00%(0.043%)
Method for UCI digit	ACC	NMI	ARI	Precision	Recall	F-score
SSC	78.02%(0.004%)	79.08%(0.007%)	71.06%(0.011%)	79.41%(0.001%)	78.02%(0.004%)	77.59%(0.001%)
LRSC	64.19%(0.001%)	68.61%(0.000%)	56.01%(0.003%)	65.54%(0.003%)	64.19%(0.001%)	63.20%(0.001%)
LMSC	74.11%(0.446%)	74.63%(0.112%)	65.19%(0.274%)	74.56%(0.591%)	74.11%(0.446%)	72.98%(0.582%)
CSMSC	83.22%(0.114%)	78.48%(0.017%)	72.29%(0.048%)	83.30%(0.182%)	83.22%(0.115%)	82.78%(0.157%)
MLRSSC	88.08%(0.374%)	85.15%(0.048%)	81.37%(0.251%)	87.71%(0.485%)	88.08%(0.374%)	87.34%(0.496%)
MLRSSC-C	89.27%(0.298%)	85.32%(0.054%)	81.81%(0.253%)	89.14%(0.399%)	89.27%(0.298%)	88.74%(0.395%)
KMLRSSC	89.35%(0.244%)	86.08%(0.033%)	82.60%(0.193%)	88.76%(0.360%)	89.35%(0.244%)	88.61%(0.342%)
KMLRSSC-C	90.97%(0.212%)	86.51%(0.033%)	84.03%(0.155%)	91.03%(0.257%)	90.97%(0.212%)	90.64%(0.284%)
MSALAA	96.07%(0.000%)	91.31%(0.001%)	91.48%(0.001%)	96.12%(0.000%)	96.07%(0.000%)	96.08%(0.000%)

4.4   Experimental Results and Analysis

We perform spectral clustering with MSALAA and six baseline methods on six
data sets. In order to verify the performance of our method, we selected six
metric criteria to measure the effect of spectral clustering: Accuracy (ACC),
Normalized Mutual Information (NMI), Adjusted Rand Index (ARI), Precision,
Recall and F-score. From the experimental results in Table 2, we can clearly
see that our method has obvious advantages over other baseline comparison
methods, validate that MSALAA can find better data self-representation. On

the data set UCI-digit, MSALAA can reach more than 90% on multiple metric criteria.

In addition, we also visualize the matrix H^v generated by the attention mechanism. Here we use t-SNE. In the visualization experiment, we use t-SNE to embed feature matrix H^v into a 2D latent feature matrix for clustering. We use t-SNE to derive a 2D latent feature matrix and depict it with a scatter plot. As shown in Fig. 2, this process is performed on the data set UCI-digit, and we can clearly see that each clustering can be easily distinguished.

5   Conclusion and Future Work

We propose a new Multi-view Subspace Adaptive Learning based on Attention and Autoencoder (MSALAA) combined with mutual subspace alignment in subspace clustering. Our method takes into account the two important factors of consistency and complementarity in multi-view learning, and also utilizes the neural network to increase the nonlinear representation ability of the model. The experiments on several real-world datasets showed that the proposed MSALAA mostly outperformed the other baseline methods, which validate that our proposed MSALAA can use self-representation of multi-views to subspace adaptive learning.

In future work, we will investigate some variants for MLRSSC. We will try to build a common self-representation matrix C^* and align C^* with all views to see the influence of performance for multi-view subspace Learning. In addition, we can improve the basic fully connected network to achieve better performance on multi-view data.

References

1. Sun, S.: A survey of multi-view machine learning. Neural Comput. Appl. **23**(7–8), 2031–2038 (2013). https://doi.org/10.1007/s00521-013-1362-6
2. Lee, D.D., Seung, H.S.: Learning the parts of objects by non-negative matrix factorization. Nature **401**(6755), 788–791 (1999). https://doi.org/10.1038/44565
3. Gao, J., Han, J., Liu, J., Wang, C.: Multi-view clustering via joint nonnegative matrix factorization. In: SDM, pp. 252–260. SIAM, Philadelphia (2013). https://doi.org/10.1137/1.9781611972832.28
4. Liu, J., Jiang, Y., Li, Z.: Partially shared latent factor learning with multiview data. IEEE Trans. Neural Netw. Learn. Syst. **29**(8), 1233–1246 (2015). https://doi.org/10.1109/TNNLS.2014.2335234
5. Wang, Z., Kong, X., Fu, H., Li, M., Zhang, Y.: Feature extraction via multi-view non-negative matrix factorization with local graph regularization. In: ICIP, pp. 3500–3504. IEEE, Quebec City (2015). https://doi.org/10.1109/ICIP.2015.7351455
6. Wang, H., Yang, Y., Li, T.: Multi-view clustering via concept factorization with local manifold regularization. In: ICDM, pp. 1245–1250. IEEE, Piscataway (2016). https://doi.org/10.1109/ICDM.2016.0167

7. Zhang, Z., Qin, Z., Li, P., Yang, Q., Shao, J.: Multi-view discriminative learning via joint non-negative matrix factorization. In: Pei, J., Manolopoulos, Y., Sadiq, S., Li, J. (eds.) DASFAA 2018. LNCS, vol. 10828, pp. 542–557. Springer, Cham (2018). https://doi.org/10.1007/978-3-319-91458-9_33

8. Elhamifar, E., Vidal, R.: Sparse subspace clustering: algorithm, theory, and applications. IEEE Trans. Pattern Anal. Mach. Intell. **35**(11), 2765–2781 (2013)

9. Vidal, R., Favaro, P.: Low rank subspace clustering (LRSC). Pattern Recogn. Lett. **43**, 47–61 (2014). https://doi.org/10.1016/j.patrec.2013.08.006

10. Luo, S., Zhang, C., Zhang, W., Cao, X.: Consistent and specific multi-view subspace clustering. In: AAAI, pp. 3730–3737. New Orleans, LA (2018)

11. Brbic, M., Kopriva, I.: Multi-view low-rank sparse subspace clustering. Pattern Recogn. **73**, 247–258 (2018). https://doi.org/10.1016/j.patcog.2017.08.024

12. Zhang, C., et al.: Generalized latent multi-view subspace clustering. IEEE Trans. Pattern Anal. Mach. Intell. **42**(1), 86–99 (2018). https://doi.org/10.1109/TPAMI.2018.2877660

13. Ji, P., Zhang, T., Li, H., Salzmann, M., Reid, I.D.: Deep subspace clustering networks. In: NIPS, pp. 24–33. Long Beach, CA (2017)

14. Cai, J.-F., Candés, E.J., Shen, Z.: A singular value thresholding algorithm for matrix completion. SIAM J. Optim. **20**(4), 1956–1982 (2010). https://doi.org/10.1137/080738970

15. Zhang, C., Hu, Q., Fu, H., Zhu, P., Cao, X.: Latent multi-view subspace clustering. In: CVPR, pp. 4333–4341. IEEE, Honolulu (2017). https://doi.org/10.1109/CVPR.2017.461

16. Hinton, G.E., Salakhutdinov, R.R.: Reducing the dimensionality of data with neural networks. Science **313**(5786), 504–507 (2006). https://doi.org/10.1126/science.1127647

17. Vaswani, A., et al.: Attention is all you need. In: NIPS, pp. 5998–6008. Long Beach, CA (2017)

Network Coding for Federated Learning Systems

Lingwei Kong, Hengtao Tao, Jianzong Wang$^{(\boxtimes)}$, Zhangcheng Huang,
and Jing Xiao

Ping An Technology (Shenzhen) Co., Ltd., Shenzhen, China
{konglingwei630,taohengtao880,wangjianzong347,huangzhangcheng624,
xiaojing661}@pingan.com.cn, jzwang@188.com

Abstract. Nowadays, artificial intelligence is limited by privacy and
security problems. Compared with the ordinary machine learning, federated learning (FL) enables multiple participants to collaboratively learn
a shared machine learning model while keeping all the training data on
local devices. However, most of the current secured federated learning
systems (FLSs) are built up with high computational and communication
costs. On the other hand, optimizing the network structure of federated
learning systems can reduce communication complexity by considering
the correlation of the transmission channels.

In this paper, we propose Network Coding Federated Learning Systems (NC-FLSs). Specifically, it considers the whole communication network by connecting all the clients and the server. Applying a linear NC
scheme to construct a linear combination of the original messages, which
is transmitted over the network instead of the messages themselves.
Based on NC-FLSs, the communication cost is halved and both data
privacy and security are improved with the imperceptibly higher computational cost. Moreover, considering that the network coding structure
is independent of the FL model, any FLSs can also be upgraded to its
corresponding NC-FLSs. We also implement differential privacy on an
NC-FLS to train an image classifier while keeping clients' local data
secure and private, which achieves superior performance and efficiency.

Keywords: Network coding · Federated learning · System security

1 Introduction

Machine learning faces two major challenges. One is that user data often exists
in the form of isolated islands. The other is that data privacy and security are
becoming increasingly important. Luckily, this requirement can be satisfied by
Federated Learning [11,14]. Federated Learning is proposed for the protection
of data privacy and decentralized machine learning, where each client's local

This paper is supported by National Key Research and Development Program of China
under grant No. 2018YFB1003500, No. 2018YFB0204400 and No. 2017YFB1401202.

data is used to train the learning models on its own device, and the data will not be exposed to the others (including the server) [10]. Thus, only the models' parameters (such as gradients and loss) can be transmitted. In addition, the transmitted parameters should be encrypted beforehand. Here are some state-of-the-art secure FLSs. [16] uses additively homomorphic encryption [2] where each client encrypts the uploaded gradients before averaging for preventing indirect leakage. In [18], differential privacy (DP) [6] is applied. Most existing FLSs are built to improve communication efficiency by reducing the communication costs (especially the uplink communication costs). These FLSs do not consider the structure of the transmission network.

On the other hand, the throughput of a network can be significantly improved by applying network coding [12]. The security of network coding was designed so that the message can be sent to the receiver without leaking any information to the eavesdropper [3]. Thanks to the security and high communication efficiency, network coding has been applied to network monitoring [8] and distributed storage systems [5].

This paper proposes secure NC-FLS, by invoking constructive results in network coding. In NC-FLSs, when a set of models parameters needs to be transmitted, the sender(the server or a client) will send a linear combination of these model parameters. These transmitted vectors are linearly independent and are transmitted over different links respectively. Thus, when the receiver (a client or the server) receives these vectors, it can obtain the original model parameters. Meanwhile, if there exists an eavesdropper who can obtain the messages from one link of the transmission network, it cannot obtain any model parameter. Compared with existing FLSs, NC-FLSs have the following three advantages:

- **Security and Privacy:** Most existing secure FLSs require high computational or communication costs. However, NC-FLSs only require negligibly higher computational and communication costs than the FLS which doesn't encrypt the transmission parameters at all. Meanwhile, NC-FLSs can also prevent information leakage, even if there is an eavesdropper who can obtain the transmitted messages from one link of the transmission network.
- **Communication Efficiency:** Existing FLSs only try to reduce the communication costs without considering the structure of the transmission network. While NC-FLSs can reduce the communication rounds typically. In the section of Related Work, we will give an example to prove that the communication rounds are halved by applying network coding.
- **Applicability (bonus):** In NC-FLSs, the construction of network coding is independent of FLS. Any other FLS can be upgraded to its NC-FLS version which has all the properties of the original FLS, and also have the properties of NC-FLS (higher security, higher privacy and higher communication efficiency).

2   Related Work

2.1   Federated Learning (FL)

The learning task in a federated learning system is solved by a loose federation of participating clients which are coordinated by a central server. Each client applies its local data to compute and sends out the model parameters with the local data kept private and secure. Based on the distribution characteristics of clients' data, [19] categorizes federated learning into horizontal, vertical and federated transfer learning.

Horizontal federated learning applies to the scenarios that clients' data has a lot of overlap on features but a little on IDs. Thus, all the clients would share the same model. Based on Asynchronous SGD [4,17], proposed a horizontal federated learning system for Google Gboard updates. Considering that the updated models may reveal information on local data, [16], based on additively homomorphic encryption, introduced an FLS to protect the privacy of these updated models. The public key is known to the server and all the clients. However, only do the clients know the secret key. NC-FLS can be applied to all of these three categories. We only consider the horizontal federated learning in this paper.

2.2   Network Coding (NC)

The basic idea of network coding [15] is to improve security and communication efficiency by mixing information in the middle of a transmission network. To improve the robustness and throughput of the transmission network, the data transmitted between devices are encoded on the sender device and decoded on the transmission destination device by network coding techniques, which can be simply algebraic algorithms. The encoding and decoding algorithms on the data can accumulate various kinds of transmission tasks. The encoded data requires fewer transmissions than sending the original data, but also need more processing at intermediary and terminal nodes. The raw idea of Network coding is proposed to improve the throughput of two-way communication through a satellite and recent works [9,13] focus on improving network coding transmission efficiency and security.

2.3   Differential Privacy (DP)

Differential privacy ensures that any change of an item of a dataset doesnot (substantially) affect the mechanism outcome information to protects privacy. A random mechanism $\mathcal{M} : \mathcal{D} \to \mathcal{R}$ satisfies (ϵ, δ)-differential privacy [7], which influences the intensity of security protection which will affect the accuracy performance and efficiency. if for all local data sets $D_1, D_2 \in \mathcal{D}$ differing on at most one element, and all $S \subset \mathcal{R}$,

$$Pr[\mathcal{M}(d) \in S] \leq e^\epsilon Pr[\mathcal{M}(d') \in S] + \delta. \tag{1}$$

In this paper, we will employ the differentially private stochastic gradient descent algorithm (DP-SGD) proposed by [1]. We propose an NC-FLS based on DP-SGD, to show that NC-FLS is compatible with the other system.

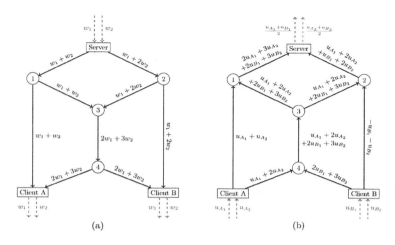

Fig. 1. Examples of NC-FLS, (a) downlink, (b) uplink.

3   A Simple NC-FLS for Two Clients

In this section, based on horizontal federated learning, we design a simple NC-FLS to show that NC-FLSs can improve data security, data privacy and transmission efficiency. In this system, each client updates the model parameters using its local data and then uploads a combination of these model parameters to the server (Uplink Transmission). After that, the server calculates the averages of the model parameters of the two clients, and then sends a combination of these averages back to both the clients (Downlink Transmission). Compared to the replicate-and-send method, network coding allows the intermediate nodes to generate output parameter by encoding previously received input parameters. Thus, two parameters can be sent out at once, i.e., half of the download transmission time could be saved by applying network coding.

3.1   Downlink Transmission

We firstly consider the downlink transmission. Although the traditional transmission scheme can be used to improve the transmission rate, it cannot be used directly, since the data security is too weak. For example, if the eavesdropper can obtain the messages from the link connecting Node 1 and Client A, then it can obtain half of the model parameters.

In order to improve the data security, we have designed a new transmission scheme, shown in Fig. 1(a). Firstly the server calculates the average of the model parameters w_1 and w_2 of all the clients, where w_1 and w_2 can be two real numbers or two-column vectors with the same length. Then it does not transmit w_1 and w_2 to Nodes 1 and 2 directly (before). Instead, the server encrypts the w_1 and w_2 beforehand. Specifically, the server sends $m_{S,1} = w_1 + w_2$ and $m_{S,2} = w_1 + 2w_2$ to Node 1 and 2 respectively.

$$\begin{bmatrix} m_{S,1} & m_{S,2} \end{bmatrix} = \begin{bmatrix} w_1 & w_2 \end{bmatrix} \begin{bmatrix} 1 & 1 \\ 1 & 2 \end{bmatrix} \tag{2}$$

For the intermediate nodes except Node 3, since the number of the messages that they receive are all 1, each of them simply needs to replicate and sends out the message it receives from upstream. For the Node 3, it receives the two messages from Node 1 and 2. It needs to send out the summation of these two messages. Finally, Client A receives two messages $m_{1,A} = w_1 + w_2$ and $m_{4,A} = 2w_1 + 3w_2$, shown in a matrix.

$$\begin{bmatrix} m_{1,A} & m_{4,A} \end{bmatrix} = \begin{bmatrix} w_1 & w_2 \end{bmatrix} \begin{bmatrix} 1 & 2 \\ 1 & 3 \end{bmatrix} \tag{3}$$

Then Client A can decrypt the messages, shown as following.

$$\begin{bmatrix} w_1 & w_2 \end{bmatrix} = \begin{bmatrix} m_{1,A} & m_{4,A} \end{bmatrix} \begin{bmatrix} 1 & 2 \\ 1 & 3 \end{bmatrix}^{-1} = \begin{bmatrix} m_{1,A} & m_{4,A} \end{bmatrix} \begin{bmatrix} 3 & -2 \\ -1 & 1 \end{bmatrix} \tag{4}$$

Similarly, Client B receives

$$\begin{bmatrix} m_{1,A} & m_{4,A} \end{bmatrix} = \begin{bmatrix} w_1 & w_2 \end{bmatrix} \begin{bmatrix} 1 & 2 \\ 2 & 3 \end{bmatrix} \tag{5}$$

From the Max-flow Min-cut Theorem, which proves there exits a minimum cut that the capacity equals to the maximum flow of the system, the information rate received by Client A or B cannot exceed 2. Thus, the transmission system has obtained the optimal communication efficiency. This system can also guarantee the security of the data. Because even if there is an eavesdropper who can obtain the message from one of the links in the transmission network, it cannot obtain w_1 or w_2.

3.2   Uplink Transmission

Now we consider the uplink transmission. Our goal is to design a system which can satisfy the following three requirements.

- **Security and Privacy.** Prevent the model parameters from revealing to the eavesdropper who can obtain the messages from one link. Also, prevent the model parameters from revealing to the server.
- **Communication Efficiency.** In order to obtain the maximum rate 2, each client needs to update two model parameters at once and the server is ensured to be able to calculate two averages.

Specifically, Client A sends out encrypted model parameters u_A^1 and u_A^2, and Client B sends out encrypted model parameters u_B^1 and u_B^2 These four model parameters mention above are all real numbers or two column vectors with the same length. Finally, the server need to calculate two averages $\frac{u_A^1 + u_B^1}{2}$ and $\frac{u_A^2 + u_B^2}{2}$.

The uplink transmission system satisfying the three requirements above is shown in Fig. 1 (b). Firstly, Client A computes the updated model parameters u_A^1 and u_A^2, and Client B computes the updated model parameters u_B^1 and u_B^2. These four parameters can be four real numbers or column vectors with the same length. Then, Client A sends $m_{A,1} = u_1^A + u_2^A$ and $m_{A,4} = u_1^A + 2u_2^A$ to Node 1 and 4 respectively, and Client B sends $m_{B,2} = -u_1^B - u_2^B$ and $m_{B,4} = 2u_1^B + 3u_2^B$ to Node 2 and 4 respectively. For the intermediate nodes except Node 4, each of them simply needs to replicate and sends out the message it receives from upstream. For the Node 4, it receives the two messages from Client A and B. It needs to send out the summation of these two messages. Finally, the server receives two messages $m_{1,S} = 2u_{A_1} + 3u_{A_2} + 2u_{B_1} + 3u_{B_2}$ and $m_{2,S} = u_{A_1} + 2u_{A_2} + u_{B_1} + 2u_{B_2}$, shown in a matrix.

$$[m_{1,S}\ m_{2,S}] = [u_{A_1}\ u_{A_2}\ u_{B_1}\ u_{B_2}] \begin{bmatrix} 2 & 1 \\ 3 & 2 \\ 2 & 1 \\ 3 & 2 \end{bmatrix} = \begin{bmatrix} \frac{u_{A_1}+u_{B_1}}{2} & \frac{u_{A_2}+u_{B_2}}{2} \end{bmatrix} \begin{bmatrix} 4 & 2 \\ 6 & 4 \end{bmatrix}. \tag{6}$$

Then we can obtain the two averages by following equation.

$$\begin{bmatrix} \frac{u_{A_1}+u_{B_1}}{2} & \frac{u_{A_2}+u_{B_2}}{2} \end{bmatrix} = [m_{1,S}\ m_{2,S}] \begin{bmatrix} 4 & 2 \\ 6 & 4 \end{bmatrix}^{-1} \tag{7}$$

Based on Eqs. 2 and 7, we have

$$\begin{aligned} [m_{S,1}\ m_{S,2}] &= [m_{1,S}\ m_{2,S}] \begin{bmatrix} 4 & 2 \\ 6 & 4 \end{bmatrix}^{-1} \begin{bmatrix} 1 & 1 \\ 1 & 2 \end{bmatrix} \\ &= [m_{1,S}\ m_{2,S}] \begin{bmatrix} \frac{1}{2} & 0 \\ \frac{-1}{2} & \frac{1}{2} \end{bmatrix} \end{aligned} \tag{8}$$

We have shown that this uplink system can be used to upload model parameters from the client to the server. Now we prove this uplink system can satisfy the three requirements.

- **Security and Privacy.** As shown in Fig. 1 (b) for any link connecting two node, the message of the link is not a model parameter. Thus, even if the eavesdropper can obtain the message from one of the links, it cannot calculate any model parameter. In this uplink system, the server cannot decrypt any model parameters. The proof is shown as followed. The server can only obtain $m_{1,S}$ and $m_{2,S}$, and the relationship between these two messages and the model parameters. When u_{A_1}, u_{A_2}, u_{B_1} and u_{B_2} are four real numbers, by solving Eq. 6, we have

$$\begin{bmatrix} u_{A_1} \\ u_{A_2} \\ u_{B_1} \\ u_{B_2} \end{bmatrix} = c_1 \begin{bmatrix} 1 \\ 0 \\ -1 \\ 0 \end{bmatrix} + c_2 \begin{bmatrix} 0 \\ 1 \\ 0 \\ -1 \end{bmatrix} + \begin{bmatrix} \eta_1 \\ \eta_2 \\ \eta_3 \\ \eta_4 \end{bmatrix}, \tag{9}$$

where $c_1, c_2, \eta_1, \eta_2, \eta_3, \eta_4 \in \mathbb{R}$, c_1, c_2 can be any two real numbers, and $(\eta_1, \eta_2, \eta_3, \eta_4)$ is one of the solutions of this equation.

– **Communication Efficiency.** At once, the server can calculate two averages. Thus the upload system has already obtained the maximum rate 2.

4   Implementation

In this section, we implement DP-SGD on a two-client NC-FLS. We follow the system design of NC-FLSs proposed in Sect. 3 using differential privacy on machine learning algorithms, which we deployed DP-SGD to protect each client's local data from the other client. The training process is shown as followed:

Algorithm 1. NC-FLS Server

Require: $\theta^{(0)}$ model initialization, the length of $\theta^{(0)}$ is $2l$

$\quad w_1^{(0)} \leftarrow \theta^{(0)}[:l]$,

$\quad w_2^{(0)} \leftarrow \theta^{(0)}[l:]$

$\quad$ Send $m_{S,1}^{(0)} \leftarrow w_1^{(0)} + w_2^{(0)}$ to Node 1

$\quad$ Send $m_{S,2}^{(0)} \leftarrow w_1^{(0)} + 2w_2^{(0)}$ to Node 2

$\quad$ **for** $t = 1, 2, \ldots$ **do**

$\quad\quad$ Wait for $m_{1,S}^{(t)}, m_{1,S}^{(t)}$ from Node 1 and 2 respectively

$\quad\quad$ Send $m_{S,1}^{(t)} \leftarrow \frac{1}{2}m_{1,S}^{(t)} - \frac{1}{2}m_{2,S}^{(t)}$ to Node 1

$\quad\quad$ Send $m_{S,2}^{(t)} \leftarrow \frac{1}{2}m_{2,S}^{(t)}$ to Node 2

$\quad$ **end for**

Step 1. The server initializes model parameters $\theta^{(0)}$, and sends out encrypted.
Step 2. Each client receives the model parameters from the server, and then decrypts it and updates the model using local data by DP-SGD. After that, each client encrypts the updated model parameters and sends them out.
Step 3. The server receives the messages from two clients, and then computes the new model parameters. After that, the new model parameters are encrypted and sent out.
Step 4. Back to Step 2.

The details are shown in Algorithms 1–2, where the detailed process on client B is similar to client A and omitted.

5   Experimental Results

We conduct experiments on the standard MNIST dataset for handwritten digit recognition. These samples are divided into two parts, allocating to Client A and Client B. Specifically, Client A has half of these data consisting of a training set of $N_A^{Tr} = 30000$ examples, and a test set of $N_A^{Te} = 5000$ samples. Client B has the remaining half, i.e., $N_B^{Tr} = 30000$ and $N_B^{Te} = 5000$. The digits have been size-normalized and centered in a fixed-size image with size 28×28. [1]

Algorithm 2. NC-FLS Client A

Require: Dataset $\mathcal{X}$
Require: Superparameters of Differentially private SGD
   for $t = 0, 1, \ldots$ **do**
      Receive and Decryption
      Wait for $m_{1,A}^{(t)}, m_{4,A}^{(t)}$ from Node 1 and 4 respectively
      $\theta^{(t)}[: l] \leftarrow 3m_{1,A}^{(t)} - m_{4,A}^{(t)}$
      $\theta^{(t)}[l :] \leftarrow -2m_{1,A}^{(t)} + m_{4,A}^{(t)}$
      Differential private SGD
      $\theta_A^{(t+1)}$ is obtained by training $\theta^{(t)}$ based on $\mathcal{X}$
      Encryption and Send
      Divide $\theta_A^{(t+1)}$ into two vectors $u_{A_1}^{(t+1)}, u_{A_2}^{(t+1)}$:
      $u_{A_1}^{(t+1)} \leftarrow \theta_A^{(t+1)}[: l]$
      $u_{A_2}^{(t+1)} \leftarrow \theta_A^{(t+1)}[l :]$
      Send $m_{A,1}^{(t+1)} \leftarrow u_{A_1}^{(t+1)} + u_{A_2}^{(t+1)}$ to Node 1
      Send $m_{A,4}^{(t+1)} \leftarrow u_{A_1}^{(t+1)} + 2u_{A_2}^{(t+1)}$ to Node 4
   end for

---

only demonstrated the training of deep neural networks with centralized data. While in this paper, we will demonstrate the training with local data and with NC-FLS respectively.

This section is divided into two parts: "without Differential Privacy" and "with Differential Privacy". In each part, we will respectively show the training with local data and with FL. There are four experiments using the same convolutional neural network (CNN). The network architecture consists of two convolutional layers followed by one fully connected layer. The first convolutional layers use 8×8 convolutions with stride 2, followed by a ReLU and 2×2 max pools, with 16 channels. The second convolutional layers use 4×4 convolutions with stride 2, followed by a ReLU and 2×2 max pools, with 32 channels. The first convolution outputs a $11 \times 11 \times 16$ tensor for each image, and the second outputs a $2 \times 2 \times 32$ tensor. The latter is fattened to a vector that gets fed into a fully connected layer with 32 units.

For each client, we set the batch size $L = 300$, so the ratio $q = L/N_A = 0.01$ and the number of epochs $E = Tq$, where T is the number of training steps.

5.1 Without Differential Privacy

Training with Local Data. We first show the training using Client A's local data without DP. The number of the training data is $N_A = 30000$. The training result is shown in Fig. 2(a). In 15 epochs (1500 steps), the testing accuracy can reach 98.45%.

Training with FLS non-NC. we show the training with all the data with federated learning with non-NC,. The number of the training data is $N_A + N_B$. The training result is shown in Fig. 2(b). In 15 epochs (1500 steps), the testing accuracy can reach 99.07%.

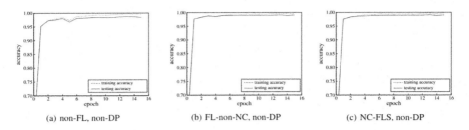

(a) non-FL, non-DP   (b) FL-non-NC, non-DP   (c) NC-FLS, non-DP

Fig. 2. Results on the accuracy for different models without differential privacy. (a) is trained using Client A's local data only (non-FL). While (b) is trained on FLS without using NC and (c) is trained by NC-FLS without uploading local data. With learning rate $\eta = 0.15$, we achieve an accuracy 98.45%, 99.07% and 99.05%, respectively.

Training with NC-FLS. Now we show the training using NC-FLS, which is depicted in Sect. 3. In this model, two clients collaborate train the same model without uploading the local data. The training result is shown in Fig. 2(c). In 15 epochs (1500 steps), the testing accuracy is improved to 99.05%, which can be considered as the same accuracy as Fig. 2(b).

5.2   With Differential Privacy

Now we consider the models with SGD-DP. Similar to the part above, we will also show the training with only local data and with NC-FLS respectively. In each experiment, we will show the results on the accuracy of different noise levels: $\sigma = 4, 2, 1$.

Training with Local Data. In this non-FL model (with SGD-DP), only is Client A's data used. The training results for different noise levels are shown in Figs. 3(a), 3(d) and 3(g).

Training with FL-non-NC. In this FL model (with SGD-DP) without using network coding, with both data on Client A and Client B. The training results for different noise levels are shown in Figs. 3(b), 3(e) and 3(h).

Training with NC-FLS. This NC-FLS (with SGD-DP) is depicted in Sect. 4. In this model, Client A trains the model with the help of Client B whose local data is kept secure and private. The training results for different noise levels are shown in Figs. 3(c), 3(f) and 3(i).

Compare the training with only local data, FL-non-NC and NC-FLS, we can observe the following results.

1. NC-FLS can achieve as good accuracy as FL-non-NC while having a more fluent raise curve of training and testing performance. The NC-FLS on SGD-DP saves about 15% training time than FLS without network coding by reducing communication costs.

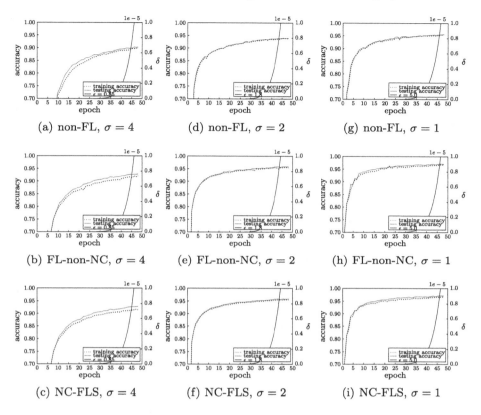

Fig. 3. Results on the accuracy for different noise levels on MNIST. (a), (d) and (g) are trained only using Client A's data (non-FL), (b), (e) and (f) are trained on FL without using network coding. While (c), (f) and (i) is trained by NC-FLS without uploading local data. For the non-FL models, with δ set to 10^{-5}, we achieve testing accuracy 89.8%, 93.8% and 95.5% with σ being 4, 2 and 1, respectively. For FL-non-NC and NC-FLS, with δ set to 10^{-5}, we achieve testing accuracy 92.7%, 95.8% and 97.1% for FL-non-NC and 92.9%, 95.8% and 97.2% for NC-FLS. In all cases, the slot size is set to 300 ($q = 0.01$) for each client.

2. When the δ is fixed, with decreasing noise level, the testing accuracy grows and the privacy of the data is better (ϵ is smaller) on all three systems.
3. When the noise level is fixed, the testing accuracy of the non-FL model is around 2% lower than systems using FL. The raise of the accuracy of the non-FL model has a greater fluctuation than NC-FLS.

Table 1. The overview experimental accuracy performance and time budget of training with SGD-DP on non-FL, FL-non-NC and NC-FLS. As shown, the NC-FLS can achieve the same accuracy while saving training time by reducing communication cost between devices.

Accuracy	$\sigma = 4$	$\sigma = 2$	$\sigma = 1$
non-FL(client A)	89.8%	93.8%	95.5%
FL-non-NC	92.9%	95.8%	97.2%
NC-FLS	92.7%	95.7%	97.1%
Traning Time (min)	$\sigma = 4$	$\sigma = 2$	$\sigma = 1$
non-FL(client A)	8.33	8.16	8.07
FL-non-NC	46.55	46.58	45.83
NC-FLS	39.58	38.78	37.25

6   Conclusion and Future Work

We propose new a federated learning system suitable for all kinds of decentralized privacy-preserving and security machine learning models, NC-FLS, which is a bridge between network coding and federated learning. NC-FLS is helpful to improve the network throughput, data privacy and data security with the imperceptible higher computational cost on the local device while saving more communication time which is shown around 15% of time-saving in our experiment. Since our approach applies directly to gradient transmission, it can be adapted to any other FLS, and the properties of the original FLS are preserved. Thus, the NC-FLS is a promising way to solve the high communication cost problem which is a bottleneck of FL. We also demonstrate a CNN with NC-FLS, and the accuracy of the model using NC-FLS is around 2% higher than the model only using local data and no accuracy loss than FLS without network coding.

This paper only considers two-party FLSs with a fixed NC scheme. For more participants, communication cost is more significant for system design and time budget. In the future, a more powerful NC-FLS matching more numbers of clients applying the random network coding scheme can be implemented.

Acknowledgement. This paper is supported by National Key Research and Development Program of China under grant No. 2018YFB1003500, No. 2018YFB0204400 and No. 2017YFB1401202.

References

1. Abadi, M., et al.: Deep learning with differential privacy. In: Proceedings of the 2016 ACM SIGSAC Conference on Computer and Communications Security, CCS '16, ACM, New York, NY, USA, pp. 308–318 (2016). https://doi.org/10.1145/2976749.2978318

2. Atayero, A.A., Feyisetan, O.: Security issues in cloud computing: the potentials of homomorphic encryption. J. Emerg. Trend Comput. Inf. Sci. **2**(10), 546–552 (2011)
3. Bhattad, K., et al.: Weakly secure network coding. NetCod, Apr 104 (2005)
4. Dean, J., et al.: Large scale distributed deep networks. In: Advances in Neural Information Processing Systems, pp. 1223–1231 (2012)
5. Dimakis, A.G., Godfrey, P.B., Wu, Y., Wainwright, M.J., Ramchandran, K.: Network coding for distributed storage systems. IEEE Trans. Inf. Theory **56**(9), 4539–4551 (2010)
6. Dwork, C.: Differential privacy. In: Encyclopedia of Cryptography and Security, pp. 338–340 (2011)
7. Dwork, C., Kenthapadi, K., McSherry, F., Mironov, I., Naor, M.: Our data, ourselves: privacy via distributed noise generation. In: Vaudenay, S. (ed.) EUROCRYPT 2006. LNCS, vol. 4004, pp. 486–503. Springer, Heidelberg (2006). https://doi.org/10.1007/11761679_29
8. Fragouli, C., Markopoulou, A.: A network coding approach to overlay network monitoring. In: Allerton Conference (2005)
9. Han, C., Yin, J., Ye, L., Ke, Y., Yang, Y.: An integrated fast data transmission scheme based on network coding. IEEE Access **7**, 112216–112228 (2019)
10. Kairouz, P., et al.: Advances and open problems in federated learning. arXiv preprint arXiv:1912.04977 (2019)
11. Konečný, J., McMahan, H.B., Ramage, D., Richtarik, P.: Federated optimization: distributed machine learning for on-device intelligence. arXiv preprint arXiv:1610.02527 (2016)
12. Li, S.Y., Yeung, R.W., Cai, N.: Linear network coding. IEEE Trans. Inf. Theory **49**(2), 371–381 (2003)
13. Martínez-Peñas, U., Kschischang, F.R.: Reliable and secure multishot network coding using linearized reed-solomon codes. IEEE Trans. Inf. Theory **65**(8), 4785–4803 (2019)
14. McMahan, B., Ramage, D.: Federated learning: collaborative machine learning without centralized training data. Google Research Blog **3**, (2017)
15. Naeem, A., Rehmani, M.H., Saleem, Y., Rashid, I., Crespi, N.: Network coding in cognitive radio networks: a comprehensive survey. IEEE Commun. Surv. Tutorials **19**(3), 1945–1973 (2017)
16. Phong, L.T., Aono, Y., Hayashi, T., Wang, L., Moriai, S.: Privacy-preserving deep learning via additively homomorphic encryption. IEEE Trans. Inf. Forensics Secur. **13**(5), 1333–1345 (2018). https://doi.org/10.1109/TIFS.2017.2787987
17. Recht, B., et al.: Hogwild: a lock-free approach to parallelizing stochastic gradient descent. In: Advances in Neural Information Processing Systems, pp. 693–701 (2011)
18. Shokri, R., Shmatikov, V.: Privacy-preserving deep learning. In: Proceedings of the 22nd ACM SIGSAC Conference on Computer and Communications Security, pp. 1310–1321. ACM (2015)
19. Yang, Q., Liu, Y., Chen, T., Tong, Y.: Federated machine learning: concept and applications. ACM Trans. Intell. Syst. Technol. (TIST) **10**(2), 12 (2019)

New Approaches to Federated XGBoost Learning for Privacy-Preserving Data Analysis

Fuki Yamamoto[1], Lihua Wang[2] (ID), and Seiichi Ozawa[1,3]([✉]) (ID)

[1] Graduate School of Engineering, Kobe University, Kobe, Japan
tamafuki929@gmail.com, ozawasei@kobe-u.ac.jp
[2] National Institute of Information and Communications Technology, Tokyo, Japan
lh-wang@nict.go.jp
[3] Center for Mathematical and Data Sciences, Kobe University, Kobe, Japan

Abstract. In this paper, we propose a new privacy-preserving machine learning algorithm called *Federated-Learning XGBoost* (FL-XGBoost), in which a federated learning scheme is introduced into XGBoost, a state-of-the-art gradient boosting decision tree model. The proposed FL-XGBoost can train a sensitive task to be solved among different entities without revealing their own data. The proposed FL-XGBoost can achieve significant reduction in the number of communications between entities by exchanging decision tree models. In our experiments, we carry out the performance comparison between FL-XGBoost and a different federated learning approach to XGBoost called FATE. The experimental results show that the proposed method can achieve high prediction accuracy with less communication even if the number of entities is increase.

Keywords: Machine learning · Federated learning · Privacy preserving · Big data analysis · Ensemble tree classifier

1 Introduction

When we analyze big data owned by multiple entities, conventional data mining technology can effectively work on conditioned that all the entities cooperatively share their own data each other. In reality, however, this condition does not always hold. On the other hand, there exist many social problems that should be cooperatively solved by sharing sensitive data among multiple entities for such as crime deterrence, medical care, and health care for elderlies. Obviously, solving such sensitive tasks could provide a big impact to our society. On the contrary, we should keep in mind that it could expose us to great danger, causing serious incidents of personal data leak. Recently, to alleviate the current difficulty in big data analysis, privacy-preserving data mining (PPDM) has attracted considerable attention.

There have been developed several PPDM approaches such as homomorphic encryption [5] and differential privacy [2]. The former allows us to conduct

© Springer Nature Switzerland AG 2020
H. Yang et al. (Eds.): ICONIP 2020, LNCS 12533, pp. 558–569, 2020.
https://doi.org/10.1007/978-3-030-63833-7_47

specific calculations (e.g., addition and multiplication) over encrypted data. The latter provides us a mechanism of adding noise to ensure a certain level of privacy from a statistical point of view. On the other hand, however, the above PPDM approaches may sacrifice accuracy in prediction or impose us some restriction in data usage. Different from the conventional PPDM approaches where data are collected and processed at a place, federated learning gives a new possibility to learn cooperatively among different entities without revealing their own data [11]; that is, each entity conduct the learning of data locally and provide only their model updates to a center server. Then, the center server distributes the whole update information to each entity. If data include confidential information, a rigorous process is generally required to follow some regulations, resulting in reducing usability and convenience. However, if only model update information is shared in learning, it would facilitate to carry out the analysis of sensitive data.

In many practical machine learning methods, Gradient Boosting Decision Tree (GBDT) [3] is applicable to many situations. Chen et al. proposed XGBoost [1] as a more scalable and accurate GBDT method, which improved the computation speed by using the sparsity-aware algorithm for sparse data and weighted quantile sketch for approximate tree learning. Unlike other black box approaches such as deep learning models, the prediction by XGBoost can give a certain level of explainability, because it gives importance score for feature conditions in decision trees. Yang et al. [10] introduces federated learning into XGBoost where gradient information on each tree nodes should be frequently communicated between a center server and entities of data owners. FATE (Federated AI Technology Enabler) [9], an open-source federated learning framework developed by WeBank, incorporated secret computation into the Yang et al.'s method. Although FATE achieves both secure and accurate computations, it generally requires frequent communications among entities to build each of decision trees. When the depth of a decision tree is d, it is estimated that about $2^d - 1$ times communications are required to carry out model update. Zhao et al. [12] introduced federated learning and differential privacy into GBDT by communicating models. Although this method can train a model with only tree-by-tree communication, the noise mechanism to ensure differential privacy often deteriorates the prediction accuracy of a model.

To solve the above-mentioned problems, we propose a new privacy-preserving machine learning algorithm called *Federated-Learning XGBoost* (FL-XGBoost). The contribution of this paper lies in the development of a practical federated learning scheme that reduces communications among entities without scarifying prediction accuracy. In the proposed FL-XGBoost, only one-time communication is necessary for keep high prediction accuracy in the model update.

This paper is organized as follow. Section 2 provides the preliminaries for the proposed federated learning approach. Then, we present the proposed FL-XGBoost in Sects. 3, In Sect. 4, after explaining experimental setups, we show the performance comparison between FATE and the proposed FL-XGBoost for some benchmark data sets. We also give a security analysis in Sect. 5, and Sect. 6 gives our conclusions and future work.

2 Preliminaries

2.1 XGBoost

XGBoost [1] is a novel GBDT method, and it is faster and more accurate than traditional methods. Since GBDT is composed of multiple decision trees, the update is performed by determining the split points and leaf weights. For XGBoost, it performs update using the gradient information of loss function, instead of the feature values of data. The following Eq. (1) shows the calculation of the gradient information, denoted by g_i, h_i.

$$g_i = \partial_{\hat{y}_i^{(k-1)}} l\left(y_i, \hat{y}_i^{(k-1)}\right), h_i = \partial^2_{\hat{y}_i^{(k-1)}} l\left(y_i, \hat{y}_i^{(k-1)}\right). \tag{1}$$

The loss function $l\left(y_i, \hat{y}_i^{(k-1)}\right)$ calculates the error between the current prediction and the true value. In the conventional GBDT methods, the split points are determined by using the gradient information as impurity and the leaf weights are determined to minimize errors. Differently, in XGBoost, the split points are determined to minimize the cost function $\mathcal{L}^{(k)}(f_k)$ as in Eq. (2), and the leaf weights $\hat{\omega}_j$ are analytically determined by quadratic approximation as shown in Eq. (3).

$$\mathcal{L}^{(k)}(f_k) = \sum_{i=1}^{n}\left[l\left(y_i, \hat{y}_i^{(k-1)}\right) + g_i f_k(x_i) + \frac{1}{2}h_i f_k^2(x_i)\right] + \Omega(f_k). \tag{2}$$

$$\hat{\omega}_j = -\frac{\Sigma_{i \in I_j} g_i}{\Sigma_{i \in I_j} h_i + \lambda}. \tag{3}$$

Furthermore, the data set in each node is divided into left and right nodes, and the gradient information of the nodes are summed up to calculate G_L, H_L and G_R, H_R, respectively. As shown in Eq. (4), these values are used to calculate *score*, and the split point is where maximizes this *score*.

$$score = \frac{G_L^2}{H_L + \lambda} + \frac{G_R^2}{H_R + \lambda} - \frac{G^2}{H + \lambda}. \tag{4}$$

2.2 Related Works

FATE [9] uses a scheme consisting of multiple data owners and a central server, and aggregates the gradients of loss functions at the central server to realize federated learning of GBDT model. This approach requires the feature values to be discretized in advance with a common criterion among the data owners. The discrete feature values are discretized candidate split points. Therefore, as the width of the discretization is reduced, the number of candidate split point increases and the loss is reduced, but the security is also reduced, correspondingly. Each data owner calculates the sum of the gradients of all the data divided at each candidate split point and sends it to the central server. The central server sums these results up and determines the split point. To prevent the central server from obtaining information from the data owners, the process to aggregate the gradient information is under encrypted form. What the algorithm depends on is not the amount of data held by each data owner, but the total amount of data from all data owners. However, this scheme requires the same times of communications as XGBoost's nodes to train the model.

Zhao et al. [12] achieved federated learning of the GBDT model in a scheme consisting of multiple data owners only, by learning a common model in turn. Specifically, each data owner updates the model with its own data set and sends the updated model to the next data owner. Since only its own data and the updated model are required for model training, the only information required for communication is the model information. However, there is still a problem that the model prediction accuracy is sacrificed due to the noise added to satisfy the differential privacy requirement.

3   Overview of FL-XGBoost

In this section, the details of the proposed FL-XGBoost are described. The proposed scheme consists of multiple data owners $U = \{u_1, u_2, u_3, ..., u_D\}$ and the central server S. Throughout this paper, we use the symbols defined in Table 1.

Table 1. The notations used in this paper.

Notation	Description
S	Central server
D	Number of data owners
U	Data owners set
u_{tr}	Data owner to train the model
T_i	The model updated for i times
$iter$	Target number of model updates
$\text{Enc}_{pk}(\cdot)$	Encryption with a public key pk
$\text{Dec}_{sk}(\cdot)$	Decryption with a secret key sk

Zhao et al. [12] realized federated learning and differential privacy by communicating models among the data owners. This method can train the model with less communication than existing federated GBDT schemes. However, in this study, the model is assumed to be shared only among data owners. Therefore, the focus should be on the performance of the model rather than differential

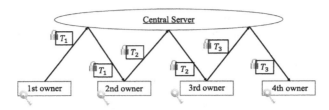

Fig. 1. The outline drawing of FL-XGBoost. T_i denotes the model that has been updated for i times.

privacy guarantees. Based on that, we propose a practical federated learning scheme for XGBoost that significantly reduces the number of communications with minimal loss. To accomplish this, we implemented a method to guarantee security without impacting the performance of the model and an algorithm to select the best data owner for the next training.

3.1   Secure Model Updates Among Multiple Data Owners

The scheme of Zhao et al. [12] consists of multiple data owners only and updates the model in a predetermined order. Therefore, it is possible to identify the data owner that updated the model. Furthermore, each data owner can easily obtain statistical information of other data owners, since each decision tree is learned by a single data owner and such trees are generally considered as statistical information.

In order to solve the above problems, we propose a scheme in which a central server S is introduced. The outline of FL-XGBoost is shown in Fig. 1. In FL-XGBoost, $u \in U$ learns T_i from its own data set, similar to the method of Zhao et al.

The difference is that the communication is done via S to prevent identifying data owner u_{tr} who updated the model. Since this u_{tr} is selected from U by S according to the algorithm described below, the other $u \in U$ cannot identify u_{tr}. The model is learned by repeatedly communicating model information between S and each u_{tr} selected by S to update the model. Here, all data owners share the encryption keys in advance and send the encrypted information to S so that S cannot obtain the information. In addition, communication is performed using a secure communication channel, and the communication content is not intercepted.

3.2   FL-XGBoost with Random and Uniform Data Owner Selection

Random selection is the simplest way for S to select u_{tr} from U. However, such selection of u_{tr} may contain a bias, impacting the performance of the trained model. Therefore, u_{tr} needs to be selected uniformly from U. Different from that, Zhao et al.'s method selects u_{tr} in a predetermined order, and each data

---

Algorithm 1: Learning algorithm of FL-XGBoost-R

1.1: **for** $i \leftarrow 1$ **to** $iter$ **do**
1.2:   **if** i is a multiple of D **then**
1.3: | S creates U_{order}.
1.4:   **end**
1.5: S selects u_{tr} according to U_{order}.
1.6: S sends $\text{Enc}_{pk}(T_{i-1})$ to u_{tr}.
1.7: u_{tr} decrypts $\text{Enc}_{pk}(T_{i-1})$ to obtain T_{i-1}.
1.8: u_{tr} updates T_{i-1} with its own data to obtain T_i.
1.9: u_{tr} encrypts T_i and obtains $\text{Enc}_{pk}(T_i)$.
1.10: u_{tr} sends $\text{Enc}_{pk}(T_i)$ to S.
1.11: **end**

---

owner is selected for the same number of times. This is risky because it is easy to obtain a set of trees trained by the same data owner.

Based on the above, we propose FL-XGBoost-R, which uses an algorithm to randomly and uniformly select the data owners. Algorithm 1 shows details. First, S creates U_{order}, where U is randomly sorted. Then, select u_{tr} according to U_{order} until all data owners are covered. Such cycle is repeated.

3.3 FL-XGBoost with Data Owner Selection Based on Prediction Confidence

Even if S selects u_{tr} from U uniformly, the model's learning may not proceed equally for all $u \in U$. To obtain an unbiased trained model for the U, we propose FL-XGBoost-G, which selects u_{tr} using $|g|$, the absolute value of g.

The gradient of the loss function is denoted by g, and $|g|$ represents the magnitude of the error between the predicted and true values. Hence, the sum of $|g|$ of the data held by $u \in U$ represents the confidence of the prediction of that data set.

Data sets with low confidence is considered to be untrained, so $u \in U$ with a small $\sum |g|$ should be selected preferentially. However, for an imbalanced dataset where the number of data per class of labels varies greatly, the data with majority class will dominate $\sum |g|$. Also, the comparison is difficult when the amount of data owned by each $u \in U$ is different, since $\sum |g|$ depends on the amount of data owned. Considering the above problems, we use G_{ave}, the mean value of $|g|$ per class of labels, to be the selection index, as shown in the following equation (5).

$$G_{ave} = \frac{G_{pos}}{pos} + \frac{G_{neg}}{neg}. \tag{5}$$

Here, pos, neg denote the amount of positive and negative data, respectively. G_{pos} and G_{neg} denote $\sum |g|$ for positive and $\sum |g|$ for negative data, respectively.

Algorithm 2: Learning algorithm of FL-XGBoost-G

2.1: **for** $i \leftarrow 1$ **to** $iter$ **do**
2.2: | **if** $i \leq D$ **then**
2.3: | | S selects u_{tr} with FL-XGBoost-R.
2.4: | **end**
2.5: | **else**
2.6: | | S selects $u \in U$ with the largest G_{ave} as u_{tr}.
2.7: | **end**
2.8: | S sends $\mathrm{Enc}_{pk}(T_{i-1})$ to u_{tr}.
2.9: | u_{tr} dectypts $\mathrm{Enc}_{pk}(T_{i-1})$ and gets T_{i-1}.
2.10: | u_{tr} updates T_{i-1} with its own data to obtain T_i.
2.11: | u_{tr} encrypts T_i and obtains $\mathrm{Enc}_{pk}(T_i)$.
2.12: | u_{tr} computes G_{ave}.
2.13: | u_{tr} sends $\mathrm{Enc}_{pk}(T_i)$ and G_{ave} to S.
2.14: **end**

Algorithm 2 shows details. FL-XGBoost-R is used for the first round. Then u_{tr} transmits $\mathrm{Enc}_{pk}(T_i)$ as well as G_{ave} to S. This implies that after the first round, S has G_{ave} of all $u \in U$. Thereafter, S selects $u \in U$ with the largest G_{ave} as u_{tr}.

4   Experiments

In this section, we describe experiments using open data sets to verify the usefulness of the proposed methods. In the scheme of the proposed methods, all elements of U own independent datasets and share the same feature space and label space. For this purpose, we divided the dataset into D subsets horizontally, and performed the experiments assuming that each subset is a dataset owned by $u \in U$. We also compared the proposed methods with the federated XGBoost implemented in FATE and the XGBoost trained on the whole dataset.

The experimental environment is UBuntu 18.04 with 64[Gb] RAM, and the programming language is Python.

4.1   Data Set

Table 2 shows information on the four binary classification datasets used in the experiment. For brevity, we named the datasets as 'Arcene' for the Arcene Data Set [6], 'Biodeg' for the QSAR biodegradation Data Set [8], 'Credit' for the Credit Card Fraud Detection [4], 'German' for the German Credit Data [7] in this paper. Basically, accuracy is the evaluation index for the models, but only for 'Credit', we use the F1-score to take account for the imbalance of the number of data per class.

Table 2. Information of the dataset used in the experiment.

Dataset	Number of data	Number of features
Arcene [6]	729	10000
Biodeg [8]	854	41
Credit [4]	230693	30
German [7]	810	20

4.2   Results

We conducted verification experiments with open data sets for the existing and proposed methods and the results of the experiments are shown in Table 3. Table 3a shows the results of XGBoost trained on the entire dataset, and Table 3b shows the results of FATE's federated XGBoost with the number of candidate split points set to 10000. The number of candidate split points is a hyperparameter, which controls the trade-off between security and prediction accuracy and was set as 10000 because the result was almost the same as Table 3a at 10000. Table 3c shows the results for FL-XGBoost-R and Table 3d shows the results for FL-XGBoost-G. The proposed methods were tested for various values of D because D is considered to affect the results.

First, we compared the results of the proposed methods. From Table 3c and Table 3d, FL-XGBoost-G shows better results for all datasets in case $D \leq 10$. However, in case $D = 15, 20$, FL-XGBoost-R shows better results depending on the dataset. This is because the effect of outliers on the value of G_{ave} increases as D increases. Some data sets have outliers that deviate from the distribution of other data, and generally such data are prone to large prediction errors. The number of outlier data is generally small and does not affect G_{ave} in large data sets, but in the case of small data sets, such as when D is large, the outlier affects G_{ave}. Therefore, the data owner with outliers is preferentially selected. This may deteriorate accuracy depending on the dataset.

Next, we compared the results of the existing methods with those of the proposed methods. FATE's federated XGBoost realized model training with little or no loss, given that the results in Tables 3a and 3b are almost identical. Comparing this with Tables 3d, FL-XGBoost-G achieves the same level of prediction accuracy when D is small. However, as the value of D increases, the prediction accuracy of the proposed methods deteriorates in all datasets.

This is because the impact of outliers on G_{ave} increases as D increases. Some data sets have outliers that deviate from the overall data distribution, and generally such outliers may lead to large prediction errors. The number of outlier data is generally small and does not affect G_{ave} in large data sets, but when D is large and the entire data set is divided into subsets containing fewer data, the outlier in a subset may have more impact on G_{ave}. Therefore, the data owner with outliers is preferentially selected. This may deteriorate accuracy depending on the dataset.

This is because the proposed methods uses a single owner's data to update the model, and therefore the amount of data available for updating is reduced.

Table 3. Experiment results of prediction accuracy for the existing and proposed methods using open data sets. (a) and (b) show the result of the existing methods, and (c) and (d) show the result of the proposed methods.

(a) The results of XGBoost trained on the entire dataset

Arcene	0.850 ± 0.00
Biodeg	0.826 ± 0.01
Credit	0.850 ± 0.00
German	0.790 ± 0.01

(b) The result of the federated XGBoost implemented in FATE

Arcene	0.850 ± 0.00
Biodeg	0.826 ± 0.01
Credit	0.850 ± 0.00
German	0.792 ± 0.02

(c) The results with FL-XGBoost-R

D	3	5	10	15	20
Arcene	0.850 ± 0.00	0.690 ± 0.00	0.740 ± 0.02	0.700 ± 0.00	0.650 ± 0.00
Biodeg	0.831 ± 0.01	0.828 ± 0.02	0.826 ± 0.02	0.842 ± 0.01	0.840 ± 0.01
Credit	0.841 ± 0.00	0.841 ± 0.00	0.838 ± 0.00	0.838 ± 0.00	0.828 ± 0.01
German	0.772 ± 0.02	0.750 ± 0.02	0.750 ± 0.01	0.724 ± 0.02	0.716 ± 0.02

(d) The results with FL-XGBoost-G

D	3	5	10	15	20
Arcene	0.850 ± 0.00	0.700 ± 0.00	0.740 ± 0.02	0.750 ± 0.00	0.650 ± 0.00
Biodeg	0.848 ± 0.01	0.830 ± 0.00	0.826 ± 0.01	0.828 ± 0.00	0.818 ± 0.01
Credit	0.861 ± 0.01	0.841 ± 0.00	0.843 ± 0.01	0.840 ± 0.01	0.812 ± 0.01
German	0.774 ± 0.02	0.764 ± 0.04	0.759 ± 0.02	0.731 ± 0.01	0.720 ± 0.01

In contrast, FATE's federated XGBoost algorithm is unaffected by the amount of data owned by each data owners as long as the total amount is the same. Improving the accuracy with large D is an issue for future works.

5   Security Analysis

In this section, with the precondition that a secure communication channel is used so that the transmitted contents are not intercepted, we provide security analysis of the proposed FL-XGBoost with the following assumptions of attackers.

– Server S being honest-but-curious
– Data owners U being honest-but-curious
– *Passive Attack* being the only attack method

Here, *Passive Attack* means that there is no collusion among the participants of the scheme, and each attacker attacks based on the information obtained through correctly following the scheme. In addition, the purpose of each attacker is as follows.

1. S aims to obtain:
   - model information of T_i,
   - personal labels and feature values, and/or
   - number of data per class of label of $u \in U$.
2. U aim to obtain:
   - u_{tr} for each update, except where the attacker data owners themselves are u_{tr},

- number of data per class of label of $u \in U$ other than the attacker data owners themselves, and/or
- labels and feature values of individuals belonging to $u \in U$ other than the attacker data owners themselves.

Here, the trained models are intended to be shared only with $u \in U$, so they do not want this information to be known by S. Also, each tree in the model is considered to be statistical information of $u \in U$, so it is necessary to prevent the identifying the data owner who trained each tree.

In addition, the information received by each attacker is assumed to be as follows.

1. Information received by S:
   - u_{tr} for each update,
   - $\text{Enc}_{pk}(T_i)(i = \{0, 1, 2, ...\})$,
   - G_{ave} (only in FL-XGBoost-G).
   So we need only consider security of FL-XGBoost-G.
2. Information received by U:
   - $T_i \ (i = \{0, 1, 2, ...\})$.

5.1   Security Against the Central Server

The G_{ave} and model information transmitted in FL-XGBoost is calculated from the gradient information of the loss function. Therefore, we first consider the information leaked from the gradient information. In the case of binary classification, when the prediction is y_{pred} and the true label is y, the gradient information g is given by following Eq. (6).

$$g = y_{pred} - y. \tag{6}$$

y_{pred} represents probability and $0 \leq y_{pred} \leq 1$. It can be inferred that the sign of g represents y, since $g \geq 0$ if $y = 0$ and $g < 0$ if $y = 1$.

No useful information from what S receives can be obtained from the encrypted information $\text{Enc}_{pk}(T_i)$ without a key. Also, since G_{ave} in equation (5) is the sum of the mean values of all classes of label of $|g|$, it is not possible to obtain the sign of g and the number of data used in the calculation from its value. Hence, S cannot get the desired information.

5.2   Security Against the Data Owners

First, we consider whether it is possible to identify u_{tr} who updated the model. Since u_{tr} is selected by S, no one other than u_{tr} and S are informed of the selected $u \in U$. However, in the case of $D = 2$, u_{tr} can be identified by excluding its own learned trees. Apart from that, in the case of $D > 2$, $u \in U$ cannot identify u_{tr} because of its indistinguishability. From the above, it is not possible to obtain information on specific data owners from the trained models. Hence, the attacker cannot even get the number of data per class of labels that $u \in U$ owns.

Next, we examine the risk of personal information leakage from the model information. Specifically, we divide the model information into thresholds and leaf weights, and consider the information that can be leaked from each.

The thresholds are determined from the feature values, and thus the values correspond to the feature values of any individual. Therefore, if a feature value is unique, we can identify whose data were used to train the model.

Since λ and h in Eq. (3) which shows the leaf weights are larger than 0, the sign of the leaf weights is determined by $\sum g$ of the data in the leaf data set. Therefore, the sign of individual g cannot be identified from $\sum g$, as long as the number of data in the leaf data set is sufficiently large. However, in an extreme case where there is only one data in the leaf data set, the sign of the leaf output can indicate the label of that individual. As a result, if such individuals can be identified, the attacker can obtain the desired information.

6   Conclusion

In this study, we propose a practical federated learning scheme for XGBoost that significantly reduces the number of communication cycles compared to existing federate approaches. The proposed method achieves the same level of prediction accuracy as existing methods when each data owner has a sufficient amount of data. However, as the amount of data owned by individual data owners decreases, the performance of the models deteriorates. It is also found that in extreme cases, there is a risk of leakage of sensitive personal information from thresholds and leaf weights. This needs to be resolved by imposing some constraints on the training of the model.

Our future work is to refine the scheme to be more practical by solving the above-mentioned problems and by implementing a method of sharing encryption keys and a secure way of utilizing G_{ave}.

Acknowledgement. We would like to thank Associate Professor Toshiaki Omori and the members of the National Institute of Information and Communications Technology (NICT) for their helpful advice and support in writing this paper. This research has been accomplished through the project "Social Implementation of Privacy-Preserving Data Analytics" (JPMJCR19F6) in the JST CREST research area "Development and Integration of Artificial Intelligence Technologies for Innovation Acceleration".

References

1. Chen, T., Guestrin, C.: Xgboost: a scalable tree boosting system. In: Proceedings of the 22nd ACM SIGKDD International Conference on Knowledge Discovery and Data Mining, pp. 785–794 (2016)
2. Dwork, C., McSherry, F., Nissim, K., Smith, A.: Calibrating noise to sensitivity in private data analysis. In: Halevi, S., Rabin, T. (eds.) TCC 2006. LNCS, vol. 3876, pp. 265–284. Springer, Heidelberg (2006). https://doi.org/10.1007/11681878_14

3. Friedman, J.: Greedy function approximation: a gradient boosting machine. Ann. Stat. pp. 1189–1232 (2001)
4. Kaggle: Credit Card Fraud Detection. https://www.kaggle.com/mlg-ulb/creditcardfraud. Accessed 14 Sep 2020
5. Paillier, P.: Public-key cryptosystems based on composite degree residuosity classes. In: Stern, J. (ed.) EUROCRYPT 1999. LNCS, vol. 1592, pp. 223–238. Springer, Heidelberg (1999). https://doi.org/10.1007/3-540-48910-X_16
6. UCI: Arcene Data Set. https://archive.ics.uci.edu/ml/datasets/Arcene. Accessed 14 Sep 2020
7. UCI: German Credit Data. https://archive.ics.uci.edu/ml/datasets/statlog+(german+credit+data. Accessed 14 Sep 2020
8. UCI: QSAR biodegradation Data Set (UCI). https://archive.ics.uci.edu/ml/datasets/QSAR+biodegradation. Accessed 14 Sep 2020
9. Webank: FATE (Federated AI Technology Enabler). https://fate.readthedocs.io/en/latest/index.html. Accessed 14 Sep 2020
10. Yang, M., Song, L., Xu, J., Li, C., Tan, G.: The tradeoff between privacy and accuracy in anomaly detection using federated XGBoost. arXiv preprint arXiv:1907.07157 (2019)
11. Yang, Q., Liu, Y., Chen, T., Tong, Y.: Federated machine learning: Concept and applications. In: ACM Transactions on Intelligent Systems and Technology (TIST), New York, NY, USA, pp. 1–19. ACM (2019)
12. Zhao, L., et al.: Inprivate digging: enabling tree-based distributed data mining with differential privacy. In: IEEE INFOCOM 2018-IEEE Conference on Computer Communications, pp. 2087–2095. IEEE (2018)

Partially Disentangled Latent Relations for Multi-label Deep Learning

Si-ming Lian, Jian-wei Liu$^{(\boxtimes)}$, and Xiong-lin Luo

Department of Automation, College of Information Science and Engineering,
China University of Petroleum, Beijing Campus (CUP), Beijing 102249, China
liujw@cup.edu.cn

Abstract. Identified the specific features from the instances belong to a certain class label is meaningful, the "purified" feature representation contains such label information can be shared with other feature learning. Besides, it is essential to distinguish the sample association relationship behind the multi-label da-tasets, which is conducive to improve the performance of the algorithm. However, most algorithms aim to capture the mapping between instances and labels, while ignoring the information about instance relations and label cor-relation hidden in the data structure. Motivated by these issues, we leverage the deep network to learn the special feature representations without aban-doning overlapped features. Meanwhile, the Euclidean metric matrices are leveraged to construct the diagonal matrix for the diffusion function, it en-sures that the results of model training by similar instance features are con-sistent. Further, considering the contributions of these feature representation are different and have influences on the final prediction results, thus the self-attention mechanism is introduced to fusion the other label specific in-stance features to build the new joint feature representation, which derive dynamic weights for multi-label prediction. Finally, experimental results of the real data sets show promising availabilities of our approach.

Keywords: Multi-label learning · Diffusion · Self-attention · Feature representation

1 Introduction

Multi-label learning has a wide range of application scenarios such as recommendation system, spam classification, etc [4,12]. Meanwhile, they also bring with them all sorts of challenges with development. One of the most basic concerns is how to find suitable implicit features of instances, which are corre-lated to corresponding multi-label, to predict the multi-label of unseen instances. Ideally, the feature representations should have the properties that contains the common features that associate with multi-label and the special features that are only related to one label [7]. It is intractable to perfect separate the special fea-tures of each label from the entire multi-label set. Hence we more prefer

© Springer Nature Switzerland AG 2020
H. Yang et al. (Eds.): ICONIP 2020, LNCS 12533, pp. 570–579, 2020.
https://doi.org/10.1007/978-3-030-63833-7_48

to learn a special feature representations, which partially contain the common representa-tion apart from the unique characteristics corresponding to the multi-label com-ponent, both consistent and diversity properties should consider in such partially disentangled way [5].

In summarize, the feature representation obtained in incompletely disentangled approach has the following advantages: (a) consistency and diversity: the learned latent features emphasized the consistency on the entire multi-label, and meanwhile the diversity of the individuals of each instance are not ignored [1,3]. (b) order independence: generally speaking, for traditional multi-label learning algorithms, rearranging the label set as a new sequential set to obtain the feature mapping may lose the certain generalization ability. However, our method maintains the generalization ability to fit those unseen test instances [6]. (c) rotation invariance: the inputs are mapped to special features of the hidden layers in deep network, then we use the Euclidean distance matrix to find the new rotation invariance representation of special features [11]. making full use of instance relations provide a new idea to tackle with the multi-label learning, which has become a worthy research directions of machine learning.

As mentioned by [8,12,13], we assume that multiple labels are sampled from one or more underlying unknown distribution, which could derive the spe-cial feature representation and united it as a new joint representation. Features learning from multi-label may play a different role during the training process. Therefore, we assign a different weight to each special feature in each task, and intuitively, these weights are the measure of the importance of features for dif-ferent multi-label components.

In our algorithm, to address the special feature representation, we utilize an in-completely disentangled method to conduct the supervised learning. First, we propose to use each class label to train the latent special features through the deep neural networks and effective increase the tolerance of variation of deep neural network. In fact, the underlying assumption reveals the special features sharing with the self-attention weights and preserving their special features from different instances, but the latent common information is retained without re-movement. They can be reappeared through self-attention from other features [14]. On that basis, we propose a Partially Disentangled Latent Relations for Mul-ti-label Deep Learning that we call PDLRMDL. In details, our method also utiliz-es a deep neural network to encode the complementary of multi-label sets, and maintain the similar instance with the similar label sets through graph-based dif-fusion function [10]. The self-attention weight associates the entire feature repre-sentation with predictive current label. Finally, we also give the experiments of PDLRMDL to verify the effectiveness of our method. The main contributions of this paper described as follows:

(1) The PDLRMDL extract the special representation without abandoning the common information and share the other latent feature representation with cur-rent feature learning. In order to reconstruct the joint feature represen-tation, we adopt the self-attention to help improve the performance of the model.

(2) Specifically, the diffusion function helps the PDLRMDL encourage the consistency of similar instances and corresponding label sets. It spread the instance information to the predicted label set, which encourage the similar in-stances have the similar label sets.

(3) We calculate the instance distance through the Euclidean distance metric to construct the symmetric adjacency matrix with zero diagonal. In this way, the back propagation in diffusion function can be ensured.

2   Problem Formulation

Given whole multi-label example set $X := \left\{x_i \in R^{1 \times d}\right\}_{i=1}^{N}$, select the first l examples as the training instance set $X_{tr} := \left\{x_i \in R^{1 \times d}\right\}_{i=1}^{l}$, and the corresponding multi-label set is $Y_{tr} : \left\{y_i = (y_i^1, y_i^2, \cdots, y_i^K)\right\}_{i=1}^{l}$, where $y_i \in \{0,1\}^K$,each training ex-ample is belong to a discrete label subset for K classes. The remaining $U = N - L$ examples $x_i, i \in \{l+1, \cdots, N\}$, denoted the testing example set $X_{te} := \left\{x_i \in R^{1 \times d}\right\}_{i=l+1}^{N}$, they are unlabeled. Our goal in the multi-label learning is to utilize the known information to train the classifier, which needs to predict the accurate mapping relations between the unseen testing instances and the corresponding multi-label.

3   Partially Disentangled Latent Relations for Multi-label Deep Learning

In this section, we present the simple neural network to perform the multi-label classification. The procedures of PDLRMDL can divide into the following stages. First, we construct multiple classifiers to fit each training class label for extracting special features; second, the Euclidean distance are adopted to keep the dis-tance information of similar latent features unchanged, which also provide the smoothness of the back propagation. Third, the diffusion function graphbased encourage the smoothness so that the nearby instances would have the similar predicted label sets. Finally, the self-attention co-regularizes the feature repre-sentation that carves up the special features into new joint features. This following section mainly elaborates the problem definition.

3.1   The PDLRMDL for Multi-label Learning

Deep neural networks have been widely adopted in various fields, and it can penetrate the multi-label classification task. We are interest in leveraging the capacity of nonlinear function approximating of deep neural networks, to learn the special feature representation with the labeled training set automatically [9]. In this way, it also can reduce the inadequacy caused by artificial design features. Most importantly, it has achieved excellent classification performance, which is superior to the existing multi-label classification algorithm in the application scenarios that meet specific conditions. The PDLRMDL is divided into the following parts, which will be discussed separately in the following sections as illustrated by Fig. 1.

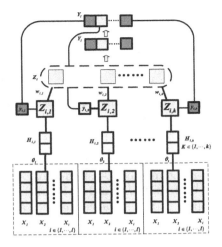

Fig. 1. The illustration of the novel Deep Networks for Multi-Label Observations that we call PDLRMDL, where $i=1,\dots\dots,$ l.

3.2 Special Representation Learning

The PDLRMDL is composed with K neural networks. Each neural network is a binary classifier. Each neural network has one input layer, three hidden layers, and one output layer. The sets of weight matrices W_i and bias vectors b_i of each neural network are denoted as parameter θ_i, and $\theta_i = \left(\theta_i^1, \theta_i^2, \cdots, \theta_i^K\right)^{\top}$, where superscript i is the index value of neural net-work, e.g., the index value of class label component. Input of each neural network is sampled from the training instance set X_{tr}, The corresponding class label of the training instance is derived from components of multi-label of the instance itself, e.g., For the k-th neural network , x_i is labeled with the value of k th component of the x_i corresponding to multi-label.

The K outputs of the first two hidden layer are denoted $H_{1,i} = (h_{1,i}^1, h_{1,i}^2, \cdots h_{1,i}^K)$, $H_{2,i} = (h_{2,i}^1, h_{2,i}^2, \cdots h_{2,i}^K)$ respectively, the K outputs $Z_i = \{z_i^1, z_i^2, \cdots z_i^K\}$ of the last hidden layer are the K special representations of instances.

Through the above methods, we learn the special representation of each instance corresponding to each class label, meanwhile preserve the effect of over-lap feature representations. The learned special feature representation must maintain complementary information as large as possible, and it will suffer extreme sensi-tivity to instance's translation and rotation during the training process. It means the algorithm we propose should maintain the instance's relations is invariance. Then, we share the special feature representation with other features corre-sponding to other multi-label components in the subsequent learning process. Therefore, it is tractable to capture instance relations and multi-label relations in multi-label classification task.

3.3   Invariant Special Representation to Translation and Rotation

In order to build the diagonal matrix that meets the requirements of diffusion function, we first get the Euclidean distance matrices, which is used to calculate matrix used in diffusion function in the subsection 3.4. In particular, the method takes K dimensional vector $\{Z_i\}_{i=1}^l$ as inputs, and each hidden layer implements the task that calculate the distance among latent features. The general idea is to produce a zeros on the diagonal symmetric initial matrix $\tilde{D}$ and then constrain the initial matrix $\tilde{D}$ through a term $\tilde{D}_i^{j,k} \in \left\| z_i^j - z_i^k \right\|_2^2$ for $i = 1, \cdots, l$, $j, k = 1, \cdots, K$ and it is equivalent to $-\frac{1}{2} W \tilde{D} W$, where $W = -\frac{1}{2} W \tilde{D} W^{\mathrm{T}}$. Next, through the above equivalent matrix transformation, a posi-tive semi-definite matrix $L \in R^{(K-1) \times (K-1)}$ of EDM $\tilde{D}$ can be obtained. Thus, the gram matrix $M \in R^{K \times K}$ corresponding to the distance matrix can be calculated by the following identity:

$$M_i^{jk} = <d_i^j, d_i^k>_2 = \frac{1}{2}(D_i^{1k} + D_i^{j1} - D_i^{jk}) \tag{1}$$

where $d_i^j = z_i^j - z_i^1, j = 1, \cdots, K, i = 1, \cdots, l$, and conversely, the distance matrix can be derived as:

$$D_i^{jk} = (M_i^{kk} + M_i^{jj} - 2M_i^{jk}) \tag{2}$$

It should be noted that the gram matrix have the specified structure with stacking the arbitrary symmetric matrix $L \in R^{(K-1) \times (K-1)}$ and zero, which can be seen in [2]. In fact, we obtain the gram matrix M and distance ma-trix D, which is derived from the output of the last hidden layers $z^{i,k}$. Hence, the symmetric matrix $\tilde{L} = \frac{1}{2}(Z_i + Z_i^{\mathrm{T}})$ is parameterized and transformed into a positive semi-definite L through the equation $L = sp(\tilde{L})$, where the $sp = g(.)$ denotes the softplus activation function, It should be noted that $sp = g(.)$ can also be derived from preserving the K-1 largest eigenvalues and explicitly setting the rest to 0. The procedures of EDMs are depicted in Fig. 2.

Through the above process, we know that the gram matrix M and distance matrix D corresponding to the special representation can be obtained from the Eq.(1) and Eq.(2) respectively. Besides, for gram matrix M we introduce a penalty term $\frac{1}{l} \sum_{i=1}^l \sum_{k=d+1}^l \lambda_k^2$ to our total objective function that drives gram matrix M towards a specific rank, and for distance matrix D we intro-duces a penalty term $\frac{1}{l} \sum_{i=1}^l \sum_{k=1}^d relu(-\mu_k)$ on negative eigenvalues [2], where μ_k and λ_k represent the eigenvalues of the matrix D and matrix M respectively. More specifically, the loss term J_1 for constraining the dimension of the latent outputs and Euclidean distance matrix respectively are written as follow:

$$J_1 = \alpha_1 \frac{1}{l} \sum_{i=1}^l \sum_{k=1}^d relu(-\mu_k) + \alpha_2 \frac{1}{l} \sum_{i=1}^l \sum_{k=d+1}^l \lambda_k^2 \tag{3}$$

In details, the network is trained by optimizing parameter θ^i with respect to $\bigtriangledown J_1$. Moreover, it is worth to note that for constraining the rank of M, we select that the dimension of the hidden layer is automatically less than dimension d, and the parameter θ^i updating process does not need to consider the rank constraint.

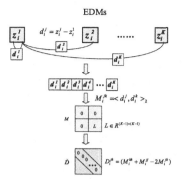

Fig. 2. The Derived Processes of the Euclidean distance matrices in PDLRMDL

3.4   The Diffusion Method for Transmitting the Instance Relations

Based on the distance matrix D, we can calculate a symmetric adjacency matrix $A = D + D^{\mathrm{T}}$, A is a $l \times K \times K$ symmetric matrix with the main diagonal elements zero, where the elements of the A are non-negative and a_i^{jk} measure the distance between the latent vec-tor z_i^j from and the latent vector z_i^k. Further, the weight Ω for the diffusion method can be calculated from the symmetrically normalized A by the formula:

$$\Omega = S^{-1/2} A S^{1/2}, S = diag(A_{1K}), \cdots, diag(A_{lK}) \tag{4}$$

where S denotes l column vectors with dimension K according to the above Eq.(4). Then the diffusion function compute latent representation with estimated label sets $\hat{Y}$ can be written as:

$$\tilde{Z} = (I - \beta\Omega)^{-1}\hat{Y} \tag{5}$$

Where $0 \leq \beta < 1$ is a parameter and $\tilde{Z} \in R^{l \times l}$. Finally, we obtain the class latent information $\tilde{Z}_i = \{\tilde{z}_i^1, \tilde{z}_i^2, \cdots \tilde{z}_i^K\}$ for the training instance $x_i, i \in (1, \cdots, l)$. So far, the final feature representation $\tilde{Z}_i = \{\tilde{z}_i^1, \tilde{z}_i^2, \cdots \tilde{z}_i^K\}$ has been obtained and the derived processes are shown in Fig. 3. We are now at the position to fusion components of feature representation $\tilde{Z}_i = \{\tilde{z}_i^1, \tilde{z}_i^2, \cdots \tilde{z}_i^K\}$ corresponding to x_i, the self-attention is used to realize this task in the followings.

3.5   The Self-Attention Layer for Multi-label Learning

The most important is we need to share the current special latent feature corresponding to each multi-label component with all the other special features corresponding to remaining multi-label components. In order to deal with such problem, we introduce the self-attention as a measure of uncertainty to alignment weight, which defined by:

$$w_i^j = \sum_{j=1}^{K} q_i^j \tanh(\sum_{k=1}^{K} v_i^k \times \tilde{z}_i^j + b_i^j) \tag{6}$$

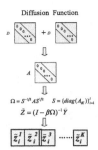

Diffusion Function

$\Omega = S^{-1/2} A S^{1/2}$ $S = \{diag(A_x)\}_{i=1}^{l}$
$\bar{Z} = (I - \beta\Omega)^{-1} \bar{Y}$

$$\boxed{\tilde{z}_i^1}\,\boxed{\tilde{z}_i^2}\,\boxed{\tilde{z}_i^3} \cdots\cdots \boxed{\tilde{z}_i^K}$$

Fig. 3. The illustration of the Derived Processes of Diffusion Function in PDLRMDL

$$w_i^j = \frac{\exp(w_i^j)}{\sum_{j=1}^{K} \exp(w_i^j)} \tag{7}$$

where $\tilde{z}_i^j$ is the special latent feature corresponding to x_i derived from the previous diffusion method, and parameter v_i^j present the alignment weight, which represent the importance of the j-th special features. Besides, other parameters q_i^j, and b_i^j are belong to the current label , not depend on the other label di-rectly, but also can be adjusted from the back propagation of the deep neural network. Given the above definitions of each latent special features $\tilde{z}_i^j$ and parameters obtained from the training datasets in an unsupervised way, the self-attention layer can catch the global connection in multi-label learning and learn the long-distance multi-label dependence, and the most important thing is that it can carry out parallel calculation. Meanwhile, in order to spread the latent instance relations to the self-attention layer in deep neural network, we leverage the diffusion function with transductive learning to transmit the latent instance relations to our self-attention layer.

3.6 The Proposed Objective Function

According to the basic principle of back propagation and gradient descent etc, we can obtain the corresponding information during the transfer process and recon-struction the distance matrix D. Besides, we consider the supervised multi-label learning and construct the loss term for each single label in multi-label of the form. There are other standard options for the loss function in multi-label classifica-tion, such as cross-entropy loss. Now, we formulate our total objective function. The objective function can be divided into three parts, supervised, unsupervised and regularization part, the fourth term constrains the diffusion function to en-sure the smoothness, and the similar instances have similar label sets.

$$J = \alpha_1 \frac{1}{l} \sum_{i=1}^{l} \sum_{k=1}^{d} relu(-\mu_k) + \alpha_2 \frac{1}{l} \sum_{i=1}^{l} \sum_{k=d+1}^{l} \lambda_k^2$$

$$+ \alpha_3 \sum_{i=1}^{l} \sum_{k=1}^{K} \left\| y_i^k - \hat{y}_i^k \right\|_2^2 + \alpha_4 \sum_{i,j=1}^{l} w_{i,j} \left\| \frac{\tilde{Z}_i}{\sqrt{d_{ii}}} - \frac{\tilde{Z}_j}{\sqrt{d_{jj}}} \right\|_2^2 + \alpha_5 \left\| \theta \right\|_2^2 \ (8)$$

For multi-label learning, our first goal is to get the special feature representation corresponding to single class label, and then according to these special feature representations, we leverage the Euclidean distance matrix to learn the new diagonal matrix with zero, and convey the instance relations to the diffusion diagonal matrix. In particular, the diffusion method maintains the consistency of the similar instances, e.g., maintains the similar instances have the similar labels. Finally, we utilize the self-attention mechanism to fusion the special features. More simply, the hidden special feature representation that we incompletely disentangle the by the deep neural network are equivalent to construct a set of the base vectors, which are corresponding to the multi-label information. Finally, the training the weight coefficient processes make full use of information corresponding training label sets.

4 Experiments and Discussion

In this section, we present the characteristic of the six real world datasets used in our experiments and discuss the effectiveness of PDLRMDL. The experimental results are listed in the Table 1. Then, it is necessary to restore the above experimental settings and details to facilitate the comparison of experiments fair comparison. We conduct the experiments to illustrate the impact of different factors involved in multi-label learning and compare with the other state of the art methods.

4.1 Datasets

Scene and Flags. The two datasets are coming from the images domain. In particular, the scene dataset consists of 2407 samples and 6 labels with distinct value 15 and density value 0.179; and the flags dataset consists of 194 samples and 7 labels with distinct value 54 and density value 0.485. We divide the datasets into the training and testing sets. The number of epoch is 300 and 100, and we repeat the experiments for 10 times. We calculate the mean error and standard deviation of the algorithm, which are shown in Table 1.

Genbase and Yeast. The two datasets are coming from the biology domain. In particular, the genbase dataset consists of 662 samples and 27 labels with distinct value 32 and density value 0.046; and the yeast dataset consists of 2417 samples and 14 labels with distinct value 198 and density value 0.303.

Medical and Emotions. The two datasets are coming from the text and music domains respectively. In particular, the medical dataset consists of 978 samples and 45 labels with distinct value 94 and density value 0.028; and the emotions dataset consists of 593 samples and 6 labels with distinct value 27 and density value 0.311.

Table 1. Experiments results for different methods(Mean±Variance).

Medical	Accuracy	Fscore	Hamming	Scene	Accuracy	Fscore	Hamming
MLKNN	41.35 ± 1.74%	2.77 ± 0.10%	1.87 ± 0.45%	MLKNN	57.48 ± 0.76%	18.10 ± 0.01%	9.68 ± 0.61%
KISAR	54.62 ± 50.02%	5.57 ± 0.52%	54.62 ± 49.93%	KISAR	31.59 ± 0.35%	17.25 ± 0.98%	68.41 ± 0.35%
MANIAC	52.17 ± 1.83%	33.24 ± 2.76%	9.27 ± 1.23%	MANIAC	60.27 ± 1.56%	17.18 ± 0.19%	28.33 ± 0.20%
C2AE	83.39 ± 0.19%	78.70 ± 0.06%	4.12 ± 4.06%	C2AE	53.51 ± 6.51%	64.79 ± 4.39%	14.99 ± 3.70%
PDLRMDL	83.72 ± 1.62%	18.56 ± 0.24%	16.29 ± 0.162%	PDLRMDL	78.41 ± 4.01%	15.27 ± 9.97%	31.00 ± 3.76%
Emotions	Accuracy	Fscore	Hamming	Flags	Accuracy	Fscore	Hamming
MLKNN	13.06 ± 1.68%	26.64 ± 0.71%	29.59 ± 1.30%	MLKNN	4.29 ± 1.27%	48.48 ± 1.09%	34.27 ± 2.09%
KISAR	51.68 ± 27.42%	37.19 ± 5.58%	48.32 ± 27.42%	KISAR	51.59 ± 22.75%	40.09 ± 27.55%	48.47 ± 22.68%
MANIAC	53.89 ± 0.11%	31.14 ± 19.86%	33.27 ± 3.23%	MANIAC	61.39 ± 3.98%	15.24 ± 2.11%	40.30 ± 0.01%
C2AE	46.70 ± 4.46%	61.00 ± 3.81%	27.20 ± 5.75%	C2AE	52.07 ± 1.67%	47.41 ± 2.77%	34.87 ± 0.67%
PDLRMDL	63.45 ± 2.46%	27.23 ± 5.27%	36.55 ± 2.45%	PDLRMDL	65.91 ± 0.16%	39.92 ± 1.40%	34.09 ± 0.16%
Yeast	Accuracy	Fscore	Hamming	Genbase	Accuracy	Fscore	Hamming
MLKNN	14.95 ± 1.04%	30.37 ± 0.01%	20.15 ± 0.89%	MLKNN	74.79 ± 1.05%	4.73 ± 0.12%	1.37 ± 0.14%
KISAR	44.81 ± 29.77%	38.45 ± 7.89%	55.19 ± 29.77%	KISAR	51.54 ± 22.71%	41.03 ± 26.44%	48.45 ± 22.71%
MANIAC	52.71 ± 1.27%	23.00 ± 0.01%	20.23 ± 3.15%	MANIAC	52.18 ± 2.20%	41.03 ± 26.44%	48.45 ± 22.71%
C2AE	46.91 ± 0.59%	40.22 ± 0.96%	24.55 ± 0.29%	C2AE	67.68 ± 12.50%	48.83 ± 4.57%	3.50 ± 1.78%
PDLRMDL	75.30 ± 1.52%	40.44 ± 3.92%	24.70 ± 1.52%	PDLRMDL	88.13 ± 1.41%	18.45 ± 1.47%	30.24 ± 0.98%

4.2   Comparison Experiments and Discussion

It can be seen from the Table 1, the PDLRMDL we proposed are comparative to several methods on six datasets in terms of the accuracy index. It also can be seen that the algorithm has achieved good results in terms of the recall on medical dataset, which indicates that our predicted results pay more attention on the nearby instance information and the diversity of features. Thus the performance of PDLRMDL on scene dataset with fewer labels and lower dis-tinct values are lower than that of the BR algorithm, but the performance is already quite competitive by contrast. Compare with other deep neural networks such as C2AE and MANIC, our method is more suitable for datasets with the abundant label relations information, even with some unrelated information.

5   Conclusion

In this paper we leverage the simple deep forward neural networks as the framework to tackle with the multi-label classification task. This idea constructs the different special feature learning classifier for each label. However, the special feature learning also contains the information should be shared with other latent feature representations. To motivate it, the self-attention gives dynamic weights for sharing other feature representation with current features. Besides, in order to ensure the consistency of the instance correlations for multi-label learning, we adopt the diffusion function to encourage similar instances contain similar label information. Taking account the back propagation into the building model, the

Euclidean distance metric is adopted to ensure the latent feature distance and provide the diffusion possible for the latter procedures. Finally, the experiment results also prove that our method is feasible.

References

1. Dou, Z., Cui, H., Wang, B.: Learning global and local consistent representations for unsupervised image retrieval via deep graph diffusion networks. CoRR, abs/2001.01284 (2020)
2. Hoffmann, M., Noé, F.: Generating valid euclidean distance matrices. CoRR, abs/1910.03131 (2019)
3. Huang, J., Li, G., Wang, S., Xue, Z., Huang, Q.: Multi-label classification by exploiting local positive and negative pair-wise label correlation. Neurocomputing **257**, 164–174 (2017)
4. Jiang, Y.-G., Zu-xuan, W., Wang, J., Xue, X., Chang, S.-F.: Exploiting feature and class relationships in video categorization with regularized deep neural networks. IEEE Trans. Pattern Anal. Mach. Intell. **40**(2), 352–364 (2018)
5. Jing, L., Shen, C., Yang, L., Yu, J., Ng, M.K.: Multi-label classification by semi-supervised singular value decomposition. IEEE Trans. Image Process. **26**(10), 4612–4625 (2017)
6. Kang, L., Wu, L., Yang, Y.H.: A novel unsupervised approach for multi-level image clustering from unordered image collection. Front. Comput. Sci. **7**(1), 69–829 (2013)
7. Ke, T., Jing, L., Lv, H., Zhang, L., Yaping, H.: Global and local learning from posi-tive and unlabeled examples. Appl. Intell. **48**(8), 2373–2392 (2018)
8. Hu, Y.-H.L.J.-H., Zhou, Y.J.Z.-H.: Towards discovering what patterns trigger what labels. In: Proceedings of the Twenty-Sixth AAAI Conference on Artificial Intelligence, 22–26 July 2012, Toronto, Ontario, Canada (2012)
9. Nam, J.: Learning Label Structures with Neural Networks for Multi-label Classification. PhD thesis, Darmstadt University of Technology, Germany (2019)
10. Seyedi, S.A., Lotfi, A., Moradi, P., Qader, N.N.: Dynamic graph-based label propagation for density peaks clustering. Expert Syst. Appl. **115**, 314–328 (2019)
11. Xiao, T., Yin-he, W., Qin-ruo, W.: A rotation and scale invariance face recognition method based on complex network and image contour. In: 12th International Conference on Control Automation Robotics & Vision, ICARCV2012, Guangzhou, China, 5–7 December 2012, pp. 371–376 (2012)
12. Fei, W., et al.: Weakly semi-supervised deep learning for multi-label image annotation. IEEE Trans. Big Data **1**(3), 109–122 (2015)
13. Guo-qiang, W., Zheng, R., Tian, Y., Liu, D.: Joint ranking SVM and binary relevance with robust low-rank learning for multi-label classification. Neural Netw. **122**, 24–39 (2020)
14. Yan, Z., Liu, W., Wen, S., Yang, Y.: Multi-label image classification by feature attention network. IEEE Access **7**, 98005–98013 (2019)

Playing Catan with Cross-Dimensional Neural Network

Quentin Gendre[1](✉) and Tomoyuki Kaneko[2]

[1] Graduate School of Interdisciplinary Information Studies,
The University of Tokyo, Tokyo, Japan
gendre@game.c.u-tokyo.ac.jp
[2] Interfaculty Initiative in Information Studies,
The University of Tokyo, Tokyo, Japan
kaneko@graco.c.u-tokyo.ac.jp
http://www.graco.c.u-tokyo.ac.jp/~kaneko/

Abstract. Catan is a strategic board game with many interesting properties, including multi-player, imperfect information, stochasticity, a complex state space structure (hexagonal board where each vertex, edge and face has its own features, cards for each player, etc.), and a large action space (including trading). Therefore, it is challenging to build AI agents by Reinforcement Learning (RL), without domain knowledge nor heuristics. In this paper, we introduce cross-dimensional neural networks to handle a mixture of information sources and a wide variety of outputs, and empirically demonstrate that the network dramatically improves RL in Catan. We also show that, for the first time, a RL agent can outperform *jsettler*, the best heuristic agent available.

Keywords: Machine learning · Board game · Catan · Imperfect information · Hexagonal grid · Reinforcement learning

1 Introduction

Among the challenges toward practical real-world AI agents, this paper focuses on three:

- learning a task with a general method and no prior domain-specific knowledge
- handling information sources of different kinds (e.g. not only images)
- acting robustly even when only a part of the world can be observed

Games have long served as testbeds for AI research, and recently AlphaZero [10] presented a general reinforcement learning method that successfully mastered chess, shogi, and Go, without human knowledge. However, these are deterministic perfect information games where agents can use Monte-Carlo tree search (MCTS), and with simple square-grid representations easily handled by standard imaging techniques (e.g. CNN). Therefore, that method may not work in harder domains with imperfect information, stochasticity, or complex state representation, where neither MCTS nor CNN are applicable. This paper focuses

H. Yang et al. (Eds.): ICONIP 2020, LNCS 12533, pp. 580–592, 2020.
https://doi.org/10.1007/978-3-030-63833-7_49

on Catan[1] – a famous Euro-style board game that has sold more than 22 million copies, and with frequent international tournaments – as a representative of such complex domains. Catan is an imperfect information and non-deterministic game, in which agents need to handle not only various observations (hexagonal board, cards for each player, etc.) but also hidden information and uncertainty depending on opponents' private cards and randomness. We integrate a standard policy gradient method in deep reinforcement learning with self-play, and introduce cross-dimensional network, a network structure supporting multiple input and output shapes in a flexible manner, that empirically outperforms the baseline *jsettler*.

2   Background and Related Work

2.1   Deep Reinforcement Learning in Two-Player Games

We follow standard notation of reinforcement learning; where an agent learns via interaction with an environment. For details, readers are referred to a textbook [11]. Usually, an environment is modeled as Markov Decision Process (MDP), (S, A, T, R, γ), though many applications of RL are not conforming Markov property in practice. At each time step t, an agent observes a state $s_t \in S$, and chooses an action $a \in A$. The environment changes its state to s_{t+1} following transition function T, and the agent receives a reward r_t. The policy $\pi : S \times A \mapsto \mathbb{R}$ of an agent is a probability distribution over actions given an observation. The (ultimate) goal of the learning is to identify the optimal policy π^* that maximizes the expected cumulative rewards $\mathbb{E}_{a \sim \pi^*}[\sum_t \gamma^{t-1} r_t]$, where $\gamma \in [0, 1]$ denotes the discount factor. The Value function $V_\pi : s \mapsto \mathbb{R}$ denotes estimated cumulative rewards, starting at state s and following policy π. In deep reinforcement learning, policy π and value function V is handled by using a (deep) neural network as a function approximator, because the state and action space, S, A, are prohibitively large in most interesting tasks.

Suppose a neural network parameterized by θ takes state s as its input and yields a probability distribution on actions $\pi(s)$ as well as an estimate of value function $v(s)$ as its output. Given a set of state transitions $\langle s_t, a_t, r_t, s_{t+1} \rangle$, (one-step) *Advantage Actor Critic* updates θ for such direction that increases the probability of a good action and moves $v(s_t)$ closer to $r_t + \gamma v(s_{t+1})$:

$$\nabla_\theta J_\pi(\theta) = \nabla_\theta \ln(\pi(a_t | s_t; \theta)) A(s_t, a_t; \theta),$$
$$\nabla_\theta J_v(\theta) = -\nabla_\theta v(s_t; \theta) (r_t + \gamma v(s_{t+1}) - v(s_t))$$

where $A(s_t, a_t)$ is advantage of taking action a_t at state s_t, and $A(s_t, a_t) = Q(s_t, a_t) - V(s_t) \approx r_t + \gamma v(s_{t+1}) - v(s_t)$. To prevent premature convergence, the entropy of the policy is often added to the objective function [7,14].

---

[1] Previously named The Settlers of Catan, renamed for the 5th Edition (2015).

Application to Two-Player Games. In typical application of RL to two-player board games, the "agent" stands for the player who is learning, and the "environment" includes both the opponent, and the rules of a game. The reward is given only at the termination of a game, as 1, 0, −1 for win, draw, loss, respectively. Given that agents are not enhanced by game-specific knowledge, the agent as well as its opponent must start as random players. AlphaZero [10] begins by gathering game records of random players, then gradually updates the agent by their experiences and periodically replace the opponent by the learn agent. Although changing the opponent along during learning makes the environment non-stationary and may introduce difficulty in training, it is effective to explore the challenging part of the state space and to improve the agent's strength. In our work, we applied reinforcement learning to two-player games in a similar way as AlphaZero.

There are several major achievements in imperfect information games, including Texas Hold'em, Marjong, and StarCraft II. However, their playing strength is supported by human game records in the target domain [5,13], or by methods based on counterfactual regret minimization [1] that is usually not applicable to games due to an intractable number of information sets growing almost exponentially along with the length of a game history.

Residual Convolutional Neural Network. A Convolutional Neural Network (CNN) is standard technique to handle images. It is also used for making RL agent in video games to understand the game screen such as Atari [6]. Residual Neural Networks [4] – or ResNet – is an enhancement for CNN to make learning efficient by adding residual path between layers. AlphaZero incorporated ResNet for RL agents in Go, Chess, or Shogi. We introduced alternative network for Catan and use ResNet as a baseline in comparison.

2.2   Rules of Two-Player Catan

The rules of Catan used for our research, as well as the naming conventions, matches the official 5th edition [2], but with only two players and no trading between them. In this game, both players compete to colonize an island represented by a board of hexagonal tiles. There are 5 resource types – Brick, Lumber, Ore, Grain, and Wool – which can be spent to make various actions. The first player to reach 10 Victory Points (VP) or more is considered the winner. VP can be acquired by various means: placing settlements (1VP) or cities (2VP) on the board, having the longest road or largest army (2VP), or special development cards (1VP).

The island of Catan is represented as a board of 19 *land* hexagonal tiles called *hexes*, randomly placed when setting up the game. Tiles can either represent a desert, or produce one of the 5 resources, in which case they will be assigned a number between 2 and 12. We will call the edge of a hex a *path*, and its corner an *intersection*.

At the beginning of the game, each player places 2 settlements, each with an adjacent road, in the following order: player A, player B, player B, player A.

Table 1. Actions in Catan. The first column denote the type of each action: 'dice' means it is mandatory and once, '+' can be performed in any order and any number of times after 'dice', '*' indicates it is allowed only once, but at any time

Type	Effect
dice	*Roll* two 6-sided dice. If the sum is 7, every player with 7 or more resources must discard half of them, and the current player moves the robber. Otherwise, every hex with the corresponding sum will produces resources, giving one resources to each settlement adjacent to it, and two for cities.
+	Buy a *Road*. Spend Brick + Lumber to place one on a path, next to another road.
+	Buy a *Settlement*. Spend Brick + Lumber + Grain + Wool to place a settlement next to a road, on an intersection surrounded by unoccupied intersections.
+	Buy a *City*. Spend 3 Ores + 2 Grains to upgrade a settlement into a city.
+	Buy a *Development Card*. Spend Ore + Grain + Wool to draw one card from the development pile, look at it, and add it to your hand at the end of your turn
+	*Trade* resources with the bank. The default ratio is four of the same resource for any one resource, but having a settlement or city on a harbor can reduce the rate to 3:1 or 2:1.
*	Use a *Development Card*. The card is revealed and consumed (see Table 2).

Settlements must be placed on intersections and can not be next to one another. During each turn, a player can take a sequence of actions under constraints, listed in Table 1. The robber is a piece located on a hex that prevents production on it. After rolling a 7 or using a Knight development card, the current player must move the robber to a new hex. If the other player has a settlement or city adjacent to this new location, the current player forcibly takes a random resource. Development cards are shuffled into a face down pile at the beginning of the game. Each has one of the effects listed in Table 2.

Table 2. Development cards

Knight card	Move the robber (see *the robber*), and increment army size
Road building	Place two roads for free
Year of Plenty	Take two resources from the bank
Monopoly	The opponent gives you all their resources of a stated type
Victory Point	Get one victory point

Challenges of Catan. There is no simple winning strategy. To obtain VP, players should aim for a stable and varied production of resources, by placing settlements next to high production hexes (with numbers around 7), and near other promising areas or harbors. What resources are produced also determines what actions can be used, so players should anticipate them (e.g. to *Buy Settlement*, a player must setup roads to a suitable intersection in advance). However, all this is very dependent on the board configuration, the random dice rolls, and the actions of the opponents (e.g. contention in acquiring empty paths and intersections), so a good player must constantly adjust his strategy.

2.3   JSettlers and Research on Catan

JSettlers [8] is an open-source Java implementation of the Catan rules. Among the many features the environment offers, it contains a hand-coded heuristic-based agent very often used as a base-line in Catan research. In this study, we used version 2.2.00 (released on the 3rd of March 2020), and kept the default agent type proportions: 30% of "smart-bots" and 70% of "fast-bots". In the rest of the paper, we will call this agent *jsettler*. We used JSettlers only for evaluation purpose (not in training) due to its slow execution speed. Note that its rules do not perfectly match the official rules (e.g. it doesn't include the 19 resources limit), but "official" agents can play in JSettlers with minor adjustments.

The earliest agent used Model Trees trained through self-play [9]. It hasn't been compared to JSettlers, but against a human, the author of the paper.

Szite et al. used Monte-Carlo Tree Search in a perfect-information variation of the game [12]. Their agent reaches 27% winrate with 1000 simulations, and 49% winrate with 10000 simulations, when playing against 3 *jsettlers*. However, this method cannot be applicable in the original (i.e. imperfect information) rule. Additionnaly, it used a hand-coded heuristic, thus domain knowledge.

We have found two papers that used Deep Reinforcement Learning, but they focused only on a subset of actions: trading. They both used a *jsettler* agent as a base and replaced its trading behavior, and compared its performance against 3 *jsettlers*: one achieved 49% winrate with Deep Q-Learning [3], the other 52% winrate with online Deep Q-Learning with LSTM [15].

In this paper, our agents do not learn trading – refusing all offers and never initiating negotiation – due to the limitation in our computational resources. We assert it is still fair as it does not introduce any advantage for our agents. We also limit the number of players to two instead of three or more. We argue that the task is still challenging, and to our best knowledge, this is the first study in which agents trained by reinforcement learning without domain knowledge successfully outperform *jsettlers*.

3   Our Approach

3.1   Training Process

Modified Advantage Actor Critic. Our agent is mostly based on Advantage Actor Critic. However, to speed up the learning and diversify experiments, some parallelism has been added. (Although there are similarities, it isn't A3C [7].)

Instead of playing one game on a single thread, experiences are acquired by 16 parallel workers, each playing 8 games at the same time. Each worker will cycle through its games, playing one move and saving the experience. Once a batch of 64 moves has been generated, the worker sends it to the trainer.

Simultaneously, another process is training the neural network on the batches it receives. After each update, the trainer propagates the weights to all workers.

Since a batch is not sent until it is full, some of the earliest experiences it contains were played with a slightly older policy. However, since this only represents a fraction of the batch, and the tardiness is of only a dozen of training steps, the off-policy aspect can be considered negligible.

Self-play Against Past Versions. Our agent is trained against a past version of itself, but each worker uses a different time stamp. Every 50 training steps ($50 \times 1000 \times 64$ moves, around an hour), the worker with the oldest opponent will update its policy to the most recent one.

This has many different advantages, and its efficiency has been shown in Fig. 10. This way, the opponents:

- change "slowly": every 50 steps, only one agent among 16 is changed.
- are varied: they have the behaviors the past trained agent had spanning over 750 steps ($\sim$18 h).
- match the level of the trained agent: the newest opponents are at a level very close to that of the trained agent.

Policy Activity Loss. In order to encourage exploration, we can add an entropy gradient as mentioned in 3.2. However entropy only affects legal actions, as the others being masked. In Catan, some actions are very rare and might be playable only once every couple of games (e.g. Monopoly). In order to prevent these actions' probabilities from drifting into near-zero during the many weights updates, we added a L2 activity loss on the policy layer. This loss is applied directly on the logits $\{p_i\}_i$, the raw output before a softmax activations maps them to probabilities. Thus, it will control the policy by pulling the average towards 0 and curbing absurdly high or low probabilities. Its empirical effect on the stability of the learning is shown in Fig. 9.

This final gradient is (with empirical hyper-parameters defined in Table 5):

$$\nabla_\theta J = \alpha_\pi \nabla_\theta J_\pi(\theta) + \alpha_v \nabla_\theta J_v(\theta) + \alpha_H \nabla_\theta \sum_a \tilde{\pi}(a|s) \ln \tilde{\pi}(a|s) + \alpha_p \nabla_\theta \sum_i p_i^2$$

3.2   Encoding and Network Structure

Brick Coordinate: Adapted CNN for Hexagonal Board. A regular board of catan contains 19 hexes, 72 paths and 54 intersections. This number being large for fully-connected networks, we would like to take advantage of the regularity of the board by using Convolutional Neural Network (CNN) layer. However, typical CNN are tailored specifically for grid-like layouts. There exist some tricks to fit a hexagonal grid into a regular grid, but the existing ones are not directly applicable in Catan as we also need to represent the paths (edges) and intersections (vertices).

Our idea, that we called "brick coordinate" (Fig. 2) was inspired by the double coordinate method (Fig. 1). By using a 5×3 kernel, the neighbors in brick coordinate considered by the CNN are very similar to the actual neighbors on the hexagonal board (Fig. 3).

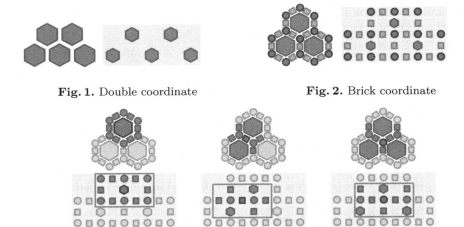

Fig. 1. Double coordinate   **Fig. 2.** Brick coordinate

Fig. 3. 5 × 3 kernel on brick coordinate

Furthermore, unlike board games like chess or go, where every position represents the same type of cell, Catan has hexes, paths and intersections that have radically different behaviors, neighbors, and features. To prevent the convolution from processing them equivalently, we separate features or actions of different types in different channels.

Cross Dimensional Neural Network. In most games where CNN can be used to efficiently process the input state, non spacial features that don't correspond to any position can be added as extra channels (e.g. turn channels in AlphaZero). However, Catan has a lot of such features, as well as actions that are completely unrelated to a position on the board (e.g. playing a development card, trading, or ending one's turn).

Intuitively, we would want to handle them using fully connected layers, but doing two networks in parallel degenerates performance. To overcome this problem, we propose using *Cross Dimensional Neural Network*. The idea is to combine two networks in parallel, each tailored for processing neurons of different dimensions, and inter-connect them to propagate information from one type into the other.

For example, in Catan, we would have one series of layers for the 2-dimensional features, one for scalar features, and interconnections between them (Fig. 4).

In order to connect features of different dimensions (i.e. brick coordinate channels and non-spacial features), we will need to either "inflate" or "deflate" them, and adjust the shape with dense layers. In this paper, we used the following:

– For inflation, each scalar value is converted to a channel filled with that value.
– For deflation, each channel is reduced to two scalars: its average and variance.

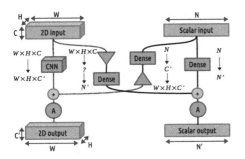

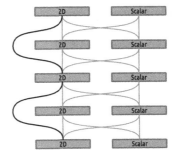

Fig. 4. Base Cross Dimensional (Xdim) layer used in our experiment: ▽ is deflation, △ is inflation, A is activation

Fig. 5. Residual Cross Dimensional layer model

Then, we can get the values for each head of the Cross dimensional neural network by summing the output of both sources and applying an activation.

As shown in Fig. 5, it is also possible to incorporate Residual paths into Cross Dimensional layers. Note that the sum is made before the activation.

Encoding of Features and Actions. A summary of how the state space and action space can be seen on Table 3 and Table 4 respectively.

We can see that Catan needs a much more complex representation for state and actions than those of Chess and Shogi used in AlphaZero. For discards (after a roll of 7), we introduced a "keep 4 resources" abstraction, using only 70 representative actions, rather than the 1 599 979 discarded resources actions (the cardinal of $\left\{b, l, o, g, w \in \{0, 1 \dots 19\}^5 \mid 3 \leq b + l + o + g + w \leq 47\right\}$, where each of b, l, o, g, w stands for a resource type). The 70 actions perfectly covers usual situations, where only four resources are kept. Even in rare cases (when holding 9 or more resources), the agent behaves robustly by randomly picking additional resources, after having saved the best four.

4   Experiments and Results

For the experiments, 3 different types of neural network architecture were used, each with 6, 8, and 10 layers (alternating *tanh* and *leaky-ReLU* activations):

- **CNNRes**, baseline, a 40-channel CNN with ResNet (without Xdim)
- **Xdim**, our method, using 15 2D-channels and 40 non-spacial neurons ($C = 15$ and $N = 40$ on Fig. 4)
- **XdimRes**, variation of our method with Residual paths (Fig. 5)

The board and positional-related actions were encoded on a 21×11 grid using brick coordinate (Figs. 2, 3). Namely, CNN layers are dimensioned $21 \times 11 \times C$.

Table 3. Input - Observable State

Board (num. of 2D channels)	17
Hexes	7
Is Desert	*1*
Production for each resource	*5*
Thief	*1*
Paths	2
Road for each player	*2*
Intersections	8
Harbors	*6*
Settlement or city for each player	*2*
Others (dim. of vector)	**45**
Self	27
Resources	*5*
Pieces left	*3*
Army size	*1*
Held development cards (new + old)	*10*
Access to each harbor	*6*
Largest Army and Longest Road	*2*
Opponent	8
Resource and Development card total	*2*
Pieces left	*3*
Army size	*1*
Largest Army and Longest Road	*2*
General	6
Bank resources	*5*
Development Card Pile	*1*
Phase	4
Has Rolled	*1*
Has development card been played	*1*
Using RoadBuilding or YearOfPlenty	*2*

Table 4. Output - Prob of Actions

Board (num. of 2D channels)	5
Hexes	2
Move thief and steal	*1*
Move thief without stealing	*1*
Paths	1
(Buy and) Place road	*1*
Intersections	2
(Buy and) Place Settlement	*1*
Buy and Place City	*1*
Others (dim. of vector)	**117**
Phase	2
Roll dice	*1*
End turn	*1*
Resources	90
Discard (4 cards to keep)	*70*
Bank trade	*20*
Development Card	22
Buy development card	*5*
Activate Knight	*5*
Activate Road Building	*1*
Activate Year of Plenty	*1*
Choose free resource	*5*
Play Monopoly (each resource)	*5*

Table 5. Hyper Parameters

Learning rate	
Initial value	3×10^{-3}
Inverse decay / training step	2×10^{-3}

Reward	
Winning reward	± 0.75
VP difference reward	± 0.02

Gradient Factor		
Policy	α_π	1×10^0
Value function	α_v	1×10^3
Entropy	α_H	1×10^{-4}
Policy activity loss	α_p	1×10^{-8}
Weight L2-regu	α_θ	1×10^{-4}

The hyper parameters used are described in Table 5. The reward is given once a game is finished, and is +0.75 for winning (resp. −0.75 for loosing) and +0.02 for every VP over the opponent's (resp. −0.02 for every VP behind).

We used Tensorflow 2.1 compiled for CUDA 10.2, and the code was run on a 32 Core CPU[2] with two GeForce GTX1080Ti 11GB GPU. In order to generate experiences quickly, we implemented a minimal environment of Catan in the Rust language, focusing on execution speed. To use it seamlessly with Tensorflow, we also turned it into a Python module using the PyO3 bindings. The code is open source and can be found at https://github.com/swynfel/rust-catan.

On the following figures, one *training* corresponds to processing 1000 batches of 64 experiences each.

4.1 Learning Curves

First, we looked at the learning curves of each model (Fig. 6).

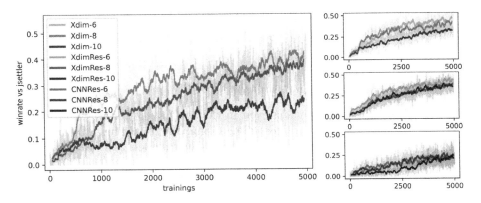

Fig. 6. Improvements of winrate against *jsettler* during early training, comparisons of architectures (left) and comparisons of layer count for each architecture (right)

Unsurprisingly, models with less layers learn faster, especially in the early steps of training. When comparing architectures of different types, *CNNRes* is lagging behind, but the other two models seems close. It is notable that even if ResNet are supposed to accelerate the early steps of training, we don't see such impact when used in conjunction with Xdim. It is even counter-productive for models with few layers. Our hypothesis is that using Xdim already introduces a sort of shortcut (information can cross from 2D values to scalar, and back to 2D). This makes ResNet not as useful for networks that aren't very deep. However for models with 10 layers, we can see its impact again.

---

[2] AMD Ryzen Threadripper 2990WX.

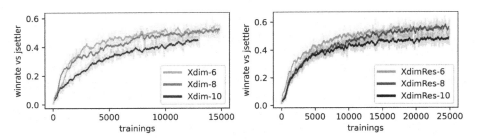

Fig. 7. Evolution of winrate against *jsettler* during long training

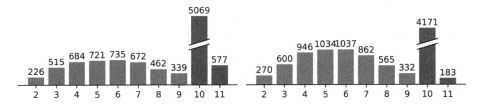

Fig. 8. Distribution of VP in 10 000 games opposing XdimRes-8 after 30 000 training steps (left) vs *jsettler* (right)

4.2   Long Term Training Results

Since CNNRes are not promising even in the early stages of training, we only kept training the models using Cross-dimensional NN. In the first 15000 training steps (around 3 weeks), we can see that the 8-layer models start catching up to the 6-layers one, and that using Residual layers does help in the long run. We can also confirm that Xdim-6, Xdim-8, XdimRes-6, and XdimRes-8 all passed over 50% win-rate (Fig. 7). When focusing on XdimRes-8 after 30000 training steps (approximately 5 weeks), we see even reached 56.5% (Fig. 8). Thus we can confidently say our agent outperforms *jsettler* in 1vs1.

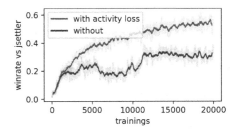

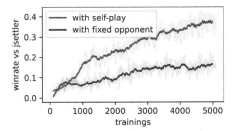

Fig. 9. Learning curves of XdimRes-8 with and without policy activity loss

Fig. 10. Learning curves with self-play (our method) and with fixed opponent

4.3 Ablation Studies

We conducted two ablation studies. Figure 9 illustrates how removing the policy activity loss makes the training unstable. Figure 10 shows the importance of self-training against various opponents. The agent trained only against a fixed "good" opponent – a copy of our agent after 10000 training steps – has trouble learning at first as the opponent is too strong, only to overfit and play poorly against *jsettler*, an unknown agent.

5 Conclusion and Future Works

In this paper, we have shown how we can successfully overcome the difficulties of Catan: The hexagonal board can be processed with CNN by using brick coordinate encoding; and the mix of positional and scalar features and actions can be handled with Cross Dimensional layers. Combining these techniques, we created a Deep RL-based agent that reached 56.5% win-rate against *jsettler* with no prior target domain specific knowledge, trained only by self-play.

We believe the methods introduced, in particular Cross-dimensional Neural Network, could be applied to real-world cases: e.g. for robots needing to combine a 2D image input with other sensory information (pressure, temperature, etc.). Thus, it would be interesting to test its effectiveness in such direction.

References

1. Bowling, M., Burch, N., Johanson, M., Tammelin, O.: Heads-up limit hold?em poker is solved. Science **347**(6218), 145–149 (2015). https://doi.org/10.1126/science.1259433
2. Catan Studio and Catan GmbH: Catan base game rules & almanac 3/4 players (5th edition). https://www.catan.com/service/game-rules (2020)
3. Cuayáhuitl, H., Keizer, S., Lemon, O.: Strategic dialogue management via deep reinforcement learning. CoRR (2015). http://arxiv.org/abs/1511.08099
4. He, K., Zhang, X., Ren, S., Sun, J.: Deep residual learning for image recognition. In: 2016 IEEE Conference on Computer Vision and Pattern Recognition (CVPR), pp. 770–778, June 2016. https://doi.org/10.1109/CVPR.2016.90
5. Li, J., et al.: Suphx: mastering mahjong with deep reinforcement learning. CoRR (2020). http://arxiv.org/abs/2003.13590
6. Mnih, V., et al.: Playing atari with deep reinforcement learning. NIPS Deep Learning Workshop (2013)
7. Mnih, V., et al.: Asynchronous methods for deep reinforcement learning. In: The 33rd International Conference on Machine Learning, pp. 1928–1937 (2016)
8. Monin, J., contributors: Jsettlers2 release-2.2.00. https://github.com/jdmonin/JSettlers2/releases/tag/release-2.2.00 (2020)
9. Pfeiffer, M.: Reinforcement learning of strategies for settlers of catan. In: Proceedings of the International Conference on Computer Games: Artificial Intelligence, Design and Education (2004)
10. Silver, D., et al.: A general reinforcement learning algorithm that masters chess, shogi, and go through self-play. Science **362**(6419), 1140–1144 (2018). https://doi.org/10.1126/science.aar6404

11. Sutton, R.S., Barto, A.G.: Introduction to Reinforcement Learning, 2nd edn. MIT Press, Cambridge, MA, USA (2018)
12. Szita, I., Chaslot, G., Spronck, P.: Monte-carlo tree search in settlers of catan. In: van den Herik, H.J., Spronck, P. (eds.) Advances in Computer Games, pp. 21–32. Springer, Berlin, Heidelberg (2010)
13. Vinyals, O., et al.: Grandmaster level in StarCraft II using multi-agent reinforcement learning. Nature **575**, 350–354 (2019)
14. Williams, R.J.: Simple statistical gradient-following algorithms for connectionist reinforcement learning. Mach. Learn. **8**(3–4), 229–256 (1992)
15. Xenou, K., Chalkiadakis, G., Afantenos, S.: Deep reinforcement learning in strategic board game environments. In: Slavkovik, M. (ed.) Multi-Agent Systems, pp. 233–248. Springer International Publishing, Cham (2019)

Port-Piece Embedding for Darknet Traffic Features and Clustering of Scan Attacks

Shintaro Ishikawa[1], Seiichi Ozawa[1,2(✉)] [iD], and Tao Ban[3]

[1] Graduate School of Engineering, Kobe University, Kobe, Japan
ozawasei@kobe-u.ac.jp
[2] Center for Mathematical and Data Sciences, Kobe University, Kobe, Japan
[3] National Institute of Information and Communications Technology, Tokyo, Japan
bantao@nict.go.jp

Abstract. With the proliferation of Internet of Things (IoT), the damage brought by cyber-attacks abusing the resources of malware-infected IoT devices is becoming more serious. Darknet monitoring, which constantly observes packets sent from malware-infected hosts to unused IP address space, has been proven effective for countermeasuring indiscriminate cyber-threats. In this paper, we presents a new machine learning scheme to track attack activities and evolving process of infected devices observed on the darknet. First, we perform feature extraction using FastText to explore the underlying correlation between targeted network services as indicated by the destination ports of scanning packets. Then, we employ a nonlinear dimension reduction technique, UMAP, to project hosts into a 2-D embedding space for a visualization purpose. Finally, we perform clustering analysis based on DBSCAN to automatically identify groups of infected hosts with similar attack behaviors. In the experiments, we use a one-month darknet traffic trace collected from a/16 darknet sensor to demonstrate the efficacy of the proposed scheme. We show that groups of Mirai variants, potentially infected by the same botnets, can be successfully detected by the proposed approach. In particular, a Mirai variant targeting vulnerabilities on TCP port 9530 are newly discovered during the observation period.

Keywords: Cybersecurity · Representation learning · Malware scan · Malware behavior analysis · Darknet analysis

1 Introduction

Cyber-attacks exploiting the vulnerabilities on IoT devices are on the rise. Compromised IoT devices are then forced to join the army of zombie devices, i.e., botnets, to perform cyber-attacks towards critical infrastructures. As a well-known example, Mirai [1], a notorious malware primarily targeting online consumer devices such as IP cameras and home routers, was first discovered in 2016. Mirai sends TCP/SYN packets to randomly IP addresses to perform a network scan searching for running services. It then intrudes the IoT devices by exploiting

© Springer Nature Switzerland AG 2020
H. Yang et al. (Eds.): ICONIP 2020, LNCS 12533, pp. 593–603, 2020.
https://doi.org/10.1007/978-3-030-63833-7_50

vulnerabilities therein and infects them to form a large-scale botnet. After the botnet is formed, DDoS attacks are performed by sending a large number of packets to a targeted server as instructed by the C&C server. The release of the Mirai source code on GitHub in September 2016 led to a burst of attacks abusing the botnet, followed by a large number of modified variants targeting other vulnerabilities. This trend is unabated and new variants are still being created so far. It is expected that early information on characterizing features such as the targeted vulnerabilities can help to reduce the damage from Mirai variants and other new malware alike.

Analyzing network traffic of infected devices to identify the cyber-threat therein belongs to the category of dynamic analysis. It is effective for obtaining the behavioral features of the malware especially when the malware program is not available. One way to collect the network traffic data is to capture the packets delivered to an unused IP address space, namely, a darknet. A darknet can capture packets which are closely related to malware infected devices, e.g., network scan packets and re-bouncing packets caused by DDoS attacks. The behavior of scanning malware can be analyzed based on these darknet packets. One of the benefits to adopt darknet analysis is its global view: darknet can observe packets coming from any host on the Internet, which reflects the trend of cyber-attacks occurring world wide. As shown in Table 1, a large number of packets (up to 327.9 billion in 2019) are observed on a darknet of 300K IP addresses operated by National Institute of Information and Communications Technology (NICT) [2]. To identify the tendency of ever-evolving attacks and detect new types of malware as early as possible from this enormous network traffic, machine learning plays an important role. Recently, assorted machine-learning based approaches have been devised to perform darknet traffic analysis [3]. In [4], the authors proposed to extract the behavior of malware as rules by using association rule analysis. In [5], IP2Vec is proposed to model the similarity between IP addresses from co-occurrence of IP addresses, destination port numbers, and protocols in packet data by using Word2Vec [6], a popular text mining method.

In this paper, we collect and analyze TCP/SYN packets observed on a darknet in order to understand the trends of port scanning from malware infected hosts. Our analysis is based on the exploration of the underlying correlation in the destination port numbers – the identifiers of the targeted network services of the scans. A quick view of the IoT related scan activities shows that destination port number pairs such as (23/TCP, 2323/TCP) and (80/TCP, 8080/TCP) are commonly spotted among related malware variants. This is caused by the cognitive habits of human beings that one tends to keep some lexical similarity in a newly assigned port numbers when replacing an existing one. Taking advantage of this convention, similarity between malware variants can be inferred from occurrence of common *sub-words*[1] in destination port numbers, and then relationships between malware variants can be modeled and evaluated. We come

---

[1] A p-piece, a.k.a., p-gram, is a sub-string of length p of a port number considered as a decimal digit string.

Table 1. Statistics of observed darknet packets per year with the NICT darknet sensor.

Year	2015	2016	2017	2018	2019
#Packets ($\times 10^8$)	545	1,281	1,504	2,121	3,279
#IP address ($\times 10^3$)	280	300	300	300	300

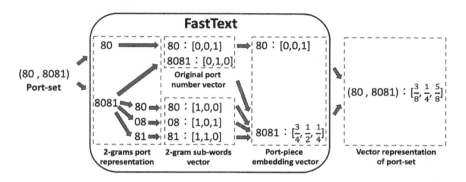

Fig. 1. Learning steps to obtain port-piece embedding vectors from a port-set using 2-gram FastText.

up with port-piece embedding with FastText to build a vector representation of port numbers that are frequently targeted by the scans, and measure the similarity between two destination-port sets based on this representation. We also perform visualization analysis based on the presentation and use it to detect the appearance of new malware variants.

2 Port-Piece Embedding for Darknet Traffic Analysis

In this section, we propose a method for quickly and accurately grasping the tendency of malware. When malware is updated, the newly assigned port numbers tend to maintain some lexical similarity due to human cognition habits. Destination port number pairs that maintain this similarity are common among related malware variants. The proposed method makes use of this convention and clusters scan activities by acquiring port-piece embedding vectors. Port-piece embedding vectors are created from the co-occurrence between port numbers and the relationship of sub-words obtained by decomposing port numbers.

2.1 Creating Port-Sets

In order to group malware-infected hosts with similar scan activities, a source IP address and a destination port number (hereafter referred to as a port) are first extracted from a TCP/SYN packet obtained by the darknet sensor. Since

Table 2. Result of clustering 9,581 port-sets including 75 Mirai-infected port-sets extracted on February 15, 2020. Port embedding vectors are obtained by FastText, TF-IDF, Word2Vec.

Cluster no.	FastText		TF-IDF		Word2Vec	
	#Port-sets	#Mirai Port-sets	#Port-sets	#Mirai Port-sets	#Port-sets	#Mirai Port-sets
1	**2,428**	**65**	671	3	4,439	0
2	494	2	27	0	237	0
3	7	0	449	0	444	0
4	40	0	87	0	23	0
5	110	0	46	0	4	0
6	9	0	1	0	3	0
Outlier	6,763	8	**8,570**	**72**	**4,701**	**75**

an individual service is assigned to a unique port, a port observed in darknet packets represents a service targeted by malware. Therefore, it is considered that a set of such ports features scan activities by a malware-infected host. We call such a set of ports a *port-set*. In the following, we try to discovery clusters of malware-infected hosts based on the similarity of port-sets.

2.2   Port-Piece Embedding Vectors

(23/TCP, 2323/TCP) and (80/TCP, 8080/TCP) are common port pairs among related malware variants that are exploiting network services hosted on similar but different ports. The similarity between the ports in these pairs are apparent due to the human convention to maintain some degree of likeness by keeping identifying morphemes – consecutive digit sequences – in related numbers. Similar malware variants can be grouped by taking advantage of this fact. By applying FastText – a popular text mining method – to the port sets, embedding vectors that take account of the identifying morphemes in port numbers can be obtained.

Figure 1 shows the procedure to obtain port-piece embedding vectors in the proposed method. To apply FastText to the port-sets, a port-set is taken as a statement in document analysis. First, port vectors, which are created from the original ports, are obtained. Then, each port is divided into n-grams (hereafter referred to as port pieces). Then, port-piece vectors, which are created for the port pieces contained in a port, are obtained. Then, the port-piece vector of a port is obtained by averaging the port vector and port-piece vectors that are contained in the port. The port-set vector for a port set is obtained by averaging the port-piece vectors of the ports that appear in the port set.

2.3   Visualization of Scan Activities

We adopt UMAP to observe the distribution of malware-infected hosts, and see if the proposed port-piece embedding gives a good interpretation for scan activities of malware. In UMAP, data vectors in a high-dimensional space are first represented as a graph. Then, the graph structure in the embedding low-dimensional space is optimized to approximate that in the high-dimensional space. This allows UMAP to perform fast and high-performance dimension reduction.

2.4   Clustering of Scan Activities

By clustering the port-set vectors, infected host groups with the same attack pattern can be identified automatically. We apply DBSCAN – a density-based clustering method – to the port-set vectors. DBSCAN consists of three steps. In the first step, DBSCAN finds the points which falls into its ϵ neighborhood for each point and identifies the points with more than k neighbors as core points. Here, ϵ is a distance threshold and k is a numerical threshold. They control the density of the points that are identified as clusters in DBSCAN. In the second step, DBSCAN finds the connected components of core points on the neighbor graph by ignoring all non-core points. In the last step, for each non-core point, it is assigned to a nearby cluster if the cluster contains points in its ϵ neighbor, otherwise it is assigned to noise. By applying DBSCAN to port-set vectors, it is possible to automatically identify groups of infected hosts which show the same attack pattern. This overcomes the difficulty of pre-determining the number of clusters in other conventional clustering algorithms.

3   Experiments

In this section, to evaluate the effectiveness of the proposed port-piece embedding, we examine the scan behaviors of the well-known IoT malware called Mirai. To assign a label to the port set exploited by a host, we check the packets from the host against the following three conditions:

- Destination IP address equals the sequence number;
- Destination port numbers include 23/TCP;
- Source port number is greater than 1024.

If more than 90% of the packets sent from the host satisfy these three conditions, we label its port set as *Mirai*.

   In the following experiments, we used 9,875,671,868 packets from February 1st, 2020 to February 29th, 2020 observed on a/16 darknet sensor operated by National Institute of Information and Communications Technology (NICT).

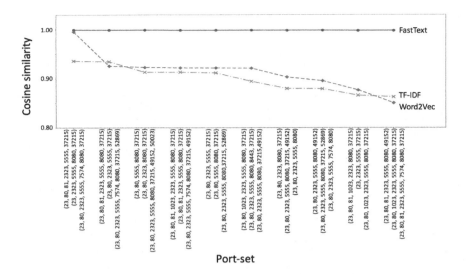

Fig. 2. Top 10 port-sets with high cosine similarity to port-set (23, 80, 2323, 5555, 8080, 37215). On each x-axis tick, the left is the FastText port-set, the center is the TF-IDF port-set, and the right is the Word2Vec port-set.

3.1 Similarity Measure of Port-Sets

In general, a good representation of an embedding space allows us to have a good interpretation when we try to understand observed phenomena. It also often leads to good performance in classification accuracy and/or meaningful clustering results. Therefore, we evaluate the proposed port-piece embedding from the perspective whether embedding vectors for Mirai-featured port-sets give us a good measure in similarity for Mirai variants.

In the following experiments, we carry out performance comparison among three feature representation methods: TF-IDF [10] features, Word2Vec embedding, and port-piece embedding. Since 62,817 ports are observed in the TCP/SYN packets from February 9th to February 15th, 2020, a TF-IDF feature is defined as a 62,817-dimensional vector. To obtain a Word2Vec embedding vector, we consider a 62,817-dimensional one-hot vector for each port set and train a skip-gram [11] network to obtain 200-dimensional Word2Vec embedding vectors. For the proposed port-piece embedding, we use FastText to set up a 65,533-dimensional one-hot vector consisting of 62,817 ports and their sub-words, and use skip-gram to reduce its dimension to 200.

To evaluate the three feature representation methods, we adopt an example of the following port-set: (23, 80, 2323, 5555, 8080, 37215). Hosts with such a port-set were observed on February 15th and their packets matched the Mirai signature. Therefore, we can consider that hosts with a port-set of similar port numbers above are likely to be infected by Mirai variants. Such similar port-sets must be represented by similar representation vectors.

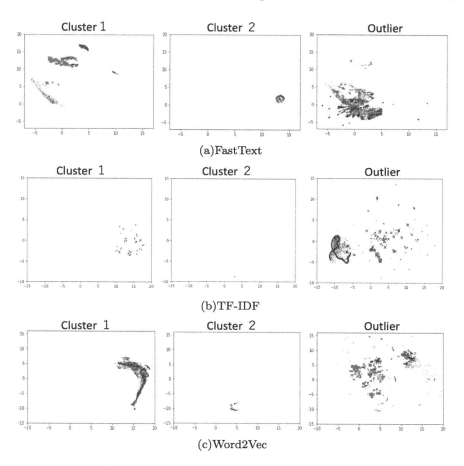

Fig. 3. Visualization of Mirai-featured (×) and Mirai-benign (○) port-sets using (a) FastText, (b) TF-IDF, and (c) Word2Vec. (Color figure online)

We calculated the cosine similarity between port sets (23, 80, 2323, 5555, 8080, 37215) and similar port sets in the embedding space obtained by the three representation methods. Figure 2 shows the top 10 port-sets with the highest cosine similarity. The port-piece embedding method yields a cosine similarity of 0.99 or higher for the top 10 port sets. It also retrieves port sets with similar port numbers in the top 10, indicating that similar port-sets will have similar representation vectors. On the other hand, the other two representation methods yield significantly degenerated cosine similarities. This result indicates that the proposed method has a favorable performance for port set representation.

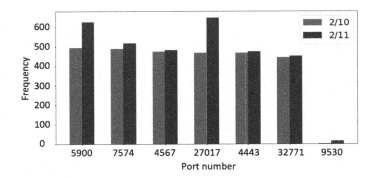

Fig. 4. Frequency of port numbers contained in port-set cluster 1 on February 10 and February 11. Ports with a frequency more than 500 as of February 10 and with a frequency increment less than 6 are omitted.

3.2 Clustering of Scan Activities

We cluster the scan activities using the same data and compared it with other methods in Sect. 3.1. To confirm the clustering accuracy of scan activities, the port-set on February 15, 2020 was labeled as Mirai. Mirai-featured port-sets were found in 75 out of the 9,581 port-sets extracted on February 15, 2020.

Table 2 shows the clustering results and Fig. 3 shows the visualization results for each method of FastText, TF-IDF, and Word2Vec. Outliers refer to points that are neither the core point nor points assigned in clusters in DBSCAN. The visualization results show only cluster 1, cluster 2, and outliers, with Mirai-featured port-sets shown in red.

The visualization result in Fig. 3(a) also shows that most Mirai-featured port-sets are contained in cluster 1. The Mirai-featured port-sets clustered in Cluster 2 contains 7547/TCP which is not included in the port-sets in Cluster 1, indicating the discovery of a major variant. An further investigation revealed that 7547/TCP is the port number targeted by a Mirai variant, which is first reported since December 2019 with a bust of increment in scanning activities. Mirai-featured port-sets assigned as outliers because they contain ports that rarely appear, e.g., 0/TCP, 654/TCP, 55717/TCP.

As for TF-IDF and Word2Vec, the visualization results in Fig. 3(b) and (c) show that the Mirai-featured port-sets are scattered into more clusters than the proposed method, with most of them assigned as outliers. This indicates TF-IDF and Word2Vec do not perform well in grouping similar port-sets into the same clusters.

The above experiment shows that the proposed method can detect Mirai-featured port-sets more appropriately than the other two referenced methods, and is therefore effective in clustering scan activities.

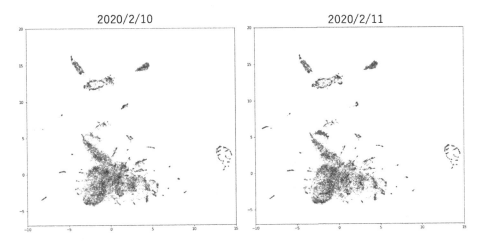

Fig. 5. Visualization of port-sets including 9530/TCP (×) and port-sets not including 9530/TCP (○) on February 10th and February 11th.

3.3 Detection of New Vulnerabilities

Due to the release of Mirai source code, the emergence of Mirai variants targeting new vulnerabilities becomes a challenging problem. By counting the frequency of ports existing in the scan activities clustered by the proposed method, we can discover the emergence of Mirai variants targeting new vulnerabilities. As a result, we were able to discover the communication to 9530/TCP which seems to be caused by a new Mirai variant started from February 11, 2020.

The port-sets on February 10 and February 11 were clustered into 8 clusters: cluster 1 to cluster 7 together with an outlier cluster. The port-sets in cluster 1 contain 23/TCP, 2323/TCP, 80/TCP, 8080/TCP with high frequency, and thus cluster 1 is assumed to mainly consist of Mirai variants. To catch the emergence of Mirai variants targeting new vulnerabilities, we count the frequencies of ports contained in this cluster and show it in Fig. 4. The frequency of 9530/TCP was only 1 as of February 10, but was increased to 16 on February 11. In the figure, ports with a frequency more than 500 and ports with a frequency increment less than 6 are omitted as they are less likely to be from new Mirai variants. The visualization results in Fig. 5 also shows that the port-sets including 9530/TCP formed compact groups and increased substantially in volume.

The increased access to 9530/TCP from February 2020 is reported to be caused by a new Mirai variant in the NICTER blog published by NICT. In order to confirm that the increased access to 9530/TCP caused by new Mirai variants is not temporary, we show the transition of the daily frequency of port-sets including 9530/TCP in Fig. 6. In the figure, we can confirm that the number of port-sets has increased since February 11. This result demonstrates the proposed method can catch the emergence of Mirai variants that encompass new vulnerabilities.

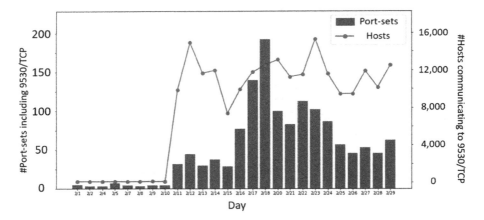

Fig. 6. Changes in the number of port-sets including 9530/TCP (lines) and changes in the number of hosts that exploit the port-sets (bars).

4   Conclusions

In this paper, we proposed a new approach to darknet traffic analysis where port-piece embedding using FastText is utilized to explore the similarity of partial patterns of the destination port numbers among malware variants. By applying FastText to TCP/SYN packets observed in the darknet, the relationship is measured based on the co-occurrence between frequent pieces of port numbers, and the clustering of scanning features is conducted in a obtained port-piece embedding space. In our experiments, darknet packet traffic collected from February 1 to 29, 2020 on a/16 network were analyzed. As a result, it is shown that the proposed method using FastText can cluster the characteristics of scanning activities of Mirai variants better than the port embedding vector obtained by TF-IDF or Word2Vec. Furthermore, the frequencies of destination port numbers were examined for the hosts in a Mirai cluster. We demonstrated that it was possible to early detect the access to 9530/TCP by the Mirai variant that started to increase from February 11, 2020. From the above results, it is expected that the proposed method works effectively to monitor malware variants which target similar port numbers.

Our future work is to improve port-piece embedding vector representation so that the accuracy of malware clustering and detection can be enhanced. In this paper, we used a Mirai signature to evaluate the accuracy of generated clusters. However, for malware other than Mirai, it is necessary to find means to accurately identify its variants and continue to verify the effectiveness of the proposed port-piece embedding representation. If malware can be correctly clustered, infected hosts acting simultaneously can be identified, leading to the detection of botnets. In addition, by monitoring the activities of such botnets, we believe that it will be possible to develop a system that can detect symptoms of new cyber-attacks.

Acknowledgement. This research is supported by the Ministry of Education, Science, Sports and Culture, Grant-in-Aid for Scientific Research (B) 16H02874 and the Commissioned Research of National Institute of Information and Communications Technology (NICT), JAPAN. The authors thank Mr. Jumpei Shimamura (clwit Inc.) for his valuable suggestions on cyber-attack monitoring.

References

1. Kolias, C., Kambourakis, G., Stavrou, A., Voas, J.: DDoS in the IoT: mirai and other botnets. IEEE Comput. **50**, 80–84 (2017)
2. National Institute of Information and Communications Technology: NICTER observation report 2019 (2019). https://www.nict.go.jp/cyber/report/NICTER_report_2019.pdf. Accessed 20 June 2020
3. Buzak, L.A., Erhan, G.: A survey of data mining and machine learning methods for cyber security intrusion detection. IEEE Commun. Surv. Tutor. **18**, 1153–1176 (2016)
4. Ozawa, S., Ban, T., Hashimoto, N., Nakazato, J., Shimamura, J.: A study of IoT malware activities using association rule learning for darknet sensor data. Int. J. Inf. Secur. **19**, 83–92 (2020). https://doi.org/10.1007/s10207-019-00439-w
5. Ring, M., Dallmann, A., Landes, D., Hotho, A.: IP2Vec: learning similarities between IP addresses. In: IEEE International Conference on Data Mining Workshops (ICDMW) (2017)
6. Mikolov, T., Chen, K., Corrado, G., Dean, J.: Efficient estimation of word representations in vector space. arXiv:1301.3781 (2013)
7. Bojanowski, P., Grave, E., Joulin, A., Mikolov, T.: Enriching word vectors with subword information. Trans. Assoc. Comput. Linguist. **5**, 135–146 (2017)
8. Bojanowski, P., Grave, E., Joulin, A., Mikolov, T.: UMAP: uniform manifold approximation and projection for dimension reduction. arXiv:1802.03426 (2018)
9. Ester, M., Kriegel, P.H., Sander, J., Xu, X.: A density-based algorithm for discovering clusters in large spatial databases with noise. In: KDD 1996: Proceedings of the Second International Conference on Knowledge Discovery and Data Mining, pp. 226–231 (1996)
10. Ramos, J.: Using TF-IDF to determine word relevance in document queries (2003)
11. Mikolov, T., Sutskever, I., Chen, K., Corrado, G., Dean, J.: Distributed representations of words and phrases and their compositionality. In: Advances in Neural Information Processing Systems, vol. 26 (2013)
12. National Institute of Information and Communications Technology: NICTER blog. https://blog.nicter.jp/. Accessed 20 June 2020

Recency-Weighted Acceleration for Continuous Control Through Deep Reinforcement Learning

Zhen Wu[1], Zongzhang Zhang[2(✉)], and Xiaofang Zhang[1]

[1] School of Computer Science and Technology, Soochow University,
Suzhou 215006, Jiangsu, China
20185227023@stu.suda.edu.cn, xfzhang@suda.edu.cn
[2] National Key Laboratory for Novel Software Technology, Nanjing University,
Nanjing 210023, China
zzzhang@nju.edu.cn

Abstract. Model-free reinforcement learning algorithms have been successfully applied to continuous control tasks. However, these algorithms suffer from severe instability and high sample complexity. Inspired by Averaged-DQN, this paper proposes a recency-weighted target estimator for actor-critic settings, which will construct a target estimator with more weight placed on recently learned value functions, obtaining a more stable and accurate value estimator. Besides, delaying policy updates with more flexible control is adopted to reduce per-update error because of value function errors. Furthermore, to improve the performance of prioritized experience replay (PER) for continuous control tasks, Phased-PER is proposed to accelerate training in different periods. Experimental results are given to demonstrate that using the same hyper-parameters and architecture the proposed algorithm is more robust and achieves better performance, surpassing the existing methods on a range of continuous control benchmark tasks.

Keywords: Deep reinforcement learning · Value estimation · Delayed policy updates · Prioritized experience replay

1 Introduction

In recent years, deep reinforcement learning (DRL) [9] has maintained a rapid development with powerful representation of deep neural networks and increasing computing ability. As a result, RL has been applied to more and more fields, such as games, natural language processing, recommendation systems, robotic control tasks and so on [14].

This work is in part supported by the Natural Science Foundation of China (61876119), the Natural Science Foundation of Jiangsu (BK20181432) and a project funded by the Priority Academic Program Development of Jiangsu Higher Education Institutions.

H. Yang et al. (Eds.): ICONIP 2020, LNCS 12533, pp. 604–615, 2020.
https://doi.org/10.1007/978-3-030-63833-7_51

As a model-free value-based RL algorithm, deep Q-network (DQN) [10] successfully combines deep neural networks with RL algorithms, using deep neural networks as value function approximators. Two novel techniques, target network and experience replay, were proposed to tackle the instability and correlation of RL problems. However, DQN still suffers from severe instability and poor sample efficiency. In order to reduce the approximation error variance in the target values, Averaged-DQN [2] averages across previously learned value estimates, leading to more stable training procedure and improved performance. Besides, PER [12] improves the sample efficiency by prioritizing replaying transitions with high learning value. However, DQN cannot handle tasks with continuous action space, which are of great importance in reality, due to the maximum operation over Q-functions.

As a model-free actor-critic RL algorithm, soft actor-critic (SAC) [5] achieves state-of-the-art performance in continuous control, but the instability and poor sample efficiency limit its further application to reality. In order to reduce target approximation error variance and make SAC more stable, we draw on the ideas underlying Averaged-DQN and further improve it by placing more weight on recently learned target values. Although the target estimates are improved, value function errors still exist. Therefore, delaying policy updates is adopted to prevent the policy deviation caused by value function errors. In addition, to improve the sample efficiency of SAC, we further propose Phased-PER, so that the algorithm can learn quickly and stably while maintaining the original distribution. With the recency-weighted acceleration (RWA) framework comprised of these improvements applied to SAC, the resulting algorithm called accelerated recency-weighted delayed soft actor-critic (ARW-DSAC) achieves higher sample efficiency and better final performance in several control tasks. The contributions of our work can be summarized as:

- The delayed recency-weighted model is proposed, where the target value estimates are calculated over previously learned target Q-networks by emphasising recently learned value estimates. Besides, the model adopts delayed policy updates in a way that allows for more flexible control.
- Adapted from PER, Phased-PER is adopted to accelerate training process, which is comprised of three periods, enabling fast learning in the first period and stable learning in the third period.
- Experiments show that our proposed algorithm exceeds existing algorithms in both sample efficiency and final performance in several continuous control tasks.

2   Related Work

2.1   Value Estimation

Value estimation is one of the most distinguished features in RL [14]. Q-learning [16] is a popular value-based RL algorithm, but it has overestimation problems. To tackle the overestimation of Q-learning, double Q-learning [6] uses a double estimator approach to determine the value of the subsequent state, but it has the underestimation problem. Weighted Double Q-learning [17] balances the overestimation in the single estimator and the underestimation in the double estimator, showing great potential for application [18]. In addition to the overestimation, value estimation often suffers from instability. Averaged-DQN stabilizes training by averaging across recent Q-networks, but it ignores relative importance which is verified necessary for actor-critic algorithms, e.g. SAC.

Twin delayed deep deterministic policy gradient (TD3) [4] demonstrated that overestimation not only existed in value-based algorithms (e.g. DQN), but also persisted in actor-critic settings, and proposed clipped double Q-learning. However, few work has been done to improve the stability of actor-critic algorithms. In addition, TD3 adopted delaying policy updates, which consists of only updating the actor and target critic networks every 2 iterations. However, this method of controlling frequency of policy updates lacks flexibility.

2.2 Experience Replay

Experience replay not only breaks the limitation of the relevance of RL problems, but also improves the sample efficiency [11]. Based on PER, distributed prioritized experience replay [8] establishes a distributed architecture, which makes full use of the advantages of parallel computing by separating exploration processes from learning processes. Hindsight experience replay [1] can automatically learn from the failed experiences of binary rewards, which can be seen as a form of implicit curriculum. However, few of them focuses on directly boosting sample efficiency of actor-critic algorithms for continuous control.

3   Preliminaries

3.1   Reinforcement Learning

RL can be described as a process where an agent interacts with an environment with the goal of maximizing the accumulative rewards. Markov decision process framework is employed to model the environment, which is defined by a tuple $(\mathcal{S}, \mathcal{A}, \mathcal{R}, \mathcal{P}, \gamma)$, where $\mathcal{S}$ is a state space, $\mathcal{A}$ is an action space, $\mathcal{R} : \mathcal{S} \times \mathcal{A} \to \mathbb{R}$ is a stochastic reward function, $\mathcal{P}(s' \mid s, a)$ is a transition function which gives distribution over next state s' given a state action tuple (s, a), and $\gamma \in [0, 1)$ is a discount factor. The goal of the agent is to search a policy $\pi(a \mid s)$ that maximizes the expected sum of discounted future rewards represented by $R_t = \sum_{i=t}^{\infty} \gamma^{i-t} r_i$. The action-value function is defined as $Q_\pi(s, a) = \mathbb{E}_{a \sim \pi(\cdot \mid s_t)}[R_t \mid s_t = s, a_t = a]$.

3.2   DQN and Variants

DQN uses deep neural networks as function approximators to represent value function $Q_\theta(s, a)$. Sampling from experience buffer, DQN updates the value function by stochastic gradient descent with the loss function constructed as below:

$$L(\theta) = \mathbb{E}_{(s,a,r,s') \sim \mathcal{B}}\left[y^{\mathrm{DQN}} - Q_\theta(s, a)\right]^2. \tag{1}$$

Here the target of DQN is $y^{\text{DQN}} = \mathbb{E}_{(s,a,r,s') \sim \mathcal{B}} [r + \gamma \max_{a' \in \mathcal{A}} Q_{\theta^-} (s', a')]$, where θ^- is the parameters of the target network, which is consistent during a fixed time interval, and $\mathcal{B}$ is the distribution from experience buffer.

Averaged-DQN stabilizes training process by averaging across the last K previously learned Q-networks. Its target estimates are calculated as below:

$$y^{\text{Averaged-DQN}} = \mathbb{E}_{(s,a,r,s') \sim \mathcal{B}} \left[r + \gamma \max_{a' \in \mathcal{A}} \frac{1}{K} \sum_{m=1}^{K} Q_{\theta^m} (s', a') \right], \qquad (2)$$

where θ^m is the m-th recently learned target value function in the past.

Instead of uniform sampling, PER prioritizes replaying transitions of high priority measured by temporal difference error. Stochastic prioritization makes sure that the probability of sampling is monotonic by the transition's priority, while guaranteeing non-zero probability even for transitions of the lowest priority. However, PER inevitably introduces bias because of the prioritization. And PER corrects this bias through importance-sampling weight $W_i = (\frac{1}{N} \cdot \frac{1}{P(i)})^\beta$, where $P(i)$ is the priority of transition i, N is the total time step, and β is the parameter controlling correction level of importance sampling. Concretely, PER linearly anneals from its initial value β_0 to 1.

3.3   Soft Actor-Critic

SAC is a state-of-the-art RL model-free algorithm based on the maximum entropy framework, where the objective of an agent is augmented with an entropy term. Therefore, the Q-function of SAC is calculated as below:

$$Q_\pi (s, a) = \sum_{t=1}^{\infty} \mathbb{E}_{a_t \sim \pi(\cdot|s_t)} [\mathcal{R} (s_t, a_t) + \alpha \mathcal{H} (\pi (\cdot \mid s_t))], \qquad (3)$$

where α is the temperature parameter that trades off exploration and exploitation and $\mathcal{H} (\pi (\cdot \mid s_t))$ is the entropy of policy π at state s_t. The specific update rule is presented in Algorithm 1. For more detail please refer to [5].

4   Recency-Weighted Acceleration Framework

In this section, the recency-weighted acceleration (RWA) framework is proposed. Firstly, we will introduce the recency-weighted model combined with delayed policy updates (short for DRW model) and present its algorithm pseudocode. Then Phased-PER is introduced to further accelerate its training process.

4.1   Delayed Recency-Weighted Model

To reduce target approximation error variance, instead of using only the last target Q-network, the DRW model makes use of last K target networks for value estimation. Moreover, instead of simply averaging over all the Q-networks,

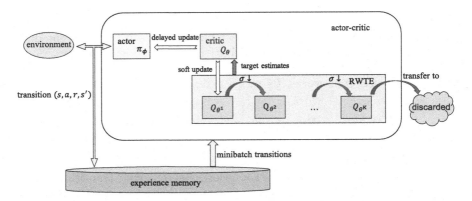

Fig. 1. The schematic diagram of the DRW model. This model has two innovations. Firstly, recency-weighted target estimator (RWTE) is made up of an FIFO structure. It maintains the recent K target networks based on the arrival time with parameters transferred to next network after one iteration, providing estimates for the critic to be updated upon. Q_{θ^1} is the most recent Q-network. Specifically, after one iteration of critics update, the parameters of Q_{θ^1} network are transferred to Q_{θ^2}, at the same time the parameters of Q_{θ^2} are transferred to Q_{θ^3}, and so on. Secondly, the DRW model adopts delaying policy updates in a more flexible approach, where d controls the ratio of the frequency of policy updates to critic updates.

this model places more weight to the more recent networks, because the latest target networks tend to have more precise estimation. The schematic diagram of the DRW model is shown in Fig. 1. As it shows that an agent adopting the actor-critic framework interacts with an environment, storing transitions (s, a, r, s') to experience buffer. Mini-batches are sampled from the experience pool to update parameters of actor and critic.

In order to put more emphasis to recently learned target estimates, recency-weighted coefficient σ is introduced to discount less recent target estimates. The weight of the most recent target network Q_{θ^1} is represented by ω_1, and that of the second is ω_2, and so on. And the weights meet the following requirements:

$$\begin{cases} \omega_{m+1} = \sigma\omega_m, \quad m = 1, 2..., K - 1, \\ \sum_{m=1}^{K} \omega_m = 1. \end{cases} \quad (4)$$

Here $\sigma \in (0, 1]$ is a constant controlling the relative importance of target Q-networks. In particular, the RWTE model will be recovered to the averaged target estimator, when σ equals to 1. With the model applied to SAC, its target values are calculated as below:

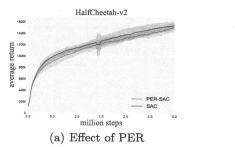

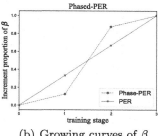

(a) Effect of PER   (b) Growing curves of β

Fig. 2. (a) PER makes little improvement on SAC and even makes it degenerate. (b) Phased-PER. The original method employed by PER to correct the distribution bias is through the importance sampling weight method, where β anneals linearly from its initial β_0 to 1. The phased update strategy for β is divided into three stages, where the growth rate of the first and third periods is relatively low, while the rate of the second period is high.

$$y^{\text{RW-DSAC}} = \mathbb{E}_{(r,s')\sim\mathcal{B}} \left[r + \gamma \sum_{m=1}^{K} \omega_m \left(\min_{i=1,2} Q_{\theta_i^m} \left(s', a'\right) - \alpha \log \left(\pi_\phi \left(a'|s'\right)\right) \right) \right]. \tag{5}$$

Delayed Policy Updates. As an off-policy actor-critic algorithm, SAC is also affected by value function errors. When the critic is inaccurate, it may lead to wrong direction of policy improvement. Here, a more general method of delaying policy updates is proposed, with d calculated as:

$$d = \frac{F_{policy}}{F_{critic}}, \tag{6}$$

where F_{policy} represents the frequency of policy updates and F_{critic} denotes the frequency of critic updates. Therefore, $d \in (0,1]$ controls the ratio of the frequency of policy updates to critic updates.

With the DRW model applied to SAC, the pseudocode of recency-weighted delayed SAC (RW-DSAC) is presented in Algorithm 1. Lines 1–3 initialize the parameters of actor, critic and their target networks. Line 4 calculates the weights of target networks. Lines 8–10 represent that the agent interacts with the environment, storing experience transitions to its replay buffer. In lines 11–14, mini-batches are sampled from replay buffer to update the critic. In line 15, $delay(t, d)$ is introduced to determine whether to update the policy based on iterations t and delaying parameter d. Specifically, $delay(t, d)$ ensures that the ratio of the frequency of policy updates to critic updates is d. Lines 16–22 update policy and target networks as well as the temperature coefficient α.

Algorithm 1: RW-DSAC

Input: Randomly initialized critic networks θ_1, θ_2, actor network ϕ

1 Initialize recency-weighted coefficient σ, and delaying parameter d;
2 Initialize array of target critic networks $(\theta_1^m, \theta_2^m) \leftarrow (\theta_1, \theta_2)$ $(m = 1, \ldots, K)$;
3 Initialize experience replay buffer $\mathcal{B}$;
4 Set weight vector ω_m so that it satisfies equation (4);
5 **for** *episode* $= 1 \to M$ *(the final episode)* **do**
6   Reset initial observation state to s_1;
7   **for** $t = 1 \to T$ **do**
8   Select action $a_t \sim \pi_\phi \left(\cdot \mid s_t \right)$;
9   Execute a_t and observe reward r_t moving to subsequent state s_{t+1};
10   Store transition (s_t, a_t, r_t, s_{t+1}) in $\mathcal{B}$;
11   Sample mini-batch of transitions (s, a, r, s') from $\mathcal{B}$;
12   Calculate recency-weighted target y^{target} using equation (5);
13   Update critic by minimizing the loss:;
14   Loss $= \mathbb{E}_{(s,a) \sim \mathcal{B}} \left[y^{target} - Q_{\theta_i} (s,a) \right]^2$, for $i \in \{1, 2\}$;
15   **if** $delay(t, d)$ **then**
16   Update actor network using:;
17              $J_\pi (\phi) = \mathbb{E}_{s \sim \mathcal{B}} \left[\mathbb{E}_{a \sim \pi_\phi (\cdot | s)} \left[\alpha \log \left(\pi_\phi (a | s) \right) - Q_{\theta_i} (s, a) \right] \right]$;
18   Update temperature parameter α using the following objective:;
19              $J(\alpha) = \mathbb{E}_{a \sim \pi_\phi} \left[-\alpha \log \pi_\phi (a | s) - \alpha \overline{\mathcal{H}} \right]$;   ▷ $\overline{\mathcal{H}}$ `represents the`
`environment-specific entropy target`
20   Transfer and update target networks:;
21              $\theta_i^m \leftarrow \theta_i^{m-1}$ $(m = 2, \ldots, k)$, for $i \in \{1, 2\}$;
22              $\theta_i^1 \leftarrow \tau \theta_i + (1 - \tau) \theta_i^1$, for $i \in \{1, 2\}$;   ▷ τ `is the smoothing`
`coefficient`
23   **end**
24   **end**
25 **end**

Output: Learned actor network π_ϕ and critic networks $Q_{\theta_1}, Q_{\theta_2}$

4.2 Phased-PER

PER is one of the two most important ingredients that lead to biggest improvement in Rainbow [7]. However as Fig. 2(a) reveals, when applied to continuous control settings, PER has no obvious improvement effect for the actor-critic algorithms. So Phased-PER is proposed to improve the performance of PER.

PER makes algorithms prioritize replaying samples with high learning value, but it changes the original sample distribution, which makes it prone to diverge. The original method employed by PER is to linearly anneal β from its initial value β_0 to 1 using importance weight. However, the linear strategy of changing β ignores the characteristics of different stages of learn process. So we modify this strategy in order to improve its effect and propose Phased-PER. As is shown in Fig. 2(b), the resulting change curve of β has lower value in the first stage and higher value in the third stage compared with the original method. Therefore,

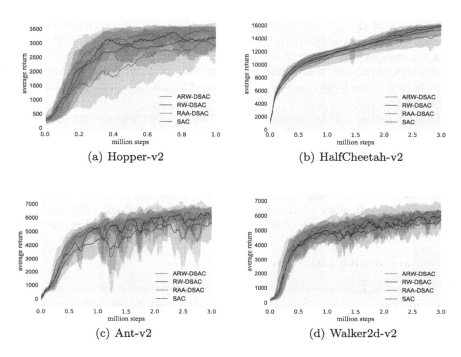

Fig. 3. Comparative results of different algorithms over continuous control tasks. Learning curves are obtained by 4 seeds where the solid curves represent the mean and the shaded regions correspond to a standard deviation. Here, RAA-DSAC is our self-implemented version of SAC applied with RAA with d equal to 0.7.

this strategy allows for fast learning in the first stage and stable training in the final stage. Specifically, the first and third periods of training account for one-eighth of the total change of β respectively, with remaining proportion of change filled by training of the second period.

5   Experiments

In order to verify the effectiveness of our proposed framework, comparative experiments are conducted in the continuous control tasks. Then empirical results are shown followed by our analysis. Finally, ablation studies are carried out to further investigate each ingredient which leads to improved performance.

5.1   Experiment Setup

We employ Gym [3] as environment interface, using MuJoCo [15] physics engine to simulate continuous control tasks. Specifically we choose Hopper, HalfCheetah, Walker and Ant as experimental environments to test our algorithms. Stability, sample efficiency and final performance are the criteria to evaluate the performance of the proposed algorithms. To make our experiments more convincing, we also choose the regularized anderson acceleration (RAA) [13] framework for comparison, which is a state-of-the-art off policy RL algorithm. It is an effective approach to accelerating the solving of fixed point problems with perturbations, which also uses previous learned target networks. For more detail please refer to [13].

Our algorithm will be compared with two baseline algorithms, one is SAC against which comparison will show the improvement of the RWA framework, the other is RAA against which comparison will reveal the utilization efficiency of multiple target Q-networks. Each task will be run for 3 million time steps except Hopper for 1 million with evaluations every 5000 time steps, where each evaluation logs the averaged return across 10 episodes. In order to solely verify the utilization efficiency of multiple target Q-networks of the DRW model, we also conduct RW-DSAC experiments without the acceleration of Phased-PER.

5.2   Comparative Evaluation

For the sake of fairness, common hyper-parameters of the four algorithms are from [5], while exclusive hyper-parameters between ARW-DSAC and RAA-DSAC stay the same. Specifically, the num of previous estimates m is set to 5. Also, RAA-DSAC adopts delayed policy updates the same as ARW-DSAC, using delaying parameter $d = 0.7$. In addition, the growth rates of the first period i_1 and third period i_3 are both 0.375, while the rate of the second period i_2 is 2.25. Hype-parameters and architectures of RW-DSAC and RAA-DSAC are the same except the calculation of target value estimates.

As shown in Fig. 3, our proposed algorithm ARW-DSAC has better sample efficiency and final performance than SAC and RAA-DSAC over all 4 tasks. Most notably in the Ant task, the ARW-DSAC only takes 1 million time steps to reach the competitive performance with SAC with less variance, exceeding both the SAC and RAA-DSAC by a wide margin. In the remaining three tasks, the proposed algorithm also has a significant improvement effect.

Compared with RAA-DSAC, RW-DSAC achieves better results demonstrating that our DRW model has better utilization efficiency of multiple target Q-networks. Compared with RW-DSAC, ARW-DSAC makes a huge progress in sample efficiency, demonstrating the acceleration effect of Phased-PER over the DRW model.

5.3   Ablation Studies

In order to analyze the contribution of each component in the RWA framework, ablation experiments are conducted to further investigate their respective effects. Due to limited space, we only choose Ant, Hopper and HalfCheetah as our experimental environments.

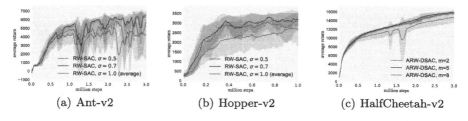

(a) Ant-v2   (b) Hopper-v2   (c) HalfCheetah-v2

Fig. 4. (a, b) Learning curves of different σ on Ant and Hopper. (c) Ablation analysis about different m on HalfCheetah.

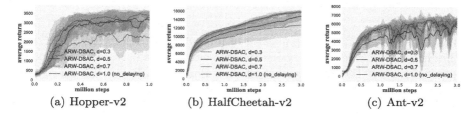

(a) Hopper-v2   (b) HalfCheetah-v2   (c) Ant-v2

Fig. 5. Performance comparison on continuous control tasks with different d. Worse performance is obtained by d of smaller value (e.g. 0.3, which means the frequency of policy updates is 0.3 times that of critics updates), while appropriate d in particular 0.7 greatly enhances the performance of the algorithm.

Recency-Weighted Coefficient. σ controls the degree of favouritism tendency of RWTE towards the more recent target networks by discounting the less recent value estimates. As shown in Fig. 4(a, b), different tasks have different sensitivities towards σ. Employment of RWTE through appropriate σ greatly improves the performance of SAC especially in the Ant task, surpassing averaged estimation ($\sigma = 1.0$) by a big margin at the same time maintaining smaller variance.

The Number of Previous Estimates. RWTE is calculated through m target networks. From Fig. 4(c), we can see that as m grows bigger, training becomes more stable and attains better final performance because of approximation error variance reduction. In consideration of increasing computing resource with larger m, we choose $m = 5$ for all algorithms.

Delayed Policy Updates. d controls the degree of delaying policy updates, with $d = 1$ meaning the algorithms without delayed policy updates. Various tasks have different sensitivities towards d. It can be concluded from the Fig. 5 that delayed policy updates further improve performance of the RWA framework, mitigating the influence of function approximation errors.

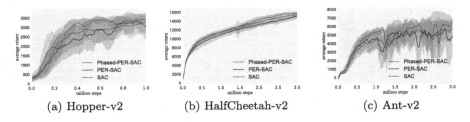

Fig. 6. Evaluations of Phased-PER. PER-SAC suffers from big variance and even has poorer performance, while Phased-PER leads to improved performance because of the adoption of phased update strategy.

Phased-PER. The acceleration effect of Phased-PER in the ARW-DSAC algorithm has been verified by previous experiments. In this part, we design experiments to compare the improvement effect of Phased-PER and PER on the SAC algorithm individually. Empirical results from Fig. 6 reveal that instead of the desired effect of accelerating training process, PER-SAC suffers from big variance and even has adverse effect, while Phased-PER obtains better results in terms of sample efficiency and final performance, which is the reason why we choose Phased-PER to be incorporated into our framework.

As ablation experiments reveal that the DRW model greatly improves performance of actor-critic algorithms. By adopting delayed policy updates, the DRW model makes further progress by becoming less prone to function approximation errors. Besides, acceleration effect caused by Phased-PER is verified.

6   Conclusion

The Instability issue not only exists in value-based control algorithms, but also persists in actor-critic settings. In order to tackle the instability issue in actor-critic settings, we draw on the idea underlying Averaged-DQN and propose the delayed recency-weighted model, which greatly reduces approximation error variance in the target value and become less prone to function approximation errors. In addition, further acceleration is attained by Phased-PER. Empirical results demonstrate that by reducing target approximation error variance and increasing sample efficiency, the RWA framework proposed in this work substantially enhances the actor-critic algorithms for continuous control.

Although the recency-weighted acceleration framework has brought vast improvement on off-policy algorithms such as SAC, but the DRW model still requires initialized parameters. It is promising to build an automatic framework that can adjust its model parameters according to its approximation errors in the training process, which is the focus of our future work.

References

1. Andrychowicz, M., et al.: Hindsight experience replay. In: Advances in Neural Information Processing Systems (NIPS), pp. 5048–5058 (2017)
2. Anschel, O., Baram, N., Shimkin, N.: Averaged-DQN: variance reduction and stabilization for deep reinforcement learning. In: International Conference on Machine Learning (ICML), pp. 176–185 (2017)
3. Brockman, G., et al.: OpenAI gym. arXiv preprint arXiv:1606.01540 (2016)
4. Fujimoto, S., Hoof, H., Meger, D.: Addressing function approximation error in actor-critic methods. In: International Conference on Machine Learning (ICML), pp. 1587–1596 (2018)
5. Haarnoja, T., et al.: Soft actor-critic algorithms and applications. arXiv preprint arXiv:1812.05905 (2018)
6. Hasselt, H.V.: Double Q-learning. In: Advances in Neural Information Processing Systems (NIPS), pp. 2613–2621 (2010)
7. Hessel, M., et al.: Rainbow: combining improvements in deep reinforcement learning. In: Thirty-Second AAAI Conference on Artificial Intelligence (AAAI) (2018)
8. Horgan, D., et al.: Distributed prioritized experience replay. arXiv preprint arXiv:1803.00933 (2018)
9. Liu, Q., et al.: A survey on deep reinforcement learning. Chin. J. Comput. **41**(1), 1–27 (2018)
10. Mnih, V., et al.: Human-level control through deep reinforcement learning. Nature **518**(7540), 529–533 (2015)
11. Munos, R., Stepleton, T., Harutyunyan, A., Bellemare, M.: Safe and efficient off-policy reinforcement learning. In: Advances in Neural Information Processing Systems (NIPS), pp. 1054–1062 (2016)
12. Schaul, T., Quan, J., Antonoglou, I., Silver, D.: Prioritized experience replay. arXiv preprint arXiv:1511.05952 (2015)
13. Shi, W., Song, S., Wu, H., Hsu, Y.C., Wu, C., Huang, G.: Regularized Anderson acceleration for off-policy deep reinforcement learning. In: Advances in Neural Information Processing Systems (NeurIPS), pp. 10231–10241 (2019)
14. Sutton, R.S., Barto, A.G.: Reinforcement Learning: An Introduction. MIT Press, Cambridge (2018)
15. Todorov, E., Erez, T., Tassa, Y.: MuJoCo: a physics engine for model-based control. In: IEEE/RSJ International Conference on Intelligent Robots and Systems (IROS), pp. 5026–5033. IEEE (2012)
16. Watkins, C.J.C.H.: Learning from delayed rewards. Ph.D. thesis, King's College, University of Cambridge (1989). https://ci.nii.ac.jp/naid/10000072699/en/
17. Zhang, Z., Pan, Z., Kochenderfer, M.J.: Weighted double Q-learning. In: International Joint Conference on Artificial Intelligence (IJCAI), pp. 3455–3461 (2017)
18. Zheng, Y., Hao, J., Zhang, Z., Meng, Z., Hao, X.: Efficient multiagent policy optimization based on weighted estimators in stochastic cooperative environments. J. Comput. Sci. Technol. **35**, 268–280 (2020). https://doi.org/10.1007/s11390-020-9967-6

Regularized Multiset Neighborhood Correlation Analysis for Semi-paired Multiview Learning

Yun-Hao Yuan[1,2], Zhaoqi Wu[1], Yun Li[1(✉)], Jipeng Qiang[1], Jianping Gou[3], and Yi Zhu[1]

[1] School of Information Engineering, Yangzhou University, Yangzhou, China
wuzhaoqiyzu@163.com, liyun@yzu.edu.cn
[2] School of Computer Science and Technology, Fudan University, Shanghai, China
[3] School of Computer Science, Jiangsu University, Zhenjiang, China

Abstract. Canonical correlation analysis (CCA) is a popular and powerful technique for two-view dimension reduction and feature extraction. But, CCA is not able to directly handle more than two view data and has a rigorous assumption that all the samples from two different views are paired. However, practical multiple view data are often semi-paired. To address this problem, we in this paper propose a novel semi-paired multiview dimension reduction approach, which takes cross-view neighborhood relationship among semi-paired data and within-view global structure information into consideration. The proposed approach can not only deal with multiview (more than two) data, but also take sufficient advantage of unpaired multiview data and then mitigate overfitting effectively caused by the limited paired data. Experimental results on two benchmark data sets demonstrate the effectiveness of our proposed method.

Keywords: Canonical correlation analysis · Multi-view learning · Neighborhood correlation · Semi-paired data

1 Introduction

In practice, one often meets such a case where an object is represented by two or more types of feature representations. For instance, a face can be depicted by color, texture, profile, and component features; a handwritten digit can be described by pixel average, moment, and morphological features. In the literature, such data with multiple representations are dubbed multiple view data [1].

Supported by the National Natural Science Foundation of China under Grant Nos. 61402203, 61703362 and 61906060, the Natural Science Foundation of Jiangsu Province of China under Grant No. BK20170513, the China Postdoctoral Science Foundation under Grant No. 2020M670995, and Yangzhou Science Project Foundation under Grant No. YZ2020173. It is also sponsored by Excellent Young Backbone Teacher (Qing Lan) Project and Scientific Innovation Project Fund of YZU under Grant No. 2017CXJ033.

© Springer Nature Switzerland AG 2020
H. Yang et al. (Eds.): ICONIP 2020, LNCS 12533, pp. 616–625, 2020.
https://doi.org/10.1007/978-3-030-63833-7_52

Multi-view data are supposed to be complementary to each other and capable of describing an object more fully. Different from single-view data learning that has been investigated for at least several decades, multiple view data learning is a young research direction. Although there have been a number of dedicated feature reduction methods for multi-view high-dimensional data, how to learn a discriminative low-dimensional representation is still a challenging problem.

In dimensionality reduction (DR), the most typical techniques are, unquestionably, principal component analysis (PCA) [3], linear discriminant analysis (LDA) [3], and locality preserving projection (LPP) [4]. PCA is the widely used unsupervised dimension reduction method. It reduces the dimension of the data by extracting the leading components and removing the redundant information as much as possible. LDA is a supervised dimension reduction method, which searches for a set of projection directions by maximizing the ratio of interclass scatter to intraclass scatter. LPP is a linearizable version derived from Laplacian eigenmap (LE) [2]. Unlike LE, LPP can yield an explicit projection matrix by computing a generalized eigenvalue problem. Despite the effectiveness of the foregoing DR methods, they are inapplicable to multi-view DR and data representation.

For multi-view data learning, canonical correlation analysis (CCA) [5] is one of the most well-known multi-view dimension reduction methods. CCA computes pairs of projection directions by maximizing the correlation between two sets of high-dimensional data. An implicit characteristic to use CCA is that multi-view training samples need to be paired. However, in practice, there is usually no shortage of unpaired data but paired are expensive. A feasible way is to make a small number of multi-view training samples paired and the rest unpaired. Such data are referred to as semi-paired data. In the semi-paired situation, it is difficult to directly employ CCA for multi-view DR.

In recent years, a number of studies have focused on semi-paired scenario. For instance, Blaschko et al. [6] proposed semi-supervised Laplacian regularization of kernel CCA (SemiLRKCCA) via considering paired and unpaired samples simultaneously, which can find highly correlated as well as high variance projection directions. Kimura et al. [7] presented a semi-paired version of CCA dubbed SemiCCA that incorporates unpaired samples into CCA model to reduce overfittng. Chen et al. [8] focused on semi-paired as well as semi-supervised scenario and developed a novel multi-view feature extraction method called S²GCA. It should be pointed out that the foregoing semi-paired CCA's variants make use of the unpaired data only in a single-view way. That is, they only take into account the intra-view information of unpaired data and do not consider the latent relationship between two views of unpaired data. To solve this problem, Zhou et al. [9] presented a novel CCA's variant named neighborhood correlation analysis (NeCA) for semi-paired two-view learning, which exploits the neighborhood relation between two-view unpaired samples.

The aforementioned methods such as SemiCCA and NeCA are only applicable to two-view scenario. When more than two views occur, they are not able to work well. To solve this problem, we propose a novel dimensionality reduction approach for semi-paired multi-view data, which incorporates between-view neighborhood relationship among semi-paired data into multiset CCA [10] and then combines it with principal component analysis (PCA). The proposed method is referred to as PCA-regularized multiset neighborhood correlation analysis (PRMNeCA). Experimental results show that PRMNeCA is encouraging.

2   Background

2.1   Multiset Canonical Correlation Analysis

Multiset canonical correlation analysis (MCCA) [10] is a statistical technique to analyze the linear relation among several sets of random variables. To be specific, suppose there are m sets of zero-mean variables $\{x^{(i)} \in R^{d_i}\}_{i=1}^m$, where d_i denotes the dimension of $x^{(i)}$. MCCA aims to find a set of projection directions $\{\omega_i \in R^{d_i}\}_{i=1}^m$ that maximize the sum of all pairwise correlations between multiset canonical variables $\{\omega_i^T x^{(i)}\}_{i=1}^m$. The optimization problem of MCCA is as follows:

$$\max_{\omega_1,\cdots,\omega_m} \sum_{\substack{i,j=1 \\ i \neq j}}^m \omega_i^T S_{ij}\omega_j$$
$$s.t. \quad \sum_{i=1}^m \omega_i^T S_{ii}\omega_i = 1, \tag{1}$$

where $S_{ij} = E(x^{(i)}x^{(j)T})$ $(i \neq j)$ is the between-set covariance matrix of $x^{(i)}$ and $x^{(j)}$, $S_{ii} = E(x^{(i)}x^{(i)T})$ is the within-set covariance matrix of $x^{(i)}$, $i = 1, 2, \cdots, m$, and $E(\cdot)$ is the expectation operator. Through the Lagrangian multiplier method, the solution to MCCA can be obtained by a generalized eigenvalue problem.

2.2   Semi-supervised Laplacian Regularization of MCCA

The multiview generalization of SemiLRKCCA has been developed in [6], which is based on the so-called kernel matrices and thus belongs to nonlinear learning category. Here, we give a linear multiview extension along the idea from SemiLRKCCA, called semi-supervised Laplacian regularization of MCCA (SemiLRMCCA). Suppose m sets (views) of training data are given as $X^{(1)}, X^{(2)}, \cdots, X^{(m)}$, where $X^{(i)} = [X_p^{(i)}, X_u^{(i)}] \in R^{d_i \times N}$ with $X_p^{(i)} = [x_1^{(i)}, x_2^{(i)}, \cdots, x_p^{(i)}] \in R^{d_i \times p}$ as paired data and $X_u^{(i)} = [x_{p+1}^{(i)}, x_{p+2}^{(i)}, \cdots, x_N^{(i)}] \in R^{d_i \times (N-p)}$ as unpaired data, d_i is the dimension of training data, p is the number of paired data, and N is the total number of paired and unpaired data in each set. A set of directions $\{\omega_i\}_{i=1}^m$ of SemiLRMCCA can be found by the following optimization problem:

$$\max_{\omega_i,\cdots,\omega_m} \sum_{\substack{i,j=1 \\ i \neq j}}^m \omega_i^T S_{ij}^p\omega_j$$
$$s.t. \quad \sum_{i=1}^m \omega_i^T \hat{S}_{ii}\omega_i = 1 \tag{2}$$

where $S_{ij}^p = X_p^{(i)} X_p^{(j)T}$ $(i \neq j)$, $\hat{S}_{ii} = X_p^{(i)} X_p^{(i)T} + \gamma_i X^{(i)} L_i X^{(i)T}$ with γ_i as the tradeoff parameter and $L_i = D^{(i)-1/2}(D^{(i)} - W^{(i)})D^{(i)-1/2}$ as the normalized graph Laplacian matrix, $W^{(i)}$ is the weight matrix of N paired and unpaired samples in i-th set, $D^{(i)}$ is a diagonal matrix whose diagonal entry as the row (column) sum of $W^{(i)}$, i.e., $[D^{(i)}]_{jj} = \sum_{k=1}^{N}[W^{(i)}]_{jk}$, $j = 1, 2, \cdots, N$.

With the Lagrange multiplier method, it is easy to show that optimization problem in (2) can be solved by the following generalized eigenvalue problem:

$$
\begin{bmatrix} 0 & S_{12}^p & \cdots & S_{1m}^p \\ S_{21}^p & 0 & \cdots & S_{2m}^p \\ \vdots & \vdots & \ddots & \vdots \\ S_{m1}^p & S_{m2}^p & \cdots & 0 \end{bmatrix} \begin{bmatrix} \omega_1 \\ \omega_2 \\ \vdots \\ \omega_m \end{bmatrix} = \lambda \begin{bmatrix} \hat{S}_{11} & & & \\ & \hat{S}_{22} & & \\ & & \ddots & \\ & & & \hat{S}_{mm} \end{bmatrix} \begin{bmatrix} \omega_1 \\ \omega_2 \\ \vdots \\ \omega_m \end{bmatrix} \tag{3}
$$

2.3 Semi-paired Learning of MCCA

In two-view scenario, SemiCCA considers the paired data as well as unpaired data under a regularization framework, which searches for pairs of projection directions by the generalized eigenvalue problem. Here, we extend SemiCCA into a multiview version that we call SemiMCCA for multi-view feature extraction, as follows.

$$A\omega = \lambda B\omega, \tag{4}$$

where $\omega^T = [\omega_1^T, \omega_2^T, \cdots, \omega_m^T]$, λ is the eigenvalue associated with eigenvector ω,

$$
A = \begin{bmatrix} 0 & S_{12}^p & \cdots & S_{1m}^p \\ S_{21}^p & 0 & \cdots & S_{2m}^p \\ \vdots & \vdots & \ddots & \vdots \\ S_{m1}^p & S_{m2}^p & \cdots & 0 \end{bmatrix} + \kappa \begin{bmatrix} S_{11} & & & \\ & S_{22} & & \\ & & \ddots & \\ & & & S_{mm} \end{bmatrix},
$$

$$
B = \begin{bmatrix} S_{11}^p & & & \\ & S_{22}^p & & \\ & & \ddots & \\ & & & S_{mm}^p \end{bmatrix} + \kappa \begin{bmatrix} I_{d_1} & & & \\ & I_{d_2} & & \\ & & \ddots & \\ & & & I_{d_m} \end{bmatrix}
$$

with κ denoting the tradeoff parameter, $S_{ii}^p = X_p^{(i)} X_p^{(i)T}$, and $I_{d_i} \in R^{d_i \times d_i}$ as the identity matrix. Clearly, when $\kappa = 0$, (4) reduces to ordinary MCCA, as described in Sect. 2.1. Note that since we build SemiMCCA directly using the eigen-form of SemiCCA, we do not give its optimization problem. In fact, the optimization model corresponding to (4) can be formulated as

$$
\max_{\omega_1, \cdots, \omega_m} \sum_{\substack{i,j=1 \\ i \neq j}}^{m} \omega_i^T S_{ij}^p \omega_j + \kappa \sum_{i=1}^{m} \omega_i^T S_{ii} \omega_i
$$

$$
s.t. \quad \sum_{i=1}^{m} \omega_i^T S_{ii}^p \omega_i + \kappa \sum_{i=1}^{m} \omega_i^T \omega_i = 1. \tag{5}
$$

3 Proposed Approach

3.1 Within-View Weight Matrix Construction

Using the locality idea from LPP [4], we are able to construct a neighborhood graph in each view $X^{(i)}$. To be specific, let $N_k(x_l^{(i)})$ denote the set of k nearest neighbors of sample point $x_l^{(i)}$. Then, the neighborhood graph $G^{(i)}$ in i-th view can be defined as $G^{(i)} = \{X^{(i)}, W^{(i)}\}$, where $X^{(i)} = [X_p^{(i)}, X_u^{(i)}] \in R^{d_i \times N}$ denotes the d_i-dimensional vertices and $W^{(i)} = [w_{lt}^{(i)}] \in R^{N \times N}$ denotes the edge weight matrix. Here, we make use of the commonly used radial basis function (RBF) to define the weight matrix, as follows:

$$w_{lt}^{(i)} = \begin{cases} \exp(-||x_l^{(i)} - x_t^{(i)}||^2/2\sigma_i^2), & x_t^{(i)} \in N_k(x_l^{(i)}) \text{ or } x_l^{(i)} \in N_k(x_t^{(i)}), \\ 0, & \text{otherwise,} \end{cases} \quad (6)$$

where $|| \cdot ||$ is the 2-norm of a vector, σ_i denotes the width parameter of RBF, and $i = 1, 2, \cdots, m$. Through (6), all the m edge weight matrices, one for each view, are able to be computed. Note that the edge weights depict the similarity of different within-view sample points in k-nearest neighborhood.

3.2 Cross-View Weight Matrix Construction

Since the samples come from multiple different views and partial pairwise information is also not available, it is difficult to directly measure the similarity among semi-paired multi-view data. Inspired with the idea from minimizing-disagreement [11], we design a simple but effective cross-view weighted graph construction to compute the similarities (weights) among semi-paired multi-view data. The central idea is to use the within-view weights as defined in Sect. 3.1 to calculate the affinity weights with the help of shared pairwise samples.

Concretely, for any two samples $x_l^{(i)}$ and $x_t^{(j)}$ in i-th and j-th views $(i \neq j)$, let us denote the paired samples they share as $\{\tilde{x}_s^{(i)}, \tilde{x}_s^{(j)}\}_{s=1}^{N_s}$, where N_s denotes the number of shared pairwise samples and $N_s \leq p$. Then, the weight of samples $x_l^{(i)}$ and $x_t^{(j)}$ is defined by

$$w_{lt}^{(ij)} = \sum_{s=1}^{N_s} w_{ls}^{(i)} w_{ts}^{(j)}, \ i, j = 1, 2, \cdots, m \text{ and } i \neq j, \quad (7)$$

where $w_{ls}^{(i)}$ ($w_{ts}^{(j)}$) denotes the affinity weight between samples $x_l^{(i)}$ and $\tilde{x}_s^{(i)}$ ($x_t^{(j)}$ and $\tilde{x}_s^{(j)}$) computed by (6), $l, t = 1, 2, \cdots, N$. Clearly, (7) reveals that if $x_l^{(i)}$ and $x_t^{(j)}$ share more paired samples, their weight $w_{lt}^{(ij)}$ should be larger. With (7), the cross-view weight matrix $W^{(ij)}$ can be formed as $W^{(ij)} = [w_{lt}^{(ij)}] \in R^{N \times N}$, $i, j = 1, 2, \cdots, m$ and $i \neq j$.

3.3   Model and Solution

According to [12], the between-set covariance matrix in MCCA can be expressed in a pairwise manner. Using (7) together with pairwise expressions, we are able to incorporate the between-view neighborhood information among semi-paired multiview data. That is, our proposed method minimizes the following

$$
\begin{aligned}
\mathcal{O} &= \sum_{\substack{i,j=1 \\ i \neq j}}^{m} \sum_{l=1}^{N} \sum_{t=1}^{N} w_{lt}^{(ij)} \| \omega_i^T x_l^{(i)} - \omega_j^T x_t^{(j)} \|^2 \\
&= \sum_{\substack{i,j=1 \\ i \neq j}}^{m} \sum_{l=1}^{N} \sum_{t=1}^{N} w_{lt}^{(ij)} (\omega_i^T x_l^{(i)} - \omega_j^T x_t^{(j)})(\omega_i^T x_l^{(i)} - \omega_j^T x_t^{(j)})^T \\
&= \sum_{\substack{i,j=1 \\ i \neq j}}^{m} (\omega_i^T X^{(i)} D_r^{(ij)} X^{(i)T} \omega_i - 2\omega_i^T X^{(i)} W^{(ij)} X^{(j)T} \omega_j \\
&\qquad + \omega_j^T X^{(j)} D_c^{(ij)} X^{(j)T} \omega_j) \\
&= \sum_{\substack{i,j=1 \\ i \neq j}}^{m} \left[\omega_i^T X^{(i)} (D_r^{(ij)} + D_c^{(ji)}) X^{(i)T} \omega_i - 2\omega_i^T X^{(i)} W^{(ij)} X^{(j)T} \omega_j \right]
\end{aligned}
\tag{8}
$$

where $D_r^{(ij)} \in R^{N \times N}$ is a diagonal matrix with l-th diagonal element as l-th row sum of weight matrix $W^{(ij)}$ and $D_c^{(ij)} \in R^{N \times N}$ is a diagonal matrix with l-th diagonal entry as l-th column sum of $W^{(ij)}$.

Let us denote

$$
\tilde{S}_{ij} = X^{(i)} W^{(ij)} X^{(j)T}
$$

$$
\tilde{S}_{ii} = X^{(i)} \left(\sum_{\substack{j=1 \\ i \neq j}}^{m} (D_r^{(ij)} + D_c^{(ji)}) \right) X^{(i)T}.
$$

If we impose $\sum_{i=1}^{m} \omega_i^T \tilde{S}_{ii} \omega_i = 1$, then minimizing the objective in (8) is equivalent to the following maximization problem:

$$
\begin{aligned}
&\max_{\omega_1, \cdots, \omega_m} \sum_{\substack{i,j=1 \\ i \neq j}}^{m} \omega_i^T \tilde{S}_{ij} \omega_j \\
&s.t. \quad \sum_{i=1}^{m} \omega_i^T \tilde{S}_{ii} \omega_i = 1
\end{aligned}
\tag{9}
$$

Similar to SemiMCCA, combining optimization model in (9) with PCA leads to the resulting optimization problem of the proposed method that we refer to as PCA-regularized multiset neighborhood correlation analysis (PRMNeCA), as follows:

$$
\begin{aligned}
&\max_{\omega_1, \cdots, \omega_m} \sum_{\substack{i,j=1 \\ i \neq j}}^{m} \omega_i^T \tilde{S}_{ij} \omega_j + \kappa \sum_{i=1}^{m} \omega_i^T S_{ii} \omega_i \\
&s.t. \quad \sum_{i=1}^{m} \omega_i^T \tilde{S}_{ii} \omega_i + \kappa \sum_{i=1}^{m} \omega_i^T \omega_i = 1,
\end{aligned}
\tag{10}
$$

where $\{S_{ii}\}_{i=1}^m$ are defined in (1) and κ is the balance parameter.

Through the Lagrange multiplier method, the optimization problem in (10) can be solved by the following generalized eigenvalue problem:

$$\tilde{A}\omega = \lambda\tilde{B}\omega, \tag{11}$$

where $\omega^T = [\omega_1^T, \omega_2^T, \cdots, \omega_m^T]$,

$$\tilde{A} = \begin{bmatrix} 0 & \tilde{S}_{12} & \cdots & \tilde{S}_{1m} \\ \tilde{S}_{21} & 0 & \cdots & \tilde{S}_{2m} \\ \vdots & \vdots & \ddots & \vdots \\ \tilde{S}_{m1} & \tilde{S}_{m2} & \cdots & 0 \end{bmatrix} + \kappa \begin{bmatrix} S_{11} & & & \\ & S_{22} & & \\ & & \ddots & \\ & & & S_{mm} \end{bmatrix},$$

$$\tilde{B} = \begin{bmatrix} \tilde{S}_{11} & & & \\ & \tilde{S}_{22} & & \\ & & \ddots & \\ & & & \tilde{S}_{mm} \end{bmatrix} + \kappa \begin{bmatrix} I_{d_1} & & & \\ & I_{d_2} & & \\ & & \ddots & \\ & & & I_{d_m} \end{bmatrix}.$$

We select the top r ($\leq \min\{d_1, \cdots, d_m\}$) generalized eigenvectors of (11) to form the projection matrices $\{P_i = [\omega_{i1}, \omega_{i2}, \cdots, \omega_{ir}] \in R^{d_i \times r}\}_{i=1}^m$ of all the views. For any given sample $x^{(i)} \in R^{d_i}$ from i-th view, its low-dimensional representation can be obtained in the form of $P_i^T x^{(i)}$, which is used to represent the original $x^{(i)}$ for classification purpose.

4   Experiments

In this section, we perform several experiments to compare the performance of PRMNeCA with MCCA, SemiLRMCCA, and SemiMCCA. We adopt two widely used data sets which are the FERET and Yale face databases. The nearest neighbor classifier is used in all our experiments.

4.1   Parameter Selection

There are several important parameters in SemiLRMCCA, SemiMCCA, and our PRMNeCA. We empirically set these parameters' values. In SemiLRMCCA and PRMNeCA, the neighborhood parameter k is chosen from the set $\{1, 2, \cdots, N_c\}$, where N_c is smaller than the number of training samples in each class, and the width parameter σ_i of RBF is set to 1. In addition, the parameter γ_i ($i = 1, 2, \cdots, m$) in SemiLRMCCA is searched from the set $\{2^{-14}, 2^{-13}, \cdots, 2^2\}$ for the best result. In SemiMCCA and PRMNeCA, the parameter κ is chosen from $\{2^{-14}, 2^{-13}, \cdots, 2^2\}$.

4.2   Results on the FERET Database

The FERET database[1] contains 14126 face images of 1199 individuals. A subset of the FERET database is used in our experiment. This subset includes 1400 face images of 200 individuals with variations in facial expression, illumination and pose. There are seven images per individual with a resolution of 80×80.

Table 1. Average accuracy (%) across ten runs of each method on FERET database.

Method	1st view	2nd view	3rd view
MCCA	35.78	38.59	38.83
SemiMCCA	43.75	43.60	44.37
SemiLRMCCA	36.57	40.18	39.30
PRMNeCA	**46.52**	**46.58**	**46.92**

In this experiment, we down-sample original face images to low-resolution images with size 40×40, and then recover them to the images with 80×80 pixels. These recovered images are taken as the first view. The wavelet transformation (i.e., Symlets wavelet) is adopted to extract facial features from original images that are regarded as the second view. The third-view data are generated by performing a 3×3 mean filter on original face images. To avoid the small sample size problem, PCA is used to reduce the dimension of each view to 120.

On this subset, four images per individual are randomly selected for training and the rest for testing. In training set, two samples per class in each view are used as paired samples and the rest as unpaired samples. Ten independent tests are run and the average results are computed for the performance evaluation. Table 1 summarizes the average recognition accuracy across ten runs of each method under nearest neighbor classifier.

From Table 1, we can see that PRMNeCA performs better in each view than MCCA, SemiMCCA, and SemiLRMCCA. SemiMCCA performs the second best and MCCA achieves the worst results. An important reason is that MCCA does not consider the unpaired data in learning low-dimensional projections. This makes MCCA possible to encounter the problem of overfitting.

4.3   Results on the Yale Database

The Yale face database[2] contains 165 grayscale images of 15 individuals. There are 11 images per subject with a resolution of 80×80, one per different facial expression or configuration: center-light, with glasses, happy, left-light, without glasses, normal, right-light, sad, sleepy, surprised, and winking.

---

[1] https://www.nist.gov/itl/products-and-services/color-feret-database.
[2] http://cvc.cs.yale.edu/cvc/projects/yalefaces/yalefaces.html.

Table 2. Average accuracy (%) across ten runs of each method with $p = 2$ on the Yale face database.

Method	Coi view	Dau view	Sym view
MCCA	58.17	59.17	59.67
SemiMCCA	63.50	64.17	63.83
SemiLRMCCA	57.33	58.50	59.00
PRMNeCA	**70.17**	**69.67**	**69.83**

Table 3. Average accuracy (%) across ten runs of each method with $p = 3$ on the Yale face database.

Method	Coi view	Dau view	Sym view
MCCA	58.50	60.00	60.83
SemiMCCA	64.17	70.17	70.17
SemiLRMCCA	59.83	59.17	59.33
PRMNeCA	**71.83**	**71.67**	**71.50**

In this experiment, we employ three kinds of different wavelet transformations, i.e., Coiflets, Daubechies, and Symlets wavelets, to extract three-view features from original face images, respectively denoted as Coi, Dau, and Sym. As used in Sect. 4.2, we also perform PCA to reduce each view's dimension to 120 before evaluating each method. In this test, seven images per individual are randomly selected for training, while the remaining four images for testing. In addition, p ($p = 2$ and 3) samples of each individual in each view are used as paired samples and the rest as unpaired samples. Ten recognition tests are independently run and the average results are computed for the performance comparison. Tables 2 and 3 list the average recognition results of each method across ten runs under nearest neighbor classifier with different values of p.

From Tables 2 and 3, we can see that our PRMNeCA method achieves the best results among all the methods with different views, irrespective of the variation of paired samples. SemiMCCA performs the second best, whether the value of p is 2 or 3. These results demonstrate again that our PRMNeCA method is effective for semi-paired multi-view dimension reduction and feature extraction.

5   Conclusion

In this paper, we have proposed a novel semi-paired multiview dimension reduction approach dubbed PRMNeCA, which not only makes full use of unpaired multiview samples, but also considers cross-view neighborhood relationship among semi-paired data and within-view global structure information. In addition, we introduce the idea from PCA to regularize our optimization model. Experimental results on two popular databases demonstrate the effectiveness of our proposed PRMNeCA method.

References

1. Li, Y., Yang, M., Zhang, Z.: A survey of multi-view representation learning. IEEE Trans. Knowl. Data Eng. **31**(10), 1863–1883 (2019)
2. Belkin, M., Niyogi, P.: Laplacian eigenmaps for dimensionality reduction and data representation. Neural Comput. **15**(6), 1373–1396 (2003)
3. Martinez, A.M., Kak, A.C.: PCA versus LDA. IEEE Trans. Pattern Anal. Mach. Intell. **23**(2), 228–233 (2001)
4. He, X., Niyogi, P.: Locality preserving projections. In: NIPS, pp. 153–160 (2003)
5. Hotelling, H.: Relations between two sets of variates. Biometrika **28**, 321–377 (1936)
6. Blaschko, M.B., Lampert, C.H., Gretton, A.: Semi-supervised laplacian regularization of kernel canonical correlation analysis. In: Daelemans, W., Goethals, B., Morik, K. (eds.) ECML PKDD 2008. LNCS (LNAI), vol. 5211, pp. 133–145. Springer, Heidelberg (2008). https://doi.org/10.1007/978-3-540-87479-9_27
7. Kimura, A., et al.: SemiCCA: efficient semi-supervised learning of canonical correlations. In: ICPR, pp. 2933–2936. IEEE, Turkey (2010)
8. Chen, X., Chen, S., Xue, H., Zhou, X.: A unified dimensionality reduction framework for semi-paired and semi-supervised multi-view data. Pattern Recognit. **45**(5), 2005–2018 (2012)
9. Zhou, X., Chen, X., Chen, S.: Neighborhood correlation analysis for semi-paired two-view data. Neural Process. Lett. **37**(3), 335–354 (2013)
10. Yuan, Y.-H., Sun, Q.-S.: Multiset canonical correlations using globality preserving projections with applications to feature extraction and recognition. IEEE Trans. Neural Networks Learn. Syst. **25**(6), 1131–1146 (2014)
11. de Sa, V.R., Gallagher, P.W., Lewis, J.M., Malave, V.L.: Multi-view kernel construction. Mach. Learn. **79**(1–2), 47–71 (2010)
12. Yuan, Y.-H., Li, Y., Shen, X.-B., Sun, Q.-S., Yang, J.-L.: Laplacian multiset canonical correlations for multiview feature extraction and image recognition. Multimed. Tools Appl. **76**(1), 731–755 (2017). https://doi.org/10.1007/s11042-015-3070-y

SAN: Sampling Adversarial Networks for Zero-Shot Learning

Chenwei Tang[1], Yangzhu Kuang[2], Jiancheng Lv[1(✉)], and Jinglu Hu[2]

[1] Sichuan University, Chengdu 610065, People's Republic of China
[2] Waseda University, Fukuoka 8080135, Japan
lvjiancheng@scu.edu.cn

Abstract. In this paper, we propose a Sampling Adversarial Networks (SAN) framework to improve Zero-Shot Learning (ZSL) by mitigating the hubness and semantic gap problem. The SAN framework incorporates a sampling model and a discriminating model, and corresponds them to the minimax two-player game. Specifically, given the semantic embedding, the sampling model samples the visual features from the training set to approach the discriminator's decision boundary. Then, the discriminator distinguishes the matching visual-semantic pairs from the sampled data. On the one hand, by the measurement of the matching degree of visual-semantic pairs and the adversarial training way, the visual-semantic embedding built by the proposed SAN decreases the intra-class distance and increases the inter-class separation. Then, the reduction of universal neighbours in the visual-semantic embedding subspace alleviates the hubness problem. On the other, the sampled rather than directly generated visual features maintain the same manifold as the real data, mitigating the semantic gap problem. Experiments show that the sampler and discriminator of the SAN framework outperform state-of-the-art methods both in conventional and generalized ZSL settings.

Keywords: Zero-Shot Learning · Sampling Adversarial Networks · Hubness problem · Semantic gap

Image classification tasks have achieved great success due to the prosperous progress of deep learning [8]. However, most deep learning methods require labeling extensive training data, which is both labor-intensive and unscalable [19]. To tackle this limitation, Zero-Shot Learning (ZSL) is proposed to recognize new categories that have never seen during training, i.e., the categories in the training and test set are disjoint [20, 22]. According to the categories included in the test, two ZSL settings are defined: conventional ZSL and Generalized ZSL (GZSL).

Thank Yangzhu Kuang for his contribution to this article. This paper is supported by the National Natural Science Fund for Distinguished Young Scholar under Grant No. 61625204 and the Key Program of National Science Foundation of China under Grant No. 61836006.

Specifically, only the unseen classes are used to evaluation in the conventional ZSL setting. The GZSL provides a more practical point of view, where both seen and unseen categories are involved for testing.

There are two fundamental challenges in ZSL: visual-semantic embedding [20] and domain adaption [5]. The knowledge can be transferred from the seen domain to the unseen domain by building a visual-semantic embedding. However, since the seen and unseen classes are different and potentially unrelated, the domain shift problem is triggered when the visual-semantic embedding is directly applied to the unseen data [5]. Thus, compared with the fully-supervised image classification tasks, the performance of ZSL is still far from perfect [3].

Most previous cross-modal embedding methods solved ZSL in two steps. First, project both visual features and semantic features to the embedding space [9]. Then, utilize nearest neighbour search in the embedding space to match the projection of visual or semantic feature vector against that of an unseen instance [28]. However, [17] proposed that there are many 'universal' neighbours, namely hubs, when performing nearest neighbour search in a high-dimensional space. They also showed that the hubness is an inherent property of data distributions in the high-dimensional vector space. Therefore, the cross-modal embedding methods always lead to the well-known hubness problem [3]. That is, a few unseen class prototypes will become the nearest neighbours of many hubs.

Recently, a new branch of methods target to ZSL by generative models [27]. They directly generate the unseen features from random noises which are conditioned by the semantic descriptions. With the generated unseen samples, zero-shot learning can be transformed to a supervised image classification task. However, since both the true and generated visual features contain intrinsic manifold structures, the manifold alignment is very challenge, especially in the high-dimension [18]. The semantic gap problem, i.e., the manifold of samples in the visual feature space is inconsistent with that of categories in the semantic space, often leads to model collapse, especially for the approaches [14] based on Generative Adversarial Networks (GAN) [6].

In this paper, we propose a novel Sampling Adversarial Networks (SAN) framework to improve ZSL by mitigating the hubness and semantic gap problem. The SAN framework proposes a new perspective for tackling ZSL by combining a sampling model and a discriminative model. The sampler aims at picking matching visual features given a semantic input. The discriminator focuses on measuring relevancy given a visual-semantic pair. These two models correspond to the minimax two-player game. On one hand, the discriminator guides the sampler to fit the underlying relevance distribution over visual features given the semantic presentation. On the other hand, the sampler tries to select visual features closing to the discriminator's decision boundary to confuse the discriminator. Our main contributes of this paper are summarized as follows:

- For the hubness problem, we construct a visual-semantic embedding by the adversarial training way. We utilize the discriminator to measure the matching degree of the visual-semantic pairs, rather than distinguish whether the visual features are real or fake. The proposed method considers the intra-class consistency and inter-class diversity, which alleviates the hubness problem.

- For the semantic gap problem, we propose to utilize the encoded attributes to sample visual features of seen classes, instead of directly generating visual features. Then, the sampled visual features space is not affected by the semantic gap between attribute and visual feature space.
- Extensive experiments demonstrate that both the sampler and discriminator of the proposed SAN framework outperform state-of-the-art methods both in the conventional and generalized ZSL setting.

1   Related Work

ZSL aims to classify images of new classes that have never been seen before, i.e., the training and test classes are disjoint. With the shared attributes annotated on class level, the ZSL is achieved by building the visual-semantic embedding to transfer the knowledge from seen classes to unseen classes [23]. GZSL is a more realistic setting, where the same information as ZSL is available at training phase, but both seen and unseen classes are classified during testing [2,26]. With the development of deep learning, many effective methods have been proposed to target to ZSL.

The cross-modal embedding models usually project either visual features or semantic features from one space to the other, or project both features into an intermediate space. Then, the compatibility function between visual and semantic features vectors is learned by using the ranking loss. ESZSL [18] learns a bilinear compatibility function between visual features, semantic features, and class labels with the square loss. LATEM [25] directly maps the visual feature to semantic space, and learns a bilinear compatibility function. SYNC [1] embeds both the visual and semantic features into another common space, and also learns a bilinear compatibility. SAE [9], following the Auto-Encoder, reconstructs the visual features in the semantic space. RethinkZSL [12] reformulates ZSL as a conditioned visual classification problem, i.e., classifying visual features based on the classifiers learned from the semantic descriptions.

The generative models reformulate ZSL as a standard fully-supervised classification task. GAZSL [29] takes noisy text descriptions about an unseen class as the input of generative model, and generates synthesized visual features for this class. f-CLSWGAN [27] synthesizes visual features conditioned on class-level semantic information, and pairs a Wasserstein GAN with a classification loss. LisGAN [24] trains a conditional Wasserstein GANs to directly generate the unseen features from random noises which are conditioned by the semantic descriptions.

Benefit from the synthesized missing features for unseen classes, the generative models achieve better results for unseen classes both in ZSL and GZSL. However, the manifold of samples in the visual feature space is inconsistent with that of categories in the semantic space. The semantic gap results in the disturbance of the generated unseen visual features to the original seen visual space.

Therefore, the generative models seem kind of "confused" for the accuracy of seen classes in the GZSL setting. Building on ideas from these many previous works, we develop a simple and effective SAN framework incorporating sampler and discriminator, and corresponding them to the minimax two-player game.

2   Sampling Adversarial Networks

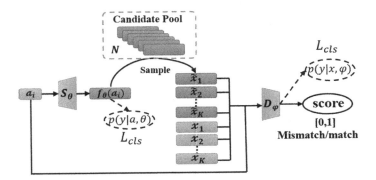

Fig. 1. The framework of the proposed SAN. a_i and x_i denote the attribute and the corresponding visual feature. $\tilde{x}_1$, $\tilde{x}_2$,..., $\tilde{x}_k$ denote the sampled visual features from all training visual instances $\mathcal{X}$. S_θ and D_φ are the sampler and discriminator, respectively.

Here we firstly introduce some notations and the problem definition. Let $\mathcal{S} = \{(x, y, a) | x \in \mathcal{X}_s, y \in \mathcal{Y}_s, a \in \mathcal{A}_s\}$ and $\mathcal{U} = \{(x, y, a) | x \in \mathcal{X}_u, y \in \mathcal{Y}_u, a \in \mathcal{A}_u\}$ where $\mathcal{S}$ and $\mathcal{U}$ denote training data of seen classes and testing data of unseen classes, respectively. $\mathcal{X}$ and $\mathcal{A}$ are the visual features and the semantic information in the form of attributes. $\mathcal{Y}_s$ and $\mathcal{Y}_u$ are the corresponding class labels. There is no overlap between seen and unseen classes, i.e., $\mathcal{Y}_s \cap \mathcal{Y}_u = \emptyset$. The goal of ZSL is to transfer the visual-semantic embedding learned in $\mathcal{S}$ to $\mathcal{U}$, and learn a classifier $f : \mathcal{X}_u \to \mathcal{Y}_u$. As for GZSL, we learn the classifier $f : \mathcal{X}_s, \mathcal{X}_u \to \mathcal{Y}_s \cup \mathcal{Y}_u$. Figure 1 shows the overview of our method. We construct the sampling model S_θ and discriminative model D_φ as follow:

Sampling Model. The sampler $\mathcal{S}_\theta : \mathcal{A} \to \mathcal{X}$ is a multi-layer perceptron. The sampling model $p_\theta(x|a)$ tries to approximate the true relevance distribution over visual features as much as possible, and confusing the discriminator's training next round. First, we utilize the sampler $\mathcal{S}_\theta$ to encode the input attribute a_i as an index vector $f_\theta(a_i)$ with the same dimension as the visual feature. Then, the sampling process of the sampling model is selecting K visual features $\{\tilde{x}_1, \tilde{x}_2, ..., \tilde{x}_K\}$ according to cosine similarity over discrete candidate pool. The candidate pool of each input semantic information is composed of N visual features randomly sampled from the whole trained seen visual features.

Discriminative Model. The discriminative model aims at distinguishing the well-matched semantic-visual features from the sampled negative data. For each attribute, K visual features of the same categories as the attribute and K fake visual features sampled by the sampler are selected. Then we can get K pairs matching semantic-visual features as the positive samples and K pairs mismatching semantic-visual features as the negative samples. In order to calculate the cosine similarity between semantic and visual information, we first encode the attribute a_i by the discriminator $\mathcal{D}_\varphi$. The discriminator $\mathcal{D}_\varphi : \mathcal{X} \times \mathcal{A} \rightarrow [0, 1]$ is also a multi-layer perceptron, where the last layer is the cosine similarity between visual and semantic features. The $\mathcal{D}_\varphi$ is designed for estimating the probability of visual feature x being relevant to the semantic information a, i.e., they belong to the same category. The discriminative score $\mathcal{D}(x, a)$ can be defined as follow:

$$\mathcal{D}(x, a) = \cos(f_\varphi(a), x). \tag{1}$$

Objective Function. Overall, as with the training procedure of GAN, the sampler $\mathcal{S}_\theta$ and discriminator $\mathcal{D}_\varphi$ of the proposed SAN framework play the two-player minimax game with the following objective function $V(\mathcal{S}, \mathcal{D})$:

$$\min_\theta \max_\varphi V(\mathcal{S}, \mathcal{D}) = \mathbb{E}_{x \sim p_{data}(x)}[\log \mathcal{D}(x, a)] + \mathbb{E}_{x \sim p_\theta(x|a)}[\log(1 - \mathcal{D}(x, a))], \tag{2}$$

where a is the attribute vector, $x \sim p_{data}(x)$ is the visual feature of the same categories as the attribute a, and $x \sim p_\theta(x|a)$ is the visual feature sampled by the sampler $\mathcal{S}_\theta$. The discriminator $\mathcal{D}_\varphi$ tries to maximize the loss, and the sampler $\mathcal{S}_\theta$ tries to minimizes it. Following a replacing trick in [6], the optimal φ^* and θ^* can be obtained as follow:

$$\varphi^* = \arg \max_\varphi (\mathbb{E}_{x \sim p_{data}(x)}[\log \mathcal{D}(x, a)] + \mathbb{E}_{x \sim p_\theta(x|a)}[\log(1 - \mathcal{D}(x, a))]). \tag{3}$$

$$\theta^* = \arg \min_\theta (\mathbb{E}_{x \sim p_{data}(x)}[\log \mathcal{D}(x, a)] + \mathbb{E}_{x \sim p_\theta(x|a)}[\log(1 - \mathcal{D}(x, a))])$$
$$= \arg \max_\theta \underbrace{\mathbb{E}_{x \sim p_\theta(x|a)}[-\log(1 - \mathcal{D}(x, a))]}_{\text{denoted as } \mathcal{J}^\mathcal{S}(a)}. \tag{4}$$

It is worth mentioning that the sampler is utilized to directly select known visual features from the candidate pool. Thus, the sampling of the visual features is discrete, which means that we cannot directly optimise the sampling model by gradient descent. Following [7], we use policy gradient [21] based on reinforcement learning to derive the gradient of $\mathcal{J}^\mathcal{S}(a)$. Given a query attribute a, the sampler is modeled as a reinforcement learning policy to sample a candidate visual feature x_n at the state, and is trained via policy gradients. The gradient of $\mathcal{J}^\mathcal{S}(a)$ can be derived as follows:

$$\nabla_\theta \mathcal{J}^\mathcal{S}(a) = \nabla_\theta \mathbb{E}_{x \sim p_\theta(x|a)}[-\log(1 - \mathcal{D}(x_n, a))]$$

$$= \sum_{n=1}^N \nabla_\theta p_\theta(x_n|a)[-\log(1 - \mathcal{D}(x, a))]$$

$$= \sum_{n=1}^N p_\theta(x_n|a) \nabla_\theta \log p_\theta(x_n|a)[-\log(1 - \mathcal{D}(x_n, a))] \qquad (5)$$

$$= \mathbb{E}_{x \sim p_\theta(x|a)}[\nabla_\theta \log p_\theta(x|a)(-\log(1 - \mathcal{D}(x, a)))]$$

$$\simeq \frac{1}{K} \sum_{k=1}^K \nabla_\theta \underbrace{\log p_\theta(x_k|a)}_{\text{the action}} \underbrace{(-\log(1 - \mathcal{D}(x_k, a)))}_{\text{the reward}},$$

where x_n denotes the n-th visual feature in the candidate pool, and x_k denotes the the k-th visual feature approximately sampled from the current version of sampler $p_\theta(x|a)$. Inspired by the reinforcement learning terminology, we use the term $\log p_\theta(x|a)$ to denote taking an action x in the environment a, and the term $(-\log(1 - \mathcal{D}(x, a)))$ to denote the reward for the policy [21].

Specifically, for the policy gradient based on reinforcement learning, we first calculate the cosine similarity between the N visual features x_n in candidate pool and corresponding index vector $f_\theta(a)$ encoded by the sampler $\mathcal{S}_\theta$. Then, the probability $p_\theta(x_n|a)$ is obtained by softmax operation on the N cosine similarity values $cos(f_\theta(a), x_n)$. After that, we choose the probability value $p_\theta(x_k|a)$ and corresponding visual feature vector x_k of the top K probability value. Then, the log value of k probability value $p_\theta(x_k|a)$ is defined as the action value $log(p_\theta(x_k|a))$, and the value of $-log(1 - \mathcal{D}(x_k, a))$ based on the discriminative score $cos(f_\varphi(a), x_k)$ is defined as the reward value. Finally, the average value of the product of the action value and reward value is the loss of the sampler.

Moreover, in order to reduce the expression differences of sampler and discriminator when they encode the attribute into the visual feature space, we introduce the classification loss $\mathcal{L}_{cls}$ based on the score function of discriminator. We apply the SoftMax classifier to both the real visual features and the sampled fake visual features. The classification loss $\mathcal{L}_{cls}$ is defined as follow:

$$\mathcal{L}_{cls} = -\sum_i y_i \ln p(y = i|x) = -\sum_i y_i \ln \frac{\exp(\mathcal{D}(x, a_i))}{\sum_{c_s} \exp(\mathcal{D}(x, a_{c_s}))}, \qquad (6)$$

where c_s denotes the number of the seen categories. $p(y = i|x)$ represents the probability that the visual feature x belongs to category i. Specifically, we utilize $\mathcal{L}_{cls}^\mathcal{S}$ and $\mathcal{L}_{cls}^\mathcal{D}$ to represent the classification loss for real visual features and sampled fake visual features, respectively.

We take the classification losses as the regularizer for enforcing the sampler to select discriminative features, and promoting the disciminator to consider both inter-class and intra-class distance. Over full objective can be derived as follows:

$$\min_\theta \max_\varphi V(\mathcal{S}, \mathcal{D}) + \alpha \mathcal{L}_{cls}^\mathcal{S} + \beta \mathcal{L}_{cls}^\mathcal{D}, \qquad (7)$$

Table 1. The details of five benchmark datasets.

Dataset	SUN	CUB	AWA1	AWA2	aPY
$\mathcal{A}$	102	312	85	85	64
$\mathcal{N}$	14340	11788	30475	37322	15339
$\mathcal{S}+\mathcal{U}$	$645+72$	$150+50$	$40+10$	$40+10$	$20+12$
$\mathcal{N}_s$	10320	7057	19832	23527	5932
$\mathcal{N}_u$	1440	2967	5685	7913	7924
$\mathcal{N}_{s \to ts}$	2580	1764	4958	5882	1483

where α and β are the hyperparameters weighting the classifiers of sampler and discriminator, respectively.

Through multiple iterations of training, both sampler $\mathcal{S}$ and discriminator $\mathcal{D}$ can be used to classification. Given the attributes of all the unseen classes, i.e., $\mathcal{A}_u$, all the test index vector $f_\theta(\mathcal{A}_u)$ can be obtained by the sampler $\mathcal{S}$. Then, we calculate the cosine similarity between any test visual feature x and the index vector $f_\theta(\mathcal{A}_u)$. For the query image feature x, the label y of $a \in \mathcal{A}_u$ with the highest compatibility score is the classification result. As for the discriminator $\mathcal{D}$, we can directly get the compatibility scores between any query image x and all the test attributes $\mathcal{A}_u$ by Eq. 1. By finding the attribute with the highest cosine value, we can get the label of the query image.

3   Experiments

3.1   Experimental Setup

Dataset. We employ the most widely-used ZSL datasets for performance evaluation, that is, SUN Attribute Database (SUN) [16], Caltech-UCSDBirds 200-2011 (CUB) [15], Animals with Attributes 1 (AWA1) [10], Animals with Attributes 2 (AWA2) [26], Attribute Pascal and Yahoo (aPY) [4]. The GBU train/test split setting proposed in [26] is adopted to evaluate both the conventional ZSL setting and GZSL setting. Table 1 shows the details of five benchmark datasets. $\mathcal{A}$ denotes the dimension of attributes. $\mathcal{S}$ and $\mathcal{U}$ are the categories numbers of seen and unseen classes. N presents the number of images. N_s and N_u are the number of images of seen and unseen classes. Note that $N_{s \to ts}$ denotes the images' number of seen classes during test in the GZSL setting.

Evaluation Metrics. To compare the performance with the existing method, we use the unified evaluation protocol, i.e., Mean Class Accuracy (MCA), proposed in [26]. MCA averages the correct predictions independently for each class before dividing the number of classes. In the GZSL setting, we adopt MCA_S on seen test classes, MCA_U on unseen test classes, and their harmonic mean $H = 2 * MCA_S * MCA_U/(MCA_S + MCA_U)$ as the evaluation metrics.

Table 2. Comparisons in conventional settings. The best results are in **bold**. SAN-D and SAN-S present the discriminator and sampler of SAN model, respectively.

Method	SUN(%)	CUB(%)	AWA1(%)	AWA2(%)	aPY(%)
ESZSL [18]	54.5	53.9	58.2	58.6	38.3
LATEM [25]	55.3	49.3	55.1	55.8	35.2
SYNC [1]	59.1	55.6	54.0	46.6	23.9
SAE [9]	40.3	33.3	53.0	54.1	8.3
RethinkZSL [12]	62.6	54.4	70.9	**71.1**	38.0
GAZSL [29]	61.3	55.8	68.2	70.2	41.1
f-CLSWAGN [27]	60.8	57.3	68.2	–	–
LisGAN [11]	61.7	**58.8**	70.6	–	43.1
SAN-D	**62.9**	57.0	**71.4**	69.7	**43.4**
SAN-S	62.7	55.7	70.3	68.1	40.3

Implementation Details. We use the 2048-dimensional top-layer pooling units with ReLU activation of the 101-layered ResNet as the visual features following the pre-trained method in [26]. The sampler and discriminator are Multi-Layer Perceptron (MLP) with ReLU activation. The dimension of the input layer of the MLP is the attribute's dimension of the corresponding dataset. For all datasets, the dimensions of the hidden layer and output layer are 1600 and 2048, respectively. We use Adam optimizer with learning rate 0.00005 to train the SAN framework. The pool size N of candidate pool is set as 100, and we sample $K = 3$ visual features from the candidate pool each time across all the datasets. We apply $\alpha = \beta = 1$ and develop our method based on PyTorch[1].

3.2 Comparisons in Conventional Setting

We compare our method with the cross-modal embedding models and generative models in the conventional ZSL setting. Table 2 shows the experimental results. In the legacy challenge of zero-shot learning, both discriminator (SAN-D) and sampler (SAN-A) provide competitive performance, i.e., **62.9%** on SUN, **71.4%** on AWA1, **43.4%** on aPY. We analyze this striking improvements owing to the visual-semantic embedding space built by the adversarial training way, which decreases the intra-class distance and increases the inter-class separation.

Compare to the cross-modal embedding models, e.g., ESZSL [18], SYNC [1], which map the visual or semantic features to the fixed anchor points in the embedding subspace, the sampler of the proposed SAN framework directly samples the unseen features. Moreover, the discriminator distinguishes whether the sampled visual features match the input attribute, rather than whether the sampled visual features are true or fake. The sampler supplies the training unseen classes, and the visual-semantic embedding built by the adversarial training way

---

[1] The source code is provided at: https://github.com/TCvivi/Zero-Shot-Learning.

decreases the intra-class distance and increases the inter-class separation. Then, the hubness problem is mitigated by reducing the universal neighbours surrounding the embedding vectors of unseen classes.

Compare to the generative models based on GAN, e.g., GAZSL [29], and f-CLSWGAN [27], we propose to utilize the encoded semantic features to sample true visual features of seen classes, instead of directly generating visual features. Thus, the sampled visual features space is not affected by the semantic gap between attribute and visual feature space. Experimental results show that the proposed method is better than the the the generative models based on GAN. Moreover, both the sampler and discriminator of the proposed SAN are able to achieve good classification results.

3.3 Comparisons in Generalized Setting

Although the generative methods have much better generalization ability than the cross-modal methods on the conventional setting, the performances of these two methods both degrade dramatically on the generalized ZSL. Table 3 shows the experimental results in the generalized ZSL setting. An interesting observation is that the cross-model methods, e.g., ESZSL [18] and SYNC [1], perform well on seen test class (S), but work poorly on the unseen test classes (U). More interesting observation can be found that the generative methods, e.g., GAZSL [29] and f-CLSWGAN [27], perform well on the unseen classes in the GZSL setting, but their classification accuracy for the seen classes is worse than the cross-model methods. From the Table 3, we can see that the proposed SAN framework performs competitive on seen classes, unseen classes, as well as harmonic mean, i.e., $H = 41.0\%$ on SUN, $H = 67.1\%$ on AWA2.

Table 3. Comparisons in generalized settings. The best results are in **bold**. $U = MCA_U$, $S = MCA_S$, and H is the harmonic mean.

Method	SUN(%)			CUB(%)			AWA1(%)			AWA2(%)			aPY(%)		
	U	S	H	U	S	H	U	S	H	U	S	H	U	S	H
ESZSL [18]	11.0	27.9	15.8	12.6	63.8	21.0	6.0	75.6	12.1	5.9	77.8	11.0	2.4	70.1	4.6
LATEM [25]	14.7	28.8	19.5	15.2	57.3	24.0	7.3	71.7	13.3	11.5	77.3	20.0	0.1	73.0	0.2
SYNC [1]	7.9	**43.3**	13.4	11.5	**70.9**	19.8	8.9	**87.3**	16.2	10.0	**90.5**	18.0	7.4	66.3	13.3
SAE [9]	8.8	18.0	11.8	7.8	54.0	13.6	1.8	77.1	3.5	1.1	82.2	2.2	0.4	**80.9**	0.9
RethinkZSL [12]	36.3	42.8	39.3	47.4	47.6	47.5	**62.7**	77.0	**69.1**	56.4	81.4	66.7	26.5	74.0	39.0
GAZSL [29]	22.1	39.3	28.3	31.7	61.3	41.8	29.6	84.2	43.8	35.4	86.9	50.3	14.2	78.6	24.0
f-CLSWAGN [27]	42.6	36.6	39.4	43.7	57.7	49.7	57.9	61.4	59.6	–	–	–	–	–	–
LisGAN [11]	42.9	37.8	40.2	46.5	57.9	**51.6**	52.6	76.3	62.3	–	–	–	**34.3**	68.2	**45.7**
SAN-D	**45.6**	37.2	**41.0**	**48.6**	49.4	49.0	61.5	76.5	68.2	**57.6**	80.4	**67.1**	32.8	68.9	44.5
SAN-S	41.3	38.5	39.8	**48.6**	46.1	48.5	58.3	76.7	66.3	55.6	79.3	65.4	31.0	68.8	42.8

The cross-modal methods transfer the knowledge learned from seen classes directly to unseen classes. Thus, the visual-semantic embedding subspace constructed by the seen classes can maintain a high supervised classification accuracy on the seen classes. Obviously, when the search space includes both seen and unseen classes, the images of unseen classes are easily divided into the seen training categories. The generative methods based on GAN build a more complete visual-semantic embedding subspace by generating the pseudo data of unseen class, so as to solve the problem of low recognition accuracy on unseen classes in the GZSL setting. However, due to the semantic gap between attribute and visual feature space, with the improvement of the subspace's ability to recognize unseen classes, it is inevitably sacrifice the classification ability on seen classes.

3.4   Model Analysis

Visualisation of the Learned Representation. To visually investigate the effectiveness of the proposed SAN framework, we adopt the t-SNE [13] approach to embed the representation of the visual features and attributes into a two-dimensional visualisation plane for the AWA1 dataset in Fig. 2. Compare to the distribution of original attributes in Fig. 2(a), the semantic representation of embedded attributes by the discriminator of our method of both seen (blue '+') and unseen (red '×') classes in Fig. 2(b) is more spatially dispersed, which proves that the proposed SAN framework considers the inter-class separation for all classes. Figure 2(c) shows the attribute features encoded by discriminator of all classes and visual features embedding of unseen classes. We can see that the discriminator is able to model the discrimination between samples from different semantic categories of unseen classes, and effectively separates the representations into several semantically clusters. It demonstrates that intra-class consistency and inter-class diversity are considered. Figure 2(d) shows the attributes and visual features embedding of all test classes. Although the proposed method is able to separates the representations of all test classes into several clusters, the distributions of seen classes and unseen classes are too close or even overlapped. It's explains that all methods is fail to achieve perfect results in GZSL setting.

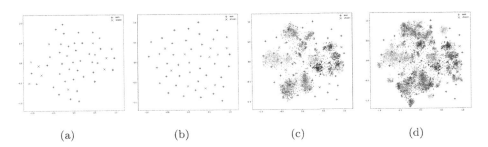

| (a) | (b) | (c) | (d) |

Fig. 2. The visualization of the learned representation of AWA1 dataset. (color figure online)

Class-Wise Accuracy. We use the confusion matrix to show the experimental result of ZSL in a more fine-grained scale. Figure 3 shows the confusion matrix of both sampler $\mathcal{S}_\theta$ (Fig. 3(a)) and discriminator $\mathcal{D}_\varphi$ (Fig. 3(b)) in the proposed SAN framework on the evaluation of AWA2 dataset. As shown in Fig. 3, sampler $\mathcal{S}_\theta$ and discriminator $\mathcal{D}_\varphi$ of the proposed method generally have better accuracy on most of the test categories. For classes such as 'seal' and 'bat', the low recognition accuracy of the both sampler $\mathcal{S}_\theta$ and discriminator $\mathcal{D}_\varphi$ mainly due to the fact that no similar categories have been seen during training. Therefore, when the visual-semantic embedding constructed in the seen classes is transferred to the novel classes, the model tends to perform poorly on these categories.

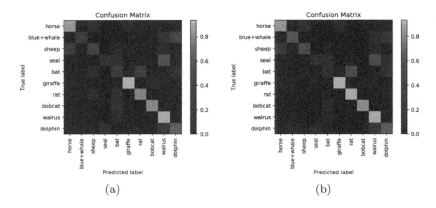

Fig. 3. The confusion matrixes on the evaluation of AWA2 dataset.

4   Conclusion

In this paper, we propose the SAN framework to improve ZSL by mitigating the hubness and semantic gap problem. For the hubness problem, we construct a visual-semantic embedding by the adversarial training way. Moreover, we utilize the discriminator to measure the matching degree of the visual-semantic pairs, rather than distinguish whether the visual features are real or fake. The proposed method considers both the intra-class consistency and inter-class diversity, which alleviates the hubness problem. For the semantic gap problem, we propose to utilize the encoded attributes to sample visual features of seen classes, instead of directly generating visual features. Then, the sampled visual features space is not affected by the semantic gap between attribute and visual feature space. Extensive experiments on five most widely-used datasets demonstrate that both the sampler and discriminator of the proposed SAN framework outperform state-of-the-art methods both in the conventional and generalized ZSL setting. In the future, we intend to perform SAN for the transductive setting as well. By adding the visual features of the unseen classes to candidate pool of the sampler, the model can learn a more comprehensive visual-semantic embedding space.

References

1. Changpinyo, S., Chao, W., Gong, B., Sha, F.: Synthesized classifiers for zero-shot learning. In: CVPR, pp. 5327–5336 (2016)
2. Chao, W.-L., Changpinyo, S., Gong, B., Sha, F.: An empirical study and analysis of generalized zero-shot learning for object recognition in the wild. In: Leibe, B., Matas, J., Sebe, N., Welling, M. (eds.) ECCV 2016. LNCS, vol. 9906, pp. 52–68. Springer, Cham (2016). https://doi.org/10.1007/978-3-319-46475-6_4
3. Dinu, G., Baroni, M.: Improving zero-shot learning by mitigating the hubness problem. In: ICLR, pp. 135–151 (2014)
4. Farhadi, A., Endres, I., Hoiem, D., Forsyth, D.A.: Describing objects by their attributes. In: CVPR, pp. 1778–1785 (2009)
5. Fu, Y., Hospedales, T.M., Xiang, T., Fu, Z., Gong, S.: Transductive multi-view embedding for zero-shot recognition and annotation. In: Fleet, D., Pajdla, T., Schiele, B., Tuytelaars, T. (eds.) ECCV 2014. LNCS, vol. 8690, pp. 584–599. Springer, Cham (2014). https://doi.org/10.1007/978-3-319-10605-2_38
6. Goodfellow, I., Pouget-Abadie, J., Mirza, M.E.A.: Generative adversarial nets. In: NIPS, pp. 2672–2680 (2014)
7. Jun, W., et al.: IRGAN: a minimax game for unifying generative and discriminative information retrieval models. In: SIGIR, pp. 515–524 (2017)
8. Kaiming, H., Xiangyu, Z., Shaoqing, R., Jian, S.: Deep residual learning for image recognition. In: CVPR, pp. 770–778 (2016)
9. Kodirov, E., Xiang, T., Gong, S.: Semantic autoencoder for zero-shot learning. In: CVPR, pp. 4447–4456 (2017)
10. Lampert, C.H., Nickisch, H., Harmeling, S.: Attribute-based classification for zero-shot visual object categorization. IEEE Trans. Pattern Anal. Mach. Intell. **36**(3), 453–465 (2014)
11. Li, J., Jing, M., Lu, K., Ding, Z., Zhu, L., Huang, Z.: Leveraging the invariant side of generative zero-shot learning. In: CVPR, pp. 7402–7411 (2019)
12. Li, K., Min, M.R., Fu, Y.: Rethinking zero-shot learning: a conditional visual classification perspective. In: ICCV, pp. 3582–3591 (2019)
13. van der Maaten, L., Hinton, G.: Visualizing data using t-SNE. J. Mach. Learn. Res. **9**(11), 2579–2605 (2008)
14. Niu, L., Cai, J., Veeraraghavan, A.: Zero-shot learning via category-specific visual-semantic mapping. IEEE Trans. Image Process. **28**(2), 965–979 (2019)
15. Welinder, P., et al.: Caltech-UCSD birds 200. In: Caltech, Technical Report CNS-TR-2010-001 (2010)
16. Patterson, G., Hays, J.: Sun attribute database: discovering, annotating, and recognizing scene attributes. In: CVPR, pp. 2751–2758 (2012)
17. Radovanovic, M., Nanopoulos, A., Ivanovic, M.: Hubs in space: popular nearest neighbors in high-dimensional data. JMLR **11**, 2487–2531 (2010)
18. Romera-Paredes, B., Torr, P.H.S.: An embarrassingly simple approach to zero-shot learning. In: ICML, pp. 2152–2161 (2015)
19. Russakovsky, O., et al.: Imagenet large scale visual recognition challenge. Int. J. Comput. Vis. **115**(3), 211–252 (2015)
20. Socher, R., Ganjoo, M., Manning, C.D., Ng, A.Y.: Zero-shot learning through cross-modal transfer. In: NIPS, pp. 935–943 (2013)
21. Sutton, R.S., et al.: Policy gradient methods for reinforcement learning with function approximation. In: NIPS, pp. 1057–1063 (2000)

22. Tang, C., Lv, J., Chen, Y., Guo, J.: An angle-based method for measuring the semantic similarity between visual and textual features. Soft Comput. **23**(12), 4041–4050 (2018). https://doi.org/10.1007/s00500-018-3051-y
23. Tang, C., Yang, X., Lv, J., He, Z.: Zero-shot learning by mutual information estimation and maximization. Knowl.-Based Syst. **194**, 105490 (2020)
24. Williams, R.J.: Leveraging the invariant side of generative zero-shot learning. Mach. Learn. **8**, 229–256 (1992)
25. Xian, Y., Akata, Z., Sharma, G., Nguyen, Q.N., Hein, M., Schiele, B.: Latent embeddings for zero-shot classification. In: CVPR, pp. 69–77 (2016)
26. Xian, Y., Lampert, C.H., Schiele, B., Akata, Z.: Zero-shot learning - a comprehensive evaluation of the good, the bad and the ugly. IEEE Trans. Pattern Anal. Mach. Intell. **41**(9), 2251–2265 (2019)
27. Xian, Y., Lorenz, T., Schiele, B., Akata, Z.: Feature generating networks for zero-shot learning. In: CVPR, pp. 5542–5551 (2018)
28. Zhang, L., Xiang, T., Gong, S.: Learning a deep embedding model for zero-shot learning. In: CVPR, pp. 3010–3019 (2017)
29. Zhu, Y., Elhoseiny, M., Liu, B., Elgammal, A.: A generative adversarial approach for zero-shot learning from noisy texts. In: CVPR, pp. 1004–1013 (2018)

Semi-supervised Classification of Data Streams Based on Adaptive Density Peak Clustering

Changjie Liu[1], Yimin Wen[1(✉)], and Yun Xue[2]

[1] Guangxi Key Laboratory of Image and Graphic Intelligent Processing,
Guilin University of Electronic Technology, Guilin, China
techat17@163.com, ymwen2004@aliyun.com
[2] School of Municipal and Surveying Engineering, Hunan City University,
Yiyang, China
yunxue1209@163.com

Abstract. In the real-world scenario of data stream classification, label scarcity is very common. More challenges are data streams always include concept drifts. To handle these challenges, an algorithm of semi-supervised classification of data streams based on adaptive density peak clustering (SSCADP) is proposed. In SSCADP, to generate concept clusters at leaves in a Hoeffding tree, a density peak clustering method and a change detection technique are combined to adaptively locate the clustering centers. Concerning concept drift detection, we argue that the change of cluster with higher density more likely reflect the change of data distribution. Hence, to detect concept drifts, an adaptive weighting method for density change detection is proposed to calculate the deviations between the history concept clusters and new ones. Experiments on synthetic and real datasets confirm the advantages of SSCADP.

Keywords: Semi-supervised classification · Data stream · Decision tree · Clustering · Concept drift

1 Introduction

The classification of data steams with concept drift is one of the main challenges of data mining [1–3]. In various real-applications, including network intrusion detection, spam filtering and credit card fraud detection[4] etc., due to labeling cost and time consuming, it is unrealistic that all instances are labeled by expert. Therefore, a semi-supervised classification algorithm which can handle concept drifts plays a critical role in addressing the issue of data stream mining.

To overcome these challenges, many researches have been reported in recent years. However, there are still some limitations. Firstly, many existing methods [5,6] commonly take clustering methods to label the unlabeled instances which should assign the number of clusters in advance and keep it constant during the

© Springer Nature Switzerland AG 2020
H. Yang et al. (Eds.): ICONIP 2020, LNCS 12533, pp. 639–650, 2020.
https://doi.org/10.1007/978-3-030-63833-7_54

processing of data streams. Secondly, many methods ignore the impact of high density clusters on concept drift detection [5,7].

Specifically in SUN [5], firstly, K-modes is used to form clusters and label the unlabeled data. However, using K-modes requires setting the number of clusters in advance and keeping it unchanged during the processing of data streams, which is unreasonable in many real-applications. Since the dynamic feature of data streams, new concept clusters may appear while old may disappear. Secondly, the average deviation between the history and new concept clusters is adopted to detect concept drifts, which ignores changes in high-density clusters which are more likely result in concept drifts.

In light of these limitations, an approach of semi-supervised classification of data streams based on adaptive density peak clustering (SSCADP) is proposed. The framework of SSCADP is the same as SUN, but there are two main differences.

Firstly, to generate concept clusters at the leaves in a Hoeffding tree, a density peak clustering method [8] and a change detection technique [9] are combined to adaptively locate the cluster centers, instead of using K-modes. Secondly, we consider that the change of clusters with higher density is more likely to reflect the change of data distribution. And hence an improved detection method based on SUN is proposed that an adaptive weighted average strategy is adopted to assign higher weight value to the higher density clusters.

The rest of this paper is organized as follows. Section 2 presents some related work. Section 3 describes the proposed algorithm in detail including adaptively locating the cluster centers, labeling the unlabeled samples, and concept drift detection method. Experiments and results are shown in Sect. 4. Finally, Sect. 5 summarizes this paper and future work.

2   Related Work

The proposed approaches for semi-supervised classification of data stream with concept drifts can be broadly divided into decision tree-based and non-decision tree-based methods. The decision tree-based methods like SUN [5] and REDLLA [7] adopt a Hoeffding tree as base classifier. During the construction of the base classifier, unlabeled data are labeled by clustering in leaves, and then added into the detection of concept drifts and the updating of the base classifier. Concept drifts are detected based on the deviation between history concept clusters and the new ones. Others like Sco-forest [10] extends Co-forest algorithm to handle evolving data streams. The concomitant ensemble is used to select samples with high classification confidence and label these samples to update the corresponding base classifier. If a concept drift is detected by Adwin2 [11], the base classifier with the worst accuracy will be discarded.

The non-decision tree-based methods usually take cluster model for data streams classification. Reasc [12] maintains an ensemble of cluster-based classifiers. When updating the ensemble, the cluster-based classifier with the worst accuracy will be removed out. SPASC [6] maintains a classifier pool which is composed of weighted clusters-based model. The weight value will be adjusted adaptively according to the correctness of classification. SCBELS [13] maintains an ensemble of cluster-based classifiers which are constructed by BIRCH [14] and incrementally updated during the classification of data stream. Local structural information of data is taken into account to deal with concept drifts. [15] proposed to dynamically maintain a set of micro-clusters, each instance is used to update the model, outdated or micro-clusters with low reliability are removed to adapt to the evolving concepts of data streams.

Other models like ECU [16], it constructs an ensemble model which combines both classifiers and clusters for classification. [17] proposed a neural network framework for streaming data classification that each layer consists of a generative network, a discriminant structure and a bridge.

3 Proposed Algorithm

3.1 The Framework of the Proposed Algorithm

In this paper, a data stream is represented as $D = \{D^0, D^1, D^2 \ldots, D^t, \ldots\}$, in which $D^t = \{x_1^t, x_2^t, \ldots, x_m^t\}$ indicates the data batch collected at the time t. SSCADP[1] is described in the algorithm 1. It employs a Hoeffding tree as its base classifier. After D^t is classified by the Hoeffding tree, each instance in it is sorted into a leaf, the corresponding statistics of the leaf are updated. If the number of instances arrived at the tree meets dp, all the labeled instances in a leaf l are utilized to label the unlabeled instances in it, and then concept drift detection is installed to detect drift at the leaf. Then, a pruning strategy is adopted on the Hoeffding tree, and if the number of instances in a new leaf meet n_{min}, the leaf attempts to split. After splitting, the updated Hoeffding tree is ready for new concept.

3.2 Adaptively Locate Cluster Centers and Label Unlabeled Instances

If a detection period is reached, a clustering method named Clustering by fast search and find of density peaks (CFSDP) [8] and a change detection technique [9] are combined to adaptively locate the cluster centers. After concept clusters at each leaf are created, graph-based label propagation [18] is installed to label the unlabeled data in each cluster. If a cluster without any labeled instance, all unlabeled instances in it will be assigned the majority label of the closest cluster.

The basic idea of CFSDP is that the clustering centers should be in the region with high data density and far away from each other. CFSDP is based on two quantities: (1) ρ_i, the local density of the i-th instance; (2) δ_i, the minimum distance between the i-th instance and the instances which have higher density than the i-th instance. ρ_i and δ_i are defined as

$$\rho_i = \sum_{j \in I_s/\{i\}} \exp\{-(d_{ij}/d_c)^2\}, \quad \delta_i = \min_{j:\rho_j > \rho_i}(d_{ij}). \quad (1)$$

---

[1] Source code: https://gitee.com/ymw12345/sscadpsrc.git.

Algorithm 1: SSCADP

Input: A data stream in the form of chunk:$D = \{D^0, D^1, D^2 ..., D^t, ...\}$ and
         parameters of n_{min}, dp, α
Output: The predicted labels of D^t
1   Initialize a leaf for tree T, $t = 0$;
2   **while** *data chunk D^t is available* **do**
3   **if** $t > 0$ **then**
4         $T.classify(D^t)$;
5   **for** *each $x_i \in D^t$* **do**
6   sort x_i into a leaf l ;
7   update the statistics of the leaf l according to it is labeled or unlabeled;
8   **if** *the number of arrived instances at T meets dp* **then**
9   **for** *each leaf l from bottom to top* **do**
10              $D_l = get\_data(l)$;
11              $labeling\_unlabeled\_data(D_l, \alpha)$;
12              $concept\_drift\_detection(l)$;
13   Installing pruning;
14   **if** *the number of arrived instances at a leaf l meets n_{min}* **then**
15   Installing split-test and growing child leaves ;
16    $t + +$;

---

where d_{ij} refers to the Euclidean distance between i-th and j-th instance. d_c represents the cutoff distance, which is set the same as CFSDP according to experience. I_s is the set of all instances indexes. Hence, a cluster center has such characteristic that ρ_i and δ_i are as large as possible. The $\rho_i - \delta_i$ plot (decision graph) can provide a visual way to determine the number of clusters.

In the function 1, $\gamma_i = \rho_i \delta_i$ is computed for each instance and then all γ_i are sorted in ascending order. CFSDP assumed that the sorted $\gamma_i(\gamma)$ follows the power-law distribution, and hence jump point of the sorted γ_i can be found.

After the cluster centers are determined, the clustering is installed, and then label propagation is conducted.

In order to find the jump point of the sorted γ_i, we refer to the idea of change detection in SAND [9], in which a change detection method is proposed to detect the location of the most significant changes in a series of values along a direction. Then, we further assume that the sorted γ_i follows a Pareto distribution. After the jump point is found, the number of clusters can be determined. As shown in the Fig. 1, the point in the red circle are jump point, the points above it are the center points.

The probability density function of Pareto distribution can be expressed as $f(x, a, k) = ak^a x^{-(a+1)}$ where a is the shape parameter and k is the proportion parameter. The corresponding logarithmic probability density function can be expressed as $\log f(x, a, k) = a \log(k) + \log(a) - (a + 1) \log(x)$. Define a random variable $\gamma \sim \text{Pareto}(a, k)$, and $\{\gamma_i\}$ is the observed value of γ. The maximum

Function 1: *labeling_unlabeled_data*

Input: Data $D = \{x_1, x_2, ..., x_n\}$; Confidence parameter α
Output: The expanded labeled data D'; Concept clusters set C

1   D'=null, labeled data are added to D';
2   Construct distance matrix on D based on d_{ij} and record it as M;
3   Calculate ρ_i and δ_i for each data x_i based on M;
4   Normalize ρ_i and δ_i, $\gamma_i = \rho_i \delta_i$ and sort the elements in γ in ascending order;
5   $JumpPoint = jump\_point\_detection(\gamma, \alpha)$, $CN = n - JumpPoint$;
6   The instances corresponding to the first CN values in γ are selected as the cluster centers. $\{C_i\}_{i=1}^{CN}$ denotes the corresponding clusters;
7   **for** *each* $x_i \in D'$ **do**
8   **if** x_i *is not a cluster center* **then**
9   x_i is assigned to the cluster the same as its nearest higher density point belongs to;

10 **for** *each* $C_i \in C$ **do**
11   Label unlabeled data at C_i by graph-based label propagation;
12   Labeled data are added to D';
13 **return** D', C;

---

likelihood estimation of the parameters are calculated as follow, where N is the total number of observed values.

$$\hat{k}_{MLE} = \min_{1 \leq i \leq N} \{\gamma_i\}, \; \hat{a}_{MLE} = N / \sum_{i=1}^{N} (\ln \gamma_i - \ln \hat{k}_{MLE}) \tag{2}$$

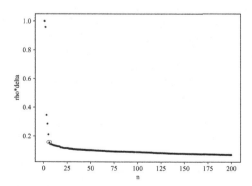

Fig. 1. Jump point detection.

To detect the jump point, in the function 2, γ is divided into two sub-windows by k from $N/2$ to $N - 3$. The k value corresponding to the maximum statistical difference between the two sub-windows is the index of the jump point.

Function 2: *jump_point_detection*

Input: Array γ with elements in ascending order; Confidence parameter α
Output: The corresponding index of the jump point e

1 $N = size(\gamma)$;
2 $e = -1, w_n = 0$;
3 **for** $k = N/2 : N - 3$ **do**
4      $m_a = \text{mean}(\gamma[1 : k])$, $m_b = \text{mean}(\gamma[k + 1 : N])$;
5   **if** $m_a < \alpha m_b$ **then**
6          $Pareto[scale_a, shape_a] < -\text{estimateParam}(\gamma[1 : k])$;
7          $Pareto[scale_b, shape_b] < -\text{estimateParam}(\gamma[k + 1 : N])$;
8          $sk = 0$;
9   **for** $i = k + 1 : N$ **do**
10             $ta = \log f(\gamma[i], scale_a, shape_a)$, $tb = \log f(\gamma[i], scale_b, shape_b)$;
11             $sk+ = \log(tb/ta)$;
12   **if** $sk > w_n$ **then**
13             $w_n = sk$, $e = k$;

14 **return** e;

3.3 Concept Drift Detection

A concept drift detection method is installed at each leaf. Before introducing the detail of concept drift detection, many variables should be defined. Respectively, r_{hist} and r_{new} denote the radius of the set of history concept clusters C_{hist} and the new ones C_{new}, n_{hist} and n_{new} represents the number of clusters in C_{hist} and C_{new}. r_k denotes the radius of a cluster and is computed as the average Euclidean distance from all instances in the cluster to its center: $r_k = \sum_{i=1}^{|C_k|} \sqrt{\sum_{j=1}^{D} (c_{kj} - x_{ij})^2} / |C_k|$, where $x_i = \{x_{i1}, x_{i2}, \ldots, x_{iD}\} \in C_k$ is the i-th instance in cluster C_k. D represents the attribute dimension. c_k refers to the cluster center of C_k and $|C_k|$ is the total number of instance in C_k. $r_{hist} = \sum_{i=1}^{n_{hist}} r_i / n_{hist}$. Similarly, r_{new} is calculated by this way. $dist$ is used to measure the average distance between these two concept cluster sets. $dist = (\sum_{i=1}^{n_{new}} \min[\sqrt{\sum_{j=1}^{D} (c_{ij} - c_{kj})^2}, c_k \in C_{hist}, 1 \le k \le n_{hist}]) / n_{new}$, where c_k and c_i denote cluster centers in C_{hist} and C_{new}, respectively. In SUN, the value of $dist$ greater than $\max(r_{hist}, r_{new})$ means concept drift.

However, in our algorithm, it is assumed that the change of points with higher density is more likely to reflect the change of data distribution. Therefore, $dist$ is redefined as $dist = \sum_{i=1}^{n_{new}} w_i dist_i$ to capture the distribution change of data more accurately. $dist_i$ and w_i are calculated by (3) and (4) respectively. $dist_i$ refers to the distance of the cluster center c_i to C_{hist}. ρ_{c_i} refers to the average density of the clusters C_i and C_k which is the closest cluster to C_i in C_{hist}, and $\rho_{c_{ij}}$ refers to the density of j-th instance belonging to C_i, n_i and n_k mean the total number of data in C_i and C_k respectively. And hence larger w_i and $dist_i$ mean concept drift. Like SUN, if the value of $dist$ is more than $\max(r_{hist}, r_{new})$, a real concept drift is considered. The process of drift detection is described in the function 3.

$$dist_i = \min[\sqrt{\sum\nolimits_{j=1}^{D} (c_{ij} - c_{kj})^2}, c_k \in C_{hist}, 1 \le k \le n_{hist}] \tag{3}$$

$$w_i = \rho_{c_i} \Big/ \sum\nolimits_{i=1}^{n_{new}} \rho_{c_i}, \rho_{c_i} = (\sum\nolimits_{j=1}^{n_i} \rho_{c_{ij}} + \sum\nolimits_{j=1}^{n_k} \rho_{c_{kj}})/(n_i + n_k). \tag{4}$$

Function 3: *concept_drift_detection*

Input: Concept clusters set C_{new} and C_{hist} saved in leaf l
Output: *Flag*

1 $Flag$ = False;
2 **if** $C_{hist} = \emptyset$ **then**
3 $\quad \lfloor\ C_{hist} = C_{new}$

4 **else**
5 $\quad$ calculate r_{hist}, r_{new} and *dist* using C_{hist} and C_{new};
6 $\quad$ **if** $dist > \max(r_{hist}, r_{new})$ **then**
7 $\quad\quad \lfloor\ Flag$ = True

8 **return** *Flag*

---

dist can be utilized to detect concept drifts caused by the change of P(x). In addition, considering concept drift can be caused by the distribution change of class labels, the class labels of history concept clusters and new ones are compared when concept drift is not detected. If the class labels of C_{hist} and C_{new} are completely opposite, it is also defined as a real concept drift.

After the bottom-up search is implemented to find all drift leaves, a pruning method is installed for adjusting the tree to cope with concept drifts. Each level of the tree will be traversed once to check concept drift of each leaf from bottom to top until the root is reached. If all child nodes of a node are detected concept drift, these child nodes are pruned and the new leaf node maybe split again.

4 Experiments

In this paper, all synthetic datasets are generated by MOA [19]. xxx-abr, xxx-gra and xxx-inc represent the concept drift types of abrupt, gradual and incremental, respectively. In the dataset with gradual drifts, it takes 5000 instances to change from one concept to another. In addition, to verify the performance of SSCADP on the datasets where the number of clusters varies apparently, we generate a Gaussian dataset with clusters change dynamically and concept changes, which are shown in Fig. 2(a), (b) and (c) represent three different distributions which

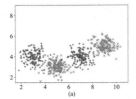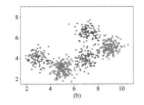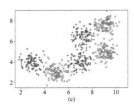

Fig. 2. Changes in clusters.

evolve along the time sequence, with the positive instances in red '+' and the negative instances in green '×', each figure contains 600 instances.

Table 1 shows the properties of all datasets. For Sea, four concepts are generated by setting $\theta = 8$, 9, 7, and 9.5. For Sine, the definition of Sine is if $a * sin(b * x_1 + \theta) + c > x_2$, the label is 0, otherwise is 1, four concepts are generated by setting $a = b = 1$ and $c = \theta = 0$, $a = b = 1$ and $c = \theta = 0$ with class labels are changed oppositely, $a = 0.3$, $b = 3\pi$, $c = 0.5$, $\theta = 0$ and $a = 0.3$, $b = 3\pi$, $c = 0.5$, $\theta = 0$ with class labels are changed oppositely. For HyperPlane-abr and HyperPlane-gra, four concepts are generated by setting $w^1 = (0, 0.5, 0.5), w^2 = (0, 1, 0), w^3 = (1, 0, 0)$ and $w^4 = (0.5, 0, 0.5)$, while for HyperPlane-inc, $d = 10$. For Agrawal, function 1, 2, 5, 6 are selected as the concepts of 1, 2, 3, 4. The Weather and Electricity dataset are used in this paper. For each synthetic data, 10 copies are randomly generated while for each real dataset labeled instances are randomly selected 10 runs.

In this paper, n_{min} refers to the minimum number of instances when a leaf attempts to do split-test and it is set to 200. dp means the detection period and $dp = 200$ empirically. $\alpha = 0.95$ is the confidence used in the clustering algorithm.

4.1   Experimental Results

SSCADP is compared with three baseline methods including SUN, SPASC, and Reasc. Three groups of experiments are conducted to evaluate the accuracy, the impact of label ratio, and concept drift tracking. In the first and last groups of the experiments, in order to simulate the situation of limited labeled data in real applications, the label ratio of all datasets are set to 0.1.

Accuracy. The accumulative accuracy is utilized to evaluate the performance of baseline methods and SSCADP. Table 2 shows the detailed results. For all datasets, each result is obtained by averaging the results of 10 runs.

It can be observed that SSCADP performs better than other baseline algorithms on almost of all datasets. The Friedman test is conducted on the results in Table 2, and the average rank is shown in Table 2, SSCADP achieves the best. Test statistic $F_F = 10.654$, the critical value for $\alpha = 0.05$ is 2.892, hence we can reject the null-hypothesis that there is no difference among the performance of

Table 1. Properties of the datasets

Datasets	Attributes	Instances	Classes	Chunk size	Concept change
Sea-abr	3	80000	2	1000	1-2-3-4-1-2-3-4
Sea-gra	3	115000	2	1000	1-2-3-4-1-2-3-4
Sine-abr	4	80000	2	1000	1-2-3-4-1-2-3-4
Sine-gra	4	115000	2	1000	1-2-3-4-1-2-3-4
HyperPlane-abr	3	80000	2	1000	1-2-3-4-1-2-3-4
HyperPlane-gra	3	115000	2	1000	1-2-3-4-1-2-3-4
HyperPlane-inc	10	80000	2	1000	Unknown
Agrawal-abr	9	80000	2	1000	1-2-3-4-1-2-3-4
Agrawal-gra	9	11500	2	1000	1-2-3-4-1-2-3-4
Gaussian	2	3600	2	600	1-2-3-1-2-3
Weather	8	18159	2	360	Unknown
Electricity	8	45312	2	1000	Unknown

all algorithms. Furthermore, in Nemenyi test $CD = 1.35$ which means SSCADP performs significantly better than SUN and Reasc. There is no significant difference in the performance between SSCADP and SPASC.

Table 2. Accumulative accuracy (%) on all datasets.

Datasets	SUN	SPASC	Reasc	SSCADP
Sea-abr	78.58 ± 2.64	83.36 ± 1.41	83.17 ± 1.58	$\mathbf{83.48 \pm 1.02}$
Sea-gra	81.16 ± 2.50	82.83 ± 1.10	82.45 ± 0.86	$\mathbf{84.45 \pm 0.51}$
Sine-abr	51.68 ± 2.37	$\mathbf{57.15 \pm 2.91}$	51.98 ± 0.31	51.92 ± 2.20
Sine-gra	53.21 ± 1.82	$\mathbf{55.35 \pm 1.76}$	51.03 ± 0.37	51.06 ± 1.40
HyperPlane-abr	64.47 ± 1.86	66.57 ± 2.29	51.38 ± 1.47	$\mathbf{67.14 \pm 0.82}$
HyperPlane-gra	65.54 ± 1.10	66.92 ± 1.99	51.17 ± 1.06	$\mathbf{68.31 \pm 1.73}$
HyperPlane-inc	74.15 ± 4.66	62.02 ± 3.38	50.06 ± 0.34	$\mathbf{74.34 \pm 7.53}$
Agrawal-abr	52.37 ± 1.03	57.78 ± 0.45	43.73 ± 0.32	$\mathbf{58.88 \pm 0.62}$
Agrawal-gra	53.36 ± 0.98	57.09 ± 0.53	44.34 ± 0.28	$\mathbf{57.86 \pm 0.46}$
Gaussian	52.64 ± 3.05	58.93 ± 5.86	50.38 ± 6.31	$\mathbf{62.17 \pm 6.37}$
Weather	67.89 ± 0.76	66.77 ± 1.21	67.95 ± 0.00	$\mathbf{68.53 \pm 0.67}$
Electricity	57.73 ± 5.04	54.75 ± 1.15	57.12 ± 1.49	$\mathbf{70.52 \pm 3.03}$
Average rank	3.00	2.25	3.41	1.33

Impact of Label Ratio. Considering the influence of labeling ratio on classification accuracy, we simulate the real scene to compare the algorithms by setting the label ratio to 0.05 and 0.2. Detailed results are shown in Table 3. In the case of 0.05, Friedman test is conducted and $F_F = 3.175$, critical value for $\alpha = 0.05$ is 2.892. This results indicate that the performance of all the algorithms is significantly different. Furthermore, in Nemenyi test $CD = 1.35$, and hence it can be concluded that SSCADP performs significantly better than SUN. These results indicate that even there are very limited labels available, SSCADP can achieve better performance, and it is suitable for semi-supervised classification of data stream.

In the case of 0.2, Friedman test is conducted and $F_F = 6.567$, critical value for $\alpha = 0.05$ is 2.892. This results indicate that the performance of all the algorithms is significantly different. Furthermore, in Nemenyi test $CD = 1.35$, and hence it can be concluded that SSCADP performs significantly better than SUN and Reasc. There is no significant difference in the performance between SSCADP and SPASC in the cases of 0.05 and 0.2.

Table 3. Impact of the label ratio on accumulative accuracy (%).

	SUN		SPASC		Reasc		SSCADP	
	0.05	0.2	0.05	0.2	0.05	0.2	0.05	0.2
Sea-abr	69.87	81.28	80.81	84.04	82.09	84.23	79.56	84.51
Sea-gra	74.07	82.53	80.89	84.78	83.13	85.18	77.33	85.22
Sine-abr	51.11	51.95	56.67	61.66	52.12	51.86	50.40	51.67
Sine-gra	52.19	52.28	57.81	54.36	53.01	53.05	52.22	50.16
H-abr	63.29	65.83	65.83	66.79	50.66	50.69	66.00	66.94
H-gra	64.22	66.97	66.04	68.32	51.39	50.65	67.28	68.86
H-inc	72.15	77.61	60.34	65.90	50.03	49.95	73.31	78.99
Agr-abr	51.69	52.98	56.87	58.82	43.69	43.80	57.63	59.30
Agr-gra	52.48	53.66	57.19	58.64	44.41	44.38	57.60	58.75
Gaussian	52.84	58.40	56.60	59.88	49.96	52.79	57.16	61.45
Weather	67.59	67.58	65.25	68.05	67.85	67.93	68.23	68.65
Electricity	59.36	70.56	53.67	56.15	59.69	57.38	68.09	70.82
Rank	3.16	3.00	2.25	2.25	2.83	3.25	1.75	1.50

Concept Drift Tracking. The drift tracking performance of all algorithms on all datasets with abrupt drift type are shown in Fig. 3. The number at bottom represent the kind of concept, and the vertical line indicate the location of concept drift. In the dataset of Sea, when new concepts arrive, the accuracy of SSCADP declined less in most cases, especially in concept 3 and 4 which are quite different. In the dataset of HyperPlane, SSCADP performed well in most cases except concept 2. In the dataset of Agrawal, the accuracy of SSCADP also declined less in most cases. In the dataset of Sine, SSCADP performed not well since it is a single model and the large differences exist between each concept. SPASC adopts ensemble model as well as can deal with recurring drifts which can achieve better performance.

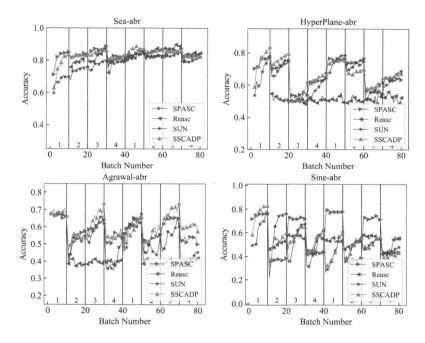

Fig. 3. Drift tracking graph.

5   Conclusions

In this paper, we propose a method of SSCADP to handle the semi-supervised classification of data streams. SSCADP can adaptively locate the cluster centers by considering both the local density and the distance between two points, and detect concept drifts by combining both the density of clusters and the distance between the historical clusters and the new ones. Experimental results illustrated that SSCADP can achieve better performance than the baseline algorithms in most datasets. In future, we will focus on how to effectively combine labeled data with unlabeled data to detect concept drift. We will also explore how to deal with recurring concept drift more effectively.

Acknowledgments. This work was partially supported by the Natural Science Foundation of Guangxi District (2018GXNSFDA138006), National Natural Science Foundation of China (61866007, 61662014), Collaborative Innovation Center of Cloud Computing and Big Data (YD16E12) and Image Intelligent Processing Project of Key Laboratory Fund (GIIP201505).

References

1. Ditzler, G., Roveri, M., Alippi, C., et al.: Learning in nonstationary environments: a survey. IEEE Comput. Intell. Mag. **10**(4), 12–25 (2015)
2. Gama, J., Žliobaitė, I., Bifet, A., et al.: A survey on concept drift adaptation. ACM Comput. Surv. **46**(4), 44–80 (2014)
3. Lu, J., Liu, A., Dong, F., et al.: Learning under concept drift: a review. IEEE Trans. Knowl. Data Eng. **31**(12), 2346–2363 (2019)
4. Sedhai, S., Sun, A.: Semi-supervised spam detection in Twitter stream. IEEE Trans. Comput. Soc. Syst. **5**(1), 169–175 (2018)
5. Wu, X., Li, P., Hu, X., et al.: Learning from concept drifting data streams with unlabeled data. Neurocomputing **92**, 145–155 (2012)
6. Hosseini, M.J., Gholipour, A., Beigy, H., et al.: An ensemble of cluster-based classifiers for semi-supervised classification of non-stationary data streams. Knowl. Inf. Syst. **46**(3), 567–597 (2016)
7. Li, P.P., Wu, X., Hu, X.: Mining recurring concept drifts with limited labeled streaming data. ACM Trans. Intell. Syst. Technol. **13**(2), 241–252 (2012)
8. Rodriguez, A., Laio, A.: Clustering by fast search and find of density peaks. Science **344**(6191), 1492–1496 (2014)
9. Haque, A., Khan, L., Baron, M., et al.: SAND: semi-supervised adaptive novel class detection and classification over data stream. In: Proceedings of the 13th AAAI Conference on Artificial Intelligence, pp. 1652–1658. AAAI, Menlo Park, CA (2016)
10. Wang, Y., Li, T.: Improving semi-supervised co-forest algorithm in evolving data streams. Appl. Intell. **48**(10), 3248–3262 (2018)
11. Bifet, A., Gavalda, R.: Learning from time-changing data with adaptive windowing. In: Proceedings of the 7th SIAM International Conference on Data Mining, pp. 443–448. SIAM, Philadelphia, PA (2007)
12. Masud, M.M., Woolam, C., Gao, J., et al.: Facing the reality of data stream classification: coping with scarcity of labeled data. Knowl. Inf. Syst. **33**(1), 213–244 (2012)
13. Wen, Y.M., Liu, S.: Semi-supervised classification of data streams by BIRCH ensemble and local structure mapping. J. Comput. Sci. Technol. **35**(2), 295–304 (2020)
14. Zhang, T., Ramakrishnan, R., Livny, M., et al.: BIRCH: an efficient data clustering method for very large databases. In: Proceedings of the 1996 ACM SIGMOD International Conference on Management of Data, pp. 103–114. ACM, New York (1996)
15. Din, S.U., Shao, J., Kumar, J., et al.: Online reliable semi-supervised learning on evolving data streams. Inf. Sci. **525**, 153–171 (2020)
16. Zhang, P., Zhu, X., Tan, J., et al.: Classifier and cluster ensembles for mining concept drifting data streams. In: Proceedings of the 10th IEEE International Conference on Data Mining, pp. 1175–1180. IEEE, Los Alamitos, CA (2010)
17. Li, Y., Wang, Y., Liu, Q., et al.: Incremental semi-supervised learning on streaming data. Pattern Recogn. **88**, 383–396 (2019)
18. Zhu X, Ghahramani Z.: Learning from labeled and unlabeled data with label propagation. Technical report CMU-CALD-02-107, Carnegie Mellon University (2002)
19. Bifet, A., Holmes, G., Kirkby, R., et al.: MOA: massive online analysis. J. Mach. Learn. Res. **11**, 1601–1604 (2010)

Stable Deep Reinforcement Learning Method by Predicting Uncertainty in Rewards as a Subtask

Kanata Suzuki[1,2] and Tetsuya Ogata[2,3(✉)]

[1] Artificial Intelligence Laboratories, Fujitsu Laboratories Ltd., Kanagawa, Japan
`suzuki.kanata@jp.fujitsu.com`
[2] School of Fundamental Science and Engineering, Waseda University, Tokyo, Japan
`ogata@waseda.jp`
[3] Artificial Intelligence Research Center, Tsukuba, Japan

Abstract. In recent years, a variety of tasks have been accomplished by deep reinforcement learning (DRL). However, when applying DRL to tasks in a real-world environment, designing an appropriate reward is difficult. Rewards obtained via actual hardware sensors may include noise, misinterpretation, or failed observations. The learning instability caused by these unstable signals is a problem that remains to be solved in DRL. In this work, we propose an approach that extends existing DRL models by adding a subtask to directly estimate the variance contained in the reward signal. The model then takes the feature map learned by the subtask in a critic network and sends it to the actor network. This enables stable learning that is robust to the effects of potential noise. The results of experiments in the Atari game domain with unstable reward signals show that our method stabilizes training convergence. We also discuss the extensibility of the model by visualizing feature maps. This approach has the potential to make DRL more practical for use in noisy, real-world scenarios.

Keywords: Deep reinforcement learning · Uncertainty · Variance branch

1 Introduction

Although deep reinforcement learning (DRL) has been shown to have high performance in various fields, some challenges remain regarding the stability of training. In applications such as games [13,19], by designing the score as a reward value, it is possible to obtain a model that obtains a performance comparable to that of a human. However, many DRL studies [11,12] only conducted experiments with simple reward signals designed by the experimenter. There is a gap between these scenarios and real environments, which often have unstable reward signals. This is an essential issue for DRL because its performance is sensitive to reward design. Therefore, a learning method that is robust to noise in the reward signals is needed.

© Springer Nature Switzerland AG 2020
H. Yang et al. (Eds.): ICONIP 2020, LNCS 12533, pp. 651–662, 2020.
https://doi.org/10.1007/978-3-030-63833-7_55

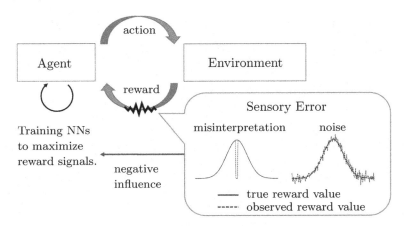

Fig. 1. There are several sources of noise in the reward of DRL in a real-world environment. If noisy signals are used as target signals, they delay convergence in training.

Noise in the DRL reward function can occur for several reasons; a typical example is the errors that occur during observation. In the real-world, reward functions are not perfect. The rewards derived from the actual environment via hardware sensors may include noise, misinterpretations, and observation failures. When misinterpretations or observation failures occur, the reward value may be calculated as an entirely different value. Another case in which noise may occur is the use of feature values as signals for training a neural network. A deep neural network extracts low-dimensional feature vectors from high-dimensional sensor information. Furthermore, we can use extracted features as the target signals of another network. For example, some studies use the feature values of images of target values as a signal to optimize robot behaviors [8,21]. When employing a reward signal to acquire advanced behavior, a variety of noise types not intended by the experimenter will occur (Fig. 1).

Among the types of unstable reward signals listed above, in this study, we focus on fine-grained noise in the signals, which has been referred to as "sensory error" in previous research [5]. In this case, we assume that the reward value is a continuous value rather than a binary signal. These unstable reward signals may inhibit DRL training; for instance, delaying convergence during training. Therefore, a DRL model needs a learning method that considers the uncertainty of the signals and updates its parameters appropriately.

Hence, we propose a stable learning method for DRL with unstable reward signals by directly estimating the variance of the rewards and adjusting the parameter update amount. We incorporate an estimation of the variance of the target signals into the model as a subtask. This makes it possible to extend the original model without significant changes to its configuration. In addition, we use an attention branch network (ABN) [6] structure that incorporates the feature map of the subtasks into the main task. This conveys the learning results of the subtask to the policy network. We evaluated our method on the Atari game domain in the Open AI Gym [3]. To verify the model can stabilize training convergence in an environment where rewards are unstable, we conducted experiments in which we added artificial noise to the reward signals. The results show that our extension to the base model improves performance. The primary contributions of this study are as follows.

1) A model is proposed to stabilize training convergence that incorporates a subtask that estimates the variance in the rewards signals.
2) An evaluation of the performance of models trained with disturbed rewards shows that the proposed approach improves performance.

2 Related Works

Several methods have been proposed to improve the convergence and stability of DRL, and they fall into two main types. One approach optimizes training, whereas the other one reduces variance. The method proposed in this study uses the latter approach.

To increase the convergence and stability of DRL, many optimization methods have been proposed. RMSprop [22] is an approach based on AdaGrad [4], which adjusts the learning rate with respect to the frequency of each parameter update and results in a rapidly decreasing learning rate. Adam [2] is a further improvement on traditional optimization methods and is used in many studies on deep learning. Furthermore, in terms of variance control, methods such as SAG [17], SDCA [18] and stochastic variance reduction (SVR) [9] have also been proposed. SVR-DQN [24] reduces the variance resulting from approximate gradient estimation. These optimizers are essential advances in the convergence and stability of DRL. However, in many studies, experimenters empirically use what is appropriate for the model.

Several previous studies consider the uncertainty of the target signals. The authors of [5] define a Markov decision process in the presence of misinterpretations of the rewards and observation failures and propose a way to deal with rewards that are not correct through sampling. However, the study remains an investigation of table-style rewards and does not consider a continuous control method. There is also a study that addresses the overestimation error in Q-learning. Double DQN uses separate Q-networks for action selection and Q-function value calculation [7]. The model is able to deal with overestimation errors that replace positive bias with negative bias. Another approach is to reduce the target approximation error. An efficient way to reduce this is to use the average of multiple models. The average DQN estimates the current action value using the previously computed K-value [1]. In contrast, an estimator has been proposed to reduce the variance of the reward signal [16]. This model updates the discount value function instead of the sampled rewards.

In this study, we propose a model extension that uses a mechanism to estimate the variance of the reward signals directly. The model estimates the mean and variance of the rewards obtained from the environment, and it can be easily integrated with the base model as a subtask. Using the estimated variance, the model updates its parameters to reduce the effects of noise. To apply the above approach, we adopt an actor-critic type network that predicts the policy and state value using the actor and critic, respectively.

3   Method

To solve the problem described above, we extend the DRL model to solve subtasks that predict the variance of reward signals. Figure 2 shows an overview of the proposed network architecture. The model consists of a base neural network model and an extended branch network. We describe the base model in Sect. 3.1 and the proposed extension in Sect. 3.2.

3.1   Base Model: ABN-A3C

As the base model, we adopt a DRL model that combines the ABN [6] and asynchronous advantage actor critic (A3C) [14]. We choose an actor-critic type network so that we may incorporate a subtask that predicts the variance of the reward signals. Moreover, the ABN enables us to visualize the focus of the subtask using a feature map. The base model, ABN-A3C, consists of a feature extractor that extracts features from the input image, a value branch that outputs state values, and a policy branch that outputs actions. The policy branch also uses the feature map $f(s_t)$ of the value branch as input. Feature map $f(s_t)$ is extracted from the current state s_t, and the value branch outputs the maximum value of feature map $f(s_t)$ using global max pooling. This emphasizes the more distinctive pixels in the feature map of a subtask when it is incorporated with the main task. The details of each model are described below.

A3C: The training of an A3C is stabilized by using various policy searches while running multiple agents in parallel. In asynchronous learning in multiple environments, there is a globally shared parameter and a separate parameter for each thread. Each worker's parameters are copied from the global network's parameters. The parameters learned by agents under different environments are reflected asynchronously in the global network. The gradient exponential moving average of RMSprop, which is used as the optimizer, is also shared globally.

A3C takes advantage of its ability to train online and updates the state value using an estimation of the reward several steps in the future as opposed to a method that estimates the reward in the next step only. As a result, learning is more stable because a more likely estimation error of the current state value is used. The value of adv, which is used to update the estimated value, is calculated by the following equation.

$$adv = \sum_{i=0}^{k-1} \gamma^i r_{t+1} + \gamma^k V(s_{t+k}) - V(s_t) \tag{1}$$

where k indicates how many future steps are used in the prediction. We decided on the prediction step that gave the best results by trying some experimental settings. In our experiment, we set the prediction step $k = 5$.

The A3C trains two models: an actor network, which represents the behavior of the agent, and a critic network, which predicts the expected rewards. An actor network is trained to predict the probability of taking action in a certain state π. The critic network is trained to predict the estimated value of state V. Because the estimated values are independent, they are easy to learn even when the action is continuous.

ABN: An ABN is a model that makes it possible to visualize the areas of focus and improve the accuracy of the network by incorporating feature maps of the subtask into the main task.

In ABN, we compute a new feature map $g'(s_t)$ from the feature map $f(s_t)$ of the value branch and the output of the feature extractor using the following residual mechanism [23]:

$$g'(s_t) = (1 + f(s_t)) * g(s_t) \qquad (2)$$

The state value in the current state is reflected in the action, and the loss of the original feature map is suppressed. The action is predicted by inputting $g'(s_t)$ to the LSTM network of the policy branch. Here, feature map $f(s_t)$ represents the features for optimizing a subtask. By visualizing the feature map overlaid on the input image, it is possible to show where the network is focusing its attention in the input image.

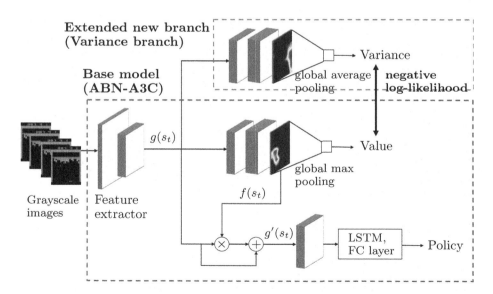

Fig. 2. Overview of the proposed network for predicting the uncertainty of signals as a subtask. The red frame indicates the base model, and the blue frame indicates the proposed extended branch. (Color figure online)

3.2   Variance Branch for Predicting Uncertainty in Rewards

To stabilize the learning convergence, we extend the base model described in the previous section. The aim is to optimize the model's parameters while ignoring the effects of reward noise. Here, the reward signal with noise is assumed to have been generated according to some probability distribution from an unknown generative model. We assume a Gaussian distribution in this study. We use the branching structure of ABN to add a new branch called the variance branch, which takes the feature map as input. The variance branch is similar to a stochastic multi-time scale recurrent neural network (SMTRNN) [15].

The SMTRNN is a type of recurrent neural network that enables the predictive learning of probability distributions based on likelihood maximization. The model extends the conventional learning method of point prediction based on squared-error minimization. It learns to minimize the negative log-likelihood and obtains the stochastic structure that underlies the target signals. To estimate the variance of the prediction of state value $V^\pi(s_t)$, the probability density $p(r_t|s_t, \theta)$ of reward r_t at time step t during an episode is dened as

$$p(r_t|s_t, \theta) = \frac{1}{\sqrt{2\pi\nu_t}} exp\left(-\frac{(V^\pi(s_t) - r_t)^2}{2\nu_t}\right) \tag{3}$$

and the log-likelihood L is defined as

$$L = \prod_{t=1}^{T} p(r_t|s_t, \theta) \tag{4}$$

where θ denotes the model parameters. This process is equivalent to minimizing the weighted prediction error by dividing the output error by predicted variance v. The model learns while ignoring errors in rewards that contain large variance, i.e., large noise. As a result, the training for the state value is stabilized.

In our method, the squared-error calculation of the state value and reward is replaced by the above function. In addition, the configuration of the variance branch is based on the value branch. To smooth the entire feature map of the variance branch, we adopt global average pooling in the final layer. The stabilization of the state value prediction caused by the variance prediction is reflected in the stability of the behavioral prediction in the policy branch. This is a result of the incorporation of the feature-map mechanism of ABN described in the previous section. Visualization of the added branch network is also possible.

4   Experiments

4.1   Model and Environment Setup

To evaluate our method, we used the model to learn to play the Atari games in the Open AI Gym [3]. We used three games: Break Out, Sea Quest, and Pong. As the input image for each game, we used 84×84 grayscale images of four time steps, and for the training, we used RMSprop [22] as the optimizer. Its learning rate is 7×10^{-4}, and the discount rate is 0.99. The number of workers in the A3C is 32. Table 1 lists the parameters of our model. We determined the above parameters by trial-and-error, choosing the parameter set that yielded the best results.

Table 1. Structures of the networks

Network	Dimensions
Feature extractor	conv@16chs - BN - Relu -
	conv@32chs - BN - Relu
Value branch	conv@32chs - BN - Relu -
	conv@64chs - BN - Relu -
	conv@1chs - BN - MaxPooling
Policy branch	Eq. (2) - conv@32chs - BN - Relu -
	LSTM@256 - FC@ActionNum
Variance branch	conv@32chs - BN - Relu -
	conv@64chs - BN - Relu -
	conv@1chs - BN - AvePooling -exp

BN: Barch normalizaion, FC: Fully connect

4.2 Evaluation Metrics

To evaluate the effectiveness of the proposed method, we added artificial noise to the reward signals in the experiments. The noise followed a Gaussian distribution of variance σ^2. In our experiments, we set σ^2 to 0.0, 0.03, and 0.05. When $\sigma^2 = 0.0$, there is no noise in the reward signals, and the noise increases as σ increases. We compared the proposed method with a base model that does not have a mechanism for estimating the variance in the reward signals. We performed experiments in each game environment five times while changing the initial weights of the model.

5 Results and Discussion

In this section, we present the results of experiments in multiple game environments to evaluate the robustness of the proposed method to noise in the reward signals. We also present a feature map of the model and analyze the points of focus in each game. Finally, we discuss the suitability of the proposed method for other deep neural network models.

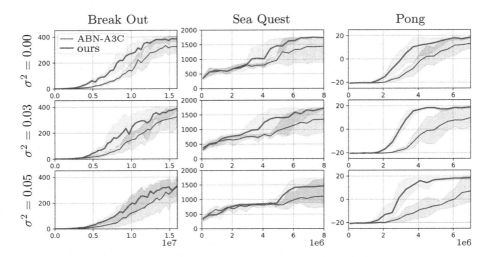

Fig. 3. Change in score during training under each condition (i.e., the type of game and level of reward signal noise). The vertical axis of each figure shows the score, and the horizontal axis shows the total number of worker epochs. Each color area shows the maximum and minimum range of the score. The red lines indicate the results of our method, and the blue lines indicate the results of the ABN-A3C base model. (Color figure online)

5.1 Atari Game Performance

The scores during the training of the proposed and base models for each game are shown in Fig. 3. The results show that the variance in the results increases with the variance in the reward signals. The time to convergence increases because the teaching signal given to the model is not stable. The proposed method converges faster than the ABN-A3C base model, regardless of the size of the variance. However, the maximum score of the proposed method is about the same as that of the base model. These results show that the proposed method predicts the mean of the reward signal and converges to the same results as the base model in less time.

The proposed method converges faster than the base model in all games, regardless of the level of noise in the reward signal. This is also true when there is no noise ($\sigma^2 = 0.0$). Although rewards are given discrete values in the standard games in the Atari game domain, the results suggest that predicting the mean of the rewards is an effective strategy. We think that was because atari's ordinary rewards include uncertainty. For example, in atari games, not all rewards given are valid. Our method may learn to ignore temporary rewards that cannot maximize cumulative rewards.

When the levels of noise are low ($\sigma^2 = 0.03$), the results of the proposed and base model differ the most. The final convergence score of the base model varies depending on the initial weight values, which are randomly chosen. In contrast, the performance of the proposed method does not depend on these values. The proposed method adequately learns the variance in the rewards and stabilizes the training of the policy network. However, when the level of reward noise increases ($\sigma^2 = 0.05$), the results of the proposed method are also worse. When the noise reaches a certain level, the training is substantially disturbed. These results demonstrate the robustness of the proposed DRL method to unstable reward signals. In our experiment, the noise added to the reward was artificially set; hence, the impact of realistic reward noise on learning needs to be considered in future work.

5.2 Visualization of the Feature Map

Next, we visualize the feature map of the proposed model to ensure that the model focuses on the appropriate areas of the feature map. The feature map for each condition superimposed on the input image is shown in Fig. 4. In Break Out, the feature map shows that the model focuses on the movement of the ball. Furthermore, when the number of blocks decreases, the area of attention moves to the blank regions above the blocks (see the results for $\sigma^2 = 0.03$). The variance branch's feature map is similarly active, focusing mainly on areas of significant change on the screen. In Sea Quest, the feature map indicates focus on agents, enemies, and the bars representing the remaining oxygen. There is less movement in the feature map than in Breakout, which may be because this is a game in which the agent employs a waiting strategy. In contrast, in Pong, the feature map is not informative in most cases, even when the obtained scores reach their upper bounds. Reviewing the gameplay after training, we found that the agent repeated a specific pattern of behavior to score points. This may be because Pong itself does not require any complex behavior.

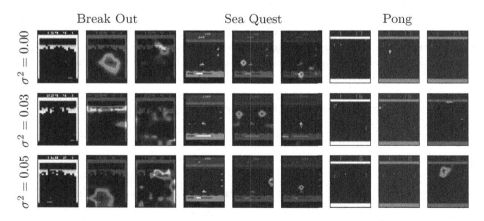

Fig. 4. Examples of visualization of feature maps for each condition. The input image, the feature map of the value branch, and the feature map of the variance branch are shown, respectively. The value of each feature map is higher as it becomes red. (Color figure online)

The above results confirm the effectiveness of our proposed method. The model paid attention to the appropriate areas on the feature map, even in environments with unstable reward signals. Furthermore, the regions of focus of the variance branch feature maps differ from those of the value branch feature maps. In other words, each branch plays a different role in the network.

5.3   Scalability

The proposed method is broadly applicable to many conventional networks because it does not require significant changes to the configuration of the original model. However, because the subtask for predicting variance requires the prediction of state values, the network to be extended should be an actor-critic type network. Because the network is extended using a branch structure, the computational complexity of the network increases; however, parallel computation is possible. Hence, the learning and prediction times should not be much different from those of the original network.

The combined method of variance prediction and feature-map visualization could be used in applications other than DRL. We are investigating an extension to recurrent neural networks for end-to-end robot control [10,20]. Robot control is a particularly promising application because it is often affected by real-world noise.

6   Conclusion

In this study, we proposed a stable reinforcement learning method for scenarios in which the reward signal contains noise. We incorporated a subtask into an actor-critic-based DRL method. The model directly estimates the variance included in the reward obtained from the environment. Moreover, we input the feature map learned by the subtask in the critic network to the actor network. We evaluated our method in the Atari game environment of the Open AI Gym. Our method enables us to stabilize the convergence of learning in an environment in which rewards are unstable. In future work, we plan to extend our method to real robot tasks.

Acknowledgment. This work was supported by JST, ACT-X Grant Number JPM-JAX190I, Japan.

References

1. Anschel, O., Baram, N., Shimkin, N.: Averaged-DQN: variance reduction and stabilization for deep reinforcement learning. In: The International Conference on Machine Learning (2017)
2. Ba, J., Kingma, D.P.: Adam: a method for stochastic optimization. In: Proceedings of International Conference on Learning Representations (2015)
3. Brockman, G., et al.: OpenAI Gym. arXiv preprint arXiv:1606.01540 (2016)

4. Duchi, J., Hazan, E., Singer, Y.: Adaptive subgradient methods for online learning and stochastic optimization. J. Mach. Learn. Res. **12**, 2121–2159 (2011)
5. Everitt, T., Krakovna, V., Orseau, L., Hutter, M., Legg, S.: Reinforcement learning with a corrupted reward channel. In: 26th International Joint Conference on Artificial Intelligence, IJCAI-17, pp. 4705–4713 (2017)
6. Fukui, H., Hirakawa, T., Yamashita, T., Fujiyoshi, H.: Attention branch network: learning of attention mechanism for visual explanation. In: International Conference on Computer Vision and Pattern Recognition (CVPR) (2019)
7. van Hasselt, H., Guez, A., Silver, D.: Deep reinforcement learning with double Q-learning. In: Proceedings of the 13th AAAI Conference on Artificial Intelligence (2016)
8. Jang, E., Devin, C., Vanhoucke, V., Levine, S.: Grasp2Vec: learning object representations from self-supervised grasping. In: Conference on Robot Learning (CoRL) (2018)
9. Johnson, R., Zhang, T.: Accelerating stochastic gradient descent using predictive variance reduction. In: Advances in Neural Information Processing Systems, pp. 315–323 (2013)
10. Kase, K., Suzuki, K., Yang, P.C., Mori, H., Ogata, T.: Put-in-box task generated from multiple discrete tasks by a humanoid robot using deep learning. In: Proceedings of the IEEE International Conference on Robots and Automation (2018)
11. Levine, S., Pastor, P., Krizhevsky, A., Quillen, D.: Learning hand-eye coordination for robotic grasping with deep learning and large-scale data collection. Int. J. Robot. Res. **37**(4–5), 421–436 (2017)
12. Levine, S., Finn, C., Darrell, T., Abbeel, P.: End-to-end training of deep visuomotor policies. J. Mach. Learn. Res. **17**(39), 1–40 (2016)
13. Mnih, V., et al.: Human-level control through deep reinforcement learning. Nature **518**, 529–533 (2015)
14. Mnih, V.: Asynchronous methods for deep reinforcement learning. In: International Conference on Machine Learning (ICML) (2016)
15. Murata, S., Namikawa, J., Arie, H., Sugano, S., Tani, J.: Learning to reproduce fluctuating time series by inferring their time-dependent stochastic properties: application in robot learning via tutoring. IEEE Trans. Auton. Ment. Dev. **5**, 298–310 (2013)
16. Romoff, J., Henderson, P., Piché, A., F-Lavet, V., Pineau, J.: Reward Estimation for Variance Reduction in Deep Reinforcement Learning. arXiv preprint arXiv:1805.03359 (2018)
17. Roux, N.L., Schmidt, M., Bach, F.: A stochastic gradient method with an exponential convergence rate for finite training sets. Adv. Neural Inf. Process. Syst. **25**, 2663–2671 (2012)
18. Shwars, S.S., Zhang, T.: Stochastic dual coordinate ascent methods for regularized loss minimization. J. Mach. Learn. Res. **14**, 567–599 (2013)
19. Silver, D., et al.: Mastering the game of Go with deep neural networks and tree search. Nature **529**, 484–489 (2016)
20. Suzuki, K., Mori, H., Ogata, T.: Motion switching with sensory and instruction signals by designing dynamical systems using deep neural network. IEEE Robot. Autom. Lett. **3**(4), 3481–3488 (2018)
21. Suzuki, K., Yokota, Y., Kanazawa, Y., Takebayashi, T.: Online self-supervised learning for object picking: detecting optimum grasping position using a metric learning approach. In: Proceedings of International Symposium on System Integrations (2020)

22. Tieleman, T., Hinton, G.: Lecture 6.5-rmsprop: divide the gradient by a running average of its recent magnitude. COURSERA: Neural Netw. Mach. Learn. **4**(2), 26–30 (2012)
23. Wang, F., et al.: Residual attention network for image classification. In: Conference on Computer Vision and Pattern Recognition (2017)
24. Zhao, W.Y., Guan, X.Y., Liu, Y., Zhao, X., Peng, J.: Stochastic Variance Reduction for Deep Q-learning. arXiv preprint arXiv:1905.08152 (2019)

STGA-LSTM: A Spatial-Temporal Graph Attentional LSTM Scheme for Multi-agent Cooperation

Huimu Wang[1,2], Zhen Liu[2(✉)], Zhiqiang Pu[1,2], and Jianqiang Yi[1,2]

[1] University of Chinese Academy of Sciences, Beijing 100049, China
[2] Institute of Automation, Chinese Academy of Sciences, Beijing 100190, China
{wanghuimu2018,liuzhen,zhiqiang.pu,jianqiang.yi}@ia.ac.cn

Abstract. Multi-agent cooperation is one of the attractive aspects in multi-agent systems. However, during the process of cooperation, communication among agents is limited by the distance or the bandwidth. Besides, the agents move around and their neighbors appear or vanish, which makes the agents hard to capture temporal dependences and to learn a stable policy. To address these issues, a Spatial-Temporal Graph Attentional Long Short-Term Memory (LSTM) Scheme (STGA-LSTM), which is composed of spatial capture network and spatiotemporal LSTM network, is proposed. The spatial capture network is designed based on graph attention network to enlarge the agents' communication range and capture the spatial structure of the multi-agent system. Based on the standard LSTM, a spatiotemporal LSTM network, which is in combination with graph convolutional network and attention mechanism, is designed to capture the temporal evolutionary patterns while keeping the spatial structure learned by spatial capture network. The results of simulations including mixed cooperative and competitive tasks indicate that the agents can learn stable and complicated strategies with STGA-LSTM.

Keywords: Multi-agent systems · Graph attention mechanism · LSTM

1 Introduction

The cooperation in a multi-agent system has shown great success in various fields, such as smart grid control [15], resource management [11]. To control such complex systems composed of many interacting components, researchers have studied multi-agent reinforcement learning (MARL) for a long time.

Recently, the advances achieved by deep reinforcement learning (DRL) [12] promote the combination between DRL and MARL to solve complex problems.

Supported by the National Key Research and Development Program of China under Grant 2018AAA0102402, and Innovation Academy for Light-duty Gas Turbine, Chinese Academy of Sciences, No. CXYJJ19-ZD-02.

H. Yang et al. (Eds.): ICONIP 2020, LNCS 12533, pp. 663–675, 2020.
https://doi.org/10.1007/978-3-030-63833-7_56

Nevertheless, when these algorithms are applied to realistic environments, there still exist several issues. First, the large number of agents results in the curse of dimensionality and the difficulty of learning a stable policy. Second, the information obtained from other agents is limited by the bandwidth and range of the communication, which affects the agents' cooperative behavior. Finally, the communication status of the agents keeps changing over time, which makes it difficult for the agents to learn dynamic strategies to adapt to this change.

Although a variety of MARL algorithms have been proposed to solve the issues above, they still suffer from different limitations. Some MARL algorithms [4,12] follow a common paradigm of centralized learning with decentralized execution to promote the cooperative behavior among the agents. These algorithms suffer from the difficulty of transferability and scalability because they directly use the state or observation in constructing critic or actor networks. The mean-field approach [18] is proposed to address the problem of scalability. However, it ignores the fact that different agent's observation has different influences. To deal with the limitation, the algorithms based on attention mechanisms [3,7,9] are proposed. They can effectively extract valuable information via the communication control. However, they are still limited by the communication bandwidth and ignore the multi-agent system's underlying structure. Considering the structure of the multi-agent system, the algorithms based on graph network [1,8,13] are proposed. Nevertheless, they do not consider the time-varying topology of the multi-systems' graph, which makes them difficult to acquire satisfying performance in the environments with dynamic graph structure.

To address the limitations mentioned above, a Spatial-Temporal Graph Attentional LSTM scheme (STGA-LSTM) is proposed. The model can be divided into two parts including a spatial capture network and a spatiotemporal LSTM network. The spatial capture network mainly focuses on learning the spatial structure among the agents and obtaining more agents' information. It is designed based on graph attention networks (GAT) [17] to capture the spatial structure and relationship among the agents and enlarge the agents' receptive field or communication field through the chain propagation characteristics of graph neural networks. The spatiotemporal LSTM network mainly focuses on the temporal dependency of dynamic graph of the multi-agent system. Based on standard LSTM, the spatiotemporal LSTM network combines with graph convolutional network and attention mechanism to overcome the limitation of ignoring spatial correlation caused by fully-connected operator within the standard LSTM. With the spatiotemporal LSTM network, the agent's features and interactions can be captured in spatial configuration and temporal evolvement.

To verify the ability of STGA-LSTM, it is evaluated in different environments including formation control, 3v1 and 5v2 predator-prey games. The simulation results demonstrate that the agents can learn complicated cooperative strategies in mixed cooperative and competitive tasks.

2   Related Works

2.1   Multi-agent Reinforcement Learning

Multi-Agent Deep Deterministic Policy Gradient (MADDPG) [12], which follows a common paradigm: centralized learning with decentralized execution, is proposed for mixed multi-agent cooperative-competitive environments. Counterfactual multi-agent (COMA) [4] also utilizes a centralized critic and computes a counterfactual advantage function which handles the problem of multi-agent credit assignment by marginalizing the effect of each agent's action. However, the centralized critic takes the observations and actions of all agents as input, which makes the algorithms more difficult to apply in large-scale environments. To better adapt to the environment with a large number of agents, the mean-field approach [18] is used to capture the interaction of agents by mean action. However, it ignores the fact that different agent's observation has different influences. Besides, [7] and [9] enable agents to obtain information effectively via attention mechanism. Nevertheless, they ignore the underlying structure of multi-agent systems and concatenate simply other agents' states and various features of the environment.

2.2   Graph Convolution Network (GCN)

Recently, graph-based methods have drawn much attention in many important real-world applications, such as social networks [10], action-recognition [16], and transportation forecasting [2], due to the effective representation of graph structure data. Graph convolution network (GCN) is a framework proposed to extract locally connected features from arbitrary graphs. Using GCN, interaction networks can reason the objects, relations and connection in complex systems, which has been proven difficult for CNNs. Earlier works such as [14,17] focus on static graph and are not designed to model temporal evolution patterns in dynamic graphs. To adapt to dynamic graphs, several methods in different areas, such as traffic forecasting [2] and human trajectory prediction [6] are proposed. But there are few interaction frameworks that have been proposed to address the dynamic graph structure in multi-agent systems. The existing methods based on the graph structure of the multi-agent systems is MAGnet [13], TRANS-FER [1] and DGN [8]. But the former two algorithms regard the graph structure as static, which is not available in realistic environments. Although DGN takes the dynamic graph into consideration, it doesn't take the communication limited by distance and bandwidth into consideration. Inspired by the graph convolutional LSTM, which is an extension of GCNs to have a recurrent architecture, STGA-LSTM is proposed to learn inherent spatiotemporal representations from the dynamic graph structure.

3   Preliminaries

3.1   Problem Definition

Let o_i^t denote the local observation of agent i including its position, velocity. There are N agents and M obstacles in this environment. We assume that at time t, the position of agent i is $p_i^t = [p_i^{t^x}, p_i^{t^y}]$, the velocity of agent i is $v_i^t = [v_i^{t^x}, v_i^{t^y}]$, the formation center position is $p_c^t = [p_c^{t^x}, p_c^{t^y}]$ and the position of obstacle j is $p_{oj}^t = [p_{oj}^{t^x}, p_{oj}^{t^y}]$. Besides, the action space for each agent is discrete. Each agent can move one step in both X and Y directions.

The connected status among the agents can be represented in an undirected graph $G = (V, E)$. Specifically, $V = \{1, \ldots, N\}$ denotes the nodes consisting of the agents. $E \subseteq V \times V$ denotes the edge set consisting of communication status among the agents where an edge from node i to node j is denoted as $(i, j) \in E$. Besides, h is a set of node features, $h = \left\{ \overrightarrow{h}_1, \overrightarrow{h}_2, \ldots, \overrightarrow{h}_N \right\}, \overrightarrow{h}_i \in \mathbb{R}^F$, where F is the number of features in each node. Moreover, N_i is a set of neighbours communicating with node i in the graph. Only when the distance between agent i and agent j is less than c, agent j belongs to the set of neighbours N_i. As indicated by (1), there is an adjacency matrix A where $a_{ij} = 1$ if $j \in N_i$ otherwise $a_{ij} = 0$. Besides, the cooperative behavior is decided not only by its neighbourhoods' information but also by its own information. Therefore, there is a self-loop for each agent.

$$a_{ij} = \begin{cases} 1 & if \ dist\,(a_i, a_j) \leq c \ or \ i = j \\ 0 & if \ dist\,(a_i, a_j) > c \end{cases} \tag{1}$$

where $dist$ is a 2-dimensional Euclidean norm to calculate the distance between agent i and agent j, and c represents the predefined communication threshold.

3.2   Partially Observable Markov Games

The environment in this paper is regarded as partially observable Markov Games which is an extension of the framework of Markov Games. It is defined by a global state S, a set of actions $A_1, \cdots, A_N$, and a set of local observations $O_1, \ldots, O_N$. To choose actions, each agent uses a learnable policy $\pi_i : O_i \to P_a(A_i)$, which produces the next state according to the state transition function $T : S \times A_1 \times \cdots \times A_N \to P_t(S')$ that defines the probability distribution over possible next states, given current states and actions for each agent. Each agent obtains rewards R_i from the environment after all agents take actions: $S \times A_1 \times \ldots \times A_N \to R$. The agents aim to learn a policy that maximizes their expected discounted returns,

$$J_i(\pi_i) = E_{a_1 \sim \pi_1, \ldots, a_N \sim \pi_N, s \sim T} \left[\sum\nolimits_{t=0}^{\infty} \gamma^t r_{it}\,(s_t, a_{1t}, \ldots, a_{Nt}) \right] \tag{2}$$

where r_{it} is the reward that agent i obtains at time t, s_t represents the global state S at time t. $\gamma \in [0, 1]$ is the discount factor that determines how much the policy favors immediate reward over long-term gain.

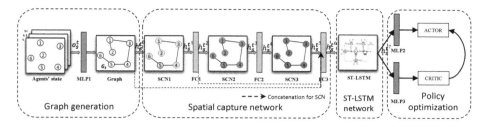

Fig. 1. STGA-LSTM scheme

4   Method

In this section, the Spatial-Temporal Graph Attentional LSTM scheme (STGA-LSTM) is designed to promote the multi-agent behavior under restricted and time-varying communication. STGA-LSTM is composed of a spatial capture network and a spatiotemporal LSTM network. As shown in Fig. 1, the historical time series data are treated as input. The inputs are embedded as graphs via Multi-Layer Perceptron (MLP). Then the spatial capture network captures the spatial structure and latent representation of the multi-agent systems and enlarges the agents' communication field by leveraging the chain propagation characteristics of GCN. The latent representation learned from the spatial structure network is feed as input to the spatiotemporal LSTM network, which captures the spatiotemporal feature and the temporal evolution of the graph. Finally, the captured node-embedding vector for each agent is subsequently used to evaluate the critic and update the actor network.

4.1   Spatial Capture Network

Usually, each agent should require all the other agents' information about their observations and actions to behave cooperatively better. However, it is not available for each agent to obtain information of all the other agents due to the limitation of communication. With the increase of the agent number, the influence caused by the limitation of communication becomes more serious.

Considering the fact that the communication status among agents can be represented as a graph naturally, the chain propagation characteristics of the graph convolution layers can be adopted to enlarge the communication range of the agents. Moreover, the attention mechanism can assign different importance to different agents. Therefore, the spatial capture network (SCN) based on graph attention network [17] is designed. It can not only enlarge the communication range of the agents by leveraging the graph convolution layers, but also handle the complex relationship to promote the cooperative behavior by utilizing attention mechanism.

We define a graph $G = (V, E)$, where each node $i \in V$ denotes an agent, and there exists an edge $e_{ij} \in E$ between agent i and agent j if they can communicate with each other. As indicated by (1), only when the distance between agent i

and agent j is less than the communication threshold c, they can communicate with each other.

For the enlarging of the communication field, SCN utilizes the chain propagation characteristics of the graph convolution. Multiple SCN layers are stacked to enlarge the agents' receptive field. In addition to enlarging the receptive fields of agent i and capturing the spatial structure, the neighbour agents need to be treated differently by agent i for promoting cooperation. Different neighbour agents have different influences on agent i. Specifically, one of the neighbour agents may be farther away from agent i than the other agents, which means agent i will be influenced differently due to the different distance among the agents. With the increase of the agent number, the relationship among the agents become more complex. Therefore, the attention mechanism in SCN layers is used to allow agent i to treat the other agents' states differently. Following an attention strategy, SCN operates on graph-structured data and computes the features of each graph node by attending over its neighbors. The hidden states of the agents are used to calculate the attention coefficients e_{ij} from agent j to agent i and its normalized form α_{ij} :

$$e_{ij} = a_G^k \left(W_G^k h_i, W_G^k h_j \right) \tag{3}$$

$$\alpha_{ij} = \text{softmax}(e_{ij}) = \frac{\exp\left(\text{LeakyReLU}\left(e_{ij}\right)\right)}{\sum_{k \in N_i} \exp\left(\text{LeakyReLU}\left(e_{ij}\right)\right)} \tag{4}$$

where a_G^k is a single-layer feedforward neural network, W_G^k is a learnable weight matrix and $LeakyReLU$ is a nonlinear activation function. [17] indicates that the import of multi-head attention is beneficial to stabilize the learning process of the attention. Moreover, each agent can extract different state representation of the nearby agents from different representation subspace with multi-head setting. Therefore, the multi-head is adopted in SCN. The output of one SCN layer with multi-head attention for node i at t is given by:

$$h_i^{t'} = \big\|_{m=1}^{K} \sigma \left(\sum_{j \in N_i} \alpha_{ij}^m W_G^m h_i^t \right) \tag{5}$$

where $\|$ represents the concatenation, K represents the number of the heads, α_{ij}^m represents the normalized attention coefficient of the m-th attention mechanism and W_G^m represents the weight matrix of the m-th linear transformation.

Furthermore, after the final SCN layer SCN3, the hidden states are concatenated, and they are fed into the fully-connected layer 3 (FC3) as shown in Fig. 1. Since the hidden state could disappear during the process of graph convolution, these hidden states are concatenated in the final layer to stabilize the training process.

$$h_s^{t^6} = \sigma \left(\left[h_s^{t^5} \big\| h_s^{t^2} \big\| h_s^{t^0} \right] W_F^3 + b_F^3 \right) \tag{6}$$

where $\|$ represents the concatenation, W_F^3 and b_F^3 are weight matrix and bias of FC3 to be learned.

4.2   Spatiotemporal LSTM Network

In the practical engineering, the location of the agents and the communication status among the agents keep changing over time, which means that the graph G formed by the agents is dynamic and evolves over time. Due to the complex time-varying graph structures, the dynamic state representation of the agents is difficult to learn.

To address the temporal sequence problem, common approach is to use a recurrent network. There are many studies which have demonstrated that LSTM [5], as a variant of RNN, has an amazing ability to model long-term temporal dependencies. However, direct use of LSTM does not help the agents to learn the dynamic state representations. The fully connected operator within LSTM ignores the spatial structure of the multi-agent systems, which causes that the spatial structure learned by SCN is useless. Considering the GAT mentioned in Sect. 4.1 has the ability to learn the structure and the traditional LSTM can handle temporal dependencies, the spatiotemporal LSTM network (ST-LSTM) is designed by combining GAT and traditional LSTM. Compared with standard LSTM, ST-LSTM can not only capture discriminative features in spatial configuration and temporal evolution, but also explore the different influences from different agents (Fig. 2).

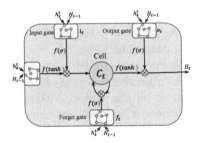

Fig. 2. ST-LSTM cell

ST-LSTM contains three gates: an input gate i_t, a forget gate f_t and an output gate o_t. These gates are obtained with the graph attention operator. The input $\mathbf{h}_t$, hidden state $\mathbf{H}_t$ and cell memory $\mathbf{C}_t$ of ST-LSTM are graph-structured data. The details of the units and the structure of ST-LSTM are illustrated in Fig. 3. Due to the graph attention operator within ST-LSTM, the cell memory $\mathbf{C}_t$ and hidden state $\mathbf{H}_t$ are able to exhibit temporal dynamics, as well as contain spatial structural information. Moreover, the graph attention operator can make the agents adaptively focus on the neighbours, which means they can obtain more effective state representations in addition to the spatial structural and temporal information. The functions of ST-LSTM cell are defined as follows:

$$
\begin{aligned}
i_t &= \sigma \left(W_{xi} *_G h_t + W_{hi} *_G H_{t-1} + b_i \right) \\
f_t &= \sigma \left(W_{xf} *_G h_t + W_{hf} *_G H_{t-1} + b_f \right) \\
o_t &= \sigma \left(W_{xo} *_G h_t + W_{ho} *_G H_{t-1} + b_o \right) \\
u_t &= \tanh \left(W_{xc} *_G h_t + W_{hc} *_G H_{t-1} + b_c \right) \\
C_t &= f_t \odot C_{t-1} + i_t \odot u_t \\
H_t &= o_t \odot \tanh(C_t)
\end{aligned}
\tag{7}
$$

where $*_G$ denotes the graph convolution operator and $\odot$ denotes the Hadamard product. $\sigma\,(\cdot)$ is the sigmoid activation function. u_t is the modulated input. H_t is an intermediate hidden state. $W_{xi} *_G X_t$ denotes a graph convolution of X_t with W_{xi}, which can be written as (5)–(7).

It is worth noting that only one graph convolution layer for the graph attention operator is used, because the function of the graph attention operator within ST-LSTM is to capture and keep the spatial structure instead of enlarging the agents receptive field. With the increase of the convolution layers, the cost of computing greatly increases and the state representation is difficult to be learned. Therefore, the number of the graph convolution layer for the graph attention operator is 1.

4.3 Policy Optimization

After the states are extracted by STGA-LSTM, they are utilized to optimize the policy of the agents. As shown in Fig. 1, all the agents' information is extracted with STGA-LSTM as $h_s^{t^7}$. As mentioned above, STGA-LSTM can let the agents obtain more agents' information. Therefore, $h_s^{t^7}$ is a function related to all the other agents' states. After $h_s^{t^7}$ is obtained, PPO is implemented in an actor-critic framework. According to the objective function of PPO, it is changed as (8) and (9) after the concatenation of all the states. Although the agents are trained with information from their nearby agents, they can obtain all the other agents' information by STGA-LSTM. Owing to STGA-LSTM, each agent can obtain more agents' information to promote the cooperation behaviour. Moreover, to scale up to more agents, the parameters sharing method is applied to train all the agents in a decentralized framework.

$$
l_t(\theta) = \frac{\pi_\theta(a_t \,|\, (h_1, h_2, \cdots, h_N))}{\pi_{\theta^k}(a_t \,|\, (h_1, h_2, \cdots, h_N))}
\tag{8}
$$

$$
\begin{aligned}
L(\theta) = E[\min(l_t(\theta) \hat{A}_t^{\theta^k}(h_1, h_2, \cdots, h_N), \\
clip(l_t(\theta), 1 - \varepsilon, 1 + \varepsilon) \hat{A}_t^{\theta^k}(h_1, h_2, \cdots, h_N)]
\end{aligned}
\tag{9}
$$

5   Simulations

In this section, the performance of STGA-LSTM is evaluated in four different scenarios as shown in Fig. 3. Scenario (a) and (b) focusing on the formation control tasks are designed to evaluate the effectiveness of STGA-LSTM. Moreover,

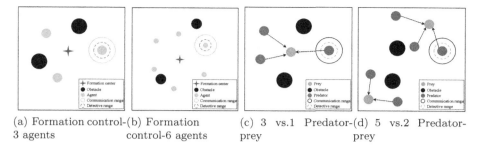

(a) Formation control-(b) Formation
3 agents   control-6 agents

(c) 3 vs.1 Predator-(d) 5 vs.2 Predator-
prey   prey

Fig. 3. The illustration of the simulation environments

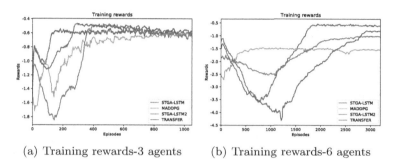

(a) Training rewards-3 agents   (b) Training rewards-6 agents

Fig. 4. The training rewards in formation control

scenario (c) and (d) focusing on the predator-prey are designed to evaluate the
decision ability of STGA-LSTM in complex environments. In the simulations,
STGA-LSTM is compared with MADDPG [12], TRANSFER [1] and STGAT-
LSTM2. The first algorithm MADDPG relies on access to the global state of
the system during training instead of the partial observation state and the communication among the agents. The second algorithm TRANSFER ignores the
temporal relationship among the agents. The final algorithm STGAT-LSTM2 is
another vision of STGA-LSTM with standard LSTM.

5.1   Formation Control

STGA-LSTM are compared with those algorithms in two different formation
control environments which include scenario (a) with 3 agents and scenario (b)
with 6 agents and larger environment size. In these scenarios, all the agents are
required to be evenly distributed around the formation center without colliding
with each other. The reward for each agent is composed of the distance reward
and the collision reward. Specifically, the distance reward is related to the distance from the agent to the formation center. Besides, if an agent collides with
another agent, the reward it obtains is −10. All the agents only observe the
formation center location and their own state. The only way to obtain the other

Table 1. Evaluating results of formation control with 3 agents

Algorithms	Success (%)	Steps	Rewards	Collision (%)
MADDPG	100	18.62	−0.73	1.25
TRANSFER	100	12.75	−0.58	0
STGA-LSTM2	100	10.68	−0.61	0.86
STGA-LSTM	**100**	**9.22**	**−0.47**	**0**

agents' states is through communication. Given the limitation of communication in reality, each agent communicates with up to two nearest neighboring agents only if their distance is less than the pre-defined threshold.

The learning curves of all the approaches in terms of mean rewards are presented in Fig. 4. In Fig. 4(a), STGA-LSTM has a similar performance with the other algorithms but has a higher convergence rate. Moreover, as shown in Fig. 4(b), STGA-LSTM obtains higher mean reward and converges faster than the other algorithms. In scenario (a), the communication status among the agents is almost fully connected, which means that the spatial graph structure of the multi-agent system can be regarded as static. Under this situation, the ability of handling the dynamic graph of STGA-LSTM is not fully reflected. On the contrast, the graph structure changes more frequently in scenario (b). Owing to the proposed scheme, the agents not only adapt to the dynamic communication status, but also enlarge their receptive field based on the chain propagation characteristics of GAT. Moreover, STGA-LSTM2 performs worse than STGA-LSTM, which means that the traditional LSTM cannot process the spatial-temporal data and even prevent the cooperative behavior because its fully connected operator disrupts the spatial structure learned by GAT.

In addition to the data of the training process, the evaluation results in Table 1 and 2 present the similar results with Fig. 4. The agents trained by STGA-LSTM have higher rewards and efficiency than MADDPG and TRANS-FER in scenario (b). The performance difference between STGA-LSTM and other algorithms demonstrates that STGA-LSTM can capture both the spatial interactions and temporal evolution of the multi-agent systems.

5.2   Predator-Prey Games

Two scenarios including 3v1 and 5v2 predator-prey games are designed to evaluate the performance of our scheme. In these scenarios, the predator moves slower and needs to capture the prey, and the prey moves faster and needs to escape from the predators. For the predators, they need to learn how to cooperate with each other. The predators can only observe their own location and velocity, and can obtain the prey's location if any prey is in their detective field. On the contrary, the preys can observe all the predators' location and velocity. If a prey is caught by one predator, the prey will obtain −10 reward. The predator can receive +20 reward when all the preys are caught. Each agent aims to maxi-

Table 2. Evaluating results of formation control with 6 agents

Algorithms	Success (%)	Steps	Rewards	Collision (%)
MADDPG	0	60	−1.52	25.80
TRANSFER	98.70	13.52	−0.63	2.60
STGA-LSTM2	97.64	12.46	−0.85	9.70
STGA-LSTM	**100**	**10.86**	**−0.42**	**0**

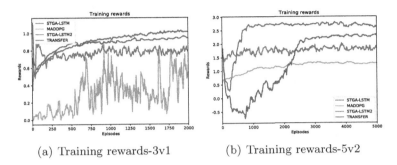

(a) Training rewards-3v1   (b) Training rewards-5v2

Fig. 5. The training rewards in predator-prey games

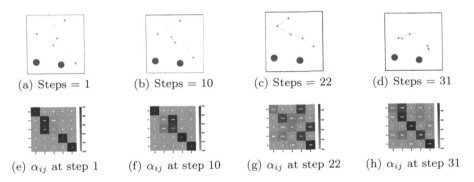

(a) Steps = 1   (b) Steps = 10   (c) Steps = 22   (d) Steps = 31

(e) α_{ij} at step 1   (f) α_{ij} at step 10   (g) α_{ij} at step 22   (h) α_{ij} at step 31

Fig. 6. The illustration of the cooperative strategy in 5V2 predator-prey games

mize their accumulated rewards, which means there is a competition between the predators and the preys, and a cooperation among the predators.

As shown in Fig. 5 and Tables 3–4, STGA-LSTM outperforms all the baselines during the training and the evaluation process. Especially in the scenario (d), SGTA-LSTM converges twice as fast as TRANSFER and obtains at least 30% rewards higher than the other methods.

To further describe our strategy, the dynamic evolution of cooperative behavior is shown in Fig. 6. In Fig. 6, five predators learn to divide themselves into two groups to chase two preys. Each predator needs to interact with the other predators to obtain the information about the preys under restricted and time-varying

Table 3. Evaluating results of 3v1 predator-prey games

Algorithms	Success (%)	Steps	Rewards	Collision (%)
MADDPG	84.2	35.26	0.54	13.60
TRANSFER	95.6	22.37	0.95	0
STGA-LSTM2	92.7	23.63	0.86	0.80
STGA-LSTM	**96.7**	**22.13**	**1.12**	**0**

Table 4. Evaluating results of 5v2 predator-prey games

Algorithms	Success (%)	Steps	Rewards	Collision (%)
MADDPG	76.3	44.18	1.25	10.24
TRANSFER	96.20	36.25	2.18	1.2
STGA-LSTM2	95.64	38.6	1.72	2.3
STGA-LSTM	**98.6**	**32.97**	**2.76**	**0.8**

communication. The final results show that the predators have learned a reasonable cooperative strategy through STGA-LSTM. Moreover, the attention value α_{ij} in (6) for different predators can be obtained in Fig. 6. For predator 1, α_{11} is 0.91 at step 1, which indicates that it mainly focuses on its own state. As the game progresses, predator 1 gets close to predator 2, and α_{12} increases from 0.09 to 0.88 at step 10, which means the communication with predator 2 becomes more important. Then predator 1 moves away from predator 2 and transfers its attention into the prey and predator 3 at step 22. α_{12} decreases from 0.88 to 0.21 and α_{13} increases from 0 to 0.16. Finally, it can be seen that α_{11} increases from 0.63 to 0.82, α_{10} increases from 0 to 0.05, and α_{13} decreases from 0.16 to 0.13, which indicates that the predator 1 pays more attention on itself, and cooperates with predator 0 and 3 to capture the prey. It can be concluded that STGA-LSTM can enhance the agents' cooperative ability through capturing the spatial structure and the temporal evolution.

6   Conclusions

In this paper, we present a novel STGA-LSTM scheme for multi-agent cooperation under restricted and time-varying topology. STGA-LSTM not only enlarges the agents' communication range, but also captures the temporal evolution while keeping the spatial structure. The scheme is shown to perform a satisfying strategy and adapt to the dynamic graph structure of the multi-agent system. Future work will take the time-delay phenomenon into consideration and conduct it in unmanned ground vehicles.

References

1. Agarwal, A., Kumar, S., Sycara, K.: Learning transferable cooperative behavior in multi-agent teams. arXiv preprint arXiv:1906.01202 (2019)
2. Cui, Z., Henrickson, K., Ke, R., Wang, Y.: Traffic graph convolutional recurrent neural network: a deep learning framework for network-scale traffic learning and forecasting. IEEE Trans. Intell. Transp. Syst. **21**, 4883–4894 (2019)
3. Das, A., et al.: TarMAC: targeted multi-agent communication. In: International Conference on Machine Learning, pp. 1538–1546 (2019)
4. Foerster, J.N., Farquhar, G., Afouras, T., Nardelli, N., Whiteson, S.: Counterfactual multi-agent policy gradients. In: 32nd AAAI Conference on Artificial Intelligence (2018)
5. Hochreiter, S., Schmidhuber, J.: Long short-term memory. Neural Comput. **9**(8), 1735–1780 (1997)
6. Huang, Y., Bi, H., Li, Z., Mao, T., Wang, Z.: STGAT: modeling spatial-temporal interactions for human trajectory prediction. In: Proceedings of the IEEE International Conference on Computer Vision, pp. 6272–6281 (2019)
7. Iqbal, S., Sha, F.: Actor-attention-critic for multi-agent reinforcement learning. In: International Conference on Machine Learning, pp. 2961–2970 (2019)
8. Jiang, J., Dun, C., Lu, Z.: Graph convolutional reinforcement learning for multi-agent cooperation. arXiv preprint arXiv:1810.09202 (2018)
9. Jiang, J., Lu, Z.: Learning attentional communication for multi-agent cooperation. In: Advances in Neural Information Processing Systems, pp. 7254–7264 (2018)
10. Kipf, T.N., Welling, M.: Semi-supervised classification with graph convolutional networks. arXiv preprint arXiv:1609.02907 (2016)
11. Li, X., Zhang, J., Bian, J., Tong, Y., Liu, T.Y.: A cooperative multi-agent reinforcement learning framework for resource balancing in complex logistics network. In: Proceedings of the 18th International Conference on Autonomous Agents and MultiAgent Systems, pp. 980–988. International Foundation for Autonomous Agents and Multiagent Systems (2019)
12. Lowe, R., Wu, Y., Tamar, A., Harb, J., Abbeel, O.P., Mordatch, I.: Multi-agent actor-critic for mixed cooperative-competitive environments. In: Advances in Neural Information Processing Systems, pp. 6379–6390 (2017)
13. Malysheva, A., Sung, T.T., Sohn, C.B., Kudenko, D., Shpilman, A.: Deep multi-agent reinforcement learning with relevance graphs. arXiv preprint arXiv:1811.12557 (2018)
14. Niepert, M., Ahmed, M., Kutzkov, K.: Learning convolutional neural networks for graphs. In: International Conference on Machine Learning, pp. 2014–2023 (2016)
15. Radhakrishnan, B.M., Srinivasan, D.: A multi-agent based distributed energy management scheme for smart grid applications. Energy **103**, 192–204 (2016)
16. Shi, L., Zhang, Y., Cheng, J., Lu, H.: Two-stream adaptive graph convolutional networks for skeleton-based action recognition. In: Proceedings of the IEEE Conference on Computer Vision and Pattern Recognition, pp. 12026–12035 (2019)
17. Veličković, P., Cucurull, G., Casanova, A., Romero, A., Liò, P., Bengio, Y.: Graph attention networks. In: International Conference on Learning Representations (2018). https://openreview.net/forum?id=rJXMpikCZ
18. Yang, Y., Luo, R., Li, M., Zhou, M., Zhang, W., Wang, J.: Mean field multi-agent reinforcement learning. In: International Conference on Machine Learning, pp. 5567–5576 (2018)

SuperConv: Strengthening the Convolution Kernel via Weight Sharing

Chuan Liu[1], Qing Ye[1], Xiaoming Huang[2], and Jiancheng Lv[1,2]($\boxtimes$)

[1] Sichuan University, Chengdu 610065, People's Republic of China
lvjiancheng@scu.edu.cn
[2] CETC Cyberspace Security Research Institute Co., Ltd., Chengdu 610041, China

Abstract. For the current neural network models, in order to improve the accuracy of the models, we need efficient plug-and-play modules. Therefore, many efficient plug-and-play operations are proposed, such as Asymmetric Convolution Block (ACB). However, the introduction of multi-branch convolution kernels in ACB increases the trainable parameters, which is an extra burden to the training of large models. In this work, SuperConv is proposed to reduce the trainable parameters while maintaining the advantages of ACB. SuperConv utilizes the method in single-path NAS to encode the convolution kernels of different sizes in multiple branches into a super-kernel, so that the convolution kernels can share some weights with each other. In addition, we introduce Super-Conv into MixConv and propose SuperMixConv (SP-MixConv). To verify the effectiveness of SP-MixConv, ACB, MixConv and SP-MixConv are inserted into the Cifar-quick model and the model with SP-MixConv gets the best accuracy on CIFAR-10 and CIFAR-100. And SuperConv and SP-MixConv will not add extra burden in inference. Simultaneously, SuperConv is very easy to implement, using existing tools such as Pytorch, and is also an interesting attempt for the design of efficient plug-and-play convolution block.

Keywords: Weight sharing · Strengthening the kernel skeletons · Super-MixConv

1 Introduction

Deep learning has made great progress and has been widely used in computer vision and natural language processing [30,31]. But with the increasing of various tasks, the design of network architecture becomes a more difficult problem. Although model architecture search has made great progress, it still needs a lot of guidance from model design experience. For example, the gradient-based

This work is supported by the Key Program of National Science Foundation of China (Grant No. 61836006) and partially supported by National Natural Science Fund for Distinguished Young Scholar (Grant No. 61625204).

© Springer Nature Switzerland AG 2020
H. Yang et al. (Eds.): ICONIP 2020, LNCS 12533, pp. 676–687, 2020.
https://doi.org/10.1007/978-3-030-63833-7_57

approaches [12,14–18], still impose some prior restrictions on the search space. Practice shows that the performance of the model is affected by many factors, such as the architecture of the model and the local receptive fields and so on. ACNet [3] utilizes BN fusion and Branch fusion mechanisms to fuse convolution kernels of different sizes. And it uses 13 and 31 non-square convolution kernels to enhance the skeletons of the 33 convolution kernel. Experiments [3] show that ACNet can improve the performance of the models and provide ideas for the design of the model. While ACNet improves the performance of the model, it also increases the trainable parameters for the model.

Single-Path NAS [2] proposes a single path model architecture search method by means of the small convolution kernel and the large convolution kernel sharing some parameters. This method is faster and has fewer trainable parameters than the multi-path model architecture search method. This fine-grained method of weight sharing can be used in many ways. In this paper, we combine it with Asymmetric Convolution Block (ACB) [3] and propose SuperConv. SuperConv utilizes weight sharing and multiple branches to enhance the skeletons of the square convolution kernel.

MixConv [4] is a very clever technique for reducing the parameters in large convolutional kernel networks and achieving an effective balance between efficiency and performance. MixConv split the input and output channels into different groups, each using different size kernel. MixConv then concats the results of the branches to get the final result. However, the results of MixConv may depend on the number of groups and how channels are partitioned. Therefore, we try to introduce SuperConv to enhance MixConv. Our experiments show that SP-MixConv improves the accuracy of the model and increases limited trainable parameters.

In this paper, we propose SuperConv, a novel method that uses weight sharing and multiple branches to enhance the skeletons of the square convolution kernel. Our key improvement is shown in Fig. 1. We find that utilizing sharing weights, SuperConv can reduce the additional trainable parameters brought by ACNet [3]. We also combine SuperConv with MixConv [4] to obtain SP-MixConv. Our experiments show that SuperConv and SP-MixConv achieving an effective balance between efficiency and performance. For details, see the experiment section.

Our contributions are as follows:

1. SuperConv: We propose a SuperConv (SP-Conv) by combining Super-Kernel [2] and ACB [3]. SuperConv takes advantage of weight sharing to reduce the trainable parameters in ACB. This way of sharing the weights in the convolution kernel is more fine-grained, which is also an interesting attempt for the design of efficient plug-and-play convolution block.

2. State-of-the-art results: Although, SuperConv is a simple improvement over ACB [3], it can also improve the accuracy of the model. In our experiments, the model with SP-MixConv gets the best accuracy on CIFAR-10 and CIFAR-100.

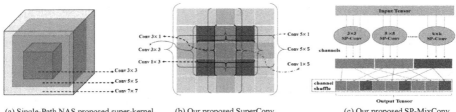

(a) Single-Path NAS proposed super-kernel   (b) Our proposed SuperConv   (c) Our proposed SP-MixConv

Fig. 1. SuperConv and Super-MixConv (best viewed in color). (a) Super-Kernel [2] encodes the small convolution kernels into the large convolution kernel by sharing weights. The choice of the size of the convolution kernel and the number of channels are determined by the threshold values. (b) we propose SuperConv by combining ACNet [3] and Super-Kernel. SuperConv includes the non-square convolution kernels and the small square convolution kernel in ACNet. But in order to reduce the trainable parameters, these convolution kernels inherit the local weights from the maximum convolution kernel. (c) we try to introduce SuperConv to enhance MixConv [4]. Experiments show that SP-MixConv not only improves the accuracy of the model, but also increases limited trainable parameters (Color figure online).

3. Inference eciency: Like SuperConv, SP-MixConv can improve the accuracy of the model, but at the same time have fewer trainable parameters. Simultaneously, SuperConv and SP-MixConv will not add extra burden in inference.

4. Reproducibility: SuperConv is very easy to implement, using existing tools such as Pytorch [24], and can be plug-and-play without having to make special adjustments to the model.

2 Related Work

SuperKernel. Single-Path NAS [2] encodes the small convolution kernel into the large convolution kernel by sharing parameters to search the neural network architecture. It proposes an novel single-path search space and the search model is state-of-the-art. Shared convolutional kernel parameters and one single-path over-parameterized ConvNet are used to reduce trainable parameters and search costs [2]. Single-path NAS can search the size of the convolution kernel and the number of channels in the Super-kernel. The search choice are determined by the threshold values. Furthermore, in Single-Path NAS, the threshold values are trainable parameters, which facilitate the search of the model structure.

ACNet. For the current neural network models, in order to improve the performance of the models, we need efficient plug-and-play modules [1,10,11,26], and high-efficiency model architectures [5–9,19,25,27–29]. [3] introduces BN fusion operation and Branch fusion operation, and through these two techniques can improve the model's inference speed. Then, [3] prove that the accuracy of the model can be improved without increasing the inference cost of the original model by increasing the branches of convolution kernel and enhancing the skeletons of ordinary convolution by using BN fusion operation and Branch fusion operation. At the same time, by enhancing the skeletons of the square convolution kernel, it is more robust to the input images. Experimental results show that ACNet have better effect on the rotated input images than the normal convolution kernel [3]. Moreover, adding BN operation [23] to the branch will also improve the accuracy of acnet model. The SuperConv we proposed also has BN operation in each branch, and we believe that BN operation plays a very important role in SuperConv.

MixConv. MixConv [4] achieves an effective balance between accuracy and efficiency by combining the different sizes of convolution kernels. [4] study the two channel partition methods: equal partition (MixConv) and exponential partition (MixConv+exp). Then [4] combine the neural network architecture search method and obtain the model with good performance and efficiency. This way of mixing the different sizes of convolution kernels has fewer parameters than single large convolution kernels, and also introduces a smaller receptive field. Moreover, the experiments prove that MixNets perform very well on other commonly used data sets [4].

3   Method

3.1   ACNet and Super-Kernel Convolution

In order to reduce the additional parameter burden caused by ACNet [3], we propose a Super-Kernel Convolution (SuperConv). SuperConv retains the advantages of ACNet while reducing trainable parameters.

BN Fusion. A normal convolution kernel can be enhanced by additional non-square convolution kernels [3]. Although the trainable parameters increased in the training stage, by combining BN fusion and Branch fusion, the prediction accuracy can be maintained without extra computations cost during the test phase. BN fusion is the fusion of BN operations [23] and convolution operations, and then processing the input. This clever approach not only increases the speed of inference, but also maintain model accuracy as before the fusion.

Branch Fusion. The Branch fusion [3], after the BN fusion, is to merge the convolution operations and BN operations [23] of the different branches. The experiments prove that similar results can be obtained by integrating the non-square convolution kernels into the large ordinary convolution kernel skeletons [3]. In this way, the precision of the model can be improved without increasing the inference cost.

Super-Kernel. Although ACNet [3] improves the model accuracy, it also increases the training cost. So we encode the small convolution kernels into the large convolution kernel, like Single-Path NAS [2]. This provides an idea for us to introduce Super-Kernel into the ACNet [3] and MixConv [4]. The Super-Kernel is shown in the Fig. 1(a). Its weights are divided into inner weights and outer weights. The small convolution kernel uses the inner weights and the large kernel convolution uses the inner and outer weights. Intuitively, this method uses the weight sharing of convolution kernels to reduce network burden.

3.2   How to Get the Weights of the Super-Kernel

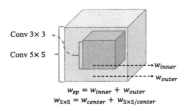

$w_{sp} = w_{inner} + w_{outer}$
$w_{5\times5} = w_{center} + w_{5\times5/center}$

Fig. 2. Weight Sharing in Super-Kernel (best viewed in color). (Color figure online)

For the update of Super-Kernel weights, we consider three ways of updating.

First, the inner weights are updated by the branch of the small convolution kernel, and the outer weights are updated by the branch of the large convolution kernel. This method of updating the weights does not reduce the trainable parameters, but only needs to store the weight parameters of Super-Kernel. For the update details of method 1, see Eq. (1), (2), and (3).

$$w_{sp} = w_{inner} + w_{outer}, \tag{1}$$

$$w_{5\times5} = w_{center} + w_{5\times5/center}, \tag{2}$$

$$update_{w_{inner}} \leftarrow update_{w_{3\times3}}, update_{w_{outer}} \leftarrow update_{w_{(5\times5/center)}}. \tag{3}$$

Where the weights of the Super-Kernel denoted as w_{sp}, consists of two parts: w_{inner} and w_{outer}. $w_{5\times5}$, $w_{3\times3}$ denote the weights of the 5×5 kernel branch, the weights of the 3×3 kernel branch, respectively. w_{center} and $w_{3\times3}$ have the same dimensions, and $w_{(5\times5/center)}$ are the outer part of the 5×5 kernel convolution.

Second, the inner weights are updated by the branch of the small convolution kernel and the branch of the large convolution kernel, and the outer weights are updated by the branch of the large convolution kernel. In order to balance the update ratio, control weight parameters are added to the branch of the large convolution kernel and the branch of the small convolution kernel. For the update details of method 2, see Eq. (4)and (5).

$$update_{w_{inner}} \leftarrow \alpha * (update_{w_{3\times3}}) + \beta * (update_{w_{center}}), \tag{4}$$

$$update_{w_{outer}} \leftarrow update_{w_{(5\times5/center)}}. \tag{5}$$

Where α and β denote the control parameter.

Third, the inner weights and outer weights are updated by the branch of the large convolution kernel, and the small convolution kernel inherit the local weights of the large convolution kernel. This method reduces trainable parameters and only needs to store the weight parameters of Super-Kernel. This way of encoding the small convolution kernel into the large convolution kernel, is easy to implement and can also get good model accuracy. Therefore, the experiments in this paper are based on this way to update the weights. For the update details of method 3, see Eq. (6) and (7).

$$update_{w_{inner}} \leftarrow update_{w_{center}}, \tag{6}$$

$$update_{w_{outer}} \leftarrow update_{w_{(5 \times 5/center)}}. \tag{7}$$

Intuitively, method one and method two, the update of weights of the Super-Kernel requires gradient back propagation of the convolution kernels in two branches. But in method three, we only need to update the weights of the large convolution kernel, and the small convolution kernel inherits the weights from the large convolution kernel. Although method three is the most coarse-grained, we adopt method three to balance the precision and the cost. While ACNet [3] belongs to a simplified version of method two, the related parameters of BN are the control parameters. The weight sharing of Super-Kernel is shown in Fig. 2. SuperConv adds non-square convolution kernels to strengthen the skeletons. The way of weight sharing is the same as Fig. 2 (Fig. 3).

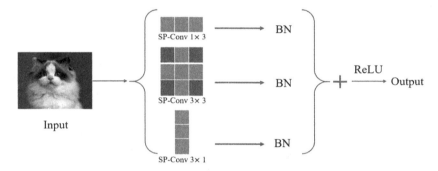

Fig. 3. Training-time SuperConv (best viewed in color). Only the weight parameters of SP-Conv 3 × 3 need to be trained. SP-Conv 1 × 3 and SP-Conv 3 × 1 inherit the local weights from SP-Conv 3 × 3. (Color figure online)

However, according to ACNet [3], it is not necessary to update the weights of the Super-Kernel at each iteration, we can update them through BN fusion [3] and Branch fusion [3] after the training is completed. Moreover, due to the weight sharing among convolution kernels of different sizes, the trainable parameters are reduced.

3.3 SuperConv

Inspired by Single-Path NAS [2], we propose SuperConv by combining ACNet [3] and Super-Kernel [2]. SuperConv has fewer trainable parameters and good performance. SuperConv is shown in Fig. 1(b).

SuperConv includes the non-square convolution kernels in ACNet [3] and the small square convolution kernels. In order to reduce the trainable parameters, these convolution kernels inherit the local weights from the maximum convolution kernel. However, SuperConv should have some requirements on the size of convolution kernel. It can be concluded that if only the 33 convolution kernels is replaced, SuperConv and ACNet have the same number of BN operations [23] and branches, so SuperConv cannot achieve similar accuracy, which is understandable. In an ideal state, by sharing the convolution kernel weights and increasing the number of branches and BN operations, a similar precision can be achieved while the parameters can be reduced. So, in some cases, we can combine SuperConv and ACNet and make some inner convolution kernels have fewer learnable parameters. The choice of the inner convolution kernels in SuperConv can also depend on the requirements of the model. In addition, we try to introduce SuperConv into MixConv [4], because it has large convolution kernels and it requires lightweight operations.

3.4 Super-MixConv

MixConv [4] is a very clever technique for reducing parameters in large convolution kernel networks and achieving an effective balance between efficiency and performance. MixConv splits the input and output channels into different groups, each using different size convolution kernels. This branch mechanism is more fine-grained, making it possible to reduce model parameters while maintaining model accuracy. However, the results of MixConv may depend on the number of groups and how channels are partitioned. Therefore, we try to introduce SuperConv to enhance MixConv. Experiments show that SP-MixConv improves the accuracy of the model, but increases limited trainable parameters. Note that when Super-Conv is inserted into MixConv, it should be in the same branch, meaning that the input should be the same. Super-MixConv is shown in Fig. 1(c).

4   Experiments[1]

4.1   CIFAR-10

The CIFAR-10 [13] dataset consists of 60k 32 × 32 color images of 10 classes, each with 6k images. It has 50k training images and 10k test images. The dataset is divided into five training batches and one test batch, each with 10k images. The test batch contains 1000 randomly selected images from each category.

ACB and SuperConv. First, we compare ACB [3] and SuperConv on CIFAR10. In experiments, we only need to replace the original convolutions with ACB and SuperConv, which is very easy to implement. For the implementation details of the experiments, please refer to the paper [3]. From the Table 1,

---

[1] For fairness and lack of computing resources, the accuracy given by our all experiments is the average of the verification accuracy of the last tenth of the whole epochs.

Table 1. Accuracy of the models with SuperConv or ACB on CIFAR10.

Models	Params	ACB	SuperConv	CIFAR10 (Ave)
Cifar-quick	117290	✗	✗	85.5720
Cifar-quick	119850 (+2.18%)	✗	✓	86.8420
Cifar-quick	137322 (+17.08%)	✓	✗	**86.8830**
WRN-16-8	10970170	✗	✗	95.5356
WRN-16-8	11013370 (+0.39%)	✗	✓	95.7450
WRN-16-8	18202138 (+65.92%)	✓	✗	**95.9206**
ResNet-56	860026	✗	✗	94.5630
ResNet-56	876282 (+1.89%)	✗	✓	94.6285
ResNet-56	1441818 (+67.65%)	✓	✗	**94.9015**

the model Cifar-quick [20] with ACB or with SuperConv achieve similar model accuracy. But the model with ACB increases about 17% more parameters, while SuperConv increases only about 2%. However, ResNet-56 [21] and WRN-16-8 [22] with SuperConv have limited improvement in model accuracy, since the size of convolution kernel we replace is 33. According to our analysis, Super-Conv has a small number of parameters, so it is a kind of weak enhancement for the skeletons and receptive field of the model. Therefore, we suggest using SuperConv on large convolution kernels or mixed convolution.

Table 2. Accuracy of the models with SP-MixConv or ACB on CIFAR10.

Models	#P	3 × 3-ACB	5 × 5-ACB	MixConv	Shuffle	CIFAR10 (Ave)
Cifar-quick	65146	✗	✗	3 × 3, 5 × 5	✓	83.296
Cifar-quick	73658	✗	1 × 5, 5 × 1	3 × 3, 5 × 5	✓	83.867
Cifar-quick	78874	1 × 3, 3 × 1	1 × 5, 5 × 1	3 × 3, 5 × 5	✓	**84.027**
Cifar-quick	78874	1 × 3, 3 × 1	1 × 5, 5 × 1	3 × 3, 5 × 5	✗	83.707

Models	Params	3 × 3-SP-Conv	5×5-SP-Conv	MixConv	Shuffle	CIFAR10(Ave)
Cifar-quick	65146	✗	✗	3 × 3, 5 × 5	✓	83.296
Cifar-quick	66426	✗	1×5, 5×1, 1× 3, 3 × 1, 3 × 3	3 × 3, 5 × 5	✓	84.046
Cifar-quick	66938	1 × 3, 3 × 1	1×5, 5×1, 1× 3, 3 × 1, 3 × 3	3 × 3, 5 × 5	✓	**84.503**

MixConv and SP-MixConv. Then, we combine SuperConv and MixConv [4] and propose SP-MixConv. In order to facilitate the experiments, we do not strictly follow the implementation details of MixConv. Instead, ordinary convolution kernels are used. See details about MixConv [4]. To verify the usefulness of SP-MixConv, we insert MixConv and SP-MixConv into the Cifar-quick model [20]. From the Table 2 and Fig. 4, SP-MixConv increases a small number of parameters relative to ACB [3] and gets the best accuracy. At the same time, it is important to note that in a MixConv model, the way that the channels are divided and the number of groups will affect the accuracy of the model [4]. For the division of channels, we use a simple experiment to illustrate, see [Ablation study].

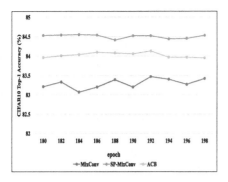
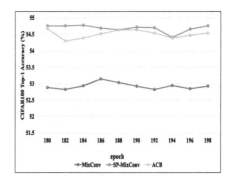

Fig. 4. MixConv, SP-MixConv and ACB performance on CIFAR10.

Fig. 5. MixConv, SP-MixConv and ACB performance on CIFAR100.

4.2   CIFAR-100

The CIFAR-100 [13] dataset is like CIFAR-10, except that it has 100 classes, each containing 600 images, each category has 500 training images and 100 test images. 100 classes are divided into 20 superclasses. We have done some simple experiments on the CIFAR-100 dataset and find that SP-MixConv could still obtain the best model accuracy. See Table 3 and Fig. 5 for the details of experimental results.

Table 3. Accuracy of the models with SP-MixConv or ACB on CIFAR100.

Models	#P	3 × 3-ACB	5 × 5-ACB	MixConv	Shuffle	CIFAR100 (Ave)
Cifar-quick	70996	✗	✗	3 × 3, 5 × 5	✓	52.924
Cifar-quick	84724	1 × 3, 3 × 1	1 × 5, 5 × 1	3 × 3, 5 × 5	✓	**54.499**
Models	#P	3 × 3-SP-Conv	5 × 5-SP-Conv	MixConv	Shuffle	CIFAR100(Ave)
Cifar-quick	70996	✗	✗	3 × 3, 5 × 5	✓	52.924
Cifar-quick	72788	1 × 3, 3 × 1	1 × 5, 5 × 1, 1 × 3, 3 × 1, 3 × 3	3 × 3, 5 × 5	✓	**54.678**

Table 4. Accuracy of the model with/without Channel shuffle on CIFAR10.

Models	Params	3 × 3-ACB	5 × 5-ACB	MixConv	Cha-shuffle	CIFAR10 (Ave)
Cifar-quick	78874	1 × 3, 3 × 1	1 × 5, 5 × 1	3 × 3, 5 × 5	✓	**84.027**
Cifar-quick	78874	1 × 3, 3 × 1	1 × 5, 5 × 1	3 × 3, 5 × 5	✗	83.707

4.3   Ablation Studies

Channel Split. In our experiments, we simply split the input and output channels of the Cifar-quick model [20]. When the number of channels divided by the number of groups has a remainder, add the remaining channels to the channel-split[0]. We can find that even if the model is small, the number of channels will still affect the accuracy of the model. So in this case, SuperConv can be used to enhance MixConv [4]. The channel-split codes come from Github[2]. In most of our other experiments, the number of channels in the 5×5 convolution kernel is channel-split[0] (Table 5).

Table 5. Accuracy of the modules with different channels-split on CIFAR10.

Models	#P	3×3-ACB	5×5-ACB	MixConv	Shuffle	Split[0]	CIFAR10 (Ave)
Cifar-quick	65146	✗	✗	$3 \times 3, 5 \times 5$	✓	5×5	83.296
Cifar-quick	64890	✗	✗	$3 \times 3, 5 \times 5$	✓	3×3	**83.381**

Channel Shuffle. A Channel shuffle operation is used in ShuffleNet [5,6] to mix channel information. We think it will be a great help to MixConv [4] as well. From Table 4, we find that adding Channel shuffle after MixConv will increase the accuracy of the model, so Channel shuffle is added by default in all MixConv experiments. We use some open codes[3].

4.4 Discussions and Future Work

Although SuperConv reduces trainable parameters, adding branches increases computation. We believe that the features obtained by different branches may be redundant. So the correct choice of the branches may reduce the calculation without reducing the accuracy of the model. In future work, we will try to make different branches share BN operation [23] and reduce branches. In addition, we are going to study whether BN fusion [3] and Branch fusion [3] are effective in the above cases.

5   Conclusion

In order to reduce the trainable parameters of ACNet, we propose SuperConv by combining ACB and Super-Kernel. In the model of large convolution kernels, the insertion of SuperConv can significantly improve the accuracy of the model. In addition, we introduce SuperConv into MixConv and propose Super-MixConv (SP-MixConv). In the experiments of SP-MixConv, the model accuracy is improved significantly when a small number of trainable parameters are added. Simultaneously, SuperConv is very easy to implement, using existing tools such as Pytorch, and can be plug-and-play without having to make special adjustments to the model. This way of sharing the weights in the convolution kernel is more fine-grained, which is also an interesting attempt for the design of efficient plug-and-play convolution block.

---

[2] https://github.com/HaiPhan1991/mixconv_pytorch.
[3] https://github.com/Randl/ShuffleNetV2-pytorch.

References

1. Li, X., Wang, W., Hu, X., Yang, J.: Selective kernel networks. In: 2019 Proceedings of the IEEE Conference on Computer Vision and Pattern Recognition, pp. 510–519 (2019)
2. Stamoulis, D., et al.: Single-Path NAS: Designing hardware-efficient ConvNets in less than 4 hours. arXiv preprint arXiv:1904.02877 (2019)
3. Ding, X., Guo, Y., Ding, G., Han, J.: ACNet: strengthening the kernel skeletons for powerful CNN via asymmetric convolution blocks. In: 2019 Proceedings of the IEEE International Conference on Computer Vision, pp. 1911–1920 (2019)
4. Tan, M., Le, Q.V.: MixConv: Mixed depthwise convolutional kernels. CoRR, abs/1907.09595 (2019)
5. Zhang, X., Zhou, X., Lin, M., Sun, J.: ShuffleNet: an extremely efficient convolutional neural network for mobile devices. In: 2018 Proceedings of the IEEE Conference on Computer Vision and Pattern Recognition, pp. 6848–6856 (2018)
6. Ma, N., Zhang, X., Zheng, H.-T., Sun, J.: ShuffleNet V2: practical guidelines for efficient CNN architecture design. In: Ferrari, V., Hebert, M., Sminchisescu, C., Weiss, Y. (eds.) Computer Vision – ECCV 2018. LNCS, vol. 11218, pp. 122–138. Springer, Cham (2018). https://doi.org/10.1007/978-3-030-01264-9_8
7. Howard, A.G., et al.: MobileNets: Efficient convolutional neural networks for mobile vision applications. arXiv preprint arXiv:1704.04861 (2017)
8. Sandler, M., Howard, A., Zhu, M., Zhmoginov, A., Chen, L.-C.: MobileNetV2: inverted residuals and linear bottlenecks. In: 2018 Proceedings of the IEEE Conference on Computer Vision and Pattern Recognition, pp. 4510–4520 (2018)
9. Howard, A., et al.: Searching for MobileNetV3. In: 2019 Proceedings of the IEEE International Conference on Computer Vision, pp. 1314–1324 (2019)
10. Hu, J., Shen, L., Sun, G.: Squeeze-and-excitation networks. In: 2018 Proceedings of the IEEE Conference on Computer Vision and Pattern Recognition, pp. 7132–7141 (2018)
11. Woo, S., Park, J., Lee, J.-Y., Kweon, I.S.: CBAM: convolutional block attention module. In: Ferrari, V., Hebert, M., Sminchisescu, C., Weiss, Y. (eds.) ECCV 2018. LNCS, vol. 11211, pp. 3–19. Springer, Cham (2018). https://doi.org/10.1007/978-3-030-01234-2_1
12. Liu, H., Simonyan, K., Yang, Y.: DARTS: Differentiable architecture search. arXiv preprint arXiv:1806.09055 (2018)
13. Krizhevsky, A., Hinton, G., et al.: Learning multiple layers of features from tiny images (2009)
14. Chen, X., Xie, L., Wu, J., Tian, Q.: Progressive differentiable architecture search: bridging the depth gap between search and evaluation. In: 2019 Proceedings of the IEEE International Conference on Computer Vision, pp. 1294–1303 (2019)
15. Li, G., Zhang, X., Wang, Z., Li, Z., Zhang, T.: StacNAS: Towards stable and consistent optimization for differentiable neural architecture search. arXiv preprint arXiv:1909.11926 (2019)
16. Xu, Y., et al.: PC-DARTS: partial channel connections for memory-efficient architecture search. In: International Conference on Learning Representations (2019)
17. Dong, X., Yang, Y.: Searching for a robust neural architecture in four GPU hours. In: 2019 Proceedings of the IEEE Conference on Computer Vision and Pattern Recognition, pp. 1761–1770 (2019)
18. Liu, C., et al.: Auto-DeepLab: hierarchical neural architecture search for semantic image segmentation. In: 2019 Proceedings of the IEEE Conference on Computer Vision and Pattern Recognition, pp. 82–92 (2019)

19. Wang, R.J., Li, X., Ling, C.X.: Pelee: a real-time object detection system on mobile devices. In: 2018 Advances in Neural Information Processing Systems, pp. 1963–1972 (2018)

20. Snoek, J., Larochelle, H., Adams, R.P.: Practical Bayesian optimization of machine learning algorithms. In: 2012 Advances in Neural Information Processing Systems, pp. 2951–2959 (2012)

21. He, K., Zhang, X., Ren, S., Sun, J.: Deep residual learning for image recognition. In: 2016 Proceedings of the IEEE Conference on Computer Vision and Pattern Recognition, pp. 770–778 (2016)

22. Zagoruyko, S., Komodakis, N.: Wide residual networks. arXiv preprint arXiv:1605.07146 (2016)

23. Ioffe, S., Szegedy, C.: Batch normalization: Accelerating deep network training by reducing internal covariate shift. arXiv preprint arXiv:1502.03167 (2015)

24. Paszke, A., et al.: Automatic differentiation in PyTorch (2017)

25. Tan, M., Le, Q.V.: EfficientNet: Rethinking model scaling for convolutional neural networks. arXiv preprint arXiv:1905.11946 (2019)

26. Chollet, F.: Xception: deep learning with depthwise separable convolutions. In: 2017 Proceedings of the IEEE Conference on Computer Vision and Pattern Recognition, pp. 1251–1258 (2017)

27. Iandola, F.N., Han, S., Moskewicz, M.W., Ashraf, K., Dally, W.J., Keutzer, K.: SqueezeNet: Alexnet-level accuracy with 50x fewer parameters and <0.5 mb model size. arXiv preprint arXiv:1602.07360 (2016)

28. Wu, B., et al.: FBNet: hardware-aware efficient ConvNet design via differentiable neural architecture search. In: 2019 Proceedings of the IEEE Conference on Computer Vision and Pattern Recognition, pp. 10 734–10 742 (2019)

29. Tan, M., et al.: MnasNet: platform-aware neural architecture search for mobile. In: 2019 Proceedings of the IEEE Conference on Computer Vision and Pattern Recognition, pp. 2820–2828 (2019)

30. Liu, D., Fu, J., Liu, P., Lv, J.: TIGS: an inference algorithm for text infilling with gradient search (2019)

31. Liu, D., Lv, J., Li, Y.: Generating style-specific Chinese Tang poetry with a simple actor-critic model. IEEE Trans. Emerg. Topics Comput. Intell. **3**, 313–321 (2019)

TAC-GAIL: A Multi-modal Imitation Learning Method

Jiacheng Zhu[✉] and Chong Jiang

School of Computer Science and Technology,
Soochow University, Suzhou 215006, China
{20185227021,20175227033}@stu.suda.edu.cn

Abstract. Imitation learning provides a family of promising frameworks that learn policies from expert demonstrations directly. However, most imitation learning methods assume that the expert demonstrations come from the same expert and have a single modality. In fact, the expert demonstrations may be generated by different experts in different modalities. Auxiliary classifier generative adversarial imitation learning (AC-GAIL) uses an auxiliary classifier to classify samples according to modalities, so that the generator can perform different actions according to different modalities, and obtain a multi-modal policy. However, we find that AC-GAIL's objective function missing a conditional entropy, and this conditional entropy cannot be calculated directly. Missing the conditional entropy can result in a decrease in the performance of the learned policy. In this paper, we propose a method that can deal with the problem of missing conditional entropy in AC-GAIL, named twin auxiliary classifiers GAIL (TAC-GAIL). Specifically, we add another auxiliary classifier to the framework of AC-GAIL, which is used to classify the generated samples. We theoretically prove the effectiveness of this method, and the experimental results on MuJoCo tasks show that TAC-GAIL can effectively improve the performance of the learned multi-modal policy.

Keywords: Imitation learning · Multi-modal · Adversarial

1 Introduction

The ability to learn from data is a key factor for agent to build decision models. In recent years, there have been many learning frameworks [2,3,8] that present promising results. Reinforcement learning (RL) [1] is one of these frameworks that learns to make decisions based on trial-and-error search in environments with a specified reward function. However, designing such an idea reward function manually is difficult, especially when the environment becomes more complex and uncertain, e.g., for autonomous driving where there is a need to balance safety, comfort and efficiency.

This work is in part supported by the Natural Science Foundation of China (61876119) and the Natural Science Foundation of Jiangsu (BK20181432) and a project funded by the Priority Academic Program Development of Jiangsu Higher Education Institutions.

© Springer Nature Switzerland AG 2020
H. Yang et al. (Eds.): ICONIP 2020, LNCS 12533, pp. 688–699, 2020.
https://doi.org/10.1007/978-3-030-63833-7_58

Imitation learning provides a method for learning decision models directly from expert demonstrations. Compared to RL, imitation learning does not require an explicit reward function, which increases its scope of application. And imitation learning has achieved remarkable successes in a wide range of problems. At present, imitation learning can be divided into two categories. The first category is behavioral cloning [21], which is a supervised learning method. It learns expert policy by directly mimicking the state-action mapping of the expert demonstrations. Behavior cloning is a relatively simple imitation learning method. When expert demonstrations can cover the entire state space, behavioral cloning can achieve good performance. However, when there are few expert demonstrations, the agent cannot learn the optimal decision in each state. Moreover, since long-term effects are not considered, subtle errors will be gradually amplified in the sequential decision-making process, resulting in compounding errors [22]. The second category is inverse reinforcement learning (IRL)[9,10]. IRL will learn a reward function so that the expert demonstrations have the highest probability of occurrence, and then use RL to learn an optimal policy based on the reward function. The policy learned by IRL has better generalization capabilities and requires fewer expert demonstrations. However, since the inner loop of IRL includes the RL process, it will increase the computational cost. Generative adversarial imitation learning (GAIL) [4] combines IRL with generative adversarial nets (GANs) [13]. GAIL is an efficient imitation learning method, and it can be extended to more complex environments.

However, most of the current imitation learning methods [5–7] assume that the expert demonstrations has only one modality. We hope that the agent can learn from multi-modal expert demonstrations and get a policy with multiple modalities. Take the example of human walking, when we are walking, we will reduce our pace and frequency, but when we are in a hurry, it is another modality, we may walk very fast, or even run.

Recently, some works [17] have emerged to solve multi-modal imitation learning tasks. Among them, auxiliary classifier GAIL (AC-GAIL) [16] can learn multi-modal policy from multi-modal expert demonstrations. It adds an auxiliary classifier to the framework of GAIL. This auxiliary classifier is used to classify state-action pairs according to modalities, so that the generator can perform different actions according to different modalities. However, we find that AC-GAIL's objective function missing a conditional entropy, and this conditional entropy cannot be calculated directly. The lack of this conditional entropy will cause a mismatch between the classification results of the classifier and the real data, affecting the performance of the learned multi-modal policy.

In this paper, inspired by Twin Auxiliary Classifiers GAN [15], we propose a method to deal with the lack of conditional entropy in AC-GAIL, named twin auxiliary classifiers GAIL (TAC-GAIL). TAC-GAIL adds an additional auxiliary classifier to the framework of AC-GAIL to classify the state-action pairs generated by the generator. In subsequent sections, we will further explain how additional auxiliary classifiers can improve the performance of the learned policy. Moreover, experimental results on several MuJoCo tasks also show that our method can outperform some existing multi-modal imitation learning methods, e.g. AC-GAIL, InfoGAIL [19].

2 Background

2.1 Preliminaries

An infinite-horizon, discounted Markov decision process (MDP) can be defined as $(S, A, P, R, \rho_0, \gamma)$, where S represents the state space, A represents the action space, $P : S \times A \times S \rightarrow [0, 1]$ denotes the transition probability distribution, $R : S \times A \rightarrow \mathbb{R}$ denotes the reward function, $\rho_0 : S \rightarrow [0, 1]$ is the distribution of the initial state s_0, and $\gamma \in [0, 1]$ is a discount factor that determines the importance of future rewards. We use $V_\pi(s_0) = \mathbb{E}_\pi \left[\sum_{t=0}^{T} \gamma^t R(s_t, a_t) | a_t \sim \pi(s_t) \right]$ to denote the cumulative reward obtained by the agent following policy π, where $a \sim \pi(s)$, and T denotes the terminal of an episode. The goal of RL is to find a policy π^* that maximizes the cumulative rewards.

2.2 Imitation Learning

Imitation learning does not require a reward function R, it directly learns how to perform a task from expert demonstrations. The set of expert demonstration trajectories is defined as $\{\tau_1, \tau_2, \cdots, \tau_N\}$, where $\tau_i = \{(s_0, a_0), (s_1, a_1), \cdots, (s_T, a_T)\}$ is a sequence of state-action pairs, $i = 1, 2, 3, \cdots, N$.

Generative Adversarial Imitation Learning. GAIL is a model-free, online imitation learning method, which can be well generalized to high-dimensional and complex environments. GAIL ignores the process of seeking reward functions in IRL, and directly extracts a policy from expert demonstrations.

By combining imitation learning with GANs, GAIL transforms the imitation learning problem into a matching problem of the state-action distributions between expert demonstrations and generated trajectories. Where the optimum is achieved when the distance between these two distributions is minimized as measured by Jensen-Shannon divergence (JSD). The formal objective of GAIL is denoted as

$$\min_{\pi_\theta} \max_{D_\omega} \mathbb{E}_{\pi_\theta}[\log(D_\omega(s, a))] + \mathbb{E}_{\pi_E}[\log(1 - D_\omega(s, a))] - \lambda H(\pi_\theta), \qquad (1)$$

where π_θ is a policy network parameterized by θ, also known as the generator, which interacts with the environment to generate trajectories. D_ω is a discriminator network parameterized by ω, which is used to discriminate state-action pairs come from the expert trajectories or the generated trajectories. $H(\pi_\theta)$ is the causal entropy of the policy π_θ. Optimization over the GAIL objective is performed by alternating a gradient step to increase Eq. 1 w.r.t. ω, and a trust region policy optimization (TRPO) [14] step to decrease Eq. 1 w.r.t. θ.

2.3   Multi-modal Imitation Learning

Recently, some works have emerged to solve multi-modal imitation learning tasks, and these works can be roughly divided into two categories. One is to learn multi-modal policies from expert demonstrations without modal labels: Info-GAIL [19] distinguishes modal information in an unsupervised manner by maximizing the mutual information between state-action pairs and latent variables; Burn-InfoGAIL [18] uses maximum mutual information from the perspective of Bayesian inference. However, such methods lack modal label information in expert demonstrations, and they may produce unexplainable behaviors. The other is to learn multi-modal policies from expert demonstrations with modal labels: VAE-GAIL [20] uses a variational autoencoder to infer modal labels to learn multi-modal policies; AC-GAIL adds an auxiliary classifier. This auxiliary classifier classifies the state-action pairs according to modalities, and is used to guide the generator to perform correct actions in each modality.

3   Twin Auxiliary Classifiers GAIL

In this section, We first explain the problems in AC-GAIL from the perspective of distribution matching. Based on our understanding of this problem, we propose TAC-GAIL to further improve the performance of multi-modal imitation learning.

3.1   Insight of AC-GAIL

In order to learn from the multi-modal expert demonstrations, AC-GAIL adds an auxiliary classifier to the framework of GAIL. The auxiliary classifier is used to classify the state-action pairs according to modalities, and its output is the probability that each state-action pair belongs to the corresponding modal label. This classifier optimizes itself based on the cross entropy between the distribution of the given data $P(c|s, a)$ and the distribution specified by the auxiliary classifier $Q(c|s, a)$. Assuming that there are K modalities, $C = \{c_1, c_2, ..., c_K\}$, AC-GAIL defines the prior distribution of these modalities as $c \sim p(c)$, and $p(c_1) = p(c_2) = ... = p(c_K)$. For simplicity, we denote the samples generated by policy π_θ in modality c as $\mathbb{E}_{c,\pi_\theta}[\cdot]$, and the samples generated by policy π_E in modality c as $\mathbb{E}_{c,\pi_E}[\cdot]$. The objective function of AC-GAIL(For brevity, we omit the policy entropy term $-\lambda_H H(\pi_\theta)$) is as follows:

$$L_{AC-GAIL} = \min_{\pi_\theta, C_\psi} \max_{D_\omega} \underbrace{\mathbb{E}_{c\sim p(c), \pi_\theta}[\log(D_\omega(s, a))] + \mathbb{E}_{c\sim p(c), \pi_E}[\log(1 - D_\omega(s, a))]}_{1}$$

$$\underbrace{-\lambda_1 \mathbb{E}_{c\sim p(c), \pi_E}[\log C_\psi(c|s, a)]}_{2} \underbrace{-\lambda_1 \mathbb{E}_{c\sim p(c), \pi_\theta}[\log C_\psi(c|s, a)]}_{3}.$$

$$(2)$$

In the above formula, λ_1 is a hyperparameter used to balance the original GAIL loss and the auxiliary classifier classification loss. Further, we can divide the objective function into 3 parts. The first part is the JSD between the expert trajectories and the generated trajectories. The second part is the cross entropy between the distribution $P_{\pi_E}(c|s,a)$ of expert demonstrations and the distribution $Q_{\pi_E}(c|s,a)$ specified by auxiliary classifier. We can intuitively think of Part 2 as minimizing the Kullback-Leibler (KL) divergence between $P_{\pi_E}(c|s,a)$ and $Q_{\pi_E}(c|s,a)$. If we add a negative conditional entropy $-H_{\pi_E}(c|s,a) = \mathbb{E}_{c\sim p(c),\pi_E}[\log P_{\pi_E}(c|s,a)]$ to the second part of the formula, we can get:

$$\mathbb{E}_{c\sim p(c),\pi_E}[\log P_{\pi_E}(c|s,a)] - \mathbb{E}_{c\sim p(c),\pi_E}[\log C_\psi(c|s,a)]$$
$$=\mathbb{E}_{c\sim p(c),\pi_E}[\log P_{\pi_E}(c|s,a)] - \mathbb{E}_{c\sim p(c),\pi_E}[\log Q_{\pi_E}(c|s,a)]$$
$$=\mathbb{E}_{c\sim p(c),\pi_E}[\log \frac{P_{\pi_E}(c|s,a)}{Q_{\pi_E}(c|s,a)}] \tag{3}$$
$$=KL(P_{\pi_E}(c|s,a)||Q_{\pi_E}(c|s,a)).$$

Minimizing the second part of the equation is equivalent to minimizing the KL divergence between $P_{\pi_E}(c|s,a)$ and $Q_{\pi_E}(c|s,a)$, because $-H_{\pi_E}(c|s,a)$ is a constant. The third part is the cross entropy between the distribution $P_{\pi_\theta}(c|s,a)$ of the generated trajectory and the distribution $Q_{\pi_\theta}(c|s,a)$ specified by auxiliary classifier. Similarly, after adding negative conditional entropy $-H_{\pi_\theta}(c|s,a) = \mathbb{E}_{c\sim p(c),\pi_\theta}[\log P_{\pi_\theta}(c|s,a)]$ to the third part of the formula, we get the following results:

$$\mathbb{E}_{c\sim p(c),\pi_\theta}[\log P_{\pi_\theta}(c|s,a)] - \mathbb{E}_{c\sim p(c),\pi_\theta}[\log C_\psi(c|s,a)]$$
$$=KL(P_{\pi_\theta}(c|s,a)||Q_{\pi_\theta}(c|s,a)). \tag{4}$$

When minimizing the third part of the equation w.r.t. the classifier C, negative conditional entropy $-H_{\pi_\theta}(c|s,a)$ can be regarded as a constant term, and therefore is equivalent to minimizing the KL divergence between $P_{\pi_\theta}(c|s,a)$ and $Q_{\pi_\theta}(c|s,a)$. However, when minimizing the third part of the equation w.r.t. the generator G, the conditional entropy $-H_{\pi_\theta}(c|s,a)$ can no longer be regarded as a constant term, because $P_{\pi_\theta}(c|s,a)$ is the conditional distribution determined by the generator G. Therefore, in the process of optimizing G, AC-GAIL ignores the conditional entropy $-H_{\pi_\theta}(c|s,a)$ and only minimizes the third part of the equation, so it cannot minimize the KL divergence between $P_{\pi_\theta}(c|s,a)$ and $Q_{\pi_\theta}(c|s,a)$.

3.2   Twin Auxiliary Classifiers GAIL (TAC-GAIL)

Based on the analysis in the previous section, we may think of adding the missing $-H_{\pi_\theta}(c|s,a)$ to the objective function to solve the problem. However, we cannot directly estimate $-H_{\pi_\theta}(c|s,a)$ because we do not know the value of $P_{\pi_\theta}(c|s,a)$. Therefore, we deal with this problem by adding a new adversarial part to the minimax game.

Its main idea is to introduce an additional auxiliary classifier C_ϕ to classify the generated data. Similar to GAIL, there is an adversarial between the generator and C_ϕ. Next, we will explain the relationship between adding classifier C_ϕ and minimizing $-H_{\pi_\theta}(c|s,a)$.

Proposition 1. *Minimizing* $-H_{\pi_\theta}(c|s,a)$ *is equivalent to minimizing the following two indicators: (1) The mutual information between data and modal labels; (2) The JSD between conditional distributions* $\{P(s,a|c = 1), ..., P(s,a|c = K)\}$.

Proof. The first is because the entropy of the modal label is a constant. The second one is as follows:

$$
\begin{aligned}
I_{\pi_\theta}(c,(s,a)) =& H(c) - H_{\pi_\theta}(c|s,a) = H_{\pi_\theta}(s,a) - H_{\pi_\theta}(s,a|c) \\
=& -\frac{1}{K}\sum_{k=1}^{K}\mathbb{E}_{c=k,\pi_\theta}[\log P(s,a)] + \frac{1}{K}\sum_{k=1}^{K}\mathbb{E}_{c=k,\pi_\theta}[\log P(s,a|c=k)] \\
=& \frac{1}{K}\sum_{k=1}^{K}KL(P(s,a|c=k)||P(s,a)) \\
=& JSD(P(s,a|c=1), ..., P(s,a|c=K)).
\end{aligned}
\tag{5}
$$

Based on the relationship between $-H_{\pi_\theta}(c|s,a)$ and the aforementioned JSD, we further expand the minimax game in AC-GAIL. We add another auxiliary classifier C_ϕ to the AC-GAIL framework to minimize the JSD, and C_ϕ is used to classify the generated data. We get the following minimax game:

$$
L_{C_\phi} = \min_{\pi_\theta}\max_{C_\phi}\mathbb{E}_{c\sim p(c),\pi_\theta}[\log(C_\phi(c|s,a))].
\tag{6}
$$

Theorem 3 can illustrate that the minimax game can effectively minimize the JSD between $\{P(s,a|c=1), ..., P(s,a|c=K)\}$.

Proposition 2. *For a fixed generator G, the optimal classifier* C_ϕ *is*

$$
C_\phi((s,a),c=k) = \frac{P(s,a|c=k)}{\sum_{k'=1}^{K}P(s,a|c=k')}.
\tag{7}
$$

The proof of Proposition 2 is in the appendix.

Theorem 3. *The global mininum of the minimax game* L_{C_ϕ} *is achieved if and only if* $P(s,a|c=1) = P(s,a|c=2) = ... = P(s,a|c=K)$.

Proof. If we add $\log K$ to the minimax game, we can obtain:

$$
\begin{aligned}
&\min_{\pi_\theta}\max_{C_\phi}\mathbb{E}_{c\sim p(c),\pi_\theta}[\log(C_\phi(c|s,a))] + \log K \\
=& \frac{1}{K}\sum_{k=1}^{K}\mathbb{E}_{c=k,\pi_\theta}[\log\frac{P(s,a|c)}{\sum_{c'=1}^{K}P(s,a|c')}] + \log K \\
=& \frac{1}{K}\sum_{k=1}^{K}\mathbb{E}_{c=k,\pi_\theta}[\log\frac{P(s,a|c)}{\frac{1}{K}\sum_{c'=1}^{K}P(s,a|c')}] \\
=& \frac{1}{K}\sum_{c=1}^{K}KL(P(s,a|c)||\frac{1}{K}\sum_{c'=1}^{K}P(s,a|c')) \\
=& JSD(P(s,a|c=1), ..., P(s,a|c=K)).
\end{aligned}
\tag{8}
$$

Then we get:

$$L_{C_\phi} = -\log K + JSD(P(s, a|c = 1), ..., P(s, a|c = K)). \tag{9}$$

Because the JSD between multiple distributions is non-negative, and is zero when multiple distributions are the same. When L_{C_ϕ} is equal to $-\log K$, this minimax game reaches the global minimum, at this point $P(s, a|c = 1) = P(s, a|c = 2) = ... = P(s, a|c = K)$.

Combining the objective function of AC-GAIL and L_{C_ϕ}, we get the objective function of TAC-GAIL:

$$L_{TAC-GAIL} = L_{AC-GAIL} + \lambda_2 L_{C_\phi}. \tag{10}$$

Because of other terms in the objective function of TAC-GAIL, L_{C_ϕ} cannot reach its global minimum. Although TAC-GAIL cannot completely remove the influence of missing conditional entropy, it can reduce the difference between the generated trajectories and the expert trajectories, and further improve the performance of the learned multi-modal policy. The optimization procedure of TAC-GAIL is shown in Algorithm 1.

Algorithm 1: Twin Auxiliary Classifier GAIL

---

input : Expert trajectories $\tau_E \sim \pi_E$, initial policy parameters θ_0, discriminator parameters ω_0, and classifier parameters ψ_0, ϕ_0.

output: Learned policy π_θ.

1 **for** $i = 0, 1, 2, ...$ **do**

2   Sample a modal-label $c_i \sim p(c)$;

3   Sample trajectories $\tau_i \sim \pi_{\theta_i}(c_i)$;

4   Sample state-action pairs $X_i \sim \tau_i$ and $X_E \sim \tau_E$ with same batch size;

5   Update the discriminator parameters from ω_i to ω_{i+1} with the gradient:
$\Delta_{\omega_i} = \mathbb{E}_{X_i} [\nabla_{\omega_i} \log(D_{\omega_i}(s, a))] + \mathbb{E}_{X_E} [\nabla_{\omega_i} \log(1 - D_{\omega_i}(s, a))]$

6   Update the discriminator parameters from ψ_i to ψ_{i+1} with the gradient:
$\Delta_{\psi_i} = \mathbb{E}_{X_i} [\nabla_{\psi_i} \log C_{\psi_i}(c|s, a)] + \mathbb{E}_{X_E} [\nabla_{\psi_i} \log C_{\psi_i}(c|s, a)]$

7   Update the discriminator parameters from ϕ_i to ϕ_{i+1} with the gradient:
$\Delta_{\phi_i} = \mathbb{E}_{X_i} [\nabla_{\phi_i} \log C_{\phi_i}(c|s, a)]$

8   Take a policy step from θ_i to θ_{i+1}, using the TRPO update rule with the following objective: $\mathbb{E}_{X_i} [\log D_{\omega_{i+1}}(s, a)] + \lambda_1 \mathbb{E}_{X_i} [\log C_{\psi_{i+1}}(c|s, a)] - \lambda_2 \mathbb{E}_{X_i} [\log C_{\phi_{i+1}}(c|s, a)]$

9 **end**

---

4   Experiments

In this section, we first introduce the experimental environment, then give the experimental steps and parameter settings, and finally compare the experimental results of various imitation learning methods.

4.1   Experimental Environments

We evaluate TAC-GAIL in a series of challenging high-dimensional simulated robotic tasks in MuJoCo. (1) Hopper is a simulation of single-legged robot jumping. Its state space size is 11, and its action space size is 3, both of which are continuous spaces; (2) HalfCheetah is a simulation of a two-legged cheetah running. Its state space size is 17 and its action space size is 6, all of which are continuous spaces. (3) Walker2d is a simulation of a two-legged robot walking. Its state space size is 17, and its action space size is 6, both of which are continuous spaces. The goal of these tasks is to move the robot forward as fast as possible.

4.2   Experimental Setup

To get expert demonstrations, we first use the true reward function defined in OpenAI gym to obtain expert policies of different modalities through RL method, and then use these policies to generate expert demonstrations of different modalities. We get 500 expert demonstration trajectories for each modality, and each trajectory contains 1000 state-action pairs. The detailed information of the experimental environments and the expert demonstrations are shown in Table 1.

Table 1. Information of the experimental environment and expert demonstrations.

Task	State space size	Action space size	Expert label 0	Expert label 1
Hopper-v2	11	3	3351	2107
HalfCheetah-v2	17	6	4184	2521
Walker2d-v2	17	6	4955	3463

In the framework of TAC-GAIL, it mainly includes the following parts: generator, discriminator and two auxiliary classifiers, all of which are implemented by multi-layer neural networks. The network structure is similar to that of AC-GAIL, with two hidden layers, each containing 100 units. We update the generator through TRPO algorithm, using Adam [12] to update the discriminator and two auxiliary classifiers. We set hyperparameters $\lambda_1 = 0.75$ and $\lambda_2 = 0.5$.

4.3   Results

In order to satisfy the demand of significance test, we set 4 random seeds for each task. In Fig. 1, we show the learning curves of AC-GAIL, TAC-GAIL and InfoGAIL over 6,000 iterations in each task.

Expert label 0 and expert label 1 represent two different modal expert demonstrations. The goal is to learn a policy with these two modalities. In Hopper and HalfCheetah, AC-GAIL and TAC-GAIL can match the expert demonstrations

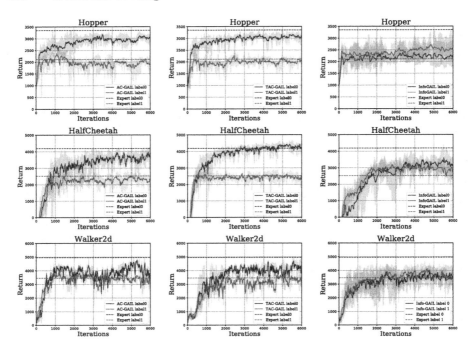

Fig. 1. The learning curves of AC-GAIL (left), TAC-GAIL (middle) and InfoGAIL (right).

in each modality. In Walker, neither AC-GAIL nor TAC-GAIL can match the expert demonstrations accurately. It is worth noting that in these three tasks, TAC-GAIL can match the expert demonstrations more accurately, which shows that the learned policy is closer to the expert policy. At the same time, the policy learned by TAC-GAIL is more stable than the policy learned by AC-GAIL. However, InfoGAIL cannot match the expert demonstrations in every task.

We use the expected error rate of the average return between the generated trajectories and the expert trajectories as the criterion. The calculation of the expected error rate is as follows:

$$Err = \sum_{c=1}^{K} p(c) \frac{|\bar{r}_{\pi_\theta}(c) - \bar{r}_{\pi_E}(c)|}{\bar{r}_{\pi_E}(c)}, \tag{11}$$

where $\bar{r}_{\pi_\theta}(c)$ represents the average return of trajectories generated by policy π_θ in modality c. For the average return of generated trajectories, we use the average of 5000–6000 iterations in the learning curve. The average error rates of TAC-GAIL and AC-GAIL in each task are shown in Table 2.

Table 2. Average error rates of TAC-GAIL and AC-GAIL.

Error rate algorithm Task	AC-GAIL	TAC-GAIL	InfoGAIL
Hopper-v2	9.52%	**5.96%**	26.86%
HalfCheetah-v2	10.56%	**2.62%**	17.37%
Walker2d-v2	11.35%	**9.68%**	15.44%

5   Conclusion

In this paper, we first theoretically analyze that the absence of conditional entropy in the objective function of AC-GAIL will affect the performance of the learned multi-modal policy. Further, we propose TAC-GAIL to deal with this problem. Its main idea is to introduce an additional auxiliary classifier in the framework of AC-GAIL to minimize the impact of missing conditional entropy. Experiments on the MuJoCo tasks also show that our method can effectively improve the performance of multi-modal imitation learning.

A   Appendix

A.1   Proof of Proposition 2

Proof. For a fixed G, Eq. 6 reduces to maximize the value function $V(G, C_\phi)$ w.r.t. $C_\phi(s, a|c = 1), ..., C_\phi(s, a|c = K)$:

$$\{C_\phi^*(s, a|c = 1), ..., C_\phi^*(s, a|c = K)\}$$

$$= \underset{C_\phi(s,a|c=1),...,C_\phi(s,a|c=K)}{\arg\max} \sum_{k=1}^{K} Q(s, a|c = k) \log(C_\phi(s, a|c = k)) \quad (12)$$

$$s.t. \sum_{k=1}^{K} C_\phi(s, a|c = k) = 1,$$

where the constraint is because C_ϕ is forced to have probability outputs that sum to 1. By applying Lagrange multipliers, we obtain the following problem:

$$\{C_\phi^*(s, a|c = 1), ..., C_\phi^*(s, a|c = K)\}$$

$$= \underset{C_\phi(s,a|c=1),...,C_\phi(s,a|c=K)}{\arg\max} \sum_{k=1}^{K} Q(s, a|c = k) \log(C_\phi(s, a|c = k)) \quad (13)$$

$$+ \lambda(\sum_{k=1}^{K} C_\phi(s, a|c = k) - 1).$$

Setting the derivative of Eq. 14 w.r.t. $C_\phi(s, a|c = k)$ to zeros, we obtain

$$C_\phi^*(s, a|c = k) = -\frac{Q(s, a|c = k)}{\lambda}. \quad (14)$$

We get λ by substituting Eq. 15 into the constraint, $\lambda = -\sum_{k=1}^{K} Q(s,a|c=k)$. Then we obtain the optimal solution

$$C_\phi^*((s,a),c=k) = \frac{P(s,a|c=k)}{\sum_{k'=1}^{K} P(s,a|c=k')} \tag{15}$$

References

1. Sutton, R.S., Barto, A.G.: Reinforcement learning: an introduction, 2nd edn. MIT Press, Cambridge (2018)
2. Silver, D., et al.: Mastering the game of Go with deep neural networks and tree search. Nature **529**(7587), 484–489 (2016)
3. Zhong, S., Liu, Q., Zhang, Z., Fu, Q.: Efficient reinforcement learning in continuous state and action spaces with Dyna and policy approximation. Front. Comput. Sci. **13**(1), 106–126 (2019)
4. Ho, J., Ermon, S.: Generative adversarial imitation learning. In: NeurIPS, pp. 4565–4573 (2016)
5. Lin, J., Zhang, Z., Jiang, C., Hao, J.: A survey of imitation learning based on generative adversarial nets. Chin. J. Comput. **43**(2), 326–351 (2020)
6. Zhu, J., et al.: Generative adversarial imitation learning from failed experiences (student abstract). In: AAAI, pp. 13997–13998 (2020)
7. Jiang, C., Zhang, Z., Chen, Z., Zhu, J., Jiang, J.: Third-person imitation learning via image difference and variational discriminator bottleneck (student abstract). In: AAAI, pp. 13819–13820 (2020)
8. Merel, J., et al.: Learning human behaviors from motion capture by adversarial imitation. arXiv preprint arXiv: 1707.02201 (2017)
9. Abbeel, P., Ng, A.Y.: Apprenticeship learning via inverse reinforcement learning. ICML 1–8 (2004)
10. Ng, A.Y., Russell, S.J.: Algorithms for inverse reinforcement learning. ICML 663–670 (2000)
11. Todorov, E., Erez, T., Tassa, Y.: Mujoco: a physics engine for model-based control. In:IROS, pp. 5026–5033 (2012)
12. Kingma, D.P., Ba, J.L.: Adam: a method for stochastic optimization. In: ICLR (2015)
13. Goodfellow, I.J., et al.: Generative adversarial nets. In: NeurIPS, pp. 2672–2680 (2014)
14. Schulman, J., Levine, S., Moritz, P., Jordan, M. Abbeel, P.: Trust region policy optimization. In: ICML, pp. 1889–1897 (2015)
15. Gong, M., Xu, Y., Li, C., Zhang, K., Batmanghelich, K.: Twin Auxilary Classifiers GAN. In: NeurIPS, pp. 1330–1339 (2019)
16. Lin, J., Zhang, Z.: ACGAIL: imitation learning about multiple intentions with auxiliary classifier GANs. In: Geng, X., Kang, B.-H. (eds.) PRICAI 2018. LNCS (LNAI), vol. 11012, pp. 321–334. Springer, Cham (2018). https://doi.org/10.1007/978-3-319-97304-3_25
17. Fei, C., et al.: Triple-GAIL: a multi-modal imitation learning framework with generative adversarial nets. In: IJCAI, pp. 2929–2935 (2020)
18. Kuefler, A., Kochenderfer, M.J.: Burn-in demonstrations for multi-modal imitation learning. In: AAMAS, pp. 1071–1078 (2018)

19. Li, Y., Song, J., Ermon, S.: InfoGAIL: Interpretable imitation learning from visual demonstrations. In: NeurIPS, pp. 3812–3822 (2017)
20. Wang, Z., Merel, J.S., Reed, S.E., de Freitas, N., Wayne, G., Heess, N.: Robust imitation of diverse behaviors. In: NeurIPS, pp. 5320–5329 (2017)
21. Ross, S., Gordon, G., Bagnell, D.: A reduction of imitation learning and structured prediction to no-regret online learning. In: AISTATS, pp. 627–635 (2011)
22. Ross, S., Bagnell, D.: Efficient reductions for imitation learning. In: AISTATS, pp. 661–668 (2010)

Top-N Recommendation in P2P Lending: A Hybrid Graph Ranking Using Investor Profile

Yuhang Liu[1], Huifang Ma[1,2,3(✉)], Yanbin Jiang[1], and Zhixin Li[2]

[1] College of Computer Science and Engineering, Northwest Normal University,
Lanzhou 730070, Gansu, China
[2] Guangxi Key Lab of Multi-source Information Mining and Security, Guangxi
Normal University, Guilin 541004, Guangxi, China
[3] Guangxi Key Laboratory of Trusted Software, Guilin University of Electronic
Technology, Guilin 541004, Guangxi, China
mahuifang@yeah.net

Abstract. The ever increasing development of P2P lending accumulates tremendous transaction data, a central question on these platforms is how to align the right products with the right investors. Most of the existing methods adapt some well-studied strategies for recommendation, we argue that an inherent drawback of such methods is that, the unique characteristics in the P2P lending scenario, such as the profile of investor, is not fully investigated. As such, the resultant recommendation may easily lead to suboptimal performance.

In this work, we propose to integrate the investor's profile into the recommendation process. We develop a new recommendation framework *enhanced Hybrid graph Ranking using Investor Profile (HRIP)*, which exploits a hybrid random walk-based recommendation via investor's profile from both the social and psychology aspects. This leads to the expressive modeling of representation of investor in investor-product hybrid graph, which can effectively deal with cold start users. Comprehensive analysis verifies the importance of the representation of investor, justifying the rationality and effectiveness of *HRIP*.

Keywords: P2P lending · Recommendation · Investor profile · Hybrid graph

1 Introduction

P2P lending is the practice of lending money to individuals or businesses through online services that match investors with borrowers. Online lending has experienced significant growth since the first P2P lending platform (i.e., Zopa[1]) was established in 2005. It is now an essential financing method for individual borrower to reach individual investor. As a fast-growing financial form, the gradual

---

[1] https://www.zopa.com.

© Springer Nature Switzerland AG 2020
H. Yang et al. (Eds.): ICONIP 2020, LNCS 12533, pp. 700–712, 2020.
https://doi.org/10.1007/978-3-030-63833-7_59

popularization of P2P lending in the industry has led to the accumulation of massive transaction data. For example, LendingClub, the world's largest P2P lending platform, can offer a plethora of historical data with over $59.2 billion by 03/31/2020 [1].

The research topic on the platform covers a variety of topics from the fields of economics, information technology and social sciences to investigate the relationship between lenders and borrowers. Among them, some researches have been devoted for guiding lenders to benefit from the P2P lending [2]. A recommender system is one solution that can alleviate the information overload problem by providing investor with personalized information [3]. Some recommendation strategies [4] have been presented to predict investors' interest by utilizing the available data information, thus helping investors find wanted projects as early as possible. In social lending context, products recommendation can help investors to find products that interest them, by aligning the right products with the right investors. However, as compared with traditional product recommendation, there are several challenges in designing recommender systems for products on social lending platforms.

On one hand, many factors may affect investors' funding decisions, the recommendation algorithm cannot simply rely on straightforward features that are directly available from the projects. On the other hand, product descriptions and investor descriptions must be analyzed so that investors' profile can be established for recommendation to address data sparsity and cold start problems.

To this end, we present a hybrid random walk-based recommendation approach to identify the potential interesting products for investor, which is capable of addressing data sparsity and cold start problem in product project recommendation. Specifically, we name our method as *Hybrid graph Ranking using Investor Profile*, i.e., *HRIP*, for this task. In experiments, we systematically evaluate our method on real-world dataset, and the experimental results demonstrate the effectiveness and robustness of *HRIP*. The contributions of this paper are summarized as follows:

- Both Profile modeling methods for product and investor are presented. Product profile is modelled via the *risk* and *return*, while the investor's profile is established not only from the perspective of his social and psychological state, but also his purchase history.
- A heterogeneous graph is constructed based on the interactions between investors and products, as well as the similarity between investors, based on which the hybrid graph ranking is performed.
- We conduct extensive experiments on real-world datasets. The experimental results clearly demonstrate the effectiveness and robustness of our approach.

The rest of the paper is organized as follows. Section 2 reviews the related works that are most relevant with our work. Section 3 describes the details of our proposed *HRIP* method. Section 4 designs experiments to evaluate our approach. Finally, Sect. 5 concludes the paper and suggests future research directions.

For the ease of description, we define terms that are interchangeably used throughout the literature and this paper – (a) we refer to a loan, a item or an investment as a product, (b) an investor is synonymous with a lender or a user.

2   Related Work

We review existing works on graph-based approaches for top-n recommendation, and recommendation systems in P2P lending, which are the most relevant works with ours.

2.1   Graph-Based Approaches for Top-N Recommendation

Top-n recommendation algorithms provide ranked lists of items tailored to the particular tastes of the users, as depicted by their past interactions within the system. There are a lot of graph-based approaches used for top-n recommendation that have been shown to have good recommended quality. The graph-based approaches perform a random walk on the user-item graph into one vector that represent user's preference for the item.

In particular, He et al. [5] developed user-item-aspect tripartite graph and used smoothness and fitting constraints on the graph for recommendation. Eksombatchai et al. [6] introduced the user-specific multi-pin transition probability on the 'pin-board' graph for recommendation. Nikolakopoulos et al. [7] proposed RecWalk, adding the item-item relationship to basic user-item bipartite graph.

Despite great success, we argue that these top-n recommendation methods are insufficient to capture the unique characteristics of the P2P lending, since these methods are general method.

2.2   Recommendation System in P2P Lending

With the burgeoning growth of P2P lending, a great deal of researches has been proposed to guide investors to recommend from the P2P lending. Here we highlight the differences with *HRIP*.

Zhao et al. [4] studied the project recommendation problem in P2P lending by managing risk through integrating portfolio theory into a personalized recommendation technique. Later, Zhao et al. [8] studied the loan recommendation problem in P2P projects from a multi-objective perspective. However, their work didn't construct product and investor profile in a comprehensive manner. In particular, they did not consider the social and psychological information of investors and failed to perform recommendation on a graph. Zhang et al. [9] studied the investor recommendation problem in P2P lending via a hybrid random walk approach, combining both collaborative filtering and content-based filtering. Although this work performed recommendation on the graph, it totally ignored investors' profile.

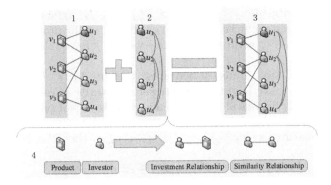

Fig. 1. An illustration of the investor-project hybrid graph construction process. The node u_1 is the target investor to provide recommendations for.

3   Hybrid Graph Ranking Method

In this section, we introduce the detail of the proposed *HRIP* model. An investment bipartite graph $G = (U, V, E)$ models the relationship between investors U and investment products V, associated with edges E connecting investors and their invested products in the past. Figure 1 illustrates the hybrid graph construction process. The investor of interest for recommendation is u_1, labeled with the red color. The first part shows the investor-project bipartite graph. The red edge is constructed if there is a purchase interaction between the investor and the product. For example, investor u_1 has invested project v_1, an edge is built between u_1 and v_1. Z_{11} represents the amount of money u_1 has invested in v_1. And the weight on the edge, denoted as τ_{ij}, is the proportion of money u_i invested in v_j divided by the total money u_i has already invested, such as $\tau_{11} = Z_{11}\big/\sum_j Z_{1j}$. The second part shows the investor similarity model that denoting the investor-investor connections. The similarity between investors is quantified on investor profile as discussed in Sect. 3.2. The third part displays the hybrid graph which is composed of the first two parts.

3.1   Product Profile

In this subsection, we show the way of assessing products' profiles on multiple characteristics. According to the previous study [10], there are three most important factors that investors take into account when deciding to invest a product: *Interest Rate, Non-Default Probability*, as well as *Fully-Funded Probability*. A natural idea is that products' profile can be portrayed with these three aspects.

Interest Rate (denoted as T_j^v). In P2P lending, borrower usually sets the product repayment rate (expressed as Re_j^v) during the auction. The T_j^v of the product can be replaced as Re_j^v, since Re_j^v are clearly given before they are released in P2P lending.

Non-Default Probability (denoted as D_j^v). The D_j^v refers to the possibility that the borrower can repay the principal or interest of the project. The following method is considered to estimate D_j^v, since we don't have relevant attributes to portray it directly.

We summarize two kinds of features: the products' features (*Amount, LendRate, Category* etc.) and the borrowers' features (*Credit, DebtToIncome* etc.). With training historical data, we can estimate D_j^v of products in V based on product features using logistic regression model. Specifically, for any $v_j \in V$, we denote its features as $\boldsymbol{v_j} = (1; v_{j,1}; v_{j,2}; \ldots; v_{j,d})$, then the D_j^v of $\boldsymbol{v_j}$ can be modeled as:

$$D_j^v = \exp(-\boldsymbol{\delta}^T \cdot v_j)/(1 + \exp(-\boldsymbol{\delta}^T \cdot v_j)) \tag{1}$$

where $\boldsymbol{\delta} = (\delta_0, \delta_1, \delta_2, \ldots, \delta_d)$ are the coefficients that to be learned. $\forall v_j \in V$, if it is paid in time by the borrower, the label of v_j on this object $y_j = 0$, else label $y_j = 1$. Given all these training data of products V, the logistic regression model learns the weight of $\boldsymbol{\delta}$ by maximum likelihood estimation.

Fully-Funded Probability (denoted as F_j^v). F_j^v suggests an estimate of the probability that the product will receive sufficient bids during the auction period. According to the '*all-or-nothing*' trading rule, the transaction will only be valid if the borrowed product receives sufficient bids during the auction period. Therefore, F_j^v is another important aspect of evaluating the product. Similar to D_j^v, we can get F_j^v through logistic regression model.

In summary, product v_j's profile can be represented as a three-element vector $\boldsymbol{P_j^v} = [T_j^v; D_j^v; F_j^v]$, where the first term is the *interest rate*, the second term is the *non-default probability*, together with the third term is the *fully-funded probability*.

3.2   Investor Profile

In this subsection, we show the way of assessing the profiles. It is generally believed that social and psychological attributes of investors will affect the investor's investment decision. Besides, the profile of the investor can also be characterized via products purchased history of the investor.

Society and Psychology-Based. Shyng et al. [11] classify personal investment as either *conservative, moderate* or *aggressive*. They established the main categories for each one and highlight the principal factors related to these types of investment, i.e. social and psychological factors. Social attributes of the investor include *gender, incomes* and *marital status* etc. Apart from that, psychological aspects include *self-esteem, emotion during risk* and so on. We adopt the same way [12] for quantifying investor's social and psychological attributes into a two-dimensional matrix. Put is another way, we construct user's profile as a two-element vector $[soc_i^u; psy_i^u]$.

Historical Purchase-Based. Besides social and psychological attributes, investor's investment preference $\boldsymbol{P_i^u} = [T_i^u; D_i^u; F_i^u]$ can be characterized through profile of products that investors have invested in history, i.e. $\boldsymbol{P_i^u}$. Generally speaking, the preference terms of $\boldsymbol{P_i^u}$ can be defined as the weighted average of the corresponding profile terms of the investor has invested: $T_i^u = \sum_{v_j \in VU_i} \tau_{ij} T_j^v$, $D_i^u = \sum_{v_j \in VU_i} \tau_{ij} D_j^v$, $F_i^u = \sum_{v_j \in VU_i} \tau_{ij} F_j^v$, where τ_{ij} is the ratio of u_i's investment in v_j, VU_i is the set of products that investor u_i has invested.

Based on these two aspects, a natural idea is that an investor's profile can be portrayed with the above two parts, i.e. $\boldsymbol{P}_i^u = [T_i^u; D_i^u; F_i^u; soc_i^u; psy_i^u]$.

3.3   Hybrid Graph Ranking Using Investor Profile

In this subsection, we introduce how *HRIP* performs the random walking process. Since the recommendations of potential products are for all investors, i.e. $u_i \in U$, *HRIP* starts from a target investor u_i at each time. For each investor u_i, *HRIP* works as follows:

At any time, *HRIP* may be at either an investor node u_i or a product node v_j. For any node position, *HRIP* has two options for the next move:

– with probability β $(0 < \beta < 1)$, *HRIP* restarts from the starting investor node u_i;
– with probability $1 - \beta$, *HRIP* continues walking.

If *HRIP* is at an investor node u_i. *HRIP* will choose to move to investors (blue lines) or products (red lines). In *HRIP*, the line selection follows a random variable R with a *Bernoulli* distribution: $R = 1$ means *HRIP* will choose blue lines, whereas $R = 0$ means *HRIP* will choose red lines. As is known different investor nodes should have different *Bernoulli* distributions which might be mainly influenced by funds held by investors. Intuitively, if an investor has few funds, the probability of moving to an investor should be higher since there are only less investor-product line options; on the contrary, for a lot of funds, it should have a higher probability of moving to a product. Consequently, the line selection probability of *HRIP* at an investor node ui can be calculated as follows:

$$p(R = 0 \,|\, u_i) = \alpha(u_i, \omega) = 2e^{-\omega M_i} / 1 + e^{-\omega M_i} \tag{2}$$

$$p(R = 1 \,|\, u_i) = 1 - \alpha(u_i, \omega) = 1 - e^{-\omega M_i} / 1 + e^{-\omega M_i} \tag{3}$$

where M_i represents funds held by u_i and $\omega \in \mathbb{R}^+$ is a parameter that can be used to adjust the value of $\alpha(.)$. When *HRIP* is at a node u_i, it will (1) move to other investors along the blue lines with probability $p(R = 1)$; (2) move to other products which has invested in v_j along the red lines with probability $p(R = 0)$.

If *HRIP* selects the blue lines when $R = 1$, the probability of moving from investor node u_i to another specific investor node u_k is calculated based on their similarity:

$$p(u_k \,|\, u_i, R = 1) = sim_{i,k} / \sum_{u_t \in U} sim_{i,t} \tag{4}$$

where $sim_{i,k}$ is the consine similarity between u_i and u_k.

By contrast, if *HRIP* selects a particular product v_j when $R = 0$, the probability of moving from u_i to v_j can be calculated as follows:

$$p(v_j \,|\, u_i) = Z_{i,j} / \sum_{v_k \in VU_i} Z_{i,k} \tag{5}$$

where $Z_{i,j}$ is the financial amount that investor u_i invests in product v_j.

If **HRIP** is at a product node v_j. HRIP will move to the investors that u_i invest in the past. The probability of selecting a particular investor u_i from product v_j is defined as similar to the move from investor node to product node:

$$p(u_i \,|v_j) = Z_{i,j} \Big/ \sum_{u_k \in UV_j} Z_{k,j} \tag{6}$$

where UV_j represents investors who invested in the product v_j.

Then the probability of HRIP visiting node u_i or node v_j can be deduced:

$$p^{(t+1)}(u_i) = \sum_{u_k \in U} p(u_i \,\big|u_k, R = 1)p^{(t)}(u_k)p(R = 1\,|u_k) + \sum_{v_j \in VU_i} p(u_i \,\big|v_j)p^{(t)}(v_j) \tag{7}$$

$$p^{(t+1)}(v_j) = \sum_{u_i \in UV_j} p(v_j \,\big|u_i, R = 0)p^{(t)}(u_i)\, p(R = 0\,|u_i) \tag{8}$$

The above formulas can be transformed into vector-matrix forms:

$$\boldsymbol{p}_u^{(t+1)} = \boldsymbol{S}(\boldsymbol{I} - \boldsymbol{A})\boldsymbol{p}_u^{(t)} + \boldsymbol{Z}_{VU}^T \boldsymbol{p}_v^{(t)} \tag{9}$$

$$\boldsymbol{p}_v^{(t+1)} = \boldsymbol{Z}_{UV}^T \boldsymbol{A}\boldsymbol{p}_u^{(t)} \tag{10}$$

where $\boldsymbol{p}_u^{(t)} = \left[p_{u_1}^{(t)}; p_{u_2}^{(t)}; \dots; p_{u_{|U|}}^{(t)}\right]$ and $\boldsymbol{p}_v^{(t)} = \left[p_{v_1}^{(t)}; p_{v_2}^{(t)}; \dots; p_{v_{|V|}}^{(t)}\right]$ represents the probability vector of visiting investor and product nodes at time t respectively. $\boldsymbol{Z}_{UV}$ is a $|U| \times |V|$ matrix, $(\boldsymbol{Z}_{UV})_{i,j} = p(v_j|u_i, R = 0)$; $\boldsymbol{Z}_{VU}$ is a $|V| \times |U|$ matrix, $(\boldsymbol{Z}_{VU})_{j,i} = p(u_i|v_j)$; In the same form, $\boldsymbol{S}$ is a $|U| \times |U|$ matrix with $(\boldsymbol{S})_{i,k} = p(u_k|u_i, R = 1)$. $\boldsymbol{A}$ is a $|U| \times |U|$ diagonal matrix with $A_{i,i} = \alpha(u_i, \omega)$. Equations.(9), (10) provide the major concepts by which HRIP approximates the result. Then we can implement the random walk process by rewriting Eqs.(9), (10), i.e.,

$$\boldsymbol{p}_u^{(t+1)} = (1 - \beta)[\boldsymbol{S}(\boldsymbol{I} - \boldsymbol{A})\boldsymbol{p}_u^{(t)} + \boldsymbol{F}_{VU}^T \boldsymbol{p}_v^{(t)}] + \beta \boldsymbol{p}_u^{(0)} \tag{11}$$

$$\boldsymbol{p}_v^{(t+1)} = (1 - \beta)\boldsymbol{F}_{UV}^T \boldsymbol{A}\boldsymbol{p}_u^{(t)} + \beta \boldsymbol{p}_v^{(0)} \tag{12}$$

HRIP performs random walks by Eqs.(11), (12), when the iteration is over and $\boldsymbol{p}_u$ reaches a stationary distribution, the stationary results $\boldsymbol{p}_u^*$ and $\boldsymbol{p}_v^*$ can be calculated. Let matrices $\boldsymbol{X}$ and $\boldsymbol{Y}$ be defined as follows:

$$\boldsymbol{X} = (\boldsymbol{I} - \boldsymbol{A}) \tag{13}$$

$$\boldsymbol{Y} = \boldsymbol{F}_{UV}^T \boldsymbol{F}_{VU}^T \boldsymbol{A} \tag{14}$$

we can obtain $\boldsymbol{p}_u^*$ and $\boldsymbol{p}_v^*$ by $\boldsymbol{p}_u^{(t+1)} = \boldsymbol{p}_u^{(t)} = \boldsymbol{p}_u^{(t-1)} = \boldsymbol{p}_u^*$ as follows:

$$\boldsymbol{p}_u^* = \beta \boldsymbol{\Psi} \boldsymbol{p}_u^0 \tag{15}$$

$$\boldsymbol{p}_v^* = \beta(1 - \beta)\boldsymbol{F}_{uv}^T \boldsymbol{A}\boldsymbol{\Psi}\boldsymbol{p}_u^0 + \beta \boldsymbol{p}_v^0 \tag{16}$$

where $\boldsymbol{\Psi} = [\boldsymbol{I} - (\boldsymbol{I} - \boldsymbol{A})(\boldsymbol{X} + \boldsymbol{Y})]^{-1}$.

Thus, the above approach can rank investors by the vector $\boldsymbol{p}_u^*$. Top-n ranked products have the highest probabilities of investing and will be recommended.

4   Experiments

In this section, we design experiments on real-world dataset to evaluate our approach. We aim to answer the following research questions:

– RQ1: How do different parameters settings (e.g.., β parameter, ω parameter) affect *HRIP*?
– RQ2: How does *HRIP* perform as compared with state-of-the-art recommendation algorithm in P2P Lending?
– RQ3: How to solve the cold start problem?

Table 1. Data Statistics of the Prosper Dataset

#products	#investors	#records	#TrR	#TeR
19,077	34,210	2,616,877	2,093,501	523,376

4.1   Experiments Setting

To evaluate the effectiveness of *HRIP*, we perform experiments on real-world dataset Prosper. We summarize the data statistics in Table 1. We choose Prosper, because Prosper has been in operation for more than 10 years, and hence, can offer a plethora of historical data that is necessary for training and testing. Other researchers who study P2P lending also use the same dataset from Propser [5], which makes it possible for us to compare our work with the current state-of-the-art techniques. We mainly use three tables of this data for our experiments.

Evaluation Metrics. For each investor in the test set, we treat all the products that the investor has not interacted with as the negative products. Then each method outputs the investor's preference scores over all the products, except the positive ones used in the training set. To evaluate the effectiveness of top-n recommendation and preference ranking, we adopt three widely-used evaluation protocols [13]: precision@n, recall@n and F1-score@n. By default, we set n=20. We report the average metrics for all investors in the test set.

Baselines. To demonstrate the effectiveness, first, the probability matrix is filled by calculating the percentage of the invest amount, then we compare our proposed *HRIP* with the following methods:

HRIP_1, *HRIP_2* and *HRIP_3* are variant methods of *HRIP*, where *HRIP_1* only performs random walk on the bipartite investment graph, and it not considers similarity between investors; *HRIP_2* executes random walk on the hybrid graph, but it only considers social and psychological attributes of the investor; *HRIP_3* runs random walk on the hybrid graph, but it only considers historical purchase relationship between investor and product.

RecWalk [7] added the item-item relationship to the basic user-item bipartite graph. *Pixie* [6] introduced the user-specific multi-pin transition probability on the 'pin-board' graph for recommendation. And *RWH* [9] is a hybrid recommendation framework tailored for loan recommendation with both collaborative filtering and content-based filtering. This method adds loan-loan routes into a loan-lender ranking framework.

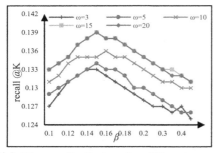 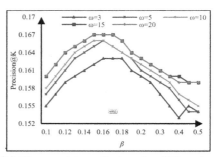

(a) The recall effect under different ω as parameter β change

(b) The precision effect under different ω as parameter β change

Fig. 2. The effect of parameters on experimental results

4.2   Parameter Affect (RQ1)

The *HRIP* method contains two important parameters, β and ω. In our approach, the β control restart probability to determine *HRIP* coming back to the starting node according to probability. And variable $R\tilde{\ }Bernoulli(\alpha(v_i, \omega))$ is introduced to determine whether *HRIP* should move to another investor node or a product node from the current investor node. This section discusses how to set these two important parameters in the experiment. The values of the parameters β and ω are estimated by the influence on the algorithm results.

Figure 2 plots the precision and recall of experimental results under different β and ω respectively on the dataset1. Specifically, Fig. 2(a) demonstrate the recall experimental result under different ω as parameter β change with $n = 20$, whereas Fig. 2(b) shows the precision experimental result under different ω as parameter β change with $n = 20$. It is clear that the performance of the algorithm increases first and then decays as β increases, and performance of the algorithm is optimal around $\beta = 0.15$. Also, when ω increases from 3 to 15, *HRIP* exhibits obvious improvements in both precision and recall. When ω continues to increase to 20, the performance of *HRIP* get steady state. For the best performance, we set $\omega = 15$ in our experiments.

Table 2. Overall Performance Comparison

	n	5	10	15	20	30	40	50
Precision	HRIP	**0.231**	**0.197**	**0.161**	**0.134**	**0.109**	**0.101**	**0.097**
	HRIP_1	0.185	0.157	0.134	0.117	0.094	0.085	0.083
	HRIP_2	0.203	0.176	0.143	0.121	0.094	0.086	0.085
	HRIP_3	0.228	0.189	0.150	0.124	0.096	0.094	0.092
	RecWalk	0.192	0.178	0.149	0.127	0.101	0.096	0.095
	RWH	0.227	0.187	0.148	0.122	0.094	0.092	0.091
	Pixie	0.083	0.078	0.069	0.052	0.042	0.036	0.034
Recall	HRIP	**0.041**	**0.081**	**0.111**	**0.139**	**0.155**	**0.161**	**0.164**
	HRIP_1	0.032	0.049	0.065	0.082	0.110	0.132	0.138
	HRIP_2	0.038	0.054	0.074	0.084	0.112	0.134	0.143
	HRIP_3	0.039	0.060	0.076	0.084	0.121	0.141	0.147
	RecWalk	0.040	0.079	0.083	0.115	0.143	0.150	0.151
	RWH	0.038	0.059	0.074	0.082	0.119	0.140	0.143
	Pixie	0.034	0.034	0.038	0.042	0.053	0.063	0.067

4.3 Performance Comparison (RQ2)

In this subsection, we mainly report the experimental results from the aspects of effectiveness. Table 2 reports the performance comparison results. We have the following observations:

The precision of the all methods decreases as top-n increases while the recall of all methods increases with the increasement of top-n.

Pixie and *HRIP_1* achieves the poorest performance across all cases. This indicates that the sparsity of the dataset is insufficient for *Pixie* to capture the complex relations between investors and products, further limiting its performance. and *RWH* consistently outperform *Pixie* and *HRIP_1* across all cases, demonstrating *RecWalk* and *RWH* have advantages over *Pixie* and *HRIP_1* by merging loan's profile into interactive relationship between lenders and loans. Moreover, *RecWalk* obtains slightly better performance than *RWH*. The reason might be that *RecWalk* defines item-item relationship in a way to enforce locality in the relations between the items, which is also easy to compute.

What's more, *RecWalk* and *RWH* achieves the better performance than *HRIP_2*. The reason might be that there are less social and psychological attribute data in the dataset. Also, *HRIP_3* consistently outperforms *RecWalk* and *RWH* in the most cases, demonstrating the importance of the characteristics of P2P lending.

Compared to other algorithms, the performance of *HRIP* incorporating P2P lending characteristics can improve the recommendation performance to some extent. Besides, *HRIP* performs better than *HRIP_1*, *HRIP_2* and *HRIP_3*. The reason might be that *HRIP* not only exploits the advantages of each of them, but also combines advantages of them.

4.4   Cold Start Affect (RQ3)

Cold Start is a common problem of recommender systems that new users or items have not yet gathered sufficient information to recommend or be recommended [14]. Here we focus on these investors and examine the performances of our model on cold-start problem of new investors.

Indeed, new investor has no interactions to be pretrained and recommender systems cannot predict investor preference. That makes many recommendation methods cannot work, especially profile-based method. However, for the *HRIP*, we can use social and psychological attributes of investors to predict investor preference. Here we test the recommendation results on 10 times from the beginning of the 0 in the interactions of the target new investor.

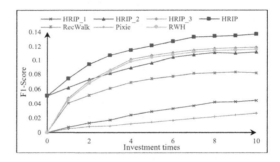

Fig. 3. Recommendation performances of new investor cold-start

The results are shown as Fig. 3. In all cases, *HRIP* and its variant methods perform better than *Recwalk* and *Pixie*. At the beginning of the investor interactions, *HRIP* and *HRIP_2* can get relatively good results because they can infer investor preferences based on the social and psychological attributes of investors. And with the increase of investors' investment times, *HRIP_2* has deteriorated to *HRIP_3* because the investor's historical the investor's historical investment records play more important role for investors' investor preferences than social and psychological attributes. Consequently, the results indicate the effectiveness for cold start investor of *HRIP* method.

From the above experimental results, we can see that our approach, i.e. *HRIP*, can provide the best results with P2P lending characteristics. In particular, *HRIP* is robust and adapted to cold start investor. This is because *HRIP* conduct a random walk on the constructed hybrid graph. The hybrid graph is constructed from investor history purchase records and investor profile, thus retaining the characteristics of the P2P lending for better recommendations.

5   Conclusions and Future Work

We have presented a *Hybrid graph Ranking using Investor Profile method* (*HRIP*) to model top-*n* recommendation for P2P lending. Particularly, we provide a consistent framework to jointly capture interactions in the investor-product graph and relationship between investors. Our experiments reveal that P2P lending characteristics plays a crucial role in the improvement of recommendation in P2P lending. Experimental results on real-world dataset show that HRIP can outperform state-of-the-art baselines.

Currently, we only incorporate the hybrid graph that combine bipartite investment graph and relationship graph between investors into recommendation, while real-world industries are associated rich other side information on items relationship. For example, products and products are associated with rich attributes. Therefore, exploring multi-hybrid network for recommendation with products relationship would be an interesting future direction.

Acknowledgment. This work is supported by the National Natural Science Foundation of China (61762078, 61363058, 61966004), Research Fund of Guangxi Key Lab of Multi-source Information Mining and Security (MIMS18-08), Major Project of Young Teachers' Scientific Research Ability Promotion Plan (NWNU-LKQN2019-2), Research Fund of Guangxi Key Laboratory of Trusted Software (kx202003) and Graduate Research Fund Project of Northwest Normal University (2019KYZZ012073).

References

1. LendingClub: 2019. LendingClub Marketplace. https://www.lendingclub.com/info/statistics.action. Accessed 6 April 2020
2. Wang, G., et al.: Product supply optimization for crowdfunding campaigns. IEEE Trans. Big Data, 1–1 (2018)
3. Ma, H., et al.: Matrix factorization recommendation algorithm fusing reliability and influence propagation. ACTA Autom. Sinica (2020)
4. Zhao, H., et al.: Investment recommendation in P2P lending: a portfolio perspective with risk management. In: 2014 IEEE International Conference on Data Mining, pp. 1109–1114. IEEE (2014)
5. He, X., et al.: Trirank: review-aware explainable recommendation by modeling aspects. In: Proceedings of the 24th ACM International on Conference on Information and Knowledge Management, pp. 1661–1670. ACM (2015)
6. Eksombatchai, C., et al.: Pixie: a system for recommending 3+ billion items to 200+ million users in real-time. In: Proceedings of the 2018 World Wide Web Conference. International World Wide Web Conferences Steering Committee, pp. 1775–1784 (2018)
7. Nikolakopoulos, A.N., Karypis, G.: Recwalk: nearly uncoupled random walks for top-n recommendation. In: Proceedings of the Twelfth ACM International Conference on Web Search and Data Mining, pp. 150–158. ACM (2019)
8. Zhao, H., et al.: Portfolio selections in P2P lending: a multi-objective perspective. In: Proceedings of the 22nd ACM SIGKDD International Conference on Knowledge Discovery and Data Mining, pp. 2075–2084. ACM (2016)

9. Zhang, H., et al.: Finding potential lenders in P2P lending: a hybrid random walk approach. Inf. Sci. **432**, 376–391 (2018)
10. Ceyhan, S., Shi, X., Leskovec, J.: Dynamics of bidding in a P2P lending service: effects of herding and predicting loan success. In: Proceedings of the 20th International Conference on World Wide Web. pp. 547–556. ACM (2011)
11. Shyng, J.Y., Shieh, H.M., Tzeng, G.H., Tzeng, S.H.: An integration method combining rough set theory with formal concept analysis for personal investment portfolios. Knowl.-Based Syst. **23**(6), 586–597 (2010)
12. Gonzalez-Carrasco, I., Colomo-Palacios, R., Lopez-Cuadrado, J.L., et al.: PB-ADVISOR: a private banking multi-investment portfolio advisor. Inf. Sci. **206**, 63–82 (2012)
13. Powers, D.M.: Evaluation: from precision, recall and F-measure to ROC, informedness, markedness and correlation. J. Mach. Learn. Technol. **1**(2), 37–63 (2011)
14. Ricci, Francesco, Rokach, Lior, Shapira, Bracha: Introduction to recommender systems handbook. In: Ricci, Francesco, Rokach, Lior, Shapira, Bracha, Kantor, Paul B. (eds.) Recommender Systems Handbook, pp. 1–35. Springer, Boston, MA (2011). https://doi.org/10.1007/978-0-387-85820-3_1

WD3-MPER: A Method to Alleviate Approximation Bias in Actor-Critic

Jiarun Cai[✉]

School of Computer Science and Technology,
Soochow University, Suzhou 215006, China
20185227073@stu.suda.edu.cn

Abstract. Deep deterministic policy gradient has been successfully applied to continuous control problems, but its function approximation errors will cause the overestimation problem and limit its performance. Existing methods alleviate the overestimation problem. However, because taking the minimum value between a pair of critics for updates, the method sometimes underestimates the values. We propose a new algorithm, which uses weighted value of two critics to alleviate the underestimation and overestimation problems caused by function approximation error. Simultaneously, in order to improve the sampling efficiency of the algorithm, we propose an improved prioritized experience replay mechanism by modifying the priority definition instead of the original random sampling. Experiments show that, compared with two state-of-the-art algorithms, our algorithm has better performance on the MuJoCo continuous control tasks.

Keywords: Reinforcement learning · Continuous control · Function approximation error · Sampling efficiency.

1 Introduction

In recent years, deep reinforcement learning has made remarkable achievements in decision-making problems and attracted a lot of attention in the field of artificial intelligence [6,7,19,22]. DeepMind uses a deep network to represent the value function, combined with Q-learning, designed the Deep Q-Network (DQN) [6], which is the first case of successfully combining deep learning and reinforcement learning. DQN performs well on discrete action space, but it performs poorly on continuous control tasks such as robot control. To solve the problem that DQN cannot handle the continuous action space, deep deterministic policy gradient algorithm (DDPG) [7] uses the Actor-Critic (AC) method, which contains a policy network to generate actions, and a Q-value network to evaluate the value of

This work is in part supported by the Natural Science Foundation of China (61876119), the Natural Science Foundation of Jiangsu (BK20181432) and a project funded by the Priority Academic Program Development of Jiangsu Higher Education Institutions.

© Springer Nature Switzerland AG 2020
H. Yang et al. (Eds.): ICONIP 2020, LNCS 12533, pp. 713–724, 2020.
https://doi.org/10.1007/978-3-030-63833-7_60

actions. DDPG differs from other actor-critic algorithms (e.g..,A3C, A2C) [5] in that it uses a deterministic policy which outputs a deterministic action [3].

The overestimation problem in reinforcement learning mainly appears in Q-learning, and the reason behind it is the maximum operation is used in the update process. Double Q-learning uses two estimators, and each estimator will use the value of the other estimator to update the value function, which alleviates the overestimation problem in the discrete action space. Fujimoto et al. prove that overestimation bias also occur in the actor-critic methods [1]. Twin Delayed Deep Deterministic policy gradient algorithm (TD3) [1] uses clipped double Q-learning based on double Q-learning. It also uses two Q-funtions. The difference is that clipped double Q-Learning takes the minimum value between the two estimates. Such an update rule minimizes the effect of overestimation bias, but it may induces an underestimation bias [13]. In this paper, we propose the Weighted Double Deep Deterministic policy gradient algorithm (WD3), which also uses two critics. Rather than taking the minimum value of two estimates, WD3 takes weighted value of two Q-estimates. This update rule greatly alleviates the underestimation and overestimation problems caused by function approximation error.

Experience replay mechanism of DDPG and TD3 is randomly sampling, which cannot more effectively use important samples to learn. Prioritized Experience Replay (PER) [2] proposes a new replay mechanism based on samples' priorities to make the algorithm more efficient. The priorities of samples is defined by TD error, which indicates how surprising or unexpected the sample is [2]. However, future rewards cannot be ignored for the importance of samples. Therefore, We modify the priority definition of the PER algorithm by considering n-step TD error and propose the Modified Prioritized Experience Replay algorithm (MPER). Finally, we combine WD3 with the improved MPER and propose Weighted Double Deep Deterministic policy gradient with Modified Prioritized Experience Replay algorithm (WD3-MPER) to alleviate the overestimation and underestimation problems and to improve sampling efficiency of experience replay.

2   Background

In this section, we introduce the basic definition of reinforcement learning, double Q-learning, deep deterministic policy gradient algorithm, twin delayed deep deterministic policy gradient algorithm and prioritized experience replay.

2.1   Reinforcement Learning

Reinforcement learning can be described as a process that an agent interacts with environment through the trial and error to maximize the cumulative reward, usually modeled as a Markov decision process (MDP) [9,20]. An MDP is represented by the tuple $M = (\mathcal{S}, \mathcal{A}, P, \gamma, \mathcal{R})$. $\mathcal{S}$ represents the state set, with $s \in \mathcal{S}$, and s_i represents the state of the i-th step. $\mathcal{A}$ represents a set of actions, with $a \in \mathcal{A}$, and a_i represents the action at step i. $P(s, a)$ represents the probability distribution over next state after the agent executes action a in state s. For example, when performing action a in state s, the probability of transition to next state s' can be expressed as $p(s'|s, a)$. $\mathcal{R}$ is the reward function. If (s, a) moves to next state s', then the reward function can be written as $r(s'|s, a)$. If the next state s' corresponding to (s, a) is unique, then the reward function can also be written as $r(s, a)$. γ is a discount factor and represents the importance of future rewards.

2.2 Double Q-Learning

The overestimation problem refers to finding the maximum value of a series of numbers first, and then averaging these numbers, usually greater than or equal to finding the average value first and then the maximum value [12].

$$\mathbb{E}\left(\max\left(X_1, X_2, \ldots, X_n\right)\right) \geq \max\left(\mathbb{E}\left(X_1\right), \mathbb{E}\left(X_2\right), \ldots, \mathbb{E}\left(X_n\right)\right) \tag{1}$$

where $X_1, X_2, \ldots, X_n$ are all sets of numbers and $\mathbb{E}(X)$ is expectation of a number set X. The overestimation problem mainly appears in Q-learning, and the reason behind it is the maximum operation is used in the update process of the following formula:

$$Q\left(s, a\right) \leftarrow Q\left(s, a\right) + \alpha\left(r + \gamma \max_a Q\left(s', a\right) - Q\left(s, a\right)\right) \tag{2}$$

where α is step-size parameter and $Q_t\left(s, a\right)$ is state action value for (s, a) at time t. The overestimation bias generated by the maximize operation will seriously affect the accuracy of value evaluation. To solve the overestimation problem, double Q-learning uses two estimators, which are represented by two functions Q^1 and Q^2, and each Q-function takes the value of another Q-function to update. Both of Q-functions must be learned from different experience buffers, but you can use two value functions at the same time to select the action to be performed.

$$Q^1(s, a) \leftarrow Q^1(s, a) + \alpha\left(r + \gamma Q^2\left(s', a^*\right) - Q^2(s, a)\right) \tag{3}$$

$$Q^2(s, a) \leftarrow Q^2(s, a) + \alpha\left(r + \gamma Q^1\left(s', a^*\right) - Q^2(s, a)\right) \tag{4}$$

where a^* is the action with the largest Q-value.

2.3 DDPG

DDPG adopts the framework of the AC method and is applied to continuous control problems. DDPG differs from other actor-critic algorithms in that it uses a deterministic policy. Deterministic policy μ means that the action performed in a state is deterministic. Stochastic policy π outputs the probability distribution of performing actions in a state. Therefore, the deterministic policy changes the output process of action, and only outputs an action instead of probability distribution of actions. After the policy becomes deterministic, the Bellman equation of the state-action value function also changes:

$$Q^{\pi}\left(s_{t}, a_{t}\right) = \mathbb{E}_{r_{t}, s_{t+1} \sim \mathcal{S}}\left[r\left(s_{t}, a_{t}\right) + \gamma \mathbb{E}_{a_{t+1} \sim \pi}\left[Q^{\pi}\left(s_{t+1}, a_{t+1}\right)\right]\right] \tag{5}$$

$$Q^{\mu}\left(s_{t}, a_{t}\right) = \mathbb{E}_{s_{t+1} \sim \mathcal{S}}\left[r\left(s_{t}, a_{t}\right) + \gamma Q^{\mu}\left(s_{t+1}, \mu\left(s_{t+1}\right)\right)\right] \tag{6}$$

where $\mu(s_{t+1})$ outputs a deterministic action. Therefore, compared to the stochastic policy, deterministic has no expectation function for actions. Due to the maximum operation of Q-learning, overestimation bias is obvious, but the presence of overestimation bias is less clear in the actor-critic setting. Fujimoto et al. proved the presence and effect of overestimation bias in actor-critic settings [1].

2.4 TD3

In double DQN, the authors propose using the target network as one of the value estimates, and obtain a policy by greedy maximization of the current value network rather than the target network [1,11].

$$Q_{\text{target}} = r + \gamma Q_{\theta'}\left(s', \pi_{\phi}\left(s'\right)\right) \tag{7}$$

where $Q_{\theta'}$ is a target critic network, Q_{target} is the learning target of action value, and π_{ϕ} is a policy network.

But in fact, due to the slow change of policy in actor-critic methods, the current and target networks are too similar to make an independent estimation. Therefore, TD3 uses original double Q-learning instead of double DQN. However, the critics are not completely independent because of TD3 uses two critics and the same replay buffer. Such a problem induces critics to overestimate Q-values. To solve this problem, clipped double Q-learning is proposed to take the minimum value between two critics.

$$Q_{\text{target}} = r + \gamma \min_{i=1,2} Q_{\theta_{i}'}\left(s', \pi_{\phi}\left(s'\right)\right) \tag{8}$$

Next, TD3 uses two tricks to alleviate the function approximation errors. One is delayed policy updates, which delays the update of the target network and policy to avoid the cumulative error during the update process. The second is target policy smoothing, which adds noise to the target action for smoothing value function. Although this setting minimizes the effect of overestimation bias, it sometimes underestimates.

2.5 Per

In order to break the association among samples, DDPG and TD3 use a replay buffer with the random sampling mechanism. However, in the case of sparse rewards, rewards are only available after multiple correct actions. There will be few samples that can incentivize agent to learn correctly. In this case, the efficiency of random sampling is very low because many samples are rewarded with 0. PER samples experience based on samples' priorities, but not just valuable experience. Since it will cause overfitting. Therefore, the lowest value also have a small probability to be sampled. In PER, the way to measure the priorities of samples is TD error. If TD error is relatively large, it means that the current Q-function is still far away from the target Q-function, and it should be trained more [11].

$$\delta_t = r + \gamma \max_a Q\left(s_{t+1}, a\right) - Q(s_t, a_t) \tag{9}$$

where δ_t is TD error. In DDPG, because of the output action is deterministic, TD error becomes the following form:

$$\delta_t' = r + \gamma Q\left(s_{t+1}, a\right) - Q(s_t, a_t) \tag{10}$$

In order to avoid overfitting, we need to ensure that the sample with TD error is equal to zero also has a small probability to be sampled. Thus, we set

$$P(i) = \frac{p_i}{\sum_j p_j} \tag{11}$$

where $p_i = |\delta_i + \epsilon|$ is the priority of transition i. ϵ is a small positive constant that prevents the sample with TD error is zero not being revisited [2].

3   Method

We propose a weighted double DDPG method to alleviate the overestimation and underestimation problems caused by function approximation error and an improved PER method to improve the sampling efficiency of the replay buffer in this section. Then we combine WD3 with the modified PER to propose the WD3-MPER algorithm.

3.1   Weighted Double DDPG

Due to the function approximation error caused by the maximum operation, Q-learning usually overestimates the action values. Double Q-learning alleviates the problem of overestimation in discrete spaces by using two independent estimators. Clipped double Q-learning uses the minimum value between a pair of critics to minimize the effects of overestimation bias in actor-critic. But clipped double Q-learning sometimes underestimates the action value. To alleviate this problem, we propose a weighted double Q-learning algorithm for actor-critic. WD3 also uses two estimators. Unlike clipped double Q-learning, we update the value function by using the weighted value of maximum value and minimum value. Therefore, WD3 alleviates the overestimation problem in Q-learning and the underestimation problem that sometimes occurs in clipped Q-learning by setting

$$Q_{\text{target}} = r + \gamma \left(w \times \min_{i=1,2} Q_{\theta_i'} \left(s', \pi_\phi \left(s' \right) \right) + \left(1 - w \right) \max_{i=1,2} Q_{\theta_i'} \left(s', \pi_\phi \left(s' \right) \right) \right) \quad (12)$$

where w is a predefined parameter, usually in the range of $[0.5, 1]$. We did not directly use the weighted values of two critics to update the value function. Instead, we first calculate the maximum value and minimum value of the two Q-values, and then calculate the weighted value of these two values to update the value network. The value calculated for updating in this way is different from the value directly weighting the two Q-values. If the latter is used for updating, the algorithm performance may be poor. In general, we pay more attention to the minimum value so that we can better avoid the overestimation problem caused by function approximation error [13, 21].

3.2   MPER

When using the one-step TD method, we update the value function at each step, which allows us to fully consider changes in the environment. However, in many cases, the environment will not change immediately. Only after a period of time the environment will change significantly. For example, the robot's behavior pattern has certain coherence between actions, and the effect of one-step update is not very good. In these cases, the n-step update usually performs better. Therefore, we propose a new method for adjusting the priorities of samples by using n-step TD error. It is worth noting that we only use n-step TD error to adjust the priority, instead of using n-step return to update the value function. n-step returns have more information than instant rewards because of the former pays more attention to future rewards. Therefore, n-step TD error also illustrates the learning degree and importance of samples. In summary, we improve the PER by using the n-step TD error return of the state action pair to adjust the priority of samples.

$$G_{t:t+n} = r_{t+1} + \gamma r_{t+2} + \cdots + \gamma^{n-1} r_{t+n} + \gamma^n Q_{t+n-1} \left(s_t, a_t \right) \quad (13)$$

$$\delta_{\text{new}} = \delta_t' + \beta \times \left(G_{t:t+n} - Q(s_t, a_t) \right) \quad (14)$$

where β is a parameter used to adjust the proportion of n-step TD error in the newly defined priority δ_{new}, and is usually set to $1/n$. $G_{t:t+n}$ is the n-step return of state-action pair at time step t.

Next, we use the number of this transition was sampled to adjust its priority. This setting is to ensure the trainings times of each transition to avoid overfitting caused by insufficient training times. Although the sample with TD error is 0 also has a lower probability to be sampled, we still hope that a strong limit can be imposed so that each sample can be trained multiple times. We map the training times of samples to an interval, and the replay buffer uses it to adjust the priorities of samples:

Algorithm 1: WD3-MPER

Input: Initialize critic networks $Q_{\theta_1}, Q_{\theta_2}$, actor network π_ϕ with random
parameters θ_1, θ_2, ϕ, and target networks $\theta_1' \leftarrow \theta_1, \theta_2' \leftarrow \theta_2, \phi' \leftarrow \phi$

Output: $Q_{\theta_1}, Q_{\theta_2}, \pi_\phi$

1 **for** $t=1{:}T$ **do**

2 $\quad$ Select action with exploration noise and observe reward r and new state s'

3 $\quad$ Set $\lambda = t + n - 1$

4 $\quad$ **if** $\lambda \geq 0$ **then**

5 $\quad\quad \Big|\ G = \sum_{i=\lambda+1}^{\min(\lambda+n,T)} \gamma^{i-\lambda-1} r_i, \beta = 1/n$

6 $\quad$ **else**

7 $\quad\quad \Big|\ G = r, \beta = 1$

8 $\quad$ **end**

9 $\quad$ Store tuple $(s, a, r, s', G, count)$ in replay buffer

10 $\quad$ Sample N tuples $(s, a, r, s', G, count)$ from replay buffer

11 $\quad$ $\tilde{a} \leftarrow \pi_{\phi'}(s') + \epsilon, \epsilon \sim \mathcal{N}(0, \sigma)$

12 $\quad$ $Q_{\text{target}} =$
$\quad\quad r + \gamma \left(w \min_{i=1,2} Q_{\theta_i'}(s', \pi_\phi(s')) + (1-w) \max_{i=1,2} Q_{\theta_i'}(s', \pi_\phi(s')) \right)$

13 $\quad$ $G = G + \gamma \left(w \min_{i=1,2} Q_{\theta_i'}(s', \pi_\phi(s')) + (1-w) \max_{i=1,2} Q_{\theta_i'}(s', \pi_\phi(s')) \right)$

14 $\quad$ $\delta_{\text{new}} = Q_{\text{target}} - Q_{\theta_1}(s, a) + \beta(G - Q_{\theta_2}(s, a))$

15 $\quad$ Use $\delta_{\text{new}}, count$ to update samples' priorities

16 $\quad$ Update critics $\theta_i \leftarrow \text{argmin}_{\theta_i} N^{-1} \sum_j \left(Q_{\text{target}}^j - Q_{\theta_i}(s_j, a_j) \right)^2$

17 $\quad$ Update ϕ by the deterministic policy gradient:

18 $\quad\quad \nabla_\phi J(\phi) = N^{-1} \sum_j \nabla_a Q_{\theta_1}(s, a) \Big|_{a_j = \pi_\phi(s_j)} \nabla_\phi \pi_\phi(s)$

19 $\quad$ Update target networks:

20 $\quad\quad \theta_i' \leftarrow \tau\theta_i + (1-\tau)\theta_i'$

21 $\quad\quad \phi' \leftarrow \tau\phi + (1-\tau)\phi'$

22 **end**

$$count \rightarrow N, N \sim [0, \rho] \qquad (15)$$

where count is the training times of samples and N is a interval in the range $[0, \rho]$. We generally set an upper limit. When the training times of a certain sample reach this upper limit, we will not give priority reward to this sample. The setting of upper limit and Eqn. 9 in PER largely guarantee minimum training times of the samples and avoid the overfitting problem.

3.3 WD3-MPER

We proposed WD3 in the previous section, which is based on TD3 and DDPG. The purpose is to alleviate the overestimation problem in Q-learning and the underestimation problem sometimes caused by taking the minimum values between the two estimators in clipped Q-learning. Next, an improved PER method is introduced to improve the sampling efficiency. WD3-MPER is summarized in Algorithm 1.

In Algorithm 1, we first initialize two critic networks and an actor network by using random parameters, and copy these parameters to the corresponding target network. In each time step, we use a policy network with exploration noise to select action and observe reward and next state given by the environment. We store tuple $(s, a, r, s', G, count)$ in replay buffer and initialize the priority of each sample to the maximum priority, which is generally 2.0. We initialize count to 0. When n-step return cannot be calculated, we simply set G to r and set β to 1. MPER degenerates into PER in this setting. When n-step return can be calculated, we set β to $1/n$. n-step TD error is generally much larger than TD error. Therefore, in order not to ignore TD error, we need to reduce the proportion of n-step TD error in δ_{new}. After calculating δ_{new}, we use δ_{new} and $count$ to update the priorities of samples. When the training number of samples is small, we give the sample a larger priority reward to encourage it to be trained again. The priority reward decreases as $count$ increases. When it exceeds the limit we set, the priority reward drops to 0. Finally, we update actor-critic networks and corresponding target network.

4   Experiments

In the first part of this section, we completely used the experimental settings in [1] to compare our algorithm with TD3 and DDPG on the MuJoCo continuous control task [14]. In the second part, we modify the hyper-parameters to show the advantages of our algorithm.

4.1   Experimental Evaluation

In order to evaluate the performance of MD3-MPER, we put it on six MuJoCo continuous control tasks of Open AI Gym [16]. Because of WD3 is improved on the basis of DDPG and TD3, we compare the performance of WD3-MPER with these algorithms.

Table 1. Comparison of the maximum average return of three algorithms running 1 million time steps on 5 random seeds. Maximum value for each task is bolded.

Environment	Algorithm		
	WD3-MPER	TD3	DDPG
Ant-v2	**5118 ± 525**	4320 ± 981	460 ± 185
HalfCheetah-v2	**10742 ± 1136**	9760 ± 920	10166 ± 985
Hopper-v2	3484 ± 173	**3499 ± 159**	1518 ± 357
Reacher-v2	−4.12 ± 0.75	**−3.95 ± 0.60**	−5.42 ± 1.27
InvertedPendulum-v2	**1000 ± 0**	**1000 ± 0**	600 ± 200
InvertedDoublePendulum-v2	**9337 ± 22**	8506 ± 840	1869 ± 65

The DDPG in experiment used the modified DDPG in [1], named OurDDPG. OurDDPG also uses a two layer feedforward neural network of 400 and 300 hidden nodes respectively, with rectified linear units between each layer. OurDDPG differs in that the critic receives both the state and action as input to the first layer, and such a setting is advantageous for DDPG to handle continuous control tasks. TD3 uses a two layer feedforward neural network of 256 and 256 hidden nodes respectively, with rectified linear units between each layer. Both network parameters are updated using Adam [8] with a learning rate of 10^{-3}.

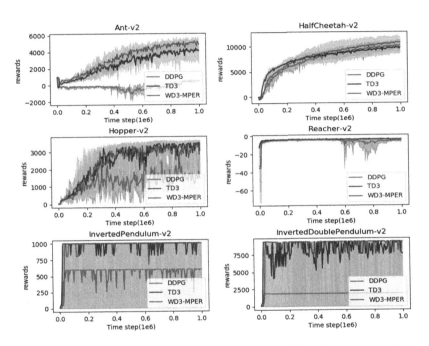

Fig. 1. Learning curves of three algorithms running 1 million time steps on 5 random seeds for the MuJoCo continuous control tasks.

We compare our algorithm with DDPG and TD3 without any parameter modification. Our results are presented in Fig. 1 and Table 1. In Table 1, we compare WD3-MPER with two benchmark algorithms by using the maximum average return. The results show that WD3-MPER outperforms DDPG in six MuJoCo continuous control tasks. WD3-MPER is significantly superior to TD3 in Ant-v2, HalfCheetah-v2, and InvertedDoublePendulum-v2, and achieves the same performance as TD3 in InvertedPendulum-v2. Although the maximum average return of WD3-MPER in Hopper-v2 and Reacher-v2 is slightly lower than TD3, the difference is very small. Figure 1 shows the comparison of learning curves between WD3-MPER and these two benchmark algorithms. In summary, WD3-MPER matches or outperforms DDPG and TD3 in both final performance and learning speed across all tasks.

4.2 Experimental Comparison

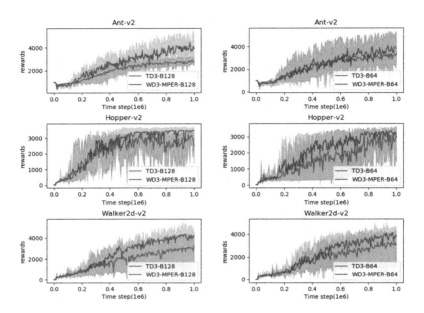

Fig. 2. The learning curves of TD3 and WD3-MPER at batch sizes of 128 and 64.

Batchsize is the number of sample sets sampled during training, which is directly related to the training times of samples. We set the batchsize to 256 in the previous part of the experiment. Under this setting, TD3 has between 0 and 450 training times for each sample, and the training times for more than half of the samples are concentrated between 200 and 300. In addition to the samples stored later, each sample can be trained enough, which makes it difficult to show the advantages of MPER. Next, we set the batchsize of WD3-MPER and TD3 to 128, 64, which reduces training times of samples. Such experimental settings can better show the performance of the algorithm.

In Fig. 2, the three graphs on the left are learning curves for TD3 and WD3-MPER with batchsize is 128. Under this setting, compared to the case of batchsize is 256, the performance of WD3-MPER has not degradation significantly, but TD3 has dropped a lot. This is because reducing the batchsize is equivalent to reducing training times of samples. As the training times of samples decrease, it means that training times of some samples may be insufficient, so random sampling has more randomness. Once the training times of important samples are small or even zero, the performance of algorithm will eventually be poor, thereby lowering the average performance of the algorithm. Figure 2 shows that when the batchsize is 64, compared to batchsize is 128, the performance gap between WD3-MPER and TD3 is narrowing. We believe that it is because we have imposed a strong limit on training times of samples in order to avoid over-fitting, thus ensuring the minimum training times of samples. Therefore, in the case where batchsize is too small, the distribution of training times of samples sampled by WD3-MPER will be close to that of TD3.

5   Summary

This paper proposes a new algorithm called weighted double deep deterministic policy gradient with modified prioritized experience replay. We propose a method between single estimator and double estimator that take minimum value between two critics, which is to use the weighted value of two Q-values. This method greatly alleviates the overestimation and underestimation problems. Then WD3 combines with improved MPER to improve sampling efficiency of experience replay. Experiments prove the effectiveness of our algorithm on the MuJoCo continuous control tasks. Next, we prove the advantage of prioritized sampling compared to random sampling by modifying experimental parameters.

In the future, in order to further improve the accuracy of function approximation and the sampling efficiency of replay buffer, we will strive to find a more accurate method of estimating Q-value and a better definition of sample's priority.

References

1. Fujimoto, S., van Hoof, H., Meger, D.: Addressing function approximation error in actor-critic methods. In: International Conference on Machine Learning (2018)
2. Schaul, T., Quan, J., Antonoglou, I., Silver, D.: Prioritized experience replay. In: International Conference on Learning Representations, Puerto Rico (2016)
3. Silver, D., Lever, G., Heess, N., Degris, T., Wierstra, D., Riedmiller, M.: Deterministic policy gradient algorithms. In: International Conference on Machine Learning, pp. 387–395 (2014)
4. Anschel, O., Baram, N., Shimkin, N.: Averaged-DQN: variance reduction and stabilization for deep reinforcement learning. In: International Conference on Machine Learning, pp. 176–185 (2017)
5. Mnih, V., et al.: Asynchronous methods for deep reinforcement learning. In: International Conference on Machine Learning, pp. 1928–1937 (2016)
6. Mnih, V., et al.: Human-level control through deep reinforcement learning. Nature 518(7540), 529–533 (2015)
7. Lillicrap, T.P., et al.: Continuous control with deep reinforcement learning. In: International Conference on Learning Representations, Puerto Rico (2016)
8. Kingma, D., Ba, J.: Adam: a method for stochastic optimization. In: The 3rd International Conference for Learning Representations, San Diego (2015)
9. Sutton, R.S., Barto, A.G.: Reinforcement learning: an introduction, vol. 1. MIT Press, Cambridge (1998)
10. Hou, Y., Liu, L., Wei, Q., Xu, X., Chen, C.: A novel DDPG method with prioritized experience replay. IEEE International Conference on Systems. IEEE (2017)
11. Van Hasselt, H., Guez, A., Silver, D.: Deep reinforcement learning with double q-learning. In: AAAI, pp. 2094–2100 (2016)

12. Van Hasselt, H.: Double q-learning. In: Advances in Neural Information Processing Systems, pp. 2613–2621 (2010)
13. Zhang, Z., Pan, Z., Kochenderfer, M.J.: Weighted double Q-learning. In: Twenty-sixth International Joint Conference on Artificial Intelligence (2017)
14. Todorov, E., Erez, T., Tassa, Y.: MuJoCo: a physics engine for model-based control. In: 2012 IEEE/RSJ International Conference on Intelligent Robots and Systems (IROS), pp. 5026–5033. IEEE (2012)
15. Barth-Maron, G., et al.: Distributed distributional deterministic policy gradients. In: International Conference on Learning Representations, Vancouver (2018)
16. Brockman, G., et al.: OpenAI Gym (2016)
17. He, Q., Hou, X.: Reducing Estimation Bias via Weighted Delayed Deep Deterministic Policy Gradient (2020)
18. Daley, B., Amato, C.: Reconciling λ-returns with experience replay. Adv. Neural Inf. Process. Syst. **32**, 1133–1142 (2019)
19. Ma, X., Driggs-Campbell, K.R., Zhang, Z., Kochenderfer, M.J.: Monte-carlo tree search for policy optimization. In: Proceedings of the 28th International Joint Conference on Artificial Intelligence (IJCAI-2019), pp. 3116–3122, Macao, China (2019)
20. Liu, Q., Zhai, J.-W., Zhang, Z.-Z., Zhong, S., Zhou, Q., Zhang, P., Xu, J.: A survey on deep reinforcement learning. Chin. J. Comput. (2018)
21. Zheng, Y., Hao, J.-Y., Zhang, Z.-Z., Meng, Z.-P., Hao, X.-T.: Efficient multiagent policy optimization based on weighted estimators in stochastic environments. J. Comput. Sci. Technol. **35**, 268–280 (2020). https://doi.org/10.1007/s11390-020-9967-6
22. Zhong, S., Liu, Q., Zhang, Z., Fu, Q.: Effcient reinforcement learning in continuous state and action spaces with Dyna and policy approximation. Front. Comput. Sci. **13**(1), 106–126 (2019)

Robotics and Control

A Novel Vascular Robotic System: Performance Evaluation

Si-Yi Wei[1], Xiao-Bo Sun[1], Xiao-Hu Zhou[2(✉)], and Zeng-Guang Hou[2]

[1] Institute of Automation, Harbin University of Science
and Technology, Harbin, China
[2] Institute of Automation, Chinese Academy of Sciences, Beijing, China
xiaohu.zhou@ia.ac.cn

Abstract. Percutaneous coronary intervention (PCI) has become a common method for the treatment of cardiovascular diseases (CVDs). However, the accumulated X-ray radiation during the procedures greatly increases the probability of medical staff suffering from cataracts and brain tumors. This study bases on an existing vascular robotic system designed in our previous work. The main component of this robotic system is a bio-inspired Dual-finger Robotic Hand (DRH), which consists of a pair of bionic thumb and forefinger to imitate the surgical manipulations of interventionalists. This study is to evaluate the performance of the robotic system through a series of experiments: advancing a guidewire at different speeds and accelerations. The mean root mean square error (RMSe) of the actual and desired axial movements is 0.72 ± 0.49 mm, demonstrating the effectiveness and robustness of the robotic system.

Keywords: Vascular robotic system · PCI · Performance evaluation

1 Introduction

Cardiovascular diseases (CVDs) are No. 1 killer of the world. In 2016, more than 17.9 million people died of CVDs, accounting for 46% among all noncommunicable diseases. The number of CVDs deaths is predicted to rise to 22.2 million by 2030 [1]. Coronary heart disease is a main CVD, accounting for 45% of CVDs deaths [2]. The main treatments for coronary heart disease are coronary artery bypass graft (CABG) and percutaneous coronary intervention (PCI). The former needs to open the patient's chest, resulting in tremendous trauma and a long recovery period. As a minimally invasive procedure, PCI has become the popular treatment for coronary heart disease.

This work was supported in part by the National Key Research and Development Program of China under Grant 2019YFB1311700; in part by the Youth Innovation Promotion Association of CAS under Grant 2020140; in part by the National Natural Science Foundation of China under Grant 61533016, Grant U1913601, and Grant 61421004; and in part by the Strategic Priority Research Program of CAS under Grant XDBS01040100.

© Springer Nature Switzerland AG 2020
H. Yang et al. (Eds.): ICONIP 2020, LNCS 12533, pp. 727–737, 2020.
https://doi.org/10.1007/978-3-030-63833-7_61

During the conventional PCI procedures, X-ray imaging is used to locate interventional devices. Hazards of being exposed to X-ray radiation include an increased risk of cataract and a possible association with the development of head and neck tumors [3]. Therefore, medical staff have to wear heavy lead aprons to prevent them from high-dose radiation. However, there are evidences showing that long hours of working by wearing lead aprons also leads to cervical and lumbar disc diseases among interventionalists [4].

To address these problems, various robotic systems are developed to assist interventionalists to deliver interventional devices during PCI procedures [5]. Beyar et al. proposed a remote navigation system (RNS), which is commercialized by Corindus Inc. [6]. Its latest generation is named CorPath GRX, which consists of two major components: an interventional cockpit and a bedside unit. Its safety and feasibility are testified through human trials and recognized by the Food and Drug Administration (FDA) [7,8]. Magellan robotic system is developed by Hansen Medical Inc., which is mainly used to conduct peripheral vascular intervention [9]. Cha et al. designed a new catheter driving system that can provide surgeons with less X-ray exposure and convenient user interface [10]. Su et al. presented a master-slave tele-operation system for percutaneous interventions under continuous magnetic resonance imaging (MRI) guidance [11]. The Niobe magnetic navigation system (Stereotaxis, MO, USA) uses a magnetic field to steer a specialized catheter with magnets at the tip [12,13]. Marcelli et al. developed a highly compact and versatile robotic system for remote navigation of standard tip-steerable electrophysiology (EP) catheters [14]. Guo et al. designed a new robotic catheter system with a master-slave structure for vascular intervention [15,16]. Tavallaei et al. presented a new compact and sterilizable telerobotic system that allows remote navigation of conventional tip-steerable catheters [17]. Bao et al. developed a novel method that provides higher operation efficiency than a previous prototype and allows for complete sterilization[18]. Kesner et al. proposed an actuated catheter tool which can compensate for the motion of heart structure by servoing a catheter guidewire inside a flexible sheath [19]. Cercenelli et al. developed a telerobotic system to remotely manipulate standard steerable EP catheter during the cardiac interventional procedures [20].

There are some problems in the current vascular intervention robotic systems. For example, the torsion of guidewire is caused by unreasonable structural design of the belt [10]. The inadequacy of grasper structure lies in the inconvenience for loading and unloading of interventional devices in clinical practice. In addition, some of them can only deliver a single catheter, not satisfying the delivery of other devices. The magnetron robotic system in [11–13] uses special magnetic devices, not suitable for patients implanted with ferruginous medical devices.

To address these problems, a bio-inspired Dual-finger Robotic Hand (DRH) was designed in our previous work [21]. Its open structure facilitates the loading and unloading of different interventional devices. Besides, it can deliver both guidewires and balloon/stent catheters.

The main contributions of this research include: 1) The effectiveness and robustness of the robotic system are evaluated. 2) Extensive comparative

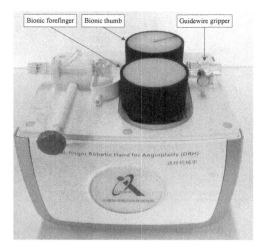

Fig. 1. The Dual-finger Robotic Hand includes a guidewire gripper, a bionic thumb, and a bionic forefinger.

experiments and statistical analysis highlights the importance of speed for high-precision delivery.

The rest of this paper is organized as follows. Section 2 describes the architecture of the DRH. Section 3 evaluates the performance of the robotic system through a series of experiments. Finally, we conclude in Sect. 4.

2   System Architecture

In PCI procedures, the guidewire is clamped by the interventionalist's right thumb and forefinger. To achieve guidewire translation, the interventionalist needs to push the guidewire by right-to-left translation of the right hand and retract the guidewire by left-to-right translation.

Inspired by the manipulations of interventionalists, the bio-inspired DRH (see Fig. 1) is developed. It mainly consists of two rollers: bionic thumb and bionic forefinger, which are controlled by two stepper motors. The bionic fingers realize the translation manipulation to deliver interventional devices by rotating around their own axes. Different from human hands, they can deliver interventional devices continuously. Meanwhile, the clamping force between two fingers can be adjusted via controlling the current of a DC motor, making it convenient to load and unload interventional devices. By fixing the guidewire with a guidewire gripper, this structure can be compatible with other interventional devices (e.g. catheter, balloon/stent) with different diameters.

Aiming to improve the accuracy and stability of delivery, a control algorithm is designed in a 32-bit ARM microcontroller (STM32F4, STMicroelectronics, USA) to control the movement of two bionic fingers. Mathematically, the trans-

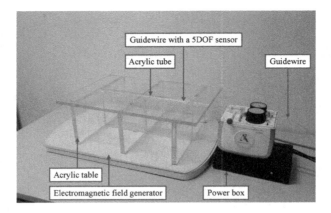

Fig. 2. The experimental setup.

lational displacement of interventional devices can be calculated by the following formula:

$$\Gamma = \frac{\pi n_p \theta_s R_{\text{roller}}}{180 i_s i_r} \tag{1}$$

where Γ is the translational displacement, n_p is the number of pulses generated by the controller, θ_s is the step angle of the stepper motor, i_s is the subdivision number of the stepper motor drivers, i_r is the reduction ratio of the gearboxes, and R_{roller} is the radius of rollers.

3   Experiments and Results

3.1   Experimental Setups

Due to the small diameter of coronary arteries, guidewire delivery requires high precision, which can reduce the contact force between the guidewire and vessel wall. Recent studies show that the high-precision delivery is critical to the success of PCI procedures [21]. In order to evaluate the performance of robotic system, extensive comparative experiments are designed at different speeds and accelerations.

As shown in Fig. 2, the experiments are performed on an electromagnetic (EM) tracking system (Aurora, Northern Digital Inc. Canada) which can offer a way to measure guidewire translation data. It includes an EM field generator, a signal processing unit, and a sensor. The EM field generator is positioned under an acrylic table. The displacement of the guidewire is acquired with a 5DOF sensor. In order to ensure the accuracy of measured data, the sensor is connected with the tip of the guidewire coaxially. Therefore, the displacement of the guidewire is equal to that of the sensor. Furthermore, an acrylic tube is fixed on the acrylic table to simulate coronary arteries.

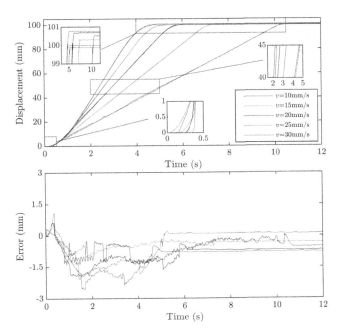

Fig. 3. The dynamic tracking trajectories of advancing the guidewire at different speeds ($a = 20\,\mathrm{mm/s^2}$). The dotted lines are desired trajectories, and the solid are actual ones.

The sample rate of the EM tracking system 40 Hz. The delay between two samplings is acceptable in a closed-loop control for real-time applications. Data acquisition program simultaneously collects the number of pulses of the robotic system and the displacement of the EM sensor. Normally, the root mean square error (RMSe), one of the most commonly used performance metrics, is selected to evaluate the performance of the robotic system. Mathematically, the RMSe between actual trajectories and desired ones can be calculated by the following formula:

$$\mathrm{RMSe} = \sqrt{\frac{\sum_{i=1}^{N} \left(Z_i - K_i\right)^2}{N}} \tag{2}$$

where Z_i is the ith actual displacement acquired by the EM sensor, K_i denotes as the ith desired displacement calculated by Eq. (1), N is the number of samples.

3.2 Experiment: Advancing the Guidewire at Different Speeds and Accelerations

In PCI procedures, the performance of the robotic system at different speeds and accelerations are the basis for high-precision delivery. In this part, different constant speeds (10–30/s, increased by 5 mm/s between two adjacent ones) are selected for advancing the guidewire. $20\,\mathrm{mm/s^2}$, $40\,\mathrm{mm/s^2}$ and $60\,\mathrm{mm/s^2}$ are

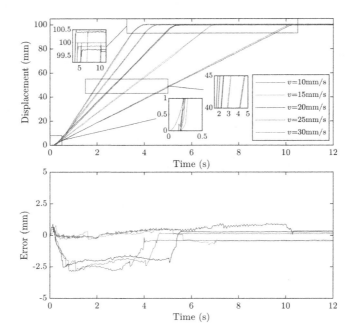

Fig. 4. The dynamic tracking trajectories of advancing the guidewire at different speeds ($a = 40\,\text{mm/s}^2$). The dotted lines are desired trajectories, and the solid are actual ones.

selected as low, medium and high accelerations, respectively. Guidewire displacement is set to 100 mm in this experiment.

The results of advancing the guidewire under different constant speeds at the low acceleration are shown in Fig. 3. Theoretically, the delivery process can be divided into three stages. The first stage is to accelerate the speed to 10 mm/s, 15 mm/s, 20 mm/s, 25 mm/s, and 30 mm/s within 0.5 s, 0.75 s, 1 s, 1.25 s, and 1.5 s, respectively. The subsequent stage is the constant-speed stage. The third stage is the deceleration stage to reduce the speed to 0 at $-20\,\text{mm/s}^2$.

Moreover, the RMSe between actual and desired trajectories are given in the Table 1, where the corresponding minimum and maximum are indicated in blue and red, respectively. According to the first row of Table 1, the minimum RMSe is 0.27 mm achieved by the speed of 10 mm/s, and the maximum is 0.49 mm yielded by the speed of 30 mm/s at the acceleration stage. For the constant-speed stage, the minimum and maximum RMSe are 0.40 mm at the speed of 15 mm/s and 0.85 mm at the speed of 10 mm/s, respectively. Finally, the minimum (0.33 mm) and maximum (3.86 mm) are implemented by the speed of 10 mm/s and 30 mm/s at the deceleration stage, respectively.

Similarly, the dynamic performance at the medium acceleration ($40\,\text{mm/s}^2$) and the high acceleration ($60\,\text{mm/s}^2$) under different constant speeds are shown in Fig. 4 and 5. Moreover, the RMSe between actual and desired trajectories are

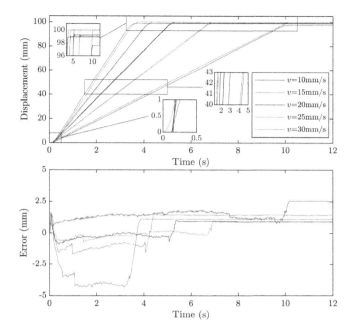

Fig. 5. The dynamic tracking trajectories of advancing the guidewire at different speeds ($a = 60\,\text{mm/s}^2$). The dotted lines are desired trajectories, and the solid are actual ones.

given in the second and third row of Table 1, respectively. Besides, the corresponding minimum and maximum are also highlighted in blue and red, respectively.

For each of the comparative experiments executed under three acceleration states (e.g. low, medium and high acceleration states), statistical significance is assessed with the Pearson correlation analysis and one-way ANOVA for quantitative features. The Pearson correlation coefficient between the speed and RMSe is shown in Figs. 6(a–c). In most cases (e.g. $20\,\text{mm/s}^2$ and $40\,\text{mm/s}^2$), the RMSe displays statistically significant difference between different speeds ($P < 0.05$). Figures 6(d–h) show no statistical significance between the acceleration and RMSe ($P > 0.05$).

The one-way ANOVA results of speed and RMSe are shown in Fig. 7 *(Left)*. According to the results, the RMSe is statistically significant difference between different speeds ($P < 0.01$). As shown in Fig. 7 *(Right)*, the one-way ANOVA results of acceleration and RMSe shows no statistical significance ($P > 0.05$).

In general, all experiments are designed after considering several practical factors:1) The first purpose is to ensure the quality of the collected data. The Pearson correlation analysis and one-way ANOVA shows that the RMSe increases with delivery speed rises linearly. 2) The second is the consideration of applications. The purpose of this study is to assist interventionalists with guidewire delivery. The movements of guidewire are usually slow.

Table 1. The RMSe (mm) of advancing the guidewire at different speeds and accelerations.

Accel-eration	Acceleration Stage					Constant-Speed Stage					Deceleration Stage				
	10	15	20	25	30	10	15	20	25	30	10	15	20	25	30
	mm/s	mm/s	mm/s	mm/s	mm/s	mm/s	mm/s	mm/s	mm/s	mm/s	mm/s	mm/s	mm/s	mm/s	mm/s
$20 mm/s^2$	**0.27**	0.29	0.36	0.37	0.49	0.85	**0.40**	0.80	0.72	0.79	**0.33**	0.73	1.49	2.29	3.86
$40 mm/s^2$	**0.27**	0.29	0.38	0.42	0.48	0.37	**0.22**	1.20	1.17	1.22	**0.05**	0.31	0.45	0.82	1.39
$60 mm/s^2$	**0.25**	0.29	0.42	0.39	0.89	1.05	**0.22**	0.26	0.62	3.19	0.27	**0.10**	0.23	0.32	0.94

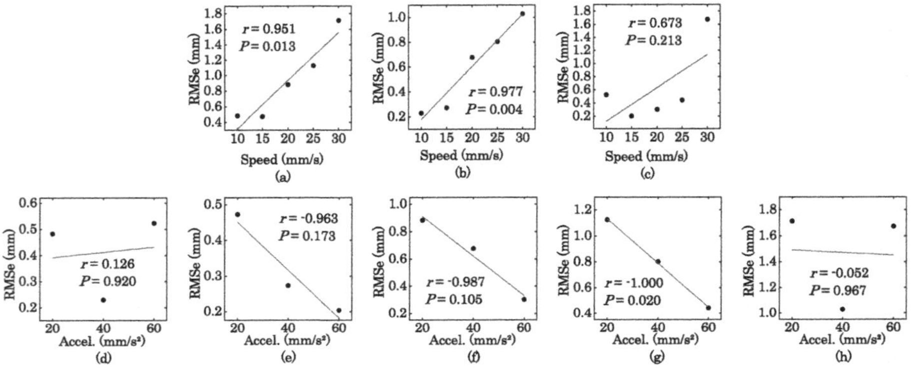

Fig. 6. The Pearson correlation coefficient of the speed and RMSe under $20\,mm/s^2$(a), $40\,mm/s^2$(b), and the $60\,mm/s^2$(c). The Pearson correlation coefficient of the acceleration and RMSe under $10\,mm/s$(d), $15\,mm/s$(e), $20\,mm/s$(f), $25\,mm/s$(g), $30\,mm/s$(h).

3.3   Discussion

The performance of the robotic system are evaluated through various comparative experiments: advancing the guidewire at different speeds and accelerations. Specifically, there are a total of three movement states are the low, medium and high acceleration states. Each states of movement is subdivided into three different forms: acceleration stage, constant-speed stage and deceleration stage. Furthermore, for the error that affected the accuracy of the translation manipulation, six practical factors are considered, including the roundness error, the sensitivity of stepper motor, friction error, vibration error, the elastic deformation of rubber sheath and the performance of the EM sensor. These factors also exist in clinical practice. Therefore, it is necessary to analyze the robotic system's delivery performance under the noise environment.

On the one hand, the RMSe between actual and desired trajectories presents a continuous upward trend at acceleration and deceleration stages with the rise of constant speed. Table 1 shows the RMSe decreases firstly and then increases with the speed climbs steadily during constant-speed stage. Moreover, the RMSe decreases with the speed climbs from $10\,mm/s$ to $15\,mm/s$, while the RMSe

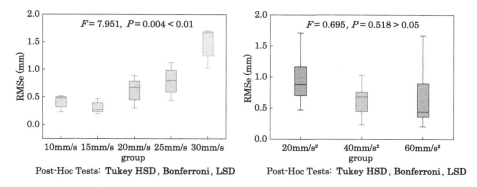

Fig. 7. The one-way ANOVA results of the speed and RMSe *(Left)*, acceleration and RMSe *(Right)*. Tukey HSD, Bonferroni and LSD are used as post-hoc tests for one-way ANOVA.

increases with the speed climbs to the 20 mm/s. After that, the smallest RMSe for each experiment is selected by comparing the changing trend on the table data. Specifically, the minimum RMSe is achieved by the speed of 10 mm/s.

In most cases, the RMSe decreases with the decline of guidewire speed ($P <$ 0.05). These promising results indicate the potential of the robotic system for facilitating the future development of PCI for coronary heart disease treatment. It is worth mentioning that the roundness error of the bionic fingers causes the RMSe to some extent during delivery. Theoretically, the bionic fingers are designed as ideal cylinders, however, there are radius errors actually. Because the delivery distance is equal to the length of the bionic finger arc. Consequently, the radius errors could directly affect delivery accuracy.

According to the Table 1, the RMSe is maximum when the guidewire is advanced at the constant speed of 30 mm/s. There are several reasons: 1) The elastic deformation of rubber sheath. After a long time of use, there are obvious wear marks on the surface of rubber sheath, resulting in losing the elasticity of the rubber sheath. 2) The accuracy of the EM sensor. Specially, the EM sensor could not accurately acquire guidewire motion data at high speeds.

Furthermore, the RMSe does not show consistent changing trends in constant-speed stage with the rise of delivery speed. The main reason for this includes: 1) The sensitivity of stepper motor. The lack of proper lubrication frequently causes overheating and excessive wear on motor bearings, resulting in that the RMSe does not change significantly with the rise of speed in the experiments. 2) The sensitivity of EM sensor. Limited by the sensitivity of the EM sensor, it is difficult to accurately measure the subtle change of guidewire movement caused by the increased speed. 3) The friction between the acrylic tube and the guidewire. The guidewire is relatively flexible, and is easily affected by the friction during translation process, which increases the RMSe to a certain extent.

In the experiment with $20\,\mathrm{mm/s^2}$, a slight vibration occurs when the guidewire moves at low speed. The vibration disappears when guidewire speed rises steadily. The main reason for this is the low pulse frequency of the stepper motor. The low pulse frequency will vibrate the stepper motor, leading to the vibration of the guidewire.

4   Conclusion

This study evaluates the effectiveness and robustness of the robotic system through a series of experiments: advancing the guidewire at different speeds and accelerations. The mean RMSe of the actual and desired axial movements is $0.72 \pm 0.49\,\mathrm{mm}$. Extensive experiments indicate that the guidewire should be advanced as slowly as possible in fragile coronary arterys. Meanwhile, high acceleration can not only reduce surgical time but also guarantee delivery accuracy, thus can decrease medical staff's exposure to ionizing radiation. In the subsequent work, several updated processing techniques will be used to reduce the radius errors of the bionic fingers. Similarly, some advanced filtering methods will be utilized to eliminate the noise in the measurement data for closed-loop control. The ergonomics, sterilization, as well as control algorithms of the robotic system will be further exploited and improved in future work. Rather than controlled and predictable setups, further research will be conducted through clinical trials.

References

1. Mendis, S.: Global Status Report on Noncommunicable Diseases 2014. World Health Organization, Geneva (2014)
2. Go, A.S., et al.: Executive summary: heart disease and stroke statistics-2014 update: a report from the American Heart Association. Circulation **129**(3), 399–410 (2014)
3. Klein, L.W., et al.: Occupational health hazards of interventional cardiologists in the current decade: results of the 2014 SCAI membership survey. Cathet. Cardiovasc. Interv. **86**(5), 913–924 (2015)
4. Pourdjabbar, A., Ang, L., Reeves, R.R., Patel, M.P., Mahmud, E.: The development of robotic technology in cardiac and vascular interventions. Rambam Maimonides Med. J. **8**(3) (2017)
5. Peters, B.S., Armijo, P.R., Krause, C., Choudhury, S.A., Oleynikov, D.: Review of emerging surgical robotic technology. Surg. Endosc. **32**(4), 1636–1655 (2018)
6. Beyar, R., et al.: Remote-control percutaneous coronary interventions: concept, validation, and first-in-humans pilot clinical trial. J. Am. Coll. Cardiol. **47**(2), 296–300 (2006)
7. Smitson, C.C., Ang, L., Pourdjabbar, A., Reeves, R., Patel, M., Mahmud, E.: Safety and feasibility of a novel, second-generation robotic-assisted system for percutaneous coronary intervention: first-in-human report. J. Invas. Cardiol. **30**(4), 152–156 (2018)

8. Almasoud, A., Walters, D., Mahmud, E.: Robotically performed excimer laser coronary atherectomy: proof of feasibility. Cathet. Cardiovasc. Intervent. **92**(4), 713–716 (2018)

9. Clements, W., Scicchitano, M., Koukounaras, J., Joseph, T., Goh, G.S.: Use of the Magellan robotic system for conventional transarterial chemoembolization (cTACE): a 6-patient case series showing safety and technical success. J. Clin. Intervent. Radiol. ISVIR **3**(02), 142–146 (2019)

10. Cha, H.-J., Yi, B.-J., Won, J.Y.: An assembly-type master-slave catheter and guidewire driving system for vascular intervention. Proc. Inst. Mech. Eng. Part H J. Eng. Med. **231**(1), 69–79 (2017)

11. Su, H., Shang, W., Li, G., Patel, N., Fischer, G.S.: An MRI-guided telesurgery system using a Fabry-Perot interferometry force sensor and a pneumatic haptic device. Ann. Biomed. Eng. **45**(8), 1917–1928 (2017)

12. Ramcharitar, S., Patterson, M.S., Van Geuns, R.J., Van Meighem, C., Serruys, P.W.: Technology insight: magnetic navigation in coronary interventions. Nat. Clin. Pract. Cardiovasc. Med. **5**(3), 148–156 (2008)

13. Patterson, M.S., et al.: Primary percutaneous coronary intervention by magnetic navigation compared with conventional wire technique. Eur. Heart J. **32**(12), 1472–1478 (2011)

14. Marcelli, E., Cercenelli, L., Plicchi, G.: A novel telerobotic system to remotely navigate standard electrophysiology catheters. In: 2008 Computers in Cardiology, pp. 137–140. IEEE (2008)

15. Guo, J., Guo, S., Shao, L., Wang, P., Gao, Q.: Design and performance evaluation of a novel robotic catheter system for vascular interventional surgery. Microsyst. Technol. **22**(9), 2167–2176 (2016)

16. Ma, X., Guo, S., Xiao, N., Yoshida, S., Tamiya, T.: Evaluating performance of a novel developed robotic catheter manipulating system. J. Micro-Bio Robot. **8**(3–4), 133–143 (2013)

17. Tavallaei, M.A., et al.: Design, development and evaluation of a compact telerobotic catheter navigation system. Int. J. Med. Robot. Comput. Assist. Surg. **12**(3), 442–452 (2016)

18. Bao, X., et al.: Operation evaluation in-human of a novel remote-controlled vascular interventional robot. Biomed. Microdevices **20**(2), 34 (2018)

19. Kesner, S.B., Howe, R.D.: Robotic catheter cardiac ablation combining ultrasound guidance and force control. Int. J. Robot. Res. **33**(4), 631–644 (2014)

20. Cercenelli, L., Marcelli, E., Plicchi, G.: Initial experience with a telerobotic system to remotely navigate and automatically reposition standard steerable EP catheters. ASAIO J. **53**(5), 523–529 (2007)

21. Feng, Z.-Q., Bian, G.-B., Xie, X.-L., Hou, Z.-G., Hao, J.-L.: Design and evaluation of a bio-inspired robotic hand for percutaneous coronary intervention. In: 2015 IEEE International Conference on Robotics and Automation (ICRA), pp. 5338–5343 (2015)

Accuracy Estimation for an Incrementally Learning Cooperative Inventory Assistant Robot

Christian Limberg[1,2(✉)] [iD], Heiko Wersing[2], and Helge Ritter[1]

[1] Research Institute for Cognition and Robotics,
Bielefeld University, Bielefeld, Germany
{climberg,helge}@techfak.uni-bielefeld.de
[2] Honda Research Institute Europe GmbH, Carl-Legien-Street 30,
63073 Offenbach am Main, Germany
heiko.wersing@honda-ri.de

Abstract. Interactive teaching from a human can be applied to extend the knowledge of a service robot according to novel task demands. This is particularly attractive if it is either inefficient or not feasible to pre-train all relevant object knowledge beforehand. Like in a normal human teacher and student situation it is then vital to estimate the learning progress of the robot in order to judge its competence in carrying out the desired task. While observing robot task success and failure is a straightforward option, there are more efficient alternatives. In this contribution we investigate the application of a recent semi-supervised confidence-based approach to accuracy estimation towards incremental object learning for an inventory assistant robot. We evaluate the approach and demonstrate its applicability in a slightly simplified, but realistic setting. We show that the configram estimation model (CGEM) outperforms standard approaches for accuracy estimation like cross-validation and interleaved test/train error for active learning scenarios, thus minimizing human training effort.

Keywords: Incremental learning · Active learning · Accuracy estimation · Household robots · Mobile object recognition

1 Introduction

The application of autonomous home service robots is nowadays still limited to constrained tasks like vacuum cleaning or lawn mowing. The main reason for this is the unresolved difficulty of robust sensory perception in dynamic environments [2]. One possible remedy for this limitation is keeping the human in the loop for supervisory control and occasional take-over of responsibility for actions or decisions of the robot. This approach can be formulated as a problem of complementary human-machine function allocation in automation [17] or viewed as a human-robot cooperation framework [10,15,16,24].

© Springer Nature Switzerland AG 2020
H. Yang et al. (Eds.): ICONIP 2020, LNCS 12533, pp. 738–749, 2020.
https://doi.org/10.1007/978-3-030-63833-7_62

An attractive means for equipping robots with the knowledge they need for successful operation within a working domain is incremental learning from a human teacher [7,8]. This differs from the standard concept of batch or offline training where basically all perceptual and action knowledge has to be extracted from exhaustive training data before its actual application on the robot. While standalone recognition performance increases with larger batch training data sets, generalization to real situated robot applications remains challenging [2, 18], due to the infinite combinatorial variations of environment and situation conditions. It may also be an explicit part of the robot task to acquire new knowledge about objects, which are not known beforehand.

If we consider human and robot as a cooperative team, they should contribute towards a joint goal with their individual competences. In order to enhance the robot competence the human can engage in incremental teaching of the robot. It is then vital to monitor the learning success during teaching, like in a human student-teacher situation. A straightforward approach would be judging the success of learning simply by observing the robot performing the planned task and counting its failure and success events. Adverse consequences of errors and inefficiency of this try and error approach, however, call for more advanced methods. Consider for example a robotic lawn mower that is trained by the owner to avoid children toys and precious flowers on the lawn while still required to mow over weeds and leaves (investigated by [13] for an online learning setting).

In this contribution we are interested in the continuous performance estimation of an incremental learning classifier that is the basis for a cooperative task scenario. We assume a typical (albeit simplified) inventory scenario as it has to be done in thousands of shops around the world on a regular basis: There are a number of different products in the shop present on displays and shelves where all instances have to be localized and counted. This is a tedious task where the support of a mobile robot that could assist in visually identifying large numbers of objects would be certainly helpful. Let us assume the shop keeper teaches the target objects based on images taken by the robot facing the shelves and display tables. In order to make this teaching as efficient as possible the following questions have to be answered:

- Since labeling takes considerable human effort, how can we provide the most efficient set of training data?
- How good is the expected performance of the robot object classifier under the constraint that a certain task error rate shall not be exceeded?

The answer to the first question has been addressed in the established research field of *active learning* [6], where most popular methods select samples with lowest classifier confidence to achieve the greatest gain per each human training input. The second question is typically answered by performing cross validation, i.e. separating a small subset of the labeled training data and evaluating the test performance on this hold-out set. It was already demonstrated, however, that for incremental learning these two approaches are incompatible [12]. The uncertainty-based selection strategy of active learning accumulates difficult labeling examples which in turn induce a much too pessimistic cross-validation error estimate. Limberg et al. [12] have developed an approach using confidence-based semi-supervised accuracy estimation (CGEM) that can resolve this issue.

In the remainder of this manuscript we review related work in Sect. 2, redefine the configram accuracy estimation approach (CGEM) from [12] in Sect. 3, investigate its application to a teachable inventory assistant robot in Sect. 4 and give our conclusions in Sect. 5.

2   Related Work

Estimating the competence of the interaction partner has been discussed as an important factor for efficient and successful human-robot cooperation. Robot errors typically have a severe impact on human trust in cooperative scenarios [22], emphasizing the need for robust performance estimators.

Incremental learning for object recognition in robots has been studied as an interesting alternative to standard batch learning approaches using offline learning, where applicability to real robot scenarios has remained rather limited [2,18]. Kirstein et al. [8] developed a vision architecture for incremental learning of multiple visual categories based on interactive in-hand object training. The architecture is capable of intuitive error-correction in real-time based on corrective speech-based user feedback. The architecture was extended by recent CNN-based feature architectures and investigated by Hasler et al. [7]. Losing et al. [13] explored the interleaved test/train error as a means for evaluating learning progress in an interactive training scenario for a garden robot.

Accuracy estimation for batch learning approaches was analyzed by Platanios et al. [19] who estimate classifier accuracy by considering the agreement rate of multiple classifiers of different types trained with independent features. Another recent approach by Donmez et al. [3] also works with a single classifier but requires the label distribution $p(y)$ for evaluating a maximum likelihood. This is applicable e.g. for medical diagnosis or handwriting recognition, where the marginal frequency of each class is known. Welinder et al. [26] showed that it is possible to estimate binary classifier precision and recall classwise by fitting a mixture model per class in a histogram of confidences and sample those mixture models

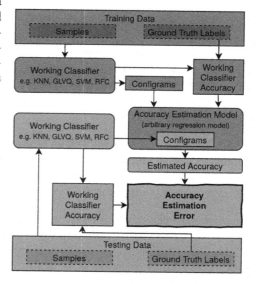

Fig. 1. Illustration of our accuracy estimation approach.

with various techniques. Aghazadeh and Carlsson [1] proposed a method to determine the quality of a train and test set by evaluating local and global moments for each class, like intra-class variation and connectivity. They evaluated this fully supervised approach via a leave one out cross validation on Pascal VOC 2007 and could predict the final mean absolute error (MAE) of the held out class with about 4–5% accuracy error.

Kreger et al. [9] apply a meta-learner using alternative features for predicting the performance in a road-terrain detection for autonomous navigation.

To unite the requirements of active and incremental learning Limberg et al. [12] have proposed a new approach to accuracy estimation based on training an estimation model using distance information from an instance based classifier. Further they demonstrated the advantage over cross-validation and interleaved test/train error on benchmark data. Building on this work, we here investigate the application of a more general approach using any classifier capable of generating confidence estimates to our scenario of a cooperative inventory robot.

3 Accuracy Estimation with CGEM

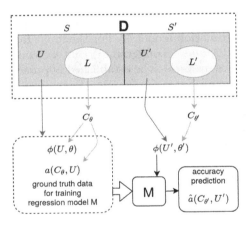

Fig. 2. A domain $\mathbf{D}$ contains data pools $\mathbf{S}, \mathbf{S'}$. Our estimation model is trained on training pool $\mathbf{S}$ to further monitor trainings on test pool $\mathbf{S'}$ or further pools $\mathbf{S''}, \mathbf{S'''}$ etc. (not depicted).

We implement our accuracy estimation approach (Fig. 1) for an incremental classification learning setting, where we use a standard incremental learning paradigm [14] for training an online classifier. The samples chosen for training can be either selected randomly or using active learning, where we select new samples for labeling based on maximal uncertainty of the current classifier [6].

Our assumption is, that training a sequence of labeled samples will improve the accuracy of the classifier over time. For improving label efficiency, we may want to stop training if the estimated accuracy of the classifier falls below a desired threshold. In an offline setting, detecting accuracy changes can be done by querying several labeled samples. This strategy, however, may be too inefficient and imprecise in our task setting, where human labels are usually expensive. Because of this, our goal is to construct an estimator that takes the usually high number of *unlabeled samples* into account.

For formal definition we denote by $\mathbf{D}$ the data set that characterizes our domain (see Fig. 2). Data pools $\mathbf{S}$ and $\mathbf{S'}$ are distinct subsets within $\mathbf{D}$, where each subset consists of an unlabeled and a labeled pool ($\mathbf{U}, \mathbf{L}$ and $\mathbf{U'}, \mathbf{L'}$ respectively).

We start training with an empty $\mathbf{L}$. A querying function is used to select samples in minibatches with size B from $\mathbf{U}$. This can be done either randomly or uncertainty-based. These minibatches are labeled by an oracle, which is typically a human annotator or another ground truth source. While training, samples from the unlabeled pool $\mathbf{U}$ become labeled and moved into the labeled pool $\mathbf{L}$. In parallel, the classifier C is trained online with the new samples. We train the classifier with a number of minibatches N.

Formally, our classifier – in the following denoted as *working classifier* – is given as a function $y = C_\theta(\mathbf{u})$ that maps unlabeled data item $\mathbf{u}$ into its class label y and $\boldsymbol{\theta}$ denotes the adaptive classifier parameters. Also, our working classifier needs to have a confidence measure $c_p(\mathbf{u}, \boldsymbol{\theta}) \in [0, 1]$ that describes how reliable input $\mathbf{u}$ can be classified, when preceding training led to classifier parameters $\boldsymbol{\theta}$.

The generation of accuracy estimator M is done for a particular working classifier *as a separate learning or regression task*. Once constructed, it can be applied to monitor trainings of various instances of the same kind of working classifier. The regression model is trained with statistics of some initial incremental training runs of a number of working classifier instances for generating the ground truth data for training the regression model M.

The statistics, denoted as $\phi(\mathbf{U}, \boldsymbol{\theta})$, have to map to the working classifiers accuracy predicting samples from pool $\mathbf{U}$. This accuracy value is then the desired output $a = M(\phi(\mathbf{U}, \boldsymbol{\theta}))$ of the accuracy estimation model M. For the input feature vector $\phi(\mathbf{U}, \boldsymbol{\theta})$ we use *confidence histograms* ("configrams") that we compute from the working classifier's confidence measure. The established approach of Platt et al. [20] consists of fitting a sigmoidal transfer function to the distribution of confidences. You may consider our approach as fitting the regression model M to the higher-dimensional space of configrams $\phi(\mathbf{U}, \boldsymbol{\theta})$.

The configrams are computed as follows: we create J-dimensional confidence histograms over "confidence bins" of width $K = 1/J$ in the confidence interval [0,1], based on sampling a "representative" subset of our data pool. The histogram count ϕ_j of bin j, j=1..J, is thus given as

$$\phi_j = \sum_{\mathbf{u} \in \mathbf{U}} D(\mathbf{u}, j) \text{ with the bin membership indicator function}$$

$$D(\mathbf{u}, j) = \begin{cases} 1 & \text{if } (j-1) \cdot K \leq c_p(\mathbf{u}, \boldsymbol{\theta}) < j \cdot K \\ 0 & \text{else} \end{cases}$$

and $\mathbf{U}$ a suitably large subset of $\mathbf{D}$. Each histogram $\phi(\mathbf{U}, \boldsymbol{\theta})$ will be a single input point for the regression model M. To determine this model, we need many such points.

Also the model should be capable of predicting working classifiers in any training state. Therefore we are not collecting a single histogram per classifier instance, but rather a sequence of n=1,2,...N histograms covering various states of training. This sequence is obtained in a similar fashion as in the later application: samples are queried in minibatches which are counted by the index n. Each new minibatch leads to an incremental update of the working classifier with changed adaptive parameters $\boldsymbol{\theta}_n$ and corresponding histogram counts ϕ_{nj} that now depend on training state n together with bin number j. In this way, we get a sequence of input feature vectors ϕ_n, component-wise given as:

$$\phi_{nj} = \sum_{\mathbf{u} \in \mathbf{U}} D_n(\mathbf{u}, j) \text{ with the bin membership indicator function}$$

$$D_n(\mathbf{u}, j) = \begin{cases} 1 & \text{if } (j-1) \cdot K \le c_p(\mathbf{u}, \boldsymbol{\theta}_n) < j \cdot K \\ 0 & \text{else} \end{cases}$$

Finally, for each training state n and sample $\mathbf{u}$ we need to know whether the prediction $\hat{y}_{\mathbf{u}} = C_{\boldsymbol{\theta}_n}(\mathbf{u})$ coincides with the true label $y_{\mathbf{u}}$ or not:

$$L_{0/1}(y, \hat{y}) = \begin{cases} 0, & \text{if } y = \hat{y} \\ 1, & \text{else} \end{cases} ; \text{ Their average } a_n(\mathbf{U}, C_{\boldsymbol{\theta}_n}) = \frac{1}{|\mathbf{U}|} \sum_{\mathbf{u} \in \mathbf{U}} 1 - L_{0/1}(y_{\mathbf{u}}, \hat{y}_{\mathbf{u}})$$

represents the ground truth accuracy that is linked to histogram vector ϕ_n. Note, that this step requires access to all ground truth labels of set $\mathbf{U}$. Finally, we are stacking all $\phi_n(\mathbf{U}, \boldsymbol{\theta})$ and $a_n(\mathbf{U}, C_{\boldsymbol{\theta}})$ into two vectors. To generate more of these configram-accuracy pairs, we are training not only one classifier instance but a classifier ensemble with Q instances $C^{1..Q}$ of the same working classifier but trained from different initializations (random queries from $\mathbf{U}$).

After training the ensemble of classifiers and collecting configram sets, they are stacked to a feature vector $\boldsymbol{\Phi}$. Analogously the ground truth accuracies are stacked to a vector $\mathbf{A}$. Our accuracy estimator M is an arbitrary regression model, trained with $(\boldsymbol{\Phi}, \mathbf{A})$ as features and target values. We tested various common regression models [4] where Nearest neighbor regression (NNR) was the most reliable approach on our data, so we select NNR to be used in our further evaluation.

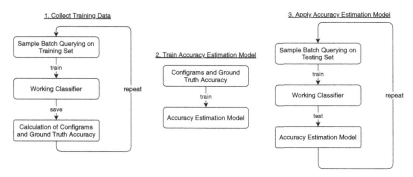

Fig. 3. Workflow of accuracy estimation

Once the Configram Estimation Model (CGEM) M has been obtained, it can be applied to working classifiers trained on pool $\mathbf{U}'$ whose statistics (configrams) may differ from pool $\mathbf{U}$, based on which M was trained. The model will then extrapolate what is has learned from $\mathbf{U}$ about the relationship between configrams and output accuracy of an incrementally trained working classifier instance and thereby permit on $\mathbf{U}'$ very quick accuracy predictions. Another advantage is that we *do not have to query any labels from* $\mathbf{U}'$ for the accuracy estimation to be applicable. However, the extrapolation assumes that the domain **D** is sufficiently homogeneous so that new pools drawn from **D** have a high likelihood to be sufficiently similar to the model training pool **S** to admit the above accuracy extrapolation through the model M. The workflow of training and applying M is shown in Fig. 3.

Also note, that we deal with two different quality measures namely the accuracy of C_θ, defined as Working Classifier Accuracy (WCA). However, more interesting for us is the error estimating with M on test set **U′**, namely Accuracy Estimation Error (AEE).

4   Experimental Evaluation

Our robot is a Scitos G5 from Metralabs [7] (see Fig. 4 left). With the attached Kinect 2 RGBD-camera, it is capable of recognizing its environment in RGB-color with additional depth information. This, together with the self-localization capabilities of the robot, can then be used for extracting global spatial coordinates of detected objects.

4.1   Setup

Let us define our inventory assistant robot setup:

i) Before the actual inventory task the CGEM accuracy estimation model is trained on a set of example object classes (set **S**) which are not necessarily used later in the task (we actually exclude them in our experiment). The purpose of this training is to learn the typical mapping from generic confidence histograms to the actual classifier accuracy for this generic object classification task.

ii) At the start of the inventory task the robot drives around and takes a few images covering a typical share of the work space containing all objects to be inventorized and thus having to be detected and classified.

Fig. 4. Left: Our service robot capturing objects in a cluttered shelf. Right: Some class representatives from our recorded real world object data set CUPSNBOTTLES. The number of a class's object samples is given in brackets.

iii) A human iteratively labels an automatically selected (active learning or random sampling) set of *object samples* (from set **U′**) for training classifier $C′$. CGEM estimates the accuracy and the human continues labeling until a desired threshold accuracy is reached.

iv) Using the trained classification model from step iii), the robot drives around the location and takes further images containing object samples (set **U′**)

covering the complete area to be scanned. Then, it detects, classifies and counts all target *object instances.*

To allow controlled and repeatable experiments we emulate the steps i)-iv) using one collected set of images (see Fig. 4 right), which we split into different subsets used in steps i)-iv). Each image contains object region proposals which we call *object samples* and which have an estimated 3D position. We want to count the number of *object instances* of a particular *object class.* To count the *class instances* in the room we cluster the *object sample* spatial coordinates into *object clusters* (see Fig. 5) and we predict a *cluster label* by taking a majority vote on the classified labels within the cluster.

By comparing the estimated number of *object instances* to the ground truth number of instances, we can calculate a *task performance* measure how well the inventory was done. Our goal is also to evaluate how the *task performance* depends on the estimated accuracy.

4.2   Experiments

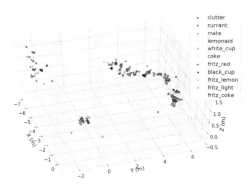

Fig. 5. Spatial positions of recorded object samples, labeled with ground truth classes. Samples are clustered only based on their spatial position and classified by an incrementally trained classifier. Each cluster's label is then defined by majority vote. The task performance is calculated by comparing the clusters labels to ground truth number of class instances.

Our evaluation data set was recorded with our service robot in a cluttered laboratory environment (see Fig. 4). We used the robot to approach five way points to record objects that were located on different kind of furniture. On each way point, the robot did a -30° followed by a 60° turn for acquiring more images from different viewing angles. All the time, the robot is patrolling its predefined path, we compute axis aligned bounding box object region proposals using the YOLO object detector [21], pre-trained on the COCO data set. From the bounding box region proposals, visual object features are calculated online by VGG19 deep convolutional net [25], which was pre-trained on imagenet data. Classification is trained incrementally based on this feature output.

Since the robot moves quite fast, there are blurry object samples within the data set. However, while the robot is turning at a waypoint, by empirical observation we make sure that all object instances are detected by YOLO and there are at most single frames where an object instance is missed. For simplifying our experiments, we include only object samples which have the YOLO-classes *cup* or *bottle*, however we consider these as meta categories which can be divided into different kind of e.g. bottles (see Fig. 4 right). Also, we exclude object samples with a YOLO-confidence with less than 20% and a distance greater than 2 m.

The total number of frames is 515 with 2179 resulting object samples in the data set. We published the data set for the community on an open access repository[1].

As mentioned earlier, our robot can locate the recognized object samples spatially in a 3d map. For instance the spatial position of the recorded object samples from the shelf (Fig. 4) can be seen in Fig. 5 at xy-coordinates (-6,-2). We cluster this spatial samples, to further count object instances, using constrained DBSCAN [23], where we add cannot-link-constrains of different object samples within the same camera frame. This represents the assumption that each present object is detected only once within a frame, which is satisfied by the YOLO output according to our parameter choices. Based on this clustering, we define the class of an actual cluster based on the majority vote of its contained samples.

Note that the region proposals generated by YOLO are not perfect and there can also be clutter. In our data set, there is a share of 5.5% remaining clutter images. Those images are exceeding the YOLO confidence threshold but they are containing other than the target objects, multiple objects or the bounding box is not centered correctly. To simulate a real world application, all algorithm (including CGEM) have to work also in the presence of clutter in the data set. Since clutter can not be avoided completely, we want to address this also in our evaluation, where clutter images are not used for classifier training (neither by C_θ or $C_{\theta'}$), since we assume that human labelers only label images of the target classes (and reject possible clutter images). However, clutter can be contained in the unsupervised training data (set $\mathbf{U}'$) for the accuracy estimator. The clutter images then contribute to the configram feature vectors, typically with lower confidences than proper object samples. We have investigated the effect on the accuracy estimation if either i) clutter samples are both contained in $\mathbf{U}$ and $\mathbf{U}'$: denoted as CC, ii) only in $\mathbf{U}$ and not in $\mathbf{U}'$: denoted as CN, iii) not in $\mathbf{U}$ and only in $\mathbf{U}'$: denoted as NC, or iv) neither in $\mathbf{U}$ nor $\mathbf{U}'$:NN. It would be reasonable to assume that training and testing conditions should match, so cases CC and NN should be expected to provide best results.

In our evaluation, we define our domains $\mathbf{S}$ and $\mathbf{S}'$ by splitting up the 10 recorded object classes into random splits of 5 classes. The clutter class was randomly split up between $\mathbf{S}$ and $\mathbf{S}'$. Another experimental dimension is the querying technique for selecting samples for human labeling. We consider the two alternatives of random sampling or active learning using maximum uncertainty sampling. For the latter we use the classifier to calculate confidence estimates for all samples within the unlabeled pool and select the sample with the least confidence for labeling/training.

For our evaluation we trained 100 single-sample mini batches by a k nearest neighbor (kNN) classifier. We chose kNN since it is a robust classifier which is flexible in incremental learning based on deep feature outputs with high accuracy [5,11]. As baseline models we choose a 5-fold cross validation (CV) and also we use interleaved test/train error with a window size of 20 samples. We choose to have $Q = 5$ classifiers in our ensemble and we repeat each experiment $W = 15$ times for averaging our results.

---

[1] https://ieee-dataport.org/open-access/cupsnbottles.

We calculate the clustering, with constrained DBSCAN as described above, on the samples from $\mathbf{S}'$. To make sure that an optimal task performance of 100% can be attainable, we check if the correct number of class instances can be found if all the samples were labeled with ground truth data.

4.3 Results

The data set evaluation results can be seen in Table 1 in condensed form.

Table 1. Experimental results of accuracy estimation with cross validation (CV), interleaved test/train error (ITT) and CGEM on our inventory object data set. For testing, 5 different objects were used, that were not in the train set. The presence of clutter (CC/NN/NC/CN) has no significant effect on the accuracy estimation quality.

Random sampling		CC	NN	NC	CN	Average
Accuracy Estimation Error	CV	0.071	0.07	0.073	0.073	0.072
	ITT	0.093	0.093	0.097	0.095	0.094
	CGEM	**0.048**	**0.045**	**0.048**	**0.045**	**0.047**
Final classifier accuracy		0.978	0.974	0.975	0.976	
Uncertainty sampling		CC	NN	NC	CN	average
Accuracy Estimation Error	CV	0.3	0.298	0.297	0.294	0.297
	ITT	0.353	0.358	0.345	0.364	0.355
	CGEM	**0.037**	**0.038**	**0.038**	**0.039**	**0.038**
Final classifier accuracy		1.0	1.0	1.0	1.0	

We see that CGEM is capable of predicting KNN's accuracy with the best precision in this realistic task compared to CV and ITT. The effect of having clutter samples in $\mathbf{S}$ or $\mathbf{S}'$ is not significant in our experiment, probably due to the low share of clutter samples (5.5%).

From Table 1 one can also clearly see that the baseline models are highly affected by active learning since they only depend on set $\mathbf{L}'$, where the most uncertain sample is selected by the uncertainty-based querying strategy. So the performance of $C_{\theta'}$ on $\mathbf{L}'$ is very different from $\mathbf{U}'$. CGEM can adapt to this since it is also taking unlabeled samples from $\mathbf{U}'$ into account.

The advantage of uncertainty-based over random sampling can also be seen by comparing the left versus the right plot of Fig. 6, which both depict training on the NC data set condition. Using uncertainty sampling, the task is done (task error below 5%) in less than 40 trained samples whereas random sampling requires nearly 60 samples to achieve the same task performance. This further motivates using an active querying of samples.

It can also be seen that the task performance directly depends on the accuracy estimate. This effect is even more visible by looking at the averaged plots. By defining a threshold for a minimum desired classifier accuracy, one can predict that the task is done with a certain performance.

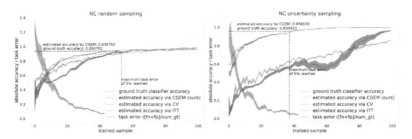

Fig. 6. Incremental classifier training using random sampling (left) and uncertainty sampling (right). Ground truth accuracy of classifier $C_{\theta'}$ on testing set $\mathbf{S'}$ together with evaluated approaches for estimating the classifier's accuracy (AEE) are displayed. The results were averaged over 15 repetitions with random object classes in the train and test splits $(\mathbf{S}, \mathbf{S'})$.

5 Conclusion

We have demonstrated the application of the CGEM accuracy estimation model towards a simplified inventory assistant robot task setting. We could show that in using a generic accuracy estimator we can predict the classification performance after incremental teaching by a human labeler using an active learning approach. This makes human teaching maximally efficient because we can limit the necessary training towards the desired accuracy of the task.

In our scenario we tested a limited number of objects and instances. We also excluded situations of strong object occlusion and crowding. A realistic application would of course naturally also contain these more difficult conditions. However, for the range of scenarios where these difficult conditions are absent even the robot assistant with the present algorithm could already be useful. Apart from that, if the general detection and recognition performance can be enhanced, the configram estimation approach can still be applied, if the classifier delivers meaningful confidences (cp. results on benchmarks data sets in [12]).

References

1. Aghazadeh, O., Carlsson, S.: Properties of datasets predict the performance of classifiers. In: Proceedings of BMVC (2013)
2. Chan, D.M., Riek, L.D.: Object proposal algorithms in the wild: are they generalizable to robot perception? In: Proceedings of IEEE/RSJ IROS. IEEE (2019)
3. Donmez, P., Lebanon, G., Balasubramanian, K.: Unsupervised supervised learning I: estimating classification and regression errors without labels. J. Mach. Learn. Res. **11**, 1323–1351 (2010)
4. Draper, N.R., Smith, H.: Applied regression analysis, vol. 326. Wiley, New York (1998)
5. Fischer, L., Hasler, S., Schrom, S., Wersing, H.: Improving online learning of visual categories by deep features. In: NIPS Workshop on Future of Interactive Learning Machines (2016)
6. Freeman, S., et al.: Active learning increases student performance in science, engineering, and mathematics. PNAS **111**(23), 8410–8415 (2014)

7. Hasler, S., Kreger, J., Bauer-Wersing, U.: Interactive incremental online learning of objects onboard of a cooperative autonomous mobile robot. In: Proceedings of ICONIP, pp. 279–290 (2018)
8. Kirstein, S., Denecke, A., Hasler, S., Wersing, H., Gross, H.M., Körner, E.: A vision architecture for unconstrained and incremental learning of multiple categories. Memetic Comput. **1**(4), 291 (2009)
9. Kreger, J., Fischer, L., Hasler, S., Weisswange, T.H., Bauer-Wersing, U.: A priori reliability prediction with meta-learning based on context information. In: Proceedings of ICANN, pp. 200–207 (2017)
10. Krüger, M., Wiebel, C., Wersing, H.: From tools towards cooperative assistants. In: Proceedings of Conference on Human Agent Interaction, HAI, pp. 287–294 (2017)
11. Limberg, C., Krieger, K., Wersing, H., Ritter, H.: Active learning for image recognition using a visualization-based user interface. In: Tetko, I.V., Kůrková, V., Karpov, P., Theis, F. (eds.) ICANN 2019. LNCS, vol. 11728, pp. 495–506. Springer, Cham (2019). https://doi.org/10.1007/978-3-030-30484-3_40
12. Limberg, C., Wersing, H., Ritter, H.J.: Efficient accuracy estimation for instance-based incremental active learning. In: European Symposium on Artificial Neural Networks (ESANN) (2018)
13. Losing, V., Hammer, B., Wersing, H.: Interactive online learning for obstacle classification on a mobile robot. In: Proceedings of IJCNN, pp. 1–8 (2015)
14. Losing, V., Hammer, B., Wersing, H.: Incremental on-line learning: a review and comparison of state of the art algorithms. Neurocomputing **275**, 1261–1274 (2018)
15. Nikolaidis, S., Nath, S., Procaccia, A.D.: Game-theoretic modeling of human adaptation in human-robot collaboration. In: Proceedings of ACM/IEEE HRI (2017)
16. Nikolaidis, S., Shah, J.: Human-robot cross-training: computational formulation, modeling and evaluation. In: Proceedings of ACM/IEEE HRI (2013)
17. Parasuraman, R., Sheridan, T.B., Wickens, C.D.: A model for types and levels of human interaction with automation. IEEE Trans. SMC A Syst. Hum. **30**(3), 286–297 (2000)
18. Pasquale, G., Ciliberto, C., Odone, F., Rosasco, L., Natale, L.: Are we done with object recognition? The iCub robot's perspective. Robot. Auton. Syst. **112**, 260–281 (2019)
19. Platanios, E., Blum, A., Mitchell, T.: Estimating accuracy from unlabeled data. In: Proceedings of the Eleventh Conference on Uncertainty in Artificial Intelligence, pp. 682–691 (2014)
20. Platt, J., et al.: Probabilistic outputs for support vector machines and comparisons to regularized likelihood methods. Adv. Large Margin Class. **10**(3), 61–74 (1999)
21. Redmon, J., Divvala, S., Girshick, R., Farhadi, A.: You only look once: unified, real-time object detection. In: Proceedings of IEEE CVPR, pp. 779–788 (2016)
22. Robinette, P., Howard, A.M., Wagner, A.R.: Effect of robot performance on human-robot trust in time-critical situations. IEEE Trans. Hum. Mach. Syst. **47**(4), 425–436 (2017)
23. Ruiz, C., Spiliopoulou, M., Menasalvas, E.: C-DBSCAN: density-based clustering with constraints. In: An, A., Stefanowski, J., Ramanna, S., Butz, C.J., Pedrycz, W., Wang, G. (eds.) RSFDGrC 2007. LNCS (LNAI), vol. 4482, pp. 216–223. Springer, Heidelberg (2007). https://doi.org/10.1007/978-3-540-72530-5_25
24. Sendhoff, B., Wersing, H.: Cooperative intelligence: a humane perspective. In: Proceedings of International Conference on Human Machine Systems (2020)
25. Simonyan, K., Zisserman, A.: Very deep convolutional networks for large-scale image recognition. In: Proceedings of CoRR (2014)
26. Welinder, P., Welling, M., Perona, P.: A lazy man's approach to benchmarking: semisupervised classifier evaluation and recalibration. In: Proceedings of IEEE CVPR, pp. 3262–3269 (2013)

Active Object Estimation for Human-Robot Collaborative Tasks

Chaoran Huang[(⊠)] and Lina Yao

School of Computer Science and Engineering, The University of New South Wales,
Sydney, NSW 2052, Australia
{chaoran.huang,lina.yao}@unsw.edu.au

Abstract. In the current exploring of interpreting human activities of daily living (ADLs), rarely we can see a specific model for training robot helpers, which in some domains has shown promising prototypes. In our proposed scenario, we aim to build a model for training a robot helper to assist human being to conduct certain activities, and for this, we are interested in (1) which objects will the subject interact with; (2) how will the subject interacts the object, or in this paper, how will the objects moving; So that in limited conditions the robotic helper can help the human conduct such interactions. The setting also includes a fixed IR based stereo camera and based on its RGB-D stream feed we utilise a generative adversarial network (GAN) for the objective movement prediction. Then object detection is applied to the produced future frame, which is compared with the last input frame, to resolve the movement of the object. IR frame is also handled, to produce the 3D distance of the object to the camera, leading to the actual 3D location of the object in the certain feature time frame. Experiment results show promising in our model.

Keywords: Future frame prediction · RGB-D video · Robotic helper · GAN

1 Introduction

Evolving from early forms of automatons and human resembling mechanisms, in the last decades, various projects have been done to mimic human beings to some extent in their appearances, actions, and speaking. Apart from the fancy androids and the fantasy of building human-like robots, there evolving one more practical kind, industrial or specialised robots, which has less restriction, at the current stage are more common. Despite their shapes and behaviours can be largely different from humans, the ultimate goals of such robots is acting as helpers and working or even living along with us.

While as the maturing of robotics technologies today, studies has emerged to bring robot helpers in various forms to homes and daily livings [22]. However, in their current stage, although there exist some interactions with human, it is still far from intimate collaboration.

© Springer Nature Switzerland AG 2020
H. Yang et al. (Eds.): ICONIP 2020, LNCS 12533, pp. 750–761, 2020.
https://doi.org/10.1007/978-3-030-63833-7_63

Nevertheless, one key part of such a system would be resolving what the subject is doing or what will happen next. Current studies on behaviour or intention prediction are promising and can handle various input sources such as environmental sensors, the status of home appliances and wearable sensors. Yet, they often targeted to get the general idea of what kind of activity would happen next, and it is still hard to have a refined prediction on what immediate move would happen. Although this may be good enough for a helper robot to finish the task by its own, an intimate collaboration with the subject can be challenging in such granularity.

Initial studies of intimate collaborative robots can be found, while most are focused on specialised services, especially for the elders or telecare. Few studies on collaborative robots can be found, while rarely they are designed for everyday activities. Hence here in this paper, we propose a general framework for predict human intention specifically for robot helpers in Activities of Daily Living (ADL) scenario. Instead of in verbatim output what the subject is doing, we aim to jointly predict/infer what object and how will the object be interacted by the subject.

On the other hand, recent popping up GAN based frame prediction has been showing considerable good results for relatively near future. And this short-term basis prediction matches exactly what is lacking in traditional activity prediction. Thus, this approach can be useful in our scenario, and specifically, we propose to adopt a fixed IR-based stereo camera, such as Microsoft Kinect or Intel RealSense Depth Camera, to capture RGB-D data, which often have built-in sensors for human skeleton recognition, i.e. human pose information. Along with recently developed rapid object recognition algorithms, specifically in our case 'Mask R-CNN' [5], we can easily list the interactable objects in the scene. Our objectives here are specifically: (1) which objects will the subject interact with; (2) how will the subject interacts the object. In the current stage, we propose to predict human action in the scenario of training robotic helper of Activities of daily living (ADL), thus the objective 2 will come down to how the objects will move, so that in limited conditions the robotic helper can help the human conduct such interactions. Based on the RGB-D stream feed from the camera, we utilise a generative adversarial network (GAN) for the prediction. Given a time range of single human-object interaction, the first step is to select the most likely object the subject will be interactive with, which can be easily discriminated by comparing the actual feature frames with the previous one when training; And the second step will be training a generator produce the following states of the object been interacted, i.e. the 3D location of the object in the certain feature time frame.

The rest of this paper is organised as follows: Sect. 2 sum up some important related works and briefly talks about some insights on how this framework is designed; Sect. 3 overviews the framework with detailed explanations; Sect. 4 introduces the evaluation dataset and setups, as well as discussion on experiment results; Sect. 5 conclude this paper with a summary and some thoughts on future works.

2   Related Works

Early studies on human intentions start with human and object recognition, often rely on pictures or 2D video clips [4,16] and focus on modelling events, while later transited to study human actions in 3D spaces [3,30]. At this stage modelling and representation of human actions emerged [18,25] along with more complex sensors such as depth cameras [23,24]. Representation for human intentions is rarely seen and limited to very definite states, often modelled by ontologies [20] and built to purposes at this stage. It has not been widely developed until recent booming of machine learning technologies [9,31]. Still, most studies are perfecting recognising human actions by various sources of input, while very limited works focus on inferring the intention of the subjects, which can be argued at least equally worthy investigating, especially in critical situations where pre-emptive actions are essential.

In the current exploring of interpreting human intentions, gaze estimation has been the interests of researchers [7,15,19]. While it is very effective for intention recognition for the moment of current inputs [28], such methods are often quite computing intensive already, which potentially can be used for intention prediction, yet that would make the model sensitive to parameter tuning, computation cost exponential and limit its application. Considering studies in Human Activity Recognition (HAR) often utilise wearable Inertial Measurement Units (IMUs) to estimate the poses of human subjects, to recognises elementary and complex activities, and achieved desirable results, we can adopt a similar approach for our objectives, and using human poses as one of our fundamental features. Yet in those studies, the objects been interacted with are usually either pre-equipped with sensors or associated with sensors to monitoring their status. And this is often not feasible and require specialisations to set up.

As for future frame prediction, it is popping up with a recent wave of neural networks. Starting from applying Deep RNN such as ConvLSTM, such prediction has been largely avoided blurry artefacts, and PredNet [10] by Lotter et al. can be one successful example. Adversarial training also proved effective in this task, and received many attentions, Vondrick et al. in [26] shows GAN can produce meaningful and vivid short video clips, and Mathieu et al. [13] further alleviate the blurry issue by introducing a gradient difference Loss. Motion constrains [2, 6,17] are also considered to produce more temporal sensible results [1,8].

3   Methodology

3.1   Overview

The general idea of our proposed work is to predict future RGB-D frames based on the stereo camera input, and compare them to the current frames in RGB-D, so that we can get the idea which object (active object) will be moved and where would it be moved to. Then the current position of the active object and its possible short-term destination can be resolved and sent to our robot helper. Figure 1 shows the idea of our framework. Specifically, our input is Human

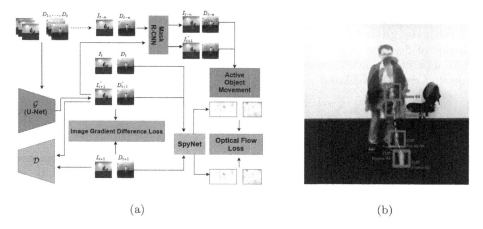

(a)   (b)

Fig. 1. The framework of our GAN for prediction (a). Here we adapt U-Net as our generator $\mathcal{G}$, and pre-trained SpyNet to calculate the optical flow for motion constrain. (b) demonstrates the outcome of our whole system in RGB Frames, as the object is detected in the prediction results following frames in sequences where the movement and the active object can be inferred. The depth position can be worked out in the same manner.

performs ADLs in RGB-D data recorded by active IR-based stereo cameras, such as Microsoft Kinect or Intel RealSense Depth Camera. Given a timestamp, the input consists of an RGB frame, an IR frame containing depth information. And in our setting, we assume the camera is fixed.

Formally, we noted Tasks in ADLs as $T = \{T_1, T_2, \ldots, T_N\}$, and each task is made up with a sequential of elementary activities $A = \{a_1, a_2, \ldots, a_M\}$. $\mathcal{I}$ is the input which according to frames have tuples $\mathcal{I} = \{(I_t, D_t) | t = 1, 2, \ldots, T\}$ where on frame t, I_t is the RGB frame and D_t is the depth inclusive IR frame. We can infer a point has the position in 3D P by combining a projected 2D position in RGB frame and IR frame.

Suppose in the scene we found a set of objects $o = o^1, \ldots o^2, \ldots, o^K$. The general objectives in this work is now to find 1)the object o^k in elementary task a_m is touched continuously by one of the hands $h \in H$ in the certain frames from t_1 to t_2, and 2) the object o^k moved from its location P_t to $P_{t+\tau}$. Here we utilize Mask R-CNN [5] for object recognition in the 2D space.

Generative adversarial networks (GAN) has been widely studied recently in short-term future frame predictions, which predicts a few future frames based on recent ones. Especially Least Squares Generative Adversarial Networks (LSGAN) by Mao et al. [12], has a modified loss function for the discriminator and been proved effective in stably generating high-quality images. Following the work by Liu et al. [8], we also adapt LSGAN with a U-Net based prediction network as our generator plus optical flow as constrain, which is illustrated in Fig. 1a. The discriminator $\mathcal{D}$ in this model is standard convolutions with fully connected layers and ReLu non-linearities as in [13].

LSGAN Discriminator Training. The Target of discriminator $\mathcal{D}$ is to distinguish the ground truth $\mathcal{I} = (I_{t+1}; D_{t+1})$ and $\hat{\mathcal{I}} = (\hat{I_{t+1}}; \hat{D_{t+1}})$. Denote the output of discriminator as $\mathcal{D}(\cdot)$, true as class 1 and false as class 0, i, j is the spatial patches indices, given the weights of generator fixed, we can have a Mean Square Error (MSE) Loss:

$$\mathcal{L}^{\mathcal{D}}(\mathcal{I}, \hat{\mathcal{I}}) = \sum_{i,j} \frac{1}{2} MSE(\mathcal{D}(\mathcal{I})_{i,j}, 1) + \sum_{i,j} \frac{1}{2} MSE(\mathcal{D}(\hat{\mathcal{I}})_{i,j}, 0) \tag{1}$$

LSGAN Generator Training. As the generator is set to fool discriminator, i.e make the $\mathcal{D}$ classify $\hat{\mathcal{I}}$ into class 1. The simple goal of training $\mathcal{G}$ can be an MSE loss as:

$$\mathcal{L}^{\mathcal{G}}{}_a(\hat{\mathcal{I}}) = \sum_{i,j} \frac{1}{2} MSE(\mathcal{D}(\hat{\mathcal{I}})_{i,j}, 1) \tag{2}$$

In practice, this adversarial pair of loss will result in an overall unstable model as the training will continue, where a regularisation is needed to constrain it. We added a $\mathcal{L}_{\mathcal{P}}$ loss in this case. As in the study by Liu et al. [8] and Mathieu et al. [13], apply Image Gradient Difference Loss as constrain can yield closer results to truth and sharpen the image, it is also combined to our model, and which is defined as:

$$\mathcal{L}_{gd}(\mathcal{I}, \hat{\mathcal{I}}) = \sum_{i,j} |||(\hat{\mathcal{I}})_{i,j} - \hat{\mathcal{I}})_{i-1,j})| - |\mathcal{I})_{i,j} - \mathcal{I})_{i-1,j}|||_1$$
$$+|||(\hat{\mathcal{I}})_{i,j} - \hat{\mathcal{I}})_{i,j-1})| - |\mathcal{I})_{i,j} - \mathcal{I})_{i,j-1}|||_1 \tag{3}$$

Although most future frame generation works can deliver meaningful and sensible results, few have motion relation between frames explicitly considered. This may not so desirable not only for critical event detection such as in [8], but also may add odd in our scenario in terms of human-robot collaboration. Given the architecture of GAN is already quite complex and costly to train, we propose to utilize a relatively lightweight and fast Spatial PYramid Network (SpyNet) [17] as the motion constrain. Even pre-trained, the SpyNet is reportedly faster than their previous works such as Flownet [2] and Flownet 2.0 [6] Denote the pre-trained SpyNet as $\mathcal{F}_{op}$, we have:

$$\mathcal{L}_{op}(\hat{\mathcal{I}_{t+1}}, \mathcal{I}_{t+1}, \mathcal{I}_t) = ||\mathcal{F}_{op}(\hat{\mathcal{I}_{t+1}}, \mathcal{I}_t) - \mathcal{F}_{op}(\mathcal{I}_{t+1}, \mathcal{I}_t)||_1 \tag{4}$$

With parameters, we can now combine the loss for training $\mathcal{G}$, as:

$$\mathcal{L}^{\mathcal{G}} = \lambda_a \mathcal{L}^{\mathcal{G}}{}_a + \lambda_p \mathcal{L}_p + \lambda_{gd} \mathcal{L}_{gd} + \lambda_{op} \mathcal{L}_{op} \tag{5}$$

4   Experiment

4.1   Dataset

We use NTU RGB+D dataset [21] to train and test our proposed framework. The dataset is rather large with 56,880 videos for 60 action classes, 40 subjects,

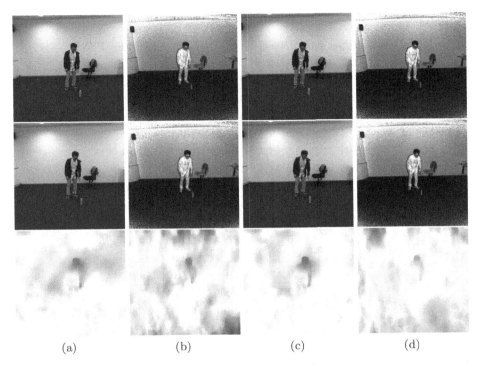

(a)   (b)   (c)   (d)

Fig. 2. Sample of visualization of RGB-D frames and optical flows in Setup 1, Camera 1, Subject 8, Repeat 2 for Action 6: (a) shows RGB frame 15, 16 and their optical flow, (b) shows IR Frames respectly; So is (c) and (b) for frame 16, 17.

and 80 viewpoints. The RGB-D video is collected by Microsoft Kinect v2, with RGB, Depth, IR and 3D human joints (skeleton) available. Since in our scenario, we assume the camera is fixed, we selected camera 001 as our viewpoint and step up number 001 as our scenario. This left us 960 pairs of RGB and IR videos, including 8 subjects, 60 actions with 2 repeats each. It is also worth mentioning that some actions may not have object involved. We also included such actions to fully evaluated our prediction networks, while in our last step, the detected objects shall have no move between frames.

4.2 Hyper-parameter Tuning

Hyper-parameter settings usually can largely affect experiment results and can be crucial to machine learning approaches. Despite relatively complex fusion in our setting for training our Generator $\mathcal{G}$, the total number of hyper-parameters are still manageable and we can safely apply Orthogonal Array Testing here. Namely we have λ_a, λ_p, λ_{gd} and λ_{op} for fusion of loss functions in training $\mathcal{G}$, Learning rate $lr_{\mathcal{D}}$, $lr_{\mathcal{G}}$ for optimizing $\mathcal{D}$ and $\mathcal{G}$ respectively. Detailed results can be found in Table 1.

Table 1. Comparison of different hyper-parameter settings.

Test #	Hyper-parameters						Metrics $PSNR$
	λ_a	λ_p	$\lambda_g d$	$\lambda_o p$	$lr_{\mathcal{G}}$	$lr_{\mathcal{D}}$	
1	0.02	0.5	0.5	1.5	0.0002	0.0005	19.23
2	0.05	0.5	0.5	1.5	0.0002	0.0005	22.30
3	0.08	0.5	0.5	1.5	0.0002	0.0005	22.18
4	0.05	1.0	0.5	1.5	0.0002	0.0005	26.41
5	0.05	1.5	0.5	1.5	0.0002	0.0005	22.30
6	0.05	1.0	1.0	1.5	0.0002	0.0005	27.77
7	0.05	1.0	1.5	1.5	0.0002	0.0005	27.60
8	0.05	1.0	1.0	2.0	0.0002	0.0005	27.31
9	0.05	1.0	1.0	2.5	0.0002	0.0005	29.13
10	0.05	1.0	1.0	2.0	0.0005	0.0005	29.84
11	0.05	1.0	1.0	2.0	0.0008	0.0005	29.82
12	0.05	1.0	1.0	2.0	0.0005	0.0010	29.33
13	0.05	1.0	1.0	2.0	0.0005	0.0015	30.10
Best	**0.05**	**1.0**	**1.0**	**2.0**	**0.0005**	**0.0010**	**30.10**

4.3   Experiment Settings

To train our prediction GAN, we cropped all video frames to the short edge size of IR video, that is 424 pixels, as the input to the U-Net in our settings. $\mathcal{D}$ is As for the input frames, we set the t to 4, i.e. using 4 previous frames to predict next one; Adam SDG is used for learning parameters, where the learning rate is set for $\mathcal{G}$ and $\mathcal{D}$ to 0.0005 and 0.001; λ_a, λ_p, λ_{gd} and λ_{op} is set to 0.05, 1, 1, and 2 respectively. Noted that in this study the RGB frame and Depth Frame is treated separately.

Table 2. Comparison of quality of produced images based on the same set of data described in Sect. 4.1.

	RGB			IR		
	PSNR	SSIM	Sharpness	PSNR	SSIM	Sharpness
ConvLSTM [29]	26.24	0.73	0.31	22.97	0.48	0.27
ConvVRNN [11]	27.80	**0.76**	0.37	21.20	0.44	**0.28**
PredNet [10]	24.60	0.71	0.34	20.61	**0.52**	0.23
Ours	**30.10**	0.74	**0.36**	**21.84**	0.40	0.19

4.4   Quantitative Evaluations

Since it is notably difficult and not quite feasible to quantitatively evaluate
robot actions, as well as the system will produce fresh results and correct itself
during the runtime, in this study, we focus on evaluating the quality of the
produced frames by our GAN module. This is critical because that the generated
frames must be clear and sensible enough for the next step, in particular, the
comparisons to the original frames to find out which object is moved and how
it is moving. Noted that we observed that in practice small runtime error can

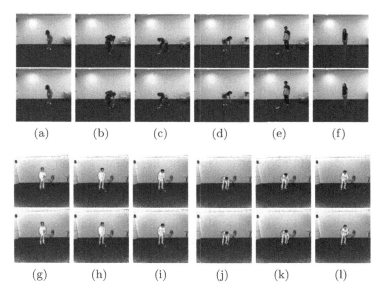

Fig. 3. Sample of ground truth and output of RGB-D frames in Setup 1, Camera 1,
Subject 8, Repeat 2 for Action 6: (a–f) shows RGB frames and (g–l) shows IR Frames.
Respectively the upper image is Ground Truth while the lower is our output.

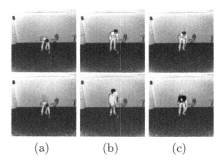

Fig. 4. Sample of output errors with ground truth. Respectively the upper image is
Ground Truth while the lower is our output.

be tolerated as the system is running lively and the inputs are continuously updating.

The quality of image predictions are usually measured with Peak Signal to Noise Ratio (PSNR) and sharpness (CPBD) [14] between the ground truth and generated ones.

$$PSNR(Y, \hat{Y}) = 10log_{10} \frac{max_{\hat{Y}}^2}{\frac{1}{N} \sum_{i=o}^{N}(Y_i - \hat{Y}_i)^2} \tag{6}$$

Also, Structure Similarity Index Measure (SSIM) by Wang et al. [27] is also quite popular.

Detailed comparison on quality of produced image frames can be found in Table 2, where the number in bold is the best in the column. It can be observed, for RGB frames, our method is, in general, the best, which slightly merit over PredNet by Cox Lab in Berkeley, while in terms of IR frames, our model produces fewer noises, but not so sharp output. By visualizing the optical flow, this was further supported, as it can be easily told that the model picked up much more 'motions' in IR frame than the RGB ones, which in our observation are essentially recording noises.

Some samples of side-by-side comparison of ground truth and output of RGB-D frames are shown in Fig. 3, where most of the outputs are quite similar to the ground truth, given minor blurry sections presented and some minor differences in the subject poses.

Noted that in our observations, there exist some outputs are different to the ground truth (see in Fig. 4), especially in IR frames: saying the groud truth is picking up objects from the ground, while the predictions showing the subject is standstill; Or there are some kind of "ghost images" presented as in the output there may have a second person partly overlapped with the primary subject. It could be caused by different subject acting patterns are changing in the training data, while further studies may be required to confirm it. These are expected limitations on our proposed method, which is rare and can be treated as outliers. The overall effectiveness is still promising.

4.5   Use Case

We combined the working model with human activity predictions system in a real-life scenario for better demonstration and evaluation. As shown in Fig. 5, here we deploy a robot arm in a scene with the proposed approach, which can move according to instructions, grip and release. Also, we employed a visual tracking based behaviour recognition system as the input for semantic intention inferring, which will be decomposed into fine procedures and carried on by the robotic arm.

In this scenario, the subject wants to make a cup of tea: Firstly, the subject walks towards the testbed, where a robot arm is set on the desk with RGB-D stereo cameras towards the scene. When the system detects the subject put down mug near the robot arm, tea bag and teaspoon with hot water, the intention of making tea is clear. Thus the system directs robot arm put a tea bag into the mug, followed by the spoon. For a better understanding of the scenario, the demonstration is available in the link[1].

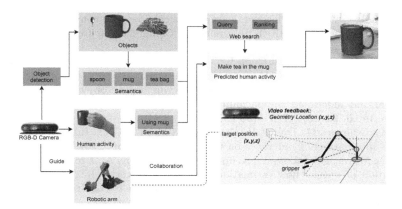

Fig. 5. Applying case of this proposed framework: Working with intention prediction in ADL to help make a tea.

5   Conclusion

In this paper, we present an RGB-D video frame prediction-based approach to estimate human promoted object movement, for robot helper in ADL scenario. The proposed framework was tested on a real-world dataset. The evaluation shows the effectiveness of RGB-D video frame prediction and the feasibility of using it for object movement estimations. The main advantage of this approach is the system is easy to set up and require no pre-installed environmental sensors or wearable sensors, which often may not desirable. While we also understand of the limitation of the proposed framework and extensive further study may be required: The testing dataset is still in a fine controlled scenario with limited objects on sight, which is quite different to the daily living scenario, especially considering part of our approach involving object detection; In the future study we are also looking to coupling optical flow for the reasoning of target object movement, to simplify the procedures of the whole system; Also the in a live test of robot helper, as a moving object itself insight, it can be influential to our optical flow module, which requires further evaluation.

Acknowledgements. This research was supported by grant ONRG NICOP N62909-19-1-2009.

---

[1] https://youtu.be/2_J9dSnvwV8.

References

1. Chen, Y., Pan, Y., Yao, T., Tian, X., Mei, T.: Mocycle-GAN: unpaired video-to-video translation. In: Proceedings of the 27th ACM International Conference on Multimedia, pp. 647–655 (2019)
2. Dosovitskiy, A., et al.: FlowNet: learning optical flow with convolutional networks. In: Proceedings of the IEEE International Conference on Computer Vision, pp. 2758–2766 (2015)
3. Grabner, H., Gall, J., Van Gool, L.: What makes a chair a chair? In: CVPR 2011, pp. 1529–1536. IEEE (2011)
4. Gupta, A., Kembhavi, A., Davis, L.S.: Observing human-object interactions: using spatial and functional compatibility for recognition. IEEE Trans. Pattern Anal. Mach. Intell. **31**(10), 1775–1789 (2009)
5. He, K., Gkioxari, G., Dollár, P., Girshick, R.: Mask R-CNN. In: Proceedings of the IEEE International Conference on Computer Vision, pp. 2961–2969 (2017)
6. Ilg, E., Mayer, N., Saikia, T., Keuper, M., Dosovitskiy, A., Brox, T.: FlowNet 2.0: evolution of optical flow estimation with deep networks. In: Proceedings of the IEEE Conference on Computer Vision and Pattern Recognition, pp. 2462–2470 (2017)
7. Kellnhofer, P., Recasens, A., Stent, S., Matusik, W., Torralba, A.: Gaze360: physically unconstrained gaze estimation in the wild. In: Proceedings of the IEEE International Conference on Computer Vision, pp. 6912–6921 (2019)
8. Liu, W., Luo, W., Lian, D., Gao, S.: Future frame prediction for anomaly detection-a new baseline. In: Proceedings of the IEEE Conference on Computer Vision and Pattern Recognition, pp. 6536–6545 (2018)
9. Liu, Z., Yao, L., Bai, L., Wang, X., Wang, C.: Spectrum-guided adversarial disparity learning. In: Proceedings of the 26th ACM SIGKDD International Conference on Knowledge Discovery & Data Mining, pp. 114–124 (2020)
10. Lotter, W., Kreiman, G., Cox, D.: Deep predictive coding networks for video prediction and unsupervised learning. arXiv preprint arXiv:1605.08104 (2016)
11. Lu, Y., Kumar, K.M., shahabeddin Nabavi, S., Wang, Y.: Future frame prediction using convolutional VRNN for anomaly detection. In: 2019 16th IEEE International Conference on Advanced Video and Signal Based Surveillance (AVSS), pp. 1–8. IEEE (2019)
12. Mao, X., Li, Q., Xie, H., Lau, R.Y., Wang, Z., Paul Smolley, S.: Least squares generative adversarial networks. In: Proceedings of the IEEE International Conference on Computer Vision, pp. 2794–2802 (2017)
13. Mathieu, M., Couprie, C., LeCun, Y.: Deep multi-scale video prediction beyond mean square error. In: 4th International Conference on Learning Representations, ICLR 2016 (2016)
14. Narvekar, N.D., Karam, L.J.: A no-reference image blur metric based on the cumulative probability of blur detection (CPBD). IEEE Trans. Image Process. **20**(9), 2678–2683 (2011)
15. Park, S., Mello, S.D., Molchanov, P., Iqbal, U., Hilliges, O., Kautz, J.: Few-shot adaptive gaze estimation. In: Proceedings of the IEEE International Conference on Computer Vision, pp. 9368–9377 (2019)
16. Pei, M., Jia, Y., Zhu, S.C.: Parsing video events with goal inference and intent prediction. In: 2011 International Conference on Computer Vision, pp. 487–494. IEEE (2011)

17. Ranjan, A., Black, M.J.: Optical flow estimation using a spatial pyramid network. In: Proceedings of the IEEE Conference on Computer Vision and Pattern Recognition, pp. 4161–4170 (2017)
18. Sadanand, S., Corso, J.J.: Action bank: a high-level representation of activity in video. In: 2012 IEEE Conference on Computer Vision and Pattern Recognition, pp. 1234–1241. IEEE (2012)
19. Sakita, K., Ogawara, K., Murakami, S., Kawamura, K., Ikeuchi, K.: Flexible cooperation between human and robot by interpreting human intention from gaze information. In: 2004 IEEE/RSJ International Conference on Intelligent Robots and Systems (IROS), vol. 1, pp. 846–851. IEEE (2004). (IEEE Cat. No. 04CH37566)
20. Schlenoff, C., Pietromartire, A., Kootbally, Z., Balakirsky, S., Foufou, S.: Ontology-based state representations for intention recognition in human-robot collaborative environments. Robot. Auton. Syst. **61**(11), 1224–1234 (2013)
21. Shahroudy, A., Liu, J., Ng, T.T., Wang, G.: NTU RGB+D: a large scale dataset for 3D human activity analysis. In: IEEE Conference on Computer Vision and Pattern Recognition (2016)
22. Sheridan, T.B.: Human-robot interaction: status and challenges. Hum. Factors **58**(4), 525–532 (2016)
23. Shotton, J., et al.: Real-time human pose recognition in parts from single depth images. In: CVPR 2011, pp. 1297–1304. IEEE (2011)
24. Sung, J., Ponce, C., Selman, B., Saxena, A.: Unstructured human activity detection from RGBD images. In: 2012 IEEE International Conference on Robotics and Automation, pp. 842–849. IEEE (2012)
25. Tang, K., Fei-Fei, L., Koller, D.: Learning latent temporal structure for complex event detection. In: 2012 IEEE Conference on Computer Vision and Pattern Recognition, pp. 1250–1257. IEEE (2012)
26. Vondrick, C., Pirsiavash, H., Torralba, A.: Generating videos with scene dynamics. In: Advances in Neural Information Processing Systems, pp. 613–621 (2016)
27. Wang, Z., Bovik, A.C., Sheikh, H.R., Simoncelli, E.P.: Image quality assessment: from error visibility to structural similarity. IEEE Trans. Image Process. **13**(4), 600–612 (2004)
28. Wei, P., Liu, Y., Shu, T., Zheng, N., Zhu, S.C.: Where and why are they looking? Jointly inferring human attention and intentions in complex tasks. In: Proceedings of the IEEE Conference on Computer Vision and Pattern Recognition, pp. 6801–6809 (2018)
29. Xingjian, S., Chen, Z., Wang, H., Yeung, D.Y., Wong, W.K., Woo, W.C.: Convolutional LSTM network: a machine learning approach for precipitation nowcasting. In: Advances in Neural Information Processing Systems, pp. 802–810 (2015)
30. Yao, B., Fei-Fei, L.: Recognizing human-object interactions in still images by modeling the mutual context of objects and human poses. IEEE Trans. Pattern Anal. Mach. Intell. **34**(9), 1691–1703 (2012)
31. Yao, L., et al.: WITS: an IoT-endowed computational framework for activity recognition in personalized smart homes. Computing **100**(4), 369–385 (2018)

Adaptive Neural CPG-Based Control for a Soft Robotic Tentacle

Marlene Hammer Jeppesen[1], Jonas Jørgensen[2],
and Poramate Manoonpong[3,4]([✉])

[1] Faculty of Engineering, University of Southern Denmark, 5230 Odense M, Denmark
mjepp15@student.sdu.dk
[2] Centre for Soft Robotics, SDU Biorobotics, The Mærsk Mc-Kinney Møller
Institute, University of Southern Denmark, Odense, Denmark
jonj@mmmi.sdu.dk
[3] Embodied AI and Neurorobotics Lab, SDU Biorobotics, The Mærsk Mc-Kinney
Møller Institute, University of Southern Denmark, Odense, Denmark
poma@mmmi.sdu.dk
[4] BRAIN Lab, School of Information Science and Technology, Vidyasirimedhi
Institute of Science and Technology, Rayong, Thailand

Abstract. Soft robotics is an area that is promising with its vast application space. One of the challenging aspects of this branch of robotics is the control of soft structures. This paper proposes a neural central pattern generator (CPG) based control architecture using an amplitude-adaptive oscillator for the movement of a low cost, pneumatically actuated soft robotic tentacle with three air chambers. The CPG is created using an $SO(2)$ oscillator that generates half-sinusoidal outputs for pneumatic control. Through the use of an adaptation mechanism, the Dual Integral Learner (DIL), the parameters of the CPG are modulated to generate oscillatory signals of larger or smaller amplitude upon external perturbations to the system. The proposed neural control is implemented on the physical system and its validity is tested through physical restriction of the pneumatic air supply to the soft robotic tentacle.

Keywords: Soft robotics · Neurorobotics · Adaptive control · Neurodynamics · Plasticity

1 Introduction

Over the past decade, the interest in soft robotics has been on the rise due to ever increasing demands of safe, human-friendly cobots [1]. Still, the control of such systems is a highly complex matter, due to the compliant nature of the materials and lack of modelling hereof. Robots consisting of soft materials can in theory exhibit infinite degrees of freedom, and are very difficult to model kinematically, leading to the need for novel approaches to traditional control theory [2].

For a rigid robot, the joint positions can be processed by forwards kinematics to determine the configuration of the robot and position of the end-effector,

© Springer Nature Switzerland AG 2020
H. Yang et al. (Eds.): ICONIP 2020, LNCS 12533, pp. 762–774, 2020.
https://doi.org/10.1007/978-3-030-63833-7_64

and inverse kinematics can be used to determine joint positions given a desired placement of the end-effector. Similar calculations are not appropriate for their soft counterparts. Additionally, soft robots made entirely of elastomers are typically under-actuated, as they also hold many passive degrees of freedom and when actuated by fluids, the soft robots are not able to fully compensate for the gravitational loading due to limited available input pressure of the fluid [2]. Just as with rigid articulated robots, accurate control of soft robots requires model-based prediction of every possible configuration. Therefore such models are complex and computationally heavy.

Currently, many soft robots are either empirically open-loop controlled or manually controlled. This is partly due to sensing in soft robotics still being a rather new field, with few studies on data interpretations, as well as current shape reconstruction algorithms being oversimplified [1]. However, efforts are made to advance simulation, modelling and control of soft robots [3–5]. Additionally, artificial neural networks and machine learning approaches are gaining increased attention to obtain data-driven models of soft robots [6–8]. Nonetheless, a common problem with data-driven algorithms is that they are domain-specific, which entails an ineffective learnt model if the domain changes slightly [9]. Furthermore, such control algorithms rely on large data sets for training the model, which, when working with a physical system, are difficult to obtain. For this reason, it is beneficial to develop a model free approach to the control of soft robots.

Towards this goal of model free learning in soft robots through the use of sensory feedback, this paper seeks to combine the rythmic motions generated by a central pattern generator (CPG) with an adaptive mechanism, known as the Dual Integral Learner (DIL)[10] for movement generation and online adaptation of a soft robotic tentacle. For the present setup, pressure sensors are used as sensory feedback and the DIL will adapt the amplitude of the oscillations of the CPG online, which directly translates to a change in the amplitude of movement in the tentacle. The adaptive control system was successfully implemented on a physical setup, with which several experiments were conducted to test the performance of the proposed adaptive control architecture.

2   Soft Robotic Tentacle

The soft robotic tentacle is inspired by soft continuum structures found in natural organisms such as the octopus' arm. In nature, the octopus arm is remarkably capable of both shortening and elongation, while having the ability to bend and twist in any direction. To mimic this compliant structure, a silicone tentacle [11] has been fabricated with three parallel volumes for pneumatic actuation as depicted in Fig. 1. The tentacle is cast in blue Ecoflex 00-30, in which fiber reinforcements in the form of braided fishing line in the outer wall of the tentacle have been implanted to prevent the parallel volumes from expanding radially. The tentacle is hung from a laboratory stand with a 3D-printed mount as depicted in Figs. 1a and 1b. To actuate the three air chambers of the robot, a control board was assembled based upon the Soft Robotics Toolkit Control Board [12], which is an open source hardware platform for controlling soft actuators. The control board contains power MOSFETs to control the on/off switching of the on-board solenoid valves. This enables the use of pulse width modulation (PWM) to control the air flow to the valves by regulating the pulses sent to the valves, opening and closing them rapidly. Varying the length of the pulses will affect the amount of time the valve is either open or closed, which in turn leads to an adjustment in chamber pressure. Three pressure sensors are placed on the control board that is interfaced with an Arduino Mega 2560 to feed back pressure sensor readings at 100 Hz. The control board alongside the soft robotic tentacle is depicted in Fig. 1a.

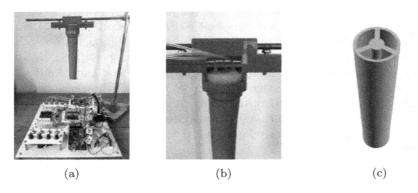

(a)   (b)   (c)

Fig. 1. The soft robotic tentacle and the test bench. (a) The setup with the blue tentacle connected with tubing to the control board. (b) The tentacle seen from the back, where the three PVC tubes supplying air for actuation are visible. (c) A 3D rendered cross-sectional view of the tentacle. (Color figure online)

3   Adaptive Neural Control and Implementation

The soft robotic tentacle is controlled by an adaptive neural CPG-based control, which utilizes feedback from the aforementioned pressure sensors to modulate the actuation amplitude generated by the CPG. An overview of the neural controller is depicted in Fig. 2. The neural controller consists of a CPG implemented as an SO(2) oscillator, highlighted in red in Fig. 2. The periodic CPG signal is post-processed to obtain a PWM-signal used for the solenoid valves controlling the air flow to the tentacle. The pressure sensors measure pressure in the tubes connected to the tentacles' three chambers during actuation. The measured pressure is post-processed to obtain a signal similar to that of the CPG signal. The error between the post-processed sensor signal and the post-processed CPG-signal is calculated, where values below a set error threshold are neglected. The error is then fed into the fast and slow learners of the adaptation mechanism called the Dual Integral Learner (DIL), highlighted in blue in Fig. 2. It is used to adapt the α-value of the CPG online and hence, the amplitude of movement of the tentacle.

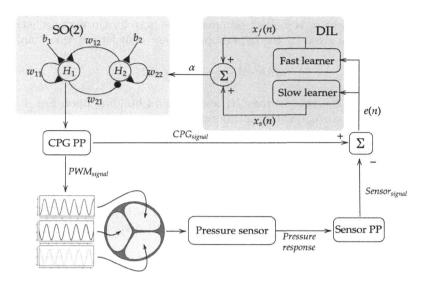

Fig. 2. Overview of the adaptive neural control system. The system combines an SO(2) oscillator (red highlighted area) and an adaptation mechanism, the DIL (blue highlighted area). The post-processed CPG signal is sent as PWM-commands to the tentacle, which pneumatically actuates the three chambers of the robot. Pressure sensors measure the pressure within the tentacle's chambers, and the post-processed sensor signal is subtracted from the CPG signal, yielding the error feedback. The error is fed into the two learners of the DIL that will adapt the α-value of the CPG. (Color figure online)

3.1 Central Pattern Generator (CPG) for Periodic Movement Generation

The CPG has been implemented as a two-neuron oscillator (SO(2)) which is a versatile recurrent neural network that can exhibit various dynamical behaviors (e.g., periodic patterns, chaotic patterns, and hysteresis effects [13,14]) by changing its synaptic weights. The dynamical behaviors of the network can also be exploited for complex movement behaviors. The SO(2)-based CPG has been depicted in Fig. 2, highlighted in red. The two neurons are fully interconnected with the synapses w_{11}, w_{12}, w_{21} and w_{22}. Each neuron sums up the weighted inputs (o_j) and the fixed bias term (b_i) and passes the final value through an activation function given by:

$$a_i(t + 1) = \sum_{j=1}^{n} w_{ij} o_j(t) + b_i \quad i = 1, ..., n, \tag{1}$$

where $a_i(t+1)$ is the activity of neuron i at time $t+1$, n is the number of inputs, w_{ij} denotes the synaptic weight associated with neuron j connected to neuron i, $o_j(t)$ refers to the input at time t, and b_i is the bias term.

The final output o_i is yielded by passing the activity through an activation function, which is given by the hyperbolic tangent transfer function (tanh):

$$tanh(a_i) = \frac{2}{1 + e^{-2a_i}} - 1. \tag{2}$$

Fully connecting the two neurons H_1 and H_2, and building upon Eqs. 1 and 2, setting $tanh(a_i) = \sigma(a_i)$, the resulting two-neuron dynamics is described by:

$$\begin{aligned} a_1(t+1) &= w_{11}\sigma(a_1(t)) + w_{12}\sigma(a_2(t)), \\ a_2(t+1) &= w_{21}\sigma(a_1(t)) + w_{22}\sigma(a_2(t)). \end{aligned} \tag{3}$$

The associated weight matrix of the SO(2)-oscillator network is given by:

$$w = \begin{pmatrix} w_{11} & w_{12} \\ w_{21} & w_{22} \end{pmatrix} = \alpha \cdot \begin{pmatrix} cos(\varphi) & sin(\varphi) \\ -sin(\varphi) & cos(\varphi) \end{pmatrix}, \tag{4}$$

where α defines the slope of the transfer function $tanh$. By varying α and φ, the amplitude and frequency of the oscillations produced by the network will be changed. According to [15], nearly sine-shaped waveforms will occur with $\alpha = 1.0 + \epsilon$ and $\epsilon \ll 1$. Increasing ε will increase the oscillation amplitude.

The initial parameters chosen for the SO(2) oscillator are $\alpha = 1.2$ and $\varphi = 0.25$, which will yield sinusoidal outputs with a phase shift of $\pi/2$ and a frequency of approximately 0.7 Hz. The output diagram is shown in Fig. 3a.

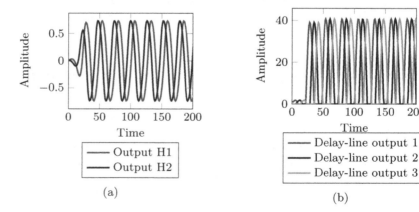

(a)

(b)

Fig. 3. Raw CPG outputs and post-processed CPG outputs. (a) The generated CPG outputs with parameters $\alpha = 1.2$ and $\varphi = 0.25$. (b) The post-processed CPG-signals to be used for PWM-control of the valves, including delay-line, zero cut-off and gain.

CPG Post-processing. After implementing the CPG to generate stable rythmic signals, a delay line has been introduced in order to apply a signal to each actuation chamber of the robot. The delay line consists of a time shifted signal of the output from the neuron H1, where the first delay is 7 timesteps and the second delay is 14 timesteps with respect to the initial output. The current actuation pattern causes the robot to move in a near-circular motion by actuating one chamber at a time with a slight overlap. Additionally, all signals driving the robot are mapped to the range of 0–100 to be used for duty cycle percentage in the PWM-signals sent to the solenoid valves. To do so, a cut-off at zero has been introduced, leading to all negative values of the CPG output being neglected. Lastly, a gain is added to the scaled signals to produce signals with an amplitude of 40 which is deemed appropriate for the given setup. The final post-processed CPG-output with delay line, zero cut-off, and gain is depicted in Fig. 3b.

3.2   Adaptation Mechanism for Online Movement Adaptation

To adapt the actuation patterns governed by the CPG in accordance with the perturbations experienced by the robot, an adaptation mechanism is needed. Thor and Manoonpong [10] proposed an adaptation method called the Dual Integral Learner (DIL), which is used for reducing tracking error between a setpoint and a system output. The rules of the DIL are given by Eqs. 5, 6, 7 and 8:

$$x_f(n) = A_f \cdot x_f(n-1) + B_f \cdot e(n) + C_f \cdot \int e(n), \tag{5}$$

$$x_s(n) = A_s \cdot x_s(n-1) + B_s \cdot e(n) + C_s \cdot \int e(n), \tag{6}$$

$$x(n) = x_f(n) + x_s(n), \tag{7}$$

$$e(n) = f(n) - x(n), \tag{8}$$

where $x_f(n)$ is the fast learner output, $x_s(n)$ is the slow learner output, $x(n)$ is the combined learner system output, $e(n)$ is the error given by the difference between the learner system output and a setpoint $f(n)$. B_f and B_s are the learning rates while A_f and A_s are the retention factors of the learners. C_f and C_s ($C_f > C_s$), are the integrator components, reducing steady state errors. The learning rates and the retention factors are chosen such that $B_f > B_s$ and $A_f < A_s$. According to this setup, the fast learner will learn quicker and forget faster and vice versa for the slow learner. The DIL is depicted in the blue highlighted area in Fig. 2.

The DIL will adapt the α-value of the CPG (Eq. 4), which in turn will adjust the amplitude of the tentacle movements. The DIL was implemented with the following, experimentally obtained parameters: $A_f = 0.7$, $A_s = 0.9$, $B_f = 0.85$, $B_s = 0.45$, $C_f = 0.01$, and $C_s = 0.005$. The given error at each timestep n is the difference between the pressure sensor feedback and the CPG-signal, and the error is used as input to the two learners, that will adjust the α-value of the CPG.

Figure 4 depicts the DIL's ability to adapt to an external sinusoidal signal with the given parameters. The signal plotted in black is the reference signal, which is generated by an SO(2) oscillator with decrementing values of α, while the signal plotted in red is obtained by using the DIL to modulate the α of a second SO(2) oscillator to mimic the signal plotted in black. As seen in Fig. 4, the DIL is able to adjust the α-value of the second SO(2) oscillator relatively fast when the amplitude of the reference signal changes, due to the online tracking error reduction executed by the DIL.

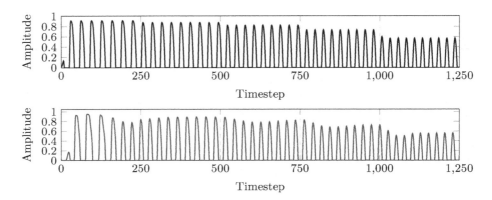

Fig. 4. CPG adaptation based on the DIL. The DIL parameters are as follows: $A_f = 0.7$, $A_s = 0.9$, $B_f = 0.85$, $B_s = 0.45$, $C_f = 0.01$, and $C_s = 0.005$. Upper plot (black) shows the reference signal where the α-value is decremented with 0.1 for every 250th timestep, lower plot (red) shows the adaptation of the CPG through the DIL to follow the change of the reference signal. (Color figure online)

Figure 5 depicts the DIL's ability to adapt to a non-periodic external signal given by rotations of one of the on-board potentiometers. The oscillations of the CPG are almost non-existent at timestep 80, but when rotating the potentiometer at this point, the DIL is able to restart the oscillations of the CPG (red highlighted area). At approximately timestep 100, the potentiometer is turned to zero, causing the DIL to lower the amplitude of the oscillations, which is visible from the plotted α-values of approximately timestep 120 to 180, and the corresponding CPG-values (blue highlighted area). This shows, that the DIL is able to adapt the amplitude of the CPG based on an external non-periodic signal. Furthermore, once the oscillatory movements generated by the CPG have ceased, it is possible to restart the oscillations by turning the potentiometer.

Sensor Signal Post-processing. The air flow to the soft robot is controlled by the PWM signals originating from the CPG. The duty cycle of the PWM signals define the relative open/close operation of the solenoid valves. Although the method is simple and efficient, it introduces noise to the measured pressure response [7]. Noisy feedback will interfere with the accuracy of the adaptation mechanism that relies on comparison of the sensor data and the CPG signal. To overcome this issue, a three-point moving average filter was applied to the measured sensor values, amplified by a factor of 4.0. After the averaging operation, it is checked whether the averaged sensor response is above or below a threshold of 7. If below, the average sensor response is set to zero, otherwise, the response remains the same. This action is performed to eliminate slight oscillations in the measured pressure (background noise), even when the robot is not actuated. To mimick the sinusoidal half-wave of the CPG-signal and to further smooth the

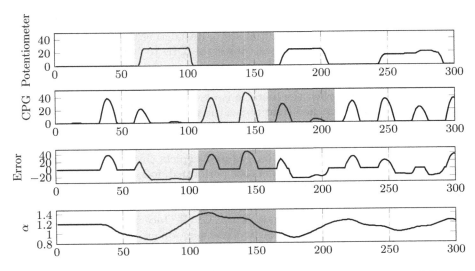

Fig. 5. Adaptation of α based on non-periodic potentiometer readings through DIL. In the highlighted red area, the DIL is able to use the reference signal of the potentiometer to restart the oscillations of the CPG. In the highlighted blue area, the DIL is able to decrease the amplitude of the CPG as a result of the potentiometer being turned to zero. (Color figure online)

sensor response, the averaged sensor response is filtered through another moving average filter and the result hereof is multiplied by an amplification gain of 1.7.

4   Experimental Results

The emergent behaviour of the tentacle when interacting with its environment has been investigated in two experiments, where the tentacle senses external perturbations through the use of the on-board pressure sensors. Both experiments are performed on the physical setup seen on Fig. 1a. The implemented neural controller generates an oscillatory movement in the tentacle through the use of the CPG, while adapting the movement amplitude, i.e. bending angle, through the DIL mechanism. All processing and control is performed online, onboard the Arduino Mega 2560.

The first experiment entails a tight grasping of the soft robotic tentacle with the hand, restricting its movement. Figure 6 depicts this experiment. Initially, two rounds of unhindered movement is performed by the tentacle, after which the tentacle is grasped tightly, generating a large pressure response in the pressure sensors, due to the lowered volume inside the air chambers of the tentacle. The restriction period is highlighted in red in the figure. After restricting the tentacle, the amplitude of the CPG increases due to the increase in α, and hence, the bending angle of the tentacle is increased. Due to lower pressure readings after the restriction period, the DIL decreases the amplitude of the CPG, which makes the tentacle return to its initial state. Restricting the tentacle once more in the second, red highlighted area causes the DIL to increase the amplitude once again.

The second experiment has been executed to better demonstrate the behaviour that occurs when the tentacle is blocked. For this experiment, the air supply tube has been pinched right above the push fitting on the control board, which generates a larger increase in pressure when the solenoid valve opens, causing a larger error in the adaptation mechanism, which in turn will cause a larger adaptation response to the stimulus. Figure 7 depicts this experiment. Initially, two periods of unhindered movements are performed, followed by the pinching of the air supply tube (highlighted red area). Figure 7 shows, that when the pressure supply is blocked by an external perturbation, the DIL tries to compensate for this by increasing the amplitude of the CPG. Once the perturbation is no longer present, the amplitude is slowly decreasing, due to the equivalently lower pressure responses. When compared to the previous experiment, this experiment shows a more definite adaptation due to the larger tracking error at the time of the perturbation.

5   Discussion

The experiments show that an adaptive behaviour in the soft robotic tentacle emerges, when pressure sensor feedback is applied. The most successful experiment hereof is the experiment of clamping the pressure supply tubes, since this action yields the largest error in the DIL mechanism. When the pressure increases, so does the movement of the robot in an effort to compensate for the error. In the experiment where the tentacle is held firmly, the pressure sensors are able to detect a slight increase in pressure due to the lowered volume inside the air chamber. This leads to an increased amplitude in the tentacle movement, somewhat reminiscent of the escaping behaviour seen in fish; when a fish is caught, it will try to wiggle to escape [16]. In that sense, when the tentacle is "caught", it will try to expand itself to be released, which is an interesting behaviour. With the proposed control, the movement of the tentacle thus increases when it bumps into an obstacle. This is also reminiscent of the reflexive behaviour seen in insects when they encounter obstacles during locomotion. Brooks' [17] reactive legged robots were inspired by this mechanism; when one of the robot's legs was blocked, it would try to swing the leg higher in order to overcome the faced obstacle.

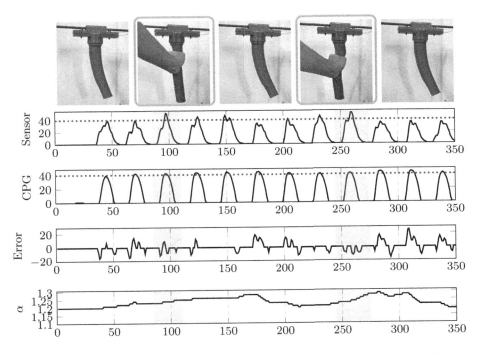

Fig. 6. Adaptation of α to restriction of the tentacle using the DIL. The red highlighted area indicates where the tentacle is restricted, causing a slightly larger pressure reading, which in turn causes the DIL to adjust the amplitude of the CPG slightly. Slowly, the tentacle returns to its initial state. Restricting the tentacle once more (second red, highlighted area) causes the amplitude to rise once again. (Color figure online)

From the experiments performed it is clear, that the sensor responses are not completely regular from one period to the next, even under the same amplitude configuration. For example in Fig. 6 a coincidental low pressure reading at timestep 180 causes the DIL to decrease the amplitude of the tentacle, which is not intended and is only due to the irregularities of the pressure response. This shows the need for a different or added sensorizing approach, such as embedding the tentacle with liquid metal strain sensors as in [4]. It is also possible, that obtaining a sensor model through the use of regression or an artificial neural network could improve the performance of the current control architecture, with the drawback of making the solution domain-specific.

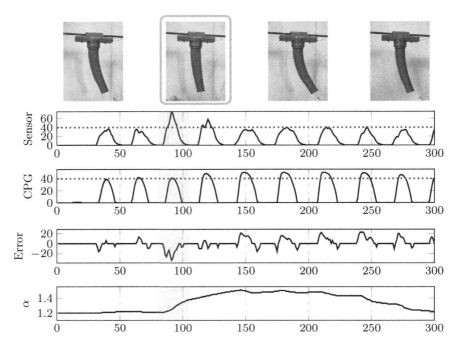

Fig. 7. Adaptation of α to blocking the pressure supply using the DIL. At timestep 80 (highlighted red), the pressure supply tube is pinched. The DIL increases the amplitude of the CPG due to the sudden high error, after which the amplitude is slowly decreased once the perturbation is removed in response to the correspondingly lower pressure readings. (Color figure online)

While the control approach shown in this study is only able to generate oscillatory movements, it can be extended to include another post processing network such as a radial basis function (RBF) network with reward-based (model free) learning for more complex pattern generation. An integrated CPG-RBF network framework with reward-based learning has recently been proposed by [19]. It is generic and can be applied to (soft) robot control. Thus, in the future we will apply the framework to the soft robotic tentacle for more complex movements towards the goal of real-world applications of human-soft robot interaction.

6   Conclusion

In this paper, a CPG implemented as an SO(2) oscillator has been applied on a physical setup consisting of a control board with an onboard Arduino Mega 2560, electrical pump and a soft robotic tentacle with three air chambers constructed from silicone. An adaptation mechanism, the DIL, was implemented that adapts the α-parameter of the SO(2) oscillator online based on an external reference signal. Sensory feedback of the tentacle was obtained from pressure sensors.

Experiments with this control architecture were performed on the physical setup, focusing on restriction of the movement of the tentacle as well as restriction of the air supply to the tentacle. The tentacle does exhibit adaptive behaviour upon external perturbations such as clamping the pressure supply tubes or restriction by holding the tentacle firmly. However, the experimental results show, that the perturbations need to be quite large for the DIL to effectively adapt α, due to the filtering of the measured pressure values.

Hence, further work includes the addition of a mapping function to the control architecture proposed in this paper. The mapping function would serve as translation from pressure sensor values to CPG-signal, and could eliminate the fluctuations of the error in the DIL mechanism, which will ultimately lead to a steadier control and adaptation mechanism. Besides this, it would be interesting to investigate the potential use of multiple (coupled) CPGs; one for each air chamber. This allows for independent adaptation of movement along the three actuation paths of the tentacle. In [18], coupled CPGs have been used for simulation of a soft robotic tentacle, where several different movements of the tentacle were achieved.

Additionally, embedding the tentacle with soft sensors such as liquid metal strain sensors could endow the soft robotic tentacle with greater proprioceptive capabilities and allow for additional experiments on emergent behaviors in confined environments. This in turn, could lead to both obstacle avoidance capabilities and an ability to reach a desired target position. Taken together, this study not only proposes an integration of online adaptive control and sensor feedback in a soft robot, but also paves the way forward for achieving motion intelligence in soft robotics in general.

References

1. Wang, H., Totaro, M., Beccai, L.: Toward perceptive soft robots: progress and challenges. Adv. Sci. **5**(1800541), 1–17 (2018). https://doi.org/10.1002/advs.201800541
2. Rus, D., Tolley, M.: Design, fabrication and control of soft robots. Nature **521**, 467–75 (2015). https://doi.org/10.1038/nature14543
3. Coevoet, E., et al.: Software toolkit for modeling, simulation, and control of soft robots. Adv. Robot. **31**, 1–17 (2017). https://doi.org/10.1080/01691864.2017.1395362
4. Tapia, J., Knoop, E., Mutný, M., Otaduy, M., Bächer, M.: MakeSense: automated sensor design for proprioceptive soft robots. Soft Robot. **3**, 332–345 (2019). https://doi.org/10.1089/soro.2018.0162
5. Case, J., White, E., Kramer, R.: Sensor enabled closed-loop bending control of soft beams. Smart Mater. Struct. **25**, 045018 (2016). https://doi.org/10.1088/0964-1726/25/4/045018
6. Wu, P., Jiangbei, W., Yanqiong, F.: The structure, design, and closed-loop motion control of a differential drive soft robot. Soft Robot. **5**(1), 71–80 (2018). https://doi.org/10.1089/soro.2017.0042
7. Elgeneidy, K., Lohse, N., Jackson, M.: Bending angle prediction and control of soft pneumatic actuators with embedded flex sensors - a data-driven approach. Mechatronics **50**, 234–247 (2017). https://doi.org/10.1016/j.mechatronics.2017.10.005

8. Zhou, Y., Ju, M., Zheng, G.: Closed-loop control of soft robot based on machine learning. In: Proceedings of the 38th Chinese Control Conference, pp. 4543–4547 (2019). https://doi.org/10.23919/ChiCC.2019.8866257
9. Zolfagharian, A., Kaynak, A., Kouzani, A.: Closed-loop 4D-printed soft robots. Mater. Des. **188**, 108411 (2019). https://doi.org/10.1016/j.matdes.2019.108411
10. Thor, M., Manoonpong, P.: A fast online frequency adaptation mechanism for CPG-based robot motion control. IEEE Robot. Autom. Lett. **4**, 3324–3331 (2019). https://doi.org/10.10007/1234567890
11. Jørgensen, J.: Constructing Soft Robot Aesthetics - Art, Sensation, and Materiality in Practice. IT University in Copenhagen, Denmark (2019). ISBN: 978-87-7949-027-7
12. Harvard Biodesign community: Soft Robotics Toolkit (2014). https://www.softroboticstoolkit.com/control-board. Accessed 24 June 2020
13. Pasemann, P.: Complex dynamics and the structure of small neural networks. Netw.: Comput. Neural Syst. **13**, 195–216 (2002). https://doi.org/10.1080/713663430
14. Steingrube, S., Timme, M., Wörgötter, F., Manoonpong, P.: Self-organized adaptation of simple neural circuits enables complex robot behavior. Nat. Phys. **6**, 224–230 (2010). https://doi.org/10.1038/nphys1508
15. Pasemann, F., Hild, M., Zahedi, K.: SO(2)-Networks as neural oscillators. In: Mira, J., Álvarez, J.R. (eds.) IWANN 2003. LNCS, vol. 2686, pp. 144–151. Springer, Heidelberg (2003). https://doi.org/10.1007/3-540-44868-3_19
16. Domenici, P., Hale, M.E.: Escape responses of fish: a review of the diversity in motor control, kinematics and behaviour. J. Exp. Biol. 222 (2019). https://doi.org/10.1242/jeb.166009
17. Brooks, R.A.: A robot that walks; emergent behaviors from a carefully evolved network. In: Proceedings, 1989 International Conference on Robotics and Automation, vol. 2, pp. 692–696 (1989). https://doi.org/10.1109/ROBOT.1989.100065
18. Tian, J., Lu, Q.: Simulation of octopus arm based on coupled CPGs. J. Robot. **2015**, 1–9 (2015). https://doi.org/10.1155/2015/529380
19. Thor, M., Kulvicius, T., Manoonpong, P.: Generic neural locomotion control framework for legged robots. IEEE Trans. Neural Netw. Learn. Syst. (2020). https://doi.org/10.1109/TNNLS.2020.3016523

Adaptive Neuromechanical Control for Robust Behaviors of Bio-Inspired Walking Robots

Carlos Viescas Huerta[1,2], Xiaofeng Xiong[1], Peter Billeschou[1], and Poramate Manoonpong[1,3(✉)]

[1] Embodied AI and Neurorobotics Lab, SDU Biorobotics,
The Mærsk Mc-Kinney Møller Institute, The University of Southern Denmark,
Odense M, Denmark
{xizi,pebil,poma}@mmmi.sdu.dk
[2] Tecnalia Research & Innovation, Donostia - San Sebastián, Spain
carlos.viescas@tecnalia.com
[3] Bio-inspired Robotics and Neural Engineering Lab, School of Information Science
and Technology, Vidyasirimedhi Institute of Science and Technology (VISTEC),
Rayong 21210, Thailand

Abstract. Walking animals show impressive locomotion. They can also online adapt their joint compliance to deal with unexpected perturbation for their robust locomotion. To emulate such ability for walking robots, we propose here adaptive neuromechanical control. It consists of two main components: Modular neural locomotion control and online adaptive compliance control. While the modular neural control based on a central pattern generator can generate basic locomotion, the online adaptive compliance control can perform online adaptation for joint compliance. The control approach was applied to a dung beetle-like robot called ALPHA. We tested the control performance on the real robot under different conditions, including impact force absorption when dropping the robot from a certain height, payload compensation during standing, and disturbance rejection during walking. We also compared our online adaptive compliance control with conventional non-adaptive one. Experimental results show that our control approach allows the robot to effectively deal with all these unexpected conditions by adapting its joint compliance online.

Keywords: Computational intelligence · Muscle models · Robot control · Bio-inspired robotics · Adaptive locomotion · Walking robots

1 Introduction

Insects show fascinating locomotion capabilities and effectively deal with unexpected conditions. Such capabilities emerge from the neuromechanical interaction between neural control and biomechanical properties [1]. However, mimicking neuromechanical control in an insect-like redundant robot is a challenging

© Springer Nature Switzerland AG 2020
H. Yang et al. (Eds.): ICONIP 2020, LNCS 12533, pp. 775–786, 2020.
https://doi.org/10.1007/978-3-030-63833-7_65

and un-fully solved problem due to neural and biomechanical complexities [2]. To address this problem, Full and Koditschek (1999) proposed a dynamic model (i.e., template) for trimming the locomotor complexities away [3]. A template is a simplified locomotor model reducing neural, muscle, and mechanical complexities. The template principles have been applied to many insect-like robots such as RHex [4]. For instance, the simplified mechanical design and coordinated controller enable the RHex robot to produce robust walking behaviors over comlpex terrains such as sands. Those robot designs and controllers are characterized by bio-inspired engineering. On the other hand, the mechanical designs and controllers are too simple to reveal locomotor principles in a robot-inspired biology perspective [5]. For example, energy-efficient insect-like walking emerges from redundant neuromusculoskeletal interactions that simplified leg design and control fail to emulate. Such emulation requires an anchor model for mimicking redundant neuromechanical control, as Full and Koditschek suggested [3]. An anchor model is an elaborate dynamic model embedded within insect-like waking behaviors. Naris et al. (2020) proposed an anchor (i.e., neuromechanical) model to explore the role of common inhibitor motor neurons in insect locomotion [6]. However, most existing anchor models are validated in neuromechanical simulations owing to physical control and design complexities. To anchor physical insect-like walking, Dürr et al. (2019) proposed a neuromechanical model for energy-efficient walking [7]. The energy-efficiency is enhanced by the muscle-like actuation of the proposed model. However, such actuation is designed in a hybrid and complicated way, therefore leading to heavy (i.e., 7.2 kg) and bulky actuators. Therefore, an elaborate (i.e., neuromechanical) model for an insect-like (i.e., redundant) robot remains a challenging problem.

To address this problem, we have developed a neuromechanical controller for emulating physical redundant insect-like walking [8]. The controller consists of a modular neural network (MNN) for coordinating 18 joint motions, and virtual agonist-antagonist muscle-like mechanisms (VAAMs) for variable compliant joint motions. As a result, such coordinated and compliant joint motions enable the insect-like robot AMOS to generate adaptive and energy-efficient walking behaviors over rough terrains such as gravels. However, a proximal-distal gradient has been applied to simplify the neuromechanical controller where proximal joint compliance gains are fixed. Such simplification leads to the trade-off between joint coordination and compliance. In this paper, we replace the VAAMs by proposing online compliance adaptation for all joints of a dung beetle-like robot. In the adaptation, the 36 joint compliance gains (K, D) are online tuned to adapt different perturbed robot behaviors. Such adaptation relies only on simple actuator position feedback, rather than complex force/torque feedback and physical compliant mechanisms (e.g., springs). Moreover, the pure software implementation can be applied to torque control of lightweight actuators, therefore greatly reducing actuator dependency and bulkiness. As a result, the modified neuromechanical controller (called here adaptive neuromechanical controller) enables the robot to achieve more robust behaviors against unexpected perturbations (such as impact force, payload, and pulling force), compared to a conventional controller. The proposed controller paves a way for implementing neuromechanical

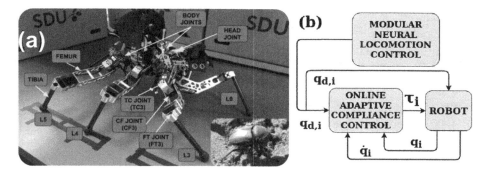

Fig. 1. (a) The dung beetle-like robot ALPHA. The inset shows the South African dung beetle (*Scarabaeus galenus*). The robot legs are numbered as L0, L1, L2, L3, L4, and L5 are the left front, left middle, left hind, right front, right middle, and right hind legs, respectively. Joints are numbered similarly starting from the left front to right hind legs (i.e., TC0, CF0, and FT0 belong to L0, until TC5, CF5, and FT5 belong to L5). (b) The block diagram of our adaptive neuromechanical control.

control in robust insect-like behavior. Taken together, the main contributions of the study are as follows: 1) an adaptive neuromechanical controller that combines modular neural locomotion control and online adaptive compliance control, 2) the compliance control, providing muscle-like function, relies only on simple joint feedback, rather than force/torque feedback and physical compliant mechanisms (e.g., spring), and 3) real robot demonstration of the control performance.

2   Dung Beetle-Like Robot ALPHA

In this work, we used the dung beetle-like robot ALPHA as our robot experimental platform (see Fig. 1 (a)). ALPHA was designed based on the South African dung beetle (*Scarabaeus galenus*) [11]. It consists of six legs. Each leg has three joints (three degrees of freedom (DOFs)). The first joint is the Thorax-Coxa (TC) joint connecting the leg to the body. It allows for forward and backward movements of the leg. The second joint is the Coxa-Femur (CF) joint. It allows for upward and downward movements of the leg. The last one is the Femur-Tibia (FT) joint. It allows for flexion and extension of the tibia part. Besides the leg joints, the robot has also additional two body joints, which provide the flexibility to the body for object manipulation and transportation tasks, and one head joint to move its head up and down for active obstacle sensing. In this study, we kept the body and head joints fixed to certain angles since we only focus on locomotion and adaptive leg compliance. All in all, ALPHA has 21 active joints which are driven by the Dynamixel motors (XM430-W350-R) producing a torque around 4.8 Nm. The motors provide joint angle and current sensory signals which are used for our adaptive compliance control and system performance monitoring, respectively. The weight of the fully equipped robot (including 21 motors, all electronic components, and battery packs) is approximately 4 kg. The

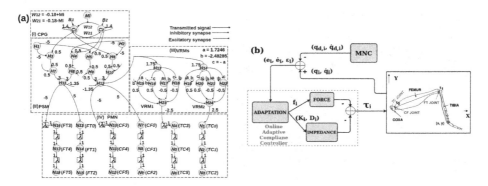

Fig. 2. (a) The modular neural locomotion control (MNC) consisting of four submodules (CPG, PSM, VRMs, and PMN). It is for locomotion generation. (b) The online adaptive compliance control (OAC). It is for joint compliance or stiffness adaptation.

robot is connected to a PC via a Dynamixel motor controller board with a USB interface.

3 Adaptive Neuromechanical Control

The adaptive neuromechanical control of ALPHA (Fig. 1(b)) is based on the control system proposed in [8]. It is divided into two main control units: Modular neural locomotion control (MNC) and online adaptive compliance control (OAC). The MNC is a biologically inspired, CPG-based neural network. It can generate different gaits and control walking direction and speed of the robot. The OAC is derived from an impedance control method [9]. It uses the desired motor positions generated by the MNC and joint angle feedback to online control and adapt leg joint stiffness. The OAC will generate final motor torque outputs to control all leg joints. Each control unit is explained in detail below.

3.1 Modular Neural Locomotion Control (MNC)

The MNC is a gait generation network (Fig. 2(a)). It has four different submodules: I) a CPG module, II) a phase switching module (PSM), III) two velocity regulation modules (VRMs), and IV) a premotor neuron module (PMN). The MNC is a fundamental control unit of the neuromechanical control, but it is not the key contribution of this paper. Therefore, it will be briefly explained. Further details of this control can be seen at [12]. All neurons of the MNC are modeled as discrete-time non-spiking neurons. The activation of each neuron is described by Eq. 1:

$$A_i(t) = \Sigma_{j=1}^{m} W_{ij} \cdot O_j(t-1) + B_i, \quad i = 1, .., m, \tag{1}$$

where m is the number of neurons, B_i an internal bias to neuron i, and W_{ij} the synaptic weight between neurons j and i. The output O_j of each neuron is calculated as the hyperbolic tangent $(tanh)$ of the activation $A_i(t)$.

The CPG, which generates basic periodic signals, is formed by two fully connected neurons, with two internal bias terms $(B_{1,2} = 0.01)$ as an initial drive of the CPG. It creates periodic signals for walking. The connection weights between the neurons can be adjusted according to a modulatory input MI while the self connection weights are fixed (see Fig. 2(a)-I). Different values of MI produce different gaits [8]. The PSM is a feed-forward network with three hidden layers (Fig. 2(a)-II). It receives the CPG signals and generates outputs to the CF and FT motor neurons. This network can reverse the CPG signals for sideways walking. The VRMs are feed-forward networks (Fig. 2(a)-III) which are used for controlling walking directions, like turning left/right or curve walking in forward and backward directions. They receive a copy of the PSM outputs and translate them to control the TC motor neurons. The PMN (Fig. 2(a)-IV) distributes the outputs of the PSM and VRMs to the different joints by means of motor neurons. The outputs of the MNC are mapped into a proper motor range of the robot and constitute the desired joint positions $q_{i,d}(t)$. The joint positions are further transmitted to the OAC which will convert them into joint torques for joint stiffness or compliance adaptation.

3.2   Online Adaptive Compliance Control (OAC)

The OAC is to generate the joint torque τ_i based on a Proportional-Derivative (PD) rule[9],

$$\tau_i = -f_i(t) - K_i(t) \cdot e_i(t) - D_i(t) \cdot \dot{e}_i(t), \quad i = 1, 2...18, \tag{2}$$

where $K_i(t)$ and $D_i(t)$ denote the compliance gains of the i-th joint of the dung beetle-like robot. $f_i(t)$ denotes the force term (gravity compensation). Finally, $e_i(t)$ and $\dot{e}_i(t)$ represent the joint position and velocity errors given by,

$$e_i(t) = q_i(t) - q_{d,i}(t), \; \dot{e}_i(t) = \dot{q}_i(t) - \dot{q}_{d,i}(t), \; \varepsilon_i(t) = e_i(t) + \beta\dot{e}_i(t), \; \beta = 0.05, \tag{3}$$

where $\varepsilon_i(t)$ denotes the joint tracking error. The desired positions $q_{d,i}(t)$ are generated by the MNC (see Fig. 2(b)). The OAC is to co-minimize the compliance efforts and motion errors of the i-th joint of the dung beetle-like robot over the time period T [9,10],

$$J_o(t) = J_c(t) + J_p(t),$$
$$J_c(t) = \frac{1}{2} \int_{t-T}^{t} (K_i(t))^2 + (D_i(t))^2, \tag{4}$$
$$J_p(t) = \frac{1}{2} \int_{t-T}^{t} V(t), V(t) = I_i(\varepsilon_i(t))^2,$$

where I_i denotes the inertial scalar of the robot i-th joint. The OAC's co-minimization leads to the robot joint compliance adaptation given by,

$$K_i(t) = f_i(t)e_i(t), \quad D_i(t) = f_i(t)\dot{e}_i(t),$$
$$f_i(t) = \frac{\varepsilon_i(t)}{\gamma_i(t)}, \quad \gamma_i(t) = \frac{a}{1 + b\varepsilon_i(t)^2}, \tag{5}$$

where $\gamma_i(t)$ is an adaptation scalar with the positive scalars $a = 0.2$ and $b = 5$. All scalars as well as the derivation of Eqs. (4) and (5) refer to our developed human-like impedance controller [9]. The values of these parameters a and b, as well as that of β in Eq. 3, were obtained from [9].

4   Experiments and Results

We performed three main experiments in order to test the performance of our adaptive neuromechanical control approach. The experiments include 1) impact force absorption when dropping the robot from a certain height, 2) payload compensation during standing, and 3) disturbance rejection during walking. We also compared the performance of our approach with non-adaptive impedance control, where we set the stiffness and damping parameters to constant values (e.g., $K = 25$ and $D = 15$).

In the first experiment, the robot was dropped from a certain height. This is to investigate the performance of the control approach that can online adjust the joint compliance to deal with impact force. The robot was set on the ground initially; then, it was picked up and suspended at a height 10 cm over the table and, finally, dropped. The results of the experiment are shown in Fig. 3. Plots (I) and (II) show stiffness and damping adaptation. Plots (III) and (IV) show the current consumption (in mA) for both the adaptive neuromechanical controller and the non-adaptive impedance controller. Here, the signals of the CF3 and FT3 joints of the right front leg (L3) are provided while the signals of other joints having similar patterns to these joints are not shown. A sequence of the experiment can be seen in the sub-figures below.

The highlighted interval corresponds to the landing moment. From the current consumption plots, we can observe that the adaptive neuromechanical controller was able to reduce the peak current of the motors and stabilize the robot posture faster when impact force was applied, compared to the non-adaptive controller. In the moment of the impact, the impedance gains $(K_i(t), D_i(t))$ of the OAC were quickly adapted to allow the robot a fast recovery to its normal state. When the robot returned to the normal state, the stiffness $(K_i(t))$ and damping $(D_i(t))$ values were automatically decreased to nearly zero (i.e., relaxation state). This effect was produced by the adaptation rule in Eq. (5). In principle, each impedance gain is adapted as a product of the force term $f_i(t)$ and the difference between desired and actual joint position/speed signals $(e_i(t), \dot{e}_i(t))$. At the steady state, the joint position and speed are almost constant. Therefore, the difference becomes nearly zero. As a consequence, the motor torque is mainly

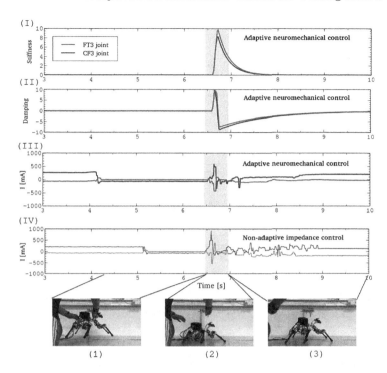

Fig. 3. Results of the first experiment. Plots show the variation of the stiffness (I), damping (II), and current consumption (III) of the joints FT3 (red) and CF3 (blue) of the adaptive neuromechanical control. Plot (IV) shows the current consumption of the non-adaptive impedance control. The video clip of this experiment can be seen at http://manoonpong.com/ICONIP2020/video1.mp4. (Color figure online)

modulated by the force term (gravity compensation, Eq. (2)), meaning that the robot joint compliance is able to adapt to the robot weight.

In the second experiment, the additional loads were incrementally applied to the robot during standing. This is to investigate the performance of the control approach that can online adjust the joint compliance to deal with different payloads. Initially, we let the robot stand. Afterwards, we incrementally added weights on it (i.e., first, the 3-kg weight and then the additional 2-kg weight). In total, the payload was 5 kg (i.e., 1.25 times body weight). The results of the experiment are shown in Fig. 4. Plots (I) and (II) show stiffness and damping adaptation. Plots (III) and (IV) show the current consumption (in mA) for both the adaptive neuromechanical controller and the non-adaptive impedance controller. Here, the signals of the CF3 and FT3 joints of the right front leg (L3) are provided while the signals of other joints having similar patterns to these joints are not shown. A sequence of the experiment can be seen in the sub-figures (1)–(4).

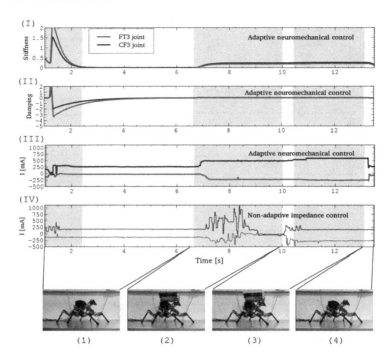

Fig. 4. Results of the second experiment. Plots show the variation of the stiffness (I), damping (II), and current consumption (III) of the joints FT3 (red) and CF3 (blue) of the adaptive neuromechanical control. Plot (IV) shows the current consumption of the non-adaptive impedance control. The video clip of this experiment can be seen at http://manoonpong.com/ICONIP2020/video2.mp4. (Color figure online)

The first highlighted interval corresponds to the own body weight adaptation. In this state, the OAC adapted the impedance gains $(K_i(t), D_i(t))$ to stabilize the robot standing posture. Once the robot had reached its stable standing posture, the impedance gains were automatically decreased to nearly zero (i.e., relaxation state) due to our adaptation rule (Eq. (5)). The second and third highlighted intervals correspond to the following phases where the 3-kg weight and then another 2-kg weight were applied. After the last interval, all weights were removed. Therefore, the robot returned to its initial stiffness (i.e., elastic recovery). The result shows that the adaptive neuromechanical control allows the robot to online adapt its joint compliance to deal with different payloads in a faster and more stable manner, compared to the non-adaptive impedance control. Note that for the non-adaptive impedance control, after adding the payloads, the robot became unstable and its motors drawn too much current (see high peak). As a consequence, the robot collapsed.

In the third experiment, the robot was disturbed during walking. This is to investigate the performance of the control approach that can online adjust the joint compliance to deal with unexpected disturbance during locomotion. Two

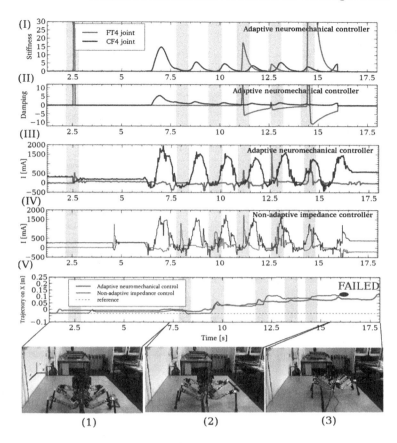

Fig. 5. Results of the third experiment (first scenario). The effect after applying disturbance to the tibia of the leg L4 during walking. Plots (I) and (II) show the stiffness and damping adaptation of the CF4 (blue) and FT4 (red) joints. Plots (III) and (IV) show the current consumption for both adaptive and non-adaptive control cases. Finally, plot (V) shows the deviation from the walking trajectory along the x direction. The adaptive neuromechanical control is represented in red, while the non-adaptive impedance control is represented in blue. The video clip of this experiment can be seen at http://manoonpong.com/ICONIP2020/video3.mp4. (Color figure online)

scenarios were performed. For the first one, we attached a string to the tibia of the middle right leg L4 (Fig. 1(a)) and horizontally pulled the leg sideways, in parallel to the ground. For the second one, we attached a string to the femur of L4 (Fig. 1(a)) and vertically pulled the leg upwards perpendicular to the ground.

The results of the experiment for the first and second scenarios are shown in Figs. 5 and 6. Each figure contains subplots. Plots (I) and (II) show stiffness and damping adaptation of the CF4 and FT4 joints of the right middle leg (L4) where the disturbance was given. Plots (III) and (IV) show the current consumption (same joints). Plot (V) of Fig. 5 shows the deviation from walking

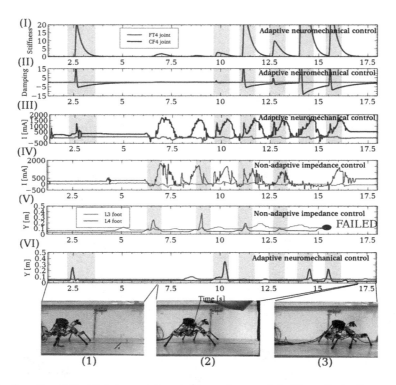

Fig. 6. Results of the third experiment (second scenario). The effect after applying disturbance to the femur of the leg L4 during walking. Plots (I) and (II) show the stiffness and damping adaptation of the CF4 (blue) and FT4 (red) joints. Plots (III) and (IV) show the current consumption for both adaptive and non-adaptive control cases. Plots (V) and (VI) show the vertical trajectory of L3 foot (red) and L4 foot (blue). The video clip of this experiment can be seen at http://manoonpong.com/ICONIP2020/video3.mp4. (Color figure online)

trajectory along the x (lateral) direction. Plots (V) and (VI) of Fig. 6 show the trajectory along the y (vertical) direction of the L3 and L4 feet. A sequence of the experiment can be seen in the sub-figures (1)–(3). The highlighted intervals correspond periods that the robot leg was pulled. Using the adaptive neuromechanical controller, the individual joints of the leg were able to produce a fast reaction similarly to muscle reflexes. This can be seen from peaks of stiffness and damping (plots (I) and (II)) of Figs. 5 and 6. In other words, the robot tried to adaptively reject the disturbance such that it can maintain its stable locomotion. In contrast, using the non-adaptive impedance controller, the robot had a difficulty to deal with the disturbance since the stiffness and damping parameters were not adapted to the spontaneous unexpected disturbance. As a consequence, the motors drawn too much current to work against the disturbance and the robot eventually collapsed (indicated by blue dots "FAILED" in plots (V) of Figs. 5 and 6.

5 Conclusions and Future Work

In this paper, we introduced a bio-inspired adaptive neuromechanical controller, applied to the dung beetle-like robot ALPHA. This controller is based on a combination of modular neural locomotion control and online adaptive compliance control. It can generate robot locomotion as well as quickly online adapt the robot joint compliance to deal with unexpected situations. We performed a series of experiments including dropping the robot from a height 10 cm above the ground, adding weights to the robots, and disturbing it during walking by pulling its leg. We also compared our control approach with a conventional non-adaptive impedance control (i.e., fixed K and D impedance gains).

The results show that the proposed control approach with the online compliance adaptation can improve the performance of the robot such that the robot can effectively deal with the unexpected situations in all experiments. Furthermore, the approach can reduce the peak current consumption of the motors (see Fig. 3) and generate fast elastic recovery and muscle reflexes, compared to the conventional one. Consequently, it will prolong the motor lifetime and prevent the robot from collapsing or sustaining damage. The results shown here also go beyond the previous works [8, 10–12] in the following aspects. Here we demonstrate all joint compliance implementation on non-uniform leg structures (i.e., all 18 joints with in total 36 joint compliance gains) while in [8, 12] joint compliance was implemented on only 12 joints (i.e., 24 joint compliance gains) where six proximal joints did not have compliance for simplification. Additionally, we use a torque control-based muscle model with fast real-time adaptation which cannot be achieved by the method in [8]. Furthermore, we demonstrate online joint compliance adaptation in more complex and multiple tasks in a real complex system (i.e., impact force absorption, payload compensation, and disturbance rejection in the 18-joint walking robot with 18 muscle models) compared to the one which was investigated in [10] (i.e., horizontal arm movements in the two-joint arm with six muscle models in simulation). Our control strategy is different from [11]. In [11], reinforcement learning was used to optimize neural control parameters for (stiff) walking without adaptive compliance in simulation. It also requires several learning trials to obtain optimal control parameters while here the controller can generate locomotion based on modular neural control and quickly adapt joint compliance based on online adaptive compliance control to deal with unexpected conditions without several learning trials. Thus, the proposed control strategy is more practical for real robot implementation with online adaptation. The work carried out for this paper is the first step toward future improvements on ALPHA. In the future, we will introduce online gait adaptation [13] to the control method for efficient locomotion on rough terrain.

Acknowledgement. This research was supported by the Human Frontier Science Program under Grant agreement no. RGP0002/2017.

References

1. Ritzmann, R., Zill, S.: Neuroethology of insect walking. Scholarpedia **8**, 30879 (2013)
2. Aoi, S., Manoonpong, P., Ambe, Y., Matsuno, F., Wörgötter, F.: Adaptive control strategies for interlimb coordination in legged robots: a review. Front. Neurorobot. **11**, 39 (2017)
3. Full, R.J., Koditschek, D.E.: Templates and anchors: neuromechanical hypotheses of legged locomotion on land. J. Exp. Biol. **202**(23), 3325–3332 (1999)
4. Wenger, G.J., Johnson, A.M., Taylor, C.J., Koditschek, D.E.: Semi-autonomous exploration of multi-floor buildings with a legged robot. In: Proceedings of Unmanned Systems Technology, vol. XVII, pp. 62–69 (2015)
5. Gravish, N., Lauder, G.V.: Robotics-inspired biology. J. Exp. Biol. **221**(7) (2018)
6. Naris, M., Szczecinski, N.S., Quinn, R.D.: A neuromechanical model exploring the role of the common inhibitor motor neuron in insect locomotion. Biol. Cybern. **114**, 23–41 (2020). https://doi.org/10.1007/s00422-019-00811-y
7. Dürr, V., et al.: Integrative biomimetics of autonomous hexapedal locomotion. Front. Neurorobot. **13**(88) (2019)
8. Xiong, X., Wörgötter, F., Manoonpong, P.: Adaptive and energy efficient walking in a hexapod robot under neuromechanical control and sensorimotor learning. IEEE Trans. Cybern. **46**(11), 2521–2534 (2016)
9. Xiong, X., Manoonpong, P.: Adaptive motor control for human-like spatial-temporal adaptation. In: Proceedings of the 2018 IEEE International Conference on Robotics and Biomimetics, pp. 2107–2112 (2018)
10. Tee, K.P., Franklin, D.W., Kawato, M., Milner, T.E., Burdet, E.: Concurrent adaptation of force and impedance in the redundant muscle system. Biol. Cybern. **102**(1), 31–44 (2010). https://doi.org/10.1007/s00422-009-0348-z
11. Pitchai, M., et al.: CPG driven RBF Network control with reinforcement learning for gait optimization of a dung beetle-like robot. In: Tetko, I.V., Kurková, V., Karpov, P., Theis, F. (eds.) ICANN 2019. LNCS, vol. 11727, pp. 698–710. Springer, Cham (2019). https://doi.org/10.1007/978-3-030-30487-4_53
12. Xiong, X., Wörgötter, F., Manoonpong, P.: Neuromechanical control for hexapedal walking on challenging surfaces and surface classification. Robot. Auton. Syst. **62**(12), 1777–1789 (2014)
13. Ngamkajornwiwat, P., Homchanthanakul, J., Teerakittikul, P., Manoonpong, P.: Bioinspired adaptive locomotion control system for online adaptation of a walking robot on complex terrains. IEEE Access (2020). https://doi.org/10.1109/ACCESS.2020.2992794

Deep Learning Based Strategy for Eye-to-Hand Robotic Tracking and Grabbing

Junwen Zhong[1,2], Weijun Sun[1,2], Qinyu Cai[1,2], Zhaowei Zhang[1,2], Zhekang Dong[1,2,3], and Mingyu Gao[1,2(✉)]

[1] School of Electronics and Information, Hangzhou Dianzi University, 1158, 2 Avenue, Hangzhou 310018, China
{jwzhong2020,weijunsun,qycai,zzw0211,englishp,mackgao}@hdu.edu.cn
[2] Zhejiang Provincial Key Lab of Equipment Electronics, 1158, 2 Avenue, Hangzhou 310018, China
[3] Department of Electrical Engineering, The Hong Kong Polytechnic University, 11 Yuk Choi Road, Hung Hom 999077, Hong Kong

Abstract. Moving target tracking and grabbing is a common task for industrial robots. Usually, industrial robot complete complex actions through programming and teaching technologies, which suffers from the limitations of complicated programming logic and low scalability. Based on this, a flexible strategy combining deep learning and Kalman filter is proposed for eye-to-hand robotic tracking and grabbing. Firstly, the classic YOLOv3 algorithm is applied for target detection, and the bounding box of the target on the conveyor belt is obtained. Secondly, the target motion model is built up to obtain the system parameter matrices. Thirdly, the prediction equations can be given by Kalman filtering, and the target prediction position can be calculated and feedback to the robotic arm for the grabbing task. Finally, the experimental results show that the proposed strategy can improve the robustness of industrial robot tracking and grabbing, and its scalability is also improved compared with traditional methods.

Keywords: Deep learning · Object detection · Kalman filter · Target tracking

1 Introduction

Industrial robots have traditionally required a precisely defined environment, with pre-planning and programming to achieve the complex movements, which means efficiency and flexibility of their work will be greatly limited by the working environment and the target objects [1,2]. While the vision-based grab control system of manipulator can automatically identify different kinds of objects and realize automatic sorting, which will liberate people from laborious and repetitive labor.

© Springer Nature Switzerland AG 2020
H. Yang et al. (Eds.): ICONIP 2020, LNCS 12533, pp. 787–798, 2020.
https://doi.org/10.1007/978-3-030-63833-7_66

In recent years, deep learning has developed rapidly and is widely used in various fields. Meanwhile, with the improvement of computer performance, it is possible to apply deep learning to robot system. K. Song et al. designed a deep neural network (DNN) approach to object recognition and combined with point cloud segmentation to enhance 3D object-pose estimation for grasping [3]. J. Zhang et al. proposed a three-dimensional positioning method based on the improved deep learning algorithm and using a monocular structured light camera to search for the target [4]. S. Wang et al. presented a vision-based robotic system to handle the picking problem involved in automatic express package dispatching. For the purpose of package recognition, the deep network framework YOLO is integrated [5].

Among the existing target detection and tracking algorithms, the detection accuracy of the deep learning based visual target detection methods has been greatly improved. However, these methods usually require expensive computing resources to achieve real-time effect due to the algorithm complexity, which hits a bottleneck for the commercialization of deep learning [6]. Especially for the target recognition and tracking in the moving scene, the recognition lag or even target loss may often occur, which may not provide stable tracking target for the manipulator. Hence, the robustness of the moving target tracking algorithm is facing challenges.

For the contradiction between real-time and accuracy in target tracking and grabbing, this paper proposes a moving target tracking and grabbing method combining deep learning target detection and Kalman filtering.

The main contributions of this work can be summarized as follows:

1) The YOLOv3 target detection algorithm is utilized to realize the moving target detection, which will make the robot system more intelligent.
2) The moving target position is filtered and estimated by Kalman filtering and so achieve more stable target tracking and motion prediction.

The remainder of this paper is organized as follows. Section 2 presents the framework of moving target tracking and grabbing strategy. Section 3 completes the target recognition based on the YOLOv3 algorithm. Section 4 implements target tracking prediction based on Kalman filter algorithm. For verification, Sect. 5 conducts a series of experiments and corresponding subjective/objective analysis.

2   Framework of Moving Target Tracking and Grabbing Strategy

Moving target tracking and grabbing has the following characteristics. Firstly, the target images collected at different angles have different postures, which may not be stably located the object. Secondly, the image sampling rate as well as the image process speed are much lower than the control period of the robot system, therefore there is a delay in tracking process.

In view of the above characteristics, this paper proposes a moving target tracking and grabbing strategy combining deep learning target detection and Kalman filter. The framework of the tracking and grabbing strategy is shown in Fig. 1. First, the YOLOv3 algorithm is used to locate the target. Then the hand-eye calibration parameters are obtained by the Eye-to-Hand calibration steps. After that, the Kalman filter is utilized to predict the possible position of the target in the image, which reduces the impact of the image delay on the real-time performance of the system. Finally, the robot start tracking according to the planed trajectory.

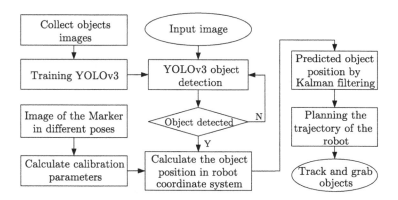

Fig. 1. Framework of tracking and grabbing strategy

3   Target Recognition Based on YOLOv3 Algorithm

The target detection networks based on deep learning mainly include SSD series, RCNN series and YOLO series [7–9]. Compared with the first two networks, the YOLOv3 algorithm takes the target detection speed and accuracy into account. Due to the high real-time requirements of moving target recognition, this paper applied YOLOv3 to realize the target recognition.

As an End-to-End target detection algorithm, YOLOv3 is mainly composed by the Darknet-53 backbone network and multi-scale fusion feature network, as shown in Fig. 2. The backbone network Darknet-53 is mainly used to extract image features. Con2d layer represents a convolutional layer, and each convolutional layer is connected to a batch normalization (BN) layer and a Leaky ReLU activation function.

The multi-scale fusion feature network makes full use of deep and shallow information by fusing feature maps of three scales (i.e., small, medium, and large scale), so that the model can achieve the target location more accurately. The network first detects the 32-fold down-sampled feature map. Because the deeper feature map has a larger receptive field, it is suitable for detecting larger objects. Secondly, the 16-fold down-sampled feature map after fusion is detected, since the feature map is in the middle of the network so that the receptive field is suitable for detecting medium-sized objects. Finally, the fused 8-fold down-sampled feature map is used to detect small objects.

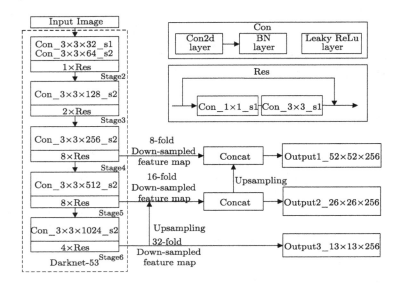

Fig. 2. Structure of YOLOv3 algorithm

YOLOv3 algorithm obtains a priori anchor box for target detection through clustering the bounding boxes in the training set. Therefore, the difficulty of the position optimization is reduced. The distance formula of K-means clustering are as follows:

$$d(b, b') = 1 - IOU(b, b') \tag{1}$$

$$IOU(b, b') = \frac{\text{area}(b \cap b')}{\text{area}(b \cup b')} \tag{2}$$

where b and b' represent the bounding box and clustering center, respectively. d denotes the distance between b and b', and IOU denotes the intersection over union of the predicted boundary box and the manual labeling boundary box. In YOLOV3 algorithm, k-means clustering algorithm generates a total of 9 prior anchor boxes, which are divided into three scales. Three larger prior anchor boxes are applied on the 13×13 feature map, three medium prior anchor boxes are applied on the 26×26 feature map, and three smaller prior anchor boxes are applied on the 52×52 feature graph.

The loss function of YOLOv3 algorithm consists of three parts: position error, confidence loss and classification loss. The expression is as follows:

$$\text{Loss} = \lambda_{coord} \sum_{i=0}^{N^2} \sum_{j=0}^{K} I_{ij}^{obj} \left[l(x_i, x_i^{'}) + l(y_i, y_i^{'}) \right]$$

$$+ \lambda_{coord} \sum_{i=0}^{N^2} \sum_{j=0}^{K} I_{ij}^{obj} \left[(\sqrt{w_i} - \sqrt{w_i^{'}})^2 + (\sqrt{h_i} - \sqrt{h_i^{'}})^2 \right]$$

$$+ \sum_{i=0}^{N^2} \sum_{j=0}^{K} I_{ij}^{obj} l(C_i, C_i^{'}) + \lambda_{noobj} \sum_{i=0}^{N^2} \sum_{j=0}^{K} I_{ij}^{noobj} l(C_i, C_i^{'}) \tag{3}$$

$$+ \sum_{i=0}^{N^2} I_i^{obj} \sum_{c \in classes} l(p_i(c), p_i^{'}(c))$$

$$l(a, a^{'}) = -a_i \log a_i^{'} + (1 - a_i) log(1 - a_i^{'}) \tag{4}$$

where parameter λ_{coord} is used to increase the weight of the bounding box position loss, and parameter λ_{noobj} is used to suppress the confidence of the bounding box of undetected objects. N^2 is the total number of the grids in the input image, and K is the number of bounding boxes predicted in each grid. I_{ij}^{obj} means that the jth bounding box in the ith grid detects the target. Meanwhile, I_{ij}^{noobj} means that the jth bounding box in the ith grid does not detect the target. $(x^{'}, y^{'}, w^{'}, h^{'})$ is the predicted coordinates of the bounding box, and (x, y, w, h) is the actual value. $p_i^{'}(c)$ is the predicted probability of the grid i belonging to the category c target. $p_i(c)$ is the actual value. $C_i^{'}$ is the prediction confidence while C_i is the actual value. The confidence calculation formula is given as follows:

$$C = \text{Pr(object)} * \text{IOU}(b, gt) \tag{5}$$

where b is the prediction bounding box, and gt is the manually marked box. When the center point of the real box falls in the prediction box, it is considered that there is a target in the prediction box, then $\text{Pr} = 1$, otherwise, $\text{Pr} = 0$.

4   Target Tracking Algorithm

The Kalman filter is widely used in visual tracking systems, which will estimate the future state of moving objects more accurately, and then guide the robot to complete the dynamic grasping task [10–12]. Due to the fact that the target detection in the video stream usually has the problems of bounding box beating, missing detection and wrong detection, etc., this paper applied Kalman filtering related theory to optimize the target tracking of the conveyor belt. Conveyor belt moving target generally performs linear motion at a constant speed. Therefore, the motion state of the target can be represented by speed and position. During the tracking process, the time interval between two adjacent frames is short, and the motion state of the target changes relatively little, so it is assumed that the target moves at a uniform speed within a unit time interval, and the speed parameter is sufficient to reflect the target's movement trend. Define the system state by a four-dimensional variable $x_k = (xs_k, ys_k, xv_k, yv_k)$, which represents the position and velocity of the target in the x and y directions. The system states should have the following relations:

$$\begin{cases} xs_k = xs_{k-1} + xv_k dt \\ ys_k = ys_{k-1} + yv_k dt \end{cases} \tag{6}$$

where the system model is established as follows, dt is the time interval between t_{k-1} and t_k:

$$\begin{pmatrix} xs_k \\ ys_k \\ xv_k \\ yv_k \end{pmatrix} = \begin{bmatrix} 1 & 0 & dt & 0 \\ 0 & 1 & 0 & dt \\ 0 & 0 & 1 & 0 \\ 0 & 0 & 0 & 1 \end{bmatrix} \begin{pmatrix} xs_{k-1} \\ ys_{k-1} \\ xv_k \\ yv_k \end{pmatrix} + w_k \tag{7}$$

$$x_k = Fx_{k-1} + w_k \tag{8}$$

In formula (8) , x_k and x_{k-1} are the optimal state estimates at time k and time $k-1$, F is the state transition matrix of the system, w_k is the process noise of the kth frame.

Because only the position of the target can be observed in the image, the observation model z_k is:

$$z_k = \begin{pmatrix} xs_k \\ ys_k \end{pmatrix} = \begin{pmatrix} 1 & 0 & 0 & 0 \\ 0 & 1 & 0 & 0 \end{pmatrix} \begin{pmatrix} xs_k \\ ys_k \\ xv_k \\ yv_k \end{pmatrix} + v_k \tag{9}$$

$$z_k = Hx_k + v_k \tag{10}$$

where x_k is the system optimal state estimates at time k, H is the observation matrix of the system, and v_k is the observation noise at time k.

Assuming that both process noise and detection noise obey Gaussian distribution, their covariance matrices are W and V, and Kalman filtering consists of two parts: prediction and update.

1) Prediction section: According to the detection bounding box of the system, when the target is detected in consecutive $f$min(threshold value of detector) frames and above, it indicates that the target is a correctly detected one. As is shown in (11), the target state is predicted through the previous state.

$$\hat{x}_k^- = F\hat{x}_{k-1} \tag{11}$$

where $\hat{x}_k^-$ and $\hat{x}_{k-1}$ is the target predicted state obtained at time k and $k-1$, the corresponding covariance matrix can be mathematically expressed by:

$$P_k^- = FP_{k-1}F^T + W \tag{12}$$

where P_k^- and P_{k-1} is the covariance matrix of $\hat{x}_k^-$ and $\hat{x}_{k-1}$.

2) Update section: When the matching relationship between the detection bounding box and the tracking bounding box is established in the system and the continuously lost detection target does not exceed the $f$max(threshold value of detector) frame, it means that the target is not really lost, and the target position and its covariance matrix need to be updated every frame.

$$K_k = P_k^- H^T (H P_k^- H^T + V)^{-1} \tag{13}$$

$$\hat{x}_k = \hat{x}_k^- + K_k(z_k - H\hat{x}_k^-) \tag{14}$$

$$P_k = (I - K_k H)P_k^- \tag{15}$$

where K_k is the Kalman gain at time k, $\hat{x}_k$ is the optimal estimated state at time k, P_k is the covariance matrix of x_k, and I is the identity matrix.

The general process of iterating according to the above prediction part and update part are as follows (Fig. 3):

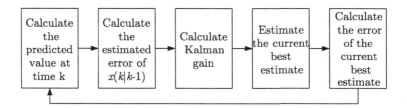

Fig. 3. Kalman filtering process

5   Experiments and Results

5.1   Train the YOLOv3 Network

The hardware and software environment of the experiment is as follows: A server with Ubuntu 16.04, GTX1080Ti GPU with 11 GB memory, deep learning framework Tensorflow 1.6.0, Python 3.6, OpenCV 3.4.1 and Darknet.

Before the network training, the target image must be collected as the training data set. Through the camera installed above the conveyor belt, video streams of target objects in different positions are collected. The video resolution is 640×480 and the frame rate is 30 fps. Finally, a total of 4200 images were extracted for model training, and 500 images were used for model testing. The images are manually annotated with Labellmg and saved in PASCAL VOC format. The marked rectangular frame is used for the classification and identification of target objects.

5.2   Analysis of Recognition Results

Figure 4 is the trend curve of Loss change when the YOLOv3 model is trained. From the figure, when the iteration exceeds 200 times, the loss value tends to stabilize, and finally drops to about 0.38.

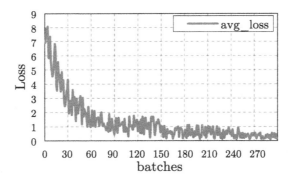

Fig. 4. Loss function curve

Figure 5(a) is the original image obtained by the camera. Figure 5(b) is the feature map extracted from the original image by the YOLOv3 network (Stage 2 to Stage 4). Based on the extracted features, the detect object bounding box is obtained, and the result is shown in Fig. 5(c).

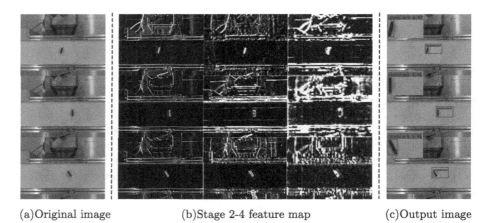

(a)Original image   (b)Stage 2-4 feature map   (c)Output image

Fig. 5. YOLOv3 detect results

5.3   Kalman Filter Target Trajectory Tracking Simulation

According Sect. 4, a Kalman filter model is established in Matlab to track the identified target position curve. As shown in Fig. 6(a), the experimental result of frame-by-frame estimation of the target position is given. The blue dots indicate the target coordinates measured by the YOLOv3. It can be seen that the discrete coordinate information is fluctuated, which may interfere with subsequent tracking and grabbing. After the Kalman filtering, the optimal estimation labeled by orange triangles is obtained, the results show that the estimated value and the measured value are basically consistent, and the estimated value is smoother than the measured value.

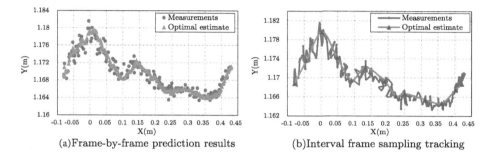

(a)Frame-by-frame prediction results   (b)Interval frame sampling tracking

Fig. 6. Kalman filter simulation results (Color figure online)

However, it takes time for the robot arm to move to the target point, so the frame-by-frame prediction does not meet the actual situation. In order to verify whether the object movement trend predicted by Kalman filtering is effective, the update rate of the Kalman filter is set to 500 ms, and the tracking result is shown in Fig. 6(b). The result shows that the estimation curve obtained by Kalman filtering tracks to the target stably, so that it can be used to predict the possible position of the target within a short time.

5.4   Experiment of Robotic Arm Tracking and Grabbing

Platforms for performing Tracking and Grabbing experiments include hardware and software. Hardware: a computer with Ubuntu 16.04, camera, and a UR3 robotic arm. Software: ROS (robot operating system), MoveIt! (motion planning library), VISP (visual servo software) [13–15]. The experimental scenario is shown in Fig. 7.

The process of tracking and grabbing according to the predicted position of the target is mainly divided into the following two parts:

1) First set the grab point according to the predicted position obtained by Kalman filtering, and next a movement trajectory that reaches the grab point is planned by MoveIt!. Then control the UR3 move to the grab point by UR3 controller, when the gripper arrived at the grab point, the movement between the gripper and target is synchronized.

Fig. 7. Grabbing experimental scenes

2) The gripper does not directly grab the target at grab point, instead of performing a synchronous displacement movement at the same speed as the target. When the gripper and the target has reached relative rest, then controls the gripper completes the grabbing action.

Finally, the actual robot control program combined with the Algorithm YOLOv3 and Kalman Filtering is developed in ROS, the actual robot tracking and grabbing experiment results are shown in Fig. 8.

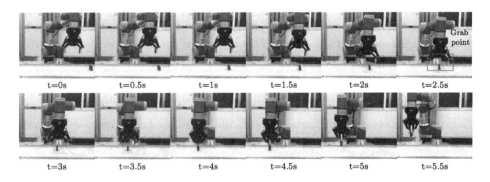

Fig. 8. Tracking and grabbing experiment

In order to illustrate the performance advantages of the proposed strategy further, this paper conducted a comparative simulation. The specific results are shown in Table 1:

Table 1. Performance comparative results

Methods	$A_q/\%$	$A_e/\%$	$A_l/\%$
Traditional method [16]	82.4	5.37	10.2
Deep learning method [17]	85.5	7.3	6.5
Proposed method	88.3	4.7	3.2

Note: A_q is the average quality of grabbing, %; A_e is the average error of tracking, %; A_l is the average target loss rate of target, %.

6   Conclusion

This paper investigates the deep learning based strategy for Eye-to-Hand robotic tracking and grabbing. Specifically, the YOLOv3 algorithm is used to identify moving targets on the conveyor belt, and the Kalman filter algorithm is used to track and predict the target position. Then through the calibration parameters of the Eye-to-Hand system and the target position predicted by the Kalman filter, the tracking and grabbing experiment based on UR3 is carried out. The experimental results show that the proposed strategy can provide more reliable data for the tracking and grabbing of moving target in the industry and help the robot industry achieve intelligence.

Acknowledgements. This work is supported by the National Natural Science Foundation of China (Grant number 61671194), Fundamental Research Funds for the Provincial Universities of Zhejiang (Grant number GK199900299012-010) and the Zhejiang Provincial Key Lab of Equipment Electronics (grant number 2019E10009).

References

1. Strandhagen, J.W., Alfnes, E., Strandhagen, J.O., Vallandingham, L.R.: The fit of Industry 4.0 applications in manufacturing logistics a multiple case study. Adv. Manuf. **5**, 344–358 (2017). https://doi.org/10.1007/s40436-017-0200-y
2. Gao, M., Yan, Y., Yang, Y., Huang, J., He, Z.: A positioning system based on monocular vision for industrial robots. In: Proceedings of the 3rd International Conference on Information Science and Control Engineering (ICISCE 2016), Beijing, pp. 784–788 (2016). https://doi.org/10.1109/ICISCE.2016.172
3. Song, K., Chang, Y., Chen, J.: 3D vision for object grasp and obstacle avoidance of a collaborative robot. In: Proceedings of the IEEE/ASME International Conference on Advanced Intelligent Mechatronics (AIM 2019), Hong Kong, China, pp. 254–258 (2019). https://doi.org/10.1109/AIM.2019.8868694
4. Zhang, J., Zhou, Z., Xing, L., Sheng, X., Wang, M.: Target recognition and Location based on deep learning. In: Proceedings of the IEEE 4th Information Technology, Networking, Electronic and Automation Control Conference (ITNEC 2020), Chongqing, China, pp. 247–250 (2020). https://doi.org/10.1109/ITNEC48623.2020.9084826

5. Wang, S., Jiang, X., Zhao, J., Wang, X., Zhou, W., Liu, Y.: Vision based picking system for automatic express package dispatching. In: Proceedings of the IEEE International Conference on Realtime Computing and Robotics (RCAR 2019), Irkutsk, Russia, pp. 797–802 (2019). https://doi.org/10.1109/RCAR47638.2019.9044094

6. LeCun, Y., Bengio, Y., Hinton, G.: Deep learning. Nature **521**, 436–444 (2015). https://doi.org/10.1038/nature14539

7. Liu, W., et al.: SSD: single shot multibox detector. In: Leibe, B., Matas, J., Sebe, N., Welling, M. (eds.) ECCV 2016. LNCS, vol. 9905, pp. 21–37. Springer, Cham (2016). https://doi.org/10.1007/978-3-319-46448-0_2

8. Girshick, R.: Fast R-CNN. In: Proceedings of the IEEE International Conference on Computer Vision, pp. 1440–1448. IEEE (2015)

9. Redmon, J., Farhadi, A.: YOLOv3: an incremental improvement [EB/OL] (2018). https://arxiv.org/abs/1804.02767

10. Smeulders, A.W.M., Chu, D.M., Cucchiara, R., Calderara, S., Dehghan, A., Shah, M.: Visual tracking: an experimental survey. IEEE Trans. Pattern Anal. Mach. Intell. **36**(7), 1442–1468 (2014). https://doi.org/10.1109/TPAMI.2013.230

11. Pallavi, S., Ramya Laxmi, K., Ramya, N., Raja, R.: Study and analysis of modified mean shift method and Kalman filter for moving object detection and tracking. In: Raju, K.S., Govardhan, A., Rani, B.P., Sridevi, R., Murty, M.R. (eds.) Proceedings of the Third International Conference on Computational Intelligence and Informatics. AISC, vol. 1090, pp. 821–828. Springer, Singapore (2020). https://doi.org/10.1007/978-981-15-1480-7_76

12. He, F., Zhen, J., Wang, Z.: Video target tracking based on adaptive Kalman filter. In: Liang, Q., Wang, W., Liu, X., Na, Z., Jia, M., Zhang, B. (eds.) Video Target Tracking Based on Adaptive Kalman Filter. LNEE, vol. 571, pp. 1198–1201. Springer, Singapore (2020). https://doi.org/10.1007/978-981-13-9409-6_141

13. Quigley, M., et al.: ROS: an open-source robot operating system. In: Proceedings of the ICRA Open-Source Software Workshop (2009)

14. Chitta, S.: MoveIt!: an introduction. In: Koubaa, A. (ed.) Robot Operating System (ROS). SCI, vol. 625, pp. 3–27. Springer, Cham (2016). https://doi.org/10.1007/978-3-319-26054-9_1

15. Marchand, E., Spindler, F., Chaumette, F.: ViSP for visual servoing: a generic software platform with a wide class of robot control skills. IEEE Robot. Autom. Mag. **12**(4), 40–52 (2005). https://doi.org/10.1109/MRA.2005.1577023

16. Qu, J., Zhang, F., Tang, Y., Fu, Y.: Dynamic visual tracking for robot manipulator using adaptive fading Kalman filter. IEEE Access **8**, 35113–35126 (2020). https://doi.org/10.1109/ACCESS.2020.2973299

17. Wan, N., Yeung, D.: Learning a deep compact image representation for visual tracking. In: Proceedings of the 26th International Conference on Neural Information Processing Systems, pp. 809–817 (2013)

Dynamical State Forcing on Central Pattern Generators for Efficient Robot Locomotion Control

Thirawat Chuthong[1], Binggwong Leung[1], Kawee Tiraborisute[1],
Potiwat Ngamkajornwiwat[3], Poramate Manoonpong[1,2,3],
and Nat Dilokthanakul[1(✉)]

[1] Bio-inspired Robotics and Neural Engineering Lab, School of Information Science
and Technology, Vidyasirimedhi Institute of Science and Technology,
Rayong 21210, Thailand
natd_pro@vistec.ac.th
[2] Embodied AI and Neurorobotics Lab, The Mærsk Mc-Kinney Møller Institute,
University of Southern Denmark, Odense, Denmark
[3] Institute of Bio-inspired Structure and Surface Engineering, College of Mechanical
and Electrical Engineering, Nanjing University of Aeronautics and Astronautics,
Nanjing, China

Abstract. Many CPG-based locomotion models have a problem known as the tracking error problem, where the mismatch between the CPG driving signal and the state of the robot can cause undesirable behaviours for legged robots. Towards alleviating this problem, we introduce a mechanism that modulates the CPG signal using the robot's interoceptive information. The key concept is to generate a driving signal that is easier for the robot to follow, yet can drive the locomotion of the robot. This can be done by nudging the CPG signal in the direction of lower tracking error, which can be analytically calculated. Unlike other reactive CPG, the proposed method does not rely on any parametric learning ability to adjust the shape of the signal, making it a unique option for a biological adaptive motor control. Our experiment results show that the proposed method successfully reduces the tracking error. We also show that the CPG signal, regulated by the proposed method, is robust to perturbation and can smoothly return back to the default pattern.

Keywords: Central Pattern Generators · Locomotion control · PID controller

1 Introduction

Central Pattern Generators (CPG) has been extensively used for robot locomotion control thanks to their ability to generate coordinated high-dimensional rhythmic patterns [1,7,9,11,21]. The pattern generated from a CPG is generally

---

T. Chuthong and B. Leung—Equal contribution.

© Springer Nature Switzerland AG 2020
H. Yang et al. (Eds.): ICONIP 2020, LNCS 12533, pp. 799–810, 2020.
https://doi.org/10.1007/978-3-030-63833-7_67

used as a high-level command signal for a low-level controller (e.g. a reference signal for a PID controller). However, it is the low-level controller that drives the robot's actuators into the correct configuration (See Fig. 1).

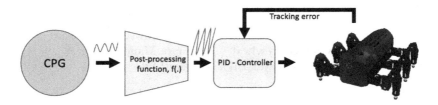

Fig. 1. The standard setup of the CPG controller consists of four main components. Firstly, the CPG neural oscillator generates a rhythmic signal. Then, the signal is transformed with the post-processing function $F(\cdot)$ to create the desired locomotion pattern. The transformed signal is then used as the setpoint for the PID controller. Finally, the PID controller drives the robot movement into the desired position or the setpoint. The robot actual movement is fed back into the PID to correct the tracking error.

One of the weaknesses of this setup is that it assumes a perfect low-level controller. However, in the real use cases, there are tracking errors between the actual system motion and the CPG driving patterns. A lack of power or an increased joint load can result in the PID being unable to adjust the robot to the target configuration. This lead to undesired behaviours such as energy-inefficient locomotion, unwanted movement and motor collapse [21]. Current solutions to this problem are to improve the responses of the low-level controller [2,22] or allowing the robot to perceive the environments and adjust the CPG driving signals accordingly [3,9,21].

In this work, we propose a novel mechanism for adjusting the high-level CPG signals with interoceptive perception. The main idea is to directly adjust the dynamics of the CPG to remove the tracking error by resetting the state of the CPG dynamics to a new position. Regardless of the dynamical state, the CPG dynamics falls back to its natural patterns due to the limit cycle behaviour. These fall-back dynamics creates a command signal that closely guides the PID controller to the desired locomotion pattern. We show in our experiment that the method can successfully provide the desired locomotion pattern while reduces the tracking error that would be large otherwise. This results in efficient and adaptive locomotion pattern.

1.1 Related Works and Contributions

According to Buchli et al. [6], there are two main characteristics of CPG: (i) Reactive CPG and (ii) Adaptive CPG. A reactive CPG [8,18] changes the dynamics of the signal only temporally while an adaptive CPG [5,13] creates lasting changes to the dynamics. This work can be categorised as a reactive CPG.

Reactive CPGs are designed such that some properties of the oscillators are temporally adjusted. Many works consider adjusting the phase of the oscillators [8,18]. Others consider adjusting the amplitude of the oscillations [15]. The most related works are the ones that consider adjusting the shape of the signal [12, 14,17,19].

The main contribution of this work is in the formulation of the dynamical state forcing CPG (DSF-CPG). The proposed DSF-CPG acts as a reactive CPG that can temporally adapts the *geometry/shape* of the CPG signal without using any learning parameters [14,17]. Instead it exploits the entrainment-like dynamics for the CPG shape adaptation. This happens via the direct perturbation of the neural activities rather than some synaptic weights. Other CPG systems that can adapt the shape of their signals typically rely on additional CPG shaping networks with learning [14,17] or multiple CPG circuits [12,19]. Our system is simple and reactive. It does not rely on any learning ability neither require high-level computational effort, which makes it a unique possible option for a biological locomotion adaptive motor control.

2   Central Pattern Generator

Central Pattern Generator (CPG) is a mathematical model of coupled neural oscillators, which is govern by coupled differential equations (continuous time) or difference equations (discrete time). There are several properties that make CPG useful for locomotion (as describes by Ijspeert (2008) [10]):

1. It exhibits limit cycle behaviour.
2. It is suitable for distributed implementation, which can be used for modular robots.
3. It allows a small number of control parameters, which can reduce a high-dimensional locomotion control problem to a low-dimensional control problem.
4. It is suited for sensory feedback integration, allowing coupling between the CPG and the robot mechanical system.

While there are many models of CPG, in this work, we use the SO(2)-network [16]. SO(2)-network can be described as a coupling of two neurons (as illustrated in Fig. 2). The activities of the neurons a_1 and a_2 follow the following discrete-time dynamics:

$$a_1(t+1) = w_{11} \tanh(a_1(t)) + w_{12} \tanh(a_2(t)), \tag{1}$$
$$a_2(t+1) = w_{21} \tanh(a_1(t)) + w_{22} \tanh(a_2(t)). \tag{2}$$

The weight matrix for an SO(2)-network is in the following form:

$$w = \alpha \cdot \begin{pmatrix} \cos(\phi) & \sin(\phi) \\ -\sin(\phi) & \cos(\phi), \end{pmatrix} \tag{3}$$

where α and ϕ are the parameters of the system dictating the oscillating frequency, amplitude and shape of the oscillation.

We utilise the signal generated from the SO(2)-network by transforming the output signal $x = \tanh(a)$ into desired motor pattern using a post-processing function. This can be done using any function $F(\cdot)$, which can be used to shape the signal pattern in the form that is suited for locomotion. In this work, we only consider an invertible function $F(\cdot)$, which is an only restriction we need for the dynamical state forcing mechanism.

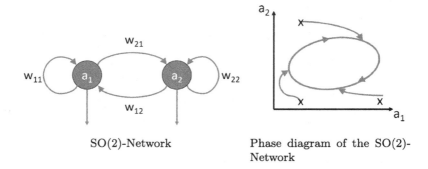

| SO(2)-Network | Phase diagram of the SO(2)-Network |

Fig. 2. SO(2)-Network is a system of coupled neurons (a). Under a certain condition of parameters W, the dynamics of the neural activities follow the limit cycle behaviour. This behaviour produces rhythmic pattern suited for driving locomotion control. The limit cycle dynamics exhibits the fallback pattern, i.e. the state fallback to the natural dynamics when it is reset to any point in the state-space (b).

3   Dynamical State Forcing on CPG

The tracking error happens when the robot actuator has not yet reached the target signal. This could be due to the lack of power, or an obstacle is pushing back the robot.

In the open-loop CPG control, the pattern keeps going, while the actual system struggles to keep up. This lead to inefficient use of energy because the proportional controller increases the gain as the tracking error increases.

Dynamical state forcing mechanism reset the current position of the CPG, i.e. the neural states of a_1 and a_2. The main idea is to reset the neural state such that it removes the tracking error. To this end, we need to find the neural state a_1 and a_2 that matches the current robot position. Let's denote r_1 and r_2 as the target robot position:

$$r_1 = F(\tanh(a_1)) \tag{4}$$
$$r_2 = F(\tanh(a_2)) \tag{5}$$

Let's denote the current robot actual position as r'_1 and r'_2. The neural activity that matches the current position can be written as:

$$a'_1 = \tanh^{-1}(F^{-1}(r'_1)) \tag{6}$$
$$a'_2 = \tanh^{-1}(F^{-1}(r'_2)) \tag{7}$$

The activities a'_1 and a'_2 map to the current robot configuration. Therefore, if the dynamical states is reset to these values, the tracking error will be zero.

However, the complete removal of tracking error could result in the actuator being in the halt state. To avoid this problem, we choose to only slightly perturb the activities of the neurons in the direction of error reduction (instead of the complete removal of error). Therefore, instead of resetting the activities, we could update the activities with the following update rule:

$$a_1(t+1) = (1-\gamma)[w_{11}\tanh(a_1(t)) + w_{12}\tanh(a_2(t))] + \gamma a'_1(t) \tag{8}$$
$$a_2(t+1) = (1-\gamma)[w_{21}\tanh(a_1(t)) + w_{22}\tanh(a_2(t))] + \gamma a'_2(t) \tag{9}$$

where $\gamma \in [0,1]$. The mixing rate γ dictates how much the update mechanism removes the tracking error. It completely removes the tracking error when $\gamma = 1$ and becomes the vanilla CPG at $\gamma = 0$.

However, this update rule leads to the continuous decreases of oscillator amplitude. The dynamics slowly converges to zero. This behaviour occurs because the mechanism would changes the system dynamics in the direction that the robot can easily reach, and that is the halt state (a fixed point). In order to overcome this problem, we introduce an outward force term that prevents the dynamics from falling to zeros.

$$a_1(t+1) = (1-\gamma)[w_{11}\tanh(a_1(t)) + w_{12}\tanh(a_2(t))] \tag{10}$$
$$+ \gamma a'_1(t) + \beta a_1(t)$$
$$a_2(t+1) = (1-\gamma)[w_{21}\tanh(a_1(t)) + w_{22}\tanh(a_2(t))] \tag{11}$$
$$+ \gamma a'_2(t) + \beta a_2(t)$$

The additional terms $\beta a_1(t)$ and $\beta a_2(t)$ increase the value of a_1 and a_2 in each update step, where β is another hyper-parameter adjust the strength of the outward force.

Figure 3 illustrates this process as a feedback from robot back to the CPG pattern generator. We show later in the experiment that this mechanism can be used to adapt the dynamics of CPG to the environment.

4   Experiment

4.1   Investigation of Dynamical State Forcing CPG on a Single Motor

We investigate the behaviour of our method on a single motor. The setup includes a 3D-printed platform (as shown in Fig. 4a), which is mounted with the motor and a rot. We implement a vanilla CPG setup and the dynamical state forcing CPG (DSF-CPG) to drive the rot in an oscillating motion.

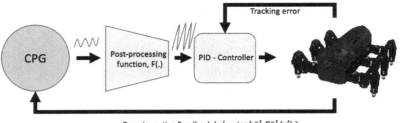

Fig. 3. The dynamical state forcing helps PID controller by propagating the error signal back to the DSF-CPG pattern generator. This mechanism creates a feedback loop, which enables the coupling between the DSF-CPG and the environment.

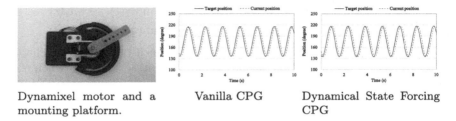

Dynamixel motor and a mounting platform.   Vanilla CPG   Dynamical State Forcing CPG

Fig. 4. Experimental Setup. A motor is mounted on a 3D-printed platform and attached to rotating rot (a). A bolt can be attached to the platform to force stop the motion of the rot. The oscillating motion can be controlled by either the SO(2)-network setup (b) or the dynamical state forcing CPG (c). For ease of comparison, we adjust the parameters such that they exhibit similar amplitude and frequency. For SO(2)-network, we set $\phi = 0.28$ and $\alpha = 1.011$. For DSF-CPG, we set $\phi = 0.45$, $\alpha = 1.00$, $\gamma = 0.3$ and $\beta = 0.2$. The signal is transformed with $F(x) = 2200x + 2000$. The parameters are selected such that the signal from CPG behaves similarly to the signal from DFS-CPG, while the oscillating pattern is slow enough for an easy intervention.

The oscillating pattern and the actual dynamics of the motor is visualised in Fig. 4b, c. We can see that both vanilla CPG and the DSF-CPG can produce similar oscillating motions. Next, we force-stop the moving rod by attaching a bolt on the platform. We can see in Fig. 5 that the actual motion of the system as read by the motor became stationary. However, for the vanilla CPG, the driving signal continues to oscillate. In contrary, the driving signal of the DSF-CPG slowly decreases as the actual system stop. After the bolt is removed, the dynamics of the driving CPG signals go back to their normal pattern. We can see that this behaviour reduces the tracking error when the actual system is forced to be in the state that is difficult for the motor.

Any motor that is controlled using current proportional to the error would, theoretically, benefit from the behaviour of the DSF-CPG because the lower error equates to lower current in these controllers. As the tracking error is reduced, the current usage would also be smaller, which potentially lead to a more energy-efficient locomotion control.

Next, we examine the behaviour of the system when the rot is forced to move to a new position. This represents a scenario when an external force is hitting a robot. The result is shown in Fig. 5. We can see that the DSF-CPG adjust itself to the perturbation. We can also see that the tracking error of DSF-CPG is significantly smaller than the error of the vanilla CPG.

4.2   Investigation of Dynamical State Forcing on a Simulated Robot

In this experiment, we demonstrate that the DSF-CPG can be used to drive robot locomotion and investigate its adaptive behaviour. We simulate the MORF hexapod robot [21] using the CoppeliaSim [20]. The locomotion of the robot is driven with a single CPG, where the output is transformed with hard-wired functions as shown in Fig. 6b.

There are 18 motors in this robot corresponding to the 18 joints (see Fig. 6b). The CPG signal is transformed with linear post-processing functions as detailed in Fig. 6b:

$$F(x) = w_i x + b_i \tag{12}$$

where $\{w_1, w_2, w_3\} = \{10, 4, 10\}$ for vanilla CPG, $\{w_1, w_2, w_3\} = \{2, 0.8, 2\}$ for DSF-CPG, and $\{b_1, b_2, b_3\} = \{1.92, -0.03, 0\}$ for both of them. Since the DSF-CPG has a higher oscillating amplitude than vanilla CPG, we scale up the signal by using larger weights in the post-processing function. The parameters of CPG are selected such that the robot can walk reliably and the parameters of DSF-CPG are chosen such that the signal generated is resemble the signal from the vanilla CPG.

As shown in Fig. 7[1], DSF-CPG is able to drive the robot with similar pattern to the normal CPG. After the heavy box is loaded onto the robot, the dynamics of DSF-CPG stops because the load is too heavy for the CF0 joint to lift the robot up. The dynamics springs back to the normal pattern after the heavy load is removed. We compare the tracking error of the DSF-CPG and the vanilla CPG in Fig. 7c.

5   Discussion

This paper introduces a novel CPG-based control method. The main benefit of this method is the reduction of tracking error by adapting the target CPG pattern to follow an unexpected condition of the system, e.g. loading condition.

---

[1] see also https://youtu.be/uMxDPPg1Q9A.

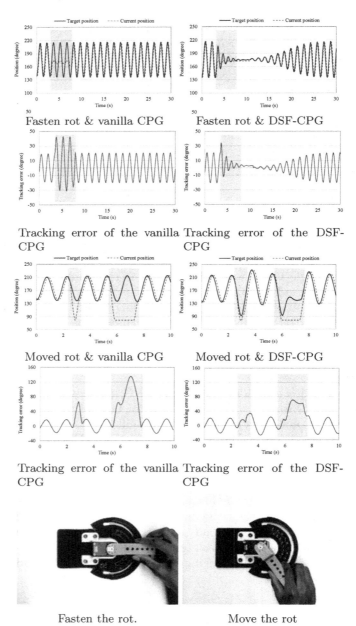

Fig. 5. Dynamics of the vanilla CPG and the DSF-CPG under influences from external forces. When the rot is fasten still, the dynamics of DSF-CPG slowly reduces to the tracking error. The driving signal goes back to the normal pattern when the rot is released. When the rot is moved from the normal position, the DSF-CPG dynamics follows the position of the rot and smoothly bring the system position back to the normal pattern. The shaded areas represent the time when the interventions happen (i.e. fasten or move the rot).

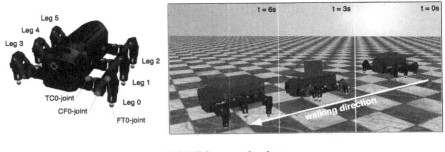

MORF hexapod robot

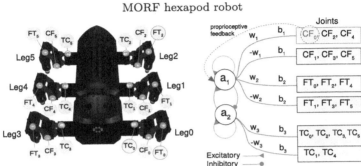

CPG locomotion controller

Fig. 6. Simulated MORF hexapod robot. We simulate a CPG-driven hexapod robot on the CoppeliaSim. The robot walks in a straight line [21]. At the middle of its path, we simulate a heavy box placed on top of the robot, which is lifted up shortly after. The propose of this experiment is to visualise the ability of DSF-CPG to generate a walking pattern and its adaptability. The feedback to the DSF-CPG comes from the configuration of the CF0 joint. CF0 joint is responsible of lifting the robot upwards and, therefore, sensitive to the carrying load of the robot.

We see that the DSF-CPG can adapt to its proprioceptive feedback, while also manage to smoothly fall-back to its original pattern. Importantly, this method can shape the geometry of the signal without any learning parameter, which makes it a unique possible option for biological adaptation mechanism.

In future work, we will explore DSF-CPG in the setup of multiple weak-coupling or decoupling CPGs. Several works propose to use multiple CPGs to drive robots' locomotions [3,4]. This multiple CPG setup removes the strong coupling restriction between each joint. This could benefit the DSF-CPG by allowing feedback to influence selectively on each joint, creating a richer repertoire of adaptive behaviours.

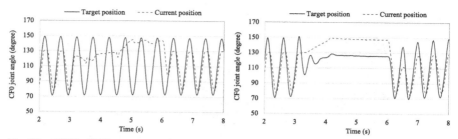

Vanilla CPG: CF0 target and real posi-   DSF-CPG: CF0 target and real positions
tions

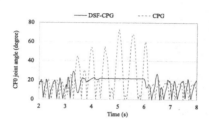

Absolute tracking error of the vanilla
CPG compared with the DSF-CPG

Fig. 7. The dynamics of the hexapod robot. We drive the MORF hexapod robot using
the vanilla CPG and the DSF-CPG. The DSF-CPG manages to drive the robot with
a similar pattern to the vanilla CPG, while also able to adapt the target pattern to
external forces reduces the tracking error. See also Fig. 7b and a video clip at https://
youtu.be/uMxDPPg1Q9A.

DSF-CPG naturally promotes robot's compliance, by changing the target
dynamics to follow external perturbations. Therefore, a careful design of the
feedback effect could potentially be used to design a novel compliant locomotion
pattern. Since an adaptive compliant control [22] has been shown to improve
energy efficiency in a robot, we hope to see whether the adaptive locomotion of
DSF-CPG can also be used to further improve in that direction.

Acknowledgement. We thank Mathias Thor for his technical support on the MORF
robot simulation and acknowledge financial support by the VISTEC research grant on
bioinspired robotics [P.M. (PI)] and in part by the NUAA Research Fund [P.M. (PI),
P.N.].

References

1. Arena, P.: The central pattern generator: a paradigm for artificial locomotion. Soft Comput. **4**(4), 251–266 (2000)
2. Åström, K.J., Hägglund, T., Astrom, K.J.: Advanced PID Control, vol. 461. ISA-The Instrumentation, Systems, and Automation Society, Research Triangle (2006)
3. Barikhan, S.S., Wörgötter, F., Manoonpong, P.: Multiple decoupled CPGs with local sensory feedback for adaptive locomotion behaviors of bio-inspired walking robots. In: del Pobil, A.P., Chinellato, E., Martinez-Martin, E., Hallam, J., Cervera, E., Morales, A. (eds.) SAB 2014. LNCS (LNAI), vol. 8575, pp. 65–75. Springer, Cham (2014). https://doi.org/10.1007/978-3-319-08864-8_7
4. Bem, T., Cabelguen, J.M., Ekeberg, Ö., Grillner, S.: From swimming to walking: a single basic network for two different behaviors. Biol. Cybern. **88**(2), 79–90 (2003)
5. Buchli, J., Ijspeert, A.J.: Distributed central pattern generator model for robotics application based on phase sensitivity analysis. In: Ijspeert, A.J., Murata, M., Wakamiya, N. (eds.) BioADIT 2004. LNCS, vol. 3141, pp. 333–349. Springer, Heidelberg (2004). https://doi.org/10.1007/978-3-540-27835-1_25
6. Buchli, J., Righetti, L., Ijspeert, A.J.: Engineering entrainment and adaptation in limit cycle systems. Biol. Cybern. **95**(6), 645 (2006)
7. Crespi, A., Ijspeert, A.J.: AmphiBot II: an amphibious snake robot that crawls and swims using a central pattern generator. In: Proceedings of the 9th International Conference on Climbing and Walking Robots (CLAWAR 2006), No. CONF, pp. 19–27 (2006)
8. Ermentrout, G.B., Kopell, N.: Inhibition-produced patterning in chains of coupled nonlinear oscillators. SIAM J. Appl. Math. **54**(2), 478–507 (1994)
9. Homchanthanakul, J., Ngamkajornwiwat, P., Teerakittikul, P., Manoonpong, P.: Neural control with an artificial hormone system for energy-efficient compliant terrain locomotion and adaptation of walking robots. In: Proceedings of the IEEE/RSJ International Conference on Intelligent Robots and Systems (IROS 2019), pp. 5475–5482 (2019)
10. Ijspeert, A.J.: Central pattern generators for locomotion control in animals and robots: a review. Neural Netw. **21**(4), 642–653 (2008)
11. Ijspeert, A.J., Crespi, A., Ryczko, D., Cabelguen, J.M.: From swimming to walking with a salamander robot driven by a spinal cord model. Science **315**(5817), 1416–1420 (2007)
12. Lu, Q., Zhang, Z., Yue, C.: The programmable CPG model based on matsuoka oscillator and its application to robot locomotion. Int. J. Model. Simul. Sci. Comput. **11**, 2050018 (2020)
13. Marbach, D., Ijspeert, A.J.: Online optimization of modular robot locomotion. In: Proceedings of the IEEE International Conference Mechatronics and Automation, vol. 1, pp. 248–253. IEEE (2005)
14. Nassour, J., Hénaff, P., Benouezdou, F., Cheng, G.: Multi-layered multi-pattern CPG for adaptive locomotion of humanoid robots. Biol. Cybern. **108**(3), 291–303 (2014)
15. Okada, M., Nakamura, D., Nakamura, Y.: On-line and hierarchical design methods of dynamics based information processing system. In: Proceedings of the IEEE/RSJ International Conference on Intelligent Robots and Systems (IROS 2003) (Cat. No.03CH37453), vol. 1, pp. 954–959. IEEE (2003)
16. Pasemann, F., Hild, M., Zahedi, K.: SO(2)-networks as neural oscillators. In: Mira, J., Álvarez, J.R. (eds.) IWANN 2003. LNCS, vol. 2686, pp. 144–151. Springer, Heidelberg (2003). https://doi.org/10.1007/3-540-44868-3_19

17. Pitchai, M., et al.: CPG driven RBF network control with reinforcement learning for gait optimization of a dung beetle-like robot. In: Tetko, I.V., Kůrková, V., Karpov, P., Theis, F. (eds.) ICANN 2019. LNCS, vol. 11727, pp. 698–710. Springer, Cham (2019). https://doi.org/10.1007/978-3-030-30487-4_53

18. Righetti, L., Buchli, J., Ijspeert, A.J.: Dynamic Hebbian learning in adaptive frequency oscillators. Physica D **216**(2), 269–281 (2006)

19. Righetti, L., Ijspeert, A.J.: Programmable central pattern generators: an application to biped locomotion control. In: Proceedings of the IEEE International Conference on Robotics and Automation, ICRA 2006, pp. 1585–1590. IEEE (2006)

20. Rohmer, E., Singh, S.P.N., Freese, M.: CoppeliaSim (formerly V-REP): a versatile and scalable robot simulation framework. In: Proceedings of the International Conference on Intelligent Robots and Systems (IROS) (2013). www.coppeliarobotics. com

21. Thor, M., Manoonpong, P.: Error-based learning mechanism for fast online adaptation in robot motor control. IEEE Trans. Neural Netw. Learn. Syst. **31**, 2042–2051 (2019)

22. Xiong, X., Wörgötter, F., Manoonpong, P.: Adaptive and energy efficient walking in a hexapod robot under neuromechanical control and sensorimotor learning. IEEE Trans. Cybern. **46**(11), 2521–2534 (2015)

GSDCN: A Customized Two-Stage Neural Network for Benthonic Organism Detection

Zhaoliang Wan[1,2], Lu Zhang[2], Hai Huang[1(✉)], and Xu Yang[2(✉)]

[1] National Key Laboratory of Science and Technology of Underwater Vehicle, Harbin Engineering University, Harbin, China
`wan.zhaoliang@icloud.com, haihus@163.com`
[2] State Key Laboratory of Management and Control for Complex System, Institute of Automation, Chinese Academy of Sciences, Beijing, China
`{zhanglu2016,xu.yang}@ia.ac.cn`

Abstract. High-quality detection of the benthonic organism is a crucial step to implement autonomous picking for the underwater robot. But there have been few studies on the underwater organism detection in recent years. Directly fine-tuning the generic object detector on an underwater dataset is limited by the domain shift, and thus cannot achieve a good performance. Then we propose a customized two-stage detector named by GSDCN and featured Guided Anchoring mechanism, Sampling Balanced strategy, and Deformable Convolutional module, which is dedicated to overcoming three challenges, i.e., geometric variations, limited underwater visibility range, and the imbalance of object samples. Extensive experiments conducted on the URPC2018 dataset (Publicly available in http://www.cnurpc.org/index.html.) show that our GSDCN improves the detection mAP of our baseline algorithm by 3.40%, and surpasses the state-of-the-art underwater object detector ROIMix [12] by a large margin to 5.39%.

Keywords: Robot perception · Underwater object detection · Autonomous robot manipulation

1 Introduction

Benthonic organisms, such as echinus, scallops, and holothurians, have high commercial value. As a multi-billion dollar industry, the development of the off-shore fisheries relies on grasping efficiency. Nevertheless, for the time being, human divers sacrifice health for picking the organisms in the deep water. Automatically capturing the benthonic organism by underwater robots would be an ideal solution in such adverse working circumstances. Undoubtedly, the precision of the detection results is of particular importance to the autonomous grasping process.

There are three main challenges in real-world underwater object detection using deep neural network (Fig. 1):

© Springer Nature Switzerland AG 2020
H. Yang et al. (Eds.): ICONIP 2020, LNCS 12533, pp. 811–820, 2020.
https://doi.org/10.1007/978-3-030-63833-7_68

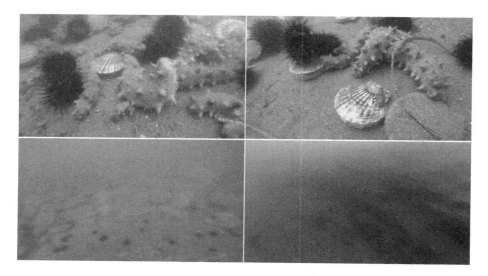

Fig. 1. Real-world underwater images of the Zhangzidao. The top images indicate the flexible holothurians, the tall holothurians, and the wide scallops. The bottom images show the imbalance of foreground and background samples.

(1) The geometric variations of holothurians and scallops. For instance, the holothurians and scallops are nonrigid, whose geometric variations may result in the low precision of these subcategories. **(2) Limited underwater visibility range.** Phytoplankton affects the visibility range of seawater in different seasons. When closing to the targets, we observe some tall or wide objects from the robot's perspective. While few images, including unduly long or wide targets in the URPC dataset, are hard to train the detector substantially. **(3) The imbalance of object samples.** The benthos lives on the sediments randomly. Due to the uneven distribution of benthos inhabiting the seabed, the serious imbalance of foreground and background samples in images collected by the robots restraints the detecting performance.

Considering the above challenges in the underwater marine organism detection, we propose a novel two-stage underwater object detector named GSDCN. Notably, by introducing the modulated deformable convolutional module, we enhance the detection accuracy of ductility holothurians and scallops. Moreover, to relive the influence of visibility range, we guide the feature maps to raise high-quality proposals, which is especially beneficial for detecting the wide or long objects in one image. Besides, we carefully distill the sampling strategy to handle the uneven distribution of benthos, which results in the sampling imbalance severely. We evaluate our detector on the challenging URPC2018 dataset, and our method beats both the on-shore and off-shore object detectors on the URPC2018 test dataset and achieving the state-of-the-art performance without bells and whistles.

2   Related Works

In sharp contrast to the flying-speed improvement of the land generic object detection, the underwater object detection makes less progress. Existing works in the literature fall into two categories: the hand-craft feature with a classifier and the CNN (Convolutional neural network) based methods. Limited by space, we focus on the CNN-based underwater detectors, while the classic underwater detectors are out of scope.

Recently, CNN-based detectors have dominated on-shore object detection. Following the typical milestone of two-stage and one-stage methods, plenty of classic on-shore detection algorithms are employed in the underwater tasks. Thanks to the success of the R-CNN [5] architecture, which combines a proposal generator with a region-wise classifier. To reduce the redundant CNN computations in the R-CNN, the Fast R-CNN [4] introduced the region-wise feature extractor. Li et al. [10] applied the Fast R-CNN to the complex underwater fish detection, acquiring the better accuracy and the shorter time than Fast R-CNN. Later, the Faster R-CNN [20] further speed-up and perform better by presenting the Region Proposal Network(RPN). Mandal [16] combine the Faster R-CNN with three classification networks for fish species detection. In [13], Faster R-CNN was employed for detecting underwater docking. What is particular, ROIMix mixed proposals of different images and achieved the state-of-the-art mAP on the URPC2018 dataset. Alternatively, considering the high efficiency, one-stage detectors have become popular. Minor modifications were based on YOLO [19] for the real-time fish detection [23], which outputs the sparse detecting results only need once forward computation. The work in [27] is based on a well-known detector SSD [14], which use multi-feature maps at different resolutions to detect various objects. The modified SSD was focused on the fish species detection in poor conditions, especially for the North and Baltic Sea.

Most underwater detection works mentioned above are based on the frameworks proposed four years ago or earlier. Otherwise, generic object detection is progressing quickly due to the emergence of key-point based detectors [15,26], center-based detectors [8,24,25], anchor refinement and matching [1,28], high-resolution representation learning [22], gradient harmonizing mechanism [9] and training from scratch [29]. However, the above methods suffer the three challenges mentioned before, their precision comparative low consequently.

3   Methodology

In this section, we first elaborate on the three introduced modules to figure out the three challenges. Then we discuss how the parts alleviate the problems. The details are presented as follows, and the overall architecture is shown in Fig. 2.

3.1   Deformable Convolutional Module

The nonrigid holothurians and scallops change their appearances dynamically according to the external stimuli. When the robot gets close to the benthic crea-

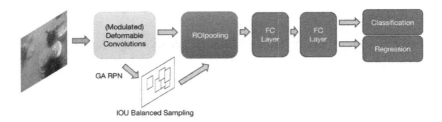

Fig. 2. Overview of the proposed GSDCN, which integrates three components: (a) Modulated DCN (b) Guided anchoring RPN and (c) IOU balanced sampling.

tures, the geometric variations of benthos influence the performance of the detector immensely. To solve the geometric variations' problem, deformable convolutional networks(DCN) [3] was proposed. Our previous work [7] firstly introduced DCN into marine objects detection. The output feature map y is calculated as

$$y(p) = \sum_{k=1}^{K} w_k \cdot x\,(p + p_k + \Delta p_k) \tag{1}$$

where w_k denotes the learnable weight, Pk is the pre-specified offset, $p_k + \Delta p_k$ is the irregular and offset sampling locations. However, as shown in the Eq.(1), the main weakness of [7] is that it lacks the spatial support, which results in the learned features that could be contaminated by irrelevant contents. Nowadays, DCN_v2 [30] performs better on the COCO dataset than DCN. Intuitively, the stronger version of DCN seems to visualize the benthos better.

We firstly introduce parts of DCN_v2 to the underwater detecting task and integrated it into our proposed detector. To enhance the spatial support, stacking more deformable convolution layers is a good solution. Intuitively, we replace the 3×3 Conv layers of the ResNet-50 to the deformable convolutions in the conv3, conv4, and conv5 stages. Besides, the modulated mechanism is necessary for sampling precisely, which is expressed as

$$y(p) = \sum_{k=1}^{K} w_k \cdot x\,(p + p_k + \Delta p_k) \cdot \Delta m_k \tag{2}$$

where Δp_k is the modulation scalar, Δm_k is the learnable offset for the k-th position, $x(p)$ is the input feature map, while $y(p)$ is the output feature map. The module is thus given the capability to adapt to the relative influence of its samples more efficiently.

3.2   Guided Anchoring Mechanism for Marine Organism

Phytoplankton affects the visibility range in different seasons. Furthermore, when seawater is muddy, the robot (camera) has to be extremely close to the target, which brings about the unduly long or wide targets in the images. Nevertheless, few images like these in the training dataset lead to terrible performance when testing in such water quality.

Let us start with the fundamental question: how to detect the tall or wide holothurians more effectively? To answer this question, we conduct experiments to find the truth behind, which introduced the Guided Anchoring Mechanism [25] into the baseline. The Guided Anchoring Mechanism consists of the anchor generation module and the feature adaptation module. The anchor generation module consists of two branches for predicting the location and shape, which applies two 1×1 convolutional layers. Meanwhile, the feature adaptation module regulates the feature in terms of the anchor shape. We firstly employ a 1×1 convolution to predict an offset and then apply a 3×3 deformable convolution to the corresponding level feature map, and thus regulates the feature map. A noticeable increase in the score of holothurian, which is listed on the ablation study part, shows that guiding scales and aspect ratios of the anchors by the semantic features can be extremely helpful in solving this problem.

3.3   Sampling Balanced Strategy with Noise Labels

Uneven distribution and reproduction of the benthos bring on one challenge, which affects the training process. When sampling the underwater images for training or testing, most images merely include few targets or none. The standard random sampling results in most of the hard positive samples that give way to the easy samples, while hard examples are extraordinary import to upgrade the detecting performance. While OHEM [21] is an alternative option yet OHEM suppresses the capability when ensembled into the baseline. In particular, underwater images and labels contain noise so that it is crucial to choose a robust method.

Our work is inspired by the Libra R-CNN [17] that proposed three components to solve the unbalanced samples problem. Contrary to intuition, we partly agree that the modules are suitable for the underwater detection task, especially the Balanced L1 loss. Consequently, this paper only extracts the IOU balanced Sampling for hard mining, which is a simple but virtual component, especially for underwater detection. Firstly, we divide the selected samples into K bins according to their IOU. Then we elect samples from each bin uniformly. The chosen probability of each sample under the introduced IOU balanced strategy is

$$p_k = \frac{N}{K} * \frac{1}{M_k}, \quad k \in [0, K) \tag{3}$$

where K is set to 3 in our experiments, and M_k is the number of candidate samples in the kth interval. N denotes the number of chosen samples.

4   Experimental Results

We analyze the impact of achieving precise detections when combining the three modules using our GSDCN detector.

4.1   Experimental Setting

If not otherwise specified, all methods listed below are based on the typical ResNet-50 backbone network, FPN [11] neck, and ROIAlign [6]. All experiments are conducted on the URPC2018 dataset, including 2901 train-Val and 800 test images. For a fair comparison with other methods, the images were resized to a maximum scale of 1300 × 800 and deleted the starfish from the dataset. We employ Mean Average Precision (mAP) metric for the evaluation.

We implement our method and experiments by PyTorch [18] and MMdetection [2]. Besides, we conduct the training processes on a GTX Titan XP GPU with 12G memory and testing on an RTX 2070 GPU with 8G memory allocated with the robot.

4.2   Results

We compare our proposed methods with the outstanding generic object detectors on the URPC *test-dev*. Through the overall pipeline elaborate design,

Table 1. Results of the outstanding detectors that were trained and tested on the URPC2018 dataset

Methods	Results			
	Holothurian	Echinus	Scallop	mAP50
SSD [14]	11.6	81.1	2.1	31.6
Faster R-CNN [20]	73.8	91.0	57.1	74
Cascaded [1]	74.1	90.2	58.6	74.3
ROIMix [12]	73.3	86.8	56.0	72.0
GHM [9]	40.7	56.2	34.9	43.9
FCOS [24]	20.4	86.2	11.8	39.5
ScratchDet [29]	57.0	87.8	48.5	64.4
Grid R-CNN [15]	71.0	88.5	55.7	71.7
Free anchor [28]	66.5	90.2	40.8	65.8
RepPoints [26]	67.1	90.8	37.1	65
FoveaBox [8]	69.7	91.4	53.5	71.5
Libra R-CNN [17]	74.9	91.0	58.9	75
Hrnet [22]	35.2	88.8	28.9	51.0
Ours	**77.3**	**91.8**	**62.9**	**77.4**

Table 2. Results of the ablation experiments

Methods	Results						
	DCN Module	GA Strategy	BA Sampling	Holothurian	Echinus	Scallop	mAP
Baseline				73.8	91.0	57.1	74.0
	✓			76.3	**92.3**	59.2	75.9
		✓		75.6	90.9	60.2	75.6
			✓	74.7	90.9	57.2	74.2
	✓	✓		76.9	91.5	60.4	76.3
		✓	✓	75.5	90.6	62.6	76.3
	✓		✓	76.7	91.5	60.8	76.3
GSDCN	✓	✓	✓	**77.3**	91.8	**62.9**	**77.4**

GSDCN achieves 77.4% mAP, which is 3.4 points higher AP than the FPN Faster R-CNN baseline. Furthermore, our method surpasses the state-of-the-art underwater object detector by a large margin to 5.4 points. Particularly, our GSDCN exceeds following outstanding generic object detectors by a distinct margin(2.4%–45.8%). The primary reason is that our GSDCN customize a detector based on the three main challenges for the specific task. Consequently, the improvement is uniquely suited for the benthonic organism detection (Table 1).

4.3   Ablation Experiments

The results of ablation experiments are shown in Table 2. First of all, we merely bring in the DCN module, GA Strategy, and Balanced Sampling, respectively. The significant improvements from the AP of holothurian and scallop validate that introducing the DCN module and GA Strategy can upgrade the baseline model, especially for detecting the nonrigid objects with geometric variations, i.e., holothurian and scallop. In particular, when we employ the Balanced Sampling on the baseline model, a slight improvement was made. Since the detector suffers from the first two challenges a lot. Afterward, after adding the GA Strategy on the baseline with the DCN module, the holothurian's AP increased by 0.6%, and scallops' rises 0.8%. Undoubtedly, attracting the GA strategy into the baseline with the DCN module further improves the capability of detecting hard categories. Last but not least, the problem of sampling unbalanced at this time is the key factor in suppressing the performance of the detector. Naturally, adding Balanced Sampling on the network mentioned before brings an increase mAP of 1.1%.

Figure 3 visualizes partial detecting results of our GSDCN detector and the baseline. As shown in Fig. 3, the baseline RPN Fast R-CNN fails to detect the wide scallops and holothurians, especially for the curving holothurian, while our GSDCN works well. The high AP of each kind, the mAP score, and the visualization of results all illustrate that GSDCN alleviates the challenges efficiently.

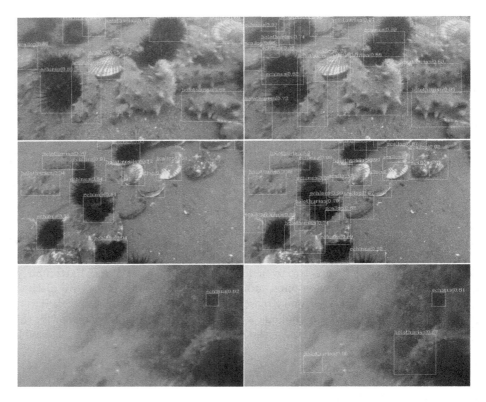

Fig. 3. Visualize the results of baseline and our GSDCN. The left column reveals the results of the baseline, and the right one indicates our outstanding performance.

5   Conclusion

This paper discusses the special challenges of underwater detection, the geometric variations of holothurians, the influence of visibility range, and the imbalance of objects' distribution. Regarding the issues discussed above, we propose a customized two-stage detector GSDCN to help underwater robots picking the holothurians, echinus, scallops autonomously. To the best of our knowledge, this is the first paper that researches the outstanding generic object detectors in the underwater conditions over the past three years. The experimental results demonstrate the promising performance of our GSDCN better than the state-of-the-art detector with a large margin to 5.39%. We believe that this paper would potentially provide new insights on designing new robust underwater object detectors. Besides, we hope this paper could spur more research works on underwater object detection.

Acknowledgements. This work is supported partly by the National Natural Science Foundation (NSFC) of China (grants 61973301, 61972020, 61633009, 51579053 and U1613213), partly by the National Key R&D Program of China (grants 2016YFC0300801 and 2017 YFB1300202), partly by Beijing Science and Technology Plan Project, partly by Field Fund (No. 61403120409) of the 13th Five-Year Plan for Equipment Pre-research Fund, and partly by Meituan Open RD Fund.

References

1. Cai, Z., Vasconcelos, N.: Cascade R-CNN: delving into high quality object detection. In: Proceedings of the IEEE Conference on Computer Vision and Pattern Recognition, pp. 6154–6162 (2018)
2. Chen, K., et al.: MMDetection: open MMLab detection toolbox and benchmark. arXiv preprint arXiv:1906.07155 (2019)
3. Dai, J., et al.: Deformable convolutional networks. In: Proceedings of the IEEE International Conference on Computer Vision, pp. 764–773 (2017)
4. Girshick, R.: Fast R-CNN. In: Proceedings of the IEEE International Conference on Computer Vision, pp. 1440–1448 (2015)
5. Girshick, R., Donahue, J., Darrell, T., Malik, J.: Rich feature hierarchies for accurate object detection and semantic segmentation. In: Proceedings of the IEEE Conference on Computer Vision and Pattern Recognition, pp. 580–587 (2014)
6. He, K., Gkioxari, G., Dollár, P., Girshick, R.: Mask R-CNN. In: Proceedings of the IEEE International Conference on Computer Vision, pp. 2961–2969 (2017)
7. Ji-Yong, L., Hao, Z., Hai, H., Xu, Y., Zhaoliang, W., Lei, W.: Design and vision based autonomous capture of sea organism with absorptive type remotely operated vehicle. IEEE Access **6**, 73871–73884 (2018)
8. Kong, T., Sun, F., Liu, H., Jiang, Y., Shi, J.: FoveaBox: beyond anchor-based object detector. arXiv preprint arXiv:1904.03797 (2019)
9. Li, B., Liu, Y., Wang, X.: Gradient harmonized single-stage detector. In: Proceedings of the AAAI Conference on Artificial Intelligence, vol. 33, pp. 8577–8584 (2019)
10. Li, X., Shang, M., Qin, H., Chen, L.: Fast accurate fish detection and recognition of underwater images with fast R-CNN. In: OCEANS 2015-MTS/IEEE Washington, pp. 1–5. IEEE (2015)
11. Lin, T.Y., Dollár, P., Girshick, R., He, K., Hariharan, B., Belongie, S.: Feature pyramid networks for object detection. In: Proceedings of the IEEE Conference on Computer Vision and Pattern Recognition, pp. 2117–2125 (2017)
12. Lin, W.H., Zhong, J.X., Liu, S., Li, T., Li, G.: RoIMix: proposal-fusion among multiple images for underwater object detection. arXiv preprint arXiv:1911.03029 (2019)
13. Liu, S., Ozay, M., Okatani, T., Xu, H., Lin, Y., Gu, H.: Learning deep representations and detection of docking stations using underwater imaging. In: 2018 OCEANS-MTS/IEEE Kobe Techno-Oceans (OTO), pp. 1–5. IEEE (2018)
14. Liu, W., et al.: SSD: single shot multibox detector. In: Leibe, B., Matas, J., Sebe, N., Welling, M. (eds.) ECCV 2016. LNCS, vol. 9905, pp. 21–37. Springer, Cham (2016). https://doi.org/10.1007/978-3-319-46448-0_2
15. Lu, X., Li, B., Yue, Y., Li, Q., Yan, J.: Grid R-CNN. In: Proceedings of the IEEE Conference on Computer Vision and Pattern Recognition, pp. 7363–7372 (2019)
16. Mandal, R., Connolly, R.M., Schlacher, T.A., Stantic, B.: Assessing fish abundance from underwater video using deep neural networks. In: Proceedings of the International Joint Conference on Neural Networks (IJCNN 2018), pp. 1–6. IEEE (2018)

17. Pang, J., Chen, K., Shi, J., Feng, H., Ouyang, W., Lin, D.: Libra R-CNN: towards balanced learning for object detection. In: Proceedings of the IEEE Conference on Computer Vision and Pattern Recognition, pp. 821–830 (2019)
18. Paszke, A., et al.: Automatic differentiation in PyTorch (2017)
19. Redmon, J., Divvala, S., Girshick, R., Farhadi, A.: You only look once: unified, real-time object detection. In: Proceedings of the IEEE Conference on Computer Vision and Pattern Recognition, pp. 779–788 (2016)
20. Ren, S., He, K., Girshick, R., Sun, J.: Faster R-CNN: towards real-time object detection with region proposal networks. In: Advances in Neural Information Processing Systems, pp. 91–99 (2015)
21. Shrivastava, A., Gupta, A., Girshick, R.: Training region-based object detectors with online hard example mining. In: Proceedings of the IEEE Conference on Computer Vision and Pattern Recognition, pp. 761–769 (2016)
22. Sun, K., Xiao, B., Liu, D., Wang, J.: Deep high-resolution representation learning for human pose estimation. arXiv preprint arXiv:1902.09212 (2019)
23. Sung, M., Yu, S.C., Girdhar, Y.: Vision based real-time fish detection using convolutional neural network. In: OCEANS 2017-Aberdeen, pp. 1–6. IEEE (2017)
24. Tian, Z., Shen, C., Chen, H., He, T.: FCOS: fully convolutional one-stage object detection. arXiv preprint arXiv:1904.01355 (2019)
25. Wang, J., Chen, K., Yang, S., Loy, C.C., Lin, D.: Region proposal by guided anchoring. In: Proceedings of the IEEE Conference on Computer Vision and Pattern Recognition, pp. 2965–2974 (2019)
26. Yang, Z., Liu, S., Hu, H., Wang, L., Lin, S.: RepPoints: point set representation for object detection. arXiv preprint arXiv:1904.11490 (2019)
27. Zhang, L., Yang, X., Liu, Z., Qi, L., Zhou, H., Chiu, C.: Single shot feature aggregation network for underwater object detection. In: Proceedings of the 24th International Conference on Pattern Recognition (ICPR 2018), pp. 1906–1911. IEEE (2018)
28. Zhang, X., Wan, F., Liu, C., Ji, R., Ye, Q.: FreeAnchor: learning to match anchors for visual object detection. In: Advances in Neural Information Processing Systems, pp. 147–155 (2019)
29. Zhu, R., et al.: ScratchDet: exploring to train single-shot object detectors from scratch. arXiv preprint arXiv:1810.08425, vol. 2 (2018)
30. Zhu, X., Hu, H., Lin, S., Dai, J.: Deformable ConvNets v2: more deformable, better results. In: Proceedings of the IEEE Conference on Computer Vision and Pattern Recognition, pp. 9308–9316 (2019)

Author Index

Printed in the United States
By Bookmasters